Cambridge Illustrated Handbook of Optoelectronics and Photonics

A one-of-a-kind illustrated encyclopedic handbook of important terms in photonics and optoelectronics, covering a wealth of material from fundamental concepts to cutting-edge applications.

Self-contained descriptions of common tools and phenomena are provided for undergraduate and graduate students, scientists, engineers and technicians in industry and laboratories.

The book strikes a balance between materials and devices-related coverage and systems-level terms, and captures key nomenclature used in the field. Equations are used where necessary, but lengthy derivations are avoided. Over 600 illustrations help to convey key concepts, creating an attractive, easy-to-use reference.

Safa Kasap is a Professor and Canada Research Chair in the Department of Electrical and Computer Engineering at the University of Saskatchewan, Canada, and has been making original research contributions to electronic and photonic materials and devices, and their applications in optical communications and x-ray imaging, for over 20 years.

Harry Ruda is Director of the Centre for Advanced Nanotechnology, and Chair Professor in Nanotechnology at the University of Toronto, Canada. His research interests focus on the fabrication and modeling of quantum functional nanostructures with applications in the fields of nanoelectronics and nanophotonics.

Yann Boucher is a Researcher at the Laboratoire RESO, ENIB (Brest National Engineering School), France. He has developed original tools for modeling optoelectronic devices, such as the Extended Transfer Matrix Formalism including sources, or so-called Couplonics, a method based on exact coupled-wave equations for investigating periodic structures.

Cambridge Illustrated Handbook of Optoelectronics and Photonics

Safa Kasap
University of Saskatchewan, Canada

Harry Ruda
University of Toronto, Canada

Yann Boucher
École Nationale d'Ingénieurs de Brest, France

CAMBRIDGE
UNIVERSITY PRESS

CAMBRIDGE UNIVERSITY PRESS
Cambridge, New York, Melbourne, Madrid, Cape Town,
Singapore, São Paulo, Delhi, Tokyo, Mexico City

Cambridge University Press
The Edinburgh Building, Cambridge CB2 8RU, UK

Published in the United States of America by Cambridge University Press, New York

www.cambridge.org
Information on this title: www.cambridge.org/9781107404236

First published 2009
First paperback edition 2012

A catalogue record for this publication is available from the British Library

Library of Congress Cataloguing in Publication Data
Kasap, S. O. (Safa O.)
Cambridge illustrated handbook of optoelectronics and photonics : an illustrated encyclopedic dictionary of significant terms, effects, phenomena, equations, and properties, from fundamental concepts to applications, from materials to devices to systems, covering optoelectronics and photonics, including optical communications / Safa
Kasap, Harry Ruda, Yann G. Boucher.
p. cm.
ISBN 978-0-521-81596-3 (hardback)
1. Optoelectronics – Encyclopedias. 2. Optoelectronic devices – Encyclopedias.
I. Ruda, Harry E., 1958– II. Boucher, Yann G. III. Title.
TA1750.K383 2009
621.381′045 – dc22 2008045090

ISBN 978-0-521-81596-3 Hardback
ISBN 978-1-107-40423-6 Paperback

Preface

The present work is an illustrated encyclopedic handbook of important terms and effects from fundamental concepts to applications in optoelectronics and photonics, including optical communications. We tried to define and describe terms from materials to devices to systems, as photonics is still in its early stages of evolution compared with microelectronics and today's billion-transistor chips, and systems on chips. It is *not* meant to be a comprehensive encyclopedia or a dictionary in this field but rather a self-contained semiquantitative description of terms and effects that frequently turn up in optoelectronics and photonics courses at the undergraduate and graduate levels. There is nothing worse than a dry reference book with no illustrations. We prepared numerous illustrations to convey the message as clearly as possible. There is an old Chinese adage that a good diagram is worth a thousand words but a bad diagram takes a thousand words to explain. Nearly all the illustrations have been prepared almost from scratch to be as self-explanatory and as clear as possible. In writing such an *encyclopedic handbook* we had to choose between short and quick definitions and definitions that encompass an extensive explanation; the choice was based on whether a term can be defined simply and still be useful, e.g. the acceptance cone of a fiber, or whether the term needs at least one page of explanations to be fair to the term and the reader, e.g. photonic crystals. In addition, we had to draw a distinction between an optics handbook and what we had in mind as a useful encyclopedic handbook for optoelectronics and photonics. We have missed many terms but we have also described many, through, undoubtedly, our own biased selection. In our own view there is not a huge difference between optoelectronics and photonics, except that in the latter there may not be any electronics involved. Today, photonics is a more fashionable term than the old optoelectronics term which usually conjures images of LEDs, solar cells, optoisolators, etc., whereas photonics is closely associated with optical communications and optical signal processing. True photonics is supposed to be a subject based on photons, or electromagnetic radiation, only; but, in practice, it is impossible to avoid electrons when photons have to be generated or detected.

"After a year's research one realises that it could have been done in a week"

Sir William Henry Bragg (1862–1942)

Numerical definitions

1/*f* noise see *excess noise.*

10Dq see *ligand field splitting parameter.*

1.24 "rule" is the equation that is widely used in converting between the photon energy E in eV and the radiation wavelength λ in microns: $\lambda(\mu m) = 1.24[E(eV)]^{-1}$. It arises from $E = h\nu = hc/\lambda$.

A

Abbe condenser is a high numerical aperture lens system that normally has two lenses designed to collect and suitably direct light to an object that is to be examined in a microscope.

Abbe diagram see *Abbe number*.

Abbe number or **constringence** of an optical medium is the inverse of its *dispersive power*, that is, it represents the relative importance of refraction and dispersion. There are two common definitions based on using different standard wavelengths. The Abbe number v_d is defined by

$$v_d = (n_d - 1)/(n_F - n_C)$$

where n_F, n_d and n_C are the refractive indices of the medium at the Fraunhofer standard wavelengths corresponding to the helium *d*-line ($\lambda_d = 587.6$ nm, yellow), hydrogen *F*-line ($\lambda_F = 486.1$ nm, blue) and hydrogen *C*-line ($\lambda_C = 656.3$ nm, red). The Abbe number v_e, on the other hand, is defined by

$$v_e = (n_e - 1)/(n_{F'} - n_{C'})$$

where n_e, $n_{F'}$ and $n_{C'}$ are refractive indices at the *e*-line (546.07 nm), *F'*-line (479.99 nm) and *C'*-line (643.85 nm) wavelengths respectively.

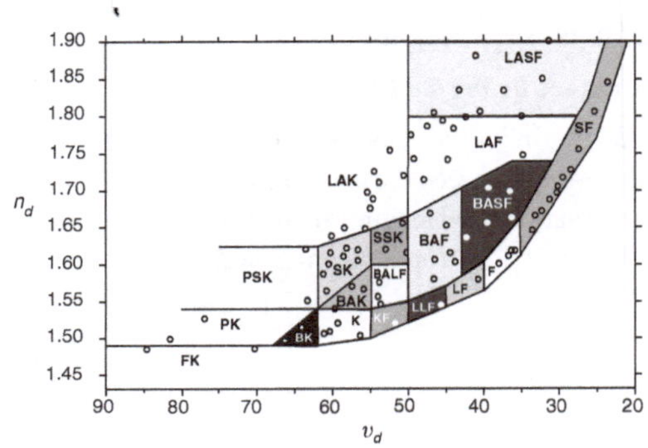

Abbe number The *Abbe diagram* is a diagram in which the refractive index n_d of glasses are plotted against their Abbe numbers in a linear n_d vs. v_d plot and, usually, with the Abbe number decreasing along the *x*-axis, rather than increasing (v_d values on the the *x*-axis have been reversed). The last letter F or K represents flint or crown glass. Other symbols are as follows: S, dense; L, light; LL, extra light; B, borosilicate; P, phosphate; LA or La, lanthanum; BA or Ba, barium. Examples: BK, dense flint; LF, light flint; LLF, extra light flint; SSK, extra dense crown; PK, phosphate crown; BAK, barium crown; LAF, lanthanum flint, etc. (Adapted from Schott Glass Website.) See *chromatic dispersion*.

Table: Abbe number PC is polycarbonate, PMMA is polymethylmethacrylate, PS is polystyrene.

Optical glass	SF11	F2	BaK1	crown glass	fused silica	PC	PMMA	PS
v_d	25.76	36.37	57.55	58.55	67.8	34	57	31

[Sources: Melles-Griot and Goodfellow websites]

Abbe-Porro prism is a prism that transmits the image through the prism laterally displaced but fully inverted, that is, flipped vertically and horizontally. It is very similar in its function to the double Porro prism. See *Porro prism*.

Abbe prism is a reflection prism that uses reflections rather than refractions to deflect (deviate) light, invert or rotate an image.

Abbe refractometer see *refractometer*.

Abbe's sine condition, discovered by Abbe and Helmholtz (1873) and also simply called the *sine condition*, specifies the condition under which arbitrary rays leaving a particular point P_o on the object are able to arrive at the same (and unique) image point P_i that is an image (or conjugate point) of P_o. Consider an image formed by a spherical surface. A point P_o on the object at a height y_o in the object space of refractive index n_o gives rise to a point P_i of height y_i in the image space of index n_i; the two media are separated by a spherical boundary with its center of curvature at C. An arbitrary ray from P_o that is incident on the spherical surface at A (an arbitrary point) is refracted, and can only pass through the required unique image point P_i if *Abbe's sine* condition is satisfied, i.e.

$$n_o y_o \sin\theta_o = n_i y_i \sin\theta_i$$

where θ_o is the angle the object ray P_oA makes with the principal ray P_oCP_i (that ray passing through the center C without refraction) and θ_i is the angle that the image ray AP_i makes with the principal ray P_oCP_i. There are no assumptions in this condition, and if an optical system is such that it satisfies the sine condition, then spherical and coma (off-axis) aberrations are eliminated since all arbitrary rays from P_o arrive at P_i. Consider a simple lens imaging system in which the object and the image are in the same refractive index media, $n_o = n_i$. The magnification is y_o/y_i, so that $y_i/y_o = \sin\theta_o/\sin\theta_i$. Both marginal and paraxial rays will produce the same magnification, that is the same P_i for a given P_o, if

$$y_i/y_o = \sin\theta_o/\sin\theta_i = \theta_{op}/\theta_{ip} = \text{constant}$$

which will then result in no spherical or coma aberration; $\sin\theta_o/\sin\theta = \theta_{op}/\theta_{ip}$ is sometimes stated as the *sine condition* for the absence of spherical and coma aberration. It has been stated that "Abbe's sine condition requires that all rays emanating from the axial object point within the incident cone must emerge in image space, where they form a converging cone toward the axial image point, at the same height at which they entered the system" (M. Born and E. Wolf, *Principles of Optics*, Cambridge University Press, 1999). The height *OA* must be the same for both the object rays and the image rays. See *aplanatic lens, aplanatic optical system, aplanatic points*.

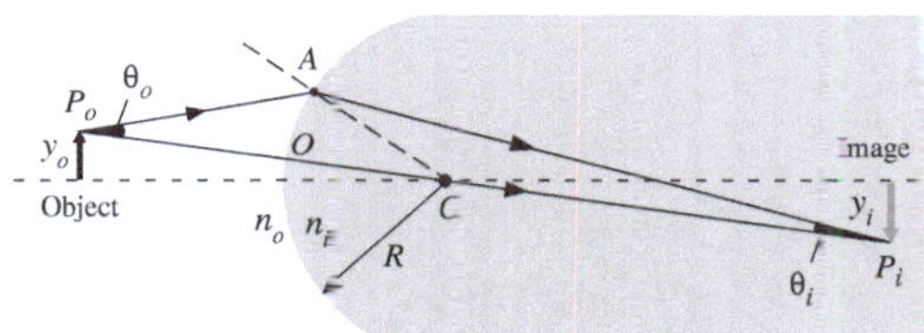

Abbe's sine condition for refraction at a curved surface. *C* is the center of curvature. Subscripts *o* and *i* refer to object and image.

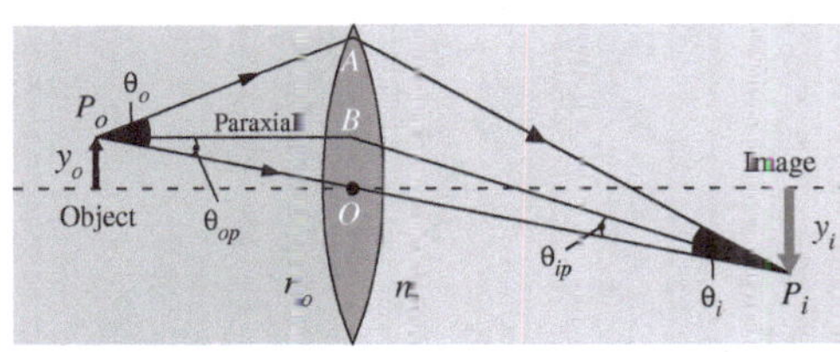

Abbe's sine condition for a lens; an arbitrary ray from P_o generates P_i.

Abbe's theory of microscope imaging refers to the case where the (thin) *phase object* is under coherent illumination, either by a laser or by light emitted from a sufficiently small source via a condenser of low aperture. The object in the Π plane acts as a diffraction phase grating, forming its

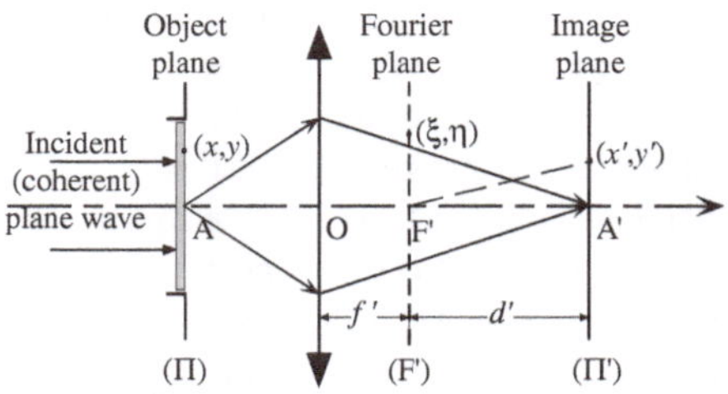

Abbe's theory of image formation. A filter conveniently located in the F′ plane enables one to select, or suppress, any given spatial frequency. (x,y) represent the coordinates of a point on the object plane, (ξ,η) the coordinates of a point on the Fourier transform plane, and (x',y') the coordinates of a point on the image plane. See also *spatial filtering*.

Ernst Abbe (1840–1905, Germany). Ernst Abbe was appointed a professor of physics and mathematics at Jena in 1870. As a result of his innovative works on optics with Carl Zeiss, Abbe became quite wealthy. (Jena Review, 1965, Zeiss Archive, courtesy AIP Emilio Segre Visual Archives, E. Scott Barr Collection.)

Fourier transform in the image focal plane F′ of the objective, where each point acts as a secondary source. The final image in the image plane Π′ is the result of the interference of all these sources. The whole theory has been summarized as follows: "The microscope image is the interference effect of diffraction phenomena."

ABCD matrix or **ray-transfer matrix** is a matrix that conveniently describes an optical operation on a ray of light; it is a direct result of the linearity of the paraxial equations. At any abscissa z along the main axis, any given optical ray propagating in a medium of refractive index n is completely determined by two numbers: the distance $\rho(z)$ to the axis and the reduced slope $\theta(z) = n(z)d\rho(z)/dz$ as illustrated in the figure.

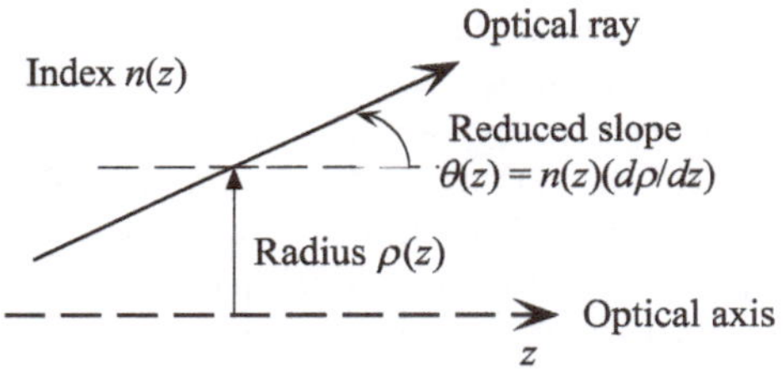

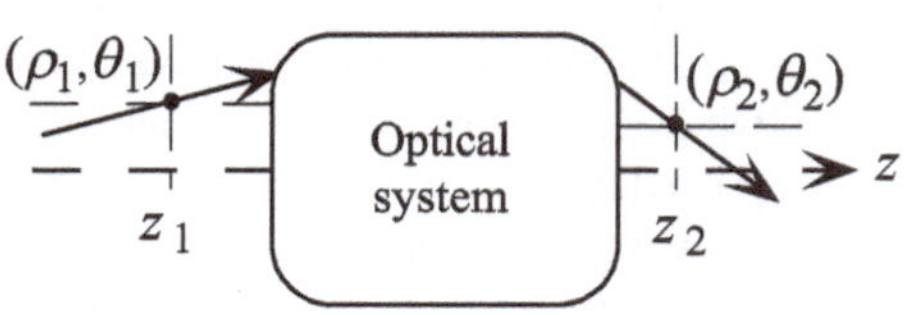

The couple $\{\rho_1,\theta_1\}$ at abscissa z_1 is transformed at abscissa z_2 into $\{\rho_2,\theta_2\}$ such that

$$\begin{pmatrix}\rho_2\\ \theta_2\end{pmatrix} = \begin{pmatrix}A & B\\ C & D\end{pmatrix}\begin{pmatrix}\rho_1\\ \theta_1\end{pmatrix},$$

with $AD-BC=1$. Some examples are reported in the following table.

Propagation in free space (length L, index n_0)	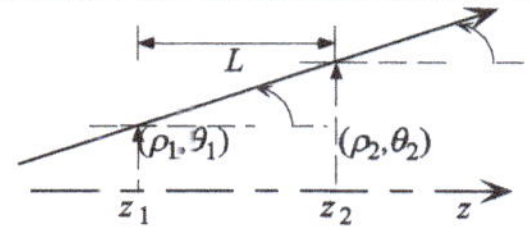	$\begin{pmatrix}\rho_2\\ \theta_2\end{pmatrix} = \begin{pmatrix}1 & L/n_0\\ 0 & 1\end{pmatrix}\begin{pmatrix}\rho_1\\ \theta_1\end{pmatrix}$
Thin lens (focal f)	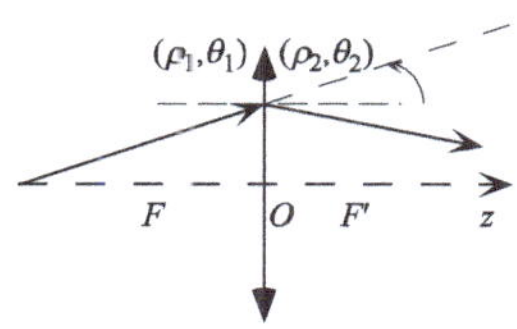	$\begin{pmatrix}\rho_2\\ \theta_2\end{pmatrix} = \begin{pmatrix}1 & 0\\ -1/f & 1\end{pmatrix}\begin{pmatrix}\rho_1\\ \theta_1\end{pmatrix}$

Any succession of optical elements, all aligned along a single axis, can therefore be associated a single ray-transfer matrix, obtained by a matrix product (in the right order) of each subcomponent. This can be successfully generalized to a linear medium exhibiting a refractive index decreasing quadratically with ρ, such as $n(\rho) = n_0(1-\gamma^2\rho^2/2)$. In this case,

$$\begin{pmatrix}\rho_2\\ \theta_2\end{pmatrix} = \begin{pmatrix}\cos(\gamma z) & \sin(\gamma z)/(n_0\gamma)\\ -(n_0\gamma)\sin(\gamma z) & \cos(\gamma z)\end{pmatrix}\begin{pmatrix}\rho_1\\ \theta_1\end{pmatrix},$$

and the optical ray is "trapped" along the axis, forced to propagate in a curved way. Graded-index quadratic fibers are based on this principle (GRIN fibers).

Two other applications of ABCD matrices should be mentioned: the stability analysis of laser cavities and the propagation of Gaussian beams generated by such a laser.

Aberrations generally refer to unwanted deviations in the optical imaging characteristics of a lens or a mirror from the ideal, based on *paraxial ray imaging*. Aberrations are usually divided into two categories. *Chromatic aberration* is due to the wavelength dependence of the refractive index and hence the focal length f, which now depends on the color of light. *Monochromatic* (or near monochromatic) *aberrations* are also called *third-order aberrations* (originally analyzed by Seidel in 1857) and are primarily *spherical*, *coma*, *astigmatism*, *field curvature*, and *distortion*, and arise as a result of the departure from the paraxial approximation when the angles of the rays in the imaging system are not sufficiently small. Aberrations typically result in the distortions of the image and in the unwanted blurring in the image quality (loss of resolution). When a light ray traveling in air is incident on a lens, it is refracted at the lens surface. It obeys Snell's law, i.e.

$$n_1 \sin\theta_1 = n_2 \sin\theta_2$$

where n_1 and n_2 are the air and glass (lens) refractive indices and θ_1 and θ_2 are the angles of incidence and refraction with respect to the normal on the lens surface at the point of incidence. There is a similar expression for the ray leaving the lens. The whole paraxial lens theory and the usual lens equations are based on assuming small angles so that $\sin\theta_1 \approx \theta_1$, and $\sin\theta_2 \approx \theta_2$, and $n_1\theta_1 \approx n_2\theta_2$. Thus, in

$$\sin\theta = \theta - \frac{\theta^3}{3!} + \frac{\theta^5}{5!} - \ldots$$

we normally neglect the third-order and higher terms, which is the paraxial approximation; small angles. The third-order term ($\theta^3/3!$) is the cause of third-order aberrations, which are sometimes called *Seidel aberrations*. Unless $\theta = 0$ (normal incidence), there will always be a third term in the above expansion (small but finite), and hence always some aberration. In practice, aberrations occur in combination which makes it difficult to cancel them exactly. Nonetheless, either by suitable lens shaping, or using two or more lenses with stops,

Table: Aberrations in optics (Summarized from various sources including J. R. Meyer-Arendt, *Introduction to Modern Optics*, Fourth Edition, Prentice Hall, 1994.)

Aberration	Cause and character	Correction
Chromatic aberration	Focal point F depends on λ (n depends on λ). Image blur	Doublet
Monochromatic aberration types		
Spherical	On-axis and off-axis blur	Lens bending; aspherical lenses; gradient index; doublet
Coma	Off-axis blur	Lens bending; spaced doublet with central stop
Astigmatism	Off-axis blur	Spaced doublet with stop
Curvature of field	Off-axis objects imaged onto a curved surface instead of a plane	Spaced doublet. Field flattener
Distortion	Off-axis distortion due to varying magnification with distance from axis	Spaced doublet with stop

it is possible to reduce aberrations to an insignificant level.

Absorption is the loss in the power of an electromagnetic radiation that is traveling in a medium. The loss is due to the conversion of light energy to other forms of energy, e.g. lattice vibrations (heat) during the polarization of the molecules of the medium, local vibrations of impurity ions, excitation of electrons from the valence band to the conduction band, etc.

Absorption broadening see *photon reabsorption broadening*.

Absorption coefficient, α, characterizes the absorption of photons as light propagates along a certain direction in a medium. It is the fractional change in the intensity of light per unit distance along the propagation direction, that is,

$$\alpha = -\frac{\delta I}{I\delta x}$$

where I is the intensity of the radiation. $1/\alpha$ is also the mean probability per unit distance that a photon is absorbed. The absorption coefficient depends on the photon energy or wavelength λ. The absorption coefficient α is a material property. Most of the photon absorption (63%) occurs over a distance $1/\alpha$ and $1/\alpha$ is called the *penetration depth* δ. The absorption coefficient depends on the radiation absorbing processes. If K is the extinction coefficient (imaginary part of the complex refractive index, $N = (n - jK)$, and λ is the free space wavelength, ω is the angular frequency ($2\pi\nu$), then

$$\alpha = 2(2\pi/\lambda)K = (2\omega/c)K.$$

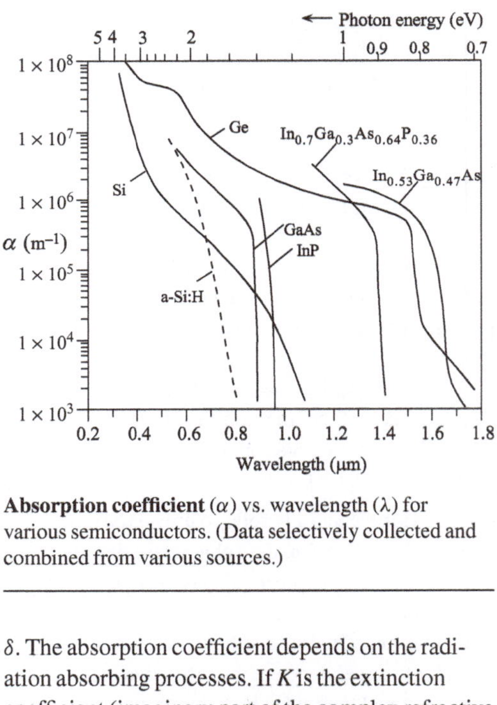

Absorption coefficient (α) vs. wavelength (λ) for various semiconductors. (Data selectively collected and combined from various sources.)

If we know the real and imaginary parts of the relative permittivity, $\varepsilon_r = \varepsilon_{r1} - j\varepsilon_{r2}$, of a medium, we can calculate its α by

$$\alpha = \frac{4\pi}{\lambda}\left[\frac{\left(\varepsilon_{r1}^2 + \varepsilon_{r2}^2\right)^{1/2} - \varepsilon_{r1}}{2}\right]^{1/2}.$$

See *cross-section, sum rules*.

Absorption cross section see *cross-section*.

Absorptivity see *emissivity*.

Acceptance angle, or the *maximum acceptance angle*, is the largest possible light launch angle from the fiber axis. Light waves within the acceptance angle that enter the fiber become guided along the fiber core. If NA is the *numerical aperture* of a step index fiber, and light is launched from a medium of refractive index n_0, then the maximum acceptance angle $\alpha_{\max}$ is given by

$$\sin\alpha_{\max} = \frac{\text{NA}}{n_0} = \frac{\left(n_1^2 - n_2^2\right)^{1/2}}{n_0}$$

where n_1 and n_2 are the refractive indices of the core and cladding of the fiber. The total acceptance angle is twice the maximum acceptance angle and is the total angle around the fiber axis within which all light rays can be launched into the fiber.

Acceptance cone is a cone with its height aligned with the fiber axis and its apex angle twice the acceptance angle so that light rays within this cone can enter the fiber and then propagate along the fiber. See *acceptance angle*.

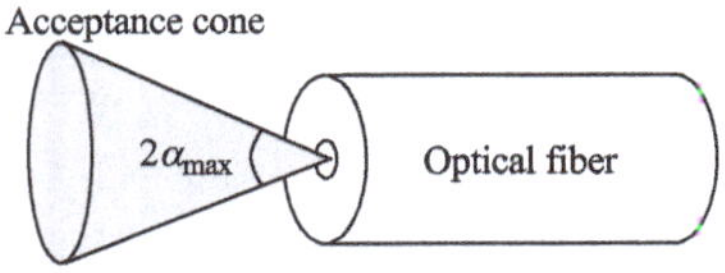

Acceptance cone.

Acceptor atoms are dopants that have one or more less valency than the host atom. They therefore accept electrons from the valence band (VB) and thereby create holes in the VB which leads to a greater hole than electron concentration, $p > n$, and hence to a p-type semiconductor.

Achromatic doublet see *chromatic dispersion*.

Achromatic lens is a composite lens (made up of two or more closely fitted lenses) so that the combination has its chromatic aberration corrected at least at two different wavelengths. Stated differently, the achromatic lens has the same focal length f at two wavelengths λ_1 and λ_2 ($>\lambda_1$) and its focal length does not vary significantly with the wavelength from λ_1 to λ_2. For example, a positive achromatic lens usually has a positive (convex) low index lens (e.g. crown glass) and a negative (concave) high index lens (e.g. flint glass) cemented together. The combination is a weaker convex lens.

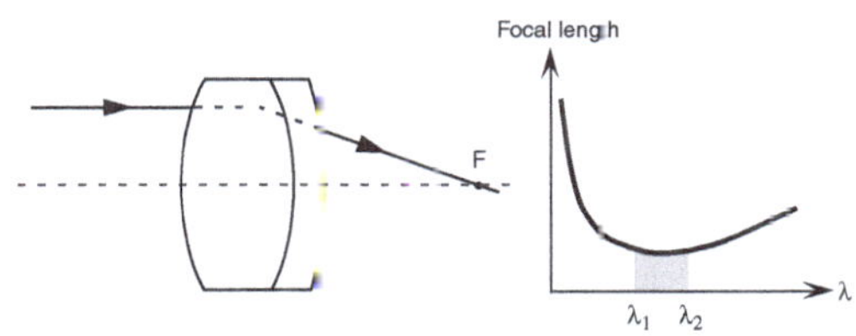

Left: **Achromatic lens** from two lenses. Right: The focal length vs. wavelength behavior.

Achromatic light, see *color*.

Achromatic triplet is a triplet lens that has three lenses to reduce chromatic dispersion. A Steinheim achromatic triplet has a symmetric convex lens of lower refractive index (crown glass) that has higher refractive index (flint) meniscus lenses cemented to its surfaces.

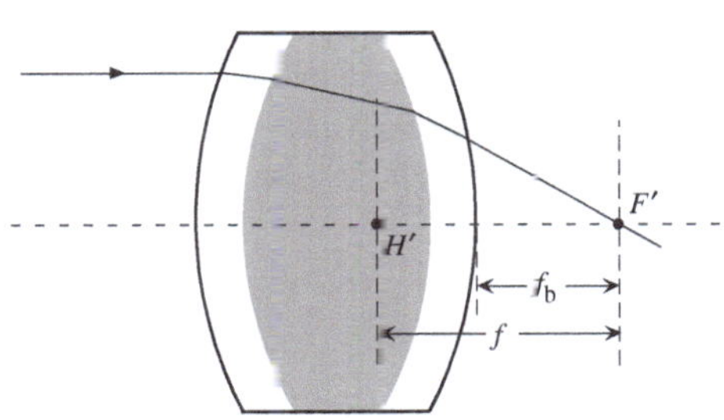

Achromatic triplet f_b is the back focal length measured from the back focal point F' on the lens axis to the lens.

Acoustic velocity see *sound velocity*.

Acoustooptic (AO) modulator makes use of the *photoelastic effect* to modulate a light beam. Suppose that we generate traveling acoustic or ultrasonic waves on the surface of a piezoelectric crystal (such as $LiNbO_3$) by attaching interdigital electrodes onto its surface and applying a modulating voltage at radio frequencies (RF). The *piezoelectric effect* is the phenomenon of generation of strain in a crystal by the application of an external electric field. The modulating voltage $V(t)$ at electrodes will therefore generate a *surface acoustic wave* (SAW) via the piezoelectric effect. These acoustic waves propagate by rarefactions and compressions of the crystal surface region which lead to a periodic variation in the density and hence a periodic variation in the refractive index in synchronization with the acoustic wave amplitude. The periodic variation in the strain S leads to a periodic variation in n owing to the *photoelastic effect*. We can simplistically view the crystal surface region as alternations in the refractive index. An incident light beam will be diffracted by this periodic variation in the refractive index. Depending on the wavelength of the optical (λ) and acoustic waves (Λ), and the length of interaction, there may be a single or multiple diffracted beams. If there are multiple diffracted beams then the AO effect is called *Raman–Nath* diffraction and if there is only the first order, it is referred to as the AO *Bragg diffraction*; the latter is easier to quantify. (See *Raman–Nath diffraction*.) Bragg diffraction occurs at high acoustic frequencies where the acoustic wavelength is short. If the acoustic wavelength is Λ, then the condition that gives the angle θ for a diffracted beam to exist is given by the *Bragg diffraction condition*,

$$2\Lambda \sin\theta = \lambda/n$$

where n is refractive index of the medium, and 2θ is the diffraction (deflection) angle inside the AO diffraction medium, within the device. The intensity of the diffracted beam depends on the power of the acoustic wave that induces the diffraction grating, and hence on the modulating voltage, a convenient way to modulate the intensity of a light beam. If the AO modulator has an essentially first-order diffracted beam, the first-order diffracted intensity I_1 is usually given by

$$\frac{I_1}{I_0} = \sin^2\left[\frac{\pi}{\lambda}\left(\frac{L}{2H}M_2 P_{\text{acoustic}}\right)^{1/2}\right] \approx \frac{\pi^2 M_2 P_{\text{acoustic}} L}{2\lambda^2 H}$$

where I_0 is the incident light intensity, P_{acoustic} is the power in the acoustic wave, M_2 is a figure of merit for the AO medium (given below), L is the acoustic beam length and H the acoustic beam height or width in a plane parallel to the propagation of incident light.

Suppose that ω is the angular frequency of the incident optical wave. The optical wave reflections occur from a moving diffraction pattern which moves with a velocity V_{acoustic}. As a result of the *Doppler effect*, the diffracted beam has either a slightly higher or slightly lower frequency depending on the direction of the traveling acoustic wave. If Ω is the frequency of the acoustic wave then the diffracted beam has a Doppler shifted frequency given by

$$\omega' = \omega \pm \Omega.$$

When the acoustic wave is traveling towards the incoming optical beam, then the diffracted optical beam frequency is up-shifted, i.e. $\omega' = \omega + \Omega$. If the acoustic wave is traveling away from the incident optical beam then the diffracted frequency is down-shifted, $\omega' = \omega - \Omega$. It is apparent that we can modulate the frequency (wavelength) of the diffracted light beam by modulating the frequency of the acoustic waves. (The diffraction angle is then also changed.)

$LiNbO_3$ is a piezoelectric crystal and hence allows the convenient generation of acoustic waves by the simple placement of interdigital electrodes;

Table: Acoustooptic modulator Figures of merit for various acoustooptic materials. n is the refractive index v is the acoustic velocity. (From the Brimstone website and others.)

Material	$LiNbO_3$	TeO_2	Ge	InP	GaAs	GaP	Fused quartz	$PbMoO_4$	Flint glass	Chalcogenide glass
λ (μm)	0.6–4.5	0.4–5	2–12	1–1.5	1–11	0.59–10	0.2–4.5	0.4–1 2	0.45–2	1.0–2.2
n	2.2	2.26	4	3.3	3.37	3.31	1.46	2.26	1.8	2.7
(at μm)	(0.633)	(0.633)	(10.6)		(1.15)	(1.15)	(6.3)	(0.633)		
v km s^{-1}	6.6	4.2	5.5	5.1	5.34	6.3	5.96	3.63	3.51	2.52
$M_2 \times 10^{-15}$ s^3/kg	7	35	180	80	104	44	1.6	50	8	164

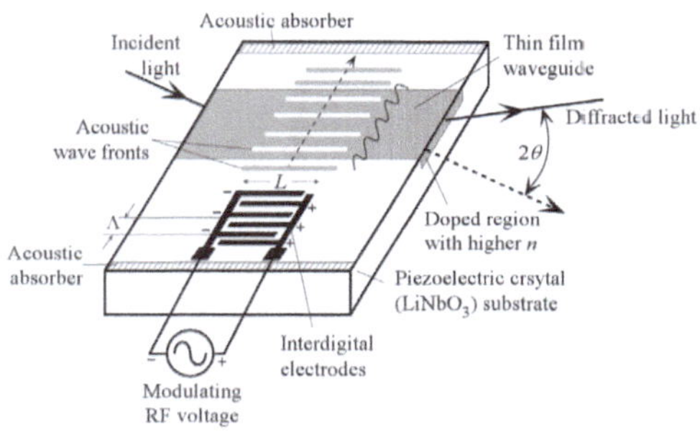

Acoustooptic modulator Traveling acoustic waves create a harmonic variation in the refractive index and thereby create a diffraction grating that diffracts the incident beam through an angle 2θ.

it is used in high-speed IR AO modulators. AO gratings can also be generated in nonpiezoelectric materials because the basic principle is the induction of a diffraction grating by an acoustic wave, which can be coupled into the material from an (external) ultrasonic transducer. The rarefactions and compressions associated with the acoustic wave generates a phase grating by virtue of the photoeleastic effect. For example, germanium is commonly used in various commercially available infrared (2–11 μm) AO devices that deflect light, modulate the light intensity, and shift the optical frequency. AO devices based on other crystals, e.g. TeO_2 (400–1000 nm), GaP (600–1000 nm), $PbMoO_4$ (400–1200 nm) and glasses, e.g. fused silica (240–450 nm), dense flint glass (440–770 nm), chalcogenide glasses (1.2–3 μm), are also available and cover a wide spectral range from UV to IR and implement various AO modulation functions. (Wavelengths in parentheses cover typical useful ranges, and not necessarily the wavelength range of optical transparency.)

The deflection efficiency of an AO modulator depends not only on the material properties such as the efficiency of the optolelastic effect but also on the power in the acoustic wave. One very simple and useful figure of merit M_2 is

$$M_2 = \frac{n^6 p^2}{\rho v^3}$$

where n is the refractive index, p is the photoelastic constant, ρ is the density and v is the acoustic velocity. The n^6 in the numerator implies that the changes in n have the largest effect on the AO modulation efficiency. The modulation speed and hence the bandwidth of an AO is determined by the transit time of the acoustic waves across the light beam waist. If the optical beam has a waist (diameter) d in the modulator, then the rise time (in digital modulation) depends on d/v, which is called the transit time across the beam diameter. Reducing v to increase the speed however results in the reduction of diffraction efficiency so there is a compromise between the speed and diffraction efficiency.

(D. Pinnow, *IEEE Journal of Quantum Electronics*, **6**, 223, 1970.)

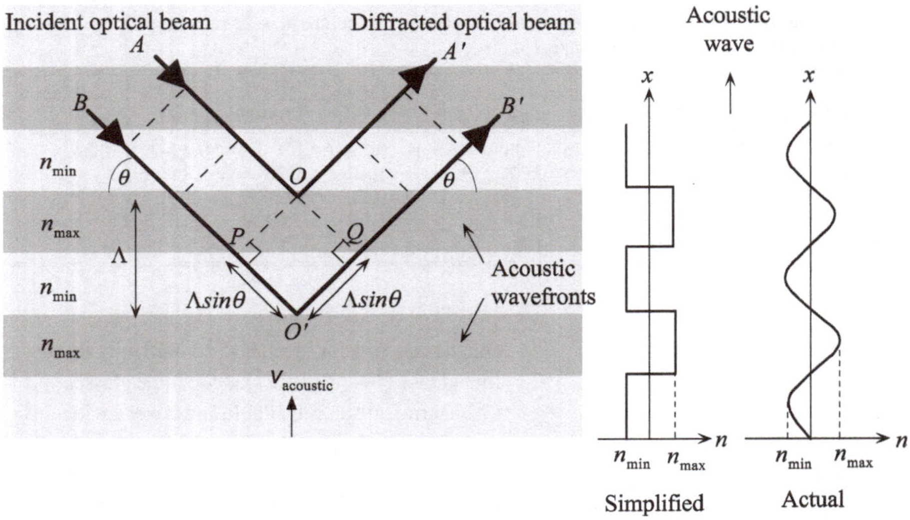

Acoustooptic modulator Consider two coherent optical waves A and B being "reflected" (strictly, scattered) from two adjacent acoustic wavefronts to become A' and B'. These reflected waves can only constitute the diffracted beam if they are in phase. The angle θ is exaggerated (typically this is a few degrees).

Activation energy is the potential energy barrier that prevents a system from changing from one state to another. For example, if two atoms A and B get together to form a product AB, the activation energy is the potential energy barrier against the formation of this product. It is the minimum energy which the reactant atom or molecule must have to be able to reach the activated state and hence be able to form the product. The probability that a system has an energy equal to the activation energy is proportional to the Boltzmann factor: $\exp(-E_A/k_BT)$, where E_A is the activation energy, k_B is the Boltzmann constant and T is the temperature (Kelvins), or it can be expressed as $\exp(\Delta H/RT)$ where ΔH would be in J mole^{-1}, and R is the gas constant.

Activator see *luminescence*.

Active device is a device that exhibits gain (current or voltage or both) and has a directional function. Transistors are active devices whereas resistors, capacitors and inductors are passive devices.

Active region is the region in a medium where direct electron-hole pair (EHP) recombination takes place. For LEDs it is the region where most EHP recombination takes place. In the laser diode it is the region where stimulated emission exceeds spontaneous emission and absorption. It is the region where coherent emission dominates.

Active matrix array (AMA) is a two-dimensional array of pixels in which each pixel has a TFT that can be externally addressed to drive a device such as an LED or a liquid crystal cell located at the pixel; or to read a signal from a sensor located at the pixel. Depending on the application, an AMA can have a few pixels or millions of pixels. The TFT AMA technology was pioneered by Peter Brody using CdSe TFTs in the early 1970s. As shown in the figure, each pixel is identical with its TFT gate

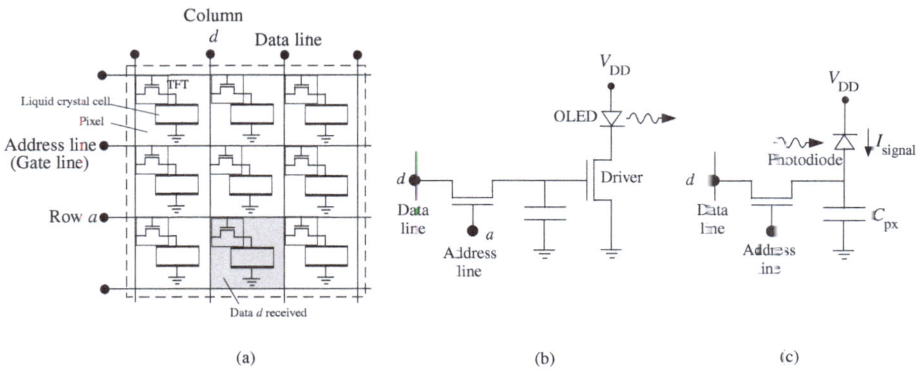

Active matrix array (a) Basic structure of an active matrix array with one TFT transistor in each pixel driving a liquid crystal cell for a liquid crystal display. The gray pixel on row *a* receives the data on data line *d*. (b) A pixel structure that drives an OLED, organic light emitting device (or some other LED with an acceptable current demand). (c) AMA used as a read out array in which a pixel has a photodiode. When light impinges on the pixel photodiode, the photocurrent generated is the signal and is read out at data line *d* by addressing row *a*.

connected to a particular address line and the source to a particular data line. The example in the figure involves an AMA driving an LCD (liquid crystal display), and each pixel has a liquid crystal cell. When a voltage is applied to the address line *a*, each TFT on row *a* is activated. The information as parallel data on the data lines is then transferred to the pixels at row *a*, e.g. column *d* is connected by the TFT to the pixel that crosses *a* and *d*, to pixel *a,d*, which receives the information. Thus, each pixel on row *a* receives its data from the data line. The data has been encoded to activate the liquid crystal cell, to turn it on or off, etc. The address and data lines have drivers, and the whole sequencing of which row is activated is done by a scanning control. One frame of an image in the LCD is generated by scanning one row after another until all the rows have been scanned, i.e. each pixel has received its input signal (data). When the pixel has an LED or an OLED (organic light emitting diode), a second TFT is incorporated as a driver. The AMA can also function to read signals from pixels. In AMA imaging sensors, each pixel has a photodiode. When light impinges on the pixel photodiode, a photocurrent is generated which charges the pixel capacitance. This charge constitutes the signal, and is read out at data line *d* by addressing row *a*. When row *a* is addressed, all signals on that row are read out, that is, the data are read out onto the data lines as *parallel data*, which can be converted into serial data. We scan the whole image row by row (line by line) by starting from the top row and sequentially activating one row after another.

Active matrix sensor see *active matrix array*.

Active pixel sensor (APS) is an active matrix array *image sensor* in which each pixel has two or more transistors, usually CMOS, that provide on-pixel amplification, either providing an actual gain or working as a source follower. In contrast, in a *passive pixel sensor* (PPS) there is no amplification at a pixel; usually a single transistor switch is used to read out the charge. See *active matrix array*, *CMOS image sensor*.

Add/drop multiplexer (ADM) see *wavelength add/drop multiplexer (WADM)*.

ADU, arbitrary digital units.

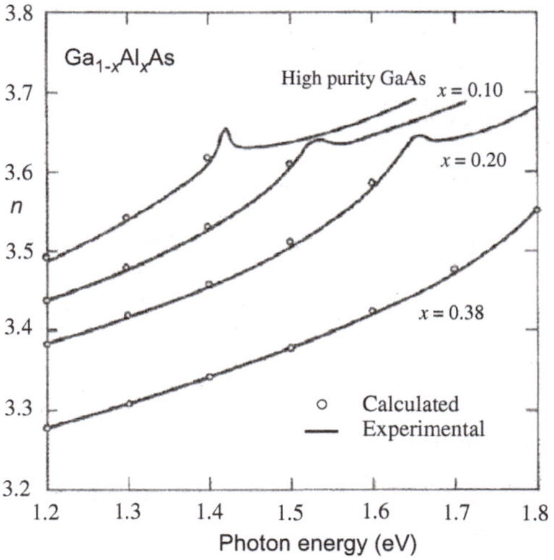

Afromowitz refractive index model Refractive index of $Ga_{1-x}Al_xAs$ as a function of photon energy above the bandgap energy for various compositions x (x is obtained experimentally). Solid curves are experimental n values and the circle points are calculated using the Afromowitz model. (Data from M. A. Afromowitz, *Sold State Communications*, **15**, 59, 1974.)

Afromowitz refractive index model provides a semiempirical method for the imaginary part ε_{r2} of the dielectric constant, and hence for the real part ε_{r1} and thus the refractive index n, of direct bandgap semiconductors as a function of the photon energy $E (= h\upsilon)$ when the photon energy is greater than the bandgap energy E_g, and includes three material-dependent constants: E_g (bandgap), E_o (resonant energy) and E_d (oscillator strength or dispersion energy). Usually E_o and E_d are determined experimentally by fitting the calculations to the experimental data. The photon absorption (ε_{r2}) is given by

$$\varepsilon_{r2} = \eta E^4 \quad \text{for } E_g < E < E_f$$

where $\eta = (\pi/2)E_d/[E_o{}^3(E_o{}^2 - E_g{}^2)]$, and $E_f = (2E_o{}^2 - E_g{}^2)^{1/2}$. The real part ε_{r1} can be obtained from the Kramers–Kronig relation as

$$\varepsilon_{r1} = 1 + M_{-1} + M_{-3}E^2 + \frac{\eta}{\pi}\ln\left[\frac{E_f^2 - E^2}{E_g^2 - E^2}\right]E^4$$

in which M_{-1} and M_{-3} are constants (called "moments" of the ε_{r2} spectrum) that depend on E, E_c and E_d as $M_{-1} = (\eta/2\pi)(E_f^4 - E_g{}^4)$ and $M_{-3} = (\eta/\pi)(E_f^2 - E_g{}^2)$. Once E_g, E_o and E_d are known (are found), ε_{r1} and hence n can be found at any photon energy in the range $E_g < E < E_f$.

Afterglow see *luminescence*.

Air Kerma see *kerma*.

Air mass see *solar radiation*.

Air-mass zero (AM0) radiation see *solar constant*.

Airy disk refers to the central white spot in the *Airy ring* pattern. *Airy rings* represent a diffraction pattern from a circular aperture illuminated by a coherent beam of light. The radius of the Airy disk corresponds to the radius of the first dark ring.

Table: Afromowitz refractive index model Afromowitz model parameters for the refractive index n. E_g represents the direct bandgap energy (sometimes denoted as E_Γ). Data from M. A. Afromowitz, *Sold State Communications*, **15**, 59, 1974.

	$Ga_{1-x}Al_xAs$	$GaAs_{1-x}P_x$	$Ga_xIn_{1-x}P$
E_o	$3.65 + 0.871x + 0.179x^2$	$3.65 + 0.721x + 0.139x^2$	$3.391 + 0.524x + 0.595x^2$
E_d	$36.1 - 2.45x$	$36.1 + 0.35x$	$28.91 + 7.54x$
E_g (or E_Γ)	$1.424 + 1.266x + 0.26x^2$	$1.441 + 1.091x + 0.21x^2$	$1.34 + 0.668x + 0.758x^2$

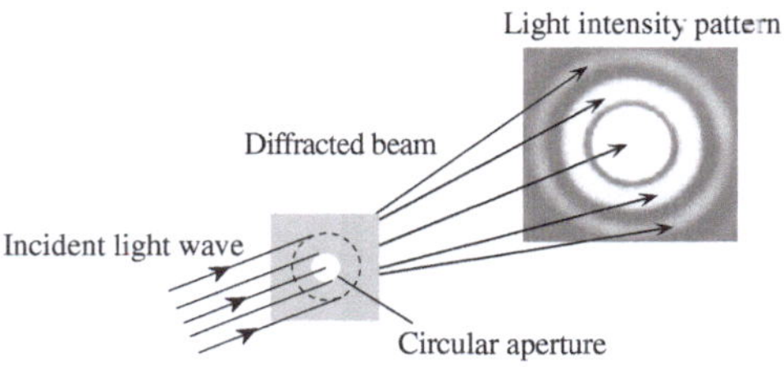

Airy disk A light beam incident on a small circular aperture becomes diffracted and its light intensity pattern after passing through the aperture is a diffraction pattern with circular bright rings (called Airy rings). If the screen is far away from the aperture, this would be a Fraunhofer diffraction pattern.

An *Airy ring pattern* is the Fraunhofer diffraction pattern produced by a circular aperture of diameter D. The circularly symmetric intensity is angularly distributed as

$$I(\theta) = I_0[2J_1(z)/z]^2,$$

where I_0 is the peak intensity, J_1 is the Bessel function of the first order, and $z = (\pi D\theta/\lambda)$. When projected into the focal plane of an imaging instrument with a focal length f, the central spot has a diameter of around $1.22\lambda f/D$. See *diffraction limited system.*

George Bidell Airy (1801–1892, England). George Airy was a Professor of Astronomy at Cambridge and then the Astromer Royal at the Royal Observatory in Greenwich, England. (C. H. Jeens, AIP Emilio Segre Visual Archives, E. Scott Barr and T. J. J. See Collections.)

Airy function is a comb-like, intensity transfer function of a *Fabry–Perot cavity.*

Airy ring pattern see *Airy disk.*

Aliasing is the wrapping or folding back of the high-frequency signal above the Nyquist frequency onto the image as a lower frequency component if the presampling signal has frequency components above the Nyquist frequency, as shown in the figure. Consider a signal $s(x)$ with a spectrum (Fourier transform) $S(f)$ that has no spectral components above f_m. Suppose $s(x)$ is sampled at a sampling frequency (rate) f_s to generate a sampled signal $s_{\text{sampled}}(x)$. The sampled signal spectrum $S_{\text{sampled}}(f)$ has the original spectrum $S(f)$ repeated at every f_s. If $f_s < 2f_m$, there exists an overlapped region in the spectrum of the sampled signal. The frequency band between $(f_s - f_m)$ and f_m is *aliased*, resulting in a distortion of the signal in this frequency band and an exact reconstruction of the signal is not possible. A higher sampling rate or an antialiasing filter can eliminate aliasing. For example, we can use an antialiasing filter to limit the higher spatial frequencies, and reduce f_m until the Nyquist criterion $f_s > 2f_m$ is satisfied and the aliasing is removed. *Aliasing noise* is the image noise associated with the aliasing phenomenon. If aliasing occurs, noise aliasing takes place, any noise component of a frequency higher than the Nyquist or critical frequency (f_c) will be replicated in frequency space and overlap with other noise frequencies in a fashion identical to the overlap of the signal frequencies. The overlapping frequency components of a digital noise power spectrum (NPS) simply add, and this effect potentially increases image noise at frequencies below f_c. Thus, aliasing noise reduces the detective quantum efficiency (DQE) at higher frequencies. See *Nyquist–Shannon theorem, detective quantum efficiency.*

Aliasing noise see *aliasing.*

Aluminum antimonide (AlSb) is an indirect bandgap III-V semiconductor crystal with zinc

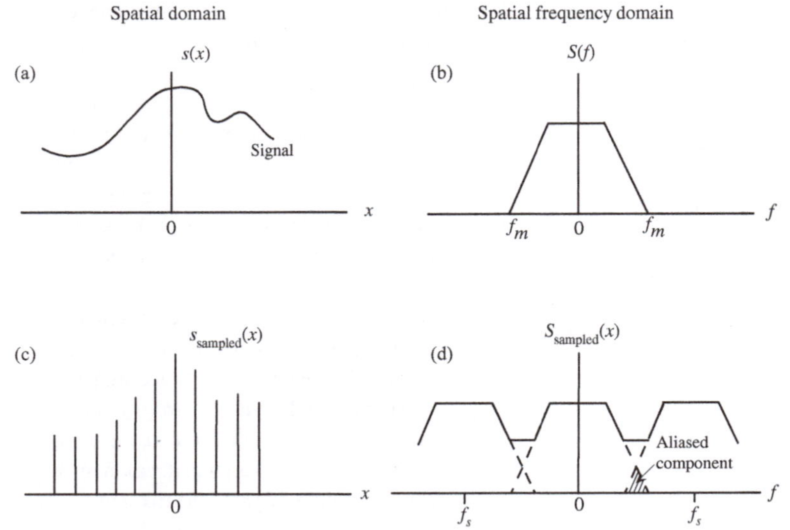

Aliasing (a) and (b): Presampling signal $s(x)$ in the spatial domain, and its Fourier transform $S(f)$ in the spatial-frequency domain, respectively. (c) and (d): Sampled signal $s_{\text{sampled}}(x)$ and its Fourier transform $S_{\text{sampled}}(f)$ in spatial and spatial-frequency domains, respectively.

blende structure having $a = 0.61355$ nm; $\rho = 4.29$ g cm^{-3}; $T_m = 1065\,°$C; $\chi = 3.53$ eV; E_g(I) = 1.615 eV (300 K) and E_g(D) = 2.30 eV (300 K); $\Delta_{\text{so}} = 0.673$ eV; $m_{eT}{}^*/m_e = 0.259$; $m_{eL}{}^*/m_e = 1.80$; $\mu_e \sim 200$ cm^2 V^{-1}s^{-1}; $\mu_h \sim 400$ cm^2 V^{-1}s^{-1}; ε_r(DC) = 12.04, $\varepsilon_r(\infty) = 10.24$. For $\lambda = 1$ to 20 μm,

$$n = 3.1340 + \frac{0.8032}{\lambda^2} - \frac{0.5166}{\lambda^4} - 0.0004\lambda^2.$$

Aluminum arsenide (AlAs) is an indirect bandgap III-V semiconductor crystal with zinc blende structure: $a = 0.566$ nm; $\rho = 3.76$ g cm^{-3}; $T_m = 1740\,°$C; $\alpha_L = 4.9\times10^{-6}$ K^{-1}; $c_s = 0.45$ J g^{-1} K^{-1}; $\kappa \approx 0.91$ W cm^{-1} K^{-1}; $\chi = 3.5$ eV; $T_{\text{Debye}} = 446$ K; E_g(I) = 2.168 eV (300K), E_g(I) = 2.230 eV (0K); $\Delta_{\text{so}} = 0.30$eV; $m_e{}^*/m_e = 0.71$; $m_{hh}{}^*/m_e = 0.76$; $m_{lh}{}^*/m_e = 0.15$; $n_i < 10$ cm^{-3}; $N_c = 1.5\times10^{19}$ cm^{-3}; $N_v = 1.65\times10^{19}$ cm^{-3}; ε_r(DC) = 10.06, $\varepsilon_r(\infty) = 8.16$; $E_{\text{breakdown}} \approx 5\times10^5$ V cm^{-1}; $\mu_e \sim 300$ cm^{2}V^{-1}s^{-1}; $\mu_h \sim 140$ cm^2 V^{-1}s^{-1}; $D_e \sim 5$ cm^2 s^{-1}; $D_h \sim 5$ cm^2 s^{-1}; $n \sim 3.3$; $\hbar\omega_{\text{op}} = 50$ meV; $B_{\text{radiative}} = 1.8\times10^{-10}$ cm^3 s^{-1}.

Aluminum gallium arsenide ($Al_xGa_{1-x}As$) is a ternary zinc blende semiconductor alloy that is near-lattice matched to GaAs over the entire compositional range with $a = 0.56533 + 0.00078x$ nm; $\rho = 5.32 - 1.56x$ g cm^{-3}; $T_m = 1240 - 58x + 558x^2\,°$C; $c_s = 0.33 + 0.12x$ J g^{-1} °C^{-1}; $\kappa = 0.55 - 2.12x + 2.48x^2$ W cm^{-1} °C^{-1}; thermal diffusivity $D_{\text{th}} = 0.31 - 1.23x + 1.46x^2$ cm^2 s^{-1}; $\alpha_L = (5.73 - 0.53x)\cdot10^{-6}$ °C^{-1}, $K = (7.55 + 0.26x)\cdot10^{11}$ dyn cm^{-2}, $G = (3.25 - 0.09x)\cdot10^{11}$dyn/cm^2, $E = (8.53 - 0.18x)\cdot10^{11}$dyn/cm^2, $v = (0.31+0.1x)$, piezoelectric constant of $e_{14} = -0.16-0.065x$ C.m^{-2}, and elastic constants of $C_{11} = (11.88 + 0.14x)\cdot10^{11}$ dyn/cm^2, $C_{12} = (5.38 + 0.32x)\cdot10^{11}$ dyn/cm^2 and $C_{44} = (5.94 - 0.05x)\cdot10^{11}$ dyn/cm^2. The material has a direct bandgap for $x < 0.5$ of E_g (D) $\approx 1.424 + 1.266x + 0.266x^2$ eV, an effective mass for electrons of $(0.063 + 0.083x)m_e$ ($x < 0.45$), density of

(a)
(b)
(c)

Aliasing (a) The original image, roof tops (Italy), where one can clearly distinguish the periodic roof tiles. (b) An undersampled image in which aliasing has introduced significant noise, and the image is not a reconstruction of the original. Many small features are unclear (noisy), edges are jagged, and there is false information. (c) Sampled image with an antialiasing filer. While the higher spatial frequencies have been lost (cut by the filter), the reconstructed image is close to the original with some blurring of very fine details. (Courtesy of Dr. A. S. Mehr, University of Saskatchewan.)

states electron effective mass of $(0.85 - 0.14x)m_e{}_o$ ($x > 0.45$), heavy-hole effective mass of $(0.51 + 0.25x)me_o$, and light-hole effective mass of $(0.082 + 0.068x)m_e$. The electron affinity is $4.07 - 1.1x$ eV for $x < 0.45$ and $3.64 - 0.14x$ eV for $x > 0.45$. $T_D = 370 + 54x + 22x^2$ K; ε_r (DC) $= 12.90 - 2.84x$, high frequency $\varepsilon_r = 10.89 - 2.73x$ and the optical phonon energy is $36.25 + 1.83x - 17.12x^2 - 5.11x^3$ meV. The valence band discontinuity for $Al_xGa_{1-x}As$ heterojunctions to GaAs is $\Delta E_v = -0.46x$ eV and that for the conduction band is $\Delta E_c = 0.79x$ eV for $x < 0.41$ and $0.475 - 0.335x + 0.143x^2$ eV for $x > 0.41$. The breakdown field strength is $(4 - 6)\times 10^5$ V cm^{-1}; $\mu_e = 8{\cdot}10^3 - 2.2{\cdot}10^4x + 10^4{\cdot}x^2$ cm^2 V^{-1} s^{-1} for $x < 0.45$ and $-255 + 1160x - 720x^2$ cm^2 V^{-1} s^{-1} for $x > 0.45$; $\mu_h = 370 - 970x + 740x^2$ cm^2 V^{-1} s^{-1}. The diffusion coefficient D_e for electrons is $200 - 550x + 250x^2$ cm^{2}s^{-1} for $x < 0.45$ and $-6.4 + 29x - 18x^2$ cm^{2}s^{-1} for $x > 0.45$; and $D_h = 9.2 - 24x + 18.5x^2$ cm^{2}s^{-1}. The refractive index n is

$$n^2 = A + \frac{B}{\lambda^2 - C} - D\lambda^2$$

where $A = 13.5 - 15.4x + 11.0x^2$, $B = 0.690 + 3.60x - 4.24x^2$, $C = 0.154 - 0.476x + 0.469x^2$, and $D = 1.84 - 8.18x + 7.00x^2$.

AMA see *active matrix array*.

Amici prism is a prism that provides a deflection through 90° and a flipping of the image.

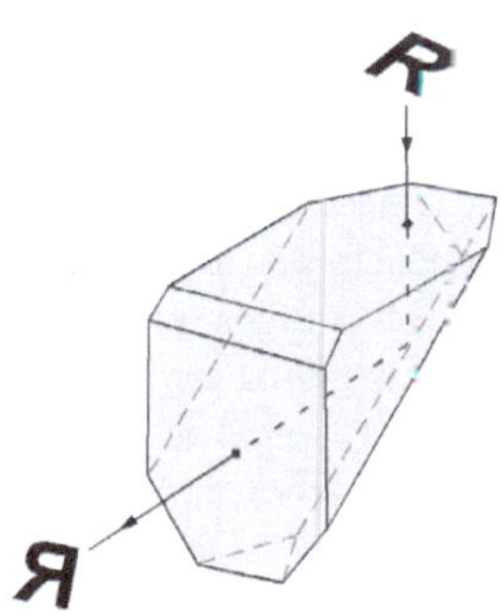

Amici prism.

Amorphous hydrogenated silicon-germanium ($a\text{-}Si_{1-x}Ge_x{:}H$) alloys are amorphous semiconductor alloy films that are prepared by various thin-film fabrication techniques such as chemical vapor deposition (CVD), etc. for use in various optoelectronic applications such as *pin* photodiodes with responsivity in the long-wavelength region; the bandgap decreases with Ge. The Tauc optical gap E_g and the refractive index n are given by

$$E_g(\text{Tauc}) = 1.20 - 0.99x + 0.65x^2$$
$$n = 3.233 + 0.177x + 0.574x^2.$$

See *amorphous semiconductor.*

Amorphous semiconductor is a type of noncrystalline solid that has no long-range periodicity in its structure (crystals have a long-range periodicity) but only short-range order that defines the local chemical binding environment – some short-range order that extends a few atomic spacings. *Hydrogenated amorphous silicon* (a-Si:H) is a noncrystalline form of silicon in which the structure has no long-range order but only short-range order as in the first figure. Each Si atom has four neighbors as in the crystal but there is no periodicity, or long-range order. Without the hydrogen, pure a-Si would have dangling bonds. In such a structure sometimes a Si atom would not be able to find a fourth neighboring Si atom to bond with and will be left with a dangling bond. The hydrogen in the structure (about ~10%) passivates (i.e. neutralizes) the unsatisfied ("dangling") bonds inherent in the noncrystalline structure and so reduces the density of dangling bonds or defects. Amorphous semiconductors do not follow typical crystalline concepts such as Bloch wavefunctions. First, due to the lack of periodicity, we cannot describe the electron as a Bloch wave. Consequently, we cannot use a wavevector k, and hence $\hbar k$, to describe the electron's motion. These semiconductors, however, do have a short-range order, and also possess an energy bandgap that separates a conduction band and a valence band. Window glass has a noncrystalline structure but also has a bandgap, which makes it transparent. Photons with energies less than the bandgap energy can pass through window glass. See *optical absorption in amorphous semiconductors, optical properties of semiconductors*.

The potential energy $V(x)$ of the electron in a noncrystalline structure fluctuates randomly from site to site. In some cases, the local changes in $V(x)$ can be quite strong, forming effective local PE wells (obviously finite wells). Such fluctuations in the PE within the solid can capture or trap electrons, that is, localize electrons at certain spatial locations. A localized electron will have a wavefunction that resembles the wavefunction in the hydrogen atom so that the probability of finding the electron is localized to the site. Such locations that can trap electrons, give them localized wavefunctions, are called *localized states*. The amorphous structure also has electrons that possess extended wavefunctions, that is, they belong to the whole solid. These extended wavefunctions are distinctly different than those in the crystal because they have a very short coherence length due to the random potential fluctuations; the electron is scattered from site to site and hence the mean free path is of the order of a few atomic spacings. The extended wavefunction has random phase fluctuations. The second figure compares localized and extended wavefunction in an amorphous semiconductor. The energy distribution of the density of states (DOS) function, $g(E)$, has defined energies E_v and E_c that separate extended states from localized states. There is a distribution of localized states, called *tail states* below E_c and above E_v. The usual *bandgap* $E_c - E_v$ is called the *mobility gap*. The reason is that there is a change in the character of charge transport, and hence in the carrier mobility, in going from extended states above E_c to localized states below E_c. The electron transport above E_c in the conduction band is dominated by scattering from random potential fluctuations arising from the disordered nature of the structure. The electrons are scattered so frequently

Amorphous semiconductor

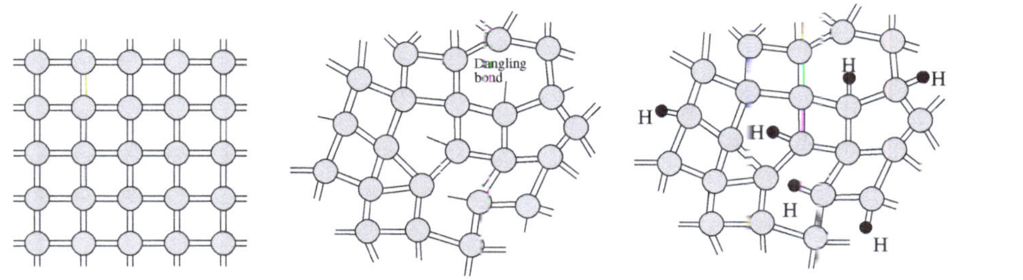

(a) Two-dimensional schematic representation of a silicon crystal

(b) Two-dimensional schematic representation of the structure of amorphous silicon. The structure has voids and dangling bonds and there is no long-range order.

(c) Two-dimensional schematic representation of the structure of hydrogenated amorphous silicon. The number of hydrogen atoms shown is exaggerated.

Amorphous semiconductor Silicon can be grown as a semiconductor crystal or as an amorphous semiconductor film. Each line represents an electron in a bond. A full covalent bond has two lines and a broken bond has one line.

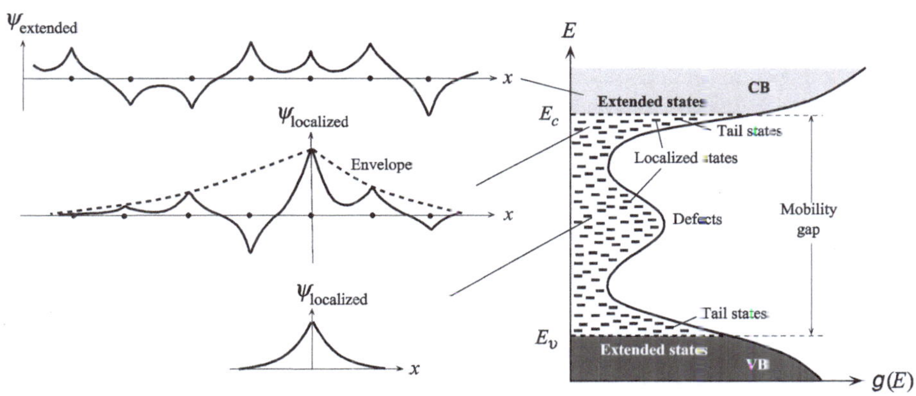

Amorphous semiconductor Schematic representation of the density of states $g(E)$ vs. energy E for an amorphous semiconductor and the associated electron wavefunctions for an electron in the extended and localized states.

that their effective mobility is much less than what it is in crystalline Si: μ_e in a-Si:H is typically 5–10 $cm^2\ V^{-1}\ s^{-1}$ whereas it is 1400 $cm^2\ V^{-1}\ s^{-1}$ in a single-crystal Si. Electron transport below E_c, on the other hand, requires an electron to jump, or hop, from one localized state to another, aided by thermal vibrations of the lattice, in an analogous way to the diffusion of an interstitial impurity in a crystal. The electron's mobility associated with this type of hopping motion among localized states is thermally activated, and its value is small. Thus, there is a change in the electron mobility across E_c, which is called the conduction band *mobility edge*. Typically, the conductivity is thermally activated, $\sigma = \sigma_0 \exp(\Delta E_\sigma / k_B T)$. The localized states (frequently simply called *traps*) between E_v and E_c have a profound effect on the overall electronic properties. The tail localized states are a direct result of the structural disorder that is inherent in noncrystalline solids, variations in the bond angles and lengths. Various prominent peaks and features in the DOS within the mobility gap have been associated with possible structural defects, such as under- and overcoordinated atoms in the structure, dangling bonds, dopants, etc. Electrons that drift in the conduction band can fall into localized states and become immobilized (trapped) for a while. Thus, electron transport in a-Si:H occurs by multiple trapping in shallow localized states. The effective electron drift mobility in a-Si:H is therefore reduced to about ~1 cm^2 V^{-1} s^{-1}. Low drift mobilities obviously prevent the use of amorphous semiconductor materials in high-speed or high-gain otpoelectronic applications. Nonetheless, low-speed electronics is just as important as high-speed electronics in such applications as flat panel displays, solar cells, image sensors, etc. A low-speed flat panel display made from hydrogenated amorphous silicon (a-Si:H) TFTs costs very roughly the same as a high-speed crystalline Si microchip that runs the CPU.

Amorphous semiconductors, optical properties see *optical absorption in amorphous semiconductors.*

Amorphous solid exhibits no crystalline structure or long-range order. It only possesses a short-range order in that the nearest neighbors of an atom are well defined by virtue of chemical bonding requirements. See *glass, amorphous semiconductor.*

Amplitude grating see *diffraction grating.*

Amplitude object (especially in microscopy) is an object that alters the amplitude, but not the phase of the transmitted wave with respect to the incident wave; e.g. an absorbing transmission grating, or any thin object with real-valued transmission function. See also *phase object.*

Amplitude-squeezed state see *squeezed states.*

AMS see *active matrix array, image sensor.*

Anamorphic lens is a lens or a lens system that has been designed for applications that involve stronger focusing in one direction than in a perpendicular direction, its extreme case being a cylindrical lens. The magnification is less in one particular lateral lens direction and more in an orthogonal direction, that is the lens squeezes the images in a particular direction. Initially the anamorphic lens was used for horizontally "squeezing" a wide-view image on a film of fixed size (e.g. a 35 mm film); a similar anamorphic lens in the projector expanded the image and restored the original aspect ratio. The anamorphic lens has now found applications in optoelectronics for shaping laser beam profiles emitted by laser diodes that are inherently noncircular. A simple combination of a spherical and a cylindrical lens, or two orthogonally oriented cylindrical lenses, can act as an anamorphic lens system.

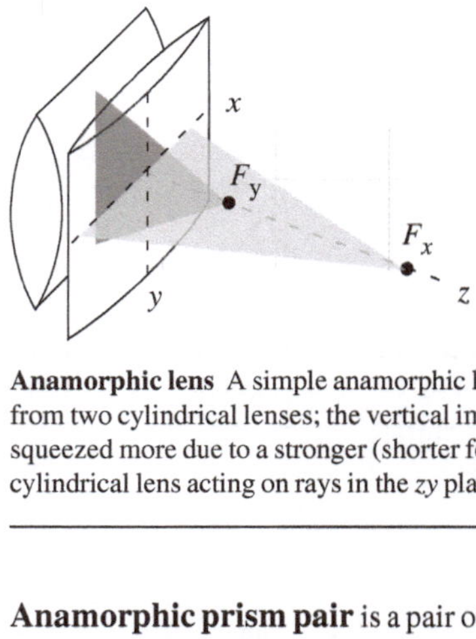

Anamorphic lens A simple anamorphic lens system from two cylindrical lenses; the vertical image (y) is squeezed more due to a stronger (shorter focal length) cylindrical lens acting on rays in the zy plane.

Anamorphic prism pair is a pair of two identical wedge-type prisms that are aligned and tilted with respect to each other in such a way that an incident beam experiences expansion only in one direction, that is, different magnifications in two orthogonal directions; the pair therefore allows a passing

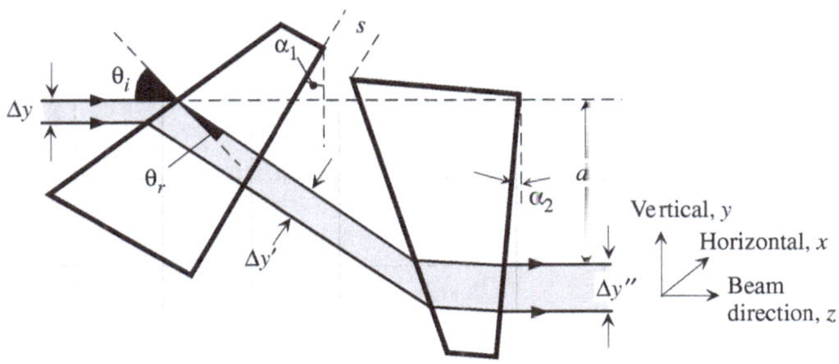

Anamorphic prism pair expands a beam from Δy to $\Delta y''$.

beam's cross-section to be shaped, for example, changed from an elliptical cross-section into circular. The emerging beam from the prism pair is parallel to the incident beam, but displaced by d, which depends on the relative orientations of the prisms and their separation; in practice, typically, d is 5–8 mm. For the prism pair shown, the beam is magnified by $\Delta y''/\Delta y$ vertically (parallel to y) but not horizontally (along x). The angles α_1 and α_2 of the flat faces with the "vertical" determine the anamorphic magnification, i.e. the *magnification aspect ratio* MAR; by changing α_1 and α_2, the anamorphic ratio can be changed from 2:1 to 6:1 quite easily, which allows a highly asymmetrical elliptical beam from a laser diode to be shaped into a circular beam cross-section. For example, for a particular pair of prisms (a Melles Griot anamorphic pair) in which each wedge prism has an apex angle of 29°26′ and, $\alpha_1 = 30.4°$, $\alpha_2 = 0.1°$, then MAR = 3:1, and $d =$ 6.4 mm for light at 830 nm. The anamorphic magnification for each prism, assuming near normal emergence from the flat face, is

$$\begin{aligned}\Delta y''/\Delta y &= \cos\theta_r / \cos\theta_i \\ &= \cos[\arcsin(n \sin\theta_i)]/\cos\theta_i\end{aligned}$$

where Δy and $\Delta y'$ are the incident and deflected beam heights (in the vertical y direction), θ_i and θ_r are the angles of incidence and refraction, and n is the prism material refractive index (normally Schott glass, e.g. SF11). The prism separation s does not affect the MAR but changes the displacement d. Anamorphic prisms are available unmounted or as a mounted pair ready for use.

Angular magnification see *magnifying power, magnification*.

Angstrom (Å) is 10^{-10} meters.

Anisotropy (optical) refers to the fact that the refractive index n of a crystal depends on the direction of the electric field in the propagating light beam. The velocity of light in a crystal depends on the direction of propagation and on the state of its polarization, i.e. the direction of the electric field Most noncrystalline materials, such as glasses and liquids, and all cubic crystals are *optically isotropic*, that is the refractive index is the same in all directions. For all classes of crystals excluding cubic structures, the refractive index depends on the propagation direction and the state of polarization. The result of optical anisotropy is that, except along certain special directions, any unpolarized light ray entering such a crystal breaks into two different rays with different polarizations and phase velocities. When we view an image through a calcite crystal, an optically anisotropic crystal, we see two images, each constituted by light of different polarization passing through the crystal, whereas there is only one image through an optically isotropic crystal. Optically anisotropic crystals are called *birefringent* because an incident light

Table: Anisotropy (optical) Classification of optical anisotropy in materials, n_o and n_e represent the ordinary and extraordinary wave refractive indices.

Classification	Crystals	Susceptibility	Refractive indices	Examples
Isotropic media	Cubic crystals, liquids, glasses	$\chi_{11} = \chi_{22} = \chi_{33}$	$n_1 = n_2 = n_3 = n_o$	Diamond, glass, fluorite
Uniaxial crystals	tetragonal, hexagonal, trigonal, hexagonal, rhombohedral	$\chi_{11} = \chi_{22} \neq \chi_{33}$ $1 + \chi_{11} = n_o^2$ $1 + \chi_{22} = n_o^2$ $1 + \chi_{33} = n_e^2$	$n_1 = n_2 = n_o$ $n_3 = n_e$ Positive uniaxial $n_e > n_o$ Negative uniaxial $n_e < n_o$	Ice, quartz, calcite, sapphire (Al_2O_3), Lithium niobate ($LiNBO_3$), zircon ($ZrSiO_4$), rutile (TiO_2), sodium nitrate ($NaNO_3$)
Biaxial crystals	orthorhombic, monoclinic, triclinic	$\chi_{11} \neq \chi_{22} \neq \chi_{33}$ $1 + \chi_{11} = n_1^2$ etc.	n_1, n_2, n_3	Mica (muscovite)

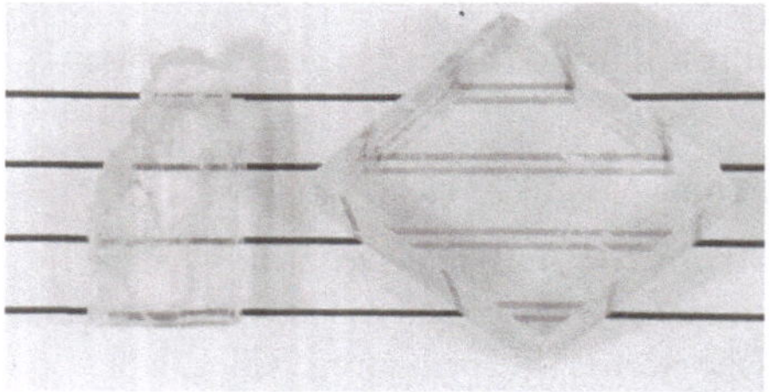

Anisotropy A line viewed through a cubic sodium chloride (halite) crystal (optically isotropic) and a calcite crystal (optically anisotropic).

beam may be doubly refracted. Experiments and theories on "most anisotropic crystals," i.e. those with the highest degree of anisotropy, show that we can describe light propagation in terms of *three* refractive indices, called *principal refractive indices* n_1, n_2 and n_3, along three mutually orthogonal directions in the crystal, say x, y and z, called the *principal axes*. These indices correspond to the polarization state of the wave along these axes. Crystals that have three distinct principal indices also have *two* optic axes and are called *biaxial crystals*. On the other hand, *uniaxial crystals* have two of their principal indices the same ($n_1 = n_2$) and only have one optic axis. Uniaxial crystals, such as quartz, that have $n_3 > n_1$ and are called *positive*, and those such as calcite that have $n_3 < n_1$, are called *negative* uniaxial crystals.

Optical anisotropy in a crystal is related to the polarization **P** in a medium not being in the same direction as the electric field **E** causing it. Suppose that the optical field **E** has components E_x, E_y and E_z along the three principal crystalline axes x, y and z. The induced polarization **P**, with components P_x, P_y and P_z along x, y and z, is then given by a tensor relationship,

$$\boldsymbol{P} = \varepsilon_0 \boldsymbol{\chi} \boldsymbol{E} \quad \text{or} \quad \begin{pmatrix} P_x \\ P_y \\ P_z \end{pmatrix} = \varepsilon_0 \begin{pmatrix} \chi_{11} & 0 & 0 \\ 0 & \chi_{22} & 0 \\ 0 & 0 & \chi_{33} \end{pmatrix} \begin{pmatrix} E_x \\ E_y \\ E_z \end{pmatrix}.$$

In tensor form, the relative permittivity is $\boldsymbol{\varepsilon}_\mathbf{r} = \mathbf{1} + \boldsymbol{\chi}$, and since $\varepsilon_r = n^2$,

$$\boldsymbol{\varepsilon}_\mathbf{r} = \begin{pmatrix} 1+\chi_{11} & 0 & 0 \\ 0 & 1+\chi_{22} & 0 \\ 0 & 0 & 1+\chi_{33} \end{pmatrix} = \begin{pmatrix} n_1^2 & 0 & 0 \\ 0 & n_2^2 & 0 \\ 0 & 0 & n_3^2 \end{pmatrix}.$$

There are three categories of optical media: isotropic media, uniaxial and biaxial crystals as summarized in the table. The *optic axis* is taken along the z-axis.

Annealing point of glass is the temperature at which the viscosity of a glass is about 10^{12} Pa s (or 10^{13} dPa s; dPa is decipascals). For example, for fused silica glass this is about 1180 °C. The viscosity of glass is strongly temperature dependent and decreases sharply as the temperature increases. At the annealing point, the viscosity is sufficient to relieve internal stresses.

Anomalous dispersion represents a dispersion behavior, refractive index, n vs. wavelength, λ, in which the refractive index decreases with the frequency of light or n increases with increasing λ. See *refractive index*, *chromatic dispersion*.

Antireflection (AR) coating is a thin dielectric layer coated on an optical device or component to reduce the reflection of light and increase the transmitted light intensity.

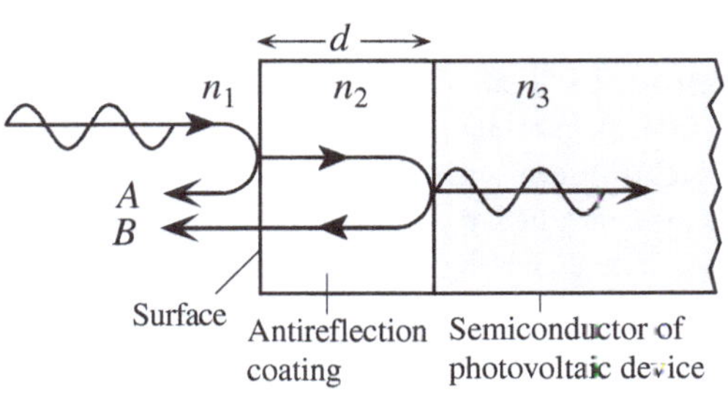

Antireflection coating Illustration of how an antireflection coating reduces the reflected light intensity.

Consider a thin layer of a dielectric material such as Si_3N_4 (silicon nitride) on the surface of a semiconductor optoelectronic device such as a solar cell as shown in the first figure. If this antireflection coating has an intermediate refractive index then the thin dielectric coating can reduce the reflected light intensity. In this case $n_1(\text{air}) = 1$, $n_2(\text{coating}) \approx 1.9$ and $n_3(\text{Si}) = 3.5$. Light is first incident on the air/coating surface and some of it becomes reflected. Suppose that this reflected wave is A. Wave A has experienced a 180° phase change on reflection as this is an *external reflection*. The wave that enters and travels in the coating then becomes reflected at the coating/semiconductor surface. This wave, say B, also suffers a 180° phase change since $n_3 > n_2$. When wave B reaches A, it has suffered a total optical path delay of traversing the thickness d of the coating twice. The phase difference is equivalent to $k_c(2d)$ where $k_c = 2\pi/\lambda_c$ is the wavevector in the coating and is given by $2\pi/\lambda_c$ where λ_c is the wavelength in the coating. Since $\lambda_c = \lambda/n_2$, where λ is the free-space wavelength, the phase difference $\Delta\phi$ between A and B is $(2\pi n_2/\lambda)(2d)$. To reduce the reflected light, A and B must interfere destructively and this requires the phase difference to be π or odd multiples of π, $m\pi$ where $m = 1,3,5,\ldots$ is an odd integer. Thus

$$\left(\frac{2\pi n_2}{\lambda}\right)2d = m\pi \quad \text{or} \quad d = m\left(\frac{\lambda}{4n_2}\right).$$

Thus, the thickness of the coating must be multiples of the *quarter wavelength* in the coating and depends on the wavelength. To obtain a good degree of destructive interference between waves A and B, the two amplitudes must be comparable. It turns out that we need $n_2 = \sqrt{n_1 n_3}$. When $n_2 = \sqrt{n_1 n_3}$ then the reflection coefficient between the air and coating is equal to that between the coating and the semiconductor. In this case we would need $\sqrt{3.5}$ or 1.87. Thus, Si_3N_4 is a good choice as an antireflection coating material on Si solar cells. Generally an AR coating operates at one or over a narrow range of wavelengths. Further, its effectiveness, the reflectance, depends on the angle of incidence

The choice of materials is not always the best for a single layer AR coating. *Double layer AR coatings* can achieve negligible small reflections at a specified wavelength. To reduce the reflection of light at the n_1/n_4 surface, two layers n_2 and n_3, each a quarter wavelength (λ/n_2 and λ/n_3) are interfaced between n_1 and n_4 as shown in the second figure.

Table: Antireflection (AR) coating Typical AR materials and their typical (approximate) refractive indices.

	MgF_2	SiO_2	Al_2O_3	CeF_3	Sb_2O_4	Si_3N_4	SiO	ZnS	TiO_2	CdS
n	1.38	1.46	1.65	1.65	1.9–2.1	1.95	2.0	2.35	2.35	2.60

The reflections A, B and C for normal incidence result in a reflectance R given by

$$R = \left[\frac{n_3^2 n_1 - n_4 n_2^2}{n_3^2 n_1 + n_4 n_2^2}\right]^2$$

and R vanishes when

$$(n_2/n_3)^2 = n_1/n_4.$$

Double layer reflectance vs. wavelength behavior usually has a V-shape, and they are called *V-coatings.* Double layer SiN and SiO_2 AR coatings are of particular interest for Si solar cells. Better AR coatings use multiple layers, or coatings that have many layers, each a quarter wave thick, whose refractive index changes from layer to layer forming an *effective graded index* AR coating.

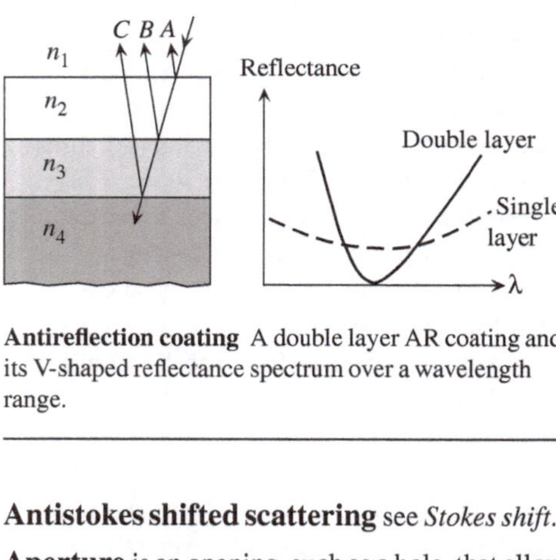

Antireflection coating A double layer AR coating and its V-shaped reflectance spectrum over a wavelength range.

Antistokes shifted scattering see *Stokes shift.*

Aperture is an opening, such as a hole, that allows light from an object to enter the optical or optoelectronic system, usually for detection or imaging.

Aperture function see *Fourier analysis of diffraction.*

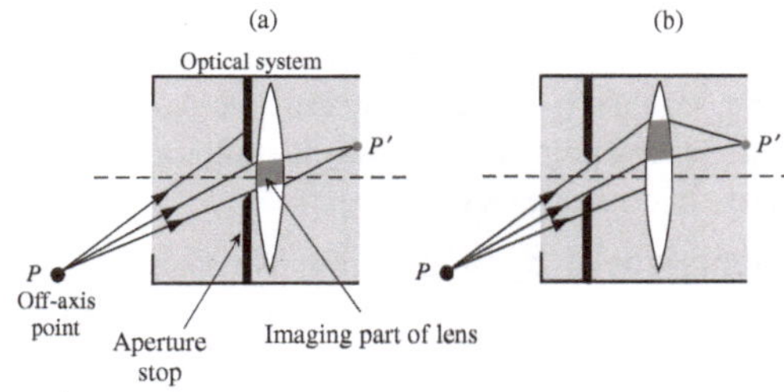

Aperture stop (a) Stop next to lens. Only the central part of the lens is used for imaging an off-axis point P to P'. (b) Stop at a distance from the lens. The outer lens region is used for imaging.

Aperture stop is the major light limiting physical constraint, such as a well-defined hole in an otherwise opaque shield (or the lens holder or retainer itself) that is usually placed on the optical axis and in front of, or behind, a lens in an optical imaging system to prevent marginal rays passing through, or to allow only light rays near the center of the lens to pass through. It is the smallest opening in a lens system, an optical system, through which light has to pass. An aperture stop whose size is adjustable is an *iris diaphragm.* Use of such aperture stops improves the optical performance, though at the expense of image illuminance (brightness); for example, spherical aberration and coma can be reduced by restricting marginal rays. Spherical aberration varies as the cube of the aperture diameter and hence there can be a significant improvement in spherical aberration by using a stop. The placement of the stop, whether right next to the lens or some distance away from the lens, is also important in optical design; the stop-to-lens separation d can affect the magnification of an off-axis object point because different parts of the lens are used for imaging, depending on d, as shown

Table: Aplanatic lens A particular aplanatic lens design that brings a collimated beam to a diffraction-limited focal spot at a back focal length of about 2 mm from the lens. (From M. Mansuripur, *Classical Optics and Its Applications*, Cambridge University Press, 2002). Lens diameter is 4 mm. D is the lens diameter. See figue for other quantities. (NA is numerical aperture. Distances in mm.)

D	t_c	f	R_1	R_2	$h = VH$	n	NA
4	1	2.6733	2.26875	3.87493	0.2894	2.49486	0.75

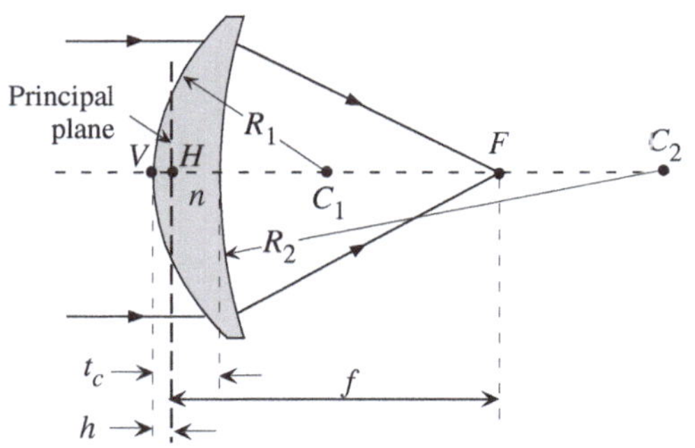

Aplanatic lens that has been designed for an object at far distance.

in the figure. Adjusting the aperture size, the iris diaphragm, in a camera changes the *depth of field*, range of object positions that can be imaged with tolerable (or negligible) blur. A smaller aperture allows a greater depth of field but at the expense of image illuminance.

Aplanatic lens is a meniscus lens (normally a positive or a converging meniscus) that has been shaped to have minimal spherical aberration and coma. Aplanatic lenses are usually aberration optimized for particular applications, such as an object at infinity, etc. since the required radii R_1 and R_2 of curvature of the two lens surfaces for aberrationless imaging depends on the object and image locations. A properly designed aplanatic lens, for example, can bring a collimated beam (parallel rays) to a diffraction-limited spot. The table lists the design characteristics of one such apalanatic lens. The aplanatic lens was invented by the Scottish astronomer Robert Blair (1748–1828). (L. C. Martin, *Proc. Phys. Soc.*, **56**, 104, 1944.)

See *aberrations, Coddington position and shape factors, coma, spherical aberration.*

Aplanatic optical system has been designed to form an image that is free of spherical (on-axis) and coma (off-axis) aberrations. The design of an aplanatic system follows Abbe's sine condition. See *Abbe's sine condition, spherical aberration, coma.*

Aplanatic points are conjugate points P and P' (i.e. object point P and its corresponding image point P') that are on the axis of an optical system, such as a converging lens, that satisfy the *Abbe sine condition*; for a lens, it means that all rays emerging from P that can be captured (and used in imaging) by the optical system converge on P'. The points P and P' are interchangeable in an optical system. The two focal points of an ellipsoidal reflecting surface form aplanatic points, P and P', because all rays from a point source placed at P will be reflected to pass through P'.

Apochromatic lens (apochromat) uses three lenses to eliminate chromatic aberration at three colors or wavelengths; three different colors are focused to the same point by the use of lenses made from three different glasses with different refractive indices. See *chromatic aberration.*

Apodization is a filtering technique mostly used under incoherent light that consists in tailoring the pupil transmittance in such a way that the secondary maxima (the "feet") of the optical transfer function are highly reduced. For instance, one can replace a uniform pupil by a Gaussian-modulated transmittance. This is often done in astronomy in order to enhance the weakest components of an image. The drawback is a slight loss in resolution.

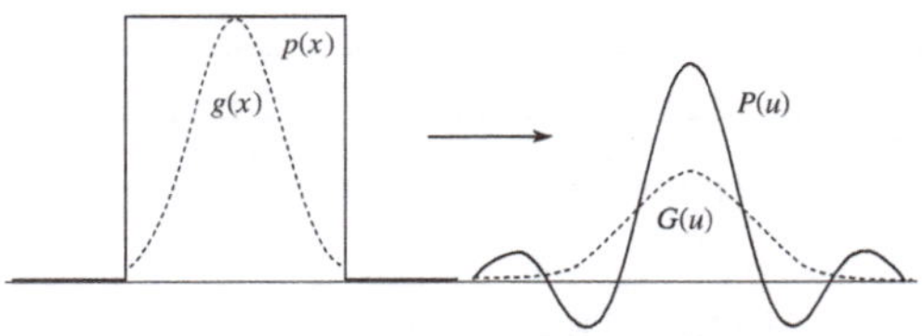

Apodization Typical example of apodization by a one-dimensional Gaussian transmittance $g(x)$, as compared to a uniform pupil $p(x)$. The Gaussian transfer function $G(u)$ appears more regular, albeit slightly larger that the sine-cardinal $P(u)$. $G(u)$ and $P(u)$ are the Fourier transforms of $g(x)$ and $p(x)$.

Arc lamp is a type of gas discharge lamp in which an electric discharge through an arc, an ionized plasma, is maintained between two electrodes. The type of gas or vapor in which the discharge is achieved characterizes the arc. For example, if the gas is xenon, it would be a xenon arc lamp; if it is mercury vapor, it is a mercury vapor arc lamp. (Apparently the first arc was demonstrated by Sir Humphry Davy in 1802 at the Royal Institution in the UK.) The electrodes should be able to withstand high temperatures and wear; typically tungsten (or impregnated tungsten) electrodes are used. Some common types of gases that are used in arc lamps are argon, xenon, neon, krypton, sodium, metal halide, and mercury. The gas between the electrodes needs to be ionized to achieve the arc.

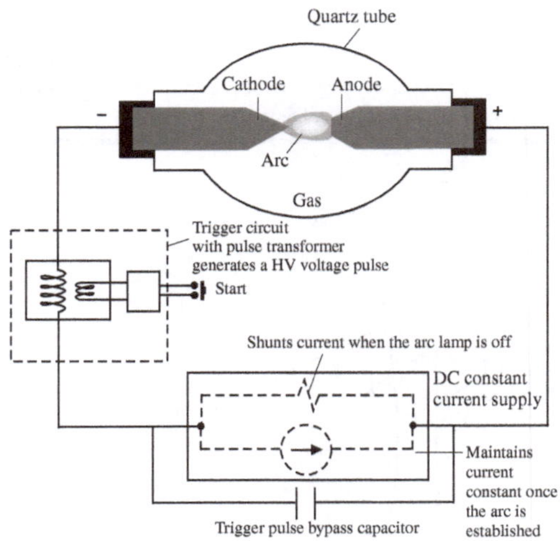

Arc lamp A highly simplified basic schematic sketch of an arc lamp. The cathode has a sharp point to create a large field, and facilitate the discharge. (Various protection electronics not shown.)

The arc is "ignited" or triggered by applying a very high-voltage pulse (several kilovolts) to start the ionization, and once the plasma is initiated and the arc is established, the actual voltage between the electrodes is substantially less than the initial trigger voltage; a circuit with a starter and a ballast controls the initial strike from a high voltage and the subsequent arc current. In steady state operation, the external circuit usually controls the current. The cathode has a sharp point to create a large field, and facilitate the discharge. The advantage of an arc

Table: Arc Lamp Some characteristics of selected arc lamps. Lm/W is the luminous efficacy and quoted values are typical; T_{color} is the color temperature of the source. The spectral ranges are typical values, which also depend on the tube, window and lens materials transmittance.

Arc lamp	Spectral range (nm)	Comment
Xe	200–2000	Bright white source, 15–45 lm/W. Broad spectrum. Light appears white with a $T_{color} \approx$ 6000–6500 K. High-intensity white light applications such as theatre and cinema projectors, stage lighting, etc.
Hg-Xe	200–2000	Similar to the Xe lamp but high UV emission
Hg	200–2000	Good efficacy, 20–50 lm/W. $T_{color} \approx$ 4000–6000 K. Roadways, landscapes, residential and commercial security, parking lots, industrial places, etc.

lamp is that it can provide a broad range of wavelengths from UV to infrared wavelengths; for example, xenon and mercury arc lamps can cover the 180–2000 nm range. The spectrum will also have numerous intense spectral peaks due to the transitions of the excited gas ions. Short arc lamps have their electrodes separated by a very short distance, a few millimeters, and are enclosed in a quartz or fused silica spherical or oblong bulb. They are very convenient in generating high-intensity light. Arc lamps are used in numerous applications as listed in the table: projectors, search lights, floodlights, street and parking lot lighting, display lighting, industrial and commercial places, security areas, UV curing of polymers, and in various scientific and medical instruments and scientific applications such as photochemistry. See *glow discharge lamp, xenon lamp*.

Argon ion laser can provide powerful CW visible coherent radiation of several watts. The laser operation is achieved as follows. The Ar atoms are ionized by electron collisions in a high current electrical discharge. Further multiple collisions with electrons excite the argon ion, Ar^+, to a group of 4*p* energy levels ~35 eV above the atomic ground state. Thus a *population inversion* forms between the 4*p* levels and the 4*s* level which is about 33.5 eV above the Ar atom ground level. Consequently, the stimulated radiation from the 4*p* levels down to the 4*s* level contains a series of wavelengths ranging from 351.1 nm to 528.7 nm. Most of the power, however, is concentrated, approximately equally, in the 488 and 514.5 nm emissions. The Ar^+ ion at the lower laser level (4*s*) returns to its neutral atomic ground state via a radiative decay to the Ar^+ ion ground state, followed by recombination with an electron to form the neutral atom. The Ar atom is then ready for "pumping" again. The Doppler broadened linewidth of the 514.5 nm radiation is about 3500 MHz ($\Delta\upsilon$) and is between the half-intensity points. Typically in the argon-ion laser the discharge tube is made of beryllia (beryllium oxide).

Typical argon ion laser parameters are (From W. T. Silfvast, *Laser Fundamentals, Second Edition*, Cambridge University Press, 2004): laser wavelengths most often used = 488.0 nm; laser transition probability = 7.8×10^7 s^{-1}; upper laser level lifetime = 1.00×10^{-8} s; stimulated emission cross-section = 2.6×10^{-16} m^2; spontaneous emission linewidth and gain bandwidth, FWHM = 2.7×10^9 Hz; population inversion density = 2×10^{15} m^{-3}; small-signal gain coefficient = 0.5 m^{-1}; laser gain-medium length (L) = 0.1–1.0 m; single-pass gain = 1.05–1.65; gas pressure < 0.1 torr; gas temperature, 1200°C; mode of operation is CW; output power = 100 mW to 50 W, TEM_{00} or multimode.

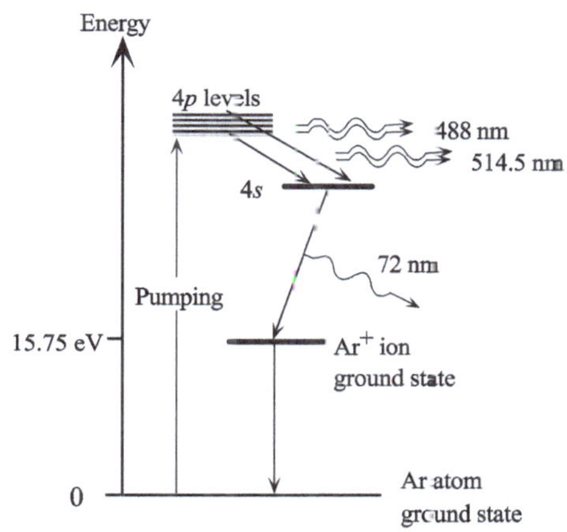

The Ar-ion laser energy diagram

Argon ion laser energy diagram.

Armoring is the protective "cover" made of metal wires or bands that is put around a cable sheath; it provides protection against harsh external environments.

Arrhenius temperature dependence implies that the rate of change in a physical or chemical process, or the particular property under observation, has an exponential temperature dependence of the form

$$\text{Rate} = C \exp\left(-\frac{\Delta H}{RT}\right) \quad \text{or}$$

$$\text{Rate} = C \exp\left(-\frac{E_A}{k_B T}\right)$$

where C is a constant, assumed to be relatively temperature independent, R is the gas constant, k_B is the Boltzmann constant, T is the temperature (in kelvins) and ΔH (or E_A) is the activation energy associated with the process, ΔH in J mol^{-1} and E_A in eV atom^{-1} if k_B is expressed in eV K^{-1} atom^{-1}. One good example of the application of the Arrhenius rate equation is the description of the *rate of failure* (R) of electronic devices, which typically follows the above Arrhenius rate equation, at least over a limited temperature range. Accelerated failure tests are carried out at elevated temperatures (e.g. measuring the breakdown voltage of a device) and then the results are extrapolated to the rate of failure at room temperature.

Aspheric lens is a lens that normally has one surface almost flat (or slightly spherical) and the other surface shaped with the right curvature to minimize spherical aberration so that the whole lens area can be utilized for optical applications without suffering significant spherical aberration; "aspherical" implies that the curved surface is not spherical. They are convenient replacements for multielement spherical lens systems that have been designed to be free of spherical aberration. The required curvature is obtained appropriately by molding the lens. In optics, they are frequently used in low *f*-number, high throughput light applications such as condenser lenses in projectors, etc. High-quality glass aspheric lenses can be used to replace multielement microscope objective lenses. Inexpensive molded optical grade plastic aspheric lenses are used in optoelectronics for obtaining collimated light from LEDs. Various aspheric lenses are available for bringing a collimated beam to a nearly diffraction limited focal point. Aspheric lenses in photonics can be used for coupling light into a fiber. Achromatic aspheric lenses correct for both spherical and chromatic aberrations. See *planoconvex lens*.

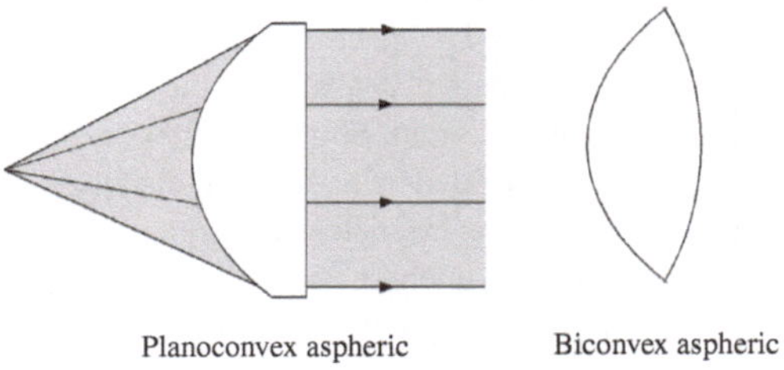

Aspheric lenses (Courtesy of Melles Griot)

Aspheric lens Planoconvex and biconvex aspherical lenses.

Astigmatism is an off-axis lens aberration in which an off-axis object point P is imaged by the lens as a blurred and distorted image due to differently inclined rays from the object not being refracted correctly, as in an ideal lens, to converge to a point (neglecting any diffraction effects). The imaging system is divided into two planes. *The tangential or meridional plane* is the plane that has the optical axis, the principal ray and the object point (P). All rays in this plane are *tangential or meridional rays*. Rays that do not lie in this plane are *skew rays*. The *sagittal plane* is a plane at right angles to the tangential plane; it has the off-axis object point, and the principal ray of the meridional plane. Astigmatism is due to the tangential rays and sagittal rays originating from the same off-axis point object P not producing the same image but rather producing a

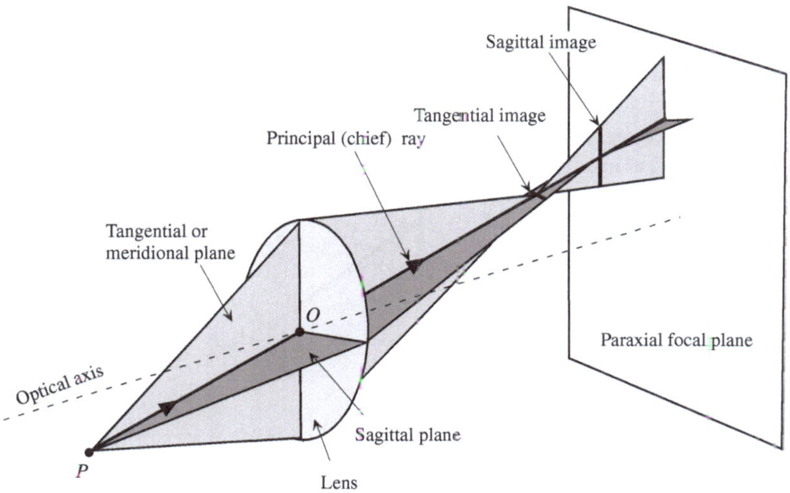

Astigmatism Illustration of astigmatic image from an off-axis point object.

tangential image (a "line" in the sagittal plane) and a sagittal image (a "line" in the tangential plane).

Asynchronous signal is not synchronized to any one time reference signal. Asynchronous transmission does not use a common clock signal between the transmitter and receiver.

Atomic force microscopy see *nanoelectronics*.

Atomic trap see *trapping of atoms*.

Attenuation is the decrease in the optical power of a traveling wave, or a guided wave in a dielectric waveguide, in the direction of propagation due to absorption and scattering. If P_o is the optical power at some location O, and if it is P at a distance L from O along the direction of propagation, then the attenuation coefficient is defined by $\alpha = 10L^{-1}\ln(P/P_o)$. In terms of attenuation measured in decibels

$$\alpha_{dB} = \frac{10}{\ln(10)}\alpha = 4.34\alpha.$$

Attenuation in optical fibers represents the loss in the optical power of a light wave as it propagates along the fiber. It is the fractional decrease in the optical power of a propagating wave per unit length of fiber as it is guided along the fiber. Suppose that the input optical power into a fiber of length L is P_{in} and the output optical power at the end is P_{out}. Then the attenuation coefficient α is

$$\alpha = -\frac{1}{L}\ln\left(\frac{P_{\text{out}}}{P_{\text{in}}}\right).$$

If we know α then we can find P_{out} from P_{in} through, $P_{\text{out}} = P_{\text{in}}\exp(-\alpha L)$. Generally optical power attenuation in a fiber is expressed in terms of decibels per unit length of fiber typically as dB per km. The attenuation of the signal in decibels per unit length is defined in terms of the logarithm to base 10 by

$$\alpha_{\text{dB}} = -\frac{1}{L}\log\left(\frac{P_{\text{out}}}{P_{\text{in}}}\right) \qquad \text{or}$$

$$\alpha_{\text{dB}} = \frac{10}{\ln(10)}\alpha = 4.34\alpha.$$

The first figure shows the attenuation in a typical silica glass-based optical fiber. The sharp increase in the attenuation at wavelengths beyond 1.6 μm in the *infrared* region is due to energy absorption by "lattice vibrations" of the constituent ions of the glass material. There is another intrinsic material absorption in the region below 500 nm, not shown in the figure, which is due to photons exciting electrons from the valence band to the conduction band of the glass. There is a marked attenuation peak centered at 1.4 μm, and a barely discernible minor peak at about 1.24 μm. These attenuation regions arise from the presence of hydroxyl ions as impurities in the glass structure inasmuch as it is difficult to remove all traces of hydroxyl (water) products during fiber production. The overtone at 1.24 μm is negligible in most modern fibers. There are two important windows in the α vs. λ behavior where the attenuation exhibits minima. The window at around 1.3 μm is the region between two neighboring OH^- absorption peaks. This window is widely used in optical communications at 1310 nm. The window at around 1.55 μm is between the first harmonic absorption of OH^- and the infrared lattice absorption tail and represents the lowest attenuation. This window is used for long-haul communications. It is important to keep the hydroxyl content in the fiber within tolerable levels. There is a background attenuation process that decreases with wavelength and is due to the Rayleigh scattering of light by the local variations in the refractive index. Random fluctuations in the refractive index give rise to light scattering and hence light attenuation along the fiber. Since a degree of structural randomness is an intrinsic property of the glass structure, this scattering process is unavoidable and represents the lowest attenuation possible through a glass medium. Rayleigh scattering process is inversely proportional to λ^4. By proper design, the attenuation window at 1.5 μm may be lowered to approach the Rayleigh scattering limit. The OH^- peak at 1.4 μm has prevented the use of the E-band in optical communications. However, recent developments in fiber manufacturing have almost totally eliminated the water peak as apparent in the α vs. λ characteristic of a newly developed low-water peak fiber in the second figure. There are also *bending losses* that arise whenever a fiber is bent. *Microbending loss* occurs typically whenever the radius of curvature of the bend is microscopically sharp, typically less than 0.1–1 mm. Bending losses are normally ignored when the bend radius is more than 100 mm or so. See *Rayleigh scattering, macrobending loss, microbending loss, reststrahlen absorption.*

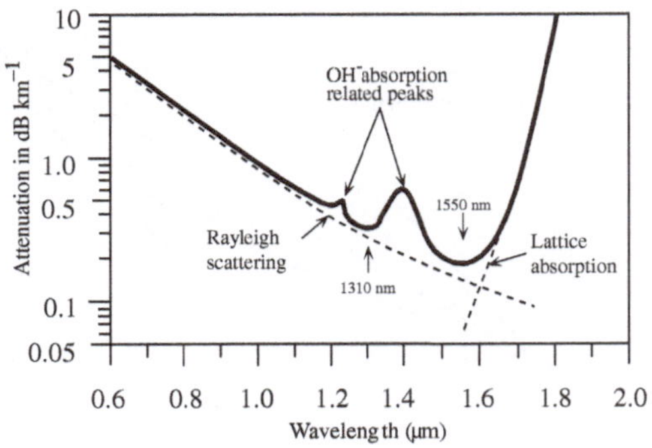

Attenuation in optical fibers Typical attenuation vs. wavelength characteristics of a silica-based optical fiber. There are two communications channels at 1310 nm and 1550 nm.

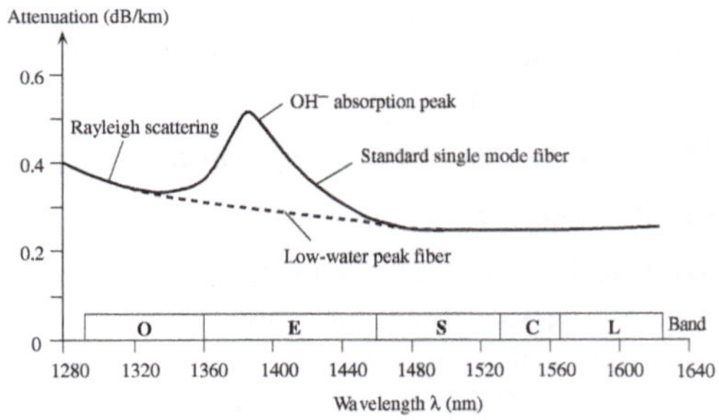

Attenuation in optical fibers Attenuation in a standard single-mode optical fiber and also in a newly developed low water peak fiber. (Data extracted from A. Lindstrom, *Photonics Spectra*, April 2002, p. 68.)

Attenuation measurement see *cut-back method, insertion loss technique.*

Auger recombination refers to a nonradiative recombination mechanism involving three carriers. Examples include the recombination of two electrons with one hole, referred to as an ehe process, or two holes with one electron, referred to as ehh process. In an Auger recombination event, the recombination energy from the recombination of an electron and a hole is transferred to the remaining carrier – thus, in the case of an ehe process, the recombination energy is transferred to another electron. The total recombination rate R_{Auger} for ehe and ehh processes can be written:

$$R_{\text{Auger}} = C_{\text{ehe}} n^2 p + C_{\text{ehh}} n p^2$$

where C_{ehe} and C_{ehh} are constants, and n and p refer to the electron and hole concentrations. Clearly, because R_{Auger} depends on the square of carrier density in a given energy band, it only becomes significant at high carrier concentrations – thus, for example, in the case of laser diodes requiring high carrier injection rates to sustain lasing this can be a very significant process. In addition, since the Auger process involves energy transfer to carriers occupying high-lying energy states in the band structure, this process tends to be very sensitive to the details of the semiconductor band structure. In the case of semiconductor heterostructures such as quantum wells, quantum wires and quantum dots, the heterostructure boundaries influence the energy and wavefunctions of carriers, actually increasing the probability of nonradiative Auger recombination. The Auger recombination lifetime τ_{A} can be found from the Auger lifetime τ_{Ai} in an intrinsic semiconductor,

$$\tau_{\text{A}} = \tau_{\text{Ai}} \left(\frac{n_{\text{i}}}{N_{\text{d}}} \right)$$

where n_{i} is the intrinsic concentration and N_{d} is the dopant concentration.

Autocorrelation function, $R(\tau)$, is a function that characterizes the correlation of two points on the same function $x(t)$ that are separated by a time interval τ by integrating the product $x(t)$ and $x(t-\tau)$ over all times

$$R(\tau) = \frac{1}{T} \int_0^T x(t)x(t-\tau)dt \quad \text{as } T \to \infty.$$

$R(\tau)$ is an even function and $R(0)$ represents the average power in the signal if $x(t)$ is a voltage or current. Autocorrelation functions are useful in signal processing, defining coherence in optics, and they can be used to conveniently obtain the spectral power densities of various quantities through the *Wiener–Khintchine theorem*. The autocorrelation function $R(t)$ and the spectral power density $S(f)$ of $x(t)$ form a Fourier transform pair. See *coherence, noise*.

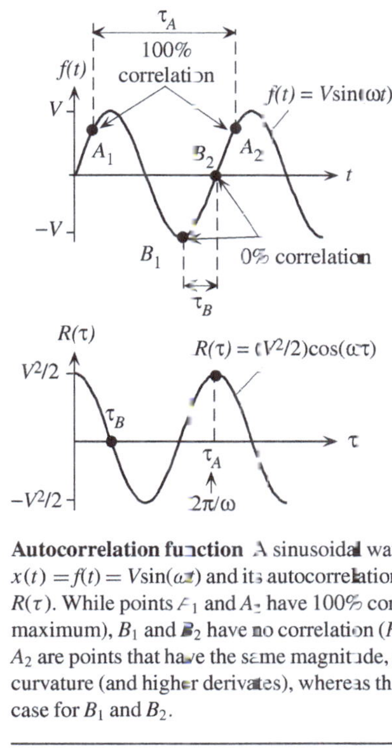

Autocorrelation function A sinusoidal waveform $x(t) = f(t) = V\sin(\omega t)$ and its autocorrelation function $R(\tau)$. While points A_1 and A_2 have 100% correlation (R is maximum), B_1 and B_2 have no correlation ($R = 0$). A_1 and A_2 are points that have the same magnitude, slope and curvature (and higher derivates), whereas this is not the case for B_1 and B_2.

Avalanche breakdown is the enormous increase in the reverse current in a *pn* junction when the applied reverse field is sufficiently high to cause the generation of electron–hole pairs by impact ionization in the space charge layer. See *impact ionization*.

Avalanche multiplication see *impact ionization*.

Avalanche multiplication factor, M, of an avalanche photodiode (APD) is defined as

$$M = \frac{\text{Multiplied photocurrent}}{\text{Primary unmultiplied photocurrent}} = \frac{I_{\text{ph}}}{I_{\text{pho}}}$$

where I_{ph} is the APD photocurrent that has been multiplied and I_{pho} is the primary or unmultiplied photocurrent.

Avalanche noise is the excess noise in the photocurrent of an avalanche photodiode due to the statistics of the multiplication process occurring in the avalanche region. The avalanche process does not occur continuously and smoothly in time but as discrete events whose frequency of occurrence fluctuates in the avalanche region about some mean value. Thus the multiplication M fluctuates about a mean value. The result of the statistics of impact ionization is an excess noise contribution, called *avalanche noise*, to the multiplied shot noise. Suppose that I_{do} is the dark current and I_{pho} is the photocurrent, both in the absence of multiplication, M is the multiplication factor and B is the bandwidth of the detector. Then the noise current in an avalanche photodiode (APD) is given by

$$i_{n-\text{APD}} = [2e(I_{\text{do}} + I_{\text{pho}})M^2FB]^{1/2}$$

where F is called the *excess noise factor* and is a function of the avalanche multiplication factor M and the impact ionization probabilities (called coefficients). Generally, F is approximated by the relationship $F \approx M^x$ where x is an index that depends on the semiconductor, the APD structure and the type of carrier that initiates the avalanche (electron or hole). For Si APDs, x is 0.3–0.5 whereas for Ge and III–V (such as InGaAs) alloys it is 0.7–1. See *avalanche photodiode, impact ionization, noise, excess noise*.

Avalanche photodiode (APD) is a photodiode with a depletion region in which the field is sufficiently large for an avalanche multiplication of photogenerated charge carriers by impact ionization. The avalanche process in the APD occurs over a limited region of the depletion layer and the photodiode design typically allows only the multiplication of one type of carrier, for example electrons for Si.

Axial or longitudinal modes are allowed electromagnetic stationary waves that exist along the optical cavity length whose electric field patterns are determined by the length L of the optical cavity, e.g.

$$m\left(\frac{\lambda}{2n}\right) = L$$

where λ is the free-space wavelength, n is the refractive index of the cavity medium and m is an integer, 1, 2, ... Longitudinal modes of an optical cavity are normally associated with transverse modes of the cavity. Each transverse mode has many longitudinal modes.

Axicons are conical lenses which image a point source into a line focus. Some authors describe it as a rotationally symmetric prism. Importantly, such lenses can generate Bessel beams – nondiffracting beams in which the transverse intensity distribution is constant along the direction of propagation. Axicons have been used for precise alignment systems for large telescopes and for scanning systems such as those used in supermarket scanners. The latter systems take advantage of the axicon's large depth of field. An axicon with a spherical lens can focus a collimated laser beam into a ring; the intensity distribution is a ring pattern around the lens axis. Other interesting applications of axicons are in realizing optical traps. The term "axicon" is due to John McLeod (1954) (J. H. McLeod, *J. Opt. Soc. Am.*, **50**, 66, 1960.)

Azimuthal mode number (l) is one of two integers that are normally used to characterize propagating modes in a *step index fiber* (or more generally in a cylindrical waveguide). It characterizes the azimuthal field distribution. In a step index fiber, there are $2l$ maxima in the light intensity around a circle centered on the fiber axis.

B

Back-tinned CCD see *charge coupled device.*

Backscattered technique see *optical time domain reflectometry.*

Ball lens see *spherical lens.*

Ballast see *fluorescent tube lamp.*

Ballistic electrons refers to electrons with an elastic mean free path l_e that considerably exceeds the dimensions of a sample, or the dimensions of some constraining channel L in which these electrons move. That is, when $l_e \gg L$, the electrons will be able to move through such a material, or in such a channel without suffering any scattering, and are said to move ballistically.

Band offset or band discontinuities can occur at a heterojunction as a result of a difference in the *electron affinities* of each of the heterojunction materials. If the electron affinities of each material are χ_A and χ_B, respectively, then the conduction band offset, ΔE_c will be given by $e(\chi_A - \chi_B)$, where e is the electron charge (1.602×10^{-19} C). Moreover, if the energy bandgaps of the two materials are E_{gA} and E_{gB}, respectively, then the valence band offset ΔE_v will be given by

$$\Delta E_v = \{|E_{gA} - E_{gB}| - \Delta E_c\}.$$

Note that the sum of the band offsets in the conduction and valence band equals the difference in the bandgaps between the two materials of the heterojunction.

Band tailing, see *band-to-band absorption.*

Band-to-band absorption, or **fundamental absorption,** is the absorption of radiation due to the photoexcitation of an electron from the valence band to the conduction band. There are two types of band-to-band absorption, which correspond to *direct* and *indirect transitions*. A direct transition is a photoexcitation process in which no phonons are involved. The photon momentum is negligible compared with the electron momentum so that when the photon is absorbed to excite an electron from the valence band (VB) to the conduction band (CB), the electron's k-vector (which represents its momentum) does not change. A direct transition on the E–k diagram is a vertical transition from an initial energy E and wavevector k in the VB to a final energy E' and wavevector k' in the CB where $k' = k$, as illustrated in the figure. The energy $(E' - E_c)$ is the kinetic energy $(\hbar k)^2/(2m_e{}^*)$ of the electron with an effective mass $m_e{}^*$, and $(E_v - E)$ is the kinetic energy $(\hbar k)^2/(2m_h{}^*)$ of the hole left behind in the VB, where $m_h{}^*$ is the hole effective mass. The ratio of the kinetic energies of the photogenerated electron and hole depends inversely on the ratio of their effective masses. The absorption coefficient α is derived from the quantum mechanical transition probability from E to E', the occupied density of states at E in the VB from which electrons are excited, and the unoccupied density of states in the CB at $E + h\upsilon$. (The calculation involves

Band-to-band absorption

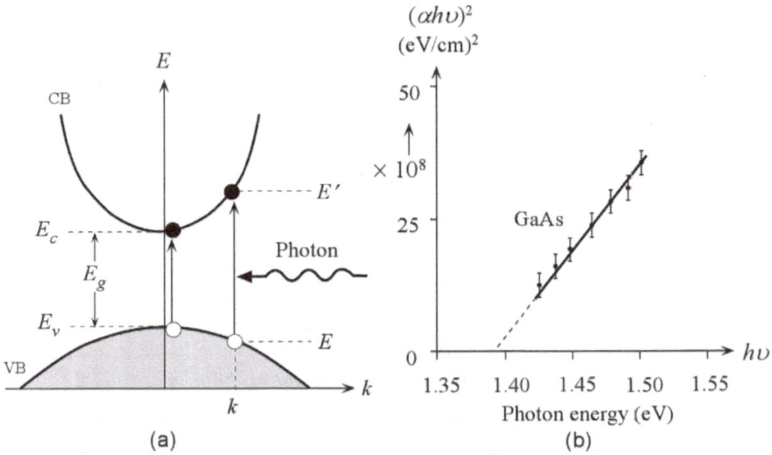

Band-to-band absorption Direct absorption: (a) A direct transition from the valence band (VB) to the conduction band (CB) by the absorption of a photon. (b) Absorption behavior represented as $(\alpha h\upsilon)^2$ vs. photon energy $h\upsilon$ near the band edge for a single crystal of *p*-type GaAs (data from I. Kudman and T. Seidel, *J. Appl. Phys.*, **33**, 771, 1962.)

Fermi's golden rule.) Thus α depends on the *joint density of states* at E and $E + h\upsilon$, and we have to suitably integrate this joint density of states. Near the band edges, the density of states can be approximated by a parabolic band, and α rises with the photon energy following

$$\alpha h\upsilon = A(h\upsilon - E_g)^{1/2} \tag{1}$$

where the constant $A \approx [(e^2/(nch^2m_e^*)](2\mu^*)^{3/2}$ in which μ^* is a reduced electron and hole effective mass, n is the refractive index, and E_g is the direct bandgap, minimum $E_c - E_v$ at the same k value. Experiments indeed show this type of behavior for photon energies above E_g and close to E_g as shown in the first figure for a GaAs crystal. The extrapolation to zero photon energy gives the direct bandgap E_g, which is about 1.40 eV for GaAs. For photon energies very close to the bandgap energy, the absorption is usually due to exciton absorption, especially at low temperatures.

In indirect bandgap semiconductors such as Si and Ge, the photon absorption for photon energies near E_g requires the absorption and emission of phonons during the absorption process as illustrated in the second figure. The absorption onset corresponds to a photon energy of $(E_g - h\vartheta)$, which represents the absorption of a phonon with energy $h\vartheta$, where ϑ is the phonon frequency. For the latter α is proportional to $[h\upsilon - (E_g - h\vartheta)]^2$. Once the photon energy reaches $(E_g + h\vartheta)$, then the photon absorption process can also occur by phonon emission for which the absorption coefficient is larger than that for phonon absorption. The absorption coefficients for the phonon absorption and emission processes are given by

$$\alpha_{\text{absorption}} = A[f_{\text{BE}}(h\vartheta)][h\upsilon - (E_g - h\vartheta)]^2; \quad h\upsilon > (E_g - h\vartheta) \tag{2}$$

and

$$\alpha_{\text{emission}} = A[(1 - f_{\text{BE}}(h\vartheta)][h\upsilon - (E_g + h\vartheta)]^2; \quad h\upsilon > (E_g + h\vartheta) \tag{3}$$

where A is a constant, and $f_{\text{BE}}(h\vartheta)$ is the Bose–Einstein distribution function at the phonon energy $h\vartheta$, i.e. $f_{\text{BE}}(h\vartheta) = [(\exp(h\vartheta/k_{\text{B}}T) - 1)]^{-1}$, where k_{B} is the Boltzmann constant and T is the

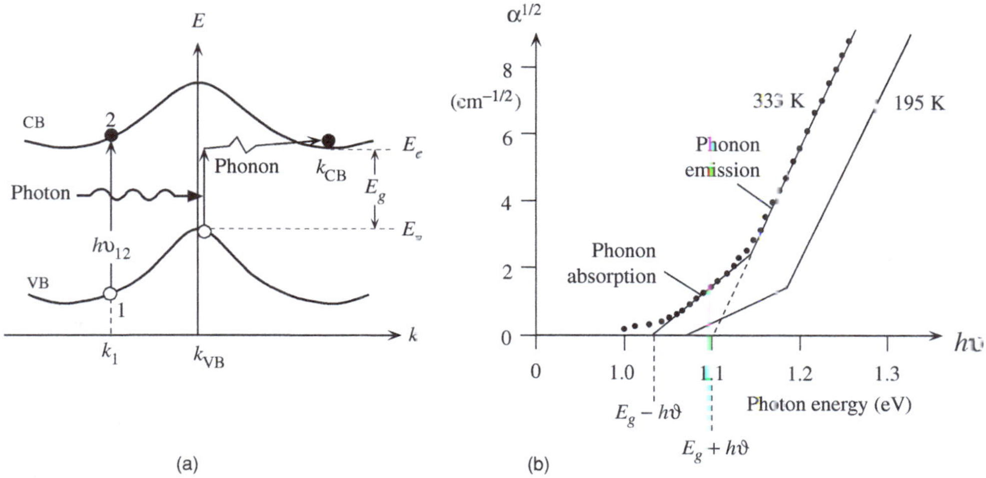

Band-to-band absorption (a) Indirect transitions across the bandgap involve phonons. Direct transitions in which *dE/dk* in the CB is parallel to *dE/dk* in the VB lead to peaks in the absorption coefficient. (b) Fundamental absorption in Si at two temperatures.

temperature. As we increase the photon energy in the range $(E_g - h\vartheta) < h\upsilon < (E_g + h\vartheta)$, the absorption is controlled by $\alpha_{\text{absorption}}$ and the plot of $\alpha^{1/2}$ vs. $h\upsilon$ has an intercept of $(E_g - h\vartheta)$.

On the other hand, for $h\upsilon > (E_g + h\vartheta)$, the overall absorption coefficient is $\alpha_{\text{absorption}} + \alpha_{\text{emission}}$, but at slightly higher photon energies than $(E_g + h\vartheta)$, α_{emission} quickly dominates over $\alpha_{\text{absorption}}$ since $[f_{\text{BE}}(h\vartheta)] \gg [1 - f_{\text{BE}}(h\vartheta)]$. At low temperatures, $f_{\text{BE}}(h\vartheta)$ is small and $\alpha_{\text{absorption}}$ decreases with decreasing temperature as is apparent in the figure. Equations (2) and (3) intersect the photon energy axis at $(E_g - h\vartheta)$ and $(E_g + h\vartheta)$, which can be used to obtain E_g.

The extinction coefficient K or ε_r'' vs. photon energy for Si shows that absorption peaks at certain photon energies corresponding to $h\upsilon \approx 3.5$ and 4.3 eV. These peaks are due to the fact that the joint density of states function peaks at these energies. The absorption coefficient peaks whenever there is a direct transition in which the E vs. **k** curve in the VB is parallel to the E vs. **k** curve in the CB as schematically illustrated in the second figure where a photon of energy $h\upsilon_{12}$ excites an electron from state 1 in the VB to state 2 in the CB in a direct transition $\mathbf{k}_1 = \mathbf{k}_2$. Such transitions where E vs. **k** curves are parallel at a photon energy $h\upsilon_{12}$ result in a peak in the absorption vs. photon energy behavior and can be represented by the condition that

$$(\nabla_k E)_{\text{CB}} - (\nabla_k E)_{\text{VB}} = 0. \tag{4}$$

The above condition is normally interpreted as the joint density of states reaching a peak value at certain points in the Brillouin zone, called *van Hove singularities*. The identification of peaks in K vs. $h\upsilon$ involves the examination of all E vs. **k** curves of a given crystal that can participate in a direct transition. Silicon's ε_r'' peaks at $h\upsilon \approx 3.5$ eV and 4.3 eV correspond to Equation (4) being satisfied at points L, along $\langle 111 \rangle$ in **k**-space, and X along $\langle 100 \rangle$ in **k**-space, at the edges of the Brillouin zone.

In *degenerate semiconductors*, the Fermi level E_F is in a band, for example, in the CB for a degenerate n-type semiconductor. Electrons in the VB can only be excited to states above E_F in the CB rather than to the bottom of the CB. The absorption coefficient then depends on the free carrier concentration since the latter determines E_F. Fundamental absorption is then said to depend on *band filling*, and there is an apparent shift in the absorption edge, called the *Burstein–Moss shift*. Furthermore, in degenerate indirect semiconductors, the indirect transition may involve a nonphonon scattering process, such as impurity or electron–electron scattering, which can change the electron's wavevector **k**. Thus in degenerate indirect bandgap semiconductors, absorption can occur without phonon assistance and the absorption coefficient becomes

$$\alpha \sim [h\upsilon - (E_g + \Delta E_F)]^2 \quad (5)$$

where ΔE_F is the energy depth of E_F into the band measured from the band edge.

Heavy doping of degenerate semiconductors normally leads to a phenomenon called bandgap narrowing and bandtailing. *Bandtailing* means that the band edges at E_v and E_c are no longer well-defined cut-off energies and there are electronic states above E_v and below E_c whose density of states falls sharply with energy away from the band edges. Consider a degenerate direct band gap *p*-type semiconductor. One can excite electrons from states below E_F in the VB where the band is nearly parabolic to tail states below E_c where the density of states decreases exponentially with energy into the bandgap, away from E_c. Such excitations lead to α depending exponentially on $h\upsilon$, a dependence that is called the *Urbach rule*

$$\alpha = A\exp[\beta(h\upsilon - E_o)] \quad (6)$$

where A, β and E_o are all constants. The Urbach rule was originally reported for alkali halides. It has been observed for many ionic crystals, degenerately doped crystalline semiconductors and almost all amorphous semiconductors. While exponential bandtailing can explain the observed *Urbach tail* of the absorption coefficient vs. photon energy, it is also possible to attribute the absorption tail behavior to strong internal fields arising, for example, from ionized dopants or defects. In the case of crystalline semiconductors that have not been heavily doped, the Urbach edge has been attributed to thermally induced disorder in the structure, and the Urbach edge has a temperature dependence that can be related to the phonon concentration. See *Urbach rule*.

Bandgap or **energy bandgap, E_g**, one of the most important nonmetalic material properties, is the least energy separation between the conduction band (where electrons are "free") and the valence band (where electrons are in bonds) in semiconductors; i.e. least energy separation between extended states that fall in the valence band and those that fall in the conduction band. In the bandgap there are no extended states and any state that has an energy in the bandgap is a localized state, reminiscent of the electron wavefunction in the H-atom. It is, however, possible to have an impurity band develop in the bandgap in heavily doped semiconductors. If the electron **k**-vector (which represents the electron's crystal momentum $\hbar\mathbf{k}$) at the bottom of the conduction band (CB) is the same as the electron **k**-vector at the top of the valence band (VB), $\mathbf{k}_{CB} = \mathbf{k}_{VB}$, then the bandgap is a *direct bandgap*, E_g(D), as in GaAs, InP, ZnTe, CdTe, etc. If the electron **k**-vector at the bottom of the CB is not the same as the electron **k**-vector at the top of the VB, $\mathbf{k}_{CB} \neq \mathbf{k}_{VB}$, then the bandgap is an *indirect bandgap*, E_g(I); Ge and Si are typical examples. Semiconductor photodetectors rely on the *photoexcitation* from the VB to the CB (as in detectors), which requires that the photon energy $h\nu > E_g$ for light absorption; the latter corresponds to *band-to-band* (fundamental) absorption. The longest wavelength λ_g that can photoexcite across E_g is the bandgap wavelength λ_g given by

$$\lambda_g(\mu\text{m}) = 1.24/E_g(\text{eV}).$$

Silicon with $E_g \approx 1.1$ eV has $\lambda_g \approx 1.1$ μm. In light-emitting diodes, the emitted photon energy is typically close to E_g and thus has a wavelength that is close to λ_g. The temperature dependence E_g is extremely important in optoelectronic device design, and has been well studied. There are various empirical and semiempirical expressions for E_g vs. T behavior such as the *Varshni equation*. E_g normally decreases with increasing temperature at a rate dE_g/dT, that is of the order of meV K^{-1}. Strain and stress dependences of the bandgap are also of considerable interest, especially since thin multilayers using different semiconductors can be strained.

Graded bandgap refers to alloy systems in which the bandgap is varied by varying the composition (i.e. fraction of a component in the alloy). If the composition is changed gradually through a structure, the bandgap changes gradually in response, leading to a so-called grading of the bandgap as a function of position in the structure. Such an approach is useful in tailoring the response of carriers within a device, as the gradient (with respect to position) of the bands corresponds to an internal electric field suitable for causing carrier drift. In addition, grading the bandgap also will alter the optical response of the structure, and is commonly used in optoelectronic devices to enhance performance. See *Varshni equation*. (R. Passler, *Phys. Stat. Sol. B*, **216**, 975, 1999.)

Bandgap electronegativity see *optical electronegativity*.

Bandgap narrowing is an effective narrowing of the bandgap as a result of the presence of a large concentration of dopants in a semiconductor. This effect is particularly important in accounting for the reduced injection efficiency of the emitter (which is heavily doped) in bipolar junction transistors. If ΔE_g is the reduction in the bandgap then for an n-type semiconductor, according to H. P. D. Lanyon and R. A. Tuft (*IEEE Trans. Elec. Dev.*, **26**, 1014, 1979),

$$\Delta E_g(\text{in meV}) = 22.5\left(\frac{n}{10^{18}}\right)^{1/2}$$

where n (in cm^{-3}) is the concentration of majority carriers which is equal to the dopant concentration if they are all ionized (for example, at room temperature). The new effective intrinsic concentration $n_{i\text{eff}}$ due to the reduced bandgap is given by

$$n_{i\text{eff}}^2 = N_c N_v \exp\left[-\frac{(E_g - \Delta E_g)}{k_B T}\right]$$
$$= n_i^2 \exp\left(\frac{\Delta E_g}{k_B T}\right)$$

where n_i is the intrinsic concentration in the absence of emitter bandgap narrowing. The equilibrium electron and hole concentrations n_{no} and p_{no} respectively obey

$$n_{no} p_{no} = n_{\text{eff}}^2$$

where $n_{no} = N_d$ since nearly all donors would be ionized at room temperature.

Bandwidth, or **optical bandwidth**, is the range of modulation frequencies up to some maximum frequency f_{op} where the optical transmission characteristic of the fiber falls by a factor of 3 dB. The optical bandwidth is very roughly the same as the bit rate capacity B. Similarly to BL, where L is the fiber length, the product $f_{op}L$, called the bandwidth-length product, remains very roughly constant and a property of the fiber. Strictly, however, BL is not constant (especially for single-mode fibers) and one should compare BL products of fibers of comparable length. Sometimes, the bandwidth-distance product is expressed as BL^{γ} = constant where γ is an index that is typically between 0.6 and 1.

Bandwidth theorem in qualitative terms refers to the fact that a signal with a finite duration Δt has spectral components (i.e. sinusoidal components) most of which are contained within a frequency width Δf such that $\Delta t\, \Delta f \sim 1$. If we take the signal as a Gaussian pulse, then its spectrum on the frequency domain will have a Gaussian distribution of frequencies, and if Δt is the full width at half maximum (FWHM) of this pulse then the spectrum of the pulse on the frequency domain will have a FWHM Δf, such that $\Delta t\, \Delta f \approx 0.88$. In photonics this theorem implies that if $\Delta\tau$ is the *coherence time* of a light pulse (such as a wave train), then the spectral width Δf in the Fourier transform of the light pulse is determined by $\Delta t\,\Delta f \sim 1$. See *coherence time*.

Bandwidth-length product see *bandwidth*.

Barium fluoride (BaF_2) is an optical material that has good transmission over a wide range, over UV and IR spectrum, from about 0.150 μm to about 13 μm (or 0.25 to 9.5 μm for 90% transmission), with a low refractive index 1.30–1.50 and $\alpha \approx 0.1\ \mathrm{cm^{-1}}$ at ~1 μm. It is used in lenses, windows, prisms and other optical applications. BaF_2 has a cubic crystal structure and cleaves on (111) planes. $\rho = 4.83\ \mathrm{g\ cm^{-3}}$; a (cubic) = 0.620 nm; $T_m \approx 1300\ ^\circ\mathrm{C}$; $v_d = 83$; $n_d = 1.4748$ (at 0.5876 μm); $E_g \approx 9.1$ eV; $\alpha_L = 16.5\text{–}19 \times 10^{-6}\ \mathrm{K^{-1}}$; $\kappa = 11.7\ \mathrm{W\ m^{-1}\ K^{-1}}$; $c_s = 410\ \mathrm{J\ kg^{-1}\ K^{-1}}$; $E = 65$ GPa; $K = 57$ GPa; $G = 25$ GPa; $\upsilon = 0.30$; $C_{11} = 90$ GPa; $C_{12} = 40$ GPa; $C_{44} = 25$ GPa; $H_{\mathrm{Knoop}} = 82\ \mathrm{kgf\ mm^{-2}}$; n vs. λ (μm) over 0.2–15 μm is

$$n^2 = 1 + \frac{0.3797\lambda^2}{\lambda^2 - (0.008674)^2} + \frac{0.7718\lambda^2}{\lambda^2 - (0.1022)^2} + \frac{4.159\lambda^2}{\lambda^2 - (48.03)^2}; \quad \lambda \text{ in } \mu\text{m}.$$

Baud is a measure of the rate of digital signal transmission, for example, voltage transitions, in terms of transitions per second.

Bayer color image sensor see *image sensor*.

Beam diameter see *Gaussian beam*.

Beam displacement see *lateral displacement*.

Beam propagation method (BPM) is a numerical technique for simulating the propagation of optical fields in materials; in particular, it is widely used to provide an accurate analysis of guided wave structures – essential for developing photonic integrated circuits. In the BPM method, the problem is usually reduced to a 1D cross-sectional index profile by defining effective indexes to various parts of the structure. Then the paraxial wave equation is solved in the resulting effective 1D index profile using an algorithm that involves a fast Fourier transform. It is a powerful technique that can simulate any arbitrary refractive index profile. This technique can take an input optical field distribution and calculate how it is modified by traveling through N longitudinal sections of a waveguide each of thickness Δz, for example, in order to determine the optical field distribution after the optical field has propagated a distance $z = N\Delta z$. The optical field is resolved into its different plane wave components and the propagation of each component is considered separately before superposing them to determine the resultant field after propagation in each short uniform section. BPM is often used to model the optical field profile in practical optical components including tapered lasers, tapered amplifiers, tapered waveguides, waveguide directional couplers, Y branch optical switches, optical filters and modulators. BPM can also simulate optical diffraction by resolving the optical field into separate plane wave components, each of which has its own wavevector. The phase changes due to the propagation of each plane wave component are calculated for each plane wave and the resultant field is obtained by superposition of the different plane waves. When implementing BPM in practice, one is concerned with keeping the step size Δz sufficiently small to ensure that the BPM remains accurate – for a typical waveguide structure the step size is of the order of a few microns. In practice, Δz is modified to check the sensitivity of the output of the BPM calculation due to changes in Δz and determining how small Δz should be for the results to become insensitive to Δz.

One approach commonly adopted in more complicated waveguide structures is the so-called *stepped taper* or *staircase concatenation method*. In this approach, the structure is split into stepped sections of uniform cross-section and thus regions within which one can define normal modes. The modes are called local normal modes since they are defined in localized region only of the taper structure. One then propagates modes across the

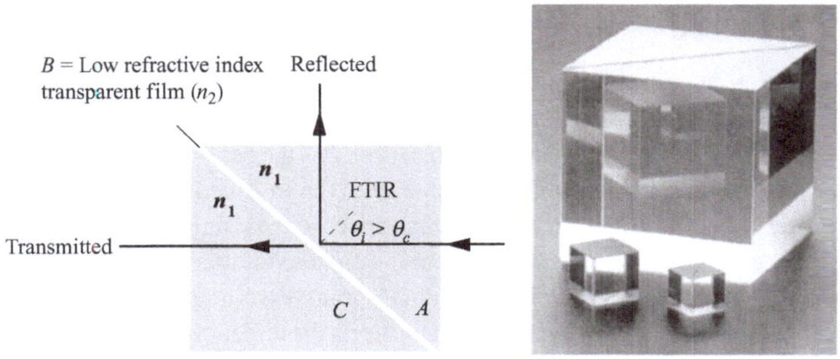

Beam splitter Left: Two prisms separated by a thin low refractive index film forming a beam splitter cube. The incident beam is split into two beams by frustrated total internal reflection (FTIR). Right: Beam splitter cubes (courtesy of Melles Griot).

interfaces by correlating the mode amplitude coefficients of modes in adjacent regions. Eventually this leads to a set of coupled mode equations. The propagation of any one mode in the set is governed by the properties of all the other contributing modes. The equations therefore must be solved simultaneously. An example of an application of this approach would be for the coupling of two modes in adjacent channel waveguides in a directional coupler.

Beam splitter is an optical device that splits an incoming beam into two beams that are traveling in different directions. Beam splitter cubes typically use frustrated total internal reflection to split an incident beam.

Beat, see *optical beating.*

Beat noise is an unwanted "signal" that arises when two or more optical signal frequencies are mixed in a nonlinear medium such as a detector.

Bend loss or **bending loss** is the attenuation in an optical fiber as a result of the bending of the fiber. See *microbending loss* and *macrobending loss.*

Bend radius is the local radius of the fiber axis when the fiber is bent, that is, the fiber axis follows a curvature. Typically attenuation increases exponentially with decreasing bend radius. Minimum bend radius is the shortest radius of curvature to which a fiber can be bent without the induced stress in the fiber exceeding the allowed stress.

BER, see *bit error rate.*

Beryllium is a light metal that is extensively used in the construction of optical mirrors, especially in the infrared where its reflectivity is above 95%; it is above 98% in 11–50 μm. It is much lighter than Al (Be's density is 1.85 g cm^{-3} vs. 2.7 g cm^{-3} for Al). Beryllium has a high thermal conductivity (2.1 W cm^{-1} K^{-1}) which makes it efficient in sinking heat away in mirror applications using high intensity laser beams. Commercial mirrors can be 5 mm–30 cm in diameter and can be very thin, and usually have a backing support or mount. Be mirrors are used in night-vision systems, space cameras for the military, high-resolution cameras for terrestrial mapping, and in optics for handling high-power lasers.

Table: Beryllium Comparison of the properties of Be and Al.

	ρ g cm^{-3}	κ W cm^{-1} K^{-1}	c_s J g^{-1} K^{-1}	E GPa	υ	α_L (K^{-1})
Be	1.85	2.16	1.93	303	0.12	11×10^{-6}
Al	2.7	1.71	1.25	69	0.33	24×10^{-6}

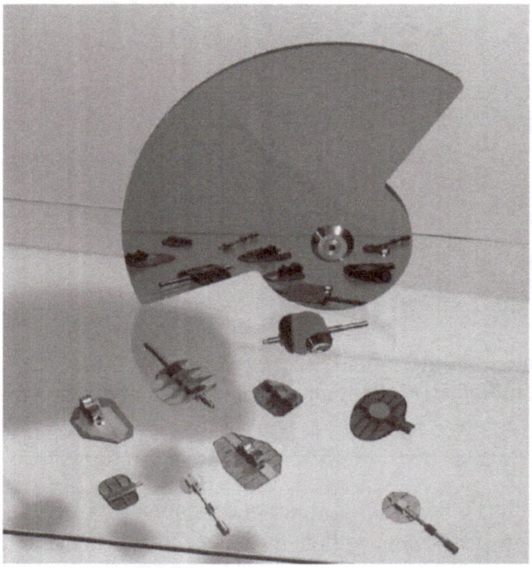

Beryllium mirrors. The camera shutter is 12 inches point-to-point. The smaller mirrors demonstrate a variety of mirror shapes as well as rib structures mounts, shafts and back side mirrors used for alignment or registration. (Courtesy of Hardric Laboratories, Inc, USA, c/o Vladimir Vudler and Peter Reachard.)

Bessel function $J_n(x)$ of order n is a mathematical function that is a solution to the differential equation

$$x^2 y'' + xy' + (x^2 - n^2)y = 0$$

where the prime represents differentiation with respect to x, i.e. $y' = dy/dx$, and n is a constant (independent of x and y). Bessel functions are either tabulated or graphed for various orders. $J_n(x)$ looks like a damped sinusoidal wave. The electric fields in cylindrical waveguides, coaxial cables, and step index optical fibers, are generally described by Bessel functions. For example, for step index fibers, zeros of the Bessel function $J_n(x)$ point to those values of the V-number at which various modes, with the exception of the fundamental mode, are cut off (cannot propagate). For example, the LP_{11} ($l = 1, m = 1$) mode is cut off when $J_0(x) = 0$ or when $V = 2.405$.

Bessel lens refers to the generation of a beam in which the intensity distribution along the direction of propagation is given by a first-order Bessel function $J_0(kBr)$, where k is the wavevector, r the radial coordinate, and B is the beam deviation angle (which is a function of the so-called *axicon* angle). Such a beam may arise from passing a uniform plane wave through an infinite-dimensional *axicon*. In this case, the transverse intensity distribution remains constant along the direction of propagation, and is referred to as a nondiffracting beam. In practice, one cannot generate a uniform plane wave or indeed an infinite-dimensional *axicon* and hence Bessel beams cannot be practically realized. One can, however, use a collimated Gaussian beam to illuminate a finite-dimensional *axicon* to generate so-called *Bessel–Gauss beams*. In this case the transverse intensity distribution is no longer invariant along the direction of propagation, but rather decreases gradually. The intensity distribution can be found from the Fresnel diffraction integral under the stationary phase approximation. See *axicon*.

Bessel–Gauss beam see *Bessel lens*.

Bias is a term used to describe the application of a continuous (dc) voltage across, or a current through, a device that puts the device in a desirable operating condition. For example, applying a reverse dc voltage across a *pin* photodiode establishes the necessary electric field and puts the *pin* into a desirable operating condition.

Biaxial crystals, see *anisotropy*.

Bimolecular recombination occurs in a semiconductor when the rate of recombination is proportional to the population of *both* excess holes and excess electrons. If R is the rate of recombination of *excess* holes and electrons, then

$$R = C_r \Delta n \Delta p$$

where Δn and Δp are the excess electron and hole concentrations (above equilibrium values). The total recombination rate R consists of the band-to-band radiative recombination rate R_r and the nonradiative recombination rates R_{nr}. Mechanisms contributing to the nonradiative recombination rate for excess injected carriers include *Auger recombination, Shockley–Read–Hall recombination* (involving recombination centers), as well as *surface state recombination*. When a direct bandgap semiconductor is illuminated by light with energy exceeding its bandgap, absorbed photons cause the creation of equal numbers of excess electrons and holes. In the inverse process, the recombination of one electron with one hole will result in the spontaneous emission of a photon of energy approximately equal to the bandgap. The instantaneous rate of such recombination events may be shown to be proportional to the product of the instantaneous electron and hole concentrations, n and p, respectively, with a constant of proportionality defined as B_{sp}. Thus, the radiative recombination rate may be written as $R_r = B_{sp}np$. Excess carriers may be created in such a semiconductor by above-bandgap illumination, as discussed above (also known as photoluminescence), by current injection (e.g. at a *pn* junction) or even by high-energy electron irradiation (also known as *cathodoluminescence*). Consider the optical injection for the two cases of when the so-called carrier injection is strong and weak for an n-type semiconductor (i.e. with $n_0 \gg p_0$, where n_0 and p_0 are the equilibrium concentrations of electrons and holes, respectively). Under strong injection, the injected concentration of carriers (i.e. the concentrations, $\Delta n = \Delta p$) far exceed n_0 and p_0 and thus $R_r \approx B_{sp}(\Delta n)^2$. Writing $R_r = -d\Delta n/dt$ provides an expression for $n(t)$,

$$n(t) = n_0/\{1 + n_0 B_{sp} t\}$$

which shows that for the case of strong injection, where the radiative recombination process is bimolecular, the excess concentration of electrons decays hyperbolically. For *weak injection*, $\Delta n = \Delta p \gg p_0$, and $R_r \approx B_{sp} n_0 \Delta n$. This gives,

$$n(t) = n_0 \exp(-n_0 B_{sp} t),$$

or an exponential decay in the photoexcited excess electron concentration, with a characteristic time constant $\tau_{sp} = 1/(n_0 B_{sp})$ for weak excitation.

Binned pixels are pixels in an imaging device that have been combined, connected in "parallel" to increase the area for photon reception and hence increase the signal. Binning pixels reduces the resolution (the pixel area is increased) but it increases the signal and hence improves the signal-to-noise ratio.

Bioluminescence see *luminescence*.

Bipolar junction transistor (BJT) is a transistor whose normal operation is based on the injection of minority carriers from the emitter into the base region and their diffusion to the collector where they give rise to a collector current. The voltage between the base and the emitter controls the collector current; this is the transistor action.

Birefringent crystals, such as calcite, are optically anisotropic which leads to an incident light beam becoming separated into ordinary (o-) and extraordinary (e-) waves with orthogonal polarizations; incident light becomes doubly refracted because these two waves experience different refractive indices n_o and n_e. There is no double refraction for light propagation along the optic axis. If we were to cut a plate from a calcite crystal so that the optic axis (along z) were parallel to two opposite faces of the plate, then a ray entering at normal incidence to one of these faces would not diverge into two separate waves. The o- and e-waves would travel in the same direction but with different speeds. The waves emerge in the same direction as well which means that we would see no double refraction. This optical arrangement is used in the construction of various optical retarders and polarizers as discussed below. If we were to cut a plate so that the optic axis was perpendicular to the plate face, then both the o and e-wave would be traveling at the same speed and in the same direction which means we would again not see any double refraction (see also *anisotropy*).

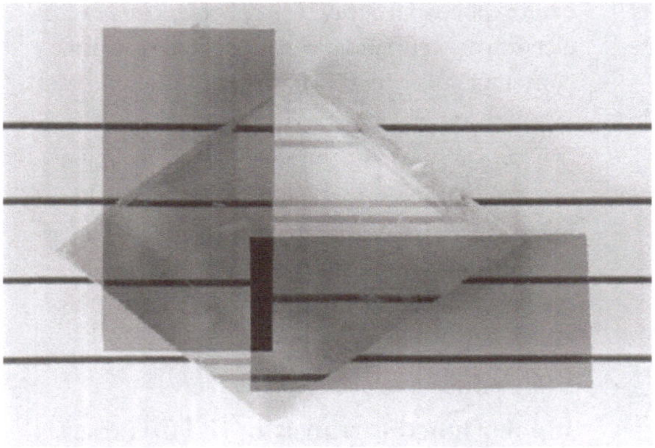

Birefringent crystals Two polaroid analyzers are placed with their transmission axes, along the long edges, at right angles to each other. The ordinary ray, undeflected, goes through the left polarizer whereas the extraordinary wave, deflected, goes through the right polarizer. The two waves therefore have orthogonal polarizations.

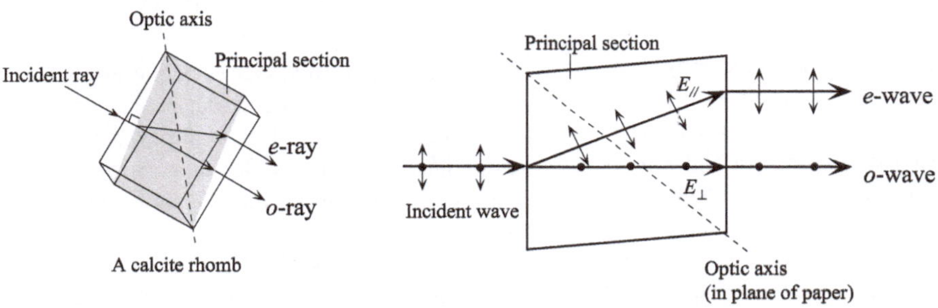

Birefringent crystals A light wave that is off the optic axis of a calcite crystal splits into two waves called ordinary and extraordinary waves. These waves have orthogonal polarizations and travel with different velocities. The *o*-wave has a polarization that is always perpendicular to the optical axis.

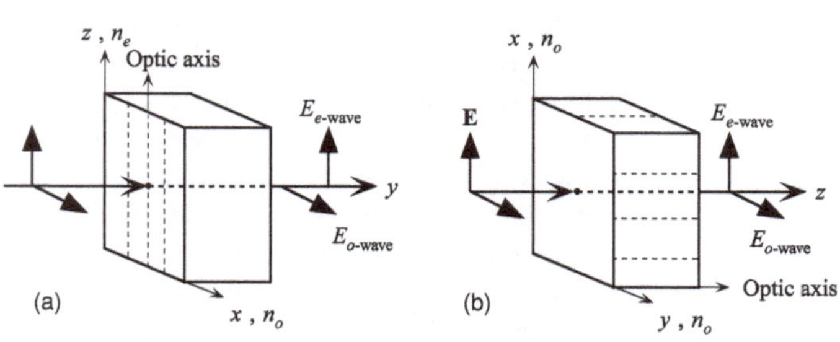

Birefringent crystals (a) A birefringent crystal plate with the optic axis parallel to the plate surfaces. (b) A birefringent crystal plate with the optic axis perpendicular to the plate surfaces.

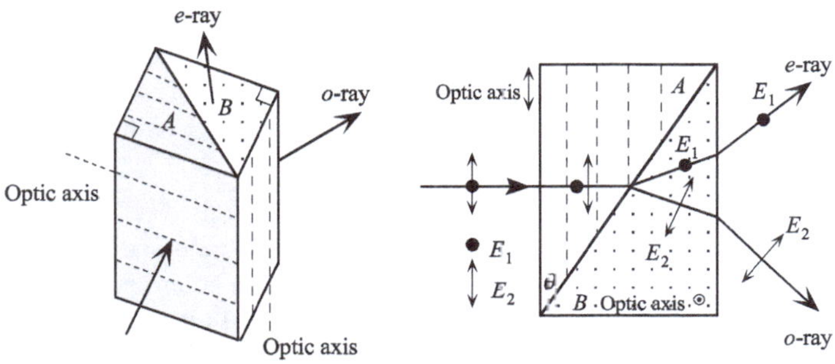

Birefringent prism The Wollaston prism is a beam polarization splitter. E_1 is orthogonal to the plane of the paper and also to the optic axis of the first prism. E_2 is in the plane of the paper and orthogonal to E_1. *e*-ray is the *extraordinary* wave and *o*-ray is the *ordinary* wave.

Birefringent prism Commercial Wollaston prisms. The actual prism is held inside a cylindrical housing (courtesy of Melles Griot).

Birefringent prisms are prisms made from birefringent crystals for producing a highly polarized light wave or polarization splitting of light. The Wollaston prism is a polarization splitter in which the split beam has orthogonal polarizations. Two calcite (or quartz) right angle prisms A and B are placed with their diagonal faces touching to form a rectangular block. Various Wollaston prisms with typical beam splitting angles of 15–45° are commercially available.

Bit is the basic unit of information in digital communication systems and corresponds to the presence or absence of pulses.

Bit error rate (BER) is the fraction of all bits that are wrong (corrupted) within a long series of bits of information received over a long period of time. Stated differently, it is the ratio of errors to the total number of bits transmitted during a long digital transmission period. The lower the BER, the better is the transmission system. Typically communication systems have BER better than 10^{-9}, that is one wrong bit in 10^9 or more bits transmitted. BER depends on the signal-to-noise ratio SNR (signal power to noise power ratio) at the detector or receiver output. If we assume that the signal amplitude has a Gaussian distribution, and the threshold level I_{th} for decision has been set to $\frac{1}{2}(I_{on} + I_{off})$, half-way, where I_{on} and I_{off} are the maximum and minimum average signal amplitudes corresponding to 1 and 0, then

$$\mathrm{BER} = \frac{1}{2}\mathrm{erfc}\left(\frac{(\mathrm{SNR})^{1/2}}{2\sqrt{2}}\right)$$

or

$$\mathrm{BER} = \frac{1}{2}\mathrm{erfc}\left(\frac{Q}{2}\right)$$

where erfc is the complementary error function, $Q = (I_{th} - I_{off})/\sigma_{off} = (I_{on} - I_{th})/\sigma_{on}$ in which σ_{on}^2 and σ_{off}^2 represent the variance of the on and off signals and hence the *noise*. The actual signal $I_{signal} = I_{on} - I_{off} = 2Q$. The significance of the above equation is that it allows the required SNR to be calculated for a given BER. For example, for high-speed SONET links, BER < 10^{-11}, SNR > 22.5 dB.

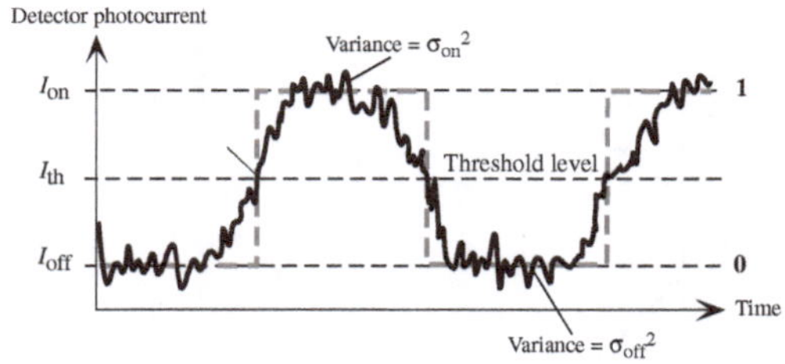

Bit error rate BER vs. SNR at detector output.

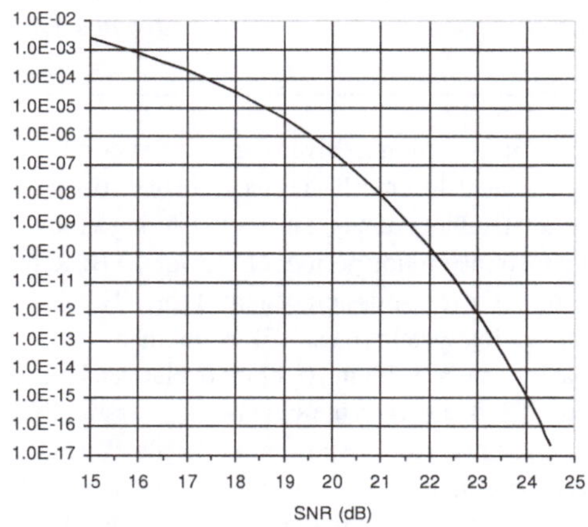

Bit error rate Noisy signal at the detector of a communications system.

Bit rate is the rate at which bits are transmitted in a digital communications system. The maximum rate at which the digital data can be transmitted along the fiber is the bit rate capacity, B (bits per second), and is directly related to the dispersion characteristics of the fiber.

Bit rate × distance (BL) is the product of the bit rate capacity and fiber length and is a measure of the information carrying capacity of a fiber. Generally, the dispersion or time spread of an optical pulse increases with distance and the bit rate capacity consequently decreases with distance. In multimode fibers, B decreases approximately linearly with L so that BL is nearly constant and a fiber characteristic. Strictly, however, BL is not constant (especially for single-mode fibers) and one should compare BL products of fibers of comparable length. See *bandwidth.*

BK7 is a borosilicate crown glass used as an optical material, in applications in visible and near infrared wavelengths, especially in transmission optics since its transmittance is in the 0.38–2.1 μm wavelength range with $n = 1.52$ at 550 nm. $\rho = 2.51$ g cm^{-3}; $\nu_d = 64.17$; $\alpha_L = 7.7 \times 10^{-6}$ K^{-1}; $\kappa = 0.011$ W cm^{-1} K^{-1}; $c_s = 0.858$ J g^{-1} K^{-1}; $E = 81.5$ GPa; $H_{\mathrm{Knoop}} = 610$ kg mm^{-2}. See *optical glasses* for refractive index.

Black body is a hypothetical (an ideal) body that completely absorbs all the electromagnetic (EM) radiation falling onto it (no reflection, no scattering) and therefore appears to be black at all wavelengths. When heated, a black body emits the maximum possible EM radiation at that temperature. A small hole in the wall of a cavity maintained at a uniform temperature emits radiation that approximately corresponds to that from a black body. At thermal equilibrium, the black body radiates as much EM energy as it absorbs, but with a spectral distribution that depends only upon its temperature. According to *Planck's law*, the *spectral energy density* (in J m^{-3} Hz^{-1}) of the black body radiation at absolute temperature T is

$$\rho(\nu) = \frac{8\pi \ h\nu^3}{c^3[\exp(h\nu/k_{\mathrm{B}}T) - 1]},$$

with k_{B} the Boltzmann constant. The spectral irradiance I_λ is the total emitted radiation irradiance (power per unit area) per unit wavelength so that $I_\lambda d\lambda$ is the irradiance (intensity) in a small range of wavelengths from λ to $\lambda + d\lambda$; it is also called the *total hemispherical emissive power.* The wavelength dependence of I_λ (in W m^{-3}) is

$$I_\lambda = \frac{2\pi hc^2}{\lambda^5[\exp(hc/\lambda k_{\mathrm{B}}T) - 1]}.$$

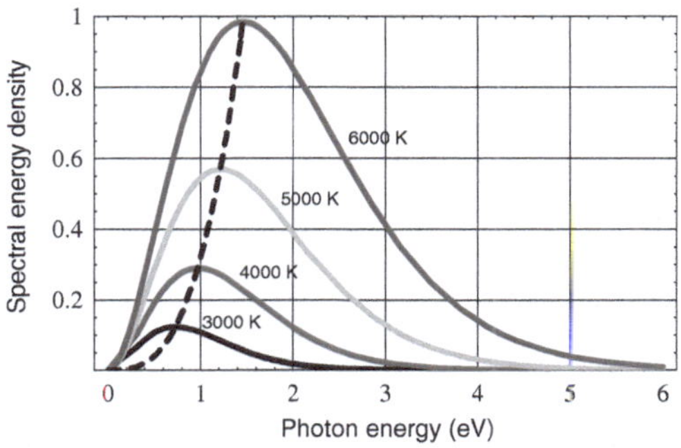

Black body Spectral energy density of a black body (in arbitrary units) as a function of $h\nu$ [eV], for $T = 3000$ K, 4000 K, 5000 K and 6000 K (from bottom to top). In dotted lines: shift of the maximum frequency, *Wien's law*.

Its spectral *luminance* (in W m^{-2} sr^{-1} Hz^{-1}) is

$$L(\nu) = \frac{2\,h\nu^3}{c^2[\exp(h\nu/k_B T) - 1]}.$$

where $L(\nu)d\nu$ is the EM power (energy per unit time) radiated per unit surface area and per unit solid angle in the small frequency interval ν to $\nu + d\nu$. Its *radiant emittance* (total power radiated per unit area) is given by the *Stefan–Boltzmann Law*,

$$M = \sigma T^4,$$

where $\sigma = 5.67 \times 10^{-8}$ W m^{-2} K^{-4} is Stefan's constant. The maximum wavelength λ_{max} is shifted with T according to *Wien's displacement law*: $\lambda_{max} T \approx 2.898 \times 10^{-3}$ m K. See *emissivity, Kirchhoff's law*. (M. Planck, *Annalen der Physik*, **4**, 553, 1901.)

Black light is the radiation from the invisible region of the electromagnetic spectrum, usually emitted by a fluorescent tube made of active mercury atoms in an inert gas. Energetic photons in the UV-B or UV-C range are absorbed by a phosphor coating at the outside of the tube that reemits photons at longer wavelengths, in the UV-A range (most visible light, if emitted at all, is blocked by the glass tube). These secondary photons can interact with any external substance containing phosphors that will in turn reemit photons in the visible range, thus glowing in the dark. Applications vary from forgery detection to entertainment to forensic sciences.

Blazed grating is a type of diffraction grating, typically a reflection grating, that is designed to maximize the diffracted energy into one order of diffraction only, with a high efficiency (e.g. around 70%). This is achieved by a special profile, usually a triangular profile, of the grating surface, each triangular groove being slanted by a *blaze angle* γ with respect to the average plane of the grating. Consider a simple example in which the incident wave is normal to the diffraction grating plane. The specular reflection from the slanted grooved surface will be at an angle γ to the normal to the slanted face. By appropriately choosing γ we can make the mth-order diffraction from the whole grating coincide with the specularly reflected beam, because the mth-order diffracted beam's direction is determined by the angle with respect to the normal to the plane of the grating (and not the normal to the slanted face) and obviously the wavelength and the grating spacing d. Thus, the magnitude of the diffraction angle has to be 2γ. The first successful blazed diffraction grating has been attributed to Robert Williams Wood.

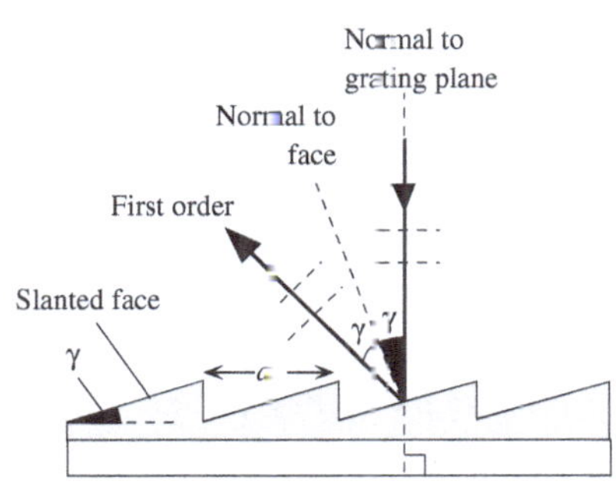

Blazed grating Triangular grooves are cut into the surface with a periodicity d. The side of a triangular groove makes an angle γ to the plane of the diffraction angle.

Blazed grating Robert Williams Wood (1868–1955) with Albert Einstein. Wood was a professor first at the University of Wisconsin and then at Johns Hopkins University. Wood was the author of many significant works, some examples of which are *Physical Optics*, 1905; *Researches in Physical Optics*, two parts, 1913–19; *Supersonics, the Science of Inaudible Sounds*,1939, and more than 200 papers and articles. (AIP Emilio Segre Visual Archives, gift of Richard Henry.)

Bloch oscillations refers to the periodic motion of electrons across the Brillouin zone, in the absence of electron scattering, being accelerated and decelerated in an applied electric field. Electrons beginning from the bottom of a given band, gain energy until they reach the top of the band, where the electron momentum reaches the Brillouin zone boundary. The electron then reflects back and continues to propagate in the opposite direction, decelerating until it reaches the bottom of the band. Then it accelerates again, and this process continues *ad infinitum* in the absence of electron scattering. In the case of a superlattice, the finite bandwidth presents an opportunity to observe such oscillations. In this case, an electron in a miniband of finite bandwidth oscillates in both real space and also momentum space, with the oscillations being Bloch oscillations. In the presence of a dc electric field E, for a superlattice with periodicity a, the frequency ν of the oscillations is given by

$$h\nu = eEa$$

where h is Planck's constant and e is the electron charge. In practice electron scattering will act to destroy these oscillations. These oscillations of charged particles (i.e. electrons) lead to the emission of radiation (typically in the THz range). See *superlattices*.

Bloch sphere see *quantum bit*.

Bloch's law see *eye*.

Blondel–Rey law of visual integration (1911, 1924) states that the human eye detection is based on the integration of the photons that arrive at the photoreceptors of the eye; the eye is a photon integrator. (A. M. Blondel and J. Rey, *C. R. Acad Sci.*, **153**, 54, 1911; **178**, 276, 1924.)

Blooming of lenses refers to the antireflection (AR) coating of lenses with a thin film of suitable material and thickness to reduce light reflections and hence increase the transmission of light; typically the AR coating is CaF_2 or MgF_2, and the lens is called a *bloomed lens*. The thickness of the AR coating is chosen to be a quarter wavelength, $\lambda/4$, where λ is the wavelength within the thin film coating, to minimize the reflections. See *antireflection coating*.

Bohr's frequency is the frequency $\upsilon_{nn'}$ associated with the energy difference $|E_n - E_{n'}|$ between any two different states (or levels) of a quantum system, such as an atom, a molecule, etc. Frequencies $\upsilon_{nn'}$ such as $|E_n - E_{n'}| = h\upsilon_{nn'}$ are called the Bohr frequencies of the system. They intuitively correspond to its oscillation frequencies. A transition between two quantum states is allowed provided that it is permitted by the *selection rules*.

Boling, Glass and Owyung rule is a semiempirical relation for predicting the second-order refractive index n_2 from the "normal" refractive index n_d at the d-line and the Abbe number v_d.

$$n_2(\text{esu}) = \frac{K(n_d - 1)(n_d^2 + 2)^2}{v_d[1.517 + (n_d^2 + 2)(n_d^2 + 1)v_d/6]^{1/2}}$$

where $K = (68–70) \times 10^{-13}$ in CGS units (esu). While the above expression is based on physical concepts and arguments, and certain simplifications, the authors also propose a much simpler empirical expression for n_2 that provides a good fit to the n_2 data as follows:

$$n_2(\text{esu}) = (391 \times 10^{-13})\frac{(n_d - 1)}{v_d^{5/4}}$$

where n_d is the refractive index at the d-line, and v_d is the Abbe number (see *Abbe number*). (N. Boling, A. J. Glass, and A. Owyoung, *IEEE J. Quantum. Electron.* **QE-14**, 601,1978; D. Milam *et al.*, *Appl. Phys. Letts.*, **31**, 822, 1977; Weber *et al.*, *Appl. Phys. Letts.*, **32**, 403, 1978; R. Adair *et al.*, *Phys Rev. B*, **39**, 3337, 1989.)

Bolometer is a thermal detector, for example a thin strip of platinum, nickel, or carbon, or a semiconductor, which absorbs and thereby transforms the incident light energy into heat, leading to a variation of its electric resistance. The bolometer element has an optically black coating to absorb all the incident radiation. It exhibits good linearity and flat detectivity for a wide range of wavelengths varying from around 0.2 μm to 10 μm, but its response time is relatively slow compared with other infrared detectors; 1–100 ms. The resistance vs. temperature characteristics, the *R/T characteristics*, are extremely important for proper detector selection and calibration. For metals, the temperature coefficient of resistivity, TCR, is positive (R increases with T) while for semiconductors TCR is negative and larger in magnitude by almost an order of magnitude. Bolometers that use a semiconductor as a sensing element are called *thermistors*; e.g. nickel-copper-cobalt-manganese oxides, vanadium oxide, silicon. Modern bolometers tend to use silicon. Thin film fabrication techniques (such as sputtering) have been used to deposit small thin film sensors to reduce the thermal capacity and shorten the response time. Recent developments in silicon micromachining have led to the use of silicon thermistors in *microbolometers* which form an array of tiny bolometers that are fabricated onto a Si chip with the microelectronic circuit. Such microbolometer arrays allow for infrared imaging. Microbolometers can have a response time 10 μs – 1 ms. For negative TCR thermistors, the *Steinhart–Hart equation* describes the R vs. T characteristics by $1/T = A + B(\ln R) + C(\ln R)^3$, where T is in kelvins, and A, B and C are thermistor-specific (not material-specific) constants. A three-point calibration allows A, B and C to be determined. Use of a look-up table, and interpolation, can provide good linearity.

Boltzmann energy distribution see *Boltzmann statistics*.

Boltzmann statistics describe the behavior of a collection of particles (*e.g.* gas atoms) in terms of their energy distribution. It specifies the number of particles, $N(E)$, with an energy E through $N(E) \propto \exp(-E/k_B T)$ where k_B is the Boltzmann constant and T is the temperature (kelvins). The

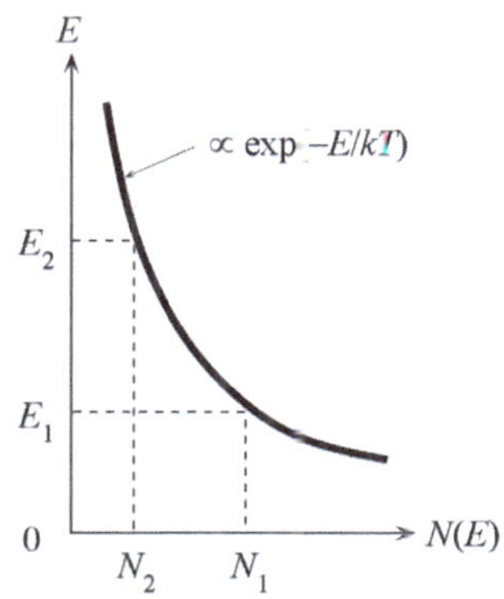

Boltzmann statistics The Boltzmann energy distribution describes the statistics of particles, e.g. electrons, when the particles do not interact with each other, i.e. when there are very few electrons compared with the number of available states.

description is nonquantum-mechanical in the sense that there is no restriction to how many particles can have the same state (the same wavefunction) with an energy E. It applies when there are only a few particles compared with the number of possible states so that the likelihood of two particles having the same state becomes negligible. This is generally the case for thermally excited electrons in the conduction band of a semiconductor where there are many more states than number of electrons. The kinetic energy distribution of gas molecules in a tank obeys the Boltzmann statistics.

Boltzmann's constant, k_B or k, is the gas constant per atom or per molecule, that is, the gas constant divided by Avogadro's number ($k_B = R/N_A$); $k_B = 1.38 \times 10^{-23}$ J K^{-1} ($^3/_2 k_B T$ is the mean kinetic energy associated with the translational motions of gas molecules in a gas cylinder at temperature T).

Born–Szigeti equation describes the dielectric response of ionic crystals and is expressed in terms of an effective charge known as the Szigeti charge, e^*:

$$\varepsilon_0 = \varepsilon_\infty + \frac{4\pi}{9\mu V}\left(\frac{e^*(\varepsilon_\infty + 2)}{e\omega_{TO}}\right)^2$$

where μ is the reduced mass of the ion pair in the crystal basis, V is the volume per ion pair, ω_{TO} is the transverse optical mode frequency, and, ε_0 and ε_∞ are the low- and high-frequency dielectric constants, respectively, and e is the free electron charge. The first term on the right-hand side describes the ultraviolet (high frequency) contribution, while the second term describes the infrared (low frequency) contribution to the polarization. This second term is a modification of the so-called Born equation which only takes into account the electronic contribution to the infrared polarization. This modified equation also takes into account the short-range interaction of electronic and atomic displacements, while the factor $[(\varepsilon_\infty + 2)/3]^2$ appears since the long-range interaction does not vanish for transverse waves. This equation is derived from the equations of motion of a biatomic lattice with associated polarization, and using the *Lyddane–Sachs–Teller* relation and the *Clausius–Mossotti* relation. Typically, the Szigeti charge ranges in value from about $0.7e$ to $0.9e$.

Boron trioxide (B_2O_3) is added to silica glass to form B-doped silica to decrease the refractive index.

Bose–Einstein condensate (BEC) see *trapping of atoms*.

Bose–Einstein statistics describe the most probable distribution of *bosons* among available energy levels. For a given energy level E, the probability, $f(E)$, that a boson has energy E is given by,

$$f(E) = 1/[A\exp(E/k_B T) - 1]$$

where A is a constant and k_B is the Boltzmann constant. The most common boson is the photon.

Boson is a quantum particle following the Bose–Einstein statistics (probability distribution) and having an integer spin or 0. Such particles have wavefunctions that are symmetric to the exchange of any pair of them, and hence do not obey the *Pauli exclusion principle*. They are called *bosons* after the statistical distribution law (*Bose–Einstein statistics*) that describes their occupation of energy states discovered in 1924–25 by Satyendra Nath Bose and Albert Einstein. Examples of bosons include photons, alpha particles and helium atoms. The mean value $\langle n\rangle_{\text{photon}}$ of the number of photons in a given mode of the electromagnetic field, at thermodynamic equilibrium at absolute temperature T, is given by

$$\langle n\rangle_{\text{photon}} = 1/[\exp(h\upsilon/k_B T) - 1].$$

Bosons are generally involved in the transmission of various forces; photons carry the electromagnetic force. Bosons are not conserved – they can be created or destroyed.

Bouguer's formula refers to light propagation in an inhomogeneous medium where the refractive index is spherically symmetric, $n = n(r)$, as approximately is the case for n in the terrestrial atmosphere. At any point M along the trajectory

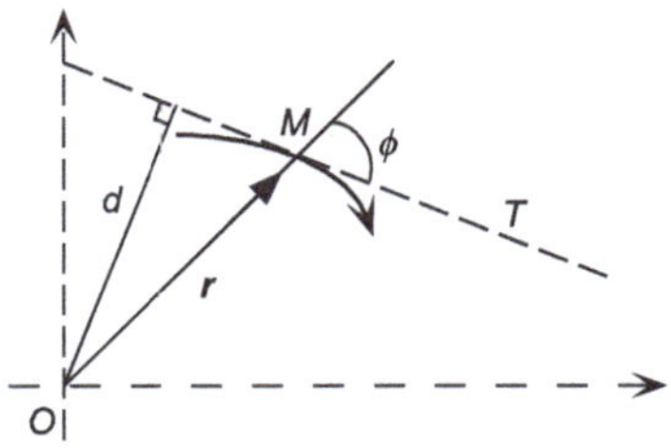

Bouguer's formula for rays propagating in a refractive medium with spherical symmetry around a center O. The distance d from O to tangent T remains constant throughout the propagation.

of a light ray, the medium can be locally described as a succession of infinitely thin concentric layers of constant index. If ϕ represents the angle between the position vector $\boldsymbol{r} = \overrightarrow{OM}$ and the tangent on the ray at point M, it can be shown that

$$n(r)r \sin\phi = \text{constant}.$$

In other words, with $d = r\sin\phi$ denoting the distance from the origin O to the tangent, Bouguer's formula can also be written as:

$$n(r)d = \text{constant}.$$

This law explains, for instance, atmospheric refraction.

Bouguer's law see *Lambert's absorption law*.

Boundary conditions in electromagnetism relate the normal and tangential components of the electric (or magnetic) field next to the boundary. The tangential component must be continuous through the boundary. If E_{t1} and E_{t2} are the tangential components of the field next to the boundary in medium 1 and 2 respectively as in the first figure, then $E_{t1} = E_{t2}$. Suppose that E_{n1} is the normal component of the field in medium 1 at the boundary and ε_{r1} is the relative permittivity in medium 1, assumed to be an isotropic medium (ε_{r1} does not depend on the direction of the field). Using a similar notation for medium 2, then the boundary condition is $\varepsilon_{r1}E_{n1} = \varepsilon_{r2}E_{n2}$. A more generalized boundary

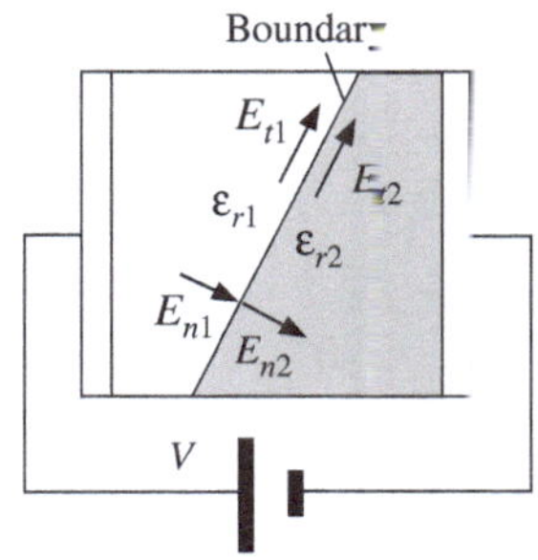

Boundary conditions between dielectrics. E_t is the tangential and E_n is the normal electric field to the boundary. Subscripts 1 and 2 refer to dielectrics 1 and 2. $\varepsilon_{r1}E_{n1} = \varepsilon_{r2}E_{n2}$ and $E_{t1} = E_{t2}$.

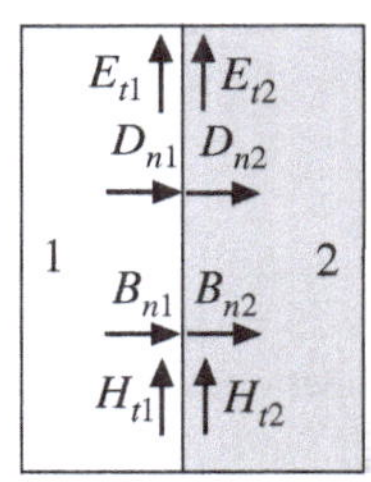

Boundary conditions General boundary condition in terms of E, D, B and H Subscripts n and t represent normal and tangential components to the boundary; there is no net space charge density nor a surface current between the two media.

condition involves four field components, the electric field E, the displacement D, the magnetic field (or flux density) B and the magnetic field intensity (or magnetizing field) H. Let subscripts n and t represent normal and tangential components to the boundary. If there is no net space charge surface density and no surface current between the two media 1 and 2, then the most general boundary condition can be expressed in simple notation as (i) $E_{t1} = E_{t2}$; (ii) $D_{n1} = D_{n2}$; (iii) $B_{n1} = B_{n2}$; (iv) $H_{t1} = H_{t2}$.

Bragg diffraction condition, see *diffraction grating*.

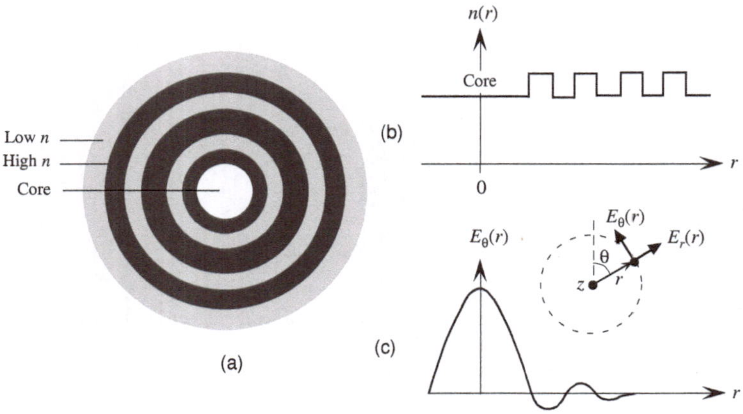

Bragg fiber (a) Concentric high and low refractive index layers clad the core. (b) The refractive index profile across the fiber. (c) The field E_θ across the fiber where the magnitude decays as the field penetrates the stratified cladding. Note: (r, θ, z) are cylindrical coordinates with z along the fiber core axis. All are schematic only.

Bragg fiber is a cylindrically symmetric fiber consisting of a low-index dielectric core surrounded with Bragg cladding, that is almost-periodically alternating layers of dielectric media with high- and low-refractive index. The mode confinement is therefore different from that of a conventional fiber, obtained by total internal reflection from a higher-index core on a lower-index cladding. Light propagation occurs in a relatively low refractive index core as a result of the minimization of power flow by using a stratified cylindrically symmetric low/high index cladding as depicted in the first figure. The cladding is normally composed of a periodic series of alternating shells of high and low index material surrounding the guiding core, in which the alternating pairs of layers are of half-wave thickness. This principle ensures that there is an exponential damping of the radial field, such that after several pairs of such cladding structure, the field decay is almost complete, serving as an approximately perfect mirror for the guided mode of interest. The concept of such structures was first introduced by Yeh *et al.* in 1978, who extended the concepts of planar waveguide structures and antireflection stacks to that of cylindrical waveguide geometries. These Bragg fibers may be considered to be a special case of a one-dimensional photonic crystal, and are usually distinguished from Holey fibers which are two-dimensional photonic crystals. An important application of Bragg fibers is to so-called air-core fibers – fibers in which light propagation occurs within air regions of a given fiber rather than, for example, in glass. The importance of light guiding in the air region is that it enables extremely low loss propagation (typically to <0.2 dB/km) and hence the possibility of raising the optical power to levels where nonlinear optical phenomena may be exploited in such systems, one of the important phenomena in this class being so-called *stimulated Raman scattering*. The first demonstration of an air-silica Bragg fiber was by G. Vienne *et al.* The second figure shows micrographs of such a fiber. Such fibers can support a single polarization nondegenerate mode, while retaining circular symmetry – for example, a single TE-polarized mode, termed a TE-Bragg fiber. In such fibers, the Bragg reflection condition is forced to coincide with the Brewster angle condition (noting that TM polarized light is

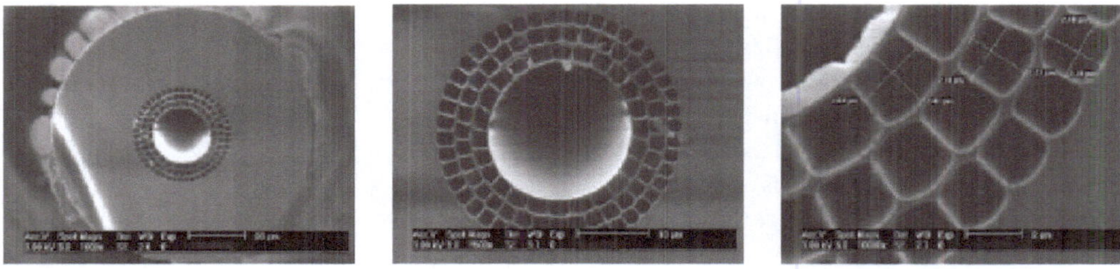

Bragg fiber SEM pictures of a 90 mm fiber, showing the entire fiber (left), the end facet (center) and the entire miscrostructure (right). (P. Yeh, A. Yariv and E. Marom, *J. Opt. Soc. Am.*, **68**, 1196, 1978; G. Vienne, Y. Xu, C. Jakobsen, H. J. Deyerl, T. P. Hansen, B. H. Larsen, J. B. Jensen, T. Sorensen, M. Terrel, Y. Huang, R. Lee, N. A. Mortensen, J. Broeng, H. Simonsen, A. Bjarklev and A. Yariv, Post Deadline Paper PDP25, *Optical Fiber Conference* 2004, Los Angeles, 22–27, Optical Society of America.)

not reflected at the Brewster angle), resulting in TM and hybrid polarization modes suffering very high confinement losses, whilst TE modes can be supported with orders of magnitude lower confinement loss. Moreover, the number of TE modes supported can be reduced to one by appropriately adjusting the core size, to furnish a single-mode air guide which potentially can eliminate both polarization and alignment problems.

Bragg fiber-optic sensor is a sensor utilizing *Bragg gratings* which act as very narrow linewidth filters which can selectively reflect a narrow band of light wavelengths – when these filters are strained, their period changes, and the reflected wavelength range shifts in proportion to the change in period. Gratings of different initial period may be distributed along a fiber and located at different points in a structure under test. In this case, since a given fiber can process in parallel these different wavelengths, one can simultaneously monitor the influence of strain at different locations. This is particularly important for monitoring the health of large structures such as bridges in a number of locations and over a period of time. Strain sensitivity, change in the Bragg wavelength per unit change in the strain ($d\lambda_B/d\varepsilon$), on the order of 1 pm/μstrain is possible, and similarly temperature and pressure changes can be measured with precision on the order of 10 pm/°C and –0.4 pm/bar, respectively. In contrast to electrical sensors, such optical sensors are immune to electromagnetic fields and the electrical conduction properties of the environment. Electromagnetic immunity and mechanical and chemical durability are the primary advantages in using fiber optic sensors. (S. W. James and R. P Tatam, *Meas. Sci. Technol.*, **14**, R49, 2003.)

Bragg wavelength is a particular wavelength λ_B of electromagnetic radiation that satisfies the Bragg diffraction condition,

$$q\frac{\lambda_B}{n} = 2\Lambda \sin\theta$$

where Λ is the periodicity of the diffracting structure, n is the refractive index of the medium diffracting the waves and θ is the diffraction angle; and q is an integer (1,2, . . .), so that the electromagnetic radiation becomes diffracted.

Bremsstrahlung radiation (German, "braking radiation") is the emission of electromagnetic (EM) waves by an accelerating or decelerating charge. According to the classical theory of electromagnetism, when a charge is accelerated (or decelerated) it will emit EM waves, the frequencies of which depend on the acceleration. When electrons are accelerated in a vacuum x-ray tube and impinge and enter a metal target, the energetic electrons interact with the columbic field of the nuclei in the metal and decelerate. This deceleration or "braking" results in the emission of EM waves, and the radiation is called "braking radiation", i.e. Bremsstrahlung radiation. The total (or integrated) Bremsstrahlung intensity I_B depends on the charge and mass of the impinging particle, *ze* (*e* is the elementary charge) and *m*, respectively, and charge of the target metal nucleus, *Ze*, as

$$I_B \propto \frac{(Ze)^2(ze)^4}{m^2}.$$

It is clear that Bremsstrahlung intensity is reduced if the impinging energetic particles are heavy, such as protons, which is the reason electrons are used in the generation of this type of radiation. Further, target metals with higher *Z* are preferred. Electrons entering the target interact with the nuclei at different distances, which means that they experience different amounts of deceleration and hence emit different frequencies of radiation, or different photon energies. The Bremsstrahlung radiation has an energy distribution that depends on the target metal and the energy of the incident electrons with its maximum E_{max} determined by the *maximum* kinetic energy KE_{max} of the incident electrons, which is the peak accelerating anode voltage, denoted as kVp (kilovolts peak), so that E_{max} is *e*kVp.

Brewster (B) is a unit used in photoelasticity to represent the induced birefringence for a difference of 10^{12} N m^{-2} in the principal stresses, i.e. 1 B = 10^{-12} m^2 N^{-1}.

Brewster's angle or **polarization angle**, θ_p, is the angle of incidence which results in the reflected wave having no electric field in the plane of incidence (plane defined by the incident ray and the normal to the surface). The electric field oscillations are in the plane perpendicular to the plane of incidence. When the angle of incidence of a light wave is equal to the polarization angle θ_p, the field in the reflected wave is then always perpendicular to the plane of incidence and hence well-defined. The reflected wave is then plane polarized. This special angle is given by

$$\tan\theta_p = \frac{n_2}{n_1}$$

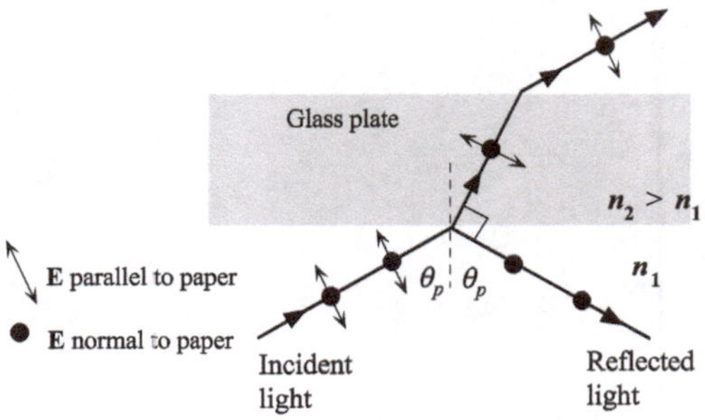

Brewster's angle When an unpolarized light wave is incident at the Brewster angle, the reflected wave is polarized with its optical field normal to the plane of incidence that is parallel to the surface of the glass plate. The angle between the refracted (transmitted) beam and the reflected beam is 90°.

Brewster's angle David Brewster (1781–1868), a British physicist, formulated the polarization law in 1815. He was a Professor of Physics at St. Andrew's University and later the Principal of the University of Edinburgh. (Courtesy of AIP Emilio Segrè Visual Archives, Zeleny Collection.)

where n_1 is the refractive index of the medium in which the incident light is traveling and n_2 is the index of the second medium that reflects the incident light. In addition, the transmitted (the refracted) wave has a greater field amplitude in the plane of incidence. By using a pile of glass plates inclined at the Brewster angle, one can construct a polarizer that provides a reasonable polarized light with the field in the plane of incidence.

Brightness see *luminance*.

Brightness sterance see *luminance*.

Brillouin scattering results from the scattering of light from high-frequency acoustic phonons ("sound") in a semiconductor. The acoustic perturbation of the material modulates the density, acting as a grating that moves at the speed of sound in the material. Light is scattered by this grating analogous to the scattering of x-rays by a crystal, and the scattering process may also be described by a form of *Bragg's law*,

$$m\lambda = 2\Lambda_{\text{phonon}} \sin\theta$$

where λ is the wavelength of incident light in the material, and Λ_{phonon} is the acoustic phonon wavelength, m is an integer (the diffraction order) and 2θ is the scattering angle. Since the dispersion relation for acoustic phonons allows for a continuum of phonon frequencies, Brillouin scattering results in a continuum of Stokes and anti-Stokes components, each being emitted in a different direction. In the former case, incident photons of energy $\hbar\omega_{\text{i}}$ give up part of their energy to generate acoustic phonons (i.e. known as phonon emission) of characteristic energy $\hbar\omega_{\text{phonon}}$, such that the scattered light will have a resultant energy, $\hbar\omega_{\text{Stokes-scattered}} = \hbar\omega_{\text{i}} - \hbar\omega_{\text{phonon}}$. Notice that this scattering process results in the emitted light being down-shifted in frequency. In the case of anti-Stokes scattered light, for a material in which the lattice is already in an excited state (e.g. the case of a material at high temperature with a large population of phonons), the scattering process can result in the emission of a photon with more energy as energy is in effect picked up from the lattice. For the anti-Stokes process $\hbar\omega_{\text{anti-Stokes-scattered}} = \hbar\omega_{\text{i}} + \hbar\omega_{\text{phonon}}$ and the resulting scattered light is up-shifted in frequency owing to phonon absorption. Since, typically, the population of phonons that can be absorbed is lower than those that can be emitted, by the factor $\exp(\hbar\omega_{\text{phonon}}/k_{\text{B}}T)$, Stokes emission lines tend to be more intense than anti-Stokes emission lines. *Stimulated Brillouin scattering* can occur under intense light illumination, resulting in the generation of coherent acoustic waves – because the light and acoustic waves are coherent, the material can act as a parametric amplifier. This case is distinct from the case of spontaneous Brillouin scattering discussed above. With stimulated scattering, the pump wave can generate an acoustic wave in the medium as a result of electrostriction. This, in turn, provides a feedback mechanism that can produce gain in the scattered intensity. The net effect is that a strong scattered signal can build up; so-called amplification. A key factor in achieving this condition is a high electrostrictive nonlinearity, becoming more pronounced with increasing pump power in the medium. Stimulated Brillouin scattering is an example of a traveling wave parametric interaction involving photons and phonons. When one of the interacting waves is a backward traveling wave, yielding spatially nonlinear distributed regeneration covering the entire medium, the system exhibits a typical backward traveling wave parametric interaction. It may oscillate with a distinct threshold. The generation of backward traveling light becomes known as optical *phase conjugation*.

Bruggeman equation, see *effective medium approximation*.

Built-in field is the internal field that exists in the depletion region of a *pn* junction due to exposed positive donor ions on the n-side, and negative acceptor ions on the p-side of the depletion region. If W_0 is the width of the depletion region (without an applied bias), for an abrupt *pn* junction the maximum field, E_0, at the metallurgical junction, is

$$E_0 = -\frac{2V_0}{W_0} = -\frac{eN_d W_n}{\varepsilon} = -\frac{eN_a W_p}{\varepsilon}$$

where V_0 is the built-in voltage, N_d and N_a are the donor and acceptor concentrations on the *n*- and *p*-sides, W_n and W_p are the widths of the depletion region on the *n*- and *p*-sides, and $\varepsilon = \varepsilon_0\varepsilon_r$ is the relative permittivity of the semiconductor. For a one-sided abrupt *pn* junction in which $N_a \gg N_d$, $E_0 = -eN_d W_0/\varepsilon$. A built-in field can also exist in an inhomogeneous semiconductor as a result of a nonuniform dopant concentration. For example, if the donor concentration in an *n*-type semiconductor varies spatially, then there will be a diffusion of electrons from high to low doped regions. This diffusion will expose positive donors, and will result in a built-in field. Suppose that the dopant concentration varies exponentially as $N_d(x) = N_0\exp(-bx)$, where *b* is a constant. Then the magnitude of the built-in potential is $|E_0| \approx k_B T/be$. See *built-in voltage, pn junction.*

Built-in voltage, V_0, is the voltage across a *pn* junction, going from a *p*- to an *n*-type semiconductor, in an open circuit; it is the potential difference across the depletion layer of a *pn* junction. It is not the voltage across the diode which is made up of V_0 as well as the contact potentials at the metal electrode to semiconductor junctions. Under open circuit or short conditions with no applied voltage or current sources, in the absence of photoexcitation, the sum of all the voltages should be zero, which means that V_0 + contact voltages at metal/*p*-type and *n*-type/metal junctions must sum to zero. If a voltage *V* is applied to a *pn* diode, this voltage drops across the depletion region and hence it directly modifies the built-in voltage V_0 by either reducing it (forward bias) or increasing it (reverse bias). The built-in voltage is responsible for the observation of a photocurrent when a *pn* junction is illuminated in a short circuit. For an abrupt *pn* junction,

$$V_0 = (k_B T/e)\ln\left(N_a N_d/n_i^2\right)$$

where N_a and N_d are the acceptor and donor concentrations on the *p*- and *n*-sides, and n_i is the intrinsic concentration. See *built-in field, pn junction.*

Bulk micro-machining see *MEMS.*

Buried double heterostructure laser diode is a double heterostructure semiconductor laser device that has its active region "buried" within the device in such a way that it is surrounded by low refractive index materials rendering the active region as a waveguide.

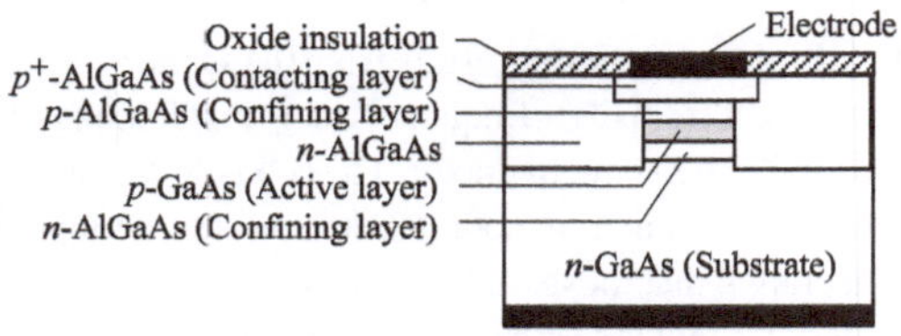

Buried double heterostructure laser diode Schematic illustration of the cross-sectional structure of a buried hetrostructure laser diode.

Burrus type device see *surface emitting LEDs.*

Burst noise see *excess noise.*

Burstein–Moss shift, see *band-to-band absorption.*

Butt jointed optical fibers are two fibers that are placed end-to-end with their flat ends as close as possible to allow light exiting one fiber to enter the other fiber.

C

Cadmium telluride (CdTe) is a direct bandgap II–VI semiconductor with the zinc blende structure. $a = 0.6482$ nm; $\rho = 5.87$ g cm^{-3}; $T_m = 1092$ °C; $c_s = 0.209$ J g^{-1} K^{-1}; $\kappa \approx 0.0628$ W cm^{-1} K^{-1}; $\alpha = 5.9 \times 10^{-6}$ K^{-1}; $\chi = 4.28$ eV; $T_{\text{Debye}} = 142$ K; $E_g = 1.49$ eV (300K), 1.61 (0 K); $\Delta_{so} = 0.80$ eV; $m_e^*/m_e = 0.090$; $m_h^*/m_e = 0.72$; $N_c = 7.3 \times 10^{17}$ cm^{-3}; $N_v = 1.7 \times 10^{19}$ cm^{-3}; $\mu_e \sim 1000$ cm^2 V^{-1}s^{-1}, $\mu_h \sim 100$ cm^2 V^{-1}s^{-1}; ε_r(DC) $= 10.20$, $\varepsilon_r(\infty) = 7.10$; $n = 2.6$ (infrared); $\hbar\omega_{op} = 17.8$ meV. For $\lambda > 0.840$ μm, and for λ in μm,

$$n^2 = 5.304 + \frac{1.899\lambda^2}{\lambda^2 - 0.573^2} - (4.188 \times 10^{-4})\lambda^2 - (2.391 \times 10^{-7})\lambda^3.$$

CdTe is also an optical material used in various applications such as lenses, wedges, prisms, beam splitters, antireflection coatings, windows, etc. operating typically in the 10–25 μm IR region, though it has a wide transmission range from 1 to 25 μm. It is used as an optical material for low power CO_2 laser applications. See *Reststrahlen absorption*.

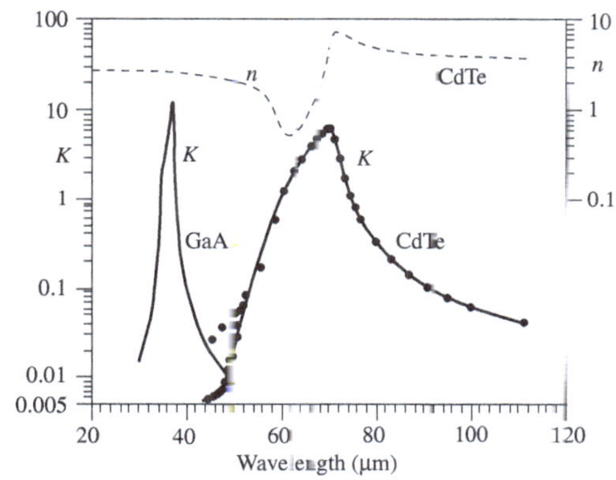

Cadmium telluride The refractive index and the extinction coefficient, n and K, of CdTe, and K for GaAs for comparison in the infrared region (reststrahlen absorption region). (From S. O. Kasap, *Principles of Electronic Materials and Devices, Third Edition*, McGraw-Hill, New York, 2006; Chapter 9.)

Calcium fluoride (CaF_2) is an optical material that has good transmission from about 0.17 to 7 μm, which includes the UV, visible and IR. Its crystal structure is fluorite-type cubic. Single crystals are grown by the vacuum Stockbarger technique. CaF_2 has been reported to be sensitive to thermal shock. $\rho = 3.18$ g cm^{-3}; $\nu_d = 94.96$; $E_g \approx 10$ eV; $T_m = 1423$ °C; $\alpha_L = 18.85 \times 10^{-6}$ (21×10^{-6} // to c-axis) K^{-1}; $\kappa = 9.71$ (21 // to c-axis) W m^{-1} K^{-1}; $c_s = 0.853$ J g^{-1} K^{-1}; $T_{\text{Debye}} = 5143$ K; $E = 75.8$ GPa; $G = 33.77$ GPa; $K = 82.71$ GPa; $H_{\text{Knoop}} = 158$ kg mm^{-2}; fracture strength $= 34$–157 MPa, depends on the surface finish, $\upsilon = 0.26$; $\varepsilon_r(1\text{ MHz}) = 6.71$; piezooptic coefficients, $q_{11} = -0.38$ pPa, $q_{12} = 1.08$ pPa, $q_{44} = 0.71$ pPa. Reststrahlen maxima at

$\lambda = 35$ μm and 25 μm; $dn/dT \approx -1.1 \times 10^{-5}$ K^{-1};

$$n^2 = 1 + \frac{0.3913\lambda^2}{\lambda^2 + (0.02911)^2} + \frac{0.6474\lambda^2}{\lambda^2 + (0.09496)^2} + \frac{3.374\lambda^2}{\lambda^2 + (32.44)^2}\ \lambda \text{ in } \mu\text{m}.$$

See *optical materials*.

Candela [cd] is a visual photometry unit (SI unit) for measuring the light *intensity* of a source which emits a luminous flux of one *lumen* in a particular direction within one unit of solid angle, i.e. 1 cd = 1 lm/sr, at a reference frequency $\upsilon = 540 \times 10^{12}$ Hz ($\lambda \approx 555$ nm), 1 cd = (1/683) W sr^{-1}. (Definition by the 16th Conférence Générale des Poids et Mesures – General Conference on Weights and Measures, 1979.) See *luminous intensity*.

Carbon dioxide (CO_2) laser is a gas laser where the optical transition takes place between molecular vibrational states of the CO_2 molecule, which can vibrate along three possible modes (each of them labeled by one quantum number): the asymmetric stretch mode ($00n_1$), the bending mode ($0n_20$), the symmetric stretch mode (n_300).

The pumping mechanism of the active medium (a gas mixture with typically $[CO_2] \approx 10\%$, $[N_2] \approx 40\%$, $[He] \approx 50\%$) is mostly indirect: an electric discharge sends N_2 molecules from their ground state to an excited metastable state; this energy is then transferred to CO_2 molecules by collisions. Two main transitions can take place between vibrational levels, from (001) to (020) or (100), at wavelengths close to 9.6 μm or 10.6 μm respectively (in the IR thermal range).

It should be noted that additional rotational modes are responsible for a lift of degeneracy, dividing each vibrational level into sublevels. This gives rise to a whole subset of possible spectral lines (with their own selection rules), separated by 20–50 GHz. Once the precise wavelength is determined (by any convenient selective mechanism), the active medium is well represented by a three-level system, with the peculiarity that the laser transition takes place between the two upper levels

Helium is used to help maintain the population inversion by favoring the relaxation to the ground level. The CO_2 laser can maintain continuous-wave operation and deliver high levels of power. From a technological point of view, the simplest CO_2 laser is a sealed tube (possibly with Brewster windows) between two reflectors.

Replacing one mirror by a diffraction grating enables one to select one out of many lines. The electric discharge can be continuous or in the radiofrequency range (≈ 80 MHz). The operation lifetime can reach a few thousand hours, limited only by CO_2 decomposition. In order to overcome this drawback, laser systems with an axial flow of gas have been developed. The CW output power can reach from 20 W to a few tens of kW, depending on the cavity geometry. Systems designed with a

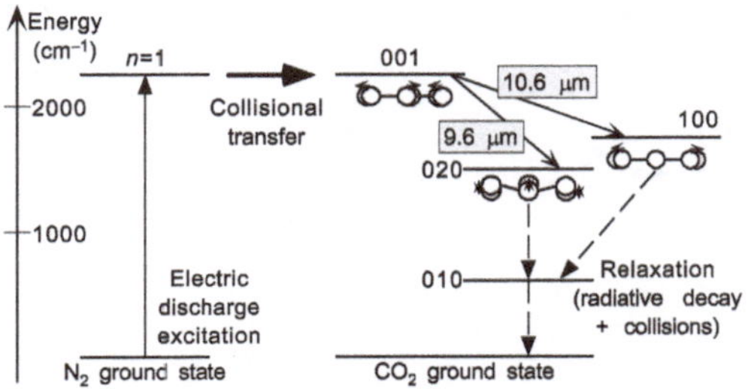

CO_2 laser Vibrational levels of the CO_2 molecule are populated through collisions with excited N_2 molecules. Rotational sublevels are not shown.

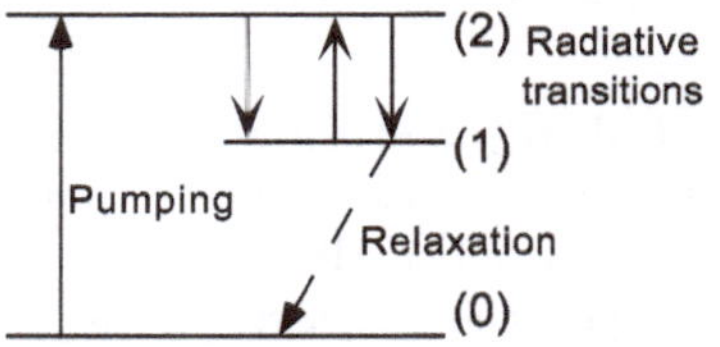

CO_2 laser Simplified three-level system for describing a CO_2 transition. The laser radiation takes place between the two upper levels.

 Cardinal points

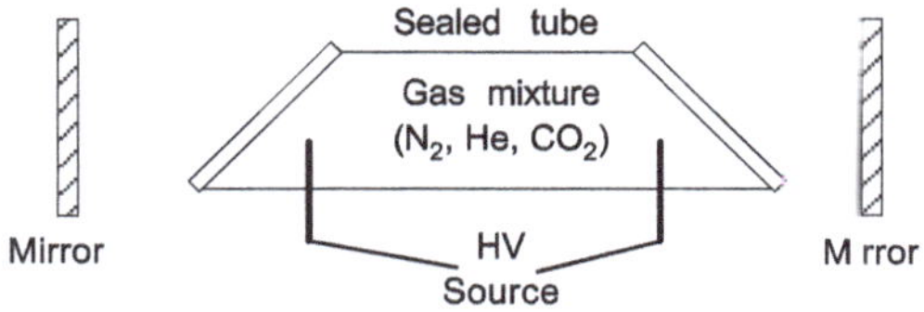

CO_2 laser Sealed tube containing a gas mixture of CO_2, N_2 and He between two reflectors. Pumping is achieved by using an electric discharge.

transverse flow (perpendicular to the electric discharge) are even more compact for the same power performance (10–20 kW). The Transversely Excited Atmospheric (TEA) CO_2 laser, with its gas mixture at atmospheric pressure, when subject to high voltage pumping pulses can deliver short optical pulses (10 ns) with a peak power higher than 10 MW at a repetition rate about 1 s^{-1}.

There are a lot of industrial applications for CO_2 lasers. The high level of available power makes it especially interesting for materials processing, such as cutting, welding, drilling, surface processing, etc. At lower levels, it is also used for medical applications.

Other gas lasers such as HCN, H_2O or CO work on exactly the same principle, except the operating wavelengths are different.

Typical carbon dioxide laser parameters are (From W. T. Silfvast, *Laser Fundamentals, Second Edition*, Cambridge University Press, 2004): laser transition probability = 0.25 s^{-1}; upper laser level lifetime = 4 s; stimulated emission cross-section = 3×10^{-22} m^2; spontaneous emission linewidth and gain bandwidth = 6×10^7 Hz; population inversion density = 3×10^{21} m^{-3}; small-signal gain coefficient = 0.9 m^{-1}; laser gain-medium length = 0.3–2.0 m; single pass gain = 1.3–6.0; gas pressure, 50–760 torr; electron temperature, 5000–10,000 K; gas temperature, 200–400°C; mode of operation may be CW or pulsed; output power, 1 W – 10 kW; TEM_{00} or multimode.

Cardinal points in *geometric optics* are characteristic points along the main axis of a centered optical system, and consist of the *focal* points (F,F′), the *principal* points (H,H′) and the *nodal* points (N,N′). Qualitatively they represent the focal, principal and nodal points of an optical system, such as a lens. A couple of points (A,A′) on the axis are said to be conjugated by the system if, to any incident ray passing through A, there corresponds an emergent ray intersecting the axis at (or in the immediate vicinity of) A′ (*approximate stigmatism*) (see *paraxial optics*). A and A′ are then called the *object* and *image*, respectively.

The *object focal point* F is such that its conjugate image point lies "at infinity" on the axis; that is, to any incident ray passing through F, there corresponds a ray emerging parallel to the axis. The *image focal point* F′ is the conjugate image of an object point lying "at infinity" on the axis; that is, to any incident ray parallel to the axis, there corresponds an emergent ray intersecting the axis at (or in the immediate vicinity of) F′. (Note: F and F′ are *not* mutually conjugated by the system.)

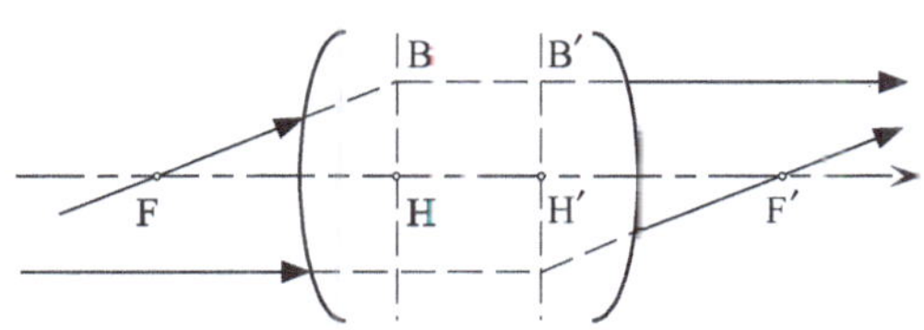

Cardinal points The *principal points* (H,H′) are defined as a couple of conjugate points such that the lateral magnification is exactly equal to unity: $\overline{H'B'}/\overline{HB} = g_y = +1$.

Geometrically, the *principal object plane* containing H is defined as the locus of the intersections between transmitted rays emerging parallel to the axis and the corresponding incident rays. The *principal image plane* containing H′ is the locus of the intersections between incident rays parallel to the axis and the corresponding emergent rays.

The *nodal points* (N,N′) are defined as a couple of conjugate points such that the angular magnification is equal to unity; $\alpha'/\alpha = g_\alpha = +1$. If the two media surrounding the system are identical, then (N,N′) are the same as (H,H′).

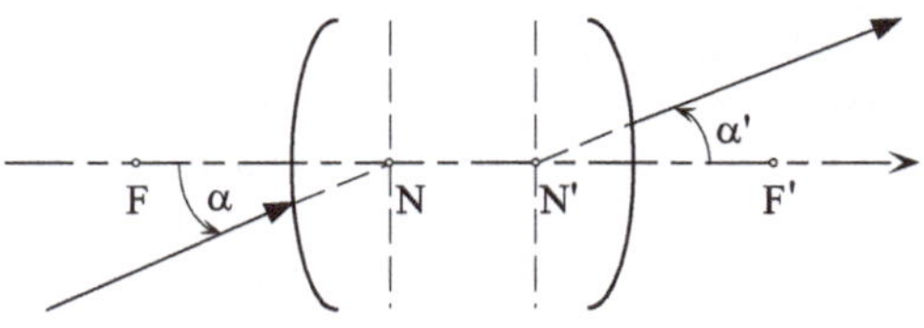

Cardinal points Definition of nodal points.

Carrier confinement in optoelectronic devices such as LEDs and laser diodes refers to the restriction of injected charge carriers to a small volume to increase the carrier concentration. The restriction of injected carriers is achieved by using a heterojunction so that there is a step change in the conduction band edge (ΔE_c) or a step change in the valence band edge (ΔE_v). The increase in the concentration of injected carriers results in a higher rate of recombination and hence greater photon emission.

Cascaded linear system model for imaging is used to characterize the performance of many imaging systems in terms of signal-transfer and noise-transfer relationships. In the cascaded linear systems model, an imaging system is described as cascades of simple and independent elementary stages. In most cases, the input and the output of each stage is a distribution of quanta. These quanta may be x-rays, light, or electrons; as an example consider x-rays. The random nature of image quanta creates statistical fluctuation in image signals contributing to image formation. Image noise is defined as random variations in image signals. The noise in the number of x-rays, or signal incident on the detector, is given by the *Poisson fluctuations*. For example, if the mean incident x-ray fluence on a detector is $\overline{\Phi}_0$ photons per unit area, the input noise power spectrum (NPS) in the number of x-rays incident on the detector is given by $S_0 = \overline{\Phi}_0$. The signal and noise are passed through various stages, such as gain, blurring, addition, and aliasing in an imaging system. At each stage the signal and NPS are calculated from the signal and NPS of the previous stage and the characteristics of the current stage.

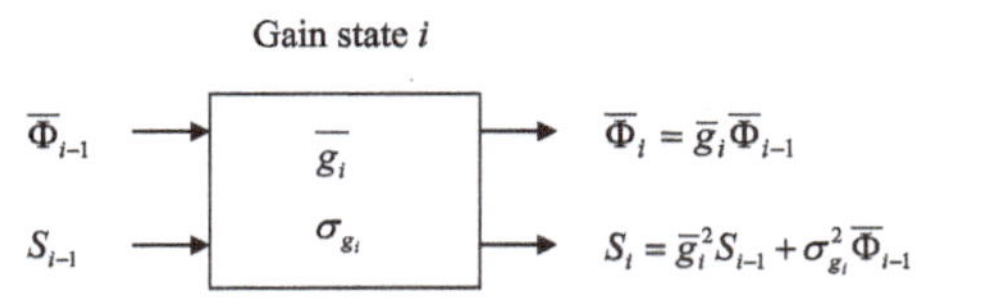

Cascaded linear system A gain state i with an average gain $\overline{g}_i$ and gain variance $\sigma_{g_i}^2$.

There are three elementary processes that have an important role in understanding signal and noise transfer: (a) stochastic amplification, (b) stochastic blurring, and (c) deterministic blurring. In stochastic amplification each quantum in a distribution is converted into g quanta, where g is a positive random variable. For the stochastic gain stage i the output mean signal quanta per unit area ($\overline{\Phi}_i$) and NPS [$S_i(f)$] are given by

$$\overline{\Phi}_i = \overline{g}_i\overline{\Phi}_{i-1}$$
$$S_i(f) = \overline{g}_i^2 S_{i-1}(f) + \sigma_{g_i}^2\overline{\Phi}_{i-1}$$

where f is the spatial frequency, $\overline{\Phi}_{i-1}$ and $S_{i-1}(f)$ are the mean number of quanta and the NPS incident on stage i, respectively, and $\overline{g}_i$ and $\sigma_{gi}{}^2$ are the mean gain and variance of the gain of the ith stage. It is seen from the above equations that the stochastic amplification increases the noise associated with the quanta in two ways. First, the quantum noise is itself amplified, and second, noise associated with the stochastic nature of the amplifying mechanism is introduced into the output.

Stochastic blurring is caused by random scattering of quanta in a distribution. That is, each quantum is randomly relocated to a new location with a certain probability. For the stochastic scattering stage i the output mean signal quanta per unit area ($\overline{\Phi}_i$)

Cascaded linear system model for imaging

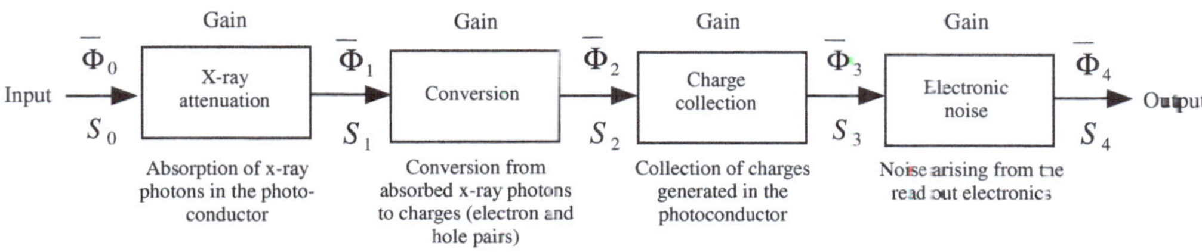

Cascaded linear system The simplest serially cascaded linear system model for a direct conversion (photoconductor) based flat-panel digital x-ray image detector.

and NPS, usually denoted by $S_i(f)$ (and not to be confused with the signal), are given by

$$\overline{\Phi}_i = \overline{\Phi}_{i-1}$$
$$S_i(f) = [S_{i-1}(f) - \overline{\Phi}_{i-1}]\mathrm{MTF}^2(f) + \overline{\Phi}_{i-1}$$

where MTF(f) is the modulation transfer function of stage i. Note that $\overline{\Phi}_{i-1}$ is the uncorrelated Poisson noise at the input of the ith stage. That is, the correlated noise component of the input noise passes through the squared MTF but the uncorrelated component does not.

Deterministic blurring can be viewed as a redistribution of the signal by weights (i.e. the signal passing through a unity-gain linear filter and is described by the convolution integral). For the deterministic blurring stage i the output mean signal quanta per unit area ($\overline{\Phi}_i$) and NPS [$S_i(f)$] are given by

$$\overline{\Phi}_i = \overline{\Phi}_{i-1}$$
$$S_i(f) = S_{i-1}(f)\,\mathrm{MTF}^2(f).$$

A very simple cascaded linear system model for a photoconductor based (a-Se) flat-panel direct conversion digital x-ray image detector describes the system by a cascade of serial elementary processes or stages as shown in the second figure, where there are four processes: x-ray attenuation and absorption; conversion of absorbed x-ray photons to charges; collection of charges, and the addition of electronic noise. While this simple serially cascaded model can be useful, a better model for the overall detector performance uses series and parallel cascading as shown in the third figure. Each of these

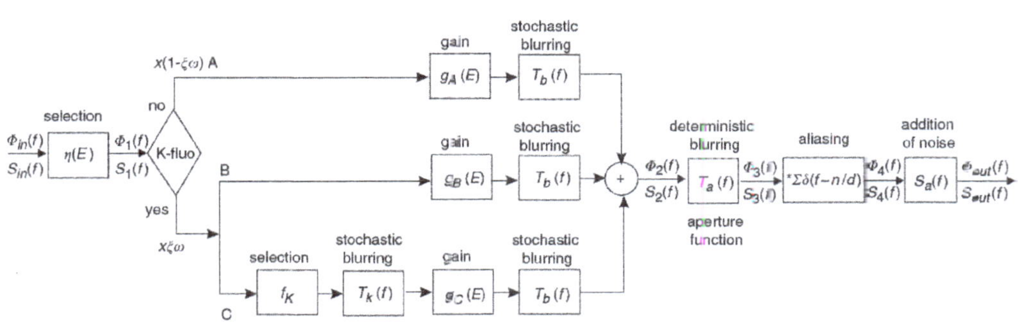

Cascaded linear system An example of a series and parallel cascaded linear system model for a direct conversion (photoconductor) based flat-panel digital x-ray image detector. The flow diagram shows the parallel and serial processes involved in the image formation. (M. Z. Kabir *et al.*, *IEE Proc. Circuits Devices Syst.*, **150**, 258, 2003.)

stages can be categorized as one of the following five processes: (1) gain or selection; (2) stochastic blurring; (3) deterministic blurring; (4) aliasing; and (5) addition.

The propagation of the signal after x-ray absorption is divided into three parallel paths: (A) when no K-fluorescence is generated, the absorbed x-ray energy is converted to charge locally; (B) when a K-fluorescent x-ray is produced, the remaining x-ray energy is converted to charge locally; and (C) when a K-fluorescent x-ray is reabsorbed remotely, its energy is converted to charge and causes image blurring. The K-fluorescence related constants include: the probability of K-shell interaction, ξ; the probability of K-fluorescence production, ω. $g_A(E)$, $g_B(E)$ and $g_C(E)$ are the conversion gains of the three parallel paths for an incident x-ray photon with energy E. $T(f)$ represents MTF of any stage. The total signal is obtained as a combination of signals from the three parallel paths. The signals are further propagated through the stages of pixel aperture function and the addition of electronic noise, as shown in the figure.

Catastrophic optical damage (COD) is a failure mode that can occur under strong injection conditions in high power cleaved-edge AlGaAs/GaAs and InGaAs/InP semiconductor laser diodes. COD arises from a feedback of energy in the optical cavity leading to thermal run-away at the facets, resulting in facet melting in the final stages of COD. Key to the feedback process is the role of surface states arising in the vicinity of the laser facets. In the case of AlGaAs-based lasers, for example, the material is very susceptible to severe levels of facet oxidation resulting in the formation of surface states which act as nonradiative recombination centres. Under high-power excitation, there is a high concentration of excess carriers, and those located within a minority carrier diffusion length of the nonradiative recombination centers (i.e. in the vicinity of the facets) will undergo nonradiative recombination – the result is the generation of heat within a thermal diffusion length of the source of the

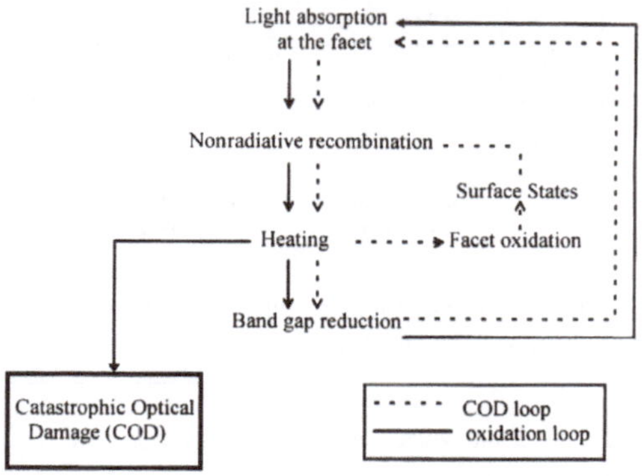

Catastrophic optical damage Feedback loops leading to catastrophic optical damage in high power semiconductor diode lasers. (Adapted from Fig. 7.16, of M. Fukuda, *Optical Semiconductor Devices*, John Wiley, 1999, who derived it from text of C. H. Henry, P. M. Petroff, R. A. Logan and F. R. Merritt, *J. Appl. Phys.*, **50**, 3721, 1979.)

heat generation. Owing to the reduction in a semiconductor's bandgap with increasing temperature, associated with the region of increased temperature, will be a region of decreased bandgap. However, since for photon energies with energy in excess of the bandgap, the absorption coefficient rises owing to a higher joint density of states, the locally hotter regions (in the vicinity of the laser facets) will act as regions with higher average absorption coefficient for the stimulated photons in the laser cavity. In the feedback cycle, the additional excess carriers produced in this location are again subject to nonradiative recombination and hence additional local heating – the thermal run-away cycle thus continues until the material melts near the facets. It should also be noted that the concentration of nonradiative surface states also increases as driven by another feedback loop – increasing the facet temperature also increases the rate of facet oxidation. Since the facets actually constitute optical mirrors (i.e. in a Fabry–Perot laser structure), damaging the facets is tantamount to damaging the mirrors and hence destroying the optical cavity responsible for optical feedback and lasing action – hence, the onset of facet melting is the onset of catastrophic optical damage. Typically, COD occurs at power densities on the order of MW/cm^2. A solution to counter the problems with COD is to use so-called nonabsorbing mirror structures, or to use a combination of chemical passivation of the facets with appropriate coatings to minimize access to the facet region by ambient species (leading to oxidation). COD is a major concern for telecommunications systems manufacturers who make use of such high-power 980 nm AlGaAs lasers as pump sources for erbium-doped fiber amplifiers used in long-haul wide-area networks.

Cathode-ray tube (CRT) is a vacuum tube device that uses an electron beam to provide an image on a luminous screen. Thermionically emitted electrons from a heated cathode at one end of the tube are accelerated by ring-shaped anodes (hollow cylinders) which form an electrostatic lens system to focus the beam. Electrons pass through the anode rings and form a beam that strikes the enlarged end of the tube. The latter is coated on the inside with a fluorescent material (a phosphor) so that the region struck by the electron beam becomes luminous. A negative voltage applied to a grid between the cathode and the anode controls the intensity of the electron beam. The cathode (including the filament), grid and the anode assembly is termed an electron gun. The electron beam from the electron gun can be deflected by x- and y-deflection plates or by magnetic coils. See *luminescence*.

Cathodoluminescence, see *luminescence*.

Cauchy dispersion equation (1836) is a useful refractive index (n) vs. wavelength of light (λ) dispersion relationship usually stated in the following short form

$$n = A + \frac{B}{\lambda^2} + \frac{C}{\lambda^4}$$

where A, B and C are material-specific constants. Typically, the Cauchy equation is used in the visible spectrum for various optical glasses and applies to normal dispersion. The third term is sometimes dropped for a simpler representation of n vs. λ behavior. The original expression was a series in light frequency,

$$n = a_0 + a_2\lambda^{-2} + a_4\lambda^{-4} + a_6\lambda^{-6} + \cdots \lambda > \lambda_h$$

or

$$n = n_0 + n_2\omega^2 + n_4\omega^4 + n_6\omega^6 + \cdots \omega < \omega_h$$

where $\hbar\omega$ is the photon energy (ω is the angular frequency of light), $\hbar\omega_h = hc/\lambda_h$ is the optical

Table: Cauchy dispersion equation Approximate Cauchy dispersion parameters for a few materials from various sources.

Material	$\hbar\omega$ (eV) Min	$\hbar\omega$ (eV) Max	n_{-2} (eV^2)	n_0	n_2 (eV^{-2})	n_4 (eV^{-4})
Diamond	0.0500	5.47	-1.0700×10^{-5}	2.3780	0.00801	1.04×10^{-4}
Si	0.0020	1.08	-2.0400×10^{-8}	3.4189	0.0815	0.0125
Ge	0.0020	0.75	-1.0000×10^{-8}	4.0030	0.2200	0.1400
AlSb	0.0620	1.24	-6.1490×10^{-4}	3.1340	0.5225	0.2185
GaP	0.0571	3.60	-1.7817×10^{-3}	3.0010	0.0784	0.0058
GaAs	0.0496	2.90	-5.9737×10^{-4}	3.3270	0.0779	0.0151
InAs	0.0496	2.40	-6.1490×10^{-4}	3.4899	0.0224	0.0284
InP	0.08	3.18	-3.0745×10^{-4}	3.0704	0.1788	
ZnSe	0.113	1.24	-4.6117×10^{-4}	2.4365	0.0316	0.0026
KDP n_o	0.62	6	-1.736×10^{-2}	1.5045	0.00181	3.3×10^{-5}
KDP n_e	0.62	6	-4.356×10^{-3}	1.4069	0.00186	2.1×10^{-5}

excitation threshold (e.g. bandgap photon energy), and $a_0, a_2, \ldots$ and $n_0, n_2, \ldots$ are constants. It has been found that a Cauchy relation of the form

$$n = n_{-2}(\hbar\omega)^{-2} + n_0 + n_2(\hbar\omega)^2 + n_4(\hbar\omega)^4$$

can be used satisfactorily over a wide range of photon energies.

Originally called the elastic-ether theory of the refractive index, the Cauchy equation has been widely used for many materials, although in recent years it has been largely replaced by the Sellmeier equation, see *refractive index, Sellmeier equation.* (A. L. Cauchy, *Bull. Sci. Math.*, **14**, 6–10, 1830); A. L. Cauchy, *Mémoire sur la Dispersion de la Lumière* (Prague: Calve, 1836); data from D. Y. Smith, M.Inokuti and W. Karstens, *J. Phys.: Condens. Matter*, **13**, 3883, 2001.)

Cauchy distribution, also known as the **Cauchy–Lorentz probability distribution,** is a *probability distribution function* that has a Lotentzian lineshape-type behavior, and has the functional form given by

$$f(x) = \frac{\gamma}{\pi[(x - x_0)^2 + \gamma^2]}$$

in which x_0 is the location of the maximum of the function, and γ is a parameter that characterizes the half width at half maximum of the distribution, and x is the variable of interest.

Caustic see *spherical aberration.*

Cavity lifetime in photonics represents an average lifetime or an overall relaxation time τ_{ph} of a photon within an optical cavity arising from all photon loss processes, e.g. losses through the end mirrors, scattering and loss through the cavity sides, free carrier absorption, etc. It is useful to represent the cavity loss lifetime τ_{ph} in terms of an attenuation coefficient α_{tot} that represents the fractional loss per unit distance, $\alpha_{tot} = (v_g \tau_{ph})^{-1}$, where v_g is the group velocity of the wave inside the cavity, which can be approximated by *c/n*, where *n* is the refractive index. For example, for a simple Fabry–Perot cavity with end reflectances R_1 and R_2, and loss coefficient γ that accounts for losses from scattering, losses from sides, free carrier absorption, etc. then the total loss coefficient α_t is

$$\alpha_t = \gamma + (1/2L)\ln[1/(R_1 R_2)] \quad \text{and}$$
$$\tau_{ph} = n/(\alpha_t c)$$

where n is the refractive index of the medium.

Cavity quantum electrodynamics (QED) in quantum computing has been used to control and manipulate *qubits.* In this case, atoms may be made to interact strongly with photons trapped between extremely reflective mirrors. It is possible to entangle the states of an atom with the light field to represent quantum logic gates.

Cavity radiation, see *black body radiation.*

CCD, see *charge coupled devices.*

Central dark ground method of observation, or **strioscopy**, see *spatial filtering.*

Centrosymmetric crystals have a unit cell with a center of symmetry. If we draw a vector **r** from point O to any charge and then reverse the vector, i.e. make it −**r**, we will find the same charge. If, on the other hand, we do *not* find the same charge when we reverse the vector, then the unit cell is *noncentrosymmetric*; **r** and −**r** point to different ions. Noncentrosymmetric crystals exhibit piezoelectricity.

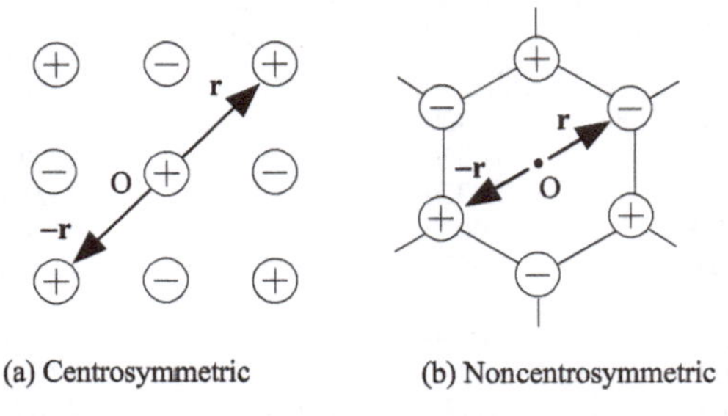

Centrosymmetric crystals (a) Centrosymmetric (b) Noncentrosymmetric.

Table: CGS system Selective physical constants in the CGS system. *e*, elementary charge; m_e, mass of electron; *c*, speed of light; *k*, Boltzmann's constant; *h*, Planck's constant; N_A, Avogadro's number; *u*, atomic mass unit; *α*, fine structure constant; r_0, Bohr's radius; μ_B, Bohr magneton; σ_S, Stefan-Boltzmann constant.

e	m_e	c	k	h	N_A	u	α	r_0	μ_B	σ_S
4.803×10^{-10} esu	9.109×10^{-28} g	2.998×10^{10} cm/s	1.381×10^{-16} erg/K	6.626×10^{-27} erg·s	6.022×10^{23} mole^{-1}	1.661×10^{-24} g	7.297×10^{-3}	5.292×10^{-9} cm	9.274×10^{-21} erg/gauss	5.670×10^{-5} erg/(s cm^2 K^4)

CGS system is based on using the centimeter, gram and second as units, hence the abbreviation CGS, and formulating certain physical laws so that they seem to appear in a simplified fashion, in particular in electrostatics and magnetism. The force is measured in dynes and energy in ergs. 1 dyne is a force that will accelerate a mass of 1 g by 1 cm s^{-2}, and is 10^{-5} N. 1 erg of energy is expanded when a force 1 dyne is displaced by 1 cm in the opposite direction to the force. In the CGS *electrostatic unit*

Table: CGS system CGS and MKS quantities compared, and Maxwell's equations in CGS and MKS/SI units.

Quantity	CGS	MKS/SI	Comment
Charge, Q	1 statcoloumb	Coulomb	
Voltage, V	statvolt or erg/esu	V	1 statvolt = 300 V
Field, **E**	statvolt/cm or dyn/esu	V/m	1 statvolt / cm = 3×10^4 V/m
Intensity/Irradiance	erg cm^{-2} s^{-1}	W m^{-2}	
Magnetic field intensity, **H**	oerstead (Oe)	A/m	1 Oe = $10^3/(4\pi)$ A/m = 79.577 A/m
Magnetic field intensity, **B**	gauss (G)	Tesla (T) or Wb m^{-2}	1 G = 10^{-4} T
Electric displacement, **D**	$\mathbf{D} = \mathbf{E} + 4\pi\mathbf{P}$	$\mathbf{D} = \varepsilon_o\mathbf{E} + \mathbf{P}$	General
Electric displacement, **D**	$\mathbf{D} = \varepsilon_r\mathbf{E}$	$\mathbf{D} = \varepsilon_r\varepsilon_o\mathbf{E}$	Displacement in linear, isotropic media
Lorentz force, **F**	$\mathbf{F} = q[\mathbf{E} + (\mathbf{v}/c)\times\mathbf{B}]$	$\mathbf{F} = q[\mathbf{E} + \mathbf{v}\times\mathbf{B}]$	
Polarization, **P**	$\mathbf{P} = \chi\mathbf{E}$	$\mathbf{P} = \varepsilon_o\chi\mathbf{E}$	Susceptibility, linear system χ = electric susceptibility
Polarizability α	$\mathbf{p} = \alpha\,\mathbf{E}_{local}$ [cm^3]	$\mathbf{p} = \alpha\,\mathbf{E}_{local}$ [C m^2 V^{-1}]	$\alpha(\text{SI}) = (4\pi\varepsilon_o 10^{-6})\alpha(\text{CGS})$
Permittivity	$\varepsilon_r = 1 + (4\pi)\chi$	$\varepsilon_r = 1 + \chi$	Linear medium ε_r = relative permittivity
Lorentz local field	$\mathbf{E}_{local} = (4\pi/3)\mathbf{P}$	$\mathbf{E}_{local} = (1/3\varepsilon_o)\mathbf{P}$	Does not apply to dipolar polarization
Clasius–Mossotti	$\dfrac{\varepsilon_r - 1}{\varepsilon_r + 2} = \dfrac{1}{3\varepsilon_o}\sum N_i\alpha_i$	$\dfrac{\varepsilon_r - 1}{\varepsilon_r + 2} = \dfrac{4\pi}{3_o}\sum N_i\alpha_i$	α_i has different units in CGS and MKS systems
Poynting vector	$\mathbf{S} = (c/4\pi)\mathbf{E}\times\mathbf{H}$	$\mathbf{S} = \mathbf{E}\times\mathbf{H}$	Flow of energy per unit area per unit time
E and **B**	$\nabla\times\mathbf{E} = -(1/c)\partial\mathbf{B}/\partial t$	$\nabla\times\mathbf{E} = -\partial\mathbf{B}/\partial t$	Maxwell's equation
H, **D** and **J**	$\nabla\times\mathbf{H} = (1/c)\partial\mathbf{D}/\partial t + (4\pi/c)\mathbf{J}$	$\nabla\times\mathbf{H} = \partial\mathbf{D}/\partial t + \mathbf{J}$	Maxwell's equation. **J** = conduction current density
B	$\nabla\cdot\mathbf{B} = 0$	$\nabla\cdot\mathbf{B} = 0$	Maxwell's equation
D and **E**	$\nabla\cdot\mathbf{D} = 4\pi\rho$ Isotropic media $\nabla\cdot\mathbf{E} = 4\pi\rho/\varepsilon_r$	$\nabla\cdot\mathbf{D} = \rho$ Isotropic media $\nabla\cdot\mathbf{E} = \rho/\varepsilon_o\varepsilon_r$	Maxwell's equation ρ = net free space charge density

(esu) *system,* Coulomb's law between two charges Q_1 and Q_2 becomes a simple equation

$$F = \frac{Q_1 Q_2}{r^2}$$

in which the units for the charge (Q_1 and Q_2) are *statcoloumbs* (not coloumbs). 1 statcoloumb is 1 esu (electrostatic unit) and has the dimensions of $[(\text{mass})^{1/2}(\text{length})^{3/2}/(\text{time})]$, i.e. $(\text{g}\cdot\text{cm}^3\cdot\text{s}^{-2})^{1/2}$. 1 statcoloumb is equivalent to 3.33564×10^{-10} C so that the elementary charge e is 4.803×10^{-10} esu. Effectively, 1 esu of charge is defined in such a way that the force between two 1 esu charges separated by 1 cm is 1 dyne. Notice that the factor $1/(4\pi\varepsilon_o)$ in the SI system has been effectively set to unity in the CGS system. One striking and convenient feature of the CGS system is that the fields **E**, **D**, **B** and **H** all have the same units, erg/cm^3. Further, in the esu system, the equation between the field generated and the current element generating the field also becomes simplified, and does not have the factor $\mu_0/4\pi$ present in the SI system.

The CGS system also has two other versions, called **Gaussian** and **electromagnetic unit (emu)** systems. The Gaussian system is similar to the esu system in terms of the Coulomb law; both have the same form, but the Gaussian system has a different proportionality constant in the expression for the magnetic field from a current element.

Chalcogenide glasses are inorganic glasses based on one or more of the chalcogen elements, S, Se, and Te, combined with other elements such as Ge, As, Sb, Ga, etc. These glasses are typically prepared by melt quenching techniques, that is, direct synthesis from elements or compounds in evacuated quartz ampoules at high temperatures where the elements or compounds are allowed to melt and mix, and then the quartz tube is quenched. They can also be prepared in thin (or thick) film form on a suitable substrate by vacuum deposition techniques such as thermal evaporation or sputtering. Chalcogenide glasses have low phonon energies (~300–500 cm^{-1}) and also generally have a wide transmission range from the visible to infrared region (up to 20 μm), but the transmission window depends on the chemical composition; for example, sulphide-based chalcogenide glasses are typically transparent in the 0.8–7 μm range, with telluride based glasses in the 2 – 12 μm range. Attenuations less than 1 dB/m have been reported for As-S fibers in the 2–3 μm wavelength range. Various potential or realized passive and active applications of chalcogenide glasses in photonics have been already demonstrated in the fabrication of IF fibers, planar waveguides, Bragg diffraction gratings, photoinduced holograms, rare-earth doped films. Further, they possess a large refractive index and a high degree of covalent bonding which increases the probability of radiative transitions compared to many other materials. Therefore chalcogenide glasses are attractive host materials for rare-earth elements, e.g. Er, Nd, Pr, etc. and numerous applications of active optical devices have been proposed. As chalcogenide glasses also have high nonlinear optical properties, they have potential for use in optical switching devices. The nonlinearity of chalcogenide glasses is about three orders of magnitude greater than that of silica. Most chalcogenide glasses exhibit a variety of *photoinduced effects* due to illumination such as photodarkening, i.e. photoinduced changes in the refractive

Table: Chalcogenide glasses Typical properties of two chalcogenide glasses.

	ρ g cm^{-3}	T_g °C	n at 1.55 μm, annealed state	dn/dT °C^{-1}	E_g eV	n_2 (W cm^{-2})$^{-1}$
As_2S_3	3.20	210	2.40	$+0.9 \times 10^{-5}$	2.4	1.7×10^{-14}
As_2Se_3	4.65	190	2.83		1.8	1.2×10^{-13}

index, absorption coefficient and bandgap, photobleaching, photoinduced dichroism, photoinduced birefringence, photoexpansion, photocrystallization, photodissolution or photodiffusion of Ag and other effects. The high sensitivity of chalcogenide glasses to photoinduced effects has been attributed to their structural flexibility given the low coordination of chalcogen elements, and to the presence of lone-pair electrons on the chalcogens that make up the valence band. Usually photoinduced effects require illumination with light of bandgap or greater photon energy, which has been shown to change the local atomic structure. Subbandgap light also causes photoinduced effects but the required intensity is much higher. Photoinduced effects can be categorized as *scalar* if they are independent of the state of polarization of the inducing, pump, beam, and *vectorial* if they depend on the polarization direction of the pump beam. Photoinduced vectorial effects are much smaller than photoinduced scalar effects. (A. Zakery and S. R. Elliott, *J. Non-Cryst. Solids*, **330**, 1, 2003).

Chalcohalides are formed by the addition of halide elements (F, Cl, Br, I) or their salts to chalcogenide glasses; they are chalcogenide-halide glasses. For example, As-S-Cl, As-S-Br, As-S-I, As-S-Cl-Br-I, etc. would be chalcohalide glasses based on the As-S chalcogenide glass and one or more halogens. The introduction of a halide disrupts the interaction of rare earth electrons with lattice vibrations of chalcogenide glass and hence substantially improves mid-infrared radiative processes (emissions) of some rare earths such as Dy; e.g. due to the addition of CsBr into the Ge-Ga-S glass, the Dy^{3+} lifetime of the 1.31 μm emission increases as much as 35 times. Chalcohalide glasses have potential for use as optical materials in such applications as low-loss fiber optics, chemical sensing, and laser or optical amplifier applications in which it serves as a host glass for rare earths.

Charge injection detector or **device** is a metal-oxide semiconductor (MOS) based detector that converts the incident light flux to collectable charges. They can be easily fabricated, usually as linear or two-dimensional sensor arrays, using the well-established PMOS, NMOS and CMOS integrated circuit technologies. See *CMOS image sensor; image sensor.*

Charge-coupled devices, CCDs, in optoelectronics are commonly used as image sensors in professional and consumer television cameras and camcorders, and as image sensors in digital still cameras. However, the term CCD in general does not imply an image sensor but a chip that is able to store and transfer signal in the form of charge. The CCD chip is an integrated circuit made from crystalline silicon and has a large number of pixels (each is a detector element); for example a 2.5 cm × 2.5 cm CCD chip may have 1024 × 1024 or 2048 × 2048 pixels on its surface. The basic pixel structure is a MOS (metal–oxide–semiconductor) or a MIS (metal–insulator–semiconductor) device. When a positive voltage is applied to the gate, the photogenerated electrons in the depletion region are collected in a layer near the interface as shown in the first figure. (With no gate voltage, photogenerated electrons and holes disappear by recombination.) These electrons are trapped inside a potential energy (PE) well, introduced by $-V$ on the gate. Their total charge is proportional to the total light exposure and is the electrical signal. The objective is to read all these charges stored at the illuminated pixels. In a *three-phase* CCD read out, there are three line voltages V_1, V_2, V_3 to which the gates are connected in an alternating fashion: G_1 to V_1, G_2 to V_2 and G_3 to V_3, G_4 to V_1, and so on. V_1, V_2 and V_3 are appropriately clocked to shift the charges from pixel to pixel to a register located at the end of the chip. If initially (time $t = t_1$) $V_1 = +V$ and $V_2 = V_3 = 0$, then photogenerated charges will be stored under G_1, G_4, etc. as Q_a, Q_b, etc. Later ($t = t_2$) we can make $V_1 = +V$, $V_2 = +V$ and $V_3 = 0$. The charge Q_a is shared between the wells under G_1 and G_2. Even later ($t = t_3$), we can bring V_1 down to zero, the charges (Q_a)

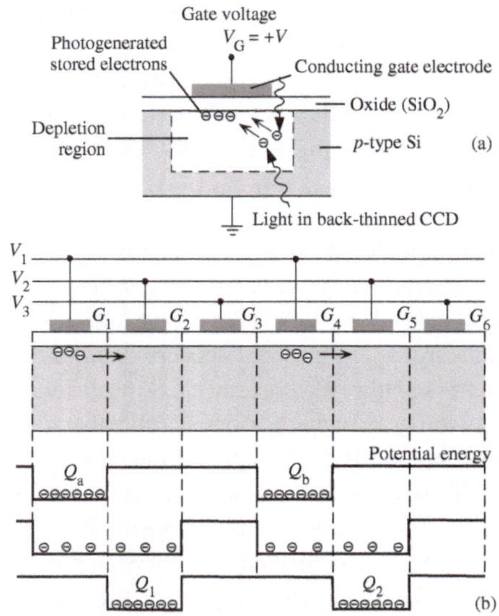

Time	V_1	V_2	V_3
t_1	$+V$	0	0
t_2	$+V$	$+V$	0
t_3	0	$+V$	0

(c)

Charge coupled device (a) One element of a CCD imaging sensor, which is a MOS (metal–oxide–semiconductor device). (b) Transfer of charge from one well to another by clocking the gate voltages. (c) The gate voltage sequences in a three phase CCD.

must go into the available potential well, which is under G_2. Thus, by toggling gate voltages, Q_a has been shifted from G_1 and G_2, and similarly Q_b from G_4 to G_5, and so on. The charges are therefore clocked progressively along the gates, from pixel to pixel, left to right, until they reach the end of the array where there is a register. The CCD readout therefore functions like a *shift register* in that clock pulses shift the information along the chain; they are often termed CCD shift registers. In back thinned CCD, light enters the depletion region not from the gate side but from the "substrate" side, which has been thinned to allow the light to pass through.

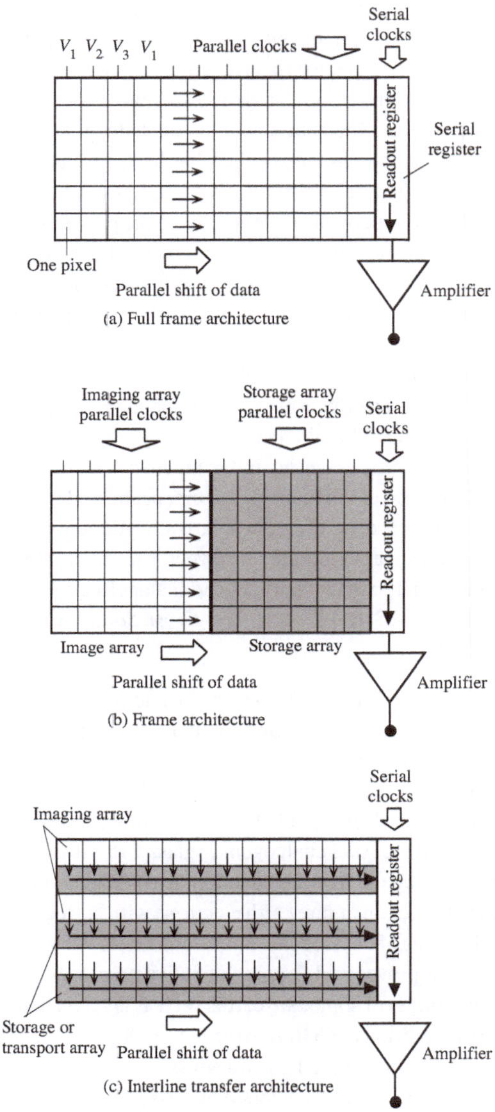

Charge coupled device CCD architectures: (a) full frame architecture; (b) frame architecture; (c) interline architecture.

Charge coupled device Kodak (KAF1001) CCD image sensor for medical and scientific imaging. Chip: 2.86 cm × 25.5 cm. Imager: 2.46 cm × 2.46 cm. Architecture: full frame CCD. Array size: 1024 × 1024. Pixel size: 24 μm × 24 μm. Fill factor: 100%. Sensitivity: 10 μV/electron.

Charge coupled device The inventors of the CCD image sensor at AT&T Bell Labs: Willard Boyle (left) and George Smith (right). The CCD was invented in 1969, the first CCD solid state camera was demonstrated in 1970, and a broadcast quality TV camera by 1975. (Courtesy of Alcatel-Lucent Bell Labs.)

There are various types of CCD read-out techniques and hence CCD architectures, e.g. full frame, frame and interline transfer architectures. In *full frame architecture* the sensor consists of parallel CCD shift registers, a serial CCD shift register as a *readout register* and an amplifier as shown in the second figure. The columns of data are fed into the serial register starting soon after the projection of the image onto the array, as the clock signals start transferring the charges pixel to pixel towards the readout register. (A short build-up time is necessary to build up the charges in the imaging arrays.) The process continues until all the columns of data have been transferred. This architecture has the highest resolution and is the densest. The *frame architecture* is similar to the full frame except that half the array is used as data storage to increase the speed. The storage array is blocked from light. The idea is to shift the data away from the image sensors quickly so that they can capture the next image; these have higher frame rates at the expense of resolution. The *interline transfer architecture* uses every other row of the CCD arrays for *storage* and *transport* right next to the imaging array. The transport register is blocked from light, and the data from the imaging array are transferred in parallel to the transport after the image has been built up in the imaging array. The data on the transport array are shifted, as in the normal CCD shift operation, to the readout register; this continues while the next image is being built up.

Charge transfer absorption is a process in which a photon is absorbed by an anion during which the photon energy releases an electron from the anion, and the electron becomes transferred to the cation. See *bandgap electronegativity*.

Chemical luminescence, see *luminescence*.

Chemical potential, μ, **or Gibbs free energy per atom**, G, of a system or a substance is the potential energy per atom for doing non-pV (nonmechanical) work, for example electrical or magnetic work. It is the change in the chemical potential (or change in Gibbs free energy per atom) that results in non-pV work being done. For example in a battery, the EMF, V, that is available is the change $\Delta\mu$. The term chemical potential (potential energy) is used because the work comes about typically through chemical changes. The chemical potential energy per electron in a semiconductor is the *Fermi energy*, E_F. When two semiconductors A and B are brought together to form a single composite material system, provided that there is no photoexcitation or an applied bias, the chemical potential difference between A and B must be zero, $\Delta\mu = 0$, since no work is done on or by the system in equilibrium which corresponds to the Fermi energy E_F being uniform throughout the system.

Chemical vapor deposition (CVD) is a chemical process by which the reaction between gaseous reactants results in a thin film of solid phase deposition. As an example consider the growth of Si using CVD. Si substrates used for preparing epitaxial devices are typically between 160 and 500 μm thick, depending on the wafer diameter. These substrates also function as mechanical supports, since the epitaxial layers grown are usually on a micron scale. The basic process for Si growth involves passing a gaseous Si compound called a precursor over a hot substrate in the presence of ultrahigh purity hydrogen carrier gas (with a flow rate of typically about 100 standard liters per minute). The most widely used Si precursor compounds are $SiCl_4$ (i.e. silicon-tetrachloride), SiH_2Cl_2 (i.e. dichloro-silane), $SiHCl_3$ (i.e. trichloro-silane) and SiH_4 (i.e. silane). For the chlorine compounds, decomposition occurs on the hot substrate by chemical reduction of the compound with hydrogen. The overall chemical reaction for the case of $SiCl_4$ may be written as:

$$SiCl_4(\text{Gas}) + H_2(\text{Gas}) \Rightarrow Si(\text{Epitaxial}) + 4HCl(\text{Gas}).$$

Notice that this reaction is reversible. If the carrier gas entering the reactor contains HCl, then the reverse reaction is favored and the Si substrate or epilayer is etched. Even if no HCl is introduced into the gas feedstock, HCl acts as an etchant for Si at typical growth temperatures. This problem is controlled by ensuring that sufficient hydrogen gas is introduced in the feedstock. Although deposition of Si can be achieved at relatively low temperatures, for growth of epitaxial layers to occur, temperatures in excess of about 1000 °C are required. This is principally dictated by the need for sufficiently high temperatures to enable Si atoms deposited on the surface to migrate to form an ordered crystal structure. Typically a growth temperature of 1250 °C is used for epitaxy, providing a growth rate of about 1 μm/minute. Doping of epilayers is accomplished by introducing controlled quantities of dopant-based hydrides into the gas stream. The most commonly used gases are PH_3 (i.e. phosphine), AsH_3 (i.e. arsine) and B_2H_6 (i.e. diborane). In practice, doping is found to depend linearly on the ratio of the supply of the dopant to that of Si-based molecules in the feedstock over a wide range of concentrations. This makes achieving controlled doping levels relatively simple.

Chief ray in a simple lens is the straight ray that travels from the object to the image, and intersects the optical axis at the lens center; it passes through the lens center, without being refracted. (Not to be confused with light traveling along the optic axis of a birefringent crystal.)

Chip is a piece (or a volume) of a semiconductor crystal which contains many integrated active and passive components to implement a circuit.

Chirp, see *optical Kerr effect, frequency chirp*.

Cholesteric phase see *liquid crystals*.

Chromatic aberration occurs because of the intrinsic *chromatic dispersion* of most optical materials. Consider for simplicity a planoconvex convergent thin lens of index $n(\lambda_0)$ and radius of curvature R in the Gaussian approximation. Its focal length f is such as $1/f = (n-1)/R$. If n varies with wavelength λ_0, so does $f = R/(n-1)$. Usually, n is a decreasing function of λ_0, so that the "blue" focal length $f_b = OF_b$ is shorter than the "red" focal length $f_r = OF_r$.

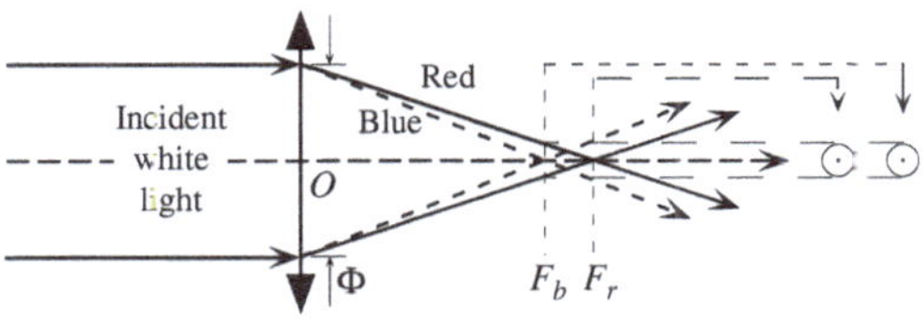

Chromatic aberration The distance $F_bF_r = f_r - f_b$ is called *longitudinal chromatic aberration.*

Chromatic dispersion of a material means that its refractive index varies with the wavelength. The dispersion of an optical glass is characterized either by its *dispersive power* $K = (n_F - n_C)/(n_D - 1)$ or by its *Abbe number* (or *constringence*) $A = 1/K$, where n_F, n_D and n_C are the refractive indices for the blue *F*, yellow *D* and red *C* reference lines ($\lambda_0 = 4861$ Å, 5875.6 Å and 6563 Å, respectively). A simplified *Sellmeier-type formula* of the type

$$n(\lambda_0) \approx A' + \left(B'/\lambda_0^2\right)$$

where the constants A' and B' depend on the material, often provides a useful approximation for *n*. Optical glasses are roughly classified into two categories: (I) Light glasses, or *crown*, with smaller index ($n_D < 1.6$) and lesser dispersion ($A > 50$), usually made of potassium or calcium silicate; one of the most widely used is potassium borosilicate BK7. (II) Heavy glasses, or *flint*, with greater index ($n_D > 1.6$) and higher dispersion ($A < 50$), usually made of potassium silicate and lead. This category also has ZnSe ($n_D \approx 2.578$ at 656 nm). An *achromatic doublet* is made of two thin lenses in contact, of focal lengths f_1, f_2 and dispersive powers K_1, K_2 such as $(K_1/f_1) + (K_2/f_2) = 0$. The blue and red focal points merge into a single point. The remaining aberration is referred to as the *secondary spectrum.*

Chromatic dispersion in optical fibers is due to the dispersion of a traveling pulse of light along an optical fiber as a result of the wavelength dependence of the propagation characteristics and waveguide properties but excluding multimode dispersion. Chromatic dispersion arises as a result of the range of wavelengths in the emission spectrum of the emitter (e.g. LED or laser diode) that are coupled into the fiber. It is the combination of *material dispersion* and *waveguide dispersion*. Chromatic dispersion leads to the temporal broadening of short pulses and hence to the reduction of the maximum rate at which information can be transmitted. This is the reason why high bit-rate telecom systems are designed for wavelengths at which fiber dispersion (expressed in ps km^{-1}nm^{-1}) is kept to a minimum. In transparent materials the refractive index *n* usually decreases with λ, whereas *active* media may exhibit *anomalous dispersion* (*n* increases with λ). Near an atomic resonance at frequency ν_0, a change of index corresponds to the sharp absorption line. This is a consequence of *Kramers–Kronig relations* between the real and the imaginary parts of the *atomic susceptibility* $\chi = -\chi_0/(\nu - \nu_0 - j\Gamma)$, where χ_0 is a scaling parameter and Γ the half width at half maximum.

If the same line is used for stimulated gain in a laser medium, both signs are inverted. In III-V

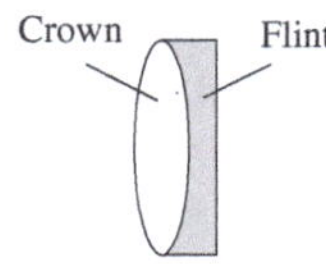

Chromatic dispersion An achromatic doublet.

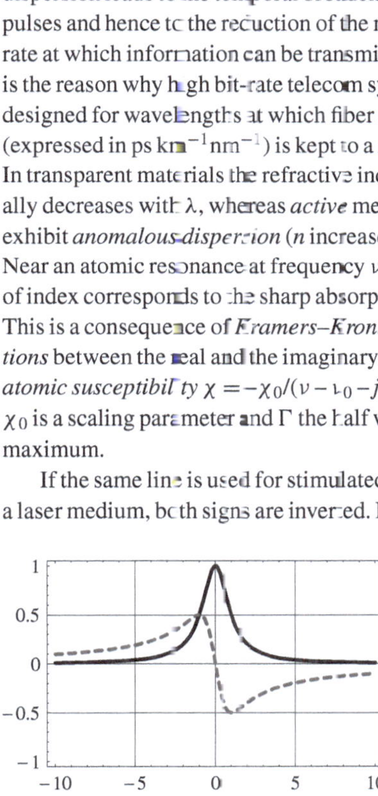

Chromatic dispersion Normalized $\mathrm{Re}[\chi/\chi_0]$ (solid line) and $|\mathrm{Im}[\chi/\chi_0]|$ (dashed line) near an atomic absorption line, as a function of the normalized detuning $X = (\nu - \nu_0)/\Gamma$.

semiconductors, n decreases with λ, with different slopes above and below the bandgap wavelength. In ternary alloys such as $Al_xGa_{1-x}As$, it also decreases with the Al concentration x. This property is used to realize multilayer structures with index steps (see *distributed Bragg reflector, vertical-cavity surface-emitting laser*). The refractive index of AlGaAs in the red and near infrared is illustrated below.

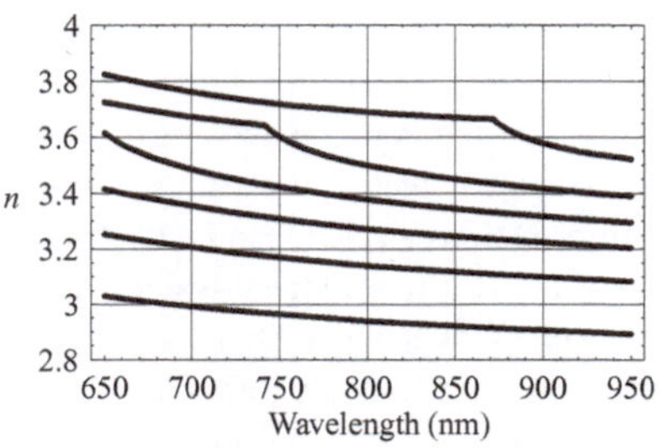

Chromatic dispersion Refractive index of $Al_xGa_{1-x}As$ at room temperature as a function of wavelength; from top to bottom, $x = 0, 0.2, 0.4, 0.6, 0.8$ and 1 (calculated values).

Apart from the intrinsic chromatic dispersion, whose origin lies in the material itself, *structural* dispersion in actual integrated systems should also be considered: the *dispersion relation* between pulsation ω and wavevector β in a waveguide shows that the *effective index* n_{eff} varies with the wavelength. See also *group index*.

Chromatic dispersion coefficient, D_{ch}, of an optical guide is the time spread of a propagating optical pulse in an optical guide per unit length and per unit spectral wavelength width due to the wavelength dependence of all the guide properties. If $\delta\lambda$ is the spread of the excitation wavelengths coupled into the fiber, $\delta\tau$ is the spread in propagation times of these different wavelengths, L is the fiber length, then,

$$D_{\text{ch}} = \frac{1}{L}\frac{\delta\tau}{\delta\lambda}.$$

The overall chromatic dispersion coefficient can be written as

$$D_{\text{ch}} = D_{\text{m}} + D_{\text{w}} + D_{\text{p}}$$

where $D_{\text{m}}, D_{\text{w}}$ and D_{p} are the material, waveguide and profile dispersion coefficients.

Chromatic resolving power see *diffraction grating*.

Chromaticity see *color*.

Chynoweth's law see *impact ionization*.

CIE chromaticity diagram see *color*.

Circle of least confusion see *spherical aberration*.

Circularly birefringent crystal is a medium in which the right- and left-handed circularly polarized waves propagate with different velocities and experience different refractive indices; n_{R} and n_{L}. Since optically active materials naturally rotate the optical field, it is not unreasonable to expect that circularly polarized light with its optical field rotating in the same sense as the optical activity will find it easier to travel through the medium. Thus, an optically active medium possesses different refractive indices for right- and left-handed circularly polarized light and exhibits circular birefringence. See *elliptically polarized light, polarization*.

Circularly polarized light has the magnitude of the field vector **E** constant but its tip at a given location on the direction of propagation, e.g. the z-direction, traces out a circle by rotating either in a clockwise sense, *right circularly polarized*, with time, as observed by the receiver of the wave, or in a counterclockwise sense, *left circularly polarized*. Consider a wave that is traveling in the z-direction. The electric field vector at any time can be resolved into two components E_x and E_y along the x- and y-directions: $E_x = A\cos(\omega t - kz)$, and $E_y = A\cos(\omega t - kz + \phi)$ where ϕ is the phase difference between the field oscillations of the x- and y-components. For $\phi = \pm\pi/2$, $E_x^2 + E_y^2 = A^2$, which represents a circle in a E_y vs. E_x plot, with radius A. Circular polarization corresponds to a phase difference of $\pm\pi/2$, between E_x and E_y and $|E_x| = |E_y|$.

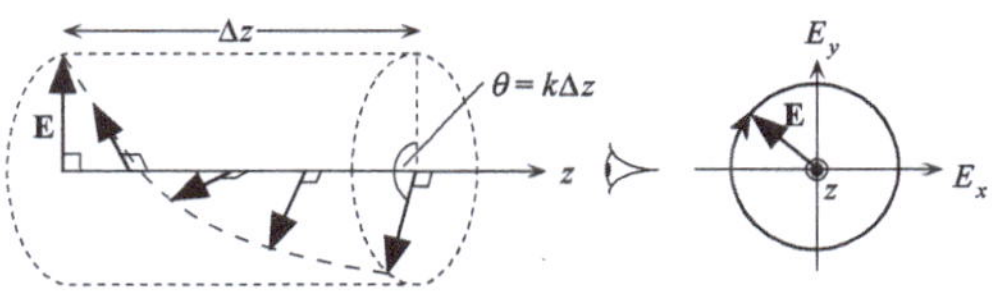

Circularly polarized light A right circularly polarized light. The field vector **E** is always at right angles to *z*, rotates clockwise around *z* with time, and traces out a full circle over one wavelength of distance propagated.

See *polarization, elliptically polarized light.*

Cladding is the dielectric layer that surrounds the dielectric core of an optical waveguide.

Cleavage plane refers to a preferred crystallographic plane for a crystalline material to fracture, resulting in a characteristic fracture surface. That is, cleaved materials exhibit atomically smooth and flat surfaces corresponding to specific atomic planes, after fracture. Typically, materials will favor the most densely packed atomic planes (or so-called close-packed planes) for cleavage, as the energy involved in moving opposing planes of atoms over each other is often lower than for other planes. This may be attributed to the relatively small displacements required for a unit of interplanar displacement on such planes. For example, the most densely packed planes for silicon are the {111} family of planes on which cleavage often occurs.

Cleaved-coupled-cavity (C^3) laser device: two different pumped laser optical cavities *L* and *D* (different lengths) are coupled in such a way that only modes that satisfy both cavities can exist for the whole system (of two cavities). The two lasers are pumped by different currents. Only those waves that can exist as modes in *both* cavities are now allowed because the system has been coupled. The modes in *L* are more closely spaced than modes in *D*. These two different set of modes coincide only at widely spaced intervals. This restriction in possible modes in the combined cavity and the wide separation between the modes results in a single-mode operation.

CMOS image sensor is an active matrix array (AMA) in which each pixel has a photodiode or a photogate and one or more CMOS transistors to read and amplify the electrical signal that has been generated by the light incident on this pixel. If the pixel has an amplifier (e.g. a source follower), then the sensor is an *active pixel sensor*. The pixel read-out operation is basically the same as that in the AMA. The pixel architecture, however, can change from application to application. CMOS is a well-established silicon fabrication technology which means that CMOS image sensors can be

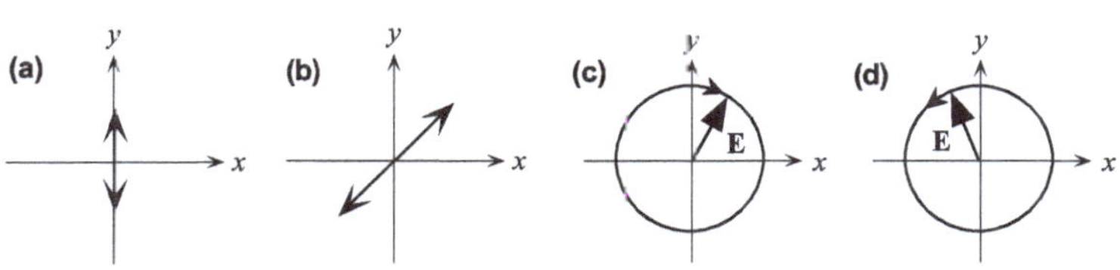

(a)	(b)	(c)	(d)
$E_{xo} = 0$	$E_{xo} = 1$	$E_{xo} = 1$	$E_{xo} = 1$
$E_{yo} = 1$	$E_{yo} = 1$	$E_{yo} = 1$	$E_{yo} = 1$
$\phi = 0$	$\phi = 0$	$\phi = \pi/2$	$\phi = -\pi/2$

Circularly polarized light Examples of linearly, (a) and (b), and circularly polarized light, (c) and (d); (c) is right circularly and (d) is left circularly polarized light (as seen when wave directly approaches the viewer).

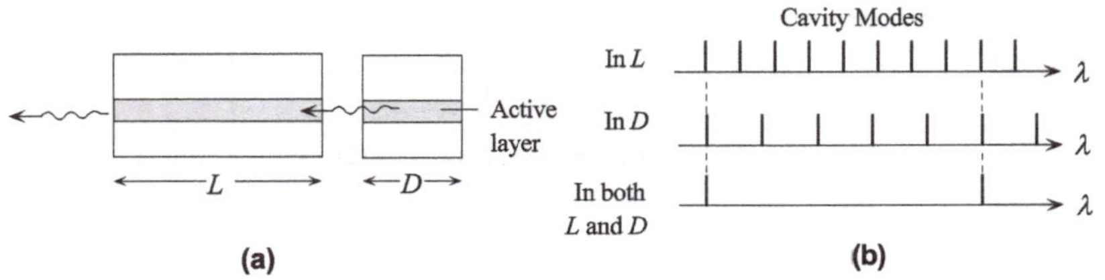

Cleaved-coupled-cavity (C^3) laser.

fabricated relatively inexpensively. If the pixel is *active* with gain, then the imager is called an *active pixel image sensor*, or *active pixel sensor*. In a *passive pixel sensor* there is only a switching transistor to read the charge out when the transistor is switched i.e. addressed. Nearly all modern CMOS imagers are active pixel sensors. One possible active pixel architecture using a photodiode is shown in the figure. T_1 is the reset transistor. T_2 works as a source follower, and T_3 is a pixel switch transistor. With T_1 off, the photocurrent charges the self-capacitance of the photodiode to a certain voltage. When the row X receives a signal, T_3 is turned on, and the signal voltage on the photodiode is transferred through the buffer T_2 to the column Y. The pixel is then reset, T_1 connects the photodiode to V_{DD} to clear the accumulated charge. CMOS imagers have a number of distinct advantages: Standard and

CMOS image sensor Micron 1.3 Megapixel CMOS active pixel digital image sensor (MI-MV13). Imager: 1.536 cm × 1.229 cm. Array size: 1280 × 1024. Pixel size: 12 μm × 12 μm. Sensitivity: 13 μV/electron. Speed: 500 fps (frames per second). Power dissipation = 150 mW at 60 fps. Electronic shutter, 2 μs – 33 ms.

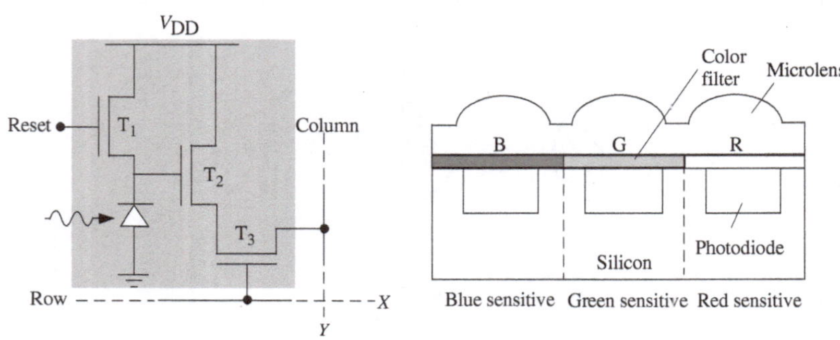

CMOS image sensor Left: The pixel architecture in a CMOS image sensor. Right: A cross-section of a CMOS imager with microlenses and color filters (B = blue, G = green, R = red) for color imaging.

well-established CMOS fabrication, and hence lower cost mean on-chip integration such as "a camera on a chip." Such a camera on a chip has a microlens at each pixel, a red, green or blue filter at each pixel for color imaging, analog signal processing after the read out of the image, and analog-to-digital conversion. Active pixel architecture can include on-pixel amplification as well, to improve the overall detective quantum efficiency. CMOS low-power consumption leads to a longer battery life. More flexible read-out technology than CCD imagers leads to more versatile applications.

CMYZ see *color*.

Coddington position factor P characterizes the relative positions of the object and its image in a lens imaging system relative to the object and image having identical distances from the lens via

$$P = (s_i - s_o)/(s_i + s_o)$$

where s_o and s_i are the object and image distances, respectively, from the lens; for a real object and real image both are positive. $P = 0$ is a symmetric object–image system i.e. $s_i = s_o$. If the object O is at infinity (incident rays are parallel), the image I is at the focal point F, and $P = -1$. If the image is at infinity, emerging light has parallel rays, $P = 1$. The *Coddington shape factor* S characterizes the extent of "bending" or deviation from symmetry in the shape of the lens; quantitatively,

$$S = (R_2 + R_1)/(R_2 - R_1)$$

where R_1 and R_2 are the radii of curvature of the two lens surfaces, and $S = 0$ is a symmetric lens. Coddington factors are useful in designing lenses with

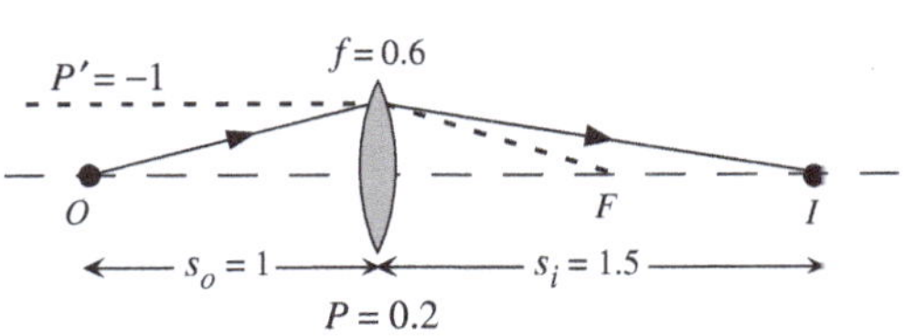

Coddington position factor P.

reduced spherical aberration, which is a minimum for a given P when

$$S = -2P(n^2 - 1)/(n + 2).$$

Clearly a particular lens shape factor for a given pair of object and image points is not the best for other object and image positions.

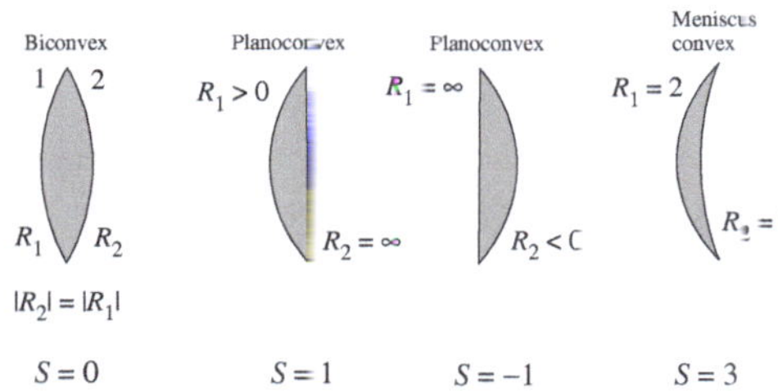

Coddington shape factors for converging lenses.

Coddington shape factor see *Coddington position factor*.

Coherence refers to the phase of a wave, or to the degree of phase relationship between different waves. Theoretically, *coherence* refers to the degree of *correlation* between the vibrations of the electromagnetic field at two points (*spatial coherence*) or two instants (*temporal coherence*). Experimentally, this notion is related to the observation of *interference fringes*. In what follows, unless specified otherwise, we will restrict ourselves to a scalar theory of light.

Let $V_1(t)$ and $V_2(t)$ be the scalar vibrations at points P_1 and P_2 in the wave field, respectively, so that the corresponding intensities are $I_i = \langle V_i V_i^* \rangle$ where $i = 1$ or 2, and the star denotes the complex conjugate. The *mutual coherence function* of the two vibrations at P_1 and P_2, between times t and $t + \tau$, is defined as

$$\Gamma_{12}(\tau) = \langle V_1(t + \tau) V_2^*(t) \rangle.$$

If $P_1 = P_2$, the quantity $\Gamma_{11}(\tau) = \langle V_1(t+\tau) V_1^*(t) \rangle$ is called the *self-coherence function* at P_1. If

$\tau = 0$, this function reduces to the usual observable intensity,

$$\Gamma_{11}(0) = I_1, \quad \Gamma_{22}(0) = I_2.$$

The *complex degree of coherence* (or *coherence factor*) $\gamma_{12}(\tau)$ is obtained by normalizing the mutual coherence Γ_{12}:

$$\gamma_{12}(\tau) = \frac{\Gamma_{12}(\tau)}{\sqrt{\Gamma_{11}(0)\Gamma_{22}(0)}}.$$

Consider the field at any other point P, distinct from P_1 and P_2. If t_1 and t_2 denote the times needed for the light to travel with no attenuation (i.e. pure phase change), from P_1 to P and from P_2 to P, then the resulting intensity at P is

$$I(P) = I_1(P) + I_2(P) + 2[I_1(P)I_2(P)]^{1/2} \times \mathrm{Re}[\gamma_{12}(t_2 - t_1)].$$

This formula, which constitutes the *general interference law for stationary optical fields*, can also be rewritten as

$$I(P) = [I_1(P) + I_2(P)] \times \{1 + m\mathrm{Re}[\gamma_{12}(t_2 - t_1)]\}$$

with $m = 2[I_1(P)I_2(P)]^{1/2}/[I_1(P) + I_2(P)]$ a modulation factor that characterizes the intensity contrast between the two sources ($m = 0$ if $I_1 I_2 = 0$, $m = 1$ for $I_1 = I_2$).

For a quasimonochromatic (wavelength λ), uniform, spatially incoherent extended source of radius ρ seen from distance D under an angular radius $\alpha = \rho/D$, the diameter of the *area of coherence* A_C is

$$\delta_C \approx 0.16\lambda/\alpha.$$

Two points P and P', illuminated by such a source, should fall inside the area of coherence in order for their vibrations to be appreciably correlated. For instance, for a screen illuminated directly by the Sun (apparent angular diameter $2\alpha \approx 0° \, 32' \approx$ 9.3 mrad), with an average wavelength $\lambda_{av} = 550$ nm, the diameter of the coherence area is $\delta_C \approx 19$ μm. In a Young-type experiment, two pinholes separated by a finite distance a and illuminated by a monochromatic, circularly extended source of area A_S located at distance d, should satisfy $a < 0.16(\lambda d)/A_S$ in order to ensure the necessary coherence between the two pinholes. (See *Young's experiment* for details.) The *coherence length* is also equal to

$$L_C = (\lambda_{0m})^2/\Delta\lambda,$$

where λ_{0m} is the mean wavelength and $\Delta\lambda$ the spectral extension. The *coherence volume* V_C is the product of the *area of coherence* and the *coherence length*:

$$V_C = A_C \times L_C.$$

Coherence length, l, is the spatial distance over which any two points P and Q on a waveform (such as the fields at P and Q) at a given instant can be correlated; one can be predicted from the other. For simplicity, consider a wave train. Suppose that a wave train has a coherence time Δt. Then its *coherence length is* $l = c\Delta t$. See *coherence, coherence time*.

Coherence time, Δt, is the longest time period over which any two points P and Q at a given location can be correlated; one can be predicted from the other. When a sine wave is chopped, truncated, so that it exists only over the time period Δt, then it is not a perfect sine wave and contains a distribution of frequencies spread around the apparent frequency of the sine wave υ_o as shown in the figure. The *spectral width* $\Delta\upsilon$ and the *temporal coherence* Δt in this particular case of a truncated sinusoidal wave are related by $\Delta\upsilon = 2/(\Delta t)$. In general, the radiation will not be a perfect truncated sine wave, and the exact relationship between $\Delta\upsilon$ and Δt will depend on the envelope function of the oscillations, which depends on the way the oscillations were generated or emitted. For example, if one takes the envelope of the oscillations to be a Gaussian so that the wave is a "Gaussian wavepacket," then the Fourier transform will also be a Gaussian, centered at the oscillation frequency υ_o; the coherence length and coherence time then usually refer to some suitable widths of the respective Gaussian envelopes and very roughly $\Delta t\Delta\upsilon \sim 1$. See *coherence, bandwidth theorem*.

 Coherent optical communication systems

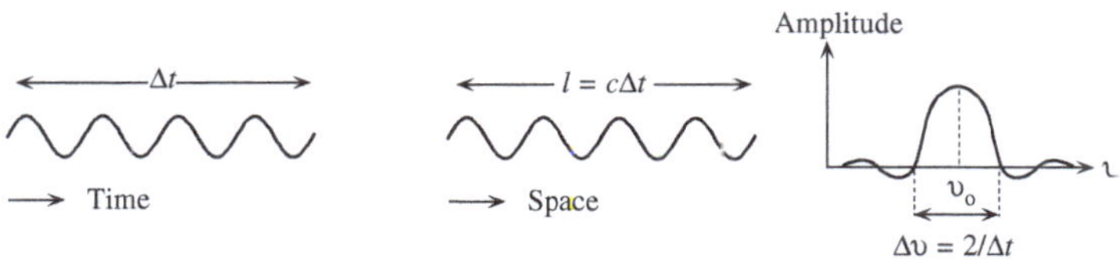

Coherence time A finite wave train lasts for a duration Δt and has a length l. Its frequency spectrum extends over $\Delta\nu = 2/\Delta t$. It has a coherence time Δt and a coherence length l.

Coherent optical communication systems exploit the coherence properties of frequency-controlled laser light in order to transmit information. Since they are based upon *interference effects* between the transmitted signal, at angular frequency ω_S, and a local laser oscillator, at ω_L, they are more elaborate than the usual – and simpler – scheme of *Intensity-Modulation Direct Detection* (IMDD), where bits of information are coded by directly modulating the laser intensity.

In a coherent system, the monochromatic signal, of power P_S and phase φ_S, is mixed with that of a spectrally pure local oscillator, of power P_L and phase φ_L, on the photodetector. The total power P_{PD} detected by the photodetector PD is written as

$$P_{\mathrm{PD}} = P_S + P_L + 2(P_S P_L)^{1/2}\langle\cos[(\omega_S - \omega_L)t + (\varphi_S - \varphi_L)]\rangle,$$

where the bracket $\langle\,\rangle$ denotes a time average over the response time τ_D of the detector. In order for the third term to be detected, the phase difference $\Delta\varphi = (\varphi_S - \varphi_L)$ as well as the frequency difference $\Delta\omega = (\omega_S - \omega_L)$ must remain extremely stable; that is what is meant by *coherence*. Obviously, the beating frequency $\Delta\omega$ must also remain much smaller than the cut-off frequency ($\sim 1/\tau_D$), in order to fall within the frequency range of the detecting system.

If $\omega_L \neq \omega_S$, the receiver is called *heterodyne*. By contrast, if $\omega_L = \omega_S$, it is called *homodyne*. The latter scheme appears somewhat trickier to implement, since the local oscillator should be tuned exactly to the signal frequency, thus requiring a feedback loop.

The most frequently met coding formats are called *Amplitude Shift Keying* (ASK), *Frequency Shift Keying* (FSK) and *Phase Shift Keying* (PSK).

Frequency modulation of the optical carrier ω_S requires rapidly tunable lasers.

Phase shift keying is obtained by modifying the absolute phase of the signal. This is realized through an external *phase modulator*; a π-shift is recorded as a sign change of the detected interference term.

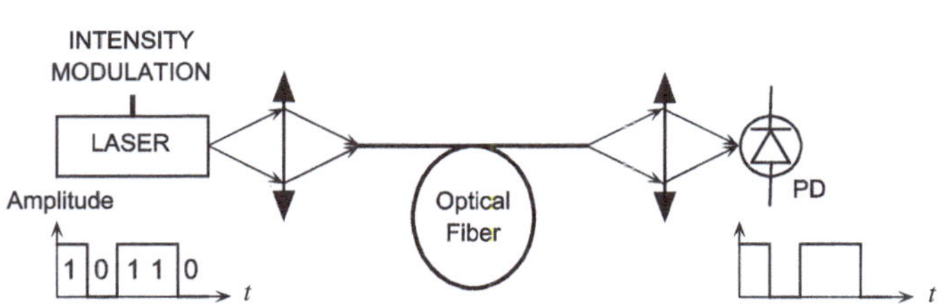

Coherent optical communication system *Noncoherent systems of transmission* are usually made of a directly modulated laser source that emits successive bits of data. After propagation inside an optical fiber, the transmitted flux is focused on a photodetector (PD), where optical bits are converted back into an electronic digital signal.

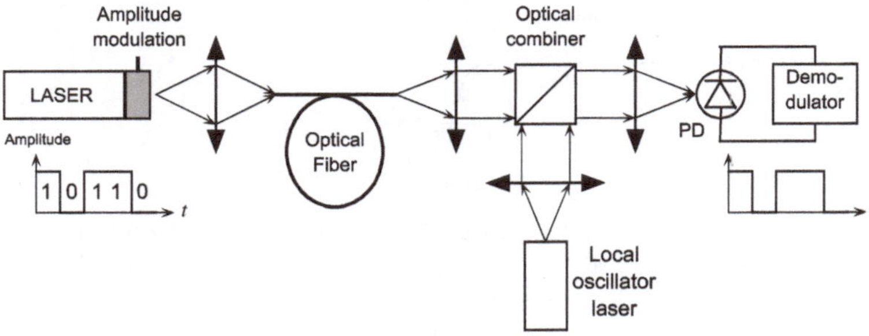

Coherent optical communication system *Amplitude shift keying*. An external modulator modulates the signal, ensuring ω_S remains stable. A demodulator, following the PD, provides an electric signal proportional to the interference term. The original data stream is reconstituted. Whereas ASK can be thought of as the mere coherent counterpart to IMDD, both FSK and PSK, involving frequency or phase, have no intensity equivalent. FSK represents a kind of frequency modulation.

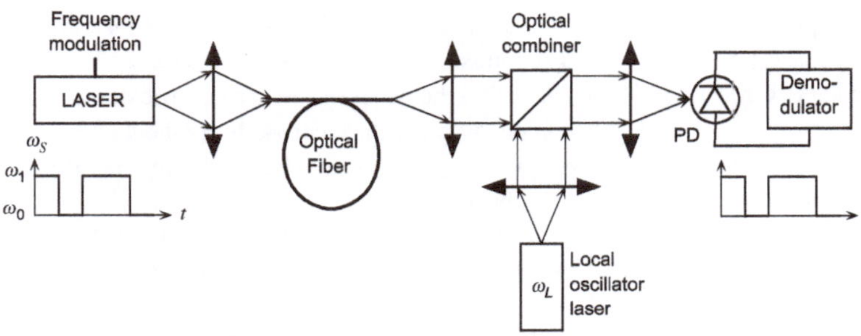

Coherent optical communication system *Frequency shift keying*. Coding is obtained by binary frequency modulation of the optical carrier ω_S (either ω_0 or ω_1). A demodulator following the PD provides an electric signal proportional to the interference term. The original data stream is reconstituted according to the beating frequency.

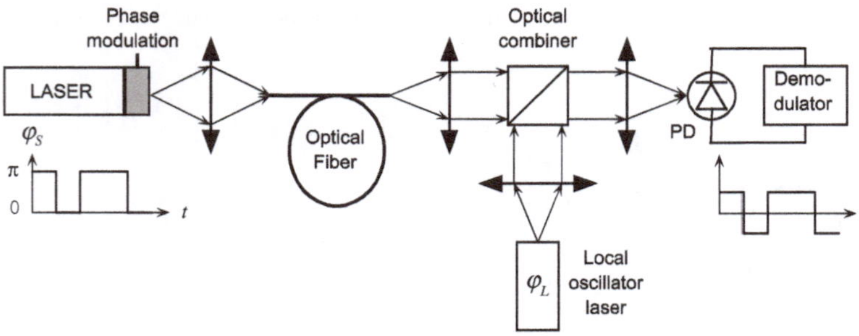

Coherent optical communication system *Phase shift keying*. Coding is obtained by shifting the phase φ_S of the signal (typically, either 0 or π). A demodulator following the PD provides an electric signal proportional to the interference term. The original data stream is reconstituted according to the sign changes of φ_S.

Although coherent optical systems appear more complex and subject to more severe stability requirements than IMDD systems, they present a number of significant advantages, such as better signal-to-noise ratio (SNR), longer haul without repeaters, increased product bandwidth × distance, reduced interchannel spacing for denser wavelength-division multiplexing (WDM) applications, etc.

Coherence volume see *coherence.*

Coherent radiation (or light) consists of waves that have the same wavelength and always have the same phase difference with respect to each other at all times.

Collimation see *laser diode output beam.*

Collision broadening is the broadening of an emission spectrum from a system of gas atoms in which collision among gas atoms interrupts the emission process and hence shortens the emission lifetime. If an atom emitting radiation is suddenly impacted by another gas atom, then the emission process becomes interrupted or perturbed, and the amplitude and phase of the emitted radiation become changed upon impact, resulting in the broadening of the spectrum (intuitively, the shortening of the emission process on the time domain increases the spectral width on the frequency domain). This type of collision broadening is called *pressure* broadening because the broadening process increases with the pressure of the gas (more collisions). The collision broadening can be soft or hard corresponding to elastic or inelastic collisions. An actual "physical" collision need not take place because, as long as the interaction between the atoms perturbs the emission process, the spectrum will be broadened. In addition, if a charged atom (ion) passes near an emitting atom, it can perturb the emission process because the field from the ion will split the energy levels in the emitting atom. Collision broadening in doped insulator type solid state lasers occurs as a result of the impacting of phonons on the emitting dopants or impurities. The impact from a phonon perturbs the emission from the dopant. Each emitting atom (or dopant) is subjected to the same broadening process, and emits at the same central frequency so that collision broadening is *homogeneous broadening,* and can be described by the Lorentzian lineshape. The emission spectrum form a CO_2 laser is usually collision broadened. Some gas discharge lamps are operated at high pressures to allow collision broadening to widen the spectrum. See *homogeneous broadening, lineshape function, Stark effect.*

Color in qualitative terms is a fundamental attribute of human vision that enables humans to distinguish between objects based on wavelength when the objects have the same extent and brightness. Color can be characterized by three main attributes: *hue, luminosity* and *saturation. Hue* characterizes the dominant wavelength(s) involved in the color, its *chromaticity.* Luminosity represents its apparent *brightness.* White and gray have the same hue (chromaticity) but gray is less bright than white. *Saturation* defines how much the color departs from white. Pink and red are perceptually different colors. Pink is obtained by adding white to red, and one identifies pink as red that has less saturation. Red, on the other hand, is fully saturated. Saturation therefore represents an extent of purity and intensity in a color. A highly saturated color tends to be vivid whereas a more "muted" color corresponds to a lower saturation. *Spectral color* denotes the sensory perception of the human eye to a monochromatic radiation. The visual spectral range is continuous from around 380 to roughly 700 nm but distinct spectral colors are arranged in *seven* "colors" as violet, blue, cyan, green, yellow, orange and red, as in the first figure; *indigo,* 445–425 nm, and magenta (color between red and blue) are not considered to be true spectral colors.

However, many combinations of different spectral components can often lead to almost the same visual effect. The perception of color experienced

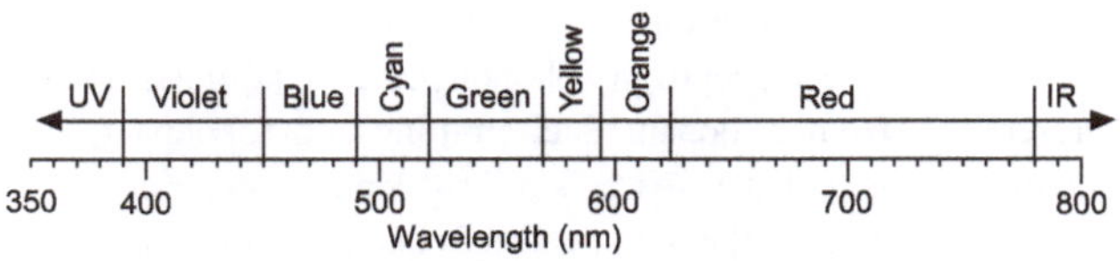

Color Arrangement of spectral colors, according to wavelength. The boundaries are somewhat arbitrary; ultraviolet (UV) and infrared (IR) mark the limits of the visible range.

by the human eye under nonmonochromatic conditions depends not only on the scene, but also on the source: at sunset, the same grass would not be seen as green as it would at midday. The color of an object is a function of the back-scattered spectrum. In white light, an object that scatters all wavelengths appears white, whereas an object that absorbs all wavelengths appears black. Note that the human visual sensitivity is not uniform across the visible range, but exhibits a peak around 555 nm.

The nature of color vision (i.e. three types of color cones in the eye with different spectral responses), and the nearly linear response of the eye, have led to development and use of *three primary colors* for matching perceived colors. Primary colors are those colors, such as *red, green* and *blue*, that cannot be generated from each other (they are mutually independent), and can be mixed in appropriate portions to generate various perceived colors. Beams of light of different primary colors can be combined to realize the *additive synthesis* of almost any color, as symbolized by the Newton color circle in the second figure. Red (R), green (G) and blue (B) are *additive primaries*. The color resulting from additive synthesis depends on the proportions of the mixture. In additive mixing equal amounts of R, G and B results in shades of gray, all the way to white in which the primary colors are fully saturated. The center of the circle corresponds to the "achromatic" case (from white to black through the achromatic grays). Note that magenta is not a pure spectral color.

On the other hand, pigments absorb certain wavelengths selectively. Nearly all printing media (e.g. colors on white paper) produce colors by *subtractive synthesis*. Subtractive primary colors are cyan (C), magenta (M) and yellow (Y). The full combination of the three should result in black but it does not generate a satisfactory black; thus black has to be added as a separate color to ensure the right

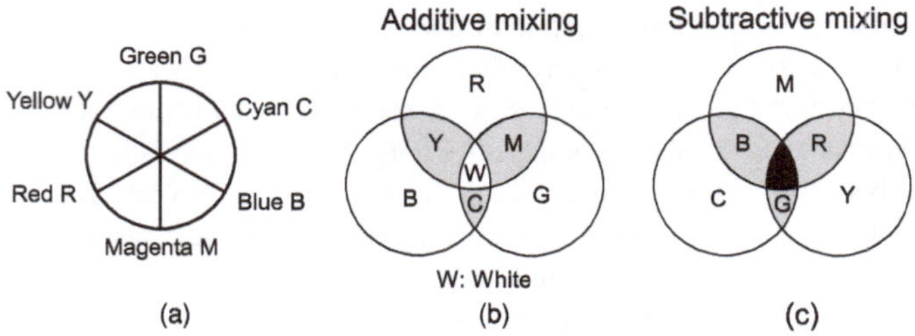

Color (a) The Newton color circle, with three primary colors: Red (R), Green (G), Blue (B) and their so-called complementary colors cyan, magenta and yellow, placed at a diametrically opposed location on the circle. (b) Additive mixing of primary colors, R, G and B. White (W) has equal amounts of R, G and B. (c) Subtractive mixing of cyan (C), magenta (M) and yellow (Y). Black has C, M and Y all subtracted.

perception of black; K, "key," is used to represent black and CMYK represents subtractive color synthesis as used in printing. Many artists use red, blue and yellow as the primary *color pigments* for their paintings. For ink color printing, *quadrichromy* is a technique based on the combination of R, B, Y and black (corresponding to CMYK).

The question of color rendering (e.g. by a television or computer screen) and color measurement has led to various standards. The most comprehensive so far is the *CEI 1931 color space*, proposed by the International Commission for Lighting (or CIE for its French acronym: Commission Internationale de l'Éclairage), in which colors are characterized by a *luminance* parameter, along with two independent coordinates on a two-dimensional map known as the *chromaticity diagram*. The *tristimulus values* are defined as the amounts of three primaries that we need to match a given color, a stimulus. The CIE *XYZ* chromaticity diagram uses three X, Y and Z tristimulus values. A color C is then written as $C = X + Y + Z$. (By choice, Y is proportional to the luminance of the color of interest.) If we normalize X, Y and Z to C, i.e. $x = X/(X + Y + Z)$, $y = Y/(X + Y + Z)$, $z = Z/(X + Y + Z)$, then $x + y + z = 1$, and we only need to plot y vs. x to represent various colors as in the third figure. One small inconvenience of the 1931 *xyz* chromaticity diagram is that the X, Y and Z are not convenient primary colors but are mathematical linear transformations of R, G and B primary colors;

$$X = 0.49R + 0.31G + 0.20B$$

$$Y = 0.17697R + 0.8124G + 0.01063B$$

$$Z = 0.00R + 0.01G + 0.99B.$$

If we know the color we wish to match, we can find its x and y coordinates, and hence its X, Y and Z tristimulus values; and then find R, G and B by the linear transformation. The 1931 CEI chromaticity diagram is by far one of the most precise among many color spaces, since all parameters are based on the spectral power distribution of the light emitted (or scattered) by an object, and corrected by the spectral sensitivity of the human eye. In addition, it encompasses not only the colors that can be synthesized by a combination of R, G and B, but also all that can be perceived by a standard observer. The primary colors X, Y and Z correspond not to any specific monochromatic colors but to three sets of color matching functions labeled, $\overline{x}(\lambda)$, $\overline{y}(\lambda)$ and $\overline{z}(\lambda)$, which have spectral dependences; $\overline{x}(\lambda)$ peaks in the blue, $\overline{y}(\lambda)$ in the green but $\overline{z}(\lambda)$ has a major peak in the red, and minor peak in the blue. Similarly R, G and B correspond to three color matching functions, $\overline{r}(\lambda)$, $\overline{g}(\lambda)$, $\overline{b}(\lambda)$. If we choose two points C_1 (x_1, y_1) and C_2 (x_2, y_2) on the CIE 1931 chromaticity diagram to represent two different colors C_1 and C_2, then the line C_1C_2 joining the two points represents all colors that can be generated by mixing C_1 and C_2. If C_1 and C_2 have the same brightness, the equal mixture of C_1 and C_2 is not at the center of the line C_1C_2, that is, the diagram cannot be used directly to determine color differences. The color spaces CIELuv and CIELab overcome this problem. Suppose that we wish to determine what gamut of colors can be generated by mixing R_1, G_1, B_1 that represent a particular set of red, green and blue light sources. The area inside the triangle $R_1G_1B_1$ represents all the colors that can be generated from these three sources as shown in the fourth figure. Colors outside the triangle cannot be formed.

A *color space* is a mathematical way of representing colors by, for example, assigning a set of three or four variables to represent a particular color perception. CIE *RGB* and *XYZ* color space are two such color spaces. If we were to construct a three-dimensional space (a three-dimensional vector) with hue along x, saturation along y and brightness along z, we would generate the *HSB color space*. See *color matching functions*, *Grassmann's law*, *primary color*.

Color centers are defects that are responsible for the coloration of transparent crystals. Color centers are created by irradiating alkali halide crystals with high-energy radiation such as x-rays or electrons. The irradiation results in the formation of defect

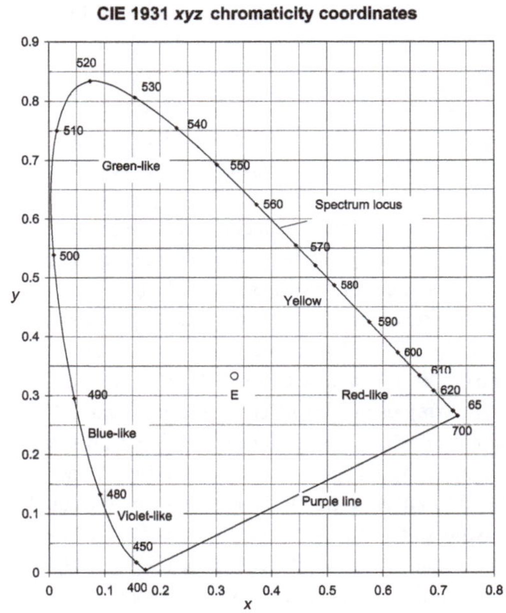

Color Approximate depiction of the CIE *xyz* chromaticity diagram. The area inside the spectrum locus represents the gamut of colors that can be perceived. Point E corresponds to *achromatic light*. Points on the outside border denote some specific wavelengths, expressed in nm.

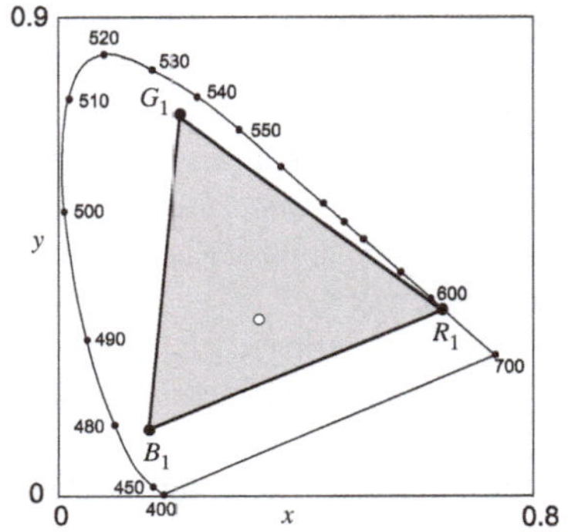

Color The gray area bounded by $R_1G_1B_1$ represents the gamut of colors that can be generated by using the three primaries based on R_1, G_1, B_1 with the coordinates shown. As is apparent, the present set of simple primaries cannot generate the colors outside the shaded gray region. The generation of the full gamut of colors needs the color matching functions, $\overline{x}(\lambda)$, $\overline{y}(\lambda)$ and $\overline{z}(\lambda)$.

energy levels lying within the energy bandgap which can absorb light within a specific wavelength band, and also as a result of electronic transitions involving these defect centers, produce characteristic light emission. In the absence of such defects these crystals would be transparent. One of the most studied defects is the so-called *F* center named after the German word for color, *Farbe*. Since the defect states form a band, tunable color center laser can be fabricated using alkali halide crystals with tunability from 0.6 to 4 μm. An important feature of color center lasers is the fact that linewidths as narrow as 4 kHz have been reported for single mode CW operation. Such lasers, for example, have application in spectroscopy and laser chemistry.

Color centers are responsible for coloration of some natural gemstones such as amethyst and topaz. Amethyst is a violet tinted gemstone consisting of crystalline quartz with iron as an added impurity, which is responsible for the violet color in amethyst by creating a color center. The color center (not the iron impurity itself but a defect in the crystal containing iron) is responsible for absorbing the "yellow" photon that makes amethyst violet. In another example, topaz, a fluoroaluminosilicate with iron as an added impurity, achieves its yellow color by means of a color center, again created by the iron impurity. On the other hand, the term "color center" is sometimes used so loosely that even transition-metal and the bandgap colorations are included.

Color matching functions (CMFs) are a set of three primary colors, e.g. $\overline{r}(\lambda)$, $\overline{g}(\lambda)$, and $\overline{b}(\lambda)$ each of which has a particular spectral distribution (hence the dependence on the wavelength) so that an appropriate combination of the three functions can be matched to a set of arbitrary colors. Three

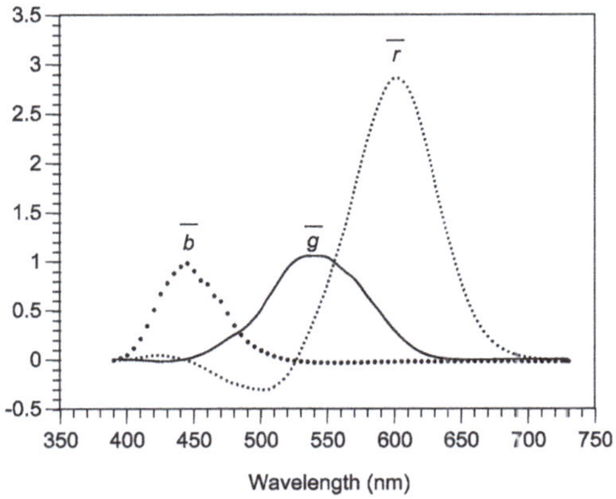

Color matching function The CIE 1931 standard observer color matching functions $\bar{r}(\lambda)$, $\bar{g}(\lambda)$, and $\bar{b}(\lambda)$. Observations within a viewing angle of 2°.

such color matching functions are the CIE 1931 standard observer color matching functions (observations within a viewing angle of 2°) denoted $\bar{r}(\lambda)$, $\bar{g}(\lambda)$, and $\bar{b}(\lambda)$. The functions are such that $\bar{r}(\lambda)$ and $\bar{g}(\lambda)$ are zero at 435.8 nm and $\bar{r}(\lambda)$ and $\bar{b}(\lambda)$ are zero 546.1 nn, and $\bar{g}(\lambda)$ and $\bar{b}(\lambda)$ are zero at 700 nm, and have the wavelength dependences shown in the first figure. A given color with a spectral power distribution $P(\lambda)$ then has R, G and B tristimulus color coordinates given by

$$R = \int P(\lambda)\bar{r}(\lambda)d\lambda, \quad G = \int P(\lambda)\bar{g}(\lambda)d\lambda,$$

$$B = \int P(\lambda)\bar{b}(\lambda)d\lambda$$

where the integration is over all wavelengths. R, G and B can be normalized to generate the chromaticity coordinates r, g and b where $r = R/(R+G+B)$, etc. and $r+g+b=1$. The CMFs $\bar{r}(\lambda)$, $\bar{g}(\lambda)$, and $\bar{b}(\lambda)$ were derived from color matching experiments carried out by Wright in 1928 and Guild in 1931 on observers who were asked to add light at 435.8 nm (blue), 546.1 nm (green) and 700 nm (red) in such a way to match the color of an object. Notice that $\bar{r}(\lambda)$ becomes negative over certain wavelengths. There are three CMFs $\bar{x}(\lambda)$, $\bar{y}(\lambda)$ and $\bar{z}(\lambda)$, shown in the second figure, that are associated with the tristimulus values Y, Y and Z used in the CIE xyz chromaticity diagram; they are defined in a similar fashion to the above equations

$$X = \int P(\lambda)\bar{x}(\lambda)d\lambda, \quad Y = \int P(\lambda)\bar{y}(\lambda)d\lambda.$$

$$Z = \int P(\lambda)\bar{z}(\lambda)d\lambda.$$

The CMFs $\bar{x}(\lambda)$, $\bar{y}(\lambda)$ and $\bar{z}(\lambda)$ are such that they are all positive over the visible wavelength range; $\bar{y}(\lambda)$ corresponds to the photopic luminous efficiency $V(\lambda)$; $\bar{z}(\lambda)$ is negligibly small above 650 nm; the constant energy white point corresponds to $X = Y = Z$.

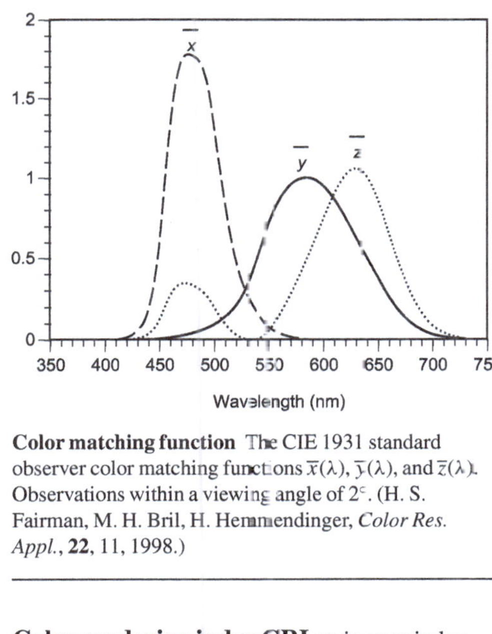

Color matching function The CIE 1931 standard observer color matching functions $\bar{x}(\lambda)$, $\bar{y}(\lambda)$, and $\bar{z}(\lambda)$. Observations within a viewing angle of 2°. (H. S. Fairman, M. H. Bril, H. Hemmendinger, *Color Res. Appl.*, **22**, 11, 1998.)

Color rendering index CRI assigns an index between 0 to 100 to a light source to represent how well it is able to render the colors of an object when the object is illuminated by this light source only. CRI is defined by comparing its effect to that due to illumination from a standard light source that has been assigned a CRI of 100. When an object is

Table: Color rendering index Typical CRI of light from various lamps

Incandescent	Fluorescent	Metal halide	White LED (GE Vio)	High pressure Na	Low pressure Na
90–100	80–85	70–90	70–85	20–65	0

illuminated by white light, its colors are well reproduced but if it is illuminated by a source with a narrow spectrum, such as a Na lamp, the color reproduction is very poor. Eight reference color samples are used to measure the CRI of a light source. The samples are illuminated by the reference light and also by the test light. For each color sample, we measure the differences observed in the three color coordinates (in uniform color space) under reference and the test illumination and then calculate a CRI for that sample using a particular equation set by CIE. Then, we average the CRIs for the eight samples. Although the actual calculation is somewhat complicated, the final result has a straightforward interpretation as representing the color reproduction ability of a light source. Although there is no well-defined cut off for CRI below which color production is unacceptable, generally a CRI below 50 gives very poor color rendering. Excellent color reproduction needs a CRI of 80–100.

Color temperature is a quantity that is expressed in kelvins to represent the spectral luminous flux from a body as if it were emitting a black body radiation at the color temperature. Higher color temperatures have more short wavelength content and hence more "blue" photons; a xenon arc lamp is bluer than an incandescent lamp. Daylight has a color temperature of about 5000 (noon)–6000 (overcast) K. Typically color temperatures above 3000 K are called "cool white" and those below 3000 K are called "warm." See *gray body*.

Colored glass filters are *filters* made from glass that selectively transmit or absorb particular wavelengths of light, and generally belong to the class of so-called *absorption filters*. Currently, these filters are available in glass or plastic-coated glass, mixed or impregnated with organic or inorganic dyes (e.g. colloidal selenides). The quality, i.e. uniformity of density and color over the whole filter, of the glass, and its stability under operating conditions, e.g. humidity, corrosive environment, are important factors. Using such filters, one can tune the color temperature of a light source such as a halogen lamp to produce the light spectrum of bright daylight or the evening sky, and also adjust contrast in photography. Color filter types include bandpass, which transmit over a band of wavelengths, and long/short pass filters – the latter transmit wavelengths starting/ending from/with a particular cut-on/cut-off wavelength. These types of color glass filters generally tend to offer lower transmittance than more expensive multilayer interference filters, as well as poorer slope characteristics (i.e. slow change in transmittance over wavelength at the beginning or end of a transmission band). In addition, most long-pass filter glasses suffer from high levels of autofluorescence. Within the category of color glass absorption filters are color compensating filters

Table: Color temperature Typical color temperatures (K)

Xenon arc lamp	Daylight	Fluorescent	Incandescent	High pressure Na	Low pressure Na	Candle flame
6500	5000–6000	3000–4000	3000	2100	1740	1800

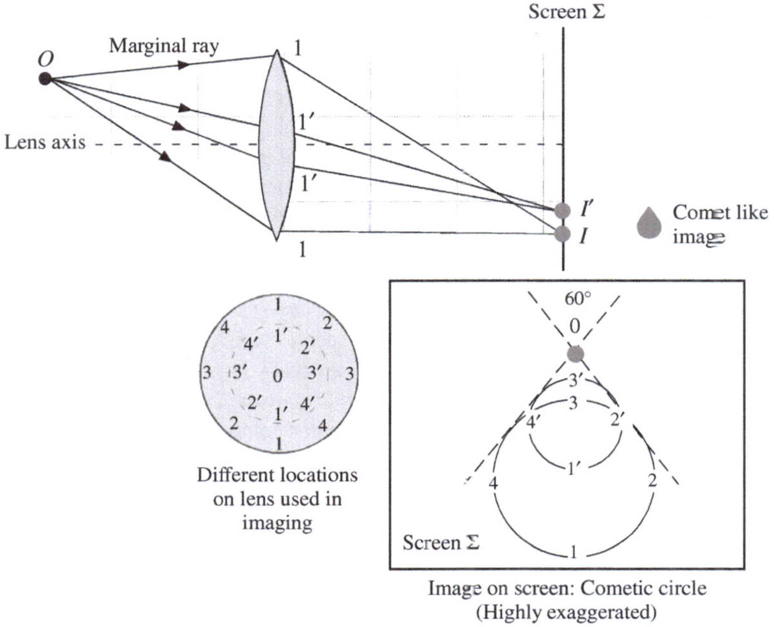

Coma is *spherical aberration* for off-axis points. The marginal rays from a point are incident on the lens at 1, 2, 3 and 4 (outer circle) and generate the image circle 1,2,3,4 on the screen Σ. The rays closer to the lens center are 1′, 2′, 3′, 4′ and generate a smaller circle on Σ. (Lower diagram adapted from Melles Griot website.)

which are designed to control color by attenuation of the red, blue and green regions of the light spectrum. By using filters with different peak transmission for one of the colors, while transmitting one or both of the other two colors untouched, subtle changes in the color balance of a light source can be produced.

Colorimetry is the science of color measurement and color specification. See *color*.

Columnar phase see *liquid crystals*.

Coma is a third-order spherical aberration that occurs for off-axis points on the object. An object point *O* that is off the lens axis is not mapped exactly at a unique image point *I* by the lens as it would be if the paraxial approximations were totally valid; instead there are various image points *I*, *I′*, etc. corresponding to rays from *O* incident on different regions of the lens, which is illustrated in the figure. Coma is an off-axis spherical aberration. Marginal rays 1, 2, 3, 4 provide a larger and a displaced image of a point object *O* than rays 1′, 2′, 3′, 4′ near the lens center. The overall image has a *comet*-like appearance, hence the name. Coma aberration can be minimized by designing the optical system in such a way that there are aperture stops and/or the design makes use of *Abbe's sine condition*. See *Abbe's sine condition, aberrations, spherical aberration*.

Comb filter is a filter that has a bandpass transmission spectrum that is periodically repeated along the frequency axis. The filter is normally obtained by taking a portion (β) of the output signal, delaying it (by τ), and then mixing it with the original input signal with the result that the sum has the required frequency transfer function. For example, if $x(t)$ is the input signal, and $y(t)$ is the output signal, then the filter has a transmission output $y = x(t) + \beta y(t-\tau)$. This is called feedback since a portion of the output is fed to the input. It is also possible to feed a portion of the input, after delay, to the output in a feedforward operation. A Fabry–Perot optical cavity has a transmission spectrum that behaves essentially like a comb filter. Fiber Bragg gratings with circulators can be used to implement comb filters as well. The separation of neighboring maxima on the frequency domain is called the *free spectral range*, FSR. In tunable comb filters, the FSR can be adjusted externally. Comb filters have applications in wavelength division multiplexing.

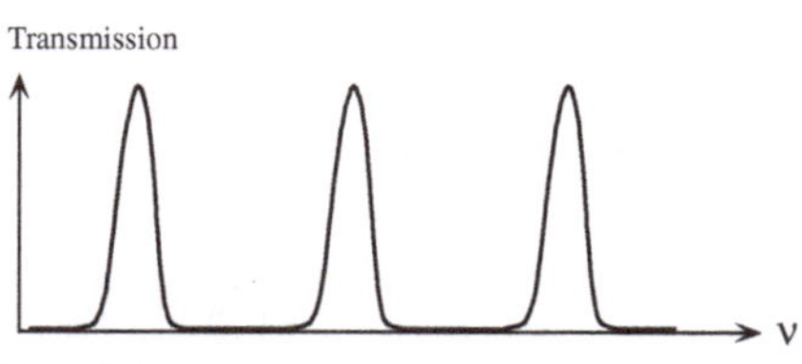

Comb filter Examples of comb filters with their transmission spectral characteristics in the frequency domain.

Compensated semiconductor contains both donors and acceptors in the same crystal region which compensate for each other's effects. For example, if there are more donors than acceptors, $N_d > N_a$, then some of the electrons released by donors are captured by acceptors and the net effect is that $N_d - N_a$ number of electrons per unit volume are left in the conduction band.

Compensating fiber is a length of fiber that is connected (spliced) to a transmission fiber whose chromatic dispersion is to be reduced. A compensating fiber has dispersion characteristics that nearly cancel the dispersion in the transmission fiber. The fiber for use is produced with some chromatic dispersion, for example +10 ps nm^{-1} km^{-1} and dispersion is allowed to build up to say $\Delta\tau_1$ over a certain length L_1. Then dispersion $\Delta\tau_1$ is "cancelled" by using a second fiber, called the *compensating fiber*, that has negative dispersion characteristics with respect to the first fiber, that is, modes that travel fast in the first fiber are slow in the second fiber. If D_t and D_c are the dispersion coefficients in the transmission and compensation fibers, and L_t is the length of the transmission fiber, then the compensating fiber has been designed such that $D_t L_t + D_c L_c$ has the right overall dispersion characteristics. The introduction of dense wavelength division multiplexing (DWDM) has led to certain requirements not only on the magnitude of dispersion but also on its dependence on the wavelength ("dispersion slope"). Consequently, dispersion management in fibers is one of the most active areas of current research development. However, while the compensating fiber enables dispersion to be controlled, one should note that the compensating fiber has higher attenuation, due to the more heavily doped core region, and needs to be of shorter length. Compensation depends on the temperature. It is more susceptible to nonlinear effects. Different cross-sections between the transmission and the compensating fiber can lead to splicing or coupling losses. They are generally used at the receiver end. Manufacturers provide the transmission fiber spliced to a compensating fiber for a well-defined D vs. λ behavior. The *inverse dispersion slope fiber* has the *inverse* dispersion of the transmission fiber, and cancels the detrimental effect of dispersion across the spectrum of communications wavelengths to allow wavelength division multiplexing to be used. See *fiber compensation and management*.

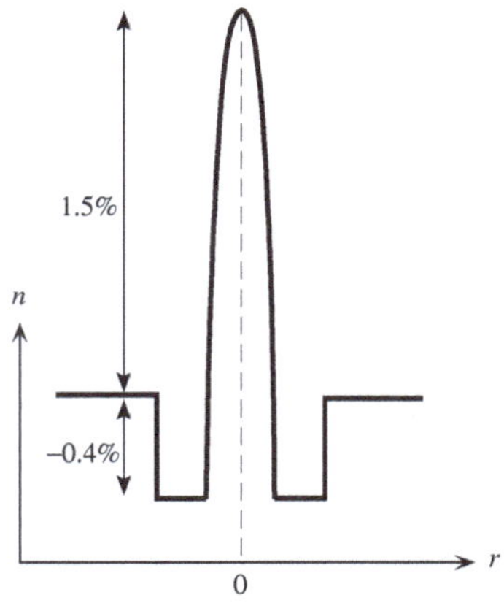

Compensating fiber The refractive index profile of a particular compensating fiber (schematic).

Complex degree of coherence or **coherence factor** see *coherence*.

Complex propagation constant, $(k' - jk'')$, describes the propagation characteristics of an electromagnetic wave that is experiencing attenuation as it travels in a lossy medium. If $k = k' - jk''$ is the complex propagation constant, then the electric field component of a plane wave traveling along z in a lossy medium can be described by

$$E = E_0 \exp(-k''z) \exp j(\omega t - k'z).$$

The amplitude decays exponentially while the wave propagates along z. The real part k' of the complex propagation constant (wavevector) describes the propagation characteristics, e.g. phase velocity $v = \omega/k$. The imaginary part k'' describes the rate of attenuation along z. The intensity I at any point along z is

$$I \propto |E|^2 \propto \exp(-2k''z)$$

so that the rate of change in the intensity with distance is

$$dI/dz = -2k''I$$

where the negative sign represents attenuation. Suppose that k_0 is the propagation constant in vacuum. This is a real quantity as a plane wave suffers no loss in free space. The *complex refractive index N* with real part n and imaginary part K is defined as the ratio of the complex propagation constant in a medium to propagation constant in free space

$$\mathrm{N} = n - jK = k/k_0 = (1/k_0)[k' - jk'']$$

$$n = k'/k_0 \quad \text{and} \quad K = k''/k_0.$$

The real part n is simply and generally called the *real refractive index* and K is called the *extinction coefficient*. In the absence of attenuation,

$$k'' = 0, \quad k = k' \quad \text{and} \quad N = n = k/k_0 = k'/k_0.$$

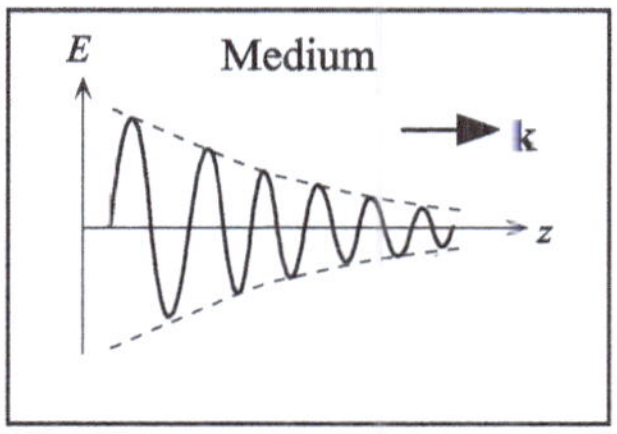

Complex propagation constant Attenuation of light in the direction of propagation.

Complex refractive index characterizes the propagation of a light wave in a medium in which there is a loss of energy, that is, the electromagnetic wave experiences attenuation, due to various loss mechanisms such as the generation of phonons (lattice waves), photogeneration, free carrier absorption, scattering, etc. If N is the complex refractive index, then $N = n - jK$ where the real part n, the refractive index, represents the effect of the medium on the phase velocity, and the imaginary part, K, called the extinction coefficient, represents the attenuation suffered by the wave as it travels along a well-defined propagation direction. N can be defined in terms of the *complex relative permittivity* $\varepsilon_r = \varepsilon_{r1} - j\varepsilon_{r2}$

$$\mathrm{N} = n - jK = \sqrt{\varepsilon_r} = \sqrt{\varepsilon_{r1} - j\varepsilon_{r2}}.$$

Thus, n and K can be directly related to ε_1 and ε_2,

$$n^2 - K^2 = \varepsilon_{r1} \quad \text{and} \quad 2nK = \varepsilon_{r2}$$

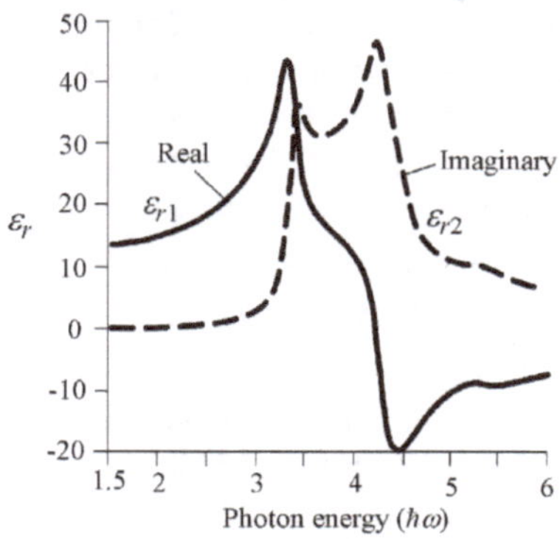

Complex relative permittivity of a silicon crystal as a function of photon energy plotted in terms of real (ε_{r1}) and imaginary (ε_{r2}) parts.

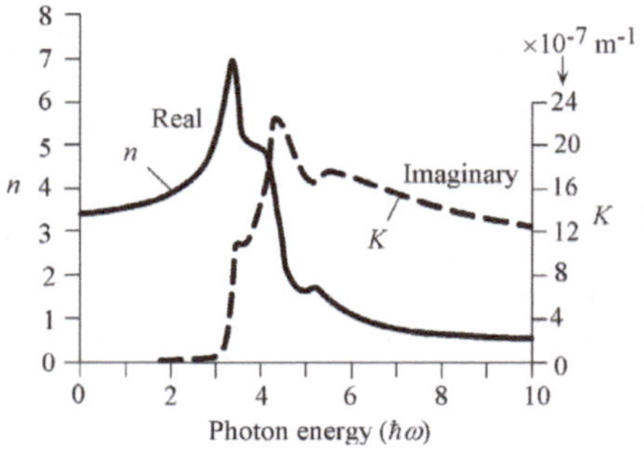

Optical properties of a silicon crystal vs. photon energy in terms of real (n) and imaginary (K) parts of the complex refractive index.

Complex refractive index Complex relative permittivity of a silicon crystal as a function of photon energy plotted in terms of real ($\varepsilon_r{}'$) and imaginary ($\varepsilon_r{}''$) parts.

---

or

$$n = \left[\frac{(\varepsilon_{r1}^2 + \varepsilon_{r2}^2)^{1/2} + \varepsilon_{r1}}{2}\right]^{1/2} \quad \text{and}$$

$$K = \left[\frac{(\varepsilon_{r1}^2 + \varepsilon_{r2}^2)^{1/2} - \varepsilon_{r1}}{2}\right]^{1/2}.$$

Optical properties of materials are typically reported either by showing the frequency dependences of n and K or ε_{r1} and ε_{r2}. The complex relative permittivity and the complex refractive index of crystalline silicon in terms of the photon energy $\hbar\omega$ (or frequency) are shown in the figure. For photon energies below the bandgap energy, both ε_{r2} and K are negligible and n is close to about 3.7. Both ε_{r2} and K increase and change strongly as the photon energy becomes greater than 3 eV, far beyond the bandgap energy (1.1 eV). Notice that both ε_{r1} and n peak around $\hbar\omega \approx 3.5$ eV.

The optical properties n and K can be determined by measuring the reflectance from the surface of a material as a function of polarization and the angle of incidence (based on Fresnel's equations). The reflection and transmission coefficients that are normally given by Fresnel's equations are based on using a real refractive index, that is, neglecting losses. We can still use the Fresnel reflection and transmission coefficients if we simply use the complex refractive index N instead of n. For example, consider a light wave traveling in free space incident on a material at normal incidence ($\theta_i = 90°$). The reflection coefficient is now,

$$r = \frac{1 - \mathrm{N}}{1 + \mathrm{N}} = \frac{n - 1 + jK}{n + 1 - jK}.$$

The reflectance is then

$$R = |r|^2 = \frac{(n-1)^2 + K^2}{(n+1)^2 + K^2}$$

which reduces to the usual forms when the extinction coefficient $K = 0$.

Complex relative permittivity, ($\varepsilon_r{}' - j\varepsilon_r{}''$ or $\varepsilon_{r1} - j\varepsilon_{r2}$), has a real part, $\varepsilon_r{}'$ or ε_{r1}, that determines the charge storage ability and an imaginary part, $\varepsilon_r{}''$ or ε_{r2}, that determines the energy losses in the material as a result of the polarization mechanism. The real part determines the capacitance through $C = \varepsilon_0\varepsilon_r{}'A/d$ and the imaginary part determines the electric power dissipation per unit volume as heat by $E^2\omega\varepsilon_0\varepsilon_r''$, where E is the electric field, ω is the angular frequency and ε_0 is the absolute permittivity.

Complex representation refers to a convenient analytical description of the electromagnetic (EM) field. Although the electric component of the EM field is real, the complex notation is particular useful in terms of manipulating time-dependent field equations. The time dependence $\exp(j\omega t)$ can be taken out as a common factor. We distinguish here the two notations by posing $\mathbf{E} = \mathrm{Re}[\boldsymbol{\mathcal{E}}]$. In real notation, a monochromatic field with a frequency ω and wavevector $\boldsymbol{k}$ can be written as

$$E_x = E_{0x}\cos(\omega t - \boldsymbol{k}.r + \varphi_x) = \mathrm{Re}[\boldsymbol{\mathcal{E}}_x]$$
$$E_y = E_{0y}\cos(\omega t - \boldsymbol{k}.r + \varphi_y) = \mathrm{Re}[\boldsymbol{\mathcal{E}}_y]$$
$$E_z = E_{0z}\cos(\omega t - \boldsymbol{k}.r + \varphi_z] = \mathrm{Re}[\boldsymbol{\mathcal{E}}_z].$$

In complex notation, the same fields become

$$\mathcal{E}_x = \boldsymbol{E}_{0x}\exp[j(\omega t - \boldsymbol{k}.r + \varphi_x)]$$
$$\mathcal{E}_y = \boldsymbol{E}_{0y}\exp[j(\omega t - \boldsymbol{k}.r + \varphi_y)]$$
$$\mathcal{E}_z = \boldsymbol{E}_{0z}\exp[j(\omega t - \boldsymbol{k}.r + \varphi_z)].$$

As a consequence, the intensity (in W m^{-2}) can be expressed as

$$I = \varepsilon_0 cn\langle|E|^2\rangle = (1/2)\varepsilon_0 cn\langle|\boldsymbol{\mathcal{E}}|^2\rangle$$
$$= (1/2)\varepsilon_0 cn(E_{0x}^2 + E_{0y}^2 + E_{0z}^2).$$

Compton effect is usually referred to the *scattering* of a high-energy photon, such as an x-ray photon, by a nearly "free" electron in a medium or by a loosely bound outer electron in an atom. When an x-ray photon interacts with an outer electron, it knocks the electron out of its orbit and imparts kinetic energy to it and, in the process, it loses energy and becomes deflected. The Compton effect is readily observed when an x-ray beam is scattered from a target that contains many conduction ("free") electrons such as a metal or graphite. When an x-ray photon interacts with a conduction electron in a

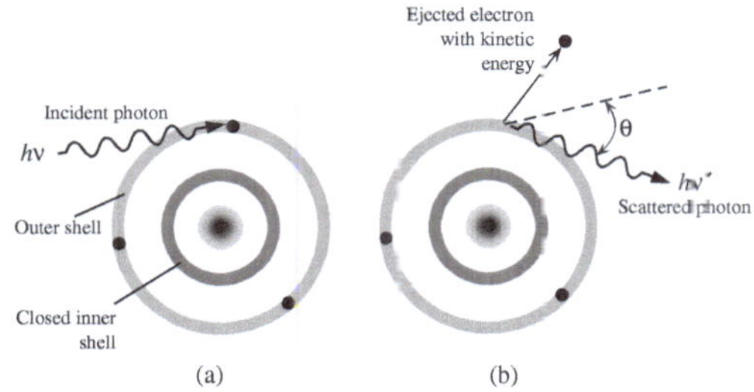

Compton effect Inelastic scattering of a high-energy photon by the outer electron of an atom. The electron becomes ejected and gains kinetic energy whereas the photon loses energy and becomes deflected.

metal it becomes deflected just as if it were a "particle" with a certain momentum p. The photon loses some of its energy to the electron and therefore suffers a reduction in the frequency from ν to ν'. From the Compton experiment, the momentum of the photon has been shown to conform to $p = h/\lambda$ (De Broglie relationship). Compton scattering is of particular interest in x-ray imaging and it is significant

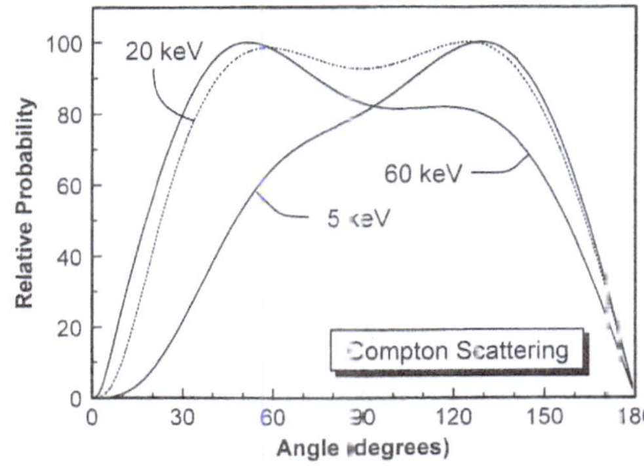

Compton effect The probability density function of Compton scattering as a function of angle for three different x-ray energies. The height of each PDF is normalized to 100%. (John M. Boone, *X-ray Production, Interaction, and Detection in Diagnostic Imaging, Handbook of Medical Imaging, Volume 1*. SPIE PM79, 2000.)

at high photon energies. If $h\nu$ is the incident photon energy and $h\nu'$ is the scattered photon energy, and θ is the angle of scattering, then *Compton scattering* follows the *Klein–Nishina equation*,

$$\frac{h\nu'}{h\nu} = \frac{1}{1+\alpha(1-\cos\theta)}; \quad \alpha = \frac{h\nu}{m_e c^2} = \frac{h\nu}{(511\,\text{keV})}$$

where $m_e c^2$ is the rest mass energy of the electron. $h\nu'$ becomes smaller as the scattering angle increases, which is significant at higher photon energies.

Compton scattering see *Compton effect*.

Conduction band (CB) is a band of energies for electrons in a semiconductor where they can gain energy from an applied field, drift, and thereby contribute to electrical conduction. The electron in the conduction band behaves as if it were a "free" particle with an effective mass m_e^*. The allowed energies for electrons in the conduction band are normally described by reference to the dependence of the electron energy E on wavevector k or the electron momentum, $\hbar k$. This is known as the dispersion relationship, and often follows the following approximate relationship,

$$E = E_c + \frac{\hbar^2 k^2}{2m_e^*}.$$

For some semiconductors, the conduction band minimum energy does not occur at $k = 0$, but rather at a particular k value, in a given direction – for example, in silicon the particular k-values are $(2\pi/a)(-0.85, 0, 0)$; $(2\pi/a)(0, -0.85, 0)$; $(2\pi/a)(0, 0, -0.85)$; $(2\pi/a)(0, -0.85, 0)$; $(2\pi/a)(0, 0, -0.85)$, where a is the lattice constant of silicon and has a value of 0.543 nm. In this case, the dispersion relationship for the conduction band is given by the equation above, but with k replaced by $k - k_{i0}$, where k_{i0} is the wavevector for minimum energy in a given direction denoted i. From the discussion above, it is clear that silicon has six equivalent k values where the electron energy in the conduction band is a minimum – when the energy for each state is identical, these states are said to be degenerate. Since the valence band maximum energy occurs at $k = 0$, a semiconductor such as silicon having its minimum energy not at $k = 0$, would require an electron to change its momentum to make an interband transition. Thus, a transition between these two bands is termed *indirect*. In the case of GaAs, the conduction band minimum occurs at $k = 0$ and no change in momentum is required to make the transition between the two bands – this semiconductor is therefore termed a direct bandgap semiconductor. The dispersion relationships discussed above may also involve a direction-dependent effective mass. For example, if the mass is the same in all directions or isotropic, and the bandgap is direct, as for GaAs, then the electron states of constant energy in the conduction band occur on the surface of a sphere having a constant radius. On the other hand, in the case of a semiconductor such as silicon, the effective mass is direction-dependent resulting in the allowed states occupying different regions of k-space, and being found on the surface of ellipsoids of revolution, characterized by transverse and longitudinal effective masses.

Confined Stark effect see *Stark effect*.

Confinement factor, see *optical mode confinement*.

Confining layer is a layer having a wider bandgap than the active layer, which is adjacent to it, resulting in the confinement of injected minority carriers in the active layer. Such an approach is commonly used in a variety of devices to increase the concentration of carriers in a particular region – this can be used to enhance the conductivity of a transistor, reduce leakage in a photodetector, or result in a critical concentration of carriers in the active region of a laser. Systems such as AlGaAs/GaAs are favored for making such heterojunctions. This is because they offer near lattice-matched interfaces free from interface states, minimizing nonradiative recombination and permitting long carrier lifetimes. Other considerations are in finding systems having the highest band-offsets to ensure large barriers are formed to the confined layer. Typically, heterojunctions result in a barrier preventing both electrons and holes from escaping from the confining layer, and are known as type I heterojunctions. However, some heterojunctions, including those formed from the antimonides, can result in the formation of a type II heterojunction, where the wider bandgap material only confines one of the carriers, while the narrow bandgap semiconductor confines the other carrier. Such heterojunctions can be used to create spatially separated confinement of electrons and holes, as distinct from type I heterojunctions, where confined electrons and holes are spatially coincident.

Confocal imaging or confocal microscopy is an imaging technique in which a point such as P in a planar section of the object is illuminated and then this point P is "imaged" (i.e. the light from this point is captured) using an objective lens and a suitably placed detector with a small pinhole to exclude light from elsewhere, as shown in the first figure. In confocal imaging, the point P to be imaged (detected) is a point in the sample on the focal plane F_1 of the objective lens system whereas the pinhole and the detector are at the other focal plane F_2. The sample is illuminated by light from a pinhole through another lens system, the condenser lens, in such a way that the illumination is focused onto P. It is apparent that the illumination, the point P that acts as the object, and the point Q that corresponds to the image of P at the detector pinhole are all the focal points of the lenses; hence the name confocal imaging. The out of focus point A in the sample cannot be detected because the light from it is prevented from entering the detector by the detector pinhole. The image Q however is only one *pixel* of the image corresponding to the point P on F_1. The sample must be stepped in the x- and y-directions so that all the points P lying on the focal plane F_1 can be detected and an image of the section F_1 of the object can be constructed by an appropriate confocal imaging software on a computer. Thus, the image is captured pixel by pixel by scanning a line at a time along y (or x). This type of imaging has distinct advantages because it discriminates against capturing light from other points while the detector is recording the light from P. For example, if the sample fluoresces, then the light emitted from other out of focus points such as A would degrade the image in normal microscopy but in confocal microscopy the light from A is excluded in building the image of the section on F_1 point by point. Obviously the section (F_1 plane) that is to be imaged must be scanned. In modern microscopes this is done most commonly not by moving the sample but by using optical scanning. In modern confocal imaging systems, a laser is used to provide the light, and a dichroic (or a dichromatic) mirror eliminates the need for using a condenser lens system as shown in the second figure. The optical scanning of the object either involves a single beam scan or a multiple beam scan. The laser is

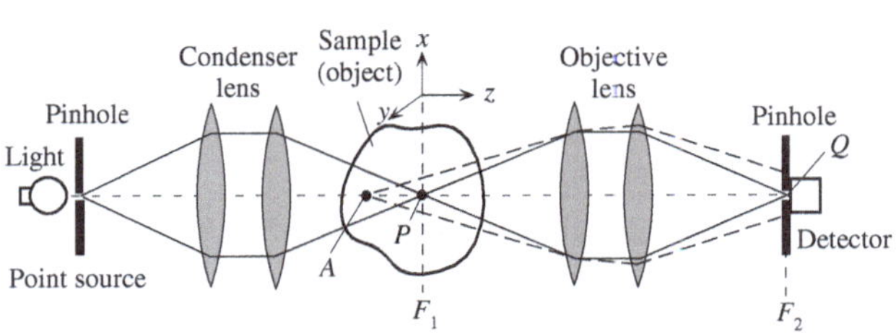

A simplified diagram illustrating **confocal imaging**.

made to scan the section of the object lying in the focal plane F_1 by using controllable mirrors. One further distinct advantage of confocal microscopy is that it is possible to build a three-dimensional (3D) image of the object by moving the focal plane F_1 across the sample (or moving the sample), in the z-direction, and, at each stage, capturing a section of the object. One can therefore build z-series (that is, z-sectioned) images of the object and hence obtain essentially 3D imaging. The principle of the confocal microscope is attributed to Marvin Minsky (US Patent 3,013467 "Microscopy Apparatus," 1957).

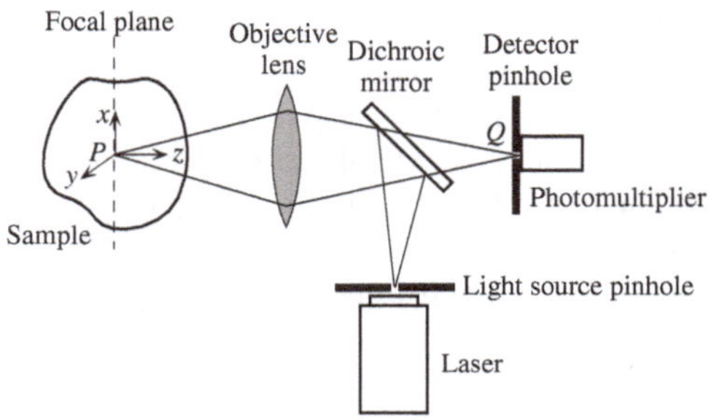

A simplified diagram of a laser **confocal imaging** system. Scanning (not shown) is done optically.

Confocal parameter see *Gaussian beam.*

Conical lens see *axicons.*

Conjugate pair ratio is the ratio of the object distance (s_o) from a lens (or other optical device) to the distance s_i of its image (object's conjugate) from the lens, i.e. s_o/s_i.

Connector (optical) is primarily a mechanical device that temporarily brings fiber ends together to couple light from one to the other. The connection is temporary because the connector can be "unplugged." The connector may contain an optical component such as a lens, or lenses, to couple the light from one fiber to the other. One of the most important characteristics of connectors and splices is their attenuation; how much light is lost through the connection. There are various reasons for the attenuation, most significant are: misalignment of fibers, mismatch of core areas, mismatch in the numerical apertures, dead space between fibers, fiber-end reflections, extraneous materials such as dirt or grease on the fibers, etc. Splices normally have much lower attenuation than connectors and are therefore preferred whenever the connection can remain permanently. An important characteristic of a connector is its *insertion loss*, that is its attenuation. Typically a splice will have an insertion loss less than 0.1 dB. On the other hand, mechanical connectors tend to have substantially higher losses that typically range from 0.2 dB to 1 dB, depending on the type of connector. An ST (straight tip) connector for a single-mode fiber would have a loss of about 0.40 dB.

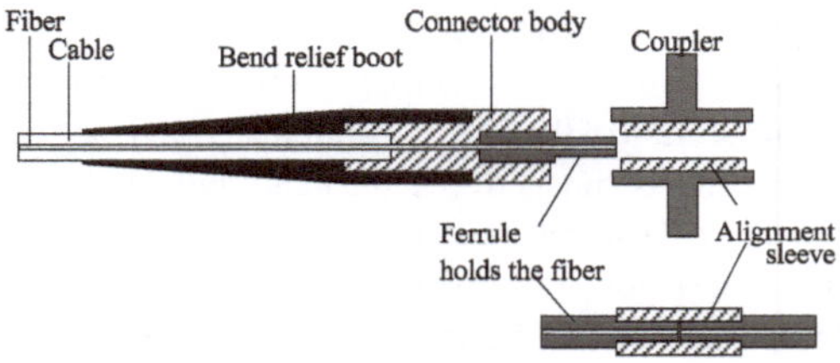

A typical **connector**.

Connector loss see *fiber-to-fiber attenuation.*

Consolidation in fiber manufacturing is a process by which the hollow soot rod obtained from the outside vapor deposition process is sintered, by passing through a furnace (1400–1600 °C), into a clear glass preform rod; this preform is next used in the fiber drawing process. See *outside vapor deposition.*

Constringence or **Abbe number**, see *chromatic dispersion.*

Constructive interference, see *interference.*

Contacting layer in a heterostructure device is a layer that is grown on a given semiconductor for the purpose of making an ohmic contact, that is for avoiding a Schottky junction. Unfortunately, contacts to a semiconductor typically result in the formation of a built-in barrier (Schottky junction type contact). Clearly, the larger the barrier, the more rectifying the contact and less ohmic the behavior – that is, the less linear is the current–voltage characteristic. Thus, the normal approach taken is to heavily dope the contact layer such that the depletion width adjacent to the metallic top contact is minimized and current transport by tunneling through the barrier is favored. It should also be noted that, where transport is controlled through tunneling, the contact resistance is proportional to $\exp(C/N_d^{1/2})$ where N_d is the dopant concentration (assuming an n-type semiconductor) and C is a constant. As an example, for heterostructure AlGaAs lasers, the optical cavity is defined by AlGaAs layers, and the contact layer is typically chosen to be a low-resistance thin heavily doped GaAs epilayer.

Continuity equation is an equation that describes the rate of change with time in the concentration of one type of carrier (i.e. either electrons or holes). In particular, the continuity equation describing the time rate of change for a process that resulted in an *excess* concentration of holes Δp, above the equilibrium concentration p_o, would include terms describing a spatial variation in the current density, generation and recombination. The current density in general has a drift contribution, arising from an electric field, and a diffusion contribution arising from a concentration gradient. Suppose that at one instant the hole concentration $p(x,t)$ as a function of distance x in a sample is as shown in the figure. Assume that there is no variation in p along y- and z-directions. Consider a small elemental volume of thickness δx. Then:

Net rate of increase in hole concentration in the elemental volume δx
= Rate of increase due to spatial change (decrease) in the current density
+ Rate of generation (G_n)
– Rate of recombination (R_n).

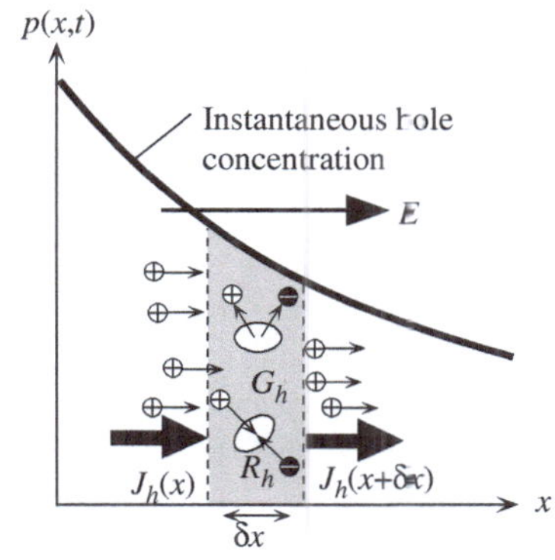

Continuity equation Arbitrary hole concentration profile $p(x,t)$. Consider an elemental volume δx at x. The net rate of change of p depends on the change in the current density J_h, generation G_h and recombination R_h processes. E is the field.

In mathematical form, the continuity equation for holes at a given point in a sample, where the current density J_h is due to the drift and diffusion of these holes, is

$$\frac{\partial \Delta p}{\partial t} = \frac{\partial p}{\partial t} = -\frac{1}{e}\frac{\partial J_h}{\partial x} + G_h - R_h$$

where p and Δp are the instantaneous and excess hole concentrations respectively, G_h and R_h refer to the *generation* and *recombination rates* for holes, e is the charge of a hole (electron charge), and J_h refers to the hole current. If drift and diffusion processes are included in the description of the hole transport equation, then the hole current density J_h can be written as,

$$J_h = e\mu_h p E - eD_h\frac{\partial p}{\partial x}$$

where μ_h, D_h are the drift mobility and diffusivity of holes, and E is the electric field, respectively. The first term on the right-hand side refers to drift, while the second term refers to diffusion (from *Fick's first law*). Consider a sample in which the electric field E is *uniform* and an excess hole concentration Δp has been created at a time $t = 0$ at some location by suitable means (e.g. by photoexcitation). Suppose that the recombination time for the excess holes is τ_h. Then the rate of recombination R_h is $\Delta p/\tau_h$ and the above two equations combine to give

$$\frac{\partial \Delta p}{\partial t} = -\mu_h E \frac{\partial \Delta p}{\partial x} + D_h \frac{\partial^2 \Delta p}{\partial x^2} - \frac{\Delta p}{\tau_h}.$$

The excess holes simply drift, diffuse and recombine with electrons, the same as indicated in the figure. If the initial excess hole concentration is a simple delta function, then the excess hole concentration is given by

$$\Delta p(x,t) = \frac{N_h}{(4\pi D_h t)^{1/2}} \exp\left[-\frac{(x-\mu_h E t)^2}{4D_h t}\right] \times \exp\left(-\frac{t}{\tau_h}\right)$$

where N_h is the total number of excess holes initially injected per unit area. The hole concentration profile shifts, that is drifts, along x with a velocity $dx/dt = v_d = \mu_h$E.

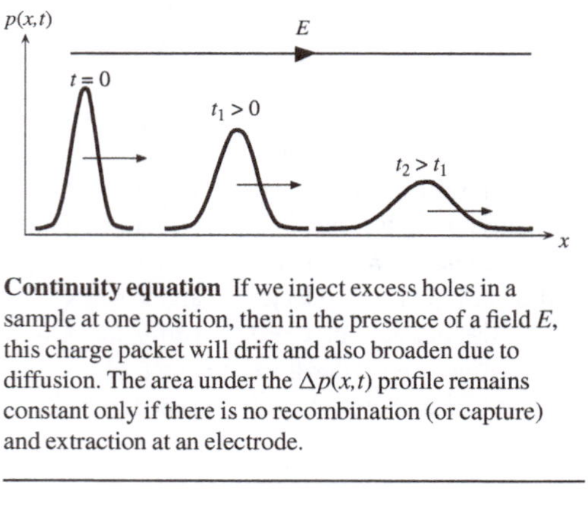

Continuity equation If we inject excess holes in a sample at one position, then in the presence of a field E, this charge packet will drift and also broaden due to diffusion. The area under the $\Delta p(x,t)$ profile remains constant only if there is no recombination (or capture) and extraction at an electrode.

Continuous wave (CW) is an untruncated sinusoidal oscillation; untruncated means over the timescale of observation as opposed to a pulsed oscillation that lasts for a short duration of time.

Contrast in an image refers to the amount of difference in the color or brightness of different regions in the image. The same qualitative definition can be applied to differences in the color and brightness of different regions of an object when these regions are inside the same field of view. One of the problems in rigorously defining and quantifying contrast is that it depends on the spatial frequency and also on the adaptation state of the eye. There are various definitions to quantify contrast. The *Michelson contrast* usually applies to periodic images. If L_{max} is the maximum luminance and L_{min} is the minimum luminance, then

$$C = \frac{L_{max} - L_{min}}{L_{max} + L_{min}}.$$

The *Weber contrast* definition is the local contrast in an image, for example, the contrast in a symbol or a small pattern, compared with its immediate background. If L_{symbol} is the luminance of the symbol and $L = L_{background}$ is the background luminance, assumed uniform, and $\Delta L = |L_{symbol} - L_{background}|$, then

$$C = \frac{\Delta L}{L}.$$

Contrast ratio, on the other hand, measures contrast by taking $L_{symbol} / L_{background}$. Root mean square contrast of an image of n pixels is

$$C_{RMS} = \left[\frac{1}{n-1}\sum_{i=1}^{n}(x_i - \overline{x})^2\right]^{1/2}$$

in which x_i is a normalized gray level of pixel i and $\overline{x}$ is the mean normalized gray level. Suppose that $F(k_x,k_y)$ is the Fourier transform of the image $f(x,y)$ with k_x and k_y the horizontal and vertical spatial frequencies. *Fourier domain contrast* is defined as

$$C_{RMS} = \frac{2A(k_x, k_y)}{A_{DC}}$$

where $A(k_x,k_y)$ is the amplitude of the Fourier transform at k_x, k_y and A_{DC} is the zero-frequency component. (E. Peli, *J. Opt. Soc. Am. A.*, **7**, 2032, 1990.)

Conventional lithography see *nanoelectronics*.

Conversion efficiency of an optoelectronic device gauges the overall efficiency of conversion from the input of electrical power to the output of optical power (or vice versa). Although this is not generally quoted in data sheets, it can be easily determined from the output power at the operating diode current and voltage. In some modern laser diodes this may be as high as 30–40%.

Convolution is a mathematical operation that takes two functions $f(x)$ and $g(x)$ and then generates a third function $h(x)$ that is also on the x-domain by the following mathematical integration procedure,

$$h(x) = f^*g = \int_{-\infty}^{\infty} f(x')g(x-x')dx'$$

where $f(x')$ (x' is the dummy variable of integration) is multiplied with "$g(x')$" that has been reversed ("flipped," x' replaced by $-x'$) and shifted by x, and the product integrated over all values of x'; that is, the overlap integral of $f(x')$ and $g(x-x')$ as a function of x. The notation f^*g designates the convolution operation. The variable x can be space (as it occurs in imaging), time or frequency. $F(\nu)$ and $G(\nu)$ are the Fourier transforms of $f(x)$ and $g(x)$. If x represents time t, ν is frequency and if x is a space coordinate, ν is the spatial frequency. The *convolution theorem* states that the product of the Fourier transforms $F(\nu)G(\nu)$ is the Fourier transform of the convolution of $f(x)$ and $g(x)$,

$$F(\nu)G(\nu) = \mathrm{FT}[f^*g].$$

The convolution integral occurs frequently in optoelectronics and photonics, and there are numerous examples. Typically, it arises when two different processes with different frequency characteristics interact either one after another or at the same time. For example, the emission spectrum from certain gas discharge lamps can be the convolution of the *Lorentzian* and *Gaussian spectra*, resulting in a *Voigt lineshape*. In imaging, one can represent an optical signal (irradiance emitted) from the object in terms of its spatial frequencies $F(\nu)$. If this is passed through an imaging system (e.g. an optical filter) with a characteristic transfer function $G(\nu)$, the resulting image will have a Fourier transform that is $F(\nu)G(\nu)$ and hence the image in real space is represented by f^*g.

Core of a optical fiber is the central region of the fiber that has a higher refractive index than the outer region. Most of the light propagates through the core region of a fiber.

Cosine fourth power law is a "law" that occurs in an optical or an imaging system, such as a camera, in which the illuminance of a small area on the image depends on the fourth power of the cosine of the angle that the small area subtends at the aperture of the imaging system; the object is assumed to be bright and extended. Consider a small area ΔS_i on the image, which is at an angle θ to the normal to the lens or the aperture of a camera. If the image distance from the lens is s_i, the aperture is D, and the luminance of the aperture is L, then the image illuminance E_i of ΔS_i is

$$E_i = \frac{\pi L \cos^4\theta}{4(s_i/D)^2}.$$

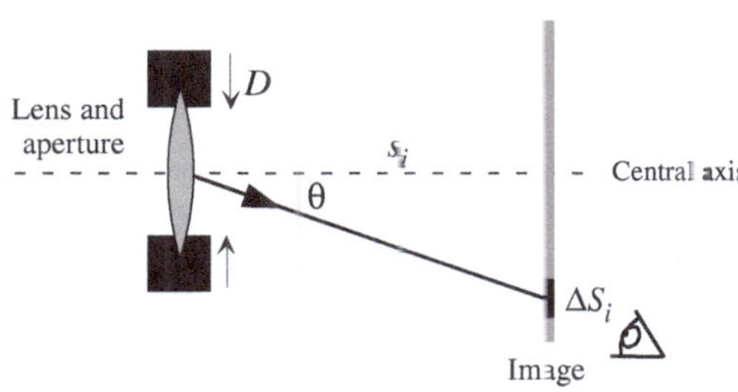

Illustration of the **cosine fourth power law** in an optical/imaging system.

Cotton–Mouton effect, also known as the **Voigt effect,** is the inducement of birefringence in a material due to the application of a magnetic field, analogous to the electrooptic Kerr effect. For example, when a magnetic field is applied to a liquid, the molecules may become aligned resulting in an optical anisotropy. The induced change in the refractive index is proportional to the square of the magnetic field.

Coupled-wave equations (CWE) describe the linear coupling between two propagating waves or, most generally, two oscillators. Under their general form, waves A_1 and A_2 are related by

$$dA_1/dz = +j\delta A_1 - j\kappa_{12}A_2,$$
$$dA_2/dz = -j\kappa_{21}A_1 - j\delta A_2,$$

where δ is the *detuning* and $(\kappa_{12}, \kappa_{21})$ the *coupling constants*. All parameters can be complex without any loss of generality. When no coupling is involved, both waves propagate without interaction. In the presence of coupling, they are solutions to the same second-order linear differential equation:

$$d^2A_p/dz^2 + [\kappa_{12}\kappa_{21} + \delta^2]A_p = 0$$

where the subscript p may be 1 or 2. They can be expressed as a linear combination of $\exp(+j\gamma z)$ and $\exp(-j\gamma z)$, where $\gamma = [\kappa_{12}\kappa_{21}+\delta^2]^{1/2}$. The coefficients are functions of boundary conditions,

$$A_1(z) = [\cos(\gamma z) + j(\delta/\gamma)\sin(\gamma z)]A_1(0) - j(\kappa_{12}/\gamma)\sin(\gamma z)A_2(0),$$
$$A_2(z) = -j(\kappa_{21}/\gamma)\sin(\gamma z)A_1(0) + [\cos(\gamma z) - j(\delta/\gamma)\sin(\gamma z)]A_2(0).$$

Coupled-wave equations are perfectly adapted to codirectional coupling between two integrated waveguides, as well as contradirectional coupling in a distributed-feedback (DFB) structure. Besides, from a formal point of view, they are exactly similar to the Schrödinger equation describing the time evolution of a two-level quantum system.

Coupled waveguides are two sufficiently closely spaced optical waveguides, say A and B, such that the electric fields associated with the propagation modes in A and B can overlap so that light can be coupled from one guide to another in a way reminiscent of frustrated total internal reflection. We can use qualitative arguments to understand the nature of light coupling between these two guides. Suppose that we launch a light wave into the guide A operating in single mode. Since the separation d of the two guides is small, some of the electric field in the evanescent wave of this mode extends into guide B, and therefore some electromagnetic energy will be transferred from guide A to B (see figure). This energy transfer will depend on the efficiency of coupling between the two guides and the nature of the modes in A and B, which in turn depend on the geometries and refractive indices of the guides and the substrate (acting as a cladding). As light in guide A travels along z, it leaks into B and, if the mode in B has the right phase, the transferred light waves build up along z as a propagating mode in B. By the same token, the light now traveling in B along z can be transferred back into A if the mode in A has the right phase. The efficient transfer of energy back and forth between the two guides A and B requires that the two modes be in phase to allow the transferred amplitude to build up along z. If the two modes are out of phase, the waves transferred into a guide do not reinforce each other and the coupling efficiency is poor. Suppose that β_A and β_B are the propagation constants of the fundamental modes in A and B. Then there is a phase mismatch per unit length along z, that is $\Delta\beta = \beta_A - \beta_B$. The efficiency of energy transfer between the two guides depends on this phase mismatch. If the phase mismatch $\Delta\beta = 0$ then full transfer of power from A to B will require a coupling distance L_0, called the *transfer distance*. This transfer distance depends on the efficiency of coupling C between the two guides A and B, which in turn depends on the refractive indices and geometries of the two guides. C depends on the extent of overlap of the mode fields E_A and E_B in A and B, respectively. The transmission length L_0 is inversely proportional to C (in fact, theory shows that $L_0 = \pi/C$). In the presence of no mismatch, $\Delta\beta = 0$, full transmission would occur over the distance L_0. However, if there is a mismatch $\Delta\beta$ then the transferred power ratio over the distance L_0 becomes a function of $\Delta\beta$. Thus, if $P_A(z)$ and $P_B(z)$ represent the light power in the guides A and B at z, then

$$\frac{P_B(L_0)}{P_A(0)} = f(\Delta\beta).$$

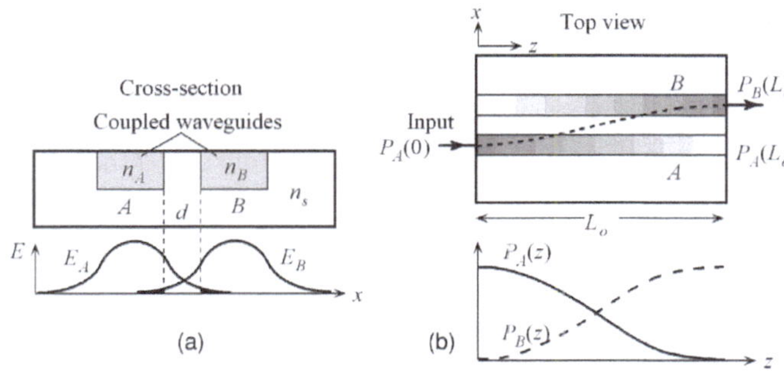

Coupled waveguides (a) Cross-section of two closely spaced waveguides *A* and *B* (separated by *d*) embedded in a substrate. The evanescent field from *A* extends into *B* and vice versa. Note: n_A and $n_B > n_s$ (= substrate index).
(b) Top view of the two guides *A* and *B* that are coupled along the *z*-direction. Light is fed into *A* at $z = 0$, and it is gradually transferred to *B* along *z*. At $z = L_0$, all the light has been transferred to *B*. Beyond this point, light begins to be transferred back to *A* in the same way.

This function has its maximum when $\Delta\beta = 0$ (no mismatch) and then decays to zero at $\Delta\beta = \pi\sqrt{3}/L_0$. If we could induce a phase mismatch of $\Delta\beta = \pi\sqrt{3}/L_0$ by applying an electric field, which modulates the refractive indices of the guides, we could then prevent the transmission of light power from *A* to *B*. Light is then *not* transferred to *B*.

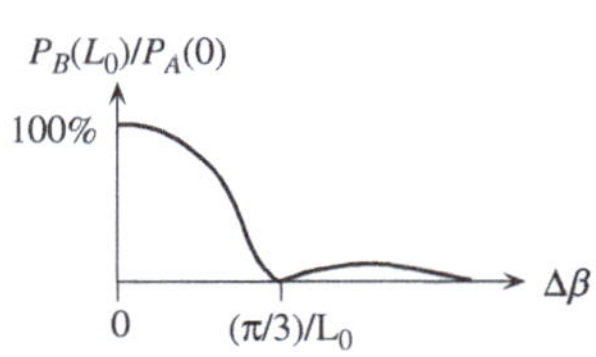

Coupled waveguides Transmission power ratio from guide *A* to guide *B* over the transmission length L_0 as a function of mismatch $\Delta\beta$.

Critical angle, θ_c, is the angle of incidence that results in a refracted wave at 90° when the incident wave is traveling in a medium of higher refractive index n_1 and is incident at a boundary with a material with a lower refractive index n_2; $\sin\theta_c = n_2/n_1$. Incidence angles greater than the critical angle result in *total internal reflection*. See *total internal reflection*.

Critical flicker frequency (CFF) represents the frequency of image sequences (frames per second) that are perceived by a human observer as a continuous flow of images just as in a movie. CFF depends on the illumination of the scene. The *Ferry–Porter law* asserts that the CFF is proportional to the logarithm of the luminance: CFF $\propto \log(L)$, where L is the illuminance in cd m^{-2}. The term *critical fusion frequency* is sometimes used to indicate the CFF. Typical CFFs range from about 40 to 80 Hz.

Critical points in the energy band structure, i.e. electron energy E vs. wavevector k, $E(k)$ of a material correspond to k-values where $|\nabla_k E_{cv}(k)| = 0$, where E_{cv} refers to the separation between the conduction and valence band energies, respectively. k-values where this condition is satisfied provide major contributions to the interband transition strength and are therefore important in describing the optical response of such materials. These critical points are also referred to as van Hove singularities and in the vicinity of such singularities E_{cv} may be given approximately by:

$$E_{cv}(k) = E_0 + \beta_1 k_1^2 + \beta_2 k_2^2 + \beta_3 k_3^2$$

where β refers to coefficients of k about the critical energy E_0. In this notation, the number of nonvanishing coefficients determines the dimensionality n_j of the jth critical point. The type of critical points (maximum, saddle point, minimum) corresponds to the possible choices of signs for the coefficients β_1, β_2 and β_3. In 3d, the critical points are labeled M_0, M_1, M_2 and M_3, whereas in 2d they are labeled D_0, D_1 and D_2, for 1d as P_0 and P_1, and for 0d as S_0. A model of the electronic structure of a material may be derived by summing the contributions from individual critical points.

Cross-phase modulation (XPM) between two waves occurs via the nonlinearity of the material in which they propagate. XPM describes the mutual interaction of one wave upon another, whereas the action of one signal upon itself is referred to as *self-phase modulation* (SPM). The injection of two signals into an active medium (optical fiber, semiconductor optical amplifier, etc.) changes the carrier concentration within the energy bands, or the spectral gain distribution, which results in a modulation of gain and refractive index. The latter modifies the phase of the signal and can be exploited in interferometric devices such as Mach–Zehnder, Michelson or Sagnac interferometers, which convert XPM into an intensity modulated signal, in various temporal and spectral switching operations.

Cross-section in photonics usually refers either to an emission or absorption cross-section, and the emission cross-section can be for stimulated or spontaneous emission. Consider the absorption or emission of radiation due to dopants or impurities in a material system, for example Er^{3+} ions in a glass host such as a silica fiber. Let N_1 (m^{-3}) be the number of dopants at energy E_1 (lower) and N_2 be the number of dopants at E_2 (higher), $E_2 > E_1$. (In equilibrium, $N_1 > N_2$.) Let α be the absorption coefficient for dopant excitations from E_1 to E_2, and W_{12} or $1/\tau_{12}$ be the rate of absorption from 1 to 2. The *absorption cross-section* σ_a can be defined as

$$\alpha = \sigma_a N_1 \qquad \text{or} \qquad W_{12} = (c/n)\sigma_a N_1$$

where n is the refractive index. By definition, $1/\alpha$ is the average distance a photon travels before absorption. Thus in the volume $(1/\alpha)\sigma_a$ there must be at least one absorber and hence $(1/\alpha)\sigma_a N_1 = 1$, which leads to the above expression. σ_a can be readily measured. If I_o and I_t are the entrance and exit light intensities through a sample of thickness L, and absorption is due to N_1 dopants, then σ_a is obtained from $\sigma_a = (N_1 L)^{-1}\ln(I_o/I_t)$. Since I_o and I_t can be measured at a particular wavelength or frequency ν, σ_a can depend on the frequency ν. Similarly, we can define an emission cross section σ_e for photon-emitting transitions from 2 to 1. If the rate of emissions is W_{21} or $1/\tau_{21}$, where τ_{21} is the radiative lifetime, then

$$W_{21} = (c/n)\sigma_e N_2.$$

The emission cross-section can be due to stimulated or spontaneous emission. Luminescence or luminescent cross-section usually refers to the cross-section associated with the luminescent (spontaneous) emission. Qualitatively, the larger the emission (or absorption) cross-section, the more efficient is the emission (or absorption). Usually it is easier to measure the absorption cross-section σ_a than the emission cross-section σ_e. Consider a system in which there are independent centers (for example Er^{3+} ions in a glass host material in which the Er^{3+} concentration is not very large) for the absorption and emission of radiation. *McCumber theory* then relates σ_a to σ_e:

$$\sigma_e(\nu) = \sigma_a(\nu)(Z_1/Z_2)\exp[(E_{ZL} - h\nu)/k_B T]$$

where $h\nu$ is the photon energy, Z_1 and Z_2 are the partition functions of the lower and upper levels, E_{ZL} is the zero-line energy, that is the energy difference between the bottom of the upper energy manifold, and the top of the bottom energy manifold. In many cases σ_a and σ_e are comparable, and depend on the type of atoms involved in the transitions and also on the host material. For example, for Er^{3+} ions, typically, σ_a and σ_e are roughly 6×10^{-21} cm^2 at 1.53 μm in various silica-germania-alumina glasses, and σ_a is $\sim 2.5 \times 10^{-21}$ at 980 nm.

See *stimulated emission*.

Crown glass is a soda-lime-silicate glass with low refractive index and low dispersion compared to various other glasses. In the Schott glass classification, "K" is used to identify a crown glass, e.g. BK is a borosilicate crown glass. See *Abbe number, chromatic dispersion, optical materials*.

Crystal is a three-dimensional periodic arrangement of atoms or molecules or ions. A characteristic property of the crystal structure is its periodicity and a degree of symmetry. For each atom the number of neighbors and their exact orientations are well defined, otherwise the periodicity will be lost. There is therefore a *long-range order* resulting from strict adherence to a well-defined bond length and relative bond angle (or exact orientation of neighbors).

Crystal field normally refers to the electric dipole field caused by the local environment of a transition metal ion. This local field causes a splitting of the energy levels of the ion, and hence gives rise to characteristic peaks in the emission or absorption of radiation associated with the particular ion. The otherwise degenerate *d*-states of the metal ion are split when subject to the so-called ligand field – namely, the field from the complexing atoms or molecules bonding to the metal ion. The magnitude of the energy splitting Δ for a given metal ion depends on the coordinating ligands. The so-called spectrochemical series lists the ligands, ranked in order of causing increasingly large Δ. An abbreviated list is:

$$I^- < Br^- < Cl^- < F^- < OH^- < H_2O < NH_3 < CN.$$

The splitting will also depend on the separation of ligands from the metal ion – for a given ligand–ion combination, the smaller the separation, the larger the splitting. Naturally, the splitting will also depend on the metal ion itself, with ions in higher oxidation states causing a larger electrostatic interaction, and therefore larger splitting. The splitting will also depend on how the ligands coordinate the ion, or the resultant symmetry, the most commonly occurring symmetries being octahedral and next, tetrahedral. Owing to weaker overlap of the *d*-orbitals for tetrahedral symmetry. Δ is lower than for octahedral. The first figure illustrates schematically, the influence of *d*-orbital orientation with respect to the ligands, on crystal field interactions for cubic, tetragonal and octahedral coordination fields, as compared with the free ion.

The second figure shows how ruby gets it characteristic red colour as a result of crystal field interactions. Part (A) (the so-called Tunabe–Sugano diagram) shows the allowed electronic excitations for Cr^{3+} in an octahedral crystal field ($^4A_2 \rightarrow {}^4T_1$ and $^4A_2 \rightarrow {}^4T_2$). The dotted vertical line shows the strength of the crystal field splitting for Cr^{3+} in Al_2O_3. The $^4A_2 \rightarrow {}^4T_1$ energy difference corresponds to the splitting between t_{2g} and e_g. The two allowed transitions for Cr^{3+} in ruby give rise to absorption in the violet and the green, as shown in (B) and (C). Since the transmission in the red is much more than the transmission in the blue, ruby crystals are red. The $^2E_1 \rightarrow {}^4A_2$ fluorescence shown in (B) and (C) add to the deep red color of a ruby.

See *ligand field splitting parameter*.

Crystal momentum of the electron in a crystal is $\hbar k$ where k is the "wavevector" of the electron in the crystal, that is $\psi_k(x) = U_k(x)\exp(-jkx)$ is a solution of the Schrödinger equation given the periodic potential energy for the particular crystal of interest. Such a solution is a traveling wave with a wavevector k and that is spatially modulated by $U_k(x)$. The rate of change of $\hbar k$ is equal to the sum of all external forces only, thus $\hbar k$ represents what we understand as the momentum of a particle. However, in this case the electron is in the crystal and subject to a periodic potential energy; hence the name crystal momentum. In general, $\hbar k$ is not the true momentum of the electron because the rate of change of $\hbar k$ is not equal to the total force, including internal forces. Nonetheless, we can treat $\hbar k$ in most cases as a "momentum."

Crystallization is a process by which crystals of a substance are formed from another phase of that substance, for example, from the melt by

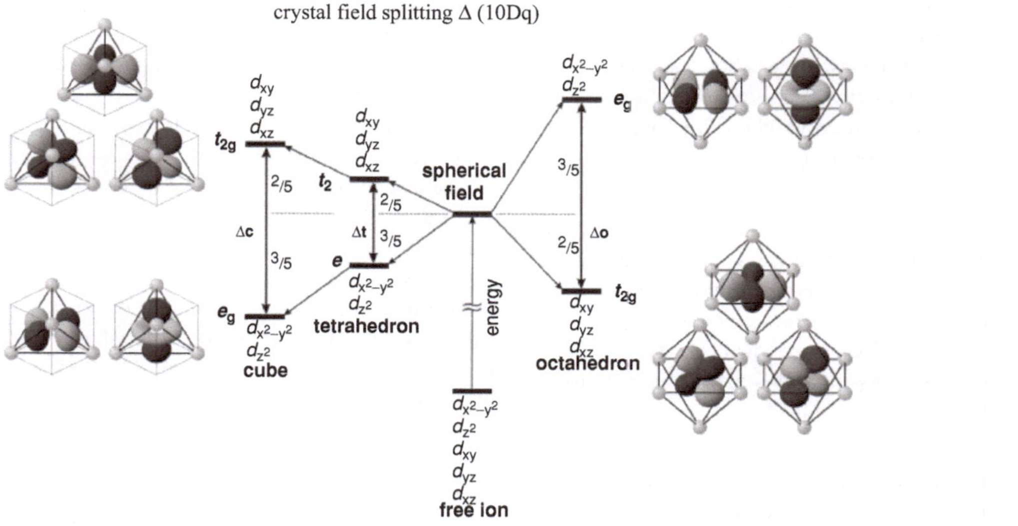

Crystal field Schematic of the influence of *d*-orbital orientation with respect to ligand positions, on crystal field interactions for cubic, tetragonal and octahedral coordination fields, as compared with the free ion. (*From:* M. Wildner, M. Andrut and C. Z. Rudowicz, *EMU Notes in Mineralogy*, Vol. **6** (2004), Ch. 3, p. 1.)

solidification just below the fusion temperature or by condensation of the molecules from the vapor phase onto a substrate. The crystallization process initially requires the formation of small nuclei of the crystal phase which contain a limited number (perhaps 10^3–10^4) of atoms or molecules of the substance. Following nucleation, the nuclei grow by atomic diffusion from the melt to the nuclei.

CTE, charge transfer efficiency in a CCD (charge coupled device).

Curie temperature see *pyroelectricity*.

Curvature of field see *field curvature*.

Cut-back method is a destructive measurement technique for determining the attenuation α of a fiber. The first part of the experiment involves measuring the optical power P_{far} coming out from the fiber at the far end as shown in the figure. Then, in the second part, keeping everything the same, the fiber is cut close to the launch or the source end. The output power P_{near} is measured at the near end from the short cut fiber. The attenuation is then given by

$$\alpha = (-10/L)\log(P_{near}/P_{far})$$

in which the L is the separation of the measurement points, the length of the cut fiber, and α is in dB per

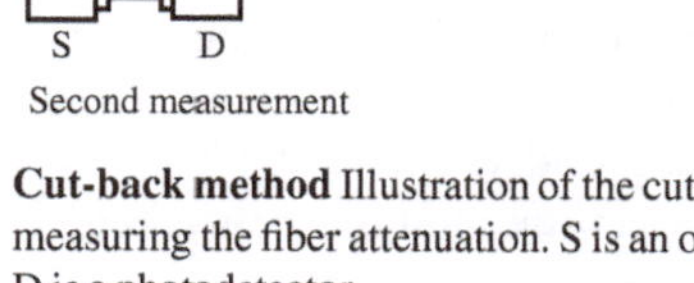

Cut-back method Illustration of the cut-back method for measuring the fiber attenuation. S is an optical source and D is a photodetector.

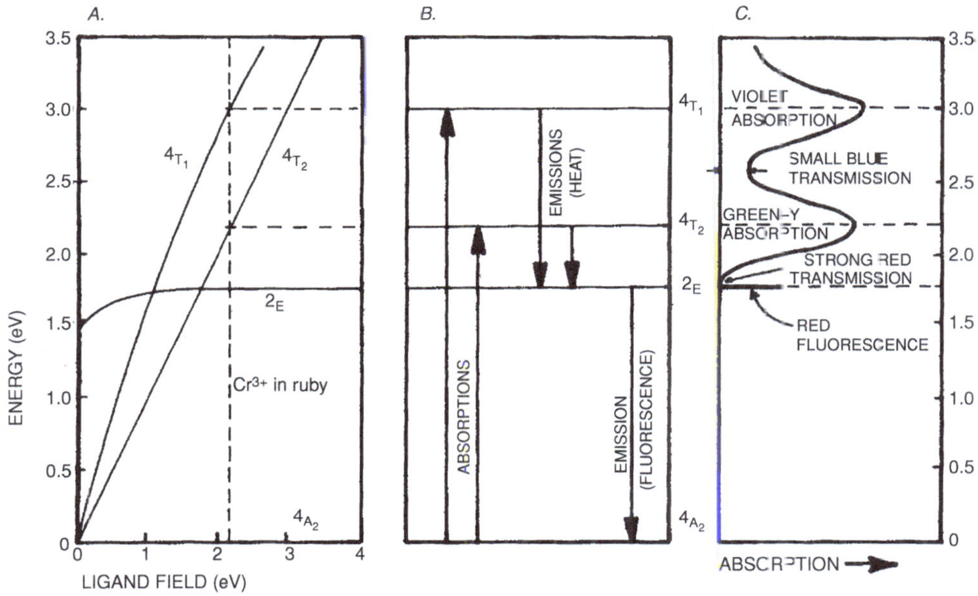

Crystal field Showing how the red colour of ruby arises from crystal field interactions (see text). (Kurt Nassau, *The Physics and Chemistry of Color: The Fifteen Causes of Color, 2nd Edition*, © Kurt Nassau 2001, Reprinted with permission of John Wiley & Sons, Inc.)

unit length. The output P_{near} from the short cut fiber in the second measurement is actually the input into the fiber under test in the first experiment. Usually a mode scrambler (mode stripper) is used for multimode fibers before the input.

Cut-off wavelength of a detector normally refers to the longest wavelength λ_c of radiation that can be detected with an acceptable signal-to-noise ratio. Typically the responsivity drops sharply as λ approaches λ_c. For a semiconductor-based detector this is normally the wavelength that corresponds to bandgap wavelength, i.e. $\lambda_c(\mu m) = 1.24[E_g(eV)]^{-1}$.

Cut-off wavelength of a mode of an optical waveguide is the critical wavelength λ_c below which the guide cannot support the propagation of this particular mode. For example, in a planar slab dielectric waveguide, when the wavelength is such that the *V-number* is less than $\pi/2$, the $m = 1$ (TE_1) mode cannot be supported and the guide can only support the $m = 0$ mode (TE_0); it becomes a single-mode waveguide.

Cylindrical lens is a lens that is shaped (or cut) as if it were part of a solid glass cylinder, which focuses light only in one plane, a plane that is perpendicular to the cylinder (long) axis. They exhibit magnification only in one plane. They are very useful in optical and optoelectronic applications because they posses a *focal line*, rather than a focal point, hence can collimate a line source or allow imaging onto a linear array of detectors. They image a point source as a line and hence focus a light point source onto a monochromator slit. Their uses include converting a circular beam to an elliptical beam, reducing astigmatism

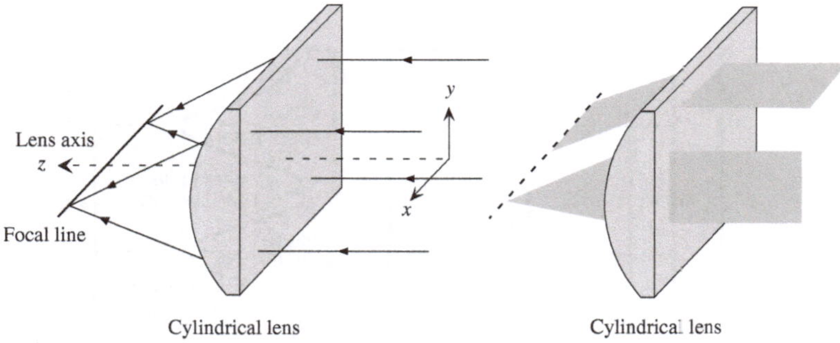

A **cylindrical lens** only focuses in one plane (the yz-plane in the above). No focusing occurs in the xz-direction.

in laser diodes, focusing onto array detectors and beam shaping in scanners, etc. See *anamorphic lens*.

Czochralski (CZ) growth is a type of melt growth technique commonly used for preparing boules of electronic materials such as silicon. The CZ growth method was first introduced by J. Czochralski in 1918, and involves dipping a single-crystal seed into the melt, and after equilibriating it with the melt, withdrawing it at a constant rate while simultaneously rotating it. In this approach, material crystallizes onto the cooler seed crystal and as the seed is slowly withdrawn further material continuously deposits onto the seed. For a given material system and operating conditions, reducing the pulling rate can be used to increase the boule diameter. Typical pulling speeds for Si are about 10–20 cm/hour using rotation rates of about 10 revolutions per minute. A great advantage of this technique is that the boule grows unconstrained (i.e. not in a crucible as would be the case for Bridgeman growth) and therefore avoids issues related to relative expansion on cooling.

D

Dark current of a photodetector is the dc current that flows in the absence of light when it is carrying a bias voltage. In a *pn* junction photodiode, the bias voltage is a reverse bias across the device, and the dark current depends strongly on the temperature. There are two contributions to the dark current which are the diffusion of minority carriers in the neutral regions and the thermal generation of electron hole pairs (EHPs) in the space charge layer (SCL). Both depend strongly on the temperature and the dark current increases exponentially with increasing temperature, typically as a thermally activated process, i.e. $I_{\text{dark}} \propto \exp(-E_g/\eta k_B T)$ where η is 1 if the reverse current is controlled by minority carrier diffusion in the neutral regions and 2 if it is controlled by thermal generation of EHPs in the SCL, E_g is the bandgap of the semiconductor material, k_B is the Boltzmann constant and T is the temperature (K). See *reverse diode current*.

Data rate is the rate at which bits are transmitted per unit time according to a well-defined information coding system used, e.g. *return-to-zero* (RZ) or *nonreturn-to-zero* (NRZ). The NRZ data rate is higher than the RZ data rate.

Data terminal equipment see *optical network*

Dawes empirical equation for the resolving power of a telescope, see *Rayleigh criterion.*

Debye equations attempt to describe the frequency response of the complex relative permittivity, $\varepsilon_r' + j\varepsilon_r''$, of a dipolar medium through the use of a single relaxation time τ to describe the sluggishness of the dipoles driven by the external ac field. If the field has an angular frequency ω, then in the presence of a dielectric loss mechanism with a single relaxation time τ, the real and imaginary dielectric constants are

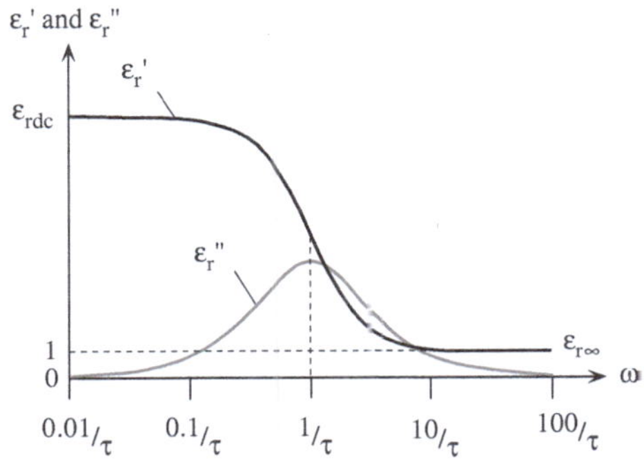

Debye equations Real and imaginary parts of the relative permittivity as a function of angular frequency for a simple relaxation process with a single time constant τ.

$$\varepsilon_r' = \varepsilon_{r\infty} + \frac{(\varepsilon_{rdc} - \varepsilon_{r\infty})}{1 + (\omega\tau)^2} \quad \text{and}$$

$$\varepsilon_r'' = \frac{(\varepsilon_{rdc} - \varepsilon_{r\infty})(\omega\tau)}{1 + (\omega\tau)^2}$$

where $\varepsilon_{r\infty}$ is the high-frequency and ε_{rdc} is the low-frequency dielectric constant. While Debye equations are highly idealized, and a single relaxation time cannot accurately describe even a simple relaxation process, there are, nonetheless, nearly Debye relaxations observed in certain solids and many liquids. For example, the dielectric properties of water can be modeled approximately by Debye equations; for water at 0.2 °C, $\varepsilon_{\mathrm{rdc}} = 87.46$, $\varepsilon_{\mathrm{r}\infty} = 4.87$ and $\tau \approx 0.017$ ns.

Decartes's law *see* Snell's law.

Decibel is a logarithmic unit of *power gain* defined as $\mathrm{dB} = 10\log(P_{\mathrm{out}}/P_{\mathrm{in}})$ where P is the power. It can be used to describe voltage gain as well in which case $\mathrm{dB} = 20\log(V_{\mathrm{out}}/V_{\mathrm{in}})$ where V is the voltage. Similarly it can also describe a current gain by $\mathrm{dB} = 20\log(I_{\mathrm{out}}/I_{\mathrm{in}})$ where I is the current.

Deep level transient spectroscopy (DLTS) is a sensitive measurement technique used to characterize traps in semiconductors. The technique is based on measuring the change in the capacitance of a reverse biased diode due to emission of carriers from traps, after the traps have been charged using a forward bias pulse. The figure below shows how the diode capacitance varies with the applied bias – in particular, the so-called recovery transient from $t = 0$ can be described by:

$$C(t) = C_0[1 - (n_t(0)/2N_\mathrm{d})\exp(-e_n t)]$$

where C_0 is the capacitance of the diode at $t = 0$, and $n_t(0)$ is the concentration of electrons in the traps at $t = 0$. Thus, by measuring the capacitance decay transients, the emission rate at a given temperature can be calculated. Using the emission rate e_n as a function of temperature, the activation energy of the trap, the capture cross-section, and the concentration of traps, can be deduced: that is, a graph of $\ln(e_n/T^2)$ versus $(1/T)$ provides the trap activation energy from its slope and the capture cross-section from the intercept. Since the emission rate increases sharply with increasing temperature, the transients will decay increasingly rapidly with increasing temperature. Thus, at low temperatures, the change in capacitance $\delta C(t)$ between two times t_1 and t_2 after the start of the transient will be relatively small since the decay curve is relatively flat; similarly, if t_1 and t_2 are relatively far apart, then at high temperatures, $\delta C(t)$ will also be small as the transient will have already saturated at t_2 as the decay was so fast. Clearly, at some intermediate temperature, for a given choice of t_1 and t_2, $\delta C(t)$ will be a maximum. An instrument called a boxcar averager is typically used to automatically measure δC as a function of temperature for a given selection of measurement times, t_1 and t_2. The resulting plot of $\delta C(T)$ vs. T is commonly referred to as a DLTS spectrum. Clearly the choice of the measurement times will determine at what temperature the DLTS spectrum has its peak. Since,

$$C(t) = (C_0 n_t(0)/2N_d)[\exp(-e_n t_2) - \exp(-e_n t_1)],$$

differentiating this expression with respect to T (noting that e_n is a function of T) and setting $\delta C(t) = 0$ at the temperature T^* when δC is a maximum, gives

$$e_n(T^*) = \ln(t_1/t_2)/(t_1 - t_2).$$

Thus a given DLTS spectrum will have a peak temperature T^* corresponding to the choice of t_1 and t_2 and hence $e_n(T^*)$. Changing the values of the boxcar times, will change the values of T^* and $e_n(T^*)$. Hence, by using a series of say 10 boxcar time settings, a series of 10 points on the Arrhenius plot of

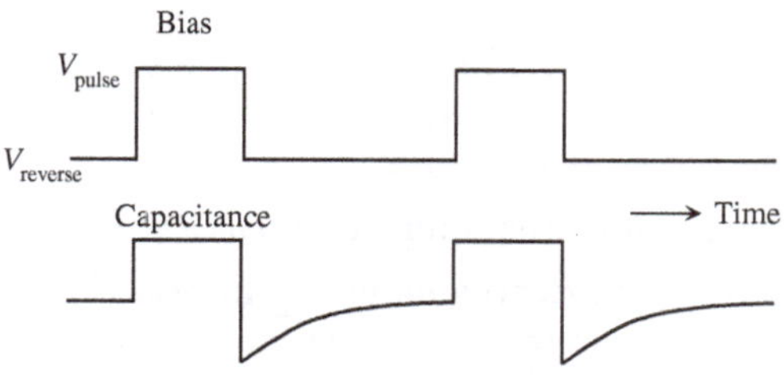

A simple schematic of a **deep level transient spectroscopy (DLTS)** experiment.

$\ln(e_n/T^2)$ versus $(1/T)$ can be found, and the trap parameters directly deduced as discussed above.

Degenerate semiconductor contains so many dopant atoms that the electron concentration in the conduction band (CB), or hole concentration in the valence band (VB), is comparable to the density of states in the respective band. Consequently, the carriers interact with each other and Fermi–Dirac statistics must be used. The Fermi level is either in the CB for an n^+ type degenerate, or in the VB for a p^+ type degenerate semiconductor. The superscript + indicates a heavily doped semiconductor. As an example, consider an n-type semiconductor with an effective density of states in the conduction band of $N_c = 2(m^* k_B T/2\pi\hbar)^{3/2}$. In general, the electron concentration in the conduction band can be written as,

$$n = (2/\pi^{1/2}) N_c \Im_{1/2}(\eta_F)$$

where η_F is the *reduced Fermi energy*, $\eta_F = (E_F - E_c)/k_B T$, giving the position of the Fermi energy with respect to the conduction band edge E_c, and $\Im_{1/2}(\eta_F)$ is the so-called Fermi–Dirac integral given by

$$\Im_{\frac{1}{2}}(\eta_F) = \int_0^\infty \frac{\eta^{\frac{1}{2}} d\eta}{1 + \exp(\eta - \eta_F)}.$$

The degenerate condition thus corresponds to $E_F > E_c$, or $\eta_F > 0$. This condition occurs at high electron concentrations, particularly for semiconductors with small effective mass and at low temperatures. The term "degenerate" is taken to mean that for $\eta_F > 0$, the occupation probability is close to unity, as distinct from the use of the word "degeneracy" to describe the occurrence of a plurality of states being available at the same energy (see the discussion of *conduction band*). One commonly used approximation for solving the Fermi–Dirac integral and for establishing η_F in the case of a degenerate semiconductor is the *Joyce–Dixon approximation*. These are given for n- and p-type semiconductors as

$$\eta_F = E_c + k_B T[(n/N_c\sqrt{8}) + \ln(n/N_c)]$$

and

$$\eta_F = E_v + k_B T[(p/N_v\sqrt{8}) + \ln(p/N_v)],$$

respectively.

Demultiplexer (DMUX) is a device that is able to separate a multiplexed signal into its original component channels. An important operation that is the cornerstone of today's optical communications is the multiplexing and demultiplexing of optical signals. In wavelength division multiplexing, N optical signals (each one a channel) at various wavelengths $\lambda_1, \lambda_2, \ldots, \lambda_N$ are multiplexed into a single fiber. At the destination, a demultiplexer has to separate out the N channels based on their wavelengths. The multiplexing procedure may be as simple as focusing the N channels into the input of a single fiber, but the demultiplexing procedure needs a "demultiplexing device" that will deflect or diffract light by an angle based on its wavelength. Prism or diffraction grating based demultiplexers are available; they can, in principle, also work as multiplexers, multiplexing N different wavelength channels into a single fiber. See *multiplexer*.

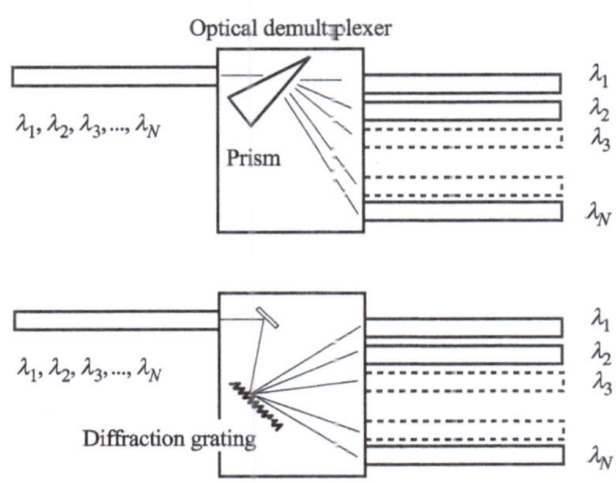

Demultiplexers using a prism and a diffraction grating.

Dense wavelength division multiplexing (DWDM) is the transmission of many channels of information that are closely spaced in wavelength along the same fiber. The channels are transmitted at wavelength $\lambda_1, \lambda_2, \ldots, \lambda_N$, which are separated in frequency by 200 GHz, or wavelength separated by 1.6 nm if the wavelengths are in the 1550 nm window. See *wavelength division multiplexing*.

Density of states, $g(E)$, for a bulk semiconductor is the number of electron states per unit energy per unit volume, so that $g(E)dE$ is the number of *distinct* states in the energy range E to $E + dE$ per unit volume. (The notation ρ(E) for $g(E)$ is also used.) The finiteness of $g(E)$ is a direct consequence of the *Pauli exclusion principle* and *Heisenberg's uncertainty principle* which establish a minimum volume of phase space that two electrons of opposite spin can occupy; for a 3D system, this gives,

$$\Delta x \Delta y \Delta z \Delta p_x \Delta p_y \Delta p_z = (2\pi\hbar)^3.$$

Invoking the *de Boglie relationship* ($p = \hbar k$), permits the phase space volume to be written as:

$$\Delta x \Delta y \Delta z \Delta k_x \Delta k_y \Delta k_z = (2\pi)^3.$$

In a 3D system, two electrons occupy a volume $8\pi^3$ of phase space, or each electron occupies a volume $4\pi^3$ of phase space. In a system of arbitrary dimensionality D, each electron occupies a volume $(2\pi)^D/2$ of phase space. We can write the volume of k-space $V_{k\text{-space}}(E)$ enclosed between two constant energy surfaces E and $E + dE$ as,

$$V_{k\text{-space}} = dE \int_{\text{Surface}} \frac{\partial k}{\partial E(k)} dS$$

where dS is an area element of the constant energy surface; in 3D k-space one can express the argument of the integral in terms of grad$_k$ and using the fact that each electron occupies a volume $4\pi^3$ of phase space, giving the density of states $g_{3\text{D}}(E)$:

$$g_{3\text{D}}(E) = \frac{1}{4\pi^3} \int_{\text{Surface}} \frac{dS}{\nabla_k E(k)}.$$

In the case of general dimensionality, the prefactor would become $2/(2\pi)^D$. The simplest $E(k)$ dispersion relationship would be that of a semiconductor with a single-valley isotropic parabolic band. For example, the conduction band in GaAs follows such a relationship near the band edge E_c to a reasonable approximation. That is, the applicable dispersion relationship would be $E(k) = E_c + h^2k^2/2m^*$. Using this dispersion relationship and the expression for the 3D density of states above, gives

$$g_{3\text{D}}(E) = \frac{1}{2\pi^3}\left[\frac{2m^*}{\hbar^2}\right]^{3/2} \sqrt{E - E_c}.$$

In cases of an anisotropic single-valley band, the dispersion relationship will be direction-dependent. For example, for the case of an ellipsoidal constant energy surface with the main axes of the ellipsoid aligning with a Cartesian coordinate system,

$$E(k) = \sum_{i=x,y,z} \frac{(\hbar k_i)^2}{2m_i^*}$$

and again using the expression for the 3D density of states in terms of grad$_k$ gives

$$g_{3\text{D}}(E) = \frac{\sqrt{2}}{\pi^2\hbar^3}\sqrt{m_x{}^*m_y{}^*m_z{}^*}.\sqrt{E - E_c}.$$

By analogy with the isotropic band case, one often defines a *density of states effective mass* $m_{\text{DOS}}{}^* = (m_x{}^*m_y{}^*m_z{}^*)^{1/3}$ recovering a similar expression to that for an isotropic band. In the case of multiple valleys, the additional states available in those valleys all contribute and one sums the states available in each valley for a given energy range E to $E + dE$. For example in the case of M_c equivalent conduction band valleys for an isotropic parabolic 3D conduction band,

$$g_{3\text{D}}(E) = \frac{\sqrt{2}M_c}{\pi^2\hbar^3}(m_{\text{DOS}}{}^*)^{3/2}\sqrt{E - E_c}.$$

In lower dimensionality systems, the dispersion relations reflect the reduced number of degrees of freedom. Thus for the case of a 2D semiconductor, i.e. a *quantum well*, having parabolic subbands, the appropriate dispersion relationship is

$$E_{2\text{D}}(k) = E_n(z) + (\hbar k_{xy})^2/2m^*;\ k_{xy}^2 = k_x^2 + k_y^2.$$

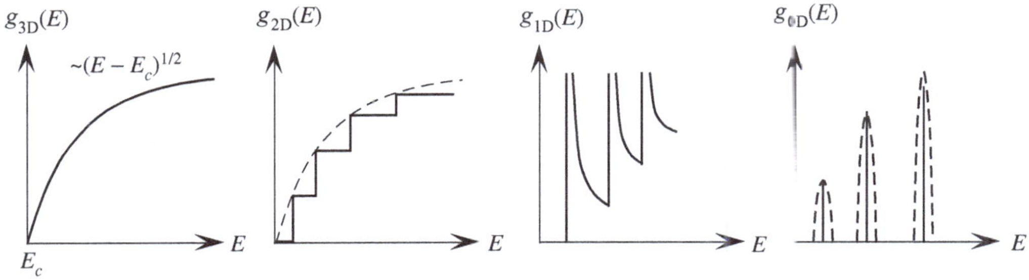

Density of states Illustration of the density of electron states $g(E)$ or number of electron energy states per unit energy shown for cases of different dimensionalities. (a) For a bulk semiconductor with no constraints on electron motion, the material is said to be 3D and follows the square root dependence. (b) For the case of a system with electrons unconfined in 2D, $g_{2D}(E)$ shows characteristic step increases as subsequent subband states are occupied. (c) In the case of a 1D system with only one degree of freedom for electron motion, when each subband defined by two quantum numbers is occupied, the density of states drops with a characteristic $E^{-1/2}$ dependence. (d) For a quantum dot having no degrees of freedom, the density of states is a series of delta functions positioned at energies determined by the corresponding three quantum numbers.

The quantum number n is introduced for the quantized energy states in the confined dimension. Using this dispersion relationship, the density of states in 2D is given by:

$$g_{2D}(E) = \sum_{\text{All subbands},n} \frac{m^*}{\pi\hbar^2}\sigma(E - E_n)$$

where $\sigma(E - E_c)$ is the *Heaviside* function.

In 1D, a so-called *quantum wire*, only one direction of motion is allowed and the 'surface' integral reduces to a single point in k-space resulting in

$$g_{1D}(E) = \frac{1}{\pi\hbar} \sum_{\text{All subbands},n} \sqrt{\frac{m^*}{2\sigma(E - E_n)}}.$$

A similar analysis for the case of 0D, a so-called *quantum dot,* gives

$$g_{0D}(E) = \sum_{\text{All subbands},n} 2\delta(E - E_n)$$

where δ is the *Kroeneker delta* function having a value of unity when $E = E_n$ and zero otherwise.

Depletion region, or **space charge layer,** SCL, is a region around the metallurgical junction where the recombination of electrons and holes has depleted this region of its large number of equilibrium majority carriers. Holes diffusing from the p-side through the junction into the n-side recombine with electrons in the n-side, and thereby expose positive donors. Holes leave behind exposed negative acceptors. Similarly, electrons diffusing from the n- to the p-side leave behind positive donors and recombine with holes in the p-side, which exposes negative acceptors. The depletion region, the SCL, therefore contains exposed negative acceptors on the p-side and exposed positive donors on the n-side. Due to the separation of positive and negative ions (charges), there is a *built-in field* and hence a *built-in voltage* across the junction. Consider an abrupt pn junction. If W_p and W_n are the widths of the SCL on the p- and n-sides, and $W_0 = (W_p + W_n)$ is the total width of the SCL, then $N_a W_p = N_d W_n$, and

$$W_0 = \left[\frac{2\varepsilon(N_a + N_d)V_o}{eN_aN_d}\right]^{1/2}$$

where N_a and N_d are the acceptor and donor concentrations on the p- and n-sides respectively, $\varepsilon = \varepsilon_0\varepsilon_r$

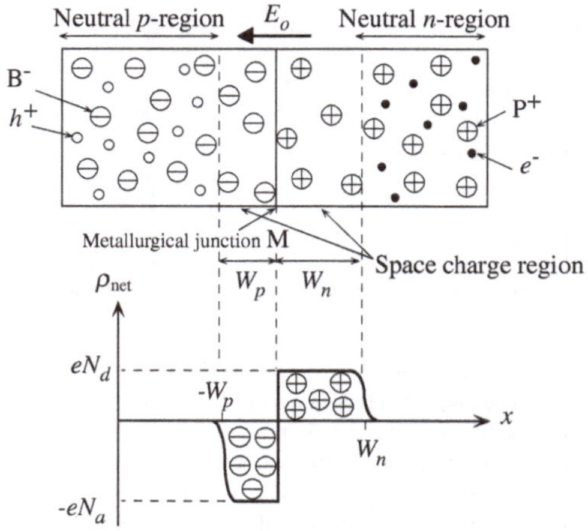

The **depletion region**, and its space charge in a *pn* junction. B^- and P^+ represent ionized boron (acceptor) and phosphorus (donor) dopants.

is the permittivity of the semiconductor and V_o is the built-in voltage; it depends on the amount of doping, N_a and N_d. When a voltage V is applied across a *pn* junction, V_o is replaced by $V_o - V$, and the depletion region width W becomes proportional to $(V_o - V)^{1/2}$; reverse bias ($V = -V_r$) increases W. The electric field is a maximum at the metallurgical junction and is given by

$$E_o = -\frac{eN_d W_n}{\varepsilon} = -\frac{eN_a W_p}{\varepsilon}.$$

Depletion region capacitance or **space charge layer capacitance,** or the **diode junction capacitance**, is the incremental capacitance (dQ/dV) due to the change in the exposed dopant charges in the depletion layer of a *pn* junction (or a Schottky junction) as a result of the change in the voltage across the *pn* junction. The depletion layer capacitance, $C_j = |dQ/dV|$, of a *pn* junction depends on the voltage V across device, and decreases with increasing reverse bias; typical values are quoted in the picofarad range but obviously depend on the *pn* junction area and doping. For a *pin* diode C_j is almost independent of the voltage. C_j can limit the high frequency operation, or the speed, of a diode. C_j is, per unit area (F m^{-2}) for an abrupt *pn* junction,

$$C_j = \left[\frac{e\varepsilon N_d N_a}{2(N_a + N_d)}\right]^{1/2} (V_o - V)^{-1/2}$$

where e is the electronic charge, N_a and N_d are the acceptor and donor concentrations on the *p*- and *n*-sides respectively, $\varepsilon = \varepsilon_o \varepsilon_r$ is the permittivity of the semiconductor, V_o is the built-in voltage and V is the applied voltage (positive for forward, negative for reverese bias). For a *linearly graded junction* in which the dopant concentration $N_{dopant}(x) = N_d - N_a = ax$, where a is a constant, and $x = 0$ is the metallurgical junction,

$$C_j = \left(\tfrac{1}{12} ea\varepsilon^2\right)^{1/3} (V_o - V)^{-1/3}.$$

Depth of field in an imaging system giving an image $A'B'$ from an object AB, describes the portion of the axis along which other objects, either in the background or in the foreground, remain as clear as $A'B'$ in the detector plane. It depends on the detector resolution as well as on the numerical aperture. The limited resolution of any detector is due to the finite size of its smallest constitutive element (cell of the retina, grain of an emulsion, pixel of a CCD, etc.). As long as the point spread function remains smaller than the size of an individual detector, $A_i'B_i'$ is as clear as $A'B'$. The depth of field is an important factor in photography, and depends on how much blurring can be tolerated. Suppose that d represents the allowable blurring diameter, that is, it is the diameter of a circle on the image plane in which blurring

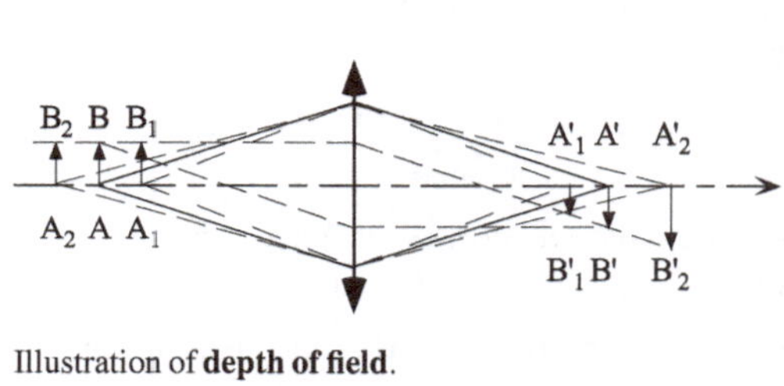

Illustration of **depth of field**.

can be tolerated. If D is the lens aperture, f is its focal length and s_o is the distance of the object from the lens, then the depth of field can be estimated from

$$\text{Depth of field} \approx \frac{2\,(d/D)\,s_o(s_o - f)f}{f^2 - (d/D)^2 s_o^2}$$

See also *depth of focus*.

Depth of focus is the maximal admissible distance between the nominal image focal plane and the actual plane of a detector in an imaging system. Any detector exhibits a finite resolution due to the finite size of its smallest constitutive element (cell of the retina, grain of an emulsion, pixel of a CCD, etc.). As long as the out-of-focus defect remains smaller than the size of an individual detector, the image quality is not affected. Consider a convergent thin lens of focal length f and diameter Φ, or aperture number N such as $\Phi = f/N$. The diameter of the emulsion grain is 2ρ. The figures illustrate the depth of focus concept: for an on-axis object point "at infinity," the image will form itself at the image focal point F'. If the plane (H) of the detector does not exactly coincide with the image focal

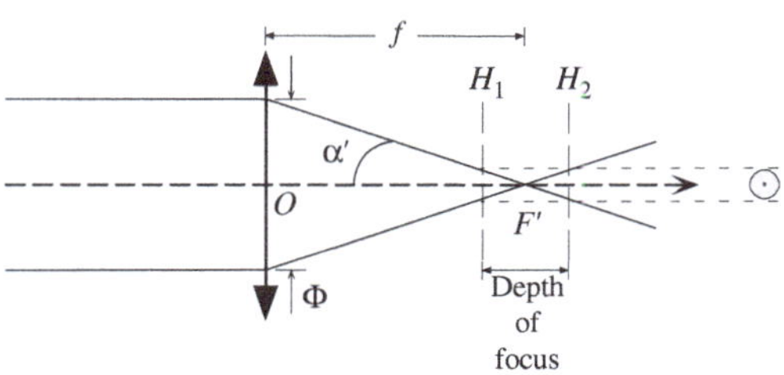

Depth of focus definition.

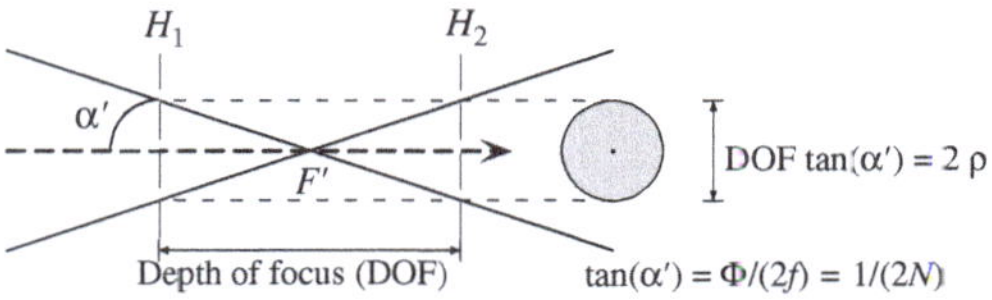

Depth of focus and spot size.

plane, then all the rays converging toward F' form a response spot of finite size instead of a point. As long as the diameter of this spot remains smaller than that of the grain, that is between planes (H_1) and (H_2), the image quality is not affected; the depth of focus DOF is

$$\text{DOF} = 2\rho/\tan(\alpha') = 4N\rho.$$

The greater the aperture number, the longer the depth of focus. The smaller the emulsion grain of diameter 2ρ, the shorter the depth of focus.

Destructive interference, see *interference*.

Detective quantum efficiency (DQE) measures the ability of a detector to transfer a signal with respect to noise from its input to its output. If the detector is perfect and "noiseless" then the signal-to-noise ratio SNR at the output would be the same as SNR at the input since both the signal and noise would have been amplified or multiplied by the same gain, and DQE would be unity. The random nature of image quanta gives rise to random fluctuations in image signals contributing to image formation and hence creates random noise. The scattering of image quanta gives rise to image blurring which is quantified by the *modulation transfer function* MTF(f). Images are partially degraded by various sources of statistical fluctuations which arise along the imaging chain. The relative increase in image noise due to an imaging system as a function of the *spatial frequency*, f, is expressed quantitatively by the spatial-frequency-dependent detective quantum efficiency, DQE(f). The DQE(f) represents signal-to-noise transfer efficiency for different frequencies of information in an image. The DQE(f) of an imaging detector is defined as

$$\text{DQE}(f) = \frac{\text{SNR}^2_{\text{out}}(f)}{\text{SNR}^2_{\text{in}}(f)}$$

where SNR_{in} and SNR_{out} are the signal-to-noise ratios at the input and output stages of the detector, respectively. DQE(f) is considered as the appropriate metric of system performance and is unity

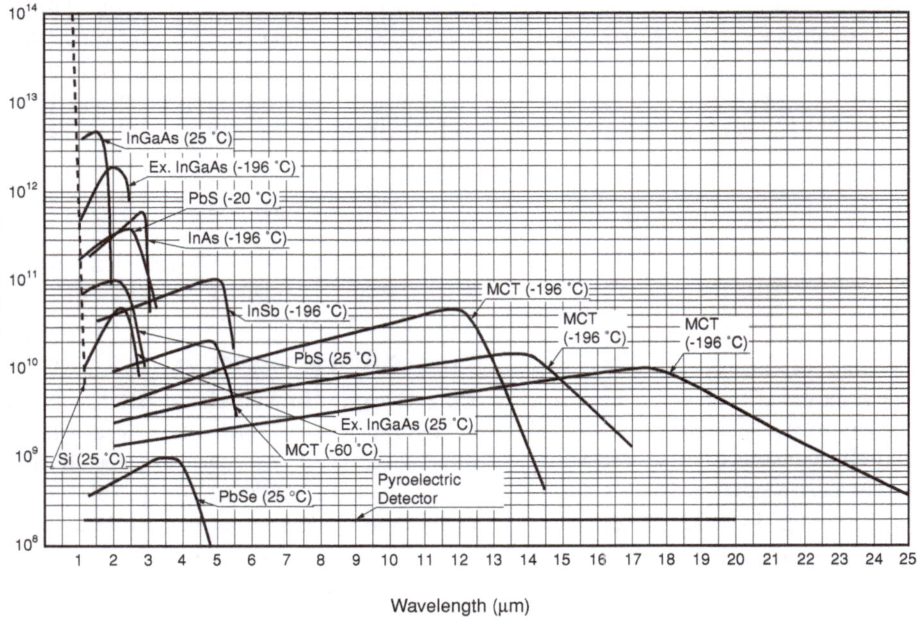

Specific **detectivity** (D^*) vs. wavelength (λ) for various infrared detectors at various temperatures. MCT is HgCdTe. (Courtesy of Hamamatsu.)

for an ideal detector. For simplicity, we are often interested in measuring the zero spatial frequency detective quantum efficiency DQE ($f = 0$) of an imaging detector. DQE(0) represents signal quality degradation due to the signal and noise transfer characteristics of the system without considering signal spreading. The relation between DQE(0) and DQE(f) is given by

$$\mathrm{DQE}(f) = \frac{\mathrm{DQE}\,(0)\,\mathrm{MTF}^2\,(f)}{\mathrm{NPS}_0\,(f)}$$

where $\mathrm{NPS}_0(f)$ is the noise power spectrum of the system normalized to unity as f approaches 0.

Detectivity is the reciprocal of *noise equivalent power* (NEP); $D = 1/\mathrm{NEP}$. It is a measure of a detector's sensitivity taking into account various noise contributions. The *specific detectivity* D^* is defined as

$$D^* = \frac{(A)^{1/2}}{\mathrm{NEP}}$$

where A is the photodetector area that is receiving the photons. Units are cm $\mathrm{Hz}^{-1/2}\,\mathrm{W}^{-1}$, or *Jones*. Typically, detectivity increases as the temperature of the detector is lowered inasmuch as the dark current and its noise both decrease as the temperature is lowered. See *noise equivalent power*.

Dextrorotatory see *optical activity*.

DFB see *distributed feedback*.

Dichroism is a phenomenon in which the optical absorption in a substance depends on the direction of propagation and the state of polarization of the light beam. A dichroic crystal is an optically anisotropic crystal in which either the ordinary wave or the extraordinary wave is heavily attenuated (absorbed).

Die is an individual integrated circuit device in a processed wafer that is cut (usually by *scribing*) from a wafer before it is packaged. It has all the necessary active and passive components fabricated into it by the fabrication process to implement the desired electronic function.

Dielectric is an insulating material in which energy can be stored by the polarization of the molecules and the conductivity is negligibly small. It is a material that increases the capacitance or charge storage ability of a capacitor. Ideally it is a nonconductor of electrical charge so that an applied field does not cause a flow of charge but instead relative displacement of opposite bound charges inside the media and hence the polarization of the medium. The energy is stored in the polarization of the medium. If E is the electric field in a small region of the deielectric, and ε_r is the relative permittivity (or the real part of the complex permittivity), then the energy stored per unit volume in this volume, the energy density, is given by $E_{vol} = (1/2)\, \varepsilon_0 \varepsilon_r\, E^2$.

Dielectric constant see *relative permittivity*.

Dielectric loss is the energy lost in the polarization process in the presence of an applied alternating electric field. The energy is absorbed from the oscillating field and converted to heat during the polarizations of the molecules. All polarization processes have some degree of energy loss, and these losses involve the conversion of energy from the applied ac field, or from an electromagnetic (EM) wave propagating in the dielectric, into unrecoverable forms of energy such as "heat" (phonons). For example, in *ionic polarization*, the oscillating field in an EM wave inside a dielectric can couple with phonons in the crystal and pass energy to phonons by forcing them to oscillate. During ionic polarization, energy is passed from the EM wave to the oscillations of the ions in the medium. Thus, when an EM wave is propagating in a dielectric, it loses energy as it propagates; it becomes attenuated. Dielectric loss should not be confused with the conduction loss σE^2 (σ is the conductivity) or V^2/R. The *imaginary* part of the relative permittivity ε_r'' gauges the extent of dielectric loss in a medium in contrast to ε_r', the real part that gauges the energy stored in the dielectric. Dielectric power loss per unit volume W_{vol} is given by

$$W_{vol} = \omega E^2 \varepsilon_0 \varepsilon_r'' = \omega E^2 \varepsilon_0 \varepsilon_r' \tan\delta$$

where ω is the angular frequency of the ac field (e.g. EM wave), and $\tan\delta$ is the loss tangent, $\tan\delta = \varepsilon_r''/\varepsilon_r'$.

See *complex relative permittivity*.

Dielectric mirror is made from alternating high and low refractive index quarter-wave thick multilayers. Such dielectric mirrors provide a high degree

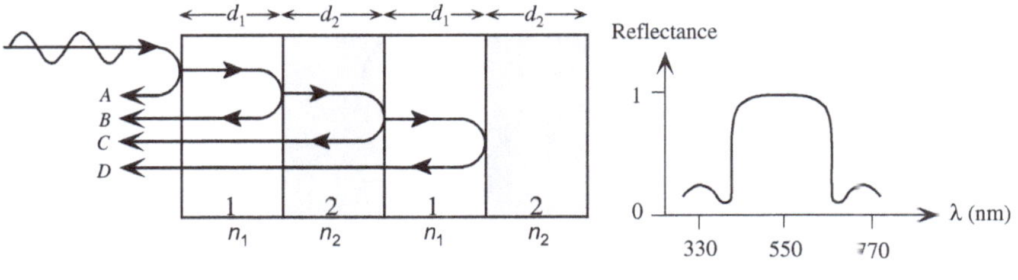

Schematic illustration of the principle of the **dielectric mirror** with many low and high refractive index layers and its reflectance (on the right). Reflected waves A, B, C, D, and so on, all interfere constructively if the layer thicknesses d_1 and d_2 satisfy the above relationship. Consider the interference between A and C to derive the above equation.

of wavelength selective reflectance at the required free space wavelength λ if the thicknesses of alternating layers d_1 and d_2 with refractive indices n_1 and n_2 are such that

$$n_1 d_1 + n_2 d_2 = \lambda/2$$

which then leads to the constructive interference of all partially reflected waves at the interfaces.

Dielectric strength is the maximum field (E_{br}) that can be sustained in a dielectric beyond which dielectric breakdown ensues, i.e. there is a large conduction current through the dielectric shorting the plates.

Diffraction is a spread or bending of a freely propagating wave when it encounters any type of obstacle. For example, sound waves become bent (deflected) around corners. When a light wave passes through an aperture (an opening in an opaque screen) it spreads out from the aperture and exhibits regions of maximum and minimum intensity as a result of interference of spherical waves emitted from every point in the aperture in accordance with the *Huygens–Fresnel principle*. The intensity distribution in the diffracted beam as registered by an observer is called the *diffraction pattern*. The light pattern of the diffracted beam does not correspond to the geometric shadow of the aperture. Diffraction phenomena are generally classified into two categories. In *Fraunhofer diffraction*, the incident light beam is a plane wave (a collimated light beam) and the observation or detection of the light intensity pattern (by placing a photographic screen, etc.) is done far away from the aperture so that the waves being received also look like plane waves. Inserting a lens between the aperture and the photographic screen enables the screen to be closer to the aperture. In *Fresnel diffraction*, the incident light beam and the received light waves are not plane waves but have significant wavefront curvatures. Typically, the light source and the photographic screen are both close to the aperture so that wavefronts are curved. Fraunhofer diffraction is by far the most important.

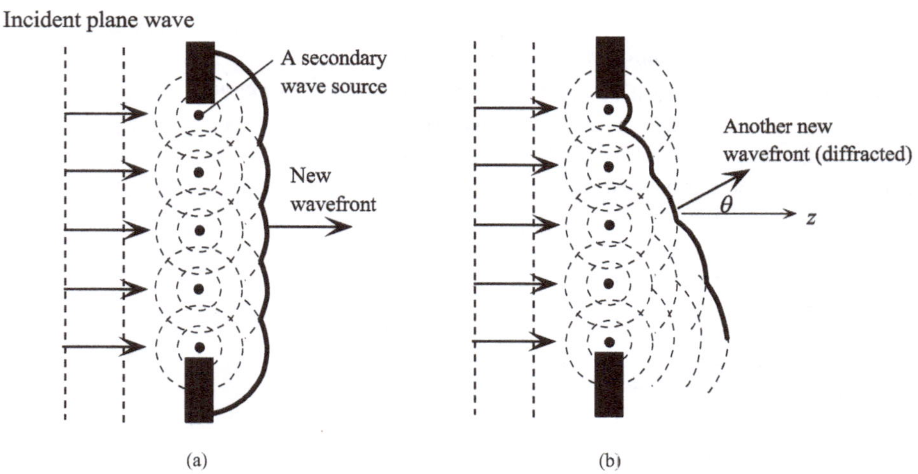

Diffraction (a) The Huygens–Fresnel principle states that each point in the aperture becomes a source of secondary waves (spherical waves). The spherical wavefronts are separated by λ. The new wavefront is the envelope of all these spherical wavefronts. (b) Another possible wavefront occurs at an angle θ to the z-direction which is a diffracted wave.

Diffraction grating is an optical device with a periodic array of slits (or grooves) for diffracting ("deflecting") an incident light beam through an angle that depends on the wavelength and the periodicity of slits in the grating. In its simplest form it has a periodic series of slits in an opaque screen. An incident beam of light is diffracted in certain well-defined directions that depend on the wavelength λ and the grating properties, that is the slit separation d,

$$d \sin\theta = m\lambda; \quad m = 0, \pm 1, \pm 2, \ldots$$

which is the well-known *grating equation*, also known as the *Bragg diffraction condition*. The value of m defines the diffraction order: $m = 0$ being zero-order, $m = 1$ being first-order, etc. Order $m = 0$ corresponds to the fraction of light unperturbed by the grating (transmitted without deviation). This basic geometrical formula does not give the amount of diffracted power in each order.

In a *transmission grating*, the incident and diffracted beams are on opposite sides of the grating. Typically parallel thin grooves on a glass plate would serve as a transmission grating. Among transmission gratings, it is customary to distinguish between *amplitude gratings* (the transmission amplitude is modulated) and so-called *phase gratings* (only the refractive index is modulated, without any losses). A *reflection grating* has the incident beam and the diffracted beams on the same side of the device. The surface of the device has a periodic reflecting structure, easily formed by etching parallel grooves in a metal film, etc. The reflecting unetched surfaces serve as synchronous secondary sources that interfere along certain directions to give diffracted beams of zero-order, first-order, etc. If θ_i is the angle of incidence, then the diffraction angle θ_m for the m-th order is given by

$$d(\sin\theta_m - \sin\theta_i) = m\lambda; \quad m = 0, \pm 1, \pm 2, \ldots$$

Some unintentional defects of periodicity can also lead to spurious diffracted beams known as *ghosts*.

A grating is not necessarily plane nor strictly periodic; volumic *holograms* can be considered as a generalized three-dimensional diffraction grating. Consider a grating with N slits. The slit width is a, and d is the periodicity. The detector is at a distance R, far away from the grating. If the incident plane wave is normal to the grating, the intensity distribution along y in the first figure is given by

$$I(y) = I_0 \left[\frac{\sin(\frac{1}{2}Nk_y d)}{\sin(\frac{1}{2}k_y d)}\right]^2 \left[\frac{\sin(\frac{1}{2}k_y a)}{\frac{1}{2}k_y a}\right]^2$$

where k_y is the scattering wavevector defined by $k_y = (2\pi/\lambda)(y/R) = (2\pi/\lambda)\sin\theta$. The first term represents the oscillations in the intensity due to interference from different slits, and the second term is the envelope of the diffraction pattern, and is the diffraction pattern of a single slit. The *resolvance* or the *resolving power* R of a diffraction grating is its ability to separate out adjacent wavelengths. If $\lambda_2 - \lambda_1 = \Delta\lambda$ is the minimum wavelength separation that can be measured, as determined by the Rayleigh criterion (the maximum of the intensity distribution at λ_1 is at the first minimum of the intensity distribution at λ_2), and λ is the average wavelength $\frac{1}{2}(\lambda_1 + \lambda_2)$ in $\Delta\lambda$, then the resolving power is defined by $R = \lambda/\Delta\lambda$. The separation $\Delta\lambda$ is also called the *spectral resolution*. If N grooves on a grating are illuminated and the order of diffraction is m, the theoretical resolving power is given simply by $R = mN$. The resolving power is also called the *chromatic resolving power* since it refers to the separation of wavelengths.

Imperfections in the grating can lead to reduced and broadened diffraction peaks or even to "ghosts" in the diffraction pattern. Random errors in the grating slit positions and widths result in a *disordered grating*. Periodic errors in the slit positions and widths also give rise to ghosts, which are spurious diffraction orders. *Rowland ghosts* are due to large-scale periodic errors in d and usually result in spurious lines symmetrically placed around a strong

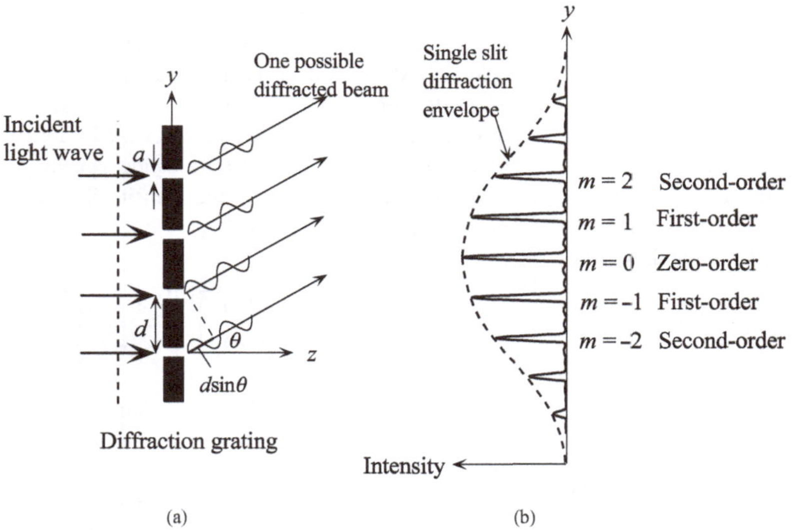

Diffraction grating (a) A diffraction grating with N slits in an opaque screen. Slit periodicity is d and slit width is a; $a \ll d$. (b) The diffracted light pattern. There are distinct beams only in certain directions (schematic).

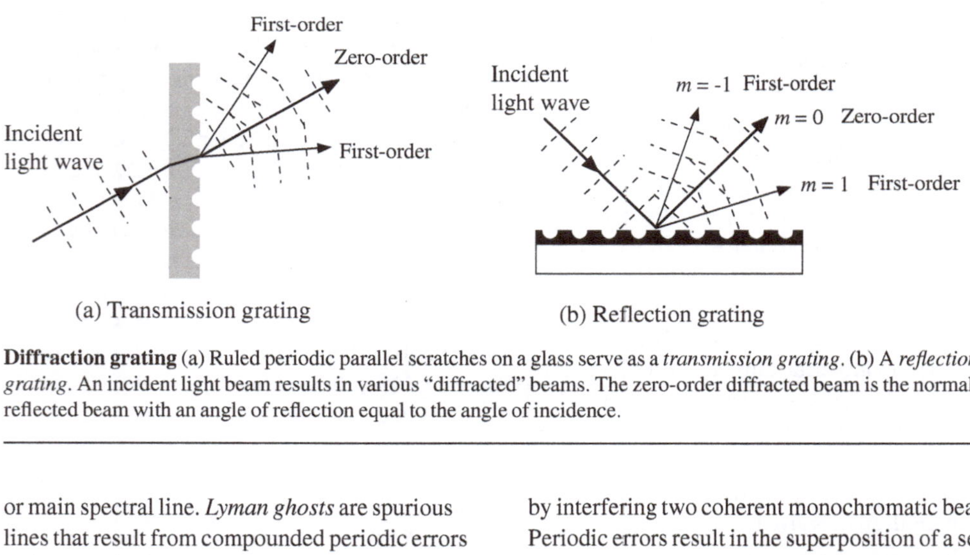

Diffraction grating (a) Ruled periodic parallel scratches on a glass serve as a *transmission grating*. (b) A *reflection grating*. An incident light beam results in various "diffracted" beams. The zero-order diffracted beam is the normal reflected beam with an angle of reflection equal to the angle of incidence.

or main spectral line. *Lyman ghosts* are spurious lines that result from compounded periodic errors and appear far from the parent spectral lines. Lyman ghosts have a period of the order of a few times the slit spacing. No such ghosts appear in gratings made by light interference techniques (e.g. those written by interfering two coherent monochromatic beams). Periodic errors result in the superposition of a secondary grating on top of the main grating, and hence generate a second diffraction pattern. In addition, scattering from the grating can also reduce the efficiency of the diffraction. See *Fourier analysis of*

Diffraction grating William Lawrence Bragg (1890–1971), Australian born British physicist, won the Nobel prize with his father William Henry Bragg for his "famous equation" when he was only 25 years old. (Courtesy of Emilio Segrè Visual Archives, AIP, Weber Collection.)

diffraction, disordered gratings (D. Rittenhouse, *Trans. Amer. Phil. Soc.* **2**, 37–42, 1786; J. F. Verill, *J. Mod. Opt., Opt. Acta*, **17**, 747, 1970; E. W. Palmer *et al. Rep. Prog. Phys.* **38,** 975, 1975; C. Palmer, *Diffraction Grating Handbook*, Sixth Edition, Newport, 2005.)

Diffraction limited system is an imaging system whose resolution is diffraction limited. Even in the case of perfect *stigmatism* the actual resolution of an imaging instrument is never perfect. For instance, a parabolic telescope, although perfectly stigmatic for the conjugation between an on-axis distant star and the focal point, exhibits a typical Airy diffraction pattern in the focal plane, with a transverse dimension inversely proportional to the diameter of the entrance pupil. From a signal-processing aspect, the Airy pattern is the response function of the apparatus. The only way to enhance the performance is to enlarge the pupil – hence the realization of so-called "very large" telescopes.

Diffuse radiation see *solar radiation air mass.*

Diffuse reflection see *reflection.*

Diffusion is a random process by which particles move from high-concentration regions to low-concentration regions. There is a flow of particles of a given species from high to low concentration regions by virtue of their random motions. Diffusion is a process by which atoms migrate by virtue of their random thermal motions. Charge particles such as electrons and holes also diffuse just like uncharged particles. The diffusion flux Γ is defined as the net flow of particles per unit time per unit area. If the concentration gradient along a particular direction x is dC/dx (C being the concentration), according to *Fick's law*, this flux Γ_x along x is proportional to the concentration gradient and is given by $\Gamma_x = -D(dC/dx)$ where D is the diffusion coefficient.

Diffusion coefficient is a measure of the rate at which atoms diffuse. It depends on the nature of the diffusion process and typically it is temperature dependent. It is defined as the diffusion flux per unit concentration gradient. Atomic diffusion in solids is a strong function of temperature and is thermally activated. The diffusion coefficient D for charge carriers is related to their drift μ mobility through the *Einstein relation* for a nondegenerate collection of carriers (those obeying Boltzmann statistics as in a nondegenerate semiconductor): $D/\mu = k_B T/e$. k_B is the Boltzmann constant, T is the absolute temperature and e is the electronic charge.

Digital communications is the transmission of information in the form of pulses from one location to another. Typically, an analog signal is sampled at fixed intervals and the value at each interval is put into a binary code containing 1s and 0s and the coded information is then transmitted as pulses (1 corresponding the presence and 0 to the absence of a pulse). Thus, the analog signal is digitized and sent by pulses.

Digital video disk see *DVD*.

Dimitrov–Sakka relationships are semiempirical relationships obtained to relate the average polarizability of an oxide ion, $\alpha(O^{2-})$, in a simple oxide solid A_pO_q to the refractive index n, the bandgap E_g and the molar volume V_m of the oxide ion. If A_pO_q represents Al_2O_3, then $p = 2$ and $q = 3$. Suppose that α_i is the polarizability of the cation. In simple terms,

$$\alpha(O^{2-}) = \frac{V_m}{2.52q}\left(1 - \sqrt{\frac{E_g}{20}}\right) - \frac{p}{q}\alpha_i$$

$$= \frac{V_m}{2.52q}\left(\frac{n^2 - 1}{n^2 + 2}\right) - \frac{p}{q}\alpha_i \text{ [esu units]}.$$

These semiempirical rules are not universal and have to be modified for certain oxides. (V. Dimitrov and S. Sakka, *J. Appl. Phys.* **79**, 1736, 1976; **79**, 1740, 1996.)

Diode (short diode) is a *pn* junction in which the neutral regions are shorter than the minority carrier diffusion lengths.

Diode pumped laser is a solid state laser such as a Nd:YAG (neodymium:yttrium aluminum garnet, $Y_3Al_5O_{12}$) laser, in which the radiation from a laser diode is used to pump the laser medium. The laser medium is usually a doped crystal such as a Nd^{3+} doped YAG crystal. The optical pumping is chosen so that the laser diode emission is matched with the 808 nm absorption band in Nd:YAG. There are efficient InGaAs laser diodes that can be used at this wavelength. The Nd:YAG crystal can be pumped from the end or from the sides. The laser crystal (Nd:YAG) is usually shaped to be a rod, disk or a slab. Most applications that require low power and high modal quality use rod shaped crystals, and usually end-pumped. Nd:YVO_4 (neodymium:yttrium orthovanadate) lasers are more efficient than Nd:YAG lasers, and can provide repetitive very short pulses as well as CW. Diode pumped lasers are compact, reliable, stable and have various industrial applications such as resistor trimming, surface marking or cleaning, materials cutting, welding, range finding, ablation, spectroscopy, nondestructive testing, and in various medical applications in dermatology, dentistry and ophthalmology, etc. They do not need special cooling arrangements. Both NdYAG and Nd:YVO_4 emit at 1.064 μm, but it is possible to double or even quadruple the frequency by using a nonlinear crystal such as lithium triborate, barium betaborate, cesium lithium borate, etc. to obtain an output at 532 nm (doubled frequency and green) or 266 nm (quadrupled frequency and UV). Typical repetition rates for optimum pulsed operation are usually less than 50 kHz.

Table: Diode pumped lasers Values from various sources including A. Kruger in *Photonics Handbook*, Laurin Publishing, Pittsfield, MA, 2002.

	Nd:YAG	Nd:YVO_4
Main lasing emission	1.0642 μm	1.0643 μm
Stimulated emission cross section	2.8×10^{-19} cm^2	15.6×10^{-19} cm^2
Pump peak wavelength	808.5	807.5
Luminescence lifetime (1 at% Nd^{3+})	0.1 μs	0.2 μs
Advantages	Compact, reliable, and stable. Pulsed or CW	Compact, reliable, and stable; more efficient than Nd:YAG, short pulses and CW; polarized output

Dioptre is a measure of the refracting power of a lens, defined by the reciprocal focal length, 1/*f*, and measured in dioptres with *f* in meters. See *prism*.

Dipolar (orientational) polarization arises when randomly oriented polar molecules in a dielectric are rotated and aligned by the application of a field so as to give rise to a net average dipole moment per molecule. In the absence of the field the dipoles (polar molecules) are randomly oriented and there is no average dipole moment per molecule. In the presence of the field the dipoles are rotated, some partially and some fully, to align with the field and hence give rise to a net dipole moment per molecule.

Dipolar relaxation equation describes the time response of the induced dipole moment per molecule in a dipolar material in the presence of a time-dependent applied field. The response of the dipoles depends on their relaxation time which is the mean time required to dissipate the stored electrostatic energy in the dipole alignment to heat through lattice vibrations or molecular collisions. Suppose that p is the induced dipole per atom or molecule at an instant t, p_0 is the steady state induced dipole moment ($p_o = \alpha_d E$, α_d is the dipolar polarizability, and E is the field). Then the rate of change in the dipole p is given by the dipolar relaxation equation,

$$\frac{dp}{dt} = -\frac{p - p_0}{\tau}$$

in which τ is a relaxation time associated with the polarization dissipation process. The above is the simplest dipolar relaxation equation and assumes that we can use a single relaxation time τ to describe the dielectric relaxation process whereas in reality there may be a distribution of relaxation times, and the polarization and its dissipation may involve long-range interactions. The solution of the above first order rate relaxation equation under an oscillating field $E_o\exp(j\omega t)$ leads to a complex relative permittivity.

Dipole matrix element see *oscillator strength*.

Dipole relaxation (dielectric resonance) occurs when the frequency of the applied ac field is such that there is maximum energy transfer from the ac voltage source to heat in the dielectric through the alternate polarization and depolarization of the molecules by the ac field. The stored electrostatic energy is dissipated through molecular collisions and lattice vibrations (in solids). The peak occurs when the angular frequency of the ac field is the reciprocal of the relaxation time.

Direct bandgap see *bandgap*.

Direct radiation see *solar radiation air mass*.

Direct recombination is a *radiative recombination* process in which an electron in the conduction band (CB) of a semiconductor recombines directly with a hole in the valence band (VB), thereby releasing a photon (electromagnetic radiation) that has an energy equal to the energy lost by the electron in the recombination process; the photon energy $h\nu$ is typically very close to the bandgap energy E_g because direct recombination typically involves electrons at the bottom of the CB, and holes at the top of the VB. The rate $R_{\text{radiative}}$ of direct recombination is proportional to both the concentration of electrons, n, and concentration of holes, p, that is, $R_{\text{radiative}} = Bnp$, where the proportionality constant B is the direct recombination coefficient.

Direct recombination capture coefficient B is a material constant that characterizes the rate of direct (radiative) recombination of excess injected carriers in a semiconductor. Suppose that excess electrons and holes have been injected into a p-side semiconductor, as would be on the p-side in a pn-junction under forward bias, and that Δn_p is the excess electron concentration and Δp_p is the excess hole concentration in the *neutral* p-side of a GaAs pn junction. Injected electron and hole concentrations would be the same to maintain charge neutrality, that is, $\Delta n_p = \Delta p_p$. Thus, at any instant,

$$n_p = n_{po} + \Delta n_p$$
$$= \text{instantaneous } \textit{minority} \text{ carrier concentration},$$

Table: Direct recombination capture coefficient Direct recombination coefficients (at 300 K) in selected semiconductors: D is direct, I is indirect. $p_{po} = 10^{17}$ cm^{-3}.
(Data combined from various sources including V. P. Varshni, *Phys. Stat. Solidi*, **19**, 459, 1967.)

Semiconductor	Si (I)	Ge (I)	GaAs (D)	GaSb (D)	GaP (I)	InAs (D)	InSb (D)	PbS (D)	PbSe (D)	PbTe (D)
B (cm^3 s^{-1})× 10^{-12}	0.0018	0.054	720	239	0.054	85	46	48	40	52
$\tau_r = 1/(2Bn_i)$ (intrinsic)	4.6 h	0.61 s		9 ms		15 μs	0.62 μs	15 μs	2 μs	2.4 μs
$\tau_r = 1/(Bp_{po})$ (extrinsic)	2.5 ms	150 μs		0.37 μs	3 ms	0.24 μs	0.12 μs	0.21 μs	0.25 μs	0.19 μs

and

$$p_p = p_{po} + \Delta n_p$$
$$= \text{instantaneous } \textit{majority} \text{ carrier concentration.}$$

where n_{po} and p_{po} are the thermal equilibrium minority and majority carrier concentrations respectively. The instantaneous recombination rate will be proportional to both the electron and hole concentrations at that instant, that is $n_p p_p$. Suppose that the thermal generation rate of electron hole pairs is G_{thermal}. The net rate of *change* of Δn_p is then,

$$\partial \Delta n_p / \partial t = -B n_p p_p + G_{\text{thermal}}$$

where B is a constant called the *direct recombination capture coefficient*. In equilibrium $\partial \Delta n_p / \partial t = 0$ and $n_p = n_{po}$ and $p_p = p_{po}$, where the subscript "*o*" refers to thermal equilibrium concentrations, so that $G_{\text{thermal}} = B n_{po} p_{po}$. Thus, the rate of change in Δn_p is

$$\partial \Delta n_p / \partial t = -B(n_p p_p - n_{po} p_{po}).$$

B values for direct bandgap semiconductors are typically of the order of 10^{13} cm^3 s^{-1} whereas B is four orders of magnitude smaller for indirect bandgap semiconductors (I). The *direct radiative recombination lifetime* τ_r for the intrinsic materials is defined by

$$\tau_r = 1/(2Bn_i)$$

where n_i is the intrinsic concentration. Shorter τ_r leads to higher direct recombination rates. In doped semiconductors, τ_r depends on whether the minority carrier injection is weak or strong. For *weak injection*, $\Delta n_p \ll p_{po}$, $\tau_r = 1/(Bp_{po})$ where p_{po} is the majority carrier concentration (i.e. N_a, acceptor concentration), and for strong injection, $\Delta n_p \gg p_{po}$, $\tau_r = 1/(B\Delta n_p)$.

See *recombination lifetime*.

Direct transitions see *band-to-band absorption*.

Directional coupler is a device that provides a convenient and fast means of switching (coupling) an optical signal from one waveguide (AA') to another (BB') by the application of an electrical signal V as depicted in the figure; light at the input into guide A is switched to B' at the output. A number of these EO directional couplers connected together can obviously switch an optical signal between various optical paths. See *directional integrated electrooptic coupler*.

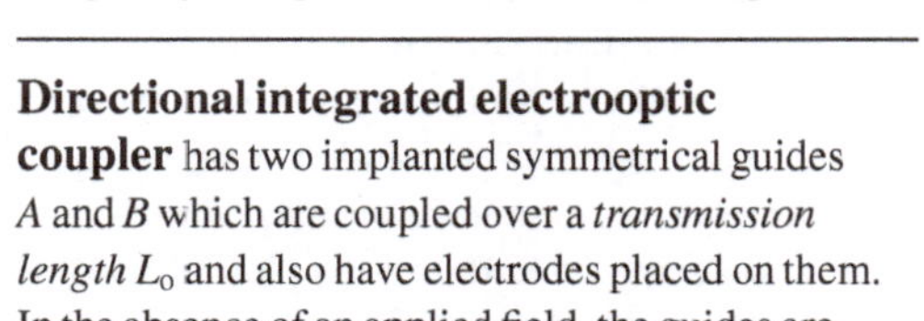

The principle of operation of a **directional coupler**.

Directional integrated electrooptic coupler has two implanted symmetrical guides A and B which are coupled over a *transmission length* L_o and also have electrodes placed on them. In the absence of an applied field, the guides are

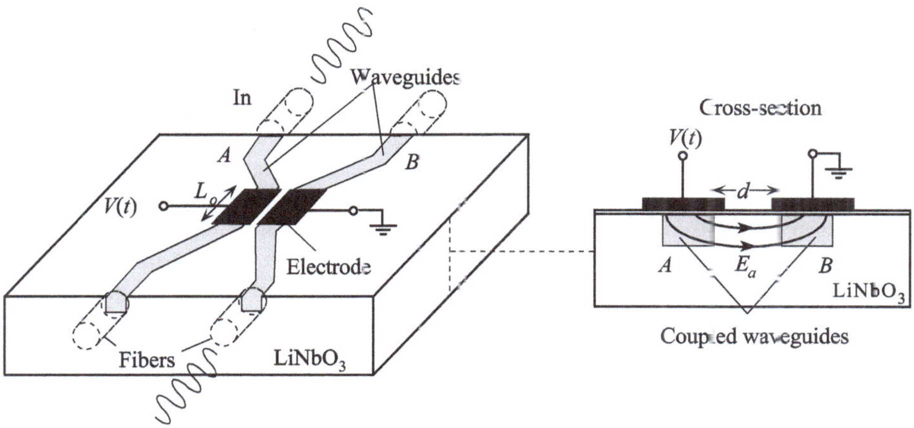

Directional integrated electrooptic coupler An integrated directional coupler. The applied field E_a alters the refractive indices of the two guides and changes the strength of coupling.

identical and the *phase mismatch*, $\Delta\beta$, is zero (no mismatch) and there is a full transmission from guide *A* to *B*. If we apply a voltage between the electrodes, the two guides experience an applied field E_a in opposite directions and hence experience opposite changes in their refractive indices. If $n\ (=n_A = n_B)$ is the refractive index of each guide, and Δn is the induced index change in each guide by the *Pockels effect*, the induced index difference Δn_{AB} between the guides is $2\Delta n$. Taking, as a first approximation, $E_a \approx V/d$, and using $\Delta\beta = \pi 3^{1/2}/L_0$ for the prevention of transfer, the corresponding switching voltage is V_o is

$$V_o = \frac{\sqrt{3}\lambda d}{2n^3 r L_0}$$

where *r* is the appropriate Pockels coefficient. Since L_0 depends inversely on the coupling efficiency *C*, for a given wavelength, V_o depends on the refractive indices and the geometry of the guides.

Dirty displacive ferroelectrics are ferroelectric materials in which each unit cell differs from every other unit cell, although an average translational symmetric is maintained. Typically for materials just below the ferroelectric phase transition temperature T_c, one can consider the ferroelectric phase to result from small displacements of the nonferroelectric phase.

Dislocation is a line imperfection within a crystal that extends over many atomic distances.

Disordered diffraction grating is a grating that deviates in some random manner from the perfect periodic structure of an ideal grating, for example, a grating in which there are random variations in the slit separations or positions and widths – each has some random distribution around a mean value. Provided the disorder is small, a disordered diffraction grating can still produce a diffraction pattern though peaks become less intense and broadened due to the slit separation having a random variation about an average, in similar fashion to the reduction and broadening of the diffraction peaks in x-ray diffraction from a crystal due to the Debye–Waller effect; in a crystal the disorder is due to uncorrelated thermal vibrations of the atoms. Deviations from perfect periodicity are usually referred to as "grating defects" and these are typically classified into *periodic errors* in the spacing and widths of the slits (or grating bars) and *random errors* in the period.

spacing, and widths of the grating slits or bars. The periodic errors give rise to what are known as *Lyman ghosts* and *Rowland ghosts* in the diffraction pattern. It can be shown that the structural uniformity of the grating is critical to the observation of high-order diffraction intensities. For example, suppose we assume identical Gaussian randomness in the leading and trailing edges of the grating slits with Gaussian variance σ. Then the ratio of the intensity of the nth order peak to that of the zero order peak is given by

$$\frac{I_n}{I_0} = \left[\frac{\sin(n\pi s/d)}{n\pi s/d}\right]^2 \exp\left[-\left(\frac{2\pi n}{d}\right)^2 \sigma^2\right]$$

where d is the average grating period and s is average slit width. It is assumed that the variations δd in d follow the Gaussian function $f_d(\delta d) = [(2\pi)^{-1/2})/\sigma]\exp[-(\delta d)^2/(2\sigma^2)]$ with a similar distribution for δs. (Grisenti *et al.*, *Phys. Rev. A*, **61**, 33608, 2000; P. Licinio *et al.*, *Am J. Phys.*, **67**, 1013, 1999.)

Dispersion compensation and management in optical fiber communications refers to the compensation of *chromatic* dispersion in the transmission fiber by inserting a *compensating fiber* that has the opposite chromatic dispersion to that in the transmission fiber. The overall *chromatic* dispersion is then managed by appropriately engineering the chromatic dispersion in the compensating fiber to meet the desired chromatic dispersion for the overall optical link; hence the term "dispersion management." Suppose that the dispersion is D_t in the transmission fiber and its length is L_t. If the dispersion in the compensating fiber is D_c and its length is L_c, the *total dispersion* is given by $D_tL_t + D_cL_c$. The dispersion D_t in the transmission fiber is normally positive, and D_c in the compensating fiber has to be negative to compensate for the dispersion in the transmission fiber, which means that waveguide dispersion in the compensating fiber has to be made dominant and appropriately shaped in terms of D_c vs. wavelength

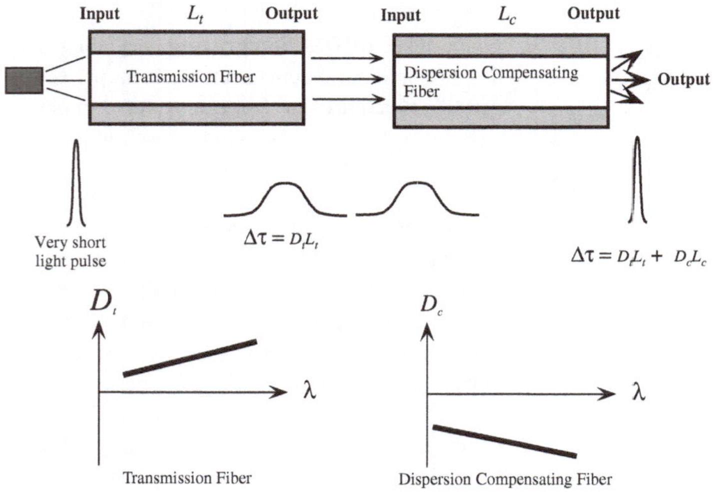

Dispersion compensation The dispersion in the transmission fiber is compensated by using a compensating fiber. The dispersion management involves engineering the right compensating fiber of the right length to produce the desired final dispersion characteristics.

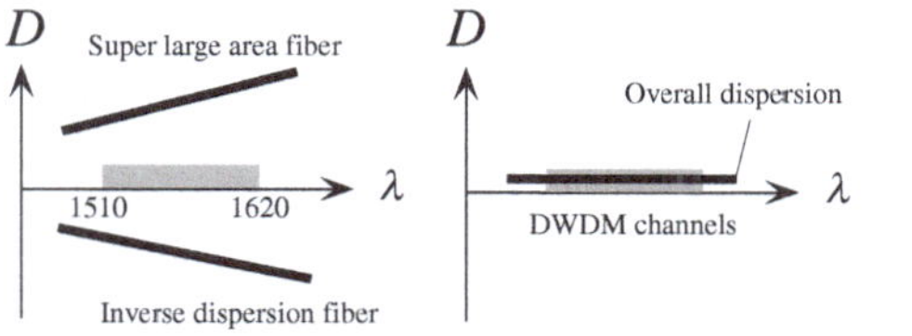

Dispersion compensation Inverse dispersion fiber enables the dispersion to be reduced and maintained flat over the communication wavelengths.

behavior by suitably engineering the fiber index profile. Dispersion compensation is also used in generating a flat dispersion over the communication wavelengths. For example, for ultralong-haul communications the so-called UltraWave fiber that is designed for dense wavelength division multiplexed (DWDM) communications has a large cross-sectional area to enable more power to be fed into the fiber – it is a super-large-area (SLA) fiber with positive dispersion. The inverse dispersion fiber (IDF) has negative dispersion slope that mirrors the dispersion slope in the SLA fiber so that the combination of SLA and IDF, the dispersion managed set, has a small and reasonably flat dispersion over the DWDM wavelengths.

See *compensating fiber*.

Dispersion diagram of a waveguide is the angular frequency, ω vs. propagation constant, β, behavior for electromagnetic waves propagating in the guide. For example, for an infinite planar dielectric waveguide, the group velocity of a mode m can be found from the dispersion diagram $\omega = f(\beta_m)$ inasmuch as

$$v_{gm} = \frac{d\omega}{d\beta_m} = \frac{d\omega}{d[\beta_m(\omega)]} = \text{function of } \omega$$

Dispersion flattened fiber is an optical fiber that has a refractive index profile that results in chromatic dispersion being small and flat over a wavelength range. Waveguide dispersion, represented by D_w, can be adjusted by changing the waveguide geometry. Waveguide dispersion arises from the group velocity v_g dependence on the wavelength

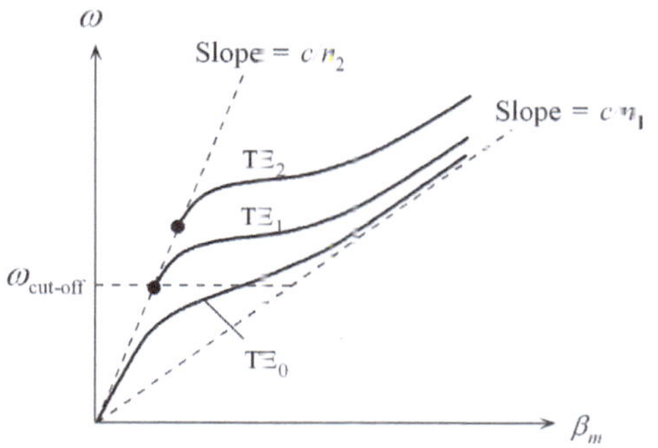

Schematic **dispersion diagram**, ω vs. β for the slab waveguide for various TE_m modes. $\omega_{cut\text{-}off}$ corresponds to $V = \pi/2$. The group velocity v_g at any ω is the slope of the ω vs. β curve at that frequency.

λ. As the wavelength increases, the field penetrates more into the cladding which changes the proportion of light energy carried by the core and the cladding and hence changes v_g. We can thus alter the waveguide geometry, that is the refractive index profile, and thereby control D_w to yield a total chromatic dispersion that is *flattened* between the

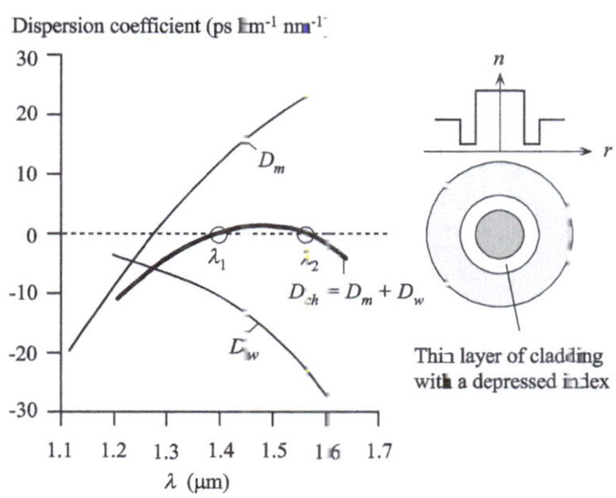

Dispersion flattened fiber example. The material dispersion coefficient (D_m) for the core material and waveguide dispersion coefficient (D_w) for the doubly clad fiber result in a flattened small chromatic dispersion between λ_1 and λ_2.

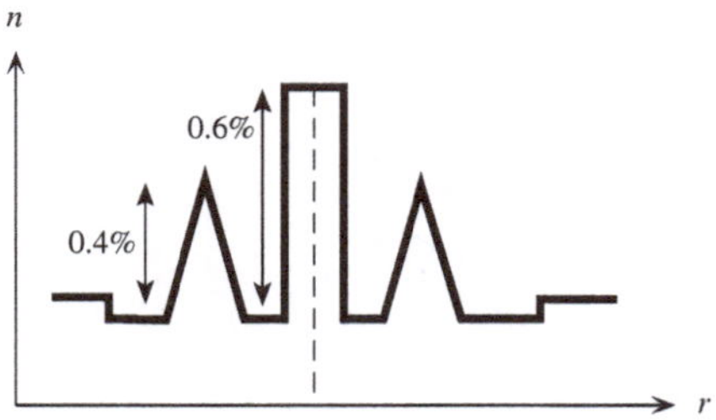

Dispersion flattened fiber The refractive index profile for a fiber that has reduced dispersion slope over the communication wavelengths.

wavelength λ_1 and λ_2. The refractive index profile of such a fiber looks like a W in which the cladding is a thin layer with a depressed refractive index; the fiber is called *doubly clad*. The simple step index fiber is singly clad. Greater control on waveguide dispersion can be obtained by using multiply clad fibers. Such fibers are more difficult to manufacture but can exhibit excellent chromatic dispersion 1–3 ps km^{-1} nm^{-1} over 1.3–1.6 μm. Low dispersion over a wavelength range, of course, allows wavelength multiplexing, e.g. using a number of wavelengths (e.g. 1.3, 1.55 μm) as communication channels.

Dispersion in fibers is the spread in time, known as temporal broadening, of an infinitesimally thin optical pulse as it propagates along the fiber. The temporal broadening is due to the different propagation characteristics of different wavelength components of light that are coupled into the fiber and propagate along the guide. The *group delay time* τ is the time it takes for very short light pulse of a particular mode of given wavelength to travel from the input to the output of the fiber; a "delay time" from the fiber's input to output. The temporal spread in τ, $\Delta\tau$, is called *dispersion*. Due to the dispersion effect there is an upper limit for the rate at which we can transmit light pulses along a fiber. For example, a single-mode step index fiber has a maximum bit rate $B = 0.59/\Delta\tau$. In single mode fibers, dispersion $\Delta\tau$ increases not only with the fiber length L but also with the spectral width $\Delta\lambda$ of the emitter (usually a laser diode) that drives the optical signals into the fiber. The dispersion coefficient D is the dispersion per unit fiber length per unit spectral width of the optical sopurce, i.e. $D = \Delta\tau\ /\ (L\Delta\lambda)$, where L is the fiber length. Many fibers are named in terms of their dispersion characteristics because dispersion ultimately controls the bandwidth for information transmission, and the

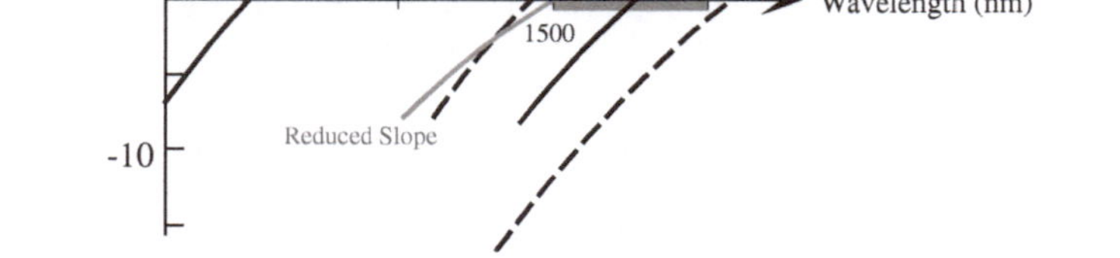

Dispersion in fibers Various fibers named after their dispersion characteristics.

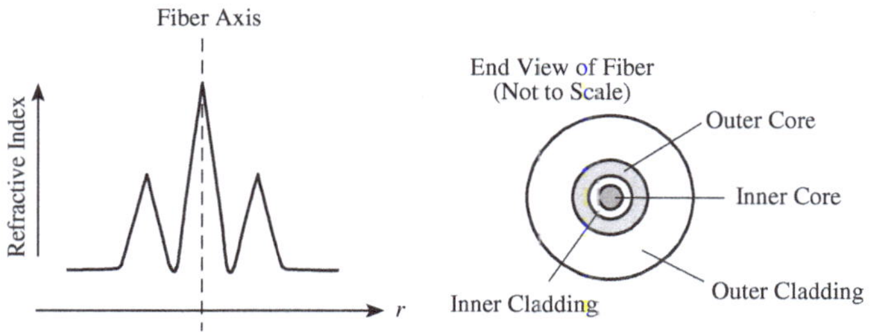

Dispersion in fibers Refractive index profile for one particular zero-dispersion shifted fiber.

way in which the fiber can be used in communications. The standard silica single-mode fiber has its chromatic dispersion zero at around 1300 nm. The fiber that has this dispersion curve shifted to have the zero in the middle of the useful 1510–1620 nm (approx.) communications channel is called the *zero dispersion shifted fiber*. The *nonzero dispersion shifted fiber* has the "zero" dispersion point outside this range. The *reduced slope fiber* has been designed to have the dispersion slope, that is D vs. wavelength slope, deduced in the same region for use in such applications as wavelength division multiplexing.

Dispersion of a light wave in optics is the separation of light into its various wavelength components, i.e. its spectral components.

Dispersion shifted fiber is a fiber in which the chromatic dispersion characteristics (dispersion vs. wavelength) has been shifted to longer wavelengths by adjusting the waveguide dispersion by appropriately changing the waveguide geometry or the refractive index profile. Zero dispersion shifted fiber has its chromatic dispersion zero at around 1550 nm. Nonzero dispersion shifted fiber is designed for use with WDM (wavelength division multiplexing) and has its chromatic dispersion outside the Er-doped amplifier band, 1525–1620 nm.

Dispersion slope specifies the rate of change in the chromatic dispersion with wavelength, $dD_{ch}/d\lambda$, over a specified wavelength range, usually over the communicating wavelengths, 1530–1565 nm. It is an important property in dense wavelength division multiplexing (DWDM) because the dispersion slope should be small over the range of channel wavelength. The absolute value of the dispersion can be compensated by using dispersion compensation See *chromatic dispersion, dispersion compensation and management.*

Dispersive medium has a refractive index n that depends on the wavelength, that is n is not a constant. The dispersion relation for the medium is the angular frequency vs. propagation constant, i.e. ω vs. β behavior in the medium. The slope of this curve at any β-value is the group velocity of the wave with that β.

Distortion is an aberration that occurs for extended objects due to the magnification of an optical system (e.g. lens) not being the same for all the points on the objects, that is, it depends on the distance from the optical axis. In distortion, off-axis points are imaged to points that do not fall onto those predicted by the paraxial approximation theory. It is therefore a third-order aberration, a failure of the paraxial approximation. Distortion normally results in vertical (horizontal) lines to bulge and exhibit a curvature which becomes more pronounced away from the optical axis. If the image lines corresponding to the perfect grid bulge inwards, distortion is "pincushion" type; if they bulge out, it is "barrel" type

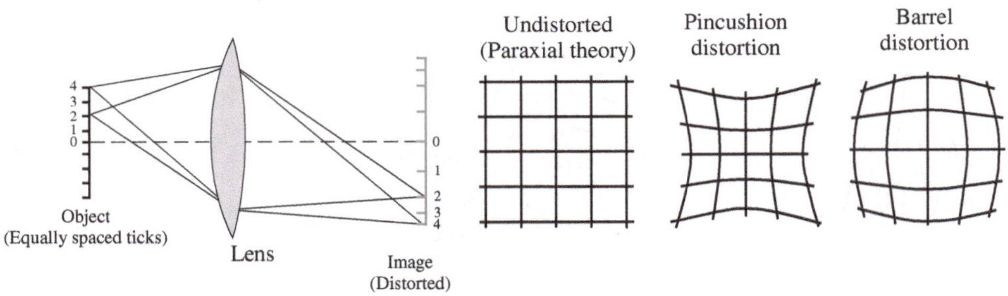

Distortion in a lens and the images.

Distributed Bragg reflector (DBR) laser diode has one of the optical cavity reflectors as a mirror that has been designed like a reflection type diffraction grating; it has a periodic corrugated structure. Intuitively, partial reflections of waves from the corrugations interfere constructively (that is reinforce each other) to give a reflected wave only when the wavelength corresponds to twice the corrugation periodicity. For example, two partially reflected waves such as *A* and *B* have an optical path difference of 2Λ where Λ is the corrugation period. They can only interfere constructively if 2Λ is a multiple of the wavelength within the medium. Each of these wavelengths is called a Bragg wavelength λ_B and given by the condition for in-phase interference,

$$q\frac{\lambda_B}{n} = 2\Lambda.$$

Distributed feedback (DFB) laser has a corrugated layer, called the *guiding layer*, next to the active layer; radiation spreads from the active layer to the guiding layer. These corrugations in the refractive index act as optical feedback over the length of the cavity by producing partial reflections. Thus optical feedback is *distributed* over the cavity length. In the DFB structure, traveling waves are reflected partially and periodically as they propagate. The left and right traveling waves can only coherently couple to set up a mode if their frequency is related to the corrugation periodicity Λ, taking into account that the medium alters the wave-amplitudes via optical gain. The allowed DFB modes are not exactly at Bragg wavelengths but are symmetrically placed about λ_B. If λ_m is an allowed

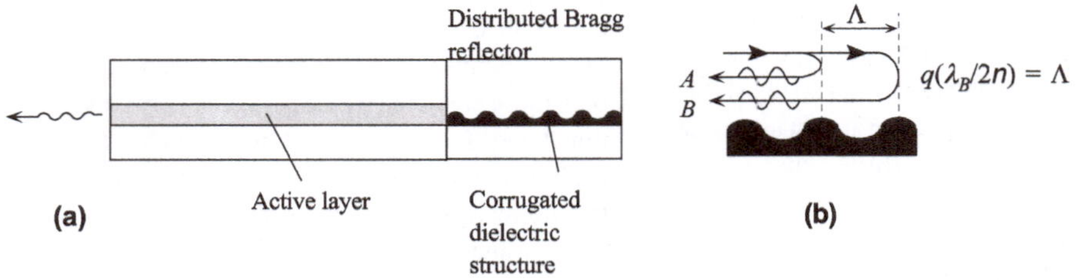

Distributed Bragg reflector laser diode (a) Distributed Bragg reflector (DBR) laser principle. (b) Partially reflected waves at the corrugations can only constitute a reflected wave when the wavelength satisfies the Bragg condition. Reflected waves *A* and *B* interfere constructively when $q(\lambda_B/2n) = \Lambda$.

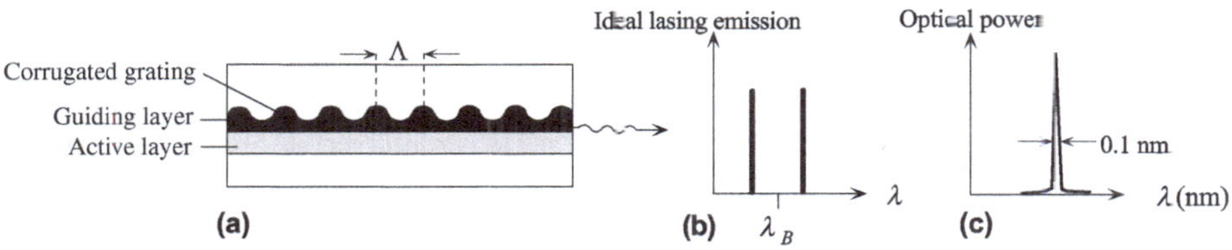

Distributed feedback (DFB) laser (a) Structure. (b) Ideal lasing emission output. (c) Typical output spectrum from a DFB laser.

DFB lasing mode then

$$\lambda_m = \lambda_B \pm \frac{\lambda_B^2}{2nL}(m+1)$$

where m is a mode integer, 0, 1, 2, . . . , and L is the effective length of the diffraction grating (corrugation length). The relative threshold gain for higher modes is so large that only the $m = 0$ mode effectively lases. A perfectly symmetric device has two equally spaced modes placed around λ_B. In reality, either inevitable asymmetry introduced by the fabrication process, or asymmetry introduced on purpose, leads to only one of the modes to appear. Further, typically the corrugation length L is so much larger than the period Λ that the second term in the above equation is very small and the emission is very close to λ_B. There are various commercially available single-mode DFB lasers on the market with spectral widths of ~0.1 nm at the communications channel of 1.55 μm.

Diverging waves have *wavevectors*, vectors that are normal to the wavefront, that are *not* parallel. The vector that is normal to a wavefront surface at a point such as P represents the direction of wave propagation (**k**) at that point P. The propagation vectors everywhere are all parallel and the plane wave propagates without the wave diverging; *the plane wave has no divergence*. The amplitude of the planar wave does not depend on the distance from a reference point, and it is the same at all points on a given plane perpendicular to **k**, i.e. independent of x and y. Moreover, as these planes extend to infinity, there is infinite energy in the plane wave. A plane wave is an idealization that is useful in analyzing many wave phenomena. In reality, however,

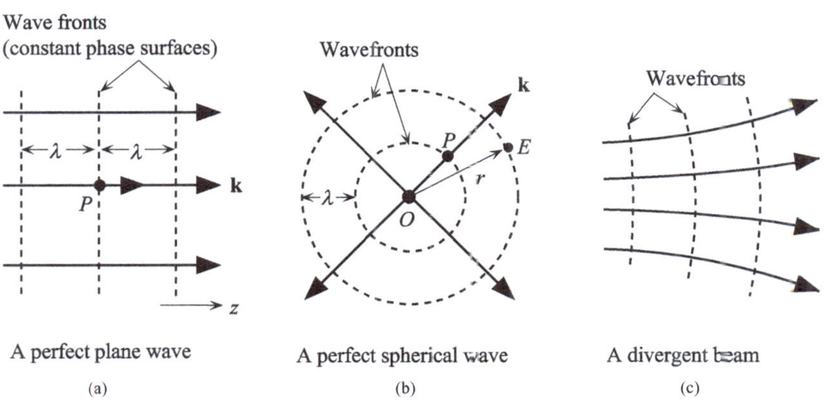

Diverging waves Examples of some possible electromagnetic waves.

the electric field in a plane at right angles to **k** does not extend to infinity since the light beam would have a finite cross-sectional area and finite power. (We would need an infinitely large electromagnetic source with infinite power to generate a perfect plane wave.)

Donor atoms are dopants in the semiconductor that have a valency one more than the host atom. They therefore donate electrons to the conduction band (CB) and thereby create electrons in the CB which leads to $n > p$ and hence to an n-type semiconductor (n is the electron concentration in the CB and p is the hole concentration in the VB).

Dopant is an element that is incorporated in trace amounts into bulk crystals to form doped wafers, or into epitaxial layers to influence their conductivity type (i.e. n- or p-type) and their resistivity. Dopants are classified as being of either acceptor or donor type.

Doped semiconductors see *extrinsic semiconductors*.

Doping is the process in which dopants are incorporated into bulk crystals or thin layers of semiconductors by using a variety of processes. These include doping during bulk material or layer fabrication (so-called *in situ* doping), and also use of diffusion and ion implantation. These approaches are all designed to permit the controlled introduction of precise amounts of an impurity to create n- or p-doped material. Using masking, implantation and diffusion processes permit the doping of only selected regions of material.

Doping, as a general term, is used to describe the addition of impurities into a material to modify its properties. For example, phosphorus dopants in a silicon crystal act as donor impurities and donate electrons into the conduction band and thereby increase the conductivity. When silica (SiO_2) is doped with germania (GeO_2), the resulting glass (Si-Ge-O) has a higher refractive index because the relative permittivity is higher.

Doppler broadened linewidth is the broadening of the emission spectrum of a collection of identical sources as a result of their random motions with respect to the observer. Since the atoms of a gas are in random motion then, due to the Doppler effect, the observer will detect a range of light frequencies or wavelengths emitted by these atoms. This leads to a Doppler broadened spectrum. When we consider the Maxwell velocity distribution of the gas atoms in the laser tube, we find that the linewidth $\Delta\upsilon_{1/2}$ between the half-intensity points (*full width at half maximum* FWHM) in the output intensity vs. frequency spectrum is given by

$$\Delta\upsilon_{1/2} = 2\upsilon_0\sqrt{\frac{2k_BT\ln(2)}{Mc^2}}$$

where M is the mass of the lasing atom or molecule and υ_0 is the center frequency of the spectrum (source emission frequency). See *inhomogeneous broadening, laser output spectrum.*

Doppler effect is the change in the measured frequency of a wave due to the motion of the source relative to the observer. If the source is moving towards the observer the frequency increases (wavelength decreases), and it decreases when the source is moving away. As a fast train approaches an observer, its whistle becomes higher in pitch (frequency) and as the train moves away from the observer, its pitch becomes lower. The true source frequency remains the same but more waves arrive at the observer per second when the source is moving toward the observer than when it is moving away. In the case of electromagnetic radiation, if v is the relative velocity of the source object toward the observer and υ_0 is the source frequency, then the measured electromagnetic wave frequency υ is $\upsilon = \upsilon_o[1 + (v/c)]$, for $(v/c) \ll 1$.

Dose represents how much energy is absorbed from an incident ionization radiation (X-rays and γ-rays) per unit mass of a body subjected to the radiation; it is measured in Grays (Gy). Suppose that when an ionizing radiation is incident on a body of mass M, a mean energy $E_{absorbed}$ is absorbed from the radiation. Then the *absorbed dose D* in units of J kg^{-3} that is in Gray or Gy is

$$D = E_{\text{absorbed}}/M.$$

Dose in *air* and *exposure* both refer to a given amount of air, and hence can be readily converted from one to the other. At STP, 1 R of exposure is equivalent to depositing 8.77×10^{-3} J into 1 kg of air. Thus 1 R = 8.77 mGy or 1 μR = 8.77 mGy; or 1 mGy = 114 mR. While exposure always involves the ionization of air, dose depends what medium has been subjected to the radiation, and thus its determination always involves calculating the actual energy absorbed by the medium per unit mass. Dose calculations require the knowledge of the spectrum of the incident radiation (photon fluence vs. photon energy) and knowledge of the mass attenuation and mass energy absorption coefficients of the constituents present in the body, quite complicated compared with exposure. See *roentgen, exposure, x-rays, mass attenuation coefficient.*

Double refraction see *birefrigence.*

Doubly clad fiber see *dispersion flattened fiber*

Dove prism rotates a passing beam by 180° and hence rotates an image – it is an image rotator. Dove prisms are typically used with collimated beams.

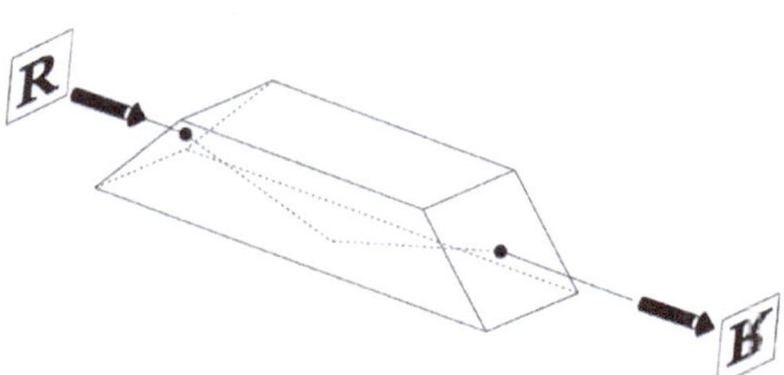

Dove prism (Courtesy of Melles Griot.)

Down-shifted light see *Stokes shift.*

Drawing in fiber manufacture is a process in which a glass preform rod is slowly fed into a hot furnace that has a hot zone around 1900–2000 °C

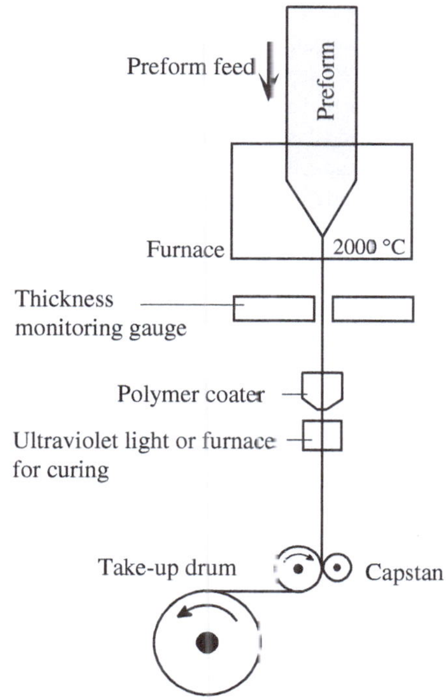

Drawing Schematic illustration of a fiber drawing tower.

where the glass flows like a viscous melt (resembling honey). As the rod reaches the hot zone and its end begins to flow, its tip is pulled, with just the right tension, to come out as a fiber and is spooled on a rotating take-up drum. The diameter of the fiber must be highly controlled to achieve the required waveguide characteristics. An optical thickness monitor gauge provides information on the changes of the fiber diameter which is used (in an automatic feedback control system) to adjust the speed of the fiber winding mechanism and the speed of the preform feeder to maintain a constant fiber diameter, typically better than 0.1%. In some cases, the preform is hollow, that is, it has a thin central hole along the rod axis. The hollow simply collapses during the drawing and does not affect the final drawn fiber. It is essential that, as soon as the fiber is drawn, it is coated with a polymeric layer (e.g. urethane acrylate) to mechanically and chemically protect the fiber surface. When a bare fiber glass surface is exposed to ambient conditions it quickly develops various microcracks on the surface and these microcracks dramatically reduce the mechanical strength (fracture strength) of the fiber. The applied polymeric coating is initially a viscous liquid and needs to be cured (hardened) which is done as the coated fiber passes through a curing oven, or ultraviolet lamps if it is UV hardenable. Sometimes two layers of polymeric coating are applied. Cladding is typically 125–150 μm and the overall diameter with the polymeric coatings is 250–500 μm. There is a thick polymeric buffer tube, or a buffer jacket, surrounding the fiber and its coating to cushion the fiber against mechanical pressure and microbending (sharp bending). Some fibers are buffered by having the fiber loose within a buffer tube. The tube may then contain a filling compound to increase the buffering ability. Single and multiple fibers are invariably used in cable form and the structure of the cable depends on the application (e.g. long-haul communications), the number of fibers carried and the cable environment (e.g. underground, underwater, overhead, etc.).

Drift mobility is the drift velocity of free charge carriers in a material per unit applied field in a direction parallel to the applied field or in the direction of applied force. If v_d is the drift velocity parallel to qE, where q is the charge of the carrier, and E is the applied electric field, then the drift mobility μ_d is defined by the equation $v_\mathrm{d} = \mu_\mathrm{d}E$. Its units are m^2 V^{-1} s^{-1}. Under low fields, charge carriers in solids drift with a constant drift mobility, that is $v_\mathrm{d} \propto E$. At high fields, for many classes of materials, the drift mobility usually depends on the field.

Drift velocity is the average velocity of electrons in the direction of an applied electrical force ($\mathbf{F} = -e\mathbf{E}$ for electrons). In the absence of an applied field, all the electrons move randomly in all directions, and the average velocity in any direction is zero. With an applied field, E_x, there is a net velocity per electron, $v_{\mathrm{d}x}$, in the opposite direction to the field where $v_{\mathrm{d}x}$ depends on E_x via $v_{\mathrm{d}x} = \mu_\mathrm{d}E_x$ where μ_d is the drift mobility. In the case of carriers moving in a weak field, the carrier distribution function remains close to the equilibrium distribution. The energy gained by the carriers from the electric field is lost to the lattice through collisions with low-energy phonons or with ionized impurities, keeping the carriers in equilibrium with the lattice. The carrier energy remains close to thermal equilibrium and so does the average speed of the electrons. In this case, the drift velocity follows directly from *Ohm's law* and is linearly proportional to the applied field, $v_{\mathrm{d}x} = \mu_\mathrm{d}E_x$ in the so-called ohmic regime. At higher fields, the efficiency of collisions in dissipating the excess carrier energy diminishes, and thus, a nonequilibrium condition is established in which the carrier system contains more energy than the lattice. It is convenient to associate the energies of the lattice and carrier systems with effective temperatures, T and T_e (i.e. for electron temperature), respectively. Thus, we can describe the effect of the high field as raising the carrier temperature over the lattice temperature, such that the carrier temperature is "hotter" than the lattice and we refer to them as *hot carriers*. For such carriers the drift velocity no longer obeys the Ohm's law form for near equilibrium, but rather becomes nonlinear in the applied field. It is not until the carriers are quite "hot" that a new pathway for energy dissipation becomes available to limit the carrier drift velocity. In this case, the dissipation mechanism is through the generation of optical phonons, which are high-energy quanta of lattice vibrations. As a result of this mechanism, the carrier velocity reaches a limiting value, the *saturation drift velocity* and no longer increases with increasing field. For Si and Ge, ohmic behavior only holds for fields of up to about 10^3 Vcm^{-1}. For higher fields, hot carrier effects prevail, until velocity saturation occurs near about 10^7 Vcm^{-1}. These effects are of great importance in the operation of electronic devices such as field effect transistors where in typical devices carriers move in relatively narrow channels and, under normal operating conditions, this corresponds to reasonably high fields. As the trend to smaller feature sizes continues for such devices, operating conditions will correspond to increasingly high applied fields.

Drude model is a primarily classical model that describes the behavior of free carriers (such as electrons) in a metal or a semiconductor in the presence of an applied field, which may be a dc or an ac field. In the Drude model (originally formulated by Paul Drude in 1900), the charge carriers are scattered with a mean scattering time τ which enters the theory as a relaxation time (i.e. $1/\tau$ is the mean probability of scattering per unit time; the mean frequency of scattering) and results in an ac conductivity σ given by

$$\sigma(\omega) = \frac{\sigma_{dc}}{1 + j\omega\tau}$$

where σ_{dc} is the dc conductivity and ω is the angular frequency of the applied field, or the electromagnetic wave that is drifting the electrons, and $j = (-1)^{1/2}$. (It is assumed that the field oscillates as $e^{j\omega t}$.) In the case of conduction electrons only contributing to conductivity, if n is the electron concentration, then $\sigma_{dc} = en\mu$ and μ is the drift mobility given by $\mu = e\tau/m_e^*$, m_e^* is the effective mass of the conduction electrons. The Drude model predicts that at high frequencies $\sigma \sim 1/\omega^2$. It turns out that for many materials, especially for semiconductors, $\sigma \sim 1/\omega^{\beta}$ where $\beta = 1.5$–3.5, which can be explained by using quantum mechanics and the appropriate scattering mechanism. The dependence of the conductivity on the frequency is usually significant at high frequencies, typically in the infrared or above. See *Drude–Smith formula, free carrier absorption.*

Drude–Smith formula is a generalization of the classical Drude equation based on an impulse response approach and using Poisson statistics. The approach introduces a new parameter c, measuring the persistence of velocity. The Drude–Smith formula for the complex conductivity σ ($= \sigma_1 + i\sigma_2$) as a function of angular frequency ω is given by

$$\sigma(\omega) = \frac{Ne^2\tau/m^*}{1 - i\omega\tau}\left[1 + \sum_{n=1}^{\infty} \frac{c_n}{(1 - i\omega\tau)^n}\right]$$

where $i = (-1)^{1/2}$, e is the elementary charge, m^* is the effective mass of the carrier, N is the free carrier concentration, τ is the relaxation time, n is an integer, and c_n is a constant that describes the persistence of the velocity after n scattering events. In practice, normally only the first term in the summation is used, such that $c_1 = c$ and $c_n = 0$ for $n > 1$. The Drude–Smith formula has often been used to describe measured THz carrier dynamics in different materials; for example, for highly polycrystalline films, or nanocrystals in an amorphous matrix, etc. In this case, c may vary between 0 and -1, corresponding to the Drude conductivity for $c = 0$, and complete carrier backscattering, or localization, for $c = -1$. For example, 200-nm thick epitaxial Si films on sapphire yields a c that is nearly zero, whereas for polynanocrystalline Si (15–25 nm domain size) films, $c \approx -0.830$, and for Si nanocrystals (nanocrystal diameters 3–7 nm, particle density $\sim 10^{18}$ cm^{-3}) in a SiO_2 film, $c \approx -0.962$ to -0.983 (N. V. Smith, *Phys. Rev. B*, **64**, 155106, 2001; D.G. Cooke *et al.*, *Phys. Rev. B*, **73**, 193311, 2006.) See *Drude model*.

Duffy relationship see *optical electronegativity*.

DVD, digital video disk, comprises a series of regions fabricated on a polycarbonate based disk of material to encode about 4.7 Gbytes of digital data that is then read using a laser-based optical readout system. Such a DVD can store up to 133 minutes of high-resolution video. The individual bytes of data are encoded as a series of embossed bumps formed in a single long spiral track on the injection molded plastic disk. The spiral data track is about 500 nm wide, and individual data bumps are at least 830 nm long and 125 nm high. The spiral tracks pass within about 1600 nm of each other as they wind around the disk. These bumps are then coated

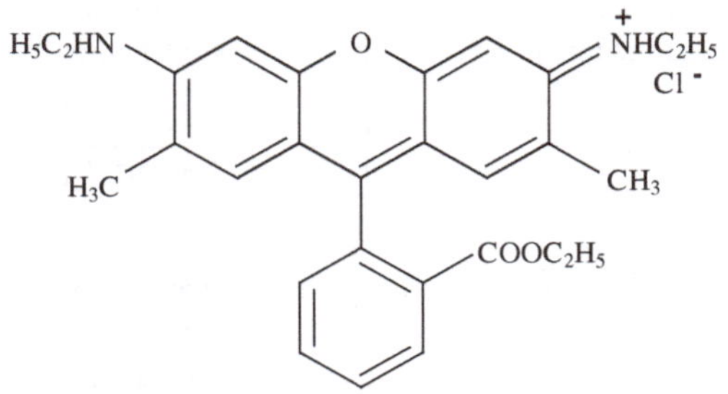

Dye laser Chemical formula of Rhodamine 6G.

with a protective aluminum layer. Data on these disks are read back by spinning the disk, typically from 200 to 500 rpm, in conjunction with an optical system. The latter is composed of a focused beam of laser light, which in conjunction with a tracking mechanism that enables the laser beam to follow the spiral track, passes through the polycarbonate layer and reflects off the bumps in the aluminum layer – the difference in height from data bits to the background level causes a difference in the signal recorded by a photodiode, registering the bits of data.

Dye laser uses organic molecules as the active medium, often in solution in water, ethanol, methanol or ethylene glycol (but also sometimes trapped in a porous *solgel*). One of the most frequently used and most widely investigated is Rhodamine 6G (or Rhodamine590) of formula $C_{28}H_{31}N_2O_3Cl$, that is benzoic acid, 2-[6(ethylamino)-3-(ethylimino)-2,7-dimethyl-3H-xanthen-9-yl]-ethyl ester, monohydrochloride. This red dye is famous for its good absorption in the blue-green, as well as its large bandwidth of fluorescence in the red part of the spectrum (560–625 nm). It has proved especially efficient for tunable continuous-wave operation (around 590 nm) or ultrashort pulse generation in earlier subpicosecond laser systems.

The major drawbacks of dye lasers lie in their chemical and thermal instability; their limited lifetime is quite prohibitive for some applications – and their potential toxicity (be it the molecule itself or the solvent). They are progressively being replaced by all-solid lasers such as titanium-sapphire ($Ti{:}Al_2O_3$) in commercial femtosecond systems.

Dynamic optical gratings are transient gratings which are generally produced as a result of appropriate illumination, and which disappear once this source is switched off. Historically, however, the first dynamic optical gratings were produced by ultrasonic waves. Indeed, such gratings are still used for light beam deflection, scanning and modulation – for example, *acoustooptical modulators*. The gratings can be produced in many solids, liquids or gases, and are detected by either the diffraction of probe beam, or by self-diffraction of the light that induced the grating. Lasers are ideally suited for generating such dynamic gratings owing to their associated properties – namely, intensity, collimation and coherence. Transient gratings are the basis of *real time holography*, *phase conjugation* and *four-wave mixing*. Normally gratings are generated by causing two TEM_{00} beams to interfere and result in plane grating formation at their intersection. The interference results in a spatially periodic light intensity and/or polarization – this changes the properties of the material in the interference region. Note that the induced changes in the material are reversible and disappear after the illumination ceases. By contrast, one example of an induced grating remaining imprinted in the material after the illumination is turned off, is a *fiber Bragg grating*. In all cases, the material response results from a nonequilibrium change in the material. For example, photoabsorption in the grating region can produce free carrier, space charge, thermal, or stress/strain gratings in the material. All of these excitations couple to a refractive index and/or absorption coefficient response. There are two classes of light-induced grating structures – quasistationary and propagating structures. In the former case, the grating maxima/minima do not oscillate or propagate and when the excitation is turned off, the grating amplitude decays without oscillation. In the latter case, there can be rapid oscillations of the grating amplitude, mostly connected with wave-like propagation. Density changes, for example, can generate acoustic phonons. Gratings are generally detected by diffraction of probe light which can be directly recorded or heterodyned with a spit-off beam from the probe light.

Dynamic range of an imaging system describes the maximum variation in signal over which the detector output represents the photon fluence without losing significant amounts of information. The dynamic range DR is defined by

$$\mathrm{DR} = \frac{X_{\max}}{X_{\mathrm{noise}}}$$

where X_{max} is the x-ray fluence corresponding to the maximum signal that the detector can accommodate and X_{noise} is the fluence that provides a signal equivalent to the quadrature sum of the detector noise and the x-ray quantum noise. A system with low dynamic range has high contrast but only over a limited portion of the image. The larger the dynamic range, the more gradual the contrast representing the range from minimal to maximal x-ray intensity. Therefore, the greater the dynamic range the better the contrast resolution. The dynamic range requirements differ among different imaging tasks. But there are some general principles for determining the requirements of each modality. If we have a mean exposure X_{mean} for a system, a multiplicative factor is necessary to get a range of exposure values above and below X_{mean}. For example, a dynamic range of 100:1 is acceptable; then, the range of exposures should be between $X_{mean}/10$ and $10\,X_{mean}$. Therefore, X_{mean} is the geometric mean of the acceptable highest and lowest exposures. The dynamic ranges of 400:1 and 100:1 are generally acceptable for mammographic and chest radiographic digital x-ray imaging systems respectively.

E

Edge emitting LEDs (ELEDs) have the emitted radiation emerging from an area on an edge of the crystal, i.e. from an area on a crystal face perpendicular to the active layer. ELEDs provide a greater intensity light and also a beam that is more collimated than surface emitting LEDs. The light is guided to the edge of the crystal by a *dielectric waveguide* formed by wider bandgap semiconductors surrounding a double heterostructure. The recombination of injected carriers occurs in the InGaAs active region which has a bandgap $E_g \approx$ 0.83 eV. Recombination is confined to this layer because the surrounding InGaAsP layers, confining layers, have a wider bandgap ($E_g \approx 1$ eV) and the InGaAsP/InGaAs/InGaAsP layers form a double heterostructure (DH). The light emitted in the active region (InGaAs) spreads into the neighboring layers (InGaAsP) which act to contain the light and guide it along the crystal to the edge. InP has a wider bandgap ($E_g \approx 1.35$ eV) and thus a lower refractive index than InGaAsP. The two InP layers adjoining the InGaAsP layers therefore act as *cladding layers* and thereby confine the light to the DH structure. Generally some kind of lens system is used to conveniently couple the emitted radiation from an ELED into a fiber. For example,

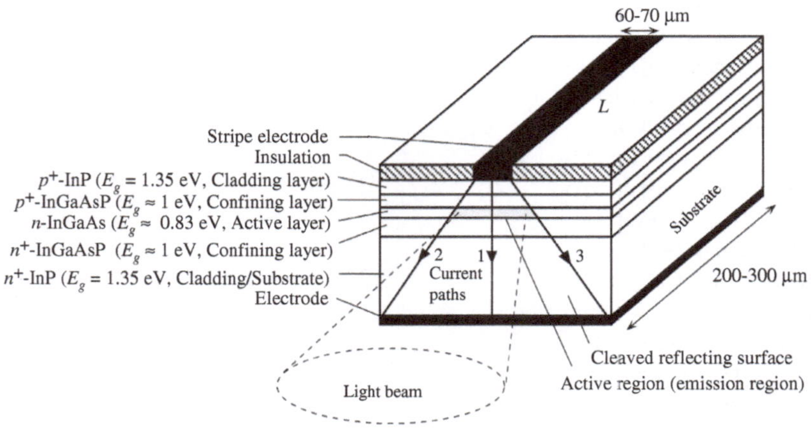

Schematic illustration of the structure of a double heterojunction stripe contact **edge emitting LED**.

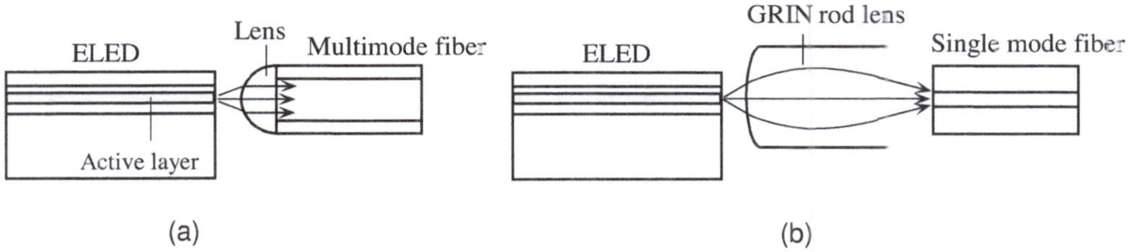

Light from an **edge emitting LED** is coupled into a fiber typically by using a lens or a GRIN rod lens.

a hemispherical lens attached to the fiber end can be used for collimating the beam into the fiber. A *graded index* (GRIN) *rod lens* is a glass rod that has a parabolic refractive index profile across its cross-section with the maximum index on the rod axis. It is like a large-diameter short-length graded index "fiber" (typical diameters are 0.5–2 mm). A GRIN rod lens can be used to focus the light from an ELED into a fiber. This coupling is particularly useful for single-mode fibers inasmuch as their core diameters are typically $\sim$10 μm. The output spectra from surface and edge emitting LEDs using the same semiconductor material is not necessarily the same. The first reason is that the active layers have different doping levels. Second is the self-absorption of some of the photons guided along the active layer as in ELED. Typically the linewidth of the output spectrum from an ELED is less than that from a SLED. In a set of experiments, for example, an InGaAsP ELED operating near 1300 nm was observed to have a linewidth of 75 nm whereas the corresponding SLED had a linewidth of 125 nm.

EDWA see *erbium doped waveguide amplifier.*

Effective density of states (N_c) at the conduction band edge is a quantity that represents all the states in the conduction band per unit volume as if they were all at E_c. Similarly N_v, which is the effective density of states at the valence band edge, is a quantity that represents all the states in the valence band per unit volume as if they were all at E_v. N_c can be calculated from

$$N_c = 2\left(\frac{2\pi m_e^* k_B T}{h^2}\right)^{3/2}$$

where m_e^* is the mean electron effective mass for the conduction band; it is called the density of states effective mass. Assuming spherical and parabolic energy bands, and for nondegenerate electron statistics, the electron concentration in the conduction band n is given by $n = N_c \exp[-(E_c - E_F)/k_B T]$, where E_F is the Fermi energy.

Effective electron mass, m_e^*, represents the inertial resistance of an electron inside a crystal against an acceleration imposed by an external force such as the applied electric field. If $F_{ext} = eE_x$ is the external applied force due to the applied field E_x then the effective mass, m_e^*, determines the acceleration a

Table: Effective density of states

	Si	Ge	GaAs	GaP	InAs	InP	β-GaN
$N_c \times 10^{19}$ cm^{-3}	2.8	1.04	0.042	4.97	0.009	0.053	1.713
$N_v \times 10^{19}$ cm^{-3}	1.04	0.6	0.88	1.46	1.51	0.63	3.82

of the electron by $eE_x = m_e{}^*a$. It takes into account the effect of the internal fields on the motion of the electron. In vacuum, where there are no internal fields, $m_e{}^*$ is obviously the mass in vacuum, m_e. The effective mass of an electron is a quantum mechanical quantity that behaves like the inertial mass in classical mechanics, $F = ma$, in that it measures the object's resistance to acceleration imposed by an external force. It relates the acceleration a of an electron in the conduction band to the applied external force F_{ext} by $F_{ext} = m_e{}^*a$. The external force is most commonly the force of an applied electric field, eE_x, and *excludes* all internal forces within the crystal. Similarly the effective mass of a hole in the valence band is its inertial mass against acceleration imposed by an external force. We should note that the internal fields depend on the location of the electron and its effect on the motion of the electron can only be determined by solving the *Schrödinger equation* with the appropriate potential energy term V. The effective mass $m_e{}^*$ depends inversely on the curvature of the electron energy vs. electron wavevector, E vs. k, behavior, i.e.

$$m_e^* = \hbar^2 \left[\frac{d^2E}{dk^2}\right]^{-1}$$

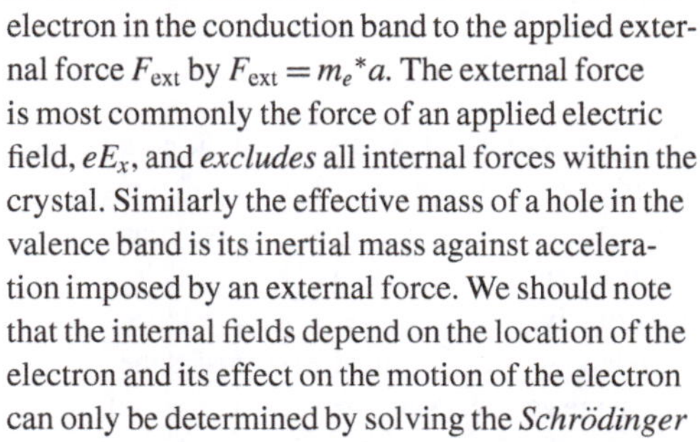

Effective electron mass n-type GaAs has the conduction electrons in the central valley at low fields. These have a light effective mass of $0.07m_e$. At sufficiently high fields some of these electrons can be scattered to the satellite valley where they have a heavier effective mass of $0.35\,m_e$.

which shows that a high curvature represents a light $m_e{}^*$ and a broad curvature represents a heavy $m_e{}^*$. For example, the GaAs E vs. k diagram for the conduction band has a central valley Γ that has a sharp curvature and a satellite valley L that has a small curvature. The electron effective mass is small in Γ but large in L. Consequently the electron drift mobility, inversely proportional to the effective mass, is high in Γ and low in L.

Effective medium approximation (EMA) is a theory that assigns *effective* optical properties to a heterogeneous or composite material system that has two or more phases (chemically and physically homogeneous material regions) by suitably combining the optical properties of the individual phases as if each phase maintained its own optical identity. EMA is not unique to optical properties, and has been applied to a wide range of properties. EMA rules for calculating the effective refractive index (or the effective relative permittivity) depend on a number of factors: the relative sizes, shapes and distributions of the constituent phases, the overall size of the mixture and the dimensions of the phases compared to the wavelength of light, disparity among the individual optical properties, etc. EMAs provide various useful mixture rules for the calculation of the effective refractive index of a composite material system. For example, the optical properties of hydrogenated microcrystalline Si has been modeled by combining the optical properties of crystalline Si and hydrogenated amorphous Si. EMA has been also applied to microscopically rough surfaces to account for optical reflections from rough surfaces. There are at least three important EMAs: *Lorentz–Lorenz, Maxwell–Garnett and Bruggeman*, all of which have the same general format, written in terms of the permittivitty, as

$$\frac{\varepsilon - \varepsilon_m}{\varepsilon + 2\varepsilon_m} = v_1\frac{\varepsilon_1 - \varepsilon_m}{\varepsilon_1 + 2\varepsilon_m} + v_2\frac{\varepsilon_2 - \varepsilon_m}{\varepsilon_2 + 2\varepsilon_m} + \cdots$$

where ε is the average or effective permittivity of the whole composite medium, ε_m is the permittivity of the matrix (host) material, $\varepsilon_1, \varepsilon_2$, ... are the permittivities of inclusions (phases) 1, 2,... in the host, $v_1, v_2, \ldots$ are the volume fractions of material (phases) 1, 2, ..., and $\Sigma v_i = v_1 + v_2 + \ldots = 0$. The fundamental assumptions are that the inclusions (phases) in the host dielectric matrix are all spherical, and that there are only dipole–dipole interactions. The origin of the above equation is due to Rayleigh (1892). The Maxwell–Garnett equation assumes that there are dispersed spherical dielectric particles of permittivity ε_d and volume fraction v_d in a host matrix with ε_m so that

$$\frac{\varepsilon - \varepsilon_m}{\varepsilon + 2\varepsilon_m} = v_d \frac{\varepsilon_d - \varepsilon_m}{\varepsilon_d + 2\varepsilon_m}$$

in which $v_d \ll 1$ is assumed. This equation has been modified by D. Polder and J. H. van Santen (*Physica*, **12**, 257, 1946) to take into account that the dispersed particles may have an ellipsoidal shape. The Bruggeman equation applies to a mixture of interdispersed phases (in contrast to the MG equation) for two mixed components 1 and 2 with permittivities ε_1 and ε_2, and with volume fractions v_1 and v_2 $(= 1 - v_1)$, even when v_1 and v_2 are comparable,

$$v_1 \frac{\varepsilon_1 - \varepsilon}{\varepsilon_1 + 2\varepsilon} + v_2 \frac{\varepsilon_2 - \varepsilon}{\varepsilon_2 + 2\varepsilon} = 1$$

where ε is the effective permittivity of the whole composite medium. Various other formulae have been proposed for calculating the dielectric properties of heterogeneous mixtures, some of which are due to H. Looyenga (*Physica*, **31**, 401, 1965), J. Monecke (*J. Phys. CM*, **6**, 907, 1994), and others. All EMA formulae have one fact in common in that they are all based on Maxwell's equations in the static limit. (J. W. Rayleigh, *Phil. Mag.*, **34**, 481, 1892; J. C. Maxwell-Garnett, *Phil. Trans. Royal Society*, **203**, 385, 1904; **205**, 237, 1906; D. A. G. Bruggeman, *Ann. Phys. (Leipzig)*, **24**, 636, 1935; D. Stroud, *Phys. Rev. B*, **12**, 3368, 1975; C. G. Granqvist and O. Hunderi, *Phys. Rev. B*, **16**, 3513, 1977.)

Eikonal equation relates the variation of the optical path length $L = \int_\Gamma n\,ds$ along a trajectory Γ of a light ray, in any linear medium, to the local value of the refractive index n; if u denotes the unit vector oriented tangentially to Γ, then the eikonal equation states

$$n\boldsymbol{u} = \nabla L$$

or, under its scalar form,

$$(\nabla L)^2 = (\partial L/\partial x)^2 + (\partial L/\partial y)^2 + (\partial L/\partial z)^2 = n^2.$$

This differential equation is useful in determining the trajectory of a light ray in a medium. In terms of curvilinear coordinate s, it can be written as

$$d(n\boldsymbol{u})/ds = \nabla n.$$

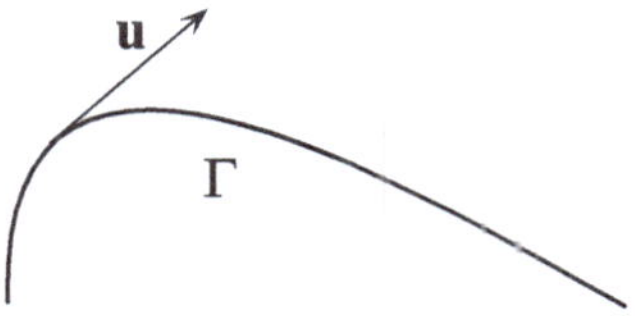

Eikonal equation **u** is the unit vector along the light trajectory Γ.

Einstein coefficients, A_{21}, B_{12} and B_{21}, are constants that are used in describing the rates of spontaneous emission, absorption and stimulated emission, respectively, between two energy levels E_1 and E_2 where $E_2 > E_1$. For example, the rate of spontaneous emission from E_2 and E_1 is proportional to the number of atoms N_2 at E_2 so that the spontaneous transition rate is $A_{21}N_2$. The rate of upward transitions from E_1 to E_2 by photon absorption is proportional to the number of atoms N_1 at E_1 and also to the number of photons with energy $h\nu = E_2 - E_1$, i.e. the absorption rate is $B_{12}N_1\rho(h\nu)$ where B_{12} is the Einstein B_{12} coefficient, and $\rho(h\nu)$ is the photon energy density per unit frequency which represents the number of photons per unit volume with an energy $h\nu$ $(= E_2 - E_1)$. The Einstein coefficients are interrelated. For example, if E_1 and E_2 have the same degeneracy, then $B_{12} = B_{21}$ and $A_{21}/B_{21} = 8\pi h\nu^3/c^3$. If E_1 and E_2 have degeneracies g_1 and g_2, then $g_1B_{12} = g_2B_{21}$. In the above expressions,

Table: Einstein coefficients

Absorption (B_{21})	Spontaneous emission (A_{21})	Stimulated emission (B_{21})
$\frac{dN_1}{dt} = -B_{12}N_1\rho(h\nu)$	$\frac{dN_2}{dt} = -A_{21}N_2 = -\frac{N_2}{\tau_{sp}}$	$\frac{dN_2}{dt} = -B_{21}N_2\rho(h\nu)$

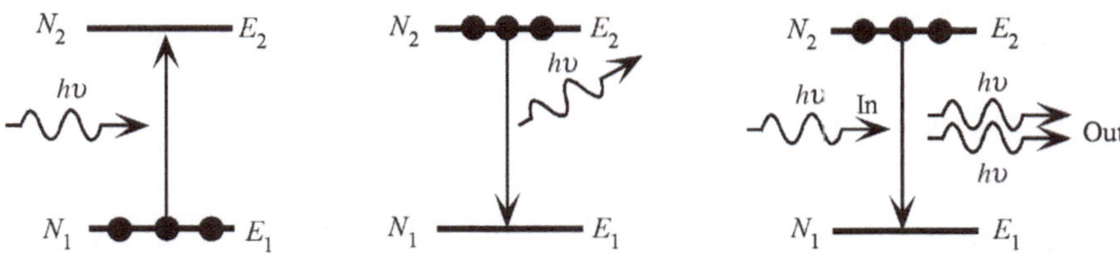

Einstein coefficients Absorption, spontaneous (random photon) emission and stimulated emission.

h is Planck's constant, ν is the radiation frequency, and c is the speed of light. $1/A_{21}$ is the spontaneous emission decay time τ_{sp}.

Einstein relation relates the diffusion coefficient, D, of a given species of charge carriers to their drift mobility, μ, via $D/\mu = k_B T/q$ where q is the electronic charge of the carrier, k_B is the Boltzmann constant and T is the temperature (in kelvin). The Einstein relation applies to a collection of particles obeying Boltzmann statistics, for example, electrons in the conduction band of a nondegenerate semiconductor. It *cannot* be applied (without extensive modification) to carriers in a degenerate semiconductor or a metal.

Elastic coefficients or constants are the elements of the elastic stiffness tensor **C** of a crystal. **C** is a tensor, an array of coefficients that relates the stress to elastic strain components for a given crystal; the stresses include both tensile and sheer stresses, and strains include both tensile and shear strains. In matrix notation, one can write the "Stress = (Elastic modulus) × Strain relationship", in mechanics as $\mathbf{T} = \mathbf{C}\,\mathbf{e}$, where **T** is the stress, **C** is the elastic stiffness and **e** is the strain matrix. **T** would have three components that are tensile, $T_1 = T_{xx}$, $T_2 = T_{yy}$, $T_3 = T_{zz}$, and three that are shear stresses, $T_4 = T_{yz}$, $T_5 = T_{xz}$, $T_6 = T_{xy}$. Similarly, **e** has three tensile, $e_1 = e_{xx}$, $e_2 = e_{yy}$, $e_3 = e_{zz}$ and three shear strains $e_4 = e_{yz}$, $e_5 = e_{xz}$, $e_6 = e_{xy}$. In general, the stress T_1 would be related to all six strains, and hence,

$$T_1 = C_{11}e_1 + C_{12}e_2 + C_{13}e_3 + C_{14}e_4 + C_{15}e_5 + C_{16}e_6.$$

In general, there are 36 elastic constants for a crystal, but only 21 independent Cs, since $C_{ij} = C_{ji}$. For cubic crystals: $C_{11} = C_{22} = C_{33}$; $C_{12} = C_{13} = C_{23}$, $C_{44} = C_{55} = C_{66}$; and the rest are zero. Further, $C_{11} = C_{12} + 2C_{44}$. C_{12} and C_{44} are called *Lamé's constants* λ and μ respectively: $C_{11} = \lambda + 2\mu$.

See *sound velocity*.

Elastic-ether theory see *Cauchy dispersion equation*.

Elastic modulus or **Young's modulus** (Y or E) indicates the ease with which a solid can be elastically deformed. The greater is Y, the more difficult it is to elastically deform the solid. When a solid of length L is subjected to a tensile *stress*, σ (force per unit area), it will extend elastically by an amount δL. $\delta L/L$ is the *strain*, ε. Stress and strain are related by $\sigma = Y\varepsilon$ so that Y is the stress needed per unit elastic strain. See *elastic coefficients*.

Table: Elastic modulus Elastic properties and their interrelationships; E = Young's modulus; B = bulk modulus ($1/B$ = compressibility); G = shear modulus; υ = Poisson's ratio; $C_{12} = \lambda$; $C_{44} = \mu$.

Cubic	$B = (^1/_3)(C_{11} + 2C_{12})$ $B = C_{12} + (^2/_3)C_{44}$	$G = (^1/_2)(C_{11} - C_{12})$ on (110) $G = C_{44}$ on (100)	$E = (C_{11} + 2C_{12})(C_{11} - C_{12})/(C_{11} + C_{12})$ $\upsilon = C_{12}/(C_{11} + C_{12})$
General	$E = 3(1 - 2\upsilon)B$	$E = 2(1 + \upsilon)G$	$E = 2(1 + \upsilon)G$
Hexagonal	$B = \frac{2}{9}[C_{11} + C_{12} + 2C_{13} + \frac{1}{2}C_{33}]$ $B = [(C_{11} + C_{12})C_{33} - 2C_{13}{}^2] / (C_{11} + C_{12} + 2C_{33} - 4C_{13})$		

Electric dipole moment exists whenever a positive charge $+Q$ is separated from a negative charge $-Q$. Even though the net charge is zero, there is nonetheless an electric dipole moment, **p**, given by $\mathbf{p} = Q\mathbf{a}$ where **a** is the distance vector from $-Q$ to $+Q$. Just as two charges exert a Coulomb force on each other, two dipoles also exert a force on each other which depends on the magnitudes of the dipoles, their separation and orientation. Consider a dipole **p** pointing in the y-direction as in the figure. The potential $V(r, \theta)$ at a point P at a distance r ($\gg a$) at an angle θ to the y-axis is given by

$$V(r, \theta) = \frac{p\cos\theta}{4\pi\varepsilon_0 r^2}.$$

The field at r, θ can be resolved into two components E_r and E_θ, one along **r** and the other perpendicular to **r** as shown in the figure:

$$E_r = \left(\frac{1}{4\pi\varepsilon_o}\right)\frac{2p\cos\theta}{r^3} \quad \text{and}$$

$$E_r = \left(\frac{1}{4\pi\varepsilon_o}\right)\frac{2p\sin\theta}{r^3}.$$

If the dipole is made to oscillate so that $p(t) = p_o\cos\omega t$, then the dipole will radiate electromagnetic (EM) radiation. The total averaged EM power radiated over a sphere enclosing the dipole is given by

$$P_{\text{average}} = \frac{p_0^2\omega^4}{12\pi\varepsilon_0 c^3}.$$

The emitted power is inversely proportional to λ^4. Shorter wavelengths or higher frequencies lead to greater EM power radiation.

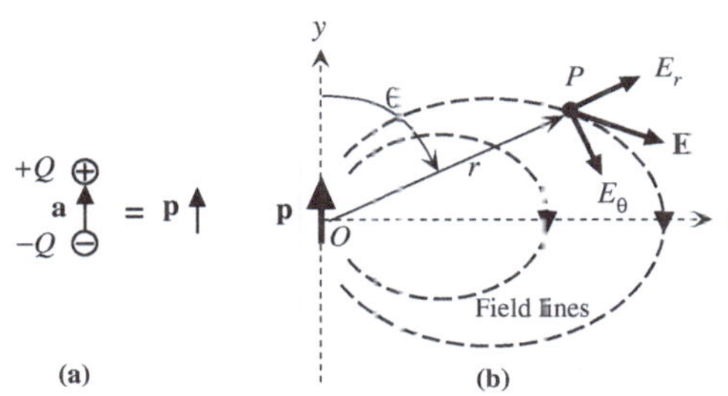

(a) Definition of an **electric dipole moment**. (b) Field put out by a dipole at a point P far away ($r \gg a$) from the dipole.

Electric susceptibility, χ, is a material quantity that measures the extent of polarization in the material per unit field. It relates the amount of polarization, **P**, at a point in the dielectric to the field, **E**, at that point via $\mathbf{P} = \chi\varepsilon_0\mathbf{E}$. If ε_r is the relative permittivity then for an isotropic medium $\chi = \varepsilon_r - 1$. The vacuum has no electric susceptibility. For an isotropic (e.g. liquids, glasses, cubic crystals) and *linear* medium, χ is a scalar quantity that does not depend on the field. In anisotropic media, the induced **P** is not in the same direction as **E**, and χ is a tensor quantity that depends on the crystal structure. In many cases the induced polarization depends on the field in a nonlinear fashion. In simple

Table: Electric susceptibility Summary of units in MKS (SI) and esu (Gaussian) systems for electric susceptibilities.

	SI or MKS	ESU or Gaussian	Conversion
P	$\mathrm{C\,m^{-2}}$	statvolt $\mathrm{cm^{-1}}$ or $(\mathrm{erg/cm^3})^{1/2}$	
E	$\mathrm{V\,m^{-1}}$	statvolt $\mathrm{cm^{-1}}$ or $(\mathrm{erg/cm^3})^{1/2}$	$E(\mathrm{MKS}) = 3 \times 10^4 E(\mathrm{esu})$
$\chi^{(1)}$	Dimensionless	Dimensionless	$\chi^{(1)}(\mathrm{MKS}) = 4\pi \times \chi^{(1)}(\mathrm{esu})$
$\chi^{(2)}$	$\mathrm{m\,V^{-1}}$	cm/statvolt or $(\mathrm{erg/cm^3})^{-1/2}$	$\chi^{(2)}(\mathrm{MKS}) = [4\pi/(3 \times 10^4)]\chi^{(2)}(\mathrm{esu})$
$\chi^{(3)}$	$\mathrm{m^2\,V^{-2}}$	$\mathrm{cm^2/statvolt^2}$ or $(\mathrm{erg/cm^3})^{-1}$	$\chi^{(3)}(\mathrm{MKS}) = [4\pi/(3 \times 10^4)^2]\chi^{(3)}(\mathrm{esu})$

terms, the dependence of the induced polarization P on the field at that point E can be written as

$$P = \chi^{(1)}\varepsilon_0 E + \chi^{(2)}\varepsilon_0 E^2 + \chi^{(3)}\varepsilon_0 E^3 + \cdots$$

where $\chi^{(1)}$, $\chi^{(2)}$, $\chi^{(3)}$, are the first-, second- and third-order susceptibilities. In Gaussian units or in esu, this equation is written normally as

$$P = \chi^{(1)} E + \chi^{(2)} E^2 + \chi^{(3)} E^3 + \cdots \text{ [esu units].}$$

The conversion between the units is not trivial and is shown in the above table.

See *polarization vector, relative permittivity, Pockels effect, optical Kerr effect.*

Electrical conductivity, σ, is a property of a material that quantifies the ease with which charges flow inside the material along or parallel to an applied electric field or a voltage gradient. It is the inverse of electrical resistivity, ρ. Since charge flow is due to a voltage gradient, σ is the rate of charge flow through unit area per unit voltage gradient; $J = \sigma E$, where J is the current density and E is the electric field. The power dissipated as heat per unit volume in a conductor of conductivity σ is given by Joule's law, that is σE^2. In a metal where the conduction electron concentration is n, and the electron drift mobility is μ, the conductivity is given by $\sigma = en\mu$, where e is the electronic charge. In a semiconductor there are both electrons in the conduction band (CB) and holes in the valence band (VB) contributing to electrical conduction and the conductivity is

$$\sigma = en\mu_e + ep\mu_h,$$

where n is the electron concentration in the CB, p is the hole concentration in the VB, μ_e is the drift mobility of electrons in the CB, and μ_h is the drift mobility of holes in the VB.

Electroabsorption is the induced absorption of light in a device as a result of applying (or changing) an electric field within the device. Such a device is an *electroabsorption modulator*. There are fundamentally three types of electroabsorption processes. In the *Franz–Keldysh* process, a strong applied field modifies the photon assisted probability of an electron tunneling from the valence band to the conduction band and thus corresponds to an effective reduction in the "bandgap energy." It was first observed for CdS in which the absorption edge was observed to shift to lower energies with the applied field, that is, photon absorption shifts to longer wavelengths with the applied field. The effect is normally quite small but nonetheless observable. In this type of electroabsorption modulation, typically, the wavelength is chosen to be slightly smaller than the bandgap wavelength so that absorption is negligible. When a field is applied, absorption is enhanced by the Franz–Keldysh effect. In *free carrier absorption*, the concentration of free carriers N in a given band is changed (modulated), for example, by an applied voltage, to change the extent of photon absorption. The absorption coefficient is proportional to N and to the light wavelength λ raised to some power, typically 2–3. In the confined Stark effect, the applied electric field modifies the energy levels in a quantum well. The energy levels are reduced by the field by an amount proportional to the square of the applied field. A multiple quantum well (MQW) *pin* device has MQWs in its intrinsic layer, between the *p* and *n* layers as shown in the figure. Without any applied bias, light with photon energy just less than the QW exciton excitation energy will not be significantly absorbed. When a field is applied, the energy levels are lowered and the incident photon energy is now sufficient to excite an electron and hole pair in the QWs. The relative transmission decreases with reverse bias V_r applied to the *pin* device. Such MQW *pin* devices are not usually very useful in the transmission mode because typically the substrate material is such that it absorbs the light (for example GaAs/AlGaAs MQW *pin* would be grown on a GaAs substrate which would absorb the radiation that excites the QWs). Thus, a reflector would be needed to reflect the light back before it reaches the substrate; such devices have indeed been fabricated.

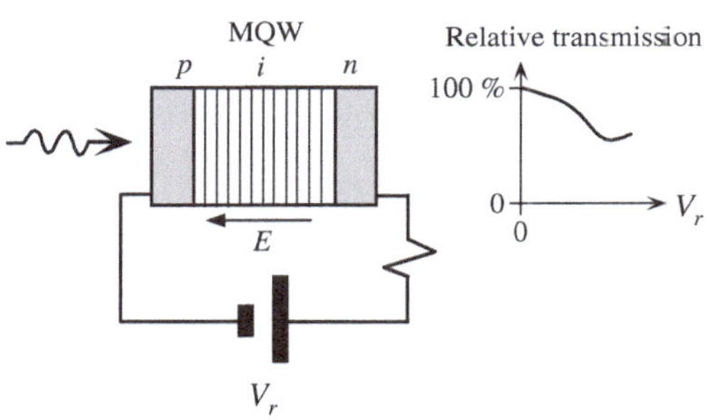

A schematic illustration of an **electroabsorption** modulator using the quantum confined Stark effect in MQWs. The *i*-region has MQWs. The transmitted light intensity can be modulated by the applied reverse bias to the *pin* device, because the electric field modifies the exciton energy in the QWs.

Electroluminescence see *luminescence.*

Electromagnetic (EM) wave is a traveling wave in which the electric displacement **D** and magnetic field **B** oscillations are perpendicular to each other and also to the direction of propagation (which is along **D** × **B**). The EM wave propagation is characterized by a *wavevector* **k** pointing in the direction of propagation whose magnitude is $2\pi/\lambda$, where λ is the wavelength (spatial periodicity of the wave) of the wave. In an EM wave, **D**, **B** and **k** are all perpendicular to each other. If the medium is isotropic and linear, then the polarization vector **P** and displacement **D** are in the same direction so that the electric field **E** is along **D** inside the medium, and **E**, **B** and **k** are all perpendicular to each other. The EM wave is a solution of Maxwell's wave equation subject to appropriate boundary conditions, which determine the form of the wave, that is, whether it is a plane, spherical, Gaussian or another type of wave. One very simple example is a *monochromatic plane polarized electromagnetic wave* that has harmonic electric and magnetic field variations with time and space, the angular frequency of the oscillations in time is well defined in terms of ω and the oscillations or periodicity in space is well defined in terms of the wavelength λ. Such a plane wave is of infinite extent. In an isotropic and linear medium, this monochromatic plane EM wave propagates with a *velocity*, called the *phase velocity*, $v = 1/(\varepsilon_0\varepsilon_r\mu_0)^{1/2}$ or $v = c/n$ where $n = \varepsilon_r^{1/2}$ is the refractive index, and ε_r is the relative permittivity.

The electric and magnetic field components in an EM wave are related. Consider a harmonic (sinusoidal) *plane* EM wave propagating along z with an electric field **E** or E_x and magnetic field **B** or B_y, a velocity v along z, and angular frequency ω. Then at all times and anywhere

$$\mathbf{k}\times\mathbf{E} = \omega\mathbf{B} \quad \text{and} \quad \mathbf{k}\times\mathbf{H} = \omega\mathbf{D}.$$

Since the wave velocity is $v = \omega/k$, more simply

$$E_x = vB_y = \frac{c}{n}B_y.$$

The two fields are simply and intimately related for an EM wave propagating in an isotropic medium. Any process that alters E_x also intimately changes B_y. The ratio of H to E is called the *impedance* Z of the dielectric medium, and given by $Z = (\mu_0/\varepsilon_0\varepsilon_r)^{1/2}$. For free-space, $Z = Z_0 = 377\ \Omega$. As the EM wave propagates in the direction of the

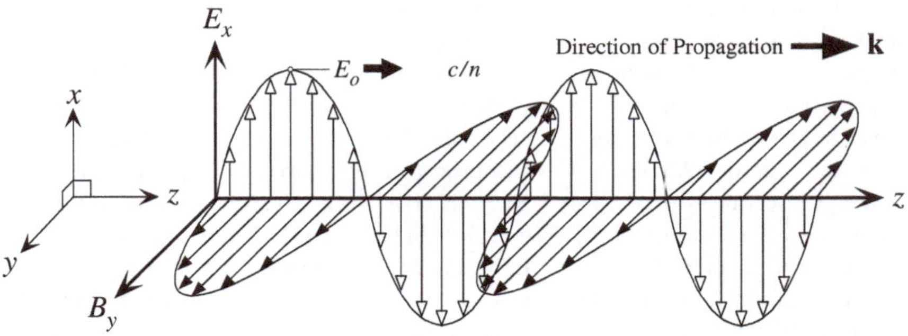

Homogeneous, isotropic and linear medium

An **electromagnetic wave** in a homogeneous, isotropic, and linear medium is a traveling wave which has time-varying electric and magnetic fields which are perpendicular to each other and the direction of propagation z. This is a snapshot at a given time of a particular EM wave in which the boundary conditions have resulted in a harmonic or a sinusoidal wave. At a time δt later, a point on the wave, such as the maximum field, would have moved a distance $v\delta t$ in the z-direction.

wavevector **k** as shown in the figure, there is an energy flow in this direction. The wave brings with it electromagnetic energy. A small region of space in which the electric field is E_x has an energy density, that is, energy per unit volume, given by $(^1/_2)\varepsilon_0\varepsilon_r E_x{}^2$. Similarly, a region of space where the magnetic field is B_y has an energy density $(^1/_2)B_y{}^2/\mu_o$. Since the two fields are related, the energy densities in the E_x and B_y fields are the same

$$\frac{1}{2}\varepsilon_0\varepsilon_r E_x^2 = \frac{1}{2\mu_o}B_y^2.$$

The total energy density in the wave is therefore $\varepsilon_0\varepsilon_r E_x{}^2$.

See *electromagnetic wave equation, plane electromagnetic wave.*

Electromagnetic (EM) wave equation is a second-order partial differential equation that is the result of combining Maxwell's equations for electromagnetism to obtain a single equation for the space and time dependence of the electric field E or the magnetic field B in a dielectric medium. Consider a nonmagnetic medium ($\mu_r = 1$) and assume (i) the medium is uniform or homogeneous, i.e. the relative permittivity ε_r is uniform (independent of x, y and z); (ii) ε_r is isotropic, i.e. the same in all directions of propagation; (iii) ε_r is independent of time; (iv) ε_r is independent of the field E, i.e. the medium is linear; (v) the conductivity $\sigma = 0$; (vi) the free charge density $\rho = 0$. Then the EM wave equation for the field E has the simplest form

$$\frac{\partial^2 E}{\partial x^2} + \frac{\partial^2 E}{\partial y^2} + \frac{\partial^2 E}{\partial z^2} - \varepsilon_0\varepsilon_r\mu_o\frac{\partial^2 E}{\partial t^2} = 0$$

or in vector form,

$$\nabla^2\mathbf{E} - \varepsilon_0\varepsilon_r\mu_o\frac{\partial^2\mathbf{E}}{\partial t^2} = 0.$$

There is a similar equation for the magnetic field **B**. The solution of the above EM wave equation with appropriate boundary conditions is usually an EM wave that propagates with a velocity $v = c/\varepsilon_r{}^{1/2}$. For fields that vary harmonically with time, $E = E(x, y, z)\exp(j\omega t)$, the wave equation becomes the *Helmholtz equation*

$$(\nabla^2 + k^2)E(x, y, z) = 0$$

where $k = \omega/v$ and represents the *wavevector* of the wave, or $k = 2\pi/\lambda$ where λ is the wavelength. If the medium has a finite conductivity, then $\mathbf{J} = \sigma\mathbf{E}$, and the new EM wave equation becomes

$$\nabla^2\mathbf{E} - \varepsilon_0\varepsilon_r\mu_0\frac{\partial^2\mathbf{E}}{\partial t^2} = \mu_0\sigma\frac{\partial\mathbf{E}}{\partial t}.$$

In the case of a wave with a harmonic time dependence, $\exp(j\omega t)$, the solution of the wave equation can be treated as in the case of a lossless medium but using a complex wavevector, complex relative permittivity $\varepsilon_r = \varepsilon_r' - j\varepsilon_r''$, and a complex refractive index, $N = n - jK$, where n and K are the real and imaginary parts. n determines the velocity of the wave through c/n and K is the extinction coefficient that determines the rate at which the field decays with distance z propagated. $K = \varepsilon_r''/2n$ depends on the imaginary part of ε_r, which is given by $\varepsilon_r'' = \sigma/\omega\varepsilon_0$.

The most general EM wave equation for $\mathbf{E}$ is quite complicated and involves source terms, that is, $\partial\mathbf{P}/\partial t$ and $\partial\mathbf{J}/\partial t$ that can change the field, i.e. in a nonmagnetic medium

$$\nabla \times (\nabla \times \mathbf{E}) + \varepsilon_0\mu_0\frac{\partial^2\mathbf{E}}{\partial t^2} = -\mu_0\frac{\partial^2\mathbf{P}}{\partial t^2} - \mu_0\frac{\partial\mathbf{J}}{\partial t}$$

the solution of which would explain the propagation of light in a polarizable medium (the $\mathbf{P}$ term) in which there are also free carriers that can drift (the $\mathbf{J}$ term). In a nonconducting medium, $\partial^2\mathbf{P}/\partial t^2$ accounts for such behavior as dispersion, refraction, double refraction, absorption, etc. To solve the general wave equation, we need to know how $\mathbf{P}$ and $\mathbf{J}$ depend on $\mathbf{E}$. In nonconducting crystals, $\mathbf{P} = \varepsilon_0\chi\mathbf{E}$ where χ is the electric susceptibility tensor. With χ given for a particular crystal, the wave equation can be solved; the solution can be a single traveling wave for $\mathbf{E}$ as in isotropic media or complicated with two solutions (two waves and hence double refraction) as in the case of anisotropic media. See *plane electromagnetic wave, spherical wave, complex refractive index, complex propagation constant, extinction coefficient*

Electromagnetically induced transparency (EIT) is a phenomenon in which a strongly absorbed light of frequency ν_{pump} causes a medium to become transparent to light of another frequency, ν_{probe}. (S. E. Harris, *Phys. Today*, **50**, 36, 1997; C. L. G. Alzar, M. A. G. Martinez, P. Nussenzveiga, *Am. J. Phys.*, **70**, 37, 2003.)

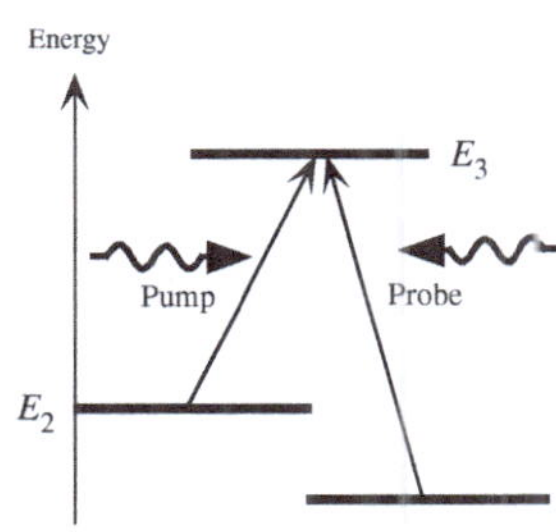

Electromagnetically induced transparency Three energy levels forming a Λ configuration can exhibit EIT.

Electron affinity, χ, is the energy required to remove an electron from the bottom of the conduction band to the vacuum level where the electron is free; it should not be confused with the work function Φ which is the average energy required to remove an electron from the semiconductor and corresponds to the energy difference between the vacuum level and the Fermi level. The electron affinity at the surface of the semiconductor may be different from in the bulk due to surface states and hence band bending that occurs near the surface.

Table: Electron affinity Electron affinities (χ) of selected semiconductors (approximate values from various sources).

	AlAs	AlSb	GaP	GaAs	GaSb	InP	InAs	InSb	Si	Ge	ZnSe	ZnTe	CdS	CdSe	CdTe
E_g (eV)	2.168	1.58	2.26	1.42	0.726	1.351	0.360	0.172	1.12	0.66	2.67	2.26	2.42	1.7	1.56
χ (eV)	3.50	3.64	4.3	4.07	4.06	4.37	4.90	4.59	4.01	4.13	4.09	3.5	4.5	4.95	4.28

Electron–hole pair (EHP) creation energy or the **ionization energy** represents the sensitivity of a high-energy particle or x-ray radiation detector in terms of the energy absorbed from the incident radiation per free electron–hole pair that is collected. Stated differently, it is the energy that the detector material absorbs from the incident beam to generate a free electron and a free hole (or one electron of collected charge in the external circuit). High sensitivity means that the amount of radiation energy required, denoted as $W_\pm$, to create a single free EHP must be as low as possible because the free (or collectable) charge ΔQ generated from an incident radiation of energy ΔE is simply $e\Delta E/W_\pm$. All experiments on examining the amount of charge generated by high-energy particles, including x-rays, have shown that this charge depends on the absorbed radiation energy, i.e. $\Delta Q \propto \Delta E$. This has led to the introduction of an EHP creation energy ($W_\pm$). The creation of EHPs by an x-ray photon first involves the generation of an energetic primary electron from an inner core shell, for example, the K-shell. As this energetic photoelectron travels in the solid it causes ionizations along its track and hence the creation of many EHPs. For many semiconductors the energy $W_\pm$ required to create an electron–hole pair has been shown to depend on the energy bandgap E_g via the *Klein rule*

$$W_\pm \approx 2.8E_g + E_{\text{phonon}}$$

where E_{phonon} is a small energy term that involves many phonons. E_{phonon} is expected to be small (0.05–0.5 eV) so that typically $W_\pm$ is close to $2.8E_g$. Further, in many crystalline semiconductors $W_\pm$ is field-independent and well-defined. This $W_\pm$ is so well-defined in crystalline semiconductors, such as high-purity Ge, ZnCdTe, for example, that they are used in spectrometers to measure the energy of the incident x-rays or high-energy charged particles. Que and Rowlands argued that if one relaxes the conservation of **k** rule as required for amorphous semiconductors then the EHP creation energy should be about $2.2E_g + E_{\text{phonon}}$ where the latter is again a small phonon energy term.

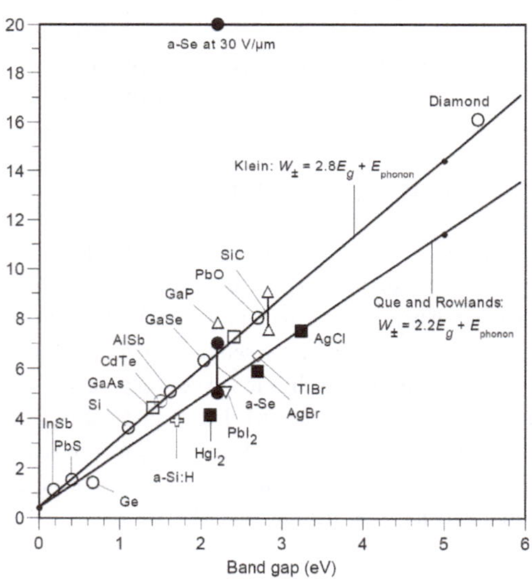

Electron–hole pair creation energy $W_\pm$ vs. energy bandgap for semiconductors (for a-Se, $W_\pm$ is field-dependent and the value plotted is extrapolated to very large fields so that $W_\pm$ is the "intrinsic" or saturated EHP creation energy).

Electronegativity is a relative measure of the ability of an atom to attract electrons to form an anion. Fluorine is the most highly electronegative atom as it most easily accepts an electron to become an anion (F^-). Typically halogens (F, Cl, Br, I) are electronegative elements. In HCl, for example, the Cl atom is more electronegative than the H atom and therefore attracts the electrons more than the H proton. Cl therefore acquires a net negative charge and the hydrogen proton becomes exposed. The molecule therefore has a permanent dipole moment.

Electronic polarization is the displacement of the electron cloud of an atom with respect to the positive nucleus. The polarization involves the displacement of electrons and hence it is effective up to very high frequencies, usually into the visible or UV range of frequencies depending on the material. If an applied oscillating field E induces, by electronic polarization, an electric dipole moment p at an atom (or a molecule), then $p = \alpha_e E$, where α_e is the *electronic polarizability* that gauges how easy it is to displace the electrons with respect to the positive nucleus. In the presence of an electromagnetic wave of frequency ω, the simplest Lorentz model predicts

$$\alpha_e(\omega) = \frac{Ze^2}{m_e(\omega_0^2 - \omega^2 + j\gamma\omega)}$$

where Z is the number of electrons in the atom that are involved in the polarization, γ is a constant that represents the losses per electron, and ω_0 is the electronic polarization resonance frequency where the losses peak. A better classical treatment would consider the electrons separately and sum their contributions to the electronic polarizability.

Electrooptic effects refer to changes in the refractive index of a material induced by the application of an external electric field, which therefore "modulates" the optical properties; the applied field is not the electric field of any light wave, but a separate external field. We can apply such an external field by placing electrodes on opposite faces of a crystal and connecting these electrodes to a battery. The presence of such a field distorts the electron motions in the atoms or molecules of the substance, or distorts the crystal structure resulting in changes in the optical properties. For example, an applied external field can cause an optically isotropic crystal such as GaAs to become birefringent. In this case, the field induces principal axes and an optic axis. Typically changes in the refractive index are small. The frequency of the applied field has to be such that the field appears static over the time-scale it takes for the medium to change its properties, that is respond, as well as for any light to cross the substance. The electrooptic effects are classified according to first- and second-order effects. If we were to take the refractive index n to be a function of the applied electric field E, that is $n = n(E)$, we can of course expand this as a Taylor series in E. The new refractive index n' is

$$n' = n + a_1 E + a_2 E^2 + \cdots$$

where the coefficients a_1 and a_2 are called the *linear* electrooptic effect and *second*-order electrooptic effect coefficients. Although we would expect even higher terms in the expansion of the above equation, these are generally very small and their effects negligible within highest practical fields. The change in n due to the first E term is called the *Pockels effect*. The change in n due to the second E^2 term is called the *Kerr effect*, and the coefficient a_2 is generally written as λK where K is called the *Kerr coefficient*. Thus, the two effects are

$$\Delta n = a_1 E$$

$$\Delta n = a_2 E^2 = (\lambda K) E^2.$$

All materials exhibit the Kerr effect. It may be thought that we will always find some (nonzero) values for a_1 for all materials but this is not true and only certain crystalline materials exhibit the Pockels effect. If we apply a field $\mathbf{E}$ in one direction to a crystal and then reverse the field and apply $-\mathbf{E}$ then Δn should change sign. If the refractive index increases for $\mathbf{E}$, it must decrease for $-\mathbf{E}$. Reversing the field should *not* lead to an identical effect (the same Δn). The structure has to respond differently to $\mathbf{E}$ and $-\mathbf{E}$. There must therefore be some *asymmetry* in the structure to distinguish between $\mathbf{E}$ and $-\mathbf{E}$. In a noncrystalline material, Δn for $\mathbf{E}$ would be the same as Δn for $-\mathbf{E}$ as all directions are equivalent in terms of dielectric properties. Thus $a_1 = 0$ for all noncrystalline materials (such as glasses and liquids). Similarly, if the crystal structure has a center of symmetry then reversing the field direction has an identical effect and a_1 is again zero. Only crystals that are noncentrosymmetric exhibit the Pockels effect. For example a NaCl crystal (centrosymmetric) exhibits no Pockels effect but a GaAs crystal (noncentrosymmetric) does. See *Kerr effect, Pockels effect, nonlinear effects.*

Ellipsometry measures changes in the polarization of light incident on a sample to sensitively characterize surfaces and thin films, for example. Interactions of incident polarized light with the sample cause a polarization change in the light which may then be measured by analyzing light reflected from the sample. For plane polarized light incident on the sample, it can be resolved into two components – the component parallel to the plane of incidence or *p*-component, and the component perpendicular to it, or *s*-component. That is, the incident light can be resolved into the components $\mathbf{E}_p$ (incident) and $\mathbf{E}_s$ (incident), where **E** refers to the respective electric field in the light wave – similarly, the reflected light can be resolved into components $\mathbf{E}_p$ (reflected) and $\mathbf{E}_s$ (reflected), resulting in the so-called Fresnel reflection coefficients, $R_p = \mathbf{E}_p$ (reflected)/$\mathbf{E}_p$ (incident) and $R_s = \mathbf{E}_s$ (reflected)/$\mathbf{E}_s$ (incident), for *p*- and *s*-components, respectively. Ellipsometry measures ρ, the ratio of these two coefficients,

$$\rho = (R_p/R_s) = e^{j\Delta}\tan(\Psi)$$

where ρ is expressed in terms of so-called ellipsometric angles Ψ ($0°\leq\Psi\leq90°$) and Δ ($0°\leq\Delta\leq360°$). These angles are defined by $\Psi = \tan^{-1}|\rho|$ and the differential phase change, $\Delta = \Delta_p - \Delta_s$. Thus, ellipsometry measures a change in polarization, expressed as Ψ and Δ, in order to characterize materials. Because ellipsometry measures the ratio of two values, it can be made to be highly accurate and reproducible. The ratio ρ is a complex number, thus it contains "phase" information, Δ, which makes the measurement very sensitive. However, establishing values for Ψ and Δ is not particularly useful in itself for sample characterization. What one really wants to determine is parameters of the sample including, for example, film thickness, optical constants, and refractive index. These characteristics can be found by using the measured values of Ψ and Δ in an appropriate model describing the interaction of light with the sample. As an example, consider light reflecting off an optically absorbing sample in air (i.e. with a refractive index of unity). The sample can be characterized by a complex refractive index $n - jk$, where n and k are the sample's refractive index and extinction coefficients at a particular wavelength; from Fresnel's equations, assuming that the light is incident at an angle Φ to the sample normal,

$$n^2 - k^2 = \sin^2(\phi)[1 + \{\tan^2(\phi)[\cos^2(2\Psi) - \sin^2(2\Psi)\sin^2(\Delta)]/ [1 + \sin(2\Psi)\cos(\Delta)]^2\}]$$

and

$$2nk = \sin^2(\phi)\tan^2(\phi)\sin(4\Psi)\sin(\Delta)/ [1 + \sin(2\Psi)\cos(\Delta)]^2.$$

Since the angle ϕ is set in the experiment, the two parameters measured from the experiment, Ψ and Δ can be used to deduce the remaining two unknown variables in the equation above – namely, n and k. For a given model (i.e. set of equations used to describe the sample), the mean squared error between the model and measured Ψ and Δ values is minimized, typically using the Marquardt–Levenberg algorithm, to quickly determine the minimum or best fit within some predetermined confidence limits. Thus, n and k values are established using this procedure. In practice, ellipsometers consist of the following basic components: a source of linearly polarized light, polarization optics and a detector. There are a number of different approaches to conducting an ellipsometry experiment, including null ellipsometers, rotating analyzer/compensator ellipsometers and spectroscopic ellipsometers. Each of these approaches is discussed briefly below.

Historically, the first ellipsometers that were developed where null ellipsometers. In this configuration, the orientation of the polarizer, compensator, and analyzer are adjusted such that the light incident on the detector is extinguished or "nulled." It should be noted that there are 32 combinations of polarizer, compensator and analyzer angles that can result in a given pair of Ψ and Δ values. However, because any two angles of the polarizer, compensator and analyzer which are 180° apart are optically equivalent, only 16 combinations of angles need to be considered if all angles are restricted to values below 180°. Even when automated, this approach is thus inherently slow and spectroscopic measurements are difficult to make. However, this configuration can be very accurate and has low systematic errors. In order to circumvent the problems of lack of measurement speed for null ellipsometers, rotating analyzer/polarizer ellipsometers were developed. In these systems, either the analyzer or polarizer is continuously rotated at a constant angular velocity (typically about 100 Hz) about the optical axis. The operating characteristics of both of these configurations are similar. However, the rotating polarizer system requires the light source to be totally unpolarized. Any residual polarization in the source results in a source of measurement error unless corrected. Similarly, a rotating analyzer system is susceptible to polarization sensitivity of the detector. However, solid state semiconductor photodetectors have extremely small polarization sensitivity. Thus, commercial systems tend to use rotating analyzer systems where residual polarization in the source is not an issue. In this case, the amplitude variation of sinusoidal variation in the detector signal output may be directly related to the ellipticity of the reflected light – Fourier analysis of this output provides values for Ψ and Δ. Such systems can provide high speed and accuracy measurements and the lack of a compensator actually improves the measurement by eliminating errors associated with these components. Spectroscopic ellipsometers extend the concepts developed for measurement at a single wavelength to measurements at multiple wavelengths – typically as many as 40 wavelengths. Being able to measure the dispersion in the optical constants with wavelength adds another dimension to the analysis providing for an unambiguous determination of material and structure parameters.

Elliptical reflecting cavity is an enclosed reflecting surface whose cross-section is an ellipse so that the light source, usually a flash tube, is placed at one focus, F_1, and the laser rod (or some other object to be illuminated) is placed at the other focus, F_2, to optically pump the laser rod. The light rays emitted from a focal point F_1 of an elliptical reflector pass through the other focal point F_2. The *elliptical reflector*, in contrast to the elliptical reflecting cavity, does not form a closed cavity and is an elliptically shaped reflecting mirror. Elliptical reflecting cavities are used in the efficient pumping of solid state laser rods: the flash light tube is placed at F_1, and the rod at F_2. The radiation intensity distribution across the rod, however, is not uniform – it is highest near the center of the rod (close to the F_2-point) and is least at its surfaces. (The exact intensity distribution depends on whether the rod

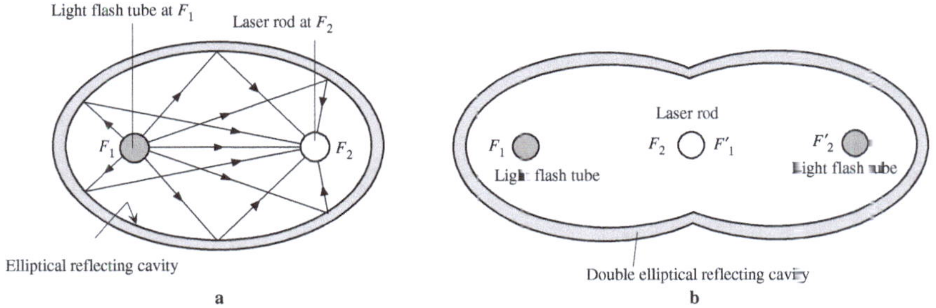

Elliptical reflecting cavity.

is weakly or strongly absorbing.) The radiation can be made more uniform across the rod thickness by using a double elliptical cavity and placing the laser rod at the shared focal point and the flash tubes at the outer focal points. Even quadruple elliptical cavities are possible.

Elliptically polarized, or **elliptic**, **light** has the tip of the electric field **E**-vector trace out an ellipse as the wave propagates through a given location in space. As in circular polarization, light can be *right* and *left elliptically polarized* depending on the clockwise or counterclockwise rotation of the **E**-vector as seen by the receiver of the wave. Consider a wave propagating along z. When the components E_x and E_y of the electric field have different amplitudes ($E_x \neq E_y$), then the resultant electric field **E** will scribe an ellipse.

We can write the x- and y-components of the field at any time generally as

$$E_x = E_{xo}\cos(\omega t - kz) \quad \text{and}$$
$$E_y = E_{yo}\cos(\omega t - kz + \phi)$$

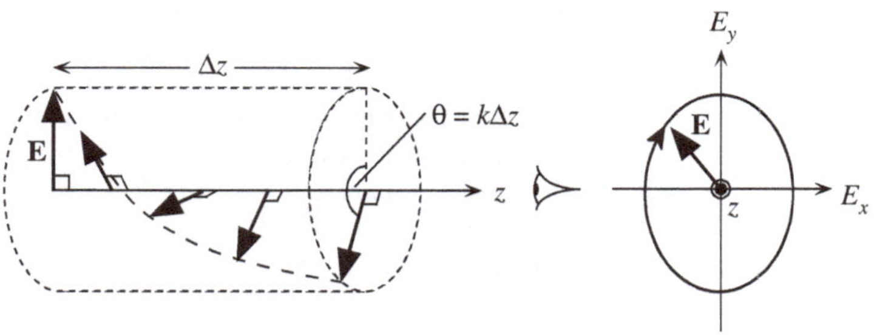

Right **elliptically polarized light** The field vector **E** is always at right angles to z (direction of propagation), rotates clockwise around z with time, and traces out a full ellipse over one wavelength of distance propagated.

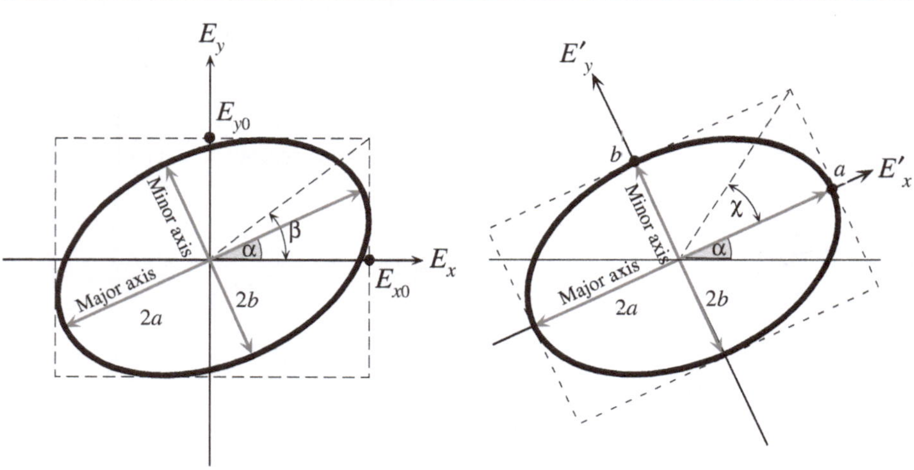

Elliptically polarized light Left: The polarization ellipse is tilted at an angle α to E_x and E_y. Right: By rotating the axes by α from E_x and E_y to E'_x and E'_y we can align the major axis to lie along E'_x.

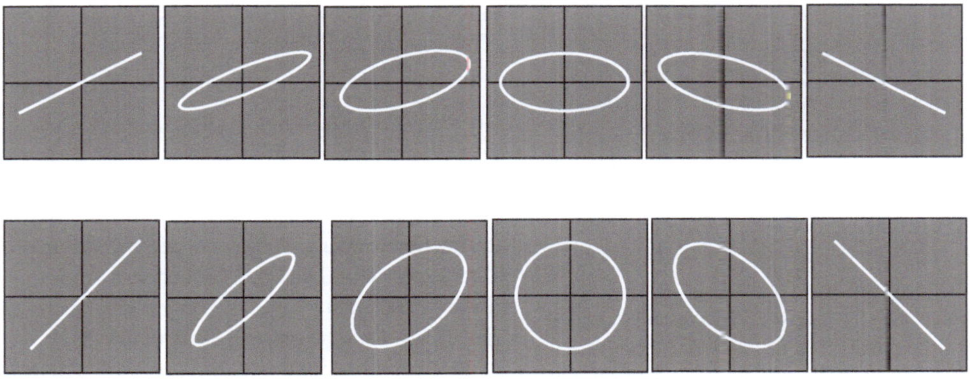

Elliptically polarized E_y vs. E_x for two ratios of $E_{xo}/E_{yo} = 2$ and 1, and various ϕ from 0 to 180°.

where E_{x0} and E_{y0} are the amplitudes of the electric fields along the x- and y-axes, k is the propagation vector along z, ωt is the time-varying component and ϕ is the phase difference between the two fields. E_x, E_y and ϕ at any instant and location then satisfy

$$\left(\frac{E_x}{E_{xo}}\right)^2 + \left(\frac{E_y}{E_{yo}}\right)^2 - 2\left(\frac{E_x}{E_{xo}}\right)\left(\frac{E_y}{E_{yo}}\right)\cos\phi = \sin^2\phi$$

which is the equation of an ellipse, called the *polarization ellipse*, on the E_y vs. E_x axes. The above equation is quadratic in E_y (or E_x), so that for every value of E_x, we can solve for E_y and then plot E_y vs. E_x.

$$\left(\frac{1}{E_{yo}}\right)^2 E_y^2 - \left[2\left(\frac{E_x}{E_{xo}}\right)\left(\frac{1}{E_{yo}}\right)\cos\phi\right]E_y + \left[\left(\frac{E_x}{E_{xo}}\right)^2 - \sin^2\phi\right] = 0.$$

The orientation α (the so-called *polarization angle*) of the polarization ellipse's major axis to E_x is given by

$$\tan(2\alpha) = \frac{2E_{xo}E_{yo}\cos\phi}{E_{xo}^2 - E_{yo}^2}.$$

The new axes E'_x and E'_y are given by $E'_x = E_x\cos\alpha + E_y\sin\alpha$ and $E'_y = E_y\cos\alpha - E_x\sin\alpha$. The length of the major axis of the polarization ellipse is $2a$ and that of the minor axis is $2b$. The ellipse must fit into a rectangular box $2a \times 2b$. The *axial ratio* or *axis-ratio* r is defined by $r = b/a$. The *ellipticity* is $\varepsilon = (a-b)/a$. The *ellipticity angle* χ is the angle of the diagonal of the rectangle $2a \times 2b$; $\tan\chi = b/a$. r is given by

$$\frac{r}{1+r^2} = \pm\frac{E_{xo}E_{yo}}{E_{xo}^2 + E_{yo}^2}\sin\phi$$

in which the sign depends on the sense of rotation of **E**. a and b are

$$\frac{1}{a^2} = \frac{\cos^2\alpha}{E_{xo}^2} + \frac{\sin^2\alpha}{E_{yo}^2} - \frac{\sin 2\alpha\cos\phi}{E_{xo}E_{yo}}$$

and

$$\frac{1}{b^2} = \frac{\sin^2\alpha}{E_{xo}^2} + \frac{\cos^2\alpha}{E_{yo}^2} + \frac{\sin 2\alpha\cos\phi}{E_{xo}E_{yo}}.$$

Emissivity usually refers to *total emissivity* which is a surface property that characterizes the relative amount of radiation that is radiated from a given surface compared with that emitted from an ideal black body at the same temperature, and over all wavelengths. (In practice, the comparison is over the same wavelength range of interest.) Consider a surface at a temperature T in an ambient temperature of T_0. If this surface were a black body then it would emit radiation energy summed over all directions at a rate $E_{\text{blackbody}} = \sigma_S T^4$ per *unit surface area* (in Watts m^{-2}), where σ_S is Stefan's constant ($= 5.67 \times 10^{-8}$ W m^{-2} K^{-4}). E is the usual symbol used in radiometry for the radiated energy per unit time per unit surface area summed over all directions (summed over all angles of emission), which is called the *total hemispherical emissive power*, though emissive power is usually understood to mean total hemispherical quantity. The same black body surface would absorb all radiation energy falling on it from the ambient. This is measured by *total hemispherical irradiation*, the rate at which energy is incident per unit area on a surface from all directions, and denoted by G. For a black body, $G_{\text{blackbody}} = \sigma_S T_0{}^4$ (in Watts m^{-2}). When the black body is in equilibrium with its environment, it emits as much radiation as it receives, $E_{\text{blackbody}} = G_{\text{blackbody}}$ and $T = T_o$. The rate of radiation energy emission from a *real* surface is characterized by the *total emissivity* ε defined by

$$E_{\text{real}} = \varepsilon E_{\text{blackbody}} = \varepsilon \sigma_S T^4.$$

The rate of absorption of radiation per unit area by a real surface is characterized by the *total absorptivity* α defined by

$$\text{Rate of radiation absorbed per unit area} = G_{\text{absobed}} = \alpha G_{\text{blackbody}} = \alpha \sigma_S T_0^4.$$

In thermal equilibrium, $T = T_0$, the surface emits as much radiation as it absorbs, $G_{\text{absorbed}} = E_{\text{real}}$, and thus $\varepsilon = \alpha$, which is *Kirchhoff's law*. If P_{radiated} is the actual *net* radiated power by a real surface then

$$P_{\text{radiated}} = \varepsilon \sigma_S S \left(T^4 - T_0^4\right)$$

where S is the surface area. If the surface is cooler than the ambient, T is below T_0, then the surface absorbs radiation energy from the ambient. *Spectral emissivity* ε_λ and *spectral absorptivity* α_λ refer to the fractional radiation power emitted and absorbed with respect to the black body radiation at a particular wavelength over a very narrow wavelength range $\delta\lambda$. If E_λ is the *spectral emissive power*, i.e. emitted radiation power summed over all directions per unit surface area and per unit wavelength, then *monochromatic emissivity*, also called the *spectral emissivity*, ε_λ, is defined as

$$E_{\lambda,\text{real}}\delta\lambda = \varepsilon_\lambda E_{\lambda,\text{blackbody}}\delta\lambda$$

i.e.

$$E_{\text{real}} = \int_0^\infty \varepsilon_\lambda(\lambda, T) E_{\lambda,\text{blackbody}} d\lambda$$

$$= 2\pi hc^2 \int_0^\infty \frac{\varepsilon_\lambda(\lambda, T) d\lambda}{\lambda^5[\exp(hc/\lambda kT) - 1]}.$$

The emissivity ε of a surface is difficult to determine exactly because it depends strongly on the surface condition of the body. For example, polished cast iron has an ε of about 0.2 whereas cast iron with surface oxidized has an ε in the range 0.6–0.7.

The total emissivity defined above is some average emissivity over all the wavelengths and

Table: Emissivity Typical emissivities. Actual values may vary substantially.

Surface	ε	Surface	ε	Surface	ε
Al, foil	0.04	Brass, dull	0.20	Iron, cast polished	0.21
Al, oxidized at 316 °C	0.05	Brick, red rough	0.93	Iron, cast oxidized	0.60–0.70
Al, anodized	0.2–0.8	Brick, building	0.45	Iron, completely rusted	0.69
Al, oxidized, black, heat sink	0.9	Chromium, polished	0.08–0.27	Iron, wrought, oxidized	0.94
Asbestos	0.93–0.96	Concrete tile	0.63	Silver, polished	0.03
Beryllium	0.16–0.30	Concrete, rough	0.94	Tungsten filament	0.35–0.39
Black paint, nonglossy	0.8–0.96	Glass	0.90	Water	0.94
Brass, polished	0.03–0.1	Gold coating	0.05–0.1	White paint, nonglossy	0.8–0.9

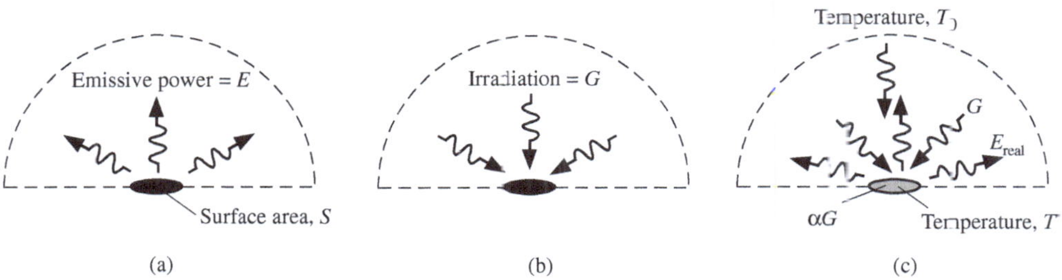

Emissivity (a) Emissive power; (b) irradiation G and (c) the real surface at a temperature T different from the ambient temperature T_o. Net emitted radiation power is $\varepsilon\sigma_S S(T^4 - T_o^4)$

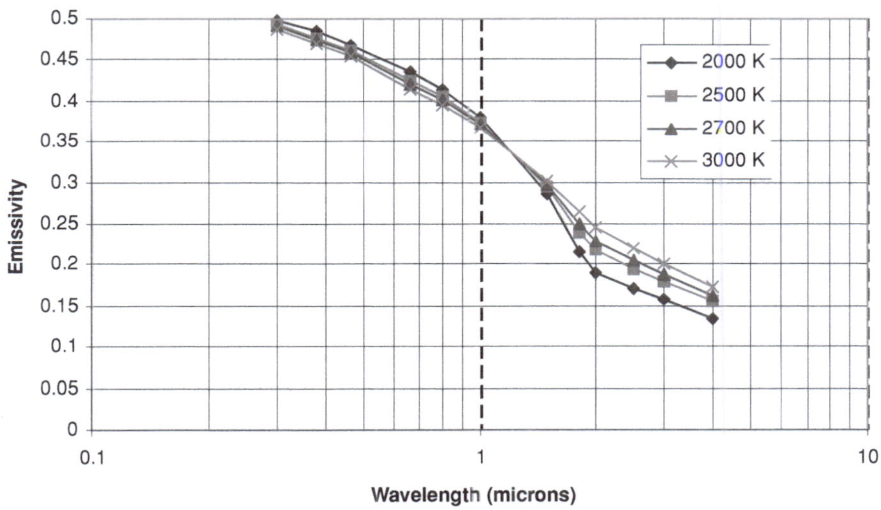

Emissivity Monochromatic emissivity vs. wavelength for a tungsten filament at various temperatures (from R. Bergman, L. Bigio and J. Ranish, "Filament Lamps", *GE Technical Information Series*, Report 98CRD027, February 1998)

at the temperature of interest. Emissivity normally depends on the temperature. This ε vs. T dependence can be quite strong. For example, polished Cr has an ε of 0.08 at 38 °C but an ε of 0.36 at 1093 °C. For tungsten, $\varepsilon \approx (8.91 \times 10^{-5})T^{1.028}$.

See *black body, Kirchhoff's law, emissive power, irradiation, radiosity, absorptivity*.

Emittance see *exitance*.

Emitter in optoelectronics and photonics refers to a device that emits electromagnetic (EM) radiation, that is optical power, such as an LED, laser diode, etc.

Emitter of a bipolar transistor is one of the two similarly doped (e.g. both *n*-type) regions that surrounds the oppositely doped (*p*-type) base and injects minority carriers into the base when the emitter–base junction is forward biased. When the emitter–base junction is forward biased, it emits minority carriers into the base, which then diffuse through the base and become collected in the collector since the base-collector is normally reverse biased.

Energy of the electron in a crystal whether in the conduction band (CB) or valence band (VB), depends on its momentum $\hbar k$ through the *E–k* behavior determined by the solutions to the Schrödinger equation for the particular crystal structure (with a particular periodic potential energy). *E–k* behavior is most conveniently represented graphically through *E–k* diagrams. For example, for an electron at the bottom of the CB, E increases as $(\hbar k)^2/(2m_e^*)$ where $\hbar k$ is the momentum and m_e^* is the effective mass of the electron which is determined from the *E–k* behavior. The quantity $\hbar k$ is called the *crystal momentum* of the electron because the rate of change of $\hbar k$ is the *external applied force* on the electron, $d(\hbar k)/dt = F_{\text{external}}$, and not the actual force which is the sum of external and internal forces. *E–k* diagrams are essentially plots of electron energy vs. crystal momentum of the electron.

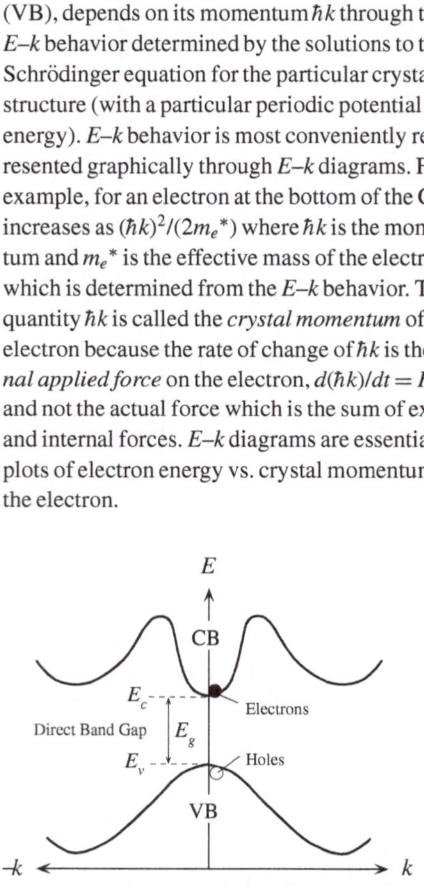

Energy of the electron in a crystal *E–k* diagram for a direct bandgap semiconductor such as GaAs. The minimum of the CB is directly above the maximum of the VB.

Epitaxial layer is typically a thin layer of near perfect crystal grown on a crystalline substrate of the same crystal structure (Greek: *epis*-layered and *taxis*-ordered). The term *epitaxy* is usually defined as the growth of a layer of single crystal on top of a single crystal substrate in such a way that the crystalline structure of the substrate is preserved in the new layer. *Homoepitaxy* refers to epitaxial growth where the layer and the substrate are of the same material, such as a layer of GaAs grown on a GaAs substrate. *Heteroepitaxy* is a method in which the layer material differs from that of the substrate material, such as the case of a InGaAsP layer grown on a GaAs substrate. The normal notation used to describe epitaxial growth is to write the epilayer chemical formula first, separated by the symbol "/" from the substrate chemical formula; for example for the growth of InGaAsP on an InP substrate we write, InGaAsP/InP. Apart from the requirement that both the epilayer and the substrate have the same crystallographic space group, the lattice parameters for both substrate and epilayer must be closely matched. This is due to the fact that any difference between the lattice parameters will require the lattice planes on either side of the interface to bend to accommodate the differences. The amount of lattice plane bending and how much of it can be tolerated will depend on the elastic stiffness of the substrate and layer, as well as the actual layer thickness grown. Thus the difference in the lattice parameters results in a strain at the interface between the layer and substrate commonly referred to as the *mismatch strain*. Beyond a critical mismatch strain, the planes at either side of the interface can no longer deform together or *coherently*, resulting in the formation of *misfit* dislocations. These defects are often highly detrimental in device applications, since they can act as nonradiative recombination centers or deep traps. For some epitaxial growth techniques, such as *molecular beam epitaxy* (MBE) *and metal-organic chemical vapour deposition* (MOCVD), growth occurs under nonequilibrium conditions, whereas for *liquid phase epitaxy* (LPE), growth occurs under near-equilibrium conditions. Nonequilibrium growth techniques such as MOCVD and MBE, can be used for the growth of single-crystal epilayers on substrates with large lattice mismatch (i.e. mismatch strain $\varepsilon > 10^{-2}$) such as GaAs/Si for example.

Equalizer for EDFA (erbium-doped fiber amplifier) is a photonic component, such as a filter, that is connected either before or after an erbium-doped fiber amplifier to flatten its gain vs. wavelength characteristics. The equalizer spectral transmission characteristics are such that the overall transmission characteristics of the EDFA and the equalizer are relatively uniform over the wavelengths of interest such as over all the channels in wavelength division multiplexing. Various techniques have been used to achieve gain flattening including Mach–Zehnder optical interferometer filters, acoustooptic filters, fiber loop mirror filters, fiber Bragg gratings, long-period fiber gratings, multilayer thin film filters, among others. Dynamic gain equalizers (DGE) use the output spectrum from the fiber amplifier in a feedback loop to a variable filter at the input of the EDFA to flatten the overall gain vs. wavelength characteristics. It is an active optical device that uses feedback to achieve the required overall spectral response. The variable filter is dynamically adjusted by tapping into the

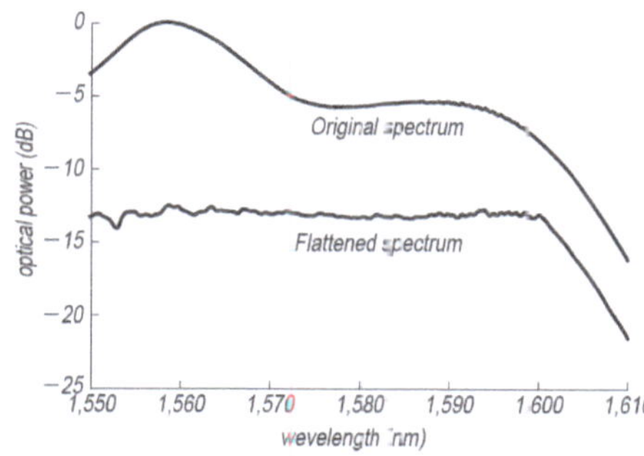

Equalizer for EDFA Overall spectral transmission characteristics of an EDFA and an EDFA with a *dynamic gain equalizer* at its input. (Data from K. Maru *et al.*, *Hitachi Cable Review*, No. 21 (August 2002) 7–10.)

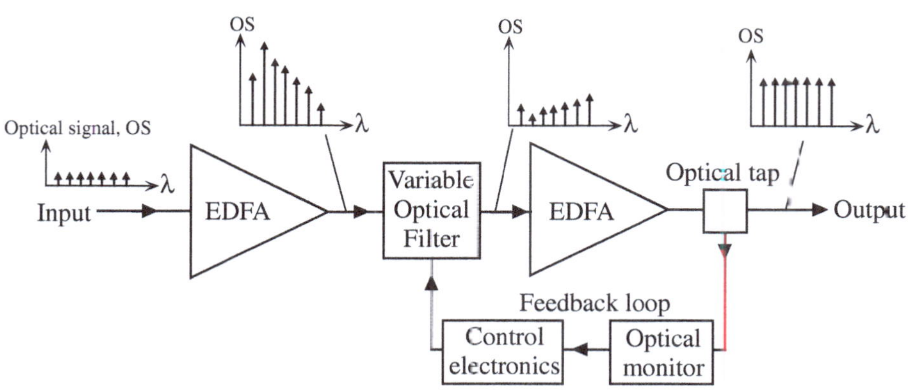

Equalizer for EDFA Schematic illustration of a dynamic gain equalizer inserted into a wavelength division multiplexed communications system to flatten the spectral response of the optical amplification system that has two erbium-doped fiber amplifiers (EDFAs). The variable optical filter modifies the input spectrum into the second EDFA so that the output from the second EDFA is flat. An optical feedback control system is used to adjust the characteristics of the variable optical filter.

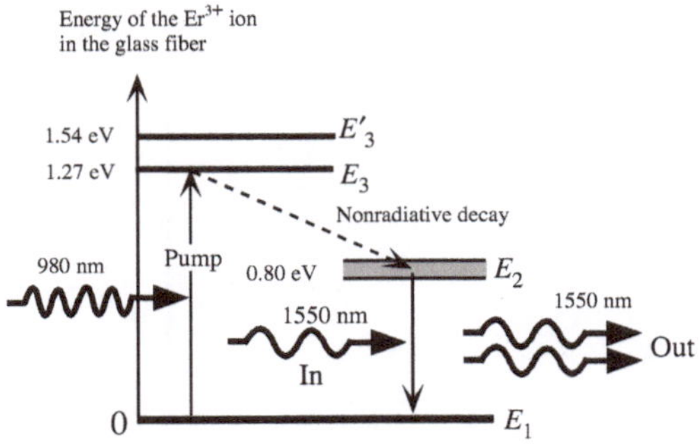

Erbium doped fiber amplifier Energy diagram for the Er^{3+} ion in the glass fiber medium and light amplification by stimulated emission from E_2 to E_1. Dashed arrows indicate radiationless transitions (energy emission by lattice vibrations). A simplified schematic illustration of an **EDFA** (optical amplifier). The erbium-ion doped fiber is pumped by feeding the light from a laser pump diode, through a coupler, into the erbium ion doped fiber.

output spectral response of the EDFA to obtain a flattened gain spectrum.

Erbium (rare earth) doped fiber amplifier (EDFA) is an optical amplifier that consists of a fiber whose core has been doped with Er^{3+} (or other rare earth ions) so that, with suitable optical pumping, the Er^{3+} ions can be population-inverted to enable the fiber to have an optical gain by virtue of stimulated emissions. In highly simplified terms, the principle of operation is as follows. When the Er^{3+} ion is implanted in the host glass material it has a ground energy of E_1, the lowest energy possible for the Er^{3+} ion. There are two convenient energy levels for optically pumping the Er^{3+} ion which are at approximately 1.27 eV and 1.54 eV above the ground energy level E_3 and E'_3, respectively. The Er^{3+} ions are optically pumped, usually from a laser diode, to excite them to E_3. The wavelength for this pumping is about 980 nm. The Er^{3+} ions decay rapidly from E'_3 and E_3 and from there to a *long-lived* energy level at E_2 which has a long lifetime of about ~10 ms (very long on the atomic scale). The decays from E'_3 to E_3 and to E_2 involve energy losses by radiationless transitions (phonon emissions) and are very rapid. Thus more and more Er^{3+} ions accumulate at E_2 which is 0.80 eV above the ground energy. The accumulation of Er^{3+} ions at E_2 leads to a population inversion between E_2 and E_1. Signal photons at 1550 nm have an energy of 0.80 eV, or $E_2 - E_1$, and give rise to *stimulated transitions* of Er^{3+} ions from E_2 to E_1. Any Er^{3+} ions left at E_1, however, will *absorb* the incoming 1550 nm photons to reach E_2. To achieve light amplification we must therefore have stimulated emission exceeding absorption. This is only possible if there are

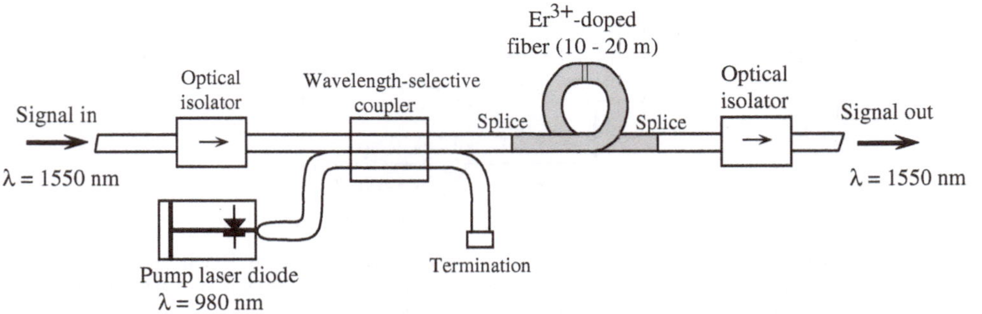

Erbium doped fiber amplifier A simplified schematic illustration of an EDFA (optical amplifier). The erbium-ion doped fiber is pumped by feeding the light from a laser pump diode, through a coupler, into the erbium-ion doped fiber.

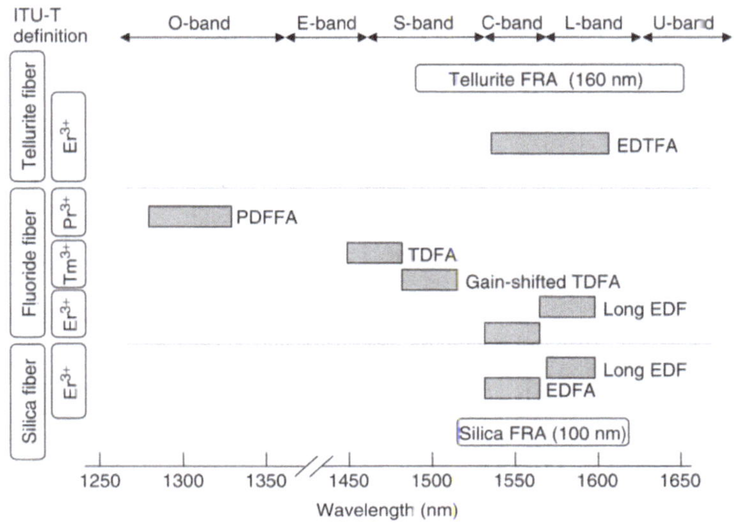

Erbium doped fiber amplifier Summary of various rare earth doped fiber amplifiers and their gain bandwidths within optical communication channels. EDTFA is a Er^{3+} doped tellurite fiber amplifier, PDFFA is a Pr^{3+} doped fluoride fiber amplifier, etc. (T. Mukai, *Amplifiers, Optical, The Optics Encyclopedia, Volume 1*, pp. 1–26, 2003. © Wiley-VCH Verlag GmbH & Co. KGaA. Reproduced with permission.)

more Er^{3+} ions at the E_2 level than at the E_1 level; if we have population inversion. If N_2 and N_1 are the number of Er^{3+} ions at E_2 and E_1 then it is clear that the difference between stimulated emission (from E_2 to E_1) and absorption (E_1 to E_2) controls the net optical gain G_{op},

$$G_{op} = \exp[L(\sigma_{em}N_2 - \sigma_{ab}N_1)]$$

where σ_{em} and σ_{ab} are the effective emission and absorption cross sections and L is the length of the pumped fiber. In practice the erbium doped fiber is inserted into the fiber communications line by splicing as shown schematically, and it is pumped from a laser diode through a coupling fiber arrangement which allows only the pumping wavelength to be coupled. Some of the Er^{3+} ions at E_2 will decay spontaneously from E_2 to E_1 which will give rise to unwanted noise in the amplified light signal.

Further, if the EDFA is not pumped at any time it will act as an attenuator as the 1550 nm photons will be absorbed by Er^{3+} ions which will become excited from E_1 to E_2. In returning back to E_1 by spontaneous emission they will emit 1550 nm photons randomly and not along the fiber axis. Although the Er^{3+} ions can also be pumped to the E'_3 level using a pumping wavelength of 810 nm, this process is much less efficient than the 980 nm pumping to the E_3 level. Optical isolators inserted at the entry and exit end of the amplifier allow only the optical signals at 1550 nm to pass in one direction and prevent the 980 pump light from propagating back or forward into the communications system. There may be another pump diode coupled at the right end of the EDFA similar to that on the left side in the figure. In addition, there is usually a photodetector system coupled to monitor the pump power or the EDFA output power. Normally there is a range of stimulated transitions from E_2 to E_1 because the level E_2 is not a single discrete level but, in reality, corresponds to closely spaced collection of energy levels that are also spread in energy. Consequently the $E_2 - E_1$ transition corresponds to a wavelength range of about 1525–1565 nm that can be amplified, an optical bandwidth of about 40 nm. Thus, the EDFA can be used as an optical amplifier in wavelength division multiplexed (WDM) systems if the wavelength range is within this optical bandwidth. However, the gain is not uniform over the whole bandwidth and special techniques must be used to "flatten" the gain. In addition, it is possible to excite the Er^{3+} ion from the "bottom" of E_1 levels to the "top" of the E_2 levels with a 1480 nm excitation which can also be used as a possible pump wavelength, though the 980 nm pumping is more efficient. The *gain efficiency* of an EDFA is the maximum optical gain achievable per unit optical pumping power and is quoted in dB/mW. Typical gain efficiencies are around 8–10 dB/mW at 980 nm pumping. A 30 dB or 10^3 gain is easily attainable with a few milliwatts of pumping at 980 nm. While Er^{3+} doped silicon fibers are quite common, there are also rare earth doped *fluoride* and *tellurite* fibers that also have optical amplification when suitably pumped.

Erbium doped waveguide amplifier (EDWA) is a waveguide that has been doped with Er^{3+} ions to enable it to operate as an optical amplifier at communication wavelengths at 1550 nm when suitably pumped, for example, at 980 nm or at 1480 nm. The waveguide may be a rib, ridge, buried, or other type of waveguide with the correct dimensions that would allow the fundamental mode to propagate at 1550 nm.

Error function is a standard mathematical function for which there are extensive tables available, which is usually written as erf(z) and defined as follows:

$$\mathrm{erf}(z) = \frac{2}{\sqrt{\pi}} \int_0^z \exp(-\varphi^2) d\varphi.$$

The error function has the following properties: – $\mathrm{erf}(z) = \mathrm{erf}(-z)$, $\mathrm{erf}(0) = 0$ and $\mathrm{erf}(\infty) = 1$. This function, and the so-called *complementary error function* $\mathrm{erfc}(z) = 1 - \mathrm{erf}(z)$, are used extensively in

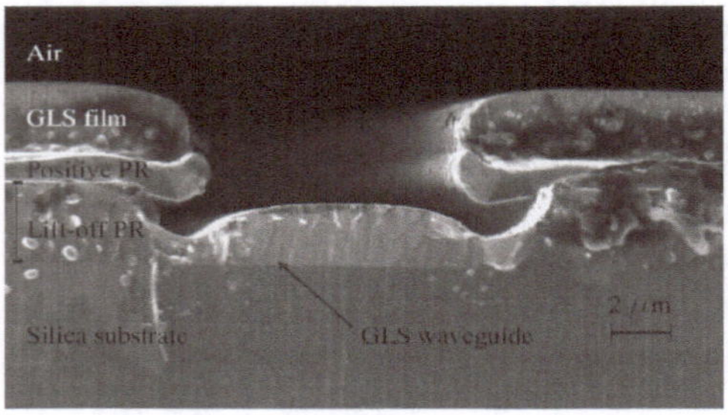

EDWA A strip-type waveguide amplifier that has a core region of Er^{3+}-doped gallium lanthanum sulfide (GLS) glass on a silica substrate. The gain is 2.7 dB/cm, 6.7 dB for the 2.4 cm experimental sample, when pumped from 1480 nm.

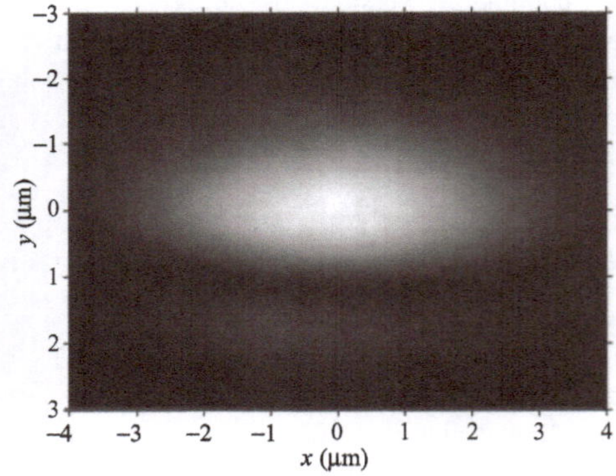

The 1550 nm output from the **EDWA**. (From J. A. Frantz, L. B. Shaw, J. S. Sanghera and I. D. Aggarwal, Optics Express, **14**, 1797, 2006.)

simplifying analytical solutions for heat and mass diffusion in solids.

ESU (electrostatic units) see *CGS units*.

Evanescent wave is the wave that propagates in a lower refractive index medium, along the boundary with a higher refractive index medium, when a plane wave traveling in the medium of higher refractive index is incident at the boundary at an angle equal to or greater than the critical angle. The amplitude of the evanescent wave decreases exponentially with distance from the boundary into the medium with a lower refractive index. The evanescent wave exists over the region of the boundary that receives the incident wave. For a plane wave with a wavevector k_i that is incident at an angle θ_i, the evanescent wave can be written as $E = E_0 \exp(-\alpha_2 y) \exp j(\omega t - k_{iz} z)$, which is a traveling wave along z (the boundary) with a wavevector $k_{iz} = k_i \sin\theta_i$, and an amplitude that decays into the second medium with an attenuation coefficient α_2 given by

$$\alpha_2 = \frac{2\pi n_2}{\lambda}\left[\left(\frac{n_1}{n_2}\right)^2 \sin^2\theta_i - 1\right]^{1/2}.$$

When light is sent along an optical fiber, there is also an evanescent wave that propagates along the cladding of the fiber.

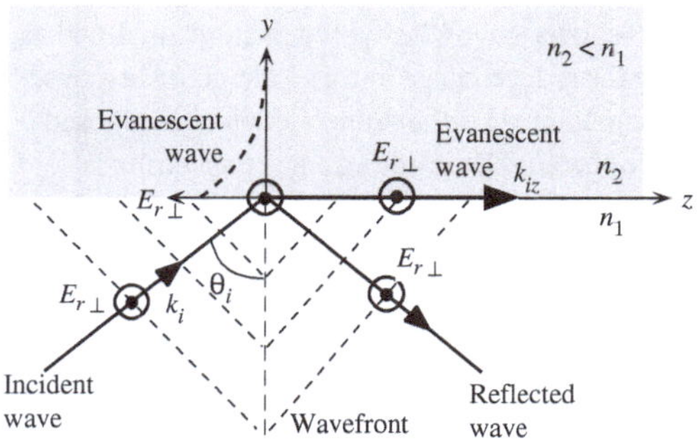

Evanescent wave When the incidence angle θ_i exceeds the critical angle, the incident plane wave is reflected and there is also an evanescent wave at the boundary propagating along z over the region of incidence.

Evanescent wave sensors are sensors that make use of the interaction of a decaying electromagnetic wave, or evanescent wave, in the vicinity of an interface, with an analyte in that location. This is a near-field phenomenon that occurs typically within about one-third of a wavelength from the interface. Sensors made using this principle often use an inverted prism arrangement such that light is reflected off the interface at an angle exceeding the critical angle, causing total internal reflection to occur. Other common arrangements include those using optical fibers where either a region of fiber cladding is removed or a microbend is introduced in a multimode fiber, causing a region with an evanescent field to occur. Sensing can be performed by measuring changes in light absorption owing to changes in the propagation of the evanescent wave in a reagent medium surrounding the sensing region, as compared with that for no reagent present. Another promising approach which may be used for sensing is one in which a high Q microring resonator is coupled to a planar waveguide, and where the transmission in the system is highly sensitive to small changes in optical path. The high Q microring can provide an ultrasensitive transducer for biochemical, gravity, pressure, acceleration and electromagnetic sensing. All of this is owing to the sensitivity of the system to very slight changes in either the mode index or the microring physical dimensions, strongly influencing transmission near the resonances of the coupled optical system.

Excess carrier concentration is the excess concentration above the thermal equilibrium value. Excess carriers are generated by an external excitation such as photogeneration or applied bias. Excess minority and majority carrier concentrations do not obey the usual mass action law that is valid for nondegenerate semiconductors; $np \neq n_i^2$, where n, p and n_i are the electron, hole and intrinsic concentrations in the excited semiconductor.

Excess noise is the noise in a device or system above the thermal noise and shot noise, both of which are white noise sources. Typical well known excess noise contributions are the *generation–recombination noise* (GR noise), *flicker* or *1/f noise* and *burst* or *popcorn noise*, that is *random telegraphic noise* (RTN). In GR noise, the conductance of a semiconductor fluctuates because of random fluctuations in the generation and recombination of carriers, which are stochastic processes. GR noise has a spectral power density that is Lorentzian. Flicker or 1/*f* noise has been given this name because the noise generating processes yield a spectral power density *S*(*f*) for noise that has a 1/*f* spectrum and the fluctuations are greater at lower frequencies (and gives the impression of "flicker" when displayed on an oscilloscope). The fluctuations are in the conductance of the material, which may be due to fluctuations in the number of carriers or their drift mobility. *Hooge's law*, a semiempirical law, has been quite useful in characterizing and comparing flicker noise in different materials because it has a relatively simple way of representing 1/*f* noise in the conductance fluctuations of a material,

$$S_I(f) = \frac{\alpha_{\mathrm{H}} I^2}{Nf}$$

where I is the dc (average) current, N is the total number of carriers in the sample, and α_{H} is Hooge's parameter, which depends on the noise process and the material; it is used for comparing the levels of flicker noise. In most cases, the 1/*f* dependence is not exact. In RTN, the current through a semiconductor device switches at random times between two or more different levels. It has been attributed to the capture and release of carriers by large traps in very small devices.

Excess noise factor, *F*, see *avalanche noise*.

Excimer stands for "excited dimer," a diatomic molecule in which the component atoms are bound in the excited state, but not in the ground state. The important gas molecules for excimer laser lines are rare gas halides including argon fluoride (ArF), krypton fluoride (KrF) and xenon chloride (XeCl). Pumped by an electron beam, they deliver pulses in the UV region, from about 190 to about 350 nm. Xenon fluoride (XeF) operates around 500 nm with a 50 nm tunability range.

Exciton in a semiconductor is a bosonic quasiparticle formed by the Coulomb interaction between an excited electron in the conduction band and a hole in the valence band. Coulomb interaction between the hole and electron can be viewed as an effective hydrogen atom with, for example, the hole establishing the coordinate reference frame about which the reduced mass electron moves. If the effective mass of the isolated electron and hole are m_{e}^* and m_{h}^*, respectively, the reduced mass m_{r}^* is given by:

$$(m_{\mathrm{r}}^*)^{-1} = (m_{\mathrm{e}}^*)^{-1} + (m_{\mathrm{h}}^*)^{-1}.$$

Note that in the case of so-called hydrogenic impurities in semiconductors (i.e. both shallow donor and acceptor impurities), the reduced mass of the nucleus takes the place of one of the terms in the equation above and hence the reduced mass is given to good approximation by the effective mass of the appropriate carrier. In the case of the exciton the carrier effective masses are comparable and hence the reduced mass is markedly lower – accordingly, the exciton binding energy is markedly lower than that for hydrogenic impurities. Analogous to shallow hydrogenic impurities, the exciton energy spectrum follows that of the hydrogen atom, modified by using the reduced mass and also dielectric response of the medium (i.e. the semiconductor, as opposed to free space). The resulting energy spectrum is:

$$E_x(n) = -(m_{\mathrm{r}}^* q^4/2h^2\varepsilon^2)/n^2 = -R_y^*/n^2$$

where n refers to the quantum number of the allowed states, with $n = 1$ being taken as the ground state; R_y^* is the so-called effective Rydberg of the exciton given by $R_y(m_{\mathrm{r}}^*/m_0)/\varepsilon^2$, for $R_y =$ 13.6 eV. For bulk GaAs, the exciton binding energy ($n = \infty$) corresponds to about 5 meV. Following from the hydrogen atom model, the extension of the excitonic wavefunction can be found from an effective Bohr radius a_{B}^* given in terms of the

Bohr radius $a_B = h^2/\pi q^2$: that is $a_B{}^* = a_B(\varepsilon/m_r{}^*)$. For GaAs this corresponds to about 12 nm or about 21 lattice constants – that is, the spherical volume of the exciton radius contains $\sim(a_B{}^*/a_0)^3$ or $\sim$9000 unit cells, where a_0 is the lattice constant of GaAs. In such a case as this when $R_y{}^* \ll E_g$ and $a_B{}^* \gg a_0$ the excitons are referred to as *Wannier or Wannier–Mott excitons.* It should be noted that the binding energy of excitons in semiconductors tends to be a strong function of the semiconductor bandgap as seen in the figure. In the case of excitons having large binding energy and correspondingly small radius (i.e. approaching a size of about one lattice parameter), the excitons become localized on a lattice site and are commonly referred to as Frenkel excitons. Unlike Wannier-based excitons which typically are dissociated at room temperature, these excitations are stable at room temperature. Moreover, there are no simple models to describe the binding of these excitations or their coupling to the underlying lattice. Indeed, such descriptions are needed in order to understand the hopping of these excitations from site to site.

When excitons dissociate they can do so radiatively emitting a series of hydrogen-like spectral lines as described by the equation above. One can estimate the binding energy of In bulk (3D) semiconductors such as GaAs; exciton lines can only be observed at low temperatures: they are easily dissociated by thermal fluctuations. On the other hand, in quantum wells and other structures of reduced dimensionality, the spatial confinement of both the electron and hole wavefunctions in the same layer ensures strong excitonic transitions a few meV below the bandgap, even at room temperature. Excitonic absorption is spectrally well located and very sensitive to optical saturation. For this reason, it plays an important role in nonlinear semiconductor devices (nonlinear Fabry–Perot resonator, nonlinear mirror, *saturable absorber*, and so on). If the valence hole is a *heavy* hole, the

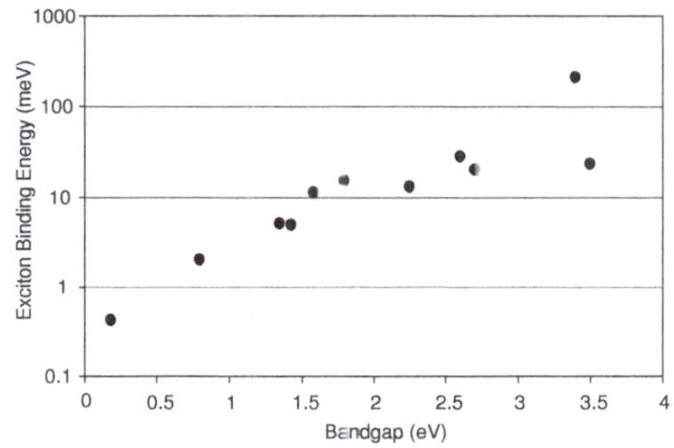

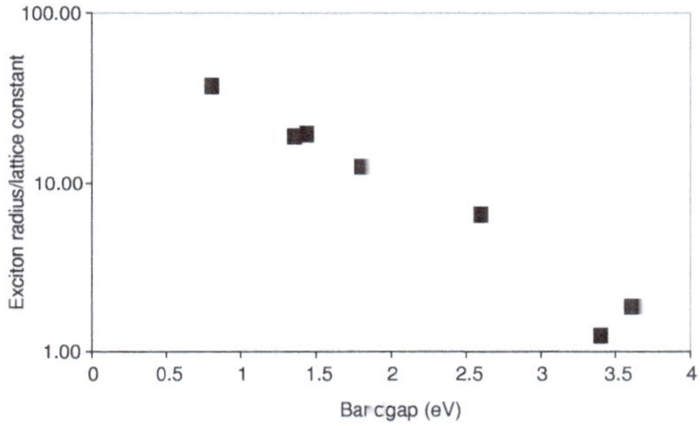

Dependence of **exciton** binding energy and size as a function of the semiconductor bandgap. Exciton binding energy increases concurrently with marked diminishing of exciton spreading as the bandgap increases. Above a bandgap of about 2 eV, the Warnier-based description is not appropriate.

exciton is called a *heavy* exciton; conversely, if the valence hole is light, the exciton is a light exciton. For practical purposes, the excitonic contribution to the overall *susceptibility* around resonance frequency ν_{ex} can be written as:

$$\chi_{exc} = -A_0 \frac{(\nu - \nu_{ex}) + j\Gamma_{ex}}{(\nu - \nu_{ex})^2 + \Gamma_{ex}{}^2(1 + S)},$$

with Γ_{ex} the linewidth and $S = I/I_S$ the *saturation parameter* of the transition. For instance, in GaAs MQW, the saturation intensity I_S is as low as

1 kW/cm^2, and Γ_{ex} ($\approx$3.55 meV at room temperature) varies with the temperature according to

$$\Gamma_{ex} = \Gamma_0 + \frac{\Gamma_1}{\exp(\hbar\omega_{LO}/kT) - 1},$$

with $\hbar\Gamma_0$ the inhomogeneous broadening ($\approx$2 meV), $\hbar\Gamma_1$ the homogeneous broadening ($\approx$5 meV), $\hbar\omega_{LO}$ the longitudinal optical phonon broadening ($\approx$36 meV). At high carrier concentrations (provided either by electrical pumping or by optical injection), an efficient mechanism saturating the excitonic line is the *screening* of the Coulomb attractive potential by free electrons and holes.

The discussion above refers to so-called free excitons formed from conduction band electrons and valence band holes. Such an excitation is able, in principle, to move through a material. It should be noted, however, that free electrons and holes move with a velocity $\hbar(dE/dk)$ where the derivative is taken for the appropriate band edge. To move through a crystal, both electron and hole must have identical translational velocities restricting the regions in k-space where these excitations can occur to those with $(dE/dk)_{electron} = (dE/dk)_{hole}$, commonly referred to as critical points.

A number of more complex pairings of carriers can also occur, which may also include fixed charges or ions. For example, for the case of three charged entities with one being an ionized donor impurity (D+) the following possibilities can occur: (D+)(+)(–), (D+)(–)(–) and (+)(+)(–) as excitonic ions, and (+)(+)(–)(–) and (D+)(+)(–)(–) as excitonic molecules. Complexity abounds in these systems as each electronic level possesses a fine structure corresponding to allowed rotational and vibrational levels. Moreover, the effective mass is often anisotropic. Note that when the exciton or exciton complex is bound to a fixed charge such as an ionized donor or acceptor center in the material, the exciton or exciton complex is referred to as a bound exciton. Indeed, bound excitons may also involve neutral fixed impurities. It is usual to relate the exciton in these cases to the center binding them – thus if an exciton is bound to a donor impurity, it is usually termed a donor-bound exciton.

See *Wannier–Mott exciton.*

Exitance or emittance, M, is the total optical power, i.e. flux (radiant or luminous), radiated, in all directions, by a unit element of a surface, measured in W m^{-2} for *radiant exitance* M_e, and in lm m^{-2} for *luminous exitance* M_v. See *emissivity, emissive power, flux.*

Exposure in photography refers to the total amount of light that is incident on the film or the image sensor. Exposure is illuminance (luminous flux incident per unit area) and duration product and is measured in lux seconds. Lux itself is lumens m^{-2}. It can also be stated in exposure values and scene luminance. Exposure is important in photography because a given film or a sensor with a given sensitivity needs a certain amount of exposure to generate the correct signal. Under- or overexposure results in images that are dark (have not received sufficient photons) or bright (have received too many photons). In a camera, the aperture controls the level of illumination, and the shutter speed controls the exposure duration. Today's digital cameras use a photometric sensor to automatically calculate the necessary exposure.

Exposure to an ionizing radiation, such as x-rays or γ-rays, is a measure of the ionization effects of the radiation in air in terms of how much charge is liberated (created) in a given amount of air that has been subjected to the ionizing radiation; it is measured in roentgens (R). One roentgen of exposure (at STP) releases 0.258 mC of charge per kg of dry air. Equivalently 1 R deposits 8.77 mJ into 1 kg of air. It should not be confused with the amount of radiation that is absorbed, which is represented by *dose.* See x-rays, γ-rays, *roentgen, dose, gray.*

Extended transfer matrix formalism in active optical systems involves 3×3 matrices (instead of traditional 2×2) to describe the action of internal sources.

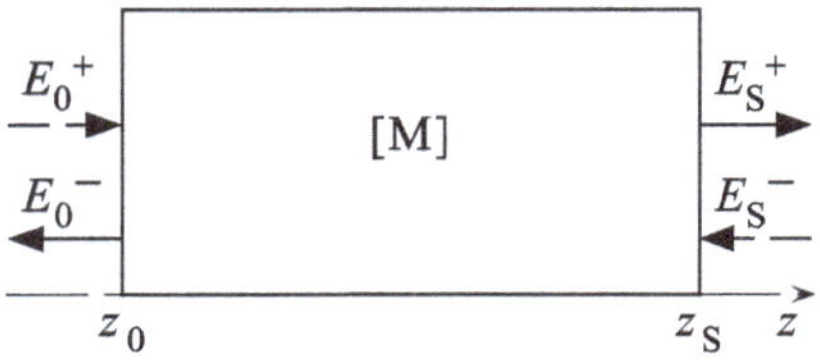

Extended transfer matrix formalism Even with an internal source, the fields at both ends (z_0, z_S) can be related by a matrix relationship, whatever the external input fields ($E_0{}^+, E_S{}^-$).

The relationship between z_0 and z_P reads:

$$\begin{pmatrix} E_0^+ \\ E_0^- \\ 1 \end{pmatrix} = \begin{pmatrix} M_{11} & M_{12} & M_{13} \\ M_{21} & M_{22} & M_{23} \\ 0 & 0 & 1 \end{pmatrix} \begin{pmatrix} E_S^+ \\ E_S^- \\ 1 \end{pmatrix}, \qquad (1a)$$

where $M_{11}, M_{12}, M_{21}, M_{22}$ are exactly that of the 2×2 transfer matrix, and M_{13}, M_{23} the source terms. The resulting "background" fields emitted at both facets by the structure when no input field is present (in which case $E_0{}^+ = E_S{}^- = 0$) are:

$$E_S^+ \equiv B_S^+ = \frac{-M_{13}}{M_{11}}, \qquad (1b)$$

$$E_0^- \equiv B_0^- = \frac{-M_{13}M_{21} + M_{23}M_{11}}{M_{11}}. \qquad (1c)$$

On the other hand, in the presence of an input field (say $E_0{}^+$, assuming $E_S{}^- = 0$):

$$E_S^+ = tE_0^+ + B_S^+, \qquad (2a)$$

$$E_0^- = rE_0^+ + B_0^-, \qquad (2b)$$

with $t = 1/M_{11}$ and $r = M_{21}/M_{11}$ the left-to-right transmittance and reflectance.

The resulting "background" spectral intensities (in W m^{-2} Hz^{-1}) are defined as:

$$I_{BS}^+ = (\varepsilon_0/2)c\, n_S \langle |B_S^+|^2 \rangle = (\varepsilon_0/2)c\, n_S \langle |M_{13}/M_{11}|^2 \rangle, \qquad (3a)$$

$$I_{B0}^- = (\varepsilon_0/2)c\, n_0 \langle |B_0^-|^2 \rangle = (\varepsilon_0/2)c\, n_0 \langle |(-M_{13}M_{21} + M_{23}M_{11})/M_{11}|^2 \rangle \qquad (3b)$$

where ε_0 is the permittivity of the vacuum, c is the speed of light, and the triangular brackets $\langle\ \rangle$ stand for a time average over the integration time of the detector under consideration. Since spontaneous emission is not coherent with respect to the signal, the different contributions will add in intensity.

Some points should be emphasized: (a) the emitted intensities coming out of the structure appear filtered through $|1/M_{11}|^2 = T(\nu)$: whatever the intrinsic emission spectrum of the source, the spontaneously emitted field is coupled preferentially towards the longitudinal "resonance modes" imposed by the structure; (b) the emitted intensities depend also (through a longitudinal overlap factor) upon the precise distribution of the emitters inside the structure, which is naturally included into the calculation; (c) the apparent asymmetry of both expressions is a consequence of the arbitrary choice for orienting the Oz-axis; in fact, the result is completely symmetrical with respect to the direction of propagation.

Localized source: Consider a source located at abscissa z_P between z_0 and z_S, and let (S_P) denote the source's own contribution to the complex amplitude of the optical field. Let [L] and [R] denote the "left" and "right" partial transfer matrices, from $z_0{}^-$ to z_P and from z_P to $z_S{}^+$, respectively, such as [L] [R] = [M].

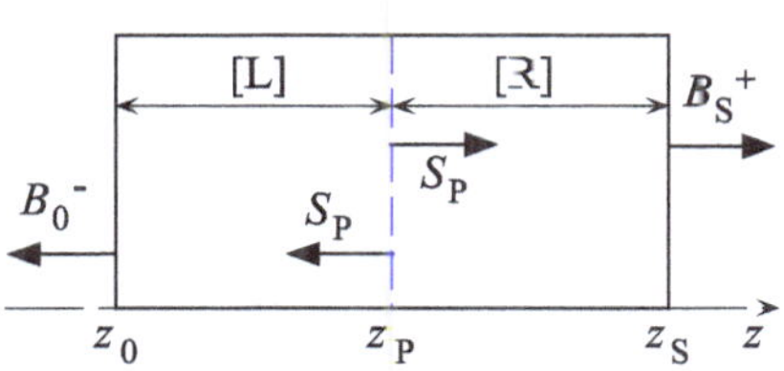

Extended transfer matrix formalism An internal source localized at abscissa z_P is responsible for an output field at both ends (z_0, z_S), $B_0{}^-$ to the left and $B_S{}^+$ to the right, even with no external input field ($E_0{}^+ = E_S{}^- = 0$).

We have $M_{13} = -(S_P)\,(L_{11} - L_{12})$ and $M_{23} = -(S_P)\,(L_{21} - L_{22})$. Although spontaneous emission is a noncoherent process, the probability of emission to the right and to the left can nevertheless interfere mutually. This interference effect is especially important if the source is localized between two reflectors, such as a quantum well in a vertical microcavity. On the other hand, as soon as the active zone exhibits a finite extension, such effects are usually washed out.

Extended source: If the source is continuously distributed along the z-axis, the total background emitted intensity can be obtained by longitudinal integration. $|(S_P)|^2$ is replaced by $|A_0(z)|^2dz$, where $|A_0|^2$ represents the local spatial density of the source, related to the spectral rate of spontaneous emission $r_{sp}(\nu)$, in s^{-1} m^{-3} Hz^{-1}, by:

$$(\varepsilon_0/2)cn\langle|A_0|^2\rangle = \beta_{sp}h\nu r_{sp}(\nu), \quad (4a)$$

with β_{sp} the fraction of spontaneous emission coupled into the mode. It is convenient to express the contribution of an active zone of length L, of transfer matrix [Z], in terms of equivalent spontaneous fields (u^+) and (u^-), which couple to the main field at both its extremities. Consider a single active zone of gain g and uniform spatial density $|A_0|^2$. If its length L is much larger than the wavelength $\lambda = \lambda_0/n$, the term $\langle(u^+)(u^-)^*\rangle$ is washed out to zero due to spatial averaging, whereas

$$\begin{aligned}\langle|(u^+)|^2\rangle &= \langle|(u^-)|^2\rangle \\ &= \langle|A_0|^2\rangle[\exp(gL) - 1]/g. \quad (4b)\end{aligned}$$

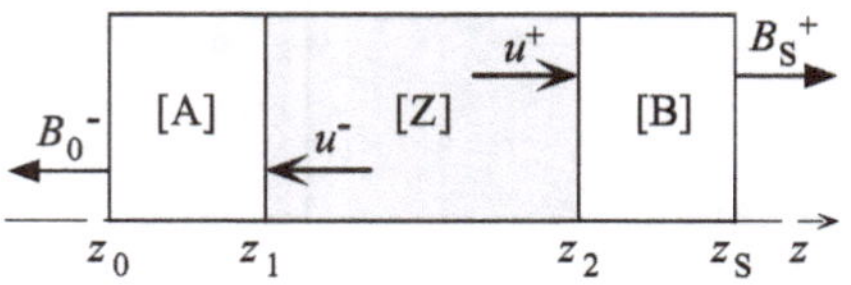

Extended transfer matrix formalism Active zone of length $L = z_2 - z_1$, of transfer matrix [Z], between two reflectors represented by [A] and [B]. Equivalent spontaneous fields (u^-) and (u^+) result from the superposition of all infinitesimal source contributions, distributed along the whole active zone.

For an active zone between two reflectors [A] and [B], the "intrinsic" intensity is $I_U(\nu) = (\varepsilon_0/2)\,c\,n\,\langle|(u^\pm)|^2\rangle$ (not to be confused with the actual value of the intensity at both extremities, which results from multiple round-trips). The emitted intensities through both mirrors read:

$$I^+_{BS}(\nu) = I_U(\nu) \times T(\nu)\left(\frac{R_A + \exp(-gL)}{T_A}\right), \quad (5a)$$

$$I^-_{B0}(\nu) = I_U(\nu) \times T(\nu)\left(\frac{R_B + \exp(-gL)}{T_B}\right). \quad (5b)$$

As pointed out before by Alfred Kastler (Nobel Prize of Physics for optical pumping), the field emitted inside a cavity is projected into its resonance modes. The emitted intensities are therefore proportional to the "intrinsic" intensity, amplified by multiple round-trips inside the resonator and spectrally filtered by $T(\nu)$. Both expressions are completely symmetrical with respect to the direction of propagation.

If several active zones coexist inside the same structure, a matrix product easily gives the total transfer matrix [M]. The source terms M_{13} and M_{23} appear as linear combinations of all the equivalent source terms $(u_a{}^+)$, $(u_a{}^-)$ coming from all active sections. Emitted intensity is obtained thanks to a simple rule: if $a \neq a'$,

$$\begin{aligned}\langle(u_a^+)(u_{a'}^-)^*\rangle &= \langle(u_a^+)(u_{a'}^+)^*\rangle \\ &= \langle(u_a^-)(u_{a'}^-)^*\rangle = 0. \quad (6)\end{aligned}$$

Formally, expressions (1)–(3) still hold. The total background intensity is simply expressed as the sum of all individual contributions.

Internal fields: The same formalism enables one to derive the spatial and spectral distribution of the field inside the cavity. The detailed calculation is tedious but straightforward. The main conclusion is that internal fields are also subject to spectral filtering by projection into the transfer function $T(\nu)$. This proves useful for nonlinear calculations when *saturation* of the active medium cannot be neglected, for instance in a semiconductor optical amplifier subject to amplified spontaneous emission (ASE). Surprisingly, it remains true of a single-mode laser above threshold, when spontaneous emission is not looked upon as mere noise, but as the very driving source of the optical field. If reasonably uniform at the scale of the active zone, the total saturating intensity I_{SAT} is obtained by longitudinal average over the active zone and spectral integration over its whole linewidth.

This saturating intensity is responsible for the saturation of the active medium: it plays an essential role in the study of intracavity light-matter interactions, threshold-crossing included. (Y. G. Boucher, *J. Europ. Opt. Soc. – Rapid Publ.*, **1**, 6027 2006; available on www.jeos.org.)

External reflection is the reflection of an electromagnetic (EM) wave initially traveling in a medium of lower refractive index n_1 at the boundary with a medium of higher refractive index n_2 ($>n_1$); a light wave traveling in n_1 is reflected at the n_1/n_2 boundary (some of the energy may, however, be transmitted). For example, external reflection occurs when a light wave traveling in air is incident on an air–water surface and becomes reflected.

Extinction coefficient, K, represents the attenuation suffered by an electromagnetic wave as it travels along a well-defined propagation direction. It is the imaginary part of the complex refractive index N which is written as $n - jK$. The amplitude of the field $|E|$ in the wave decreases as $|E| \propto \exp(-k_0 K z)$ where k_0 is the propagation constant in vacuum.

Extraordinary wave is a linearly polarized light wave that propagates in an anisotropic crystal without necessarily having the electric field **E** oscillations orthogonal to the wavevector **k**; in ordinary light **E** and **k** are orthogonal. The **E** oscillations do not necessarily lie in the wavefront plane. The energy flow direction (the *Poynting vector*) is not along the wavevector direction and this phenomenon is called the Poynting vector "walk-off." The extraordinary and the ordinary waves have orthogonal polarizations.

Extrinsic semiconductor is a semiconductor that has been doped so that the concentration of one type of charge carriers far exceeds that of the other. Adding donor impurities releases electrons into the conduction band and the concentration of electrons (n) in the conduction band (CB) far exceeds the concentration of holes (p) in the valence band (VB) and the semiconductor becomes n-type. When acceptor impurities are added, these acceptors accept electrons from the VB and thereby create holes in the VB so that the hole concentration (p) far exceeds the electron concentration (n); the semiconductor becomes p-type. The mass action law, $np = n_i^2$, remains valid in extrinsic semiconductors as long as the semiconductor is in the dark and in thermal equilibrium. Thus, adding donors to increase n concomitantly also decreases p by the same ratio.

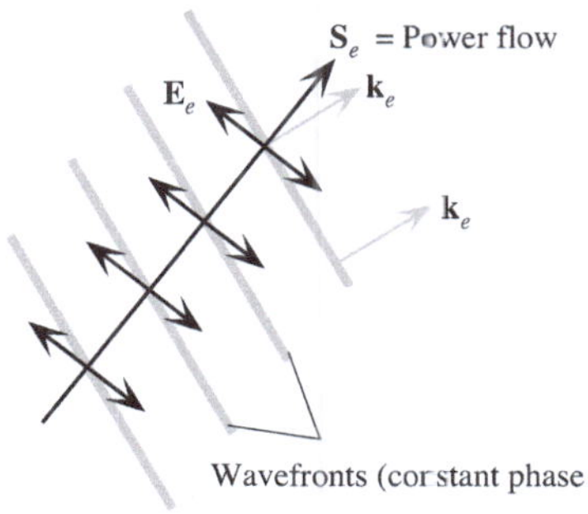

An **extraordinary wave** in an anisotropic crystal with a wavevector $\mathbf{k}_e$ at an angle to the optic axis The electric field $\mathbf{E}_e$ is *not* normal to $\mathbf{k}_e$. The energy flow (group velocity) is along $\mathbf{S}_e$ (the Poynting vector) which is different than $\mathbf{k}_e$.

Eye (human) is a biological optical instrument that enables humans to capture images The basic structure of the eye, which resembles a camera, is shown in the first figure. The *sclera* is a tough outside coating (normally white) that covers and protects the eye. The *cornea* is a curved transparent portion in the front of the eye with a fixed optical power that helps in focusing the light. The *anterior chamber* is the space between the lens and the cornea, and is filled with a clear watery fluid called the *aqueous humor*. The *pupil*, an aperture in the middle of the iris, behind the cornea, allows the light to pass

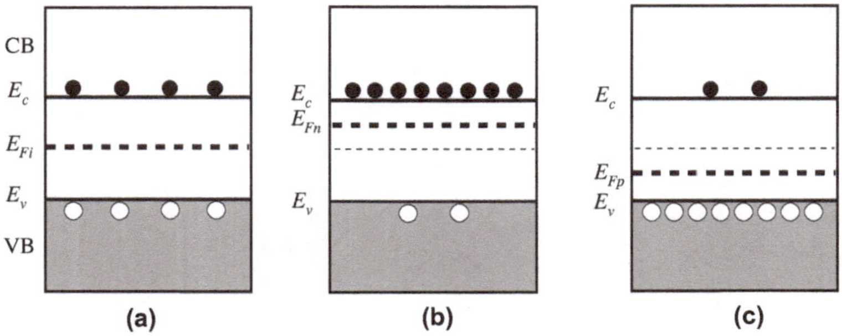

Extrinsic semiconductor Energy band diagrams for (a) intrinsic (b) n-type and (c) p-type semiconductors. In all cases, $np = n_i^2$. Note that donor and acceptor energy levels are not shown. CB = conduction band, VB = valence band, E_c = CB edge, E_v = VB edge, E_F = Fermi level in intrinsic semiconductor, E_{Fn} = Fermi level in n-type semiconductor, E_{Fp} = Fermi level in p-type semiconductor.

through towards the lens. The pupil size (2–8 mm) acts as an adjustable aperture. The pupil size is controlled by the iris muscles. The *lens* is a crystalline biconvex lens with a refractive index that decreases from the center to the surface, i.e. a GRIN-type lens. The *ciliary muscles* adjust the lens shape (radii of curvature) and hence adjust the focus; this is called *accommodation*, the ability of the eye to change its effective focal length to focus images. Very roughly, two-thirds of refraction are provided by the cornea and the aqueous humor, and the lens provides a further one-third of the refraction. The focused light forms an image on the *retina* of the eye, the inside layer that has millions of photoreceptors for detecting the light. The *posterior chamber* is the space between the lens and the retina. It is filled with a clear water-based liquid, called the *vitreous humor*, to provide support for the eye structure.

Photons passing through the pupil are focused by the cornea and the lens on to the retina of the eye and are detected by two types of photosensitive cells, called *rods* and *cones*. Rods are highly sensitive photoreceptors with a peak response at the wavelength at about 507 nm (roughly cyan). They do not register color and are responsible for our vision under dimmed light conditions, what is termed *scotopic vision*. Cones are responsible for our color perception and daytime vision, called *photopic vision*.

There are three types of cone photoreceptors which are sensitive to blue, green and red at wavelengths 430 nm, 535 nm and 575 nm. All three cones have an overall peak response at 555 nm (green), which represents the peak response of an average daylight adapted eye or in our photopic vision. Various experiments, the most well-known being by S. Hecht *et al.*, *J. Opt. Soc. America*, **38**, 196, 1942, have tested the threshold sensitivity of the dark adapted eye and have estimated that visual perception requires a minimum of roughly 90 photons to be incident onto the cornea in front of the eye's pupil and within 1/10 seconds. Taking 90 incident photons every 100 ms for threshold visual perception, the eye needs 900 photons per second, and the required optical power is $\sim 4 \times 10^{-16}$ W or 2000 eV/s. Not all photons incident on the eye make it to the actual photoreceptors in the retina. It has been estimated that only 1 in 10 photons arriving at the eye's cornea actually make it to rod photoreceptors, due to various reflections and absorptions in the eye, and other loss mechanisms. Thus, only nine photons make it to photoreceptors on the retina. (Sometimes

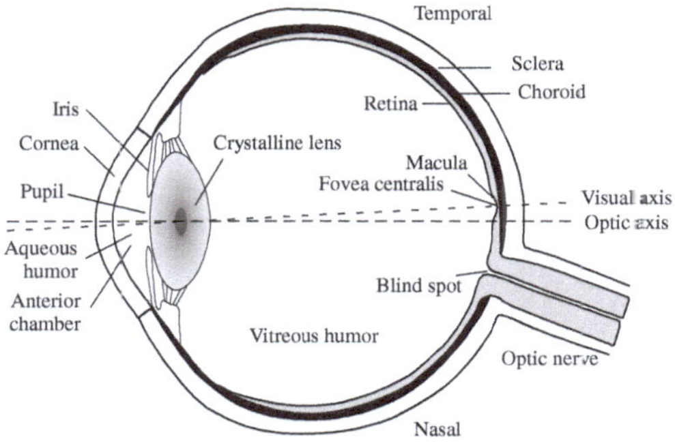

Eye The basic structure of the eye (schematic).

one comes across a statement that the eye can detect a single photon. While a rod photoreceptor can indeed sense a single photon, or, put differently, a photon can activate a single rod, the overall human visual perception needs roughly nine photons at around 507 nm to consciously register a visual sensation.) It is estimated that the nine test photons fall randomly onto a circular area of about 0.0025 mm^2 but there are 150,000 rods mm^{-2} in this area of the eye, which means that there are 350 or more rods in the spot area. Thus, a rod must be able to *sense* a single photon but it takes nine excited rods, somehow summed up by the visual system, to generate the visual sensation. It is estimated that at least 200,000 photons per second must be incident on the eye to generate a color sensation by exciting the cones.

The eye integrates, i.e. takes a temporal summary, of the photons that are received on the photoreceptors to register a signal. The integration time (the storage time) τ is of the order of 100–300 ms for rods and 10–50 ms for cones. τ determines the temporal resolution of the eye; its ability to distinguish light flashes occurring one after another. The value of τ balances between being able to distinguish two time separated events, and the summation of photons required to construct detectable signal. For stimuli shorter than 100 ms, the exposure and intensity seem to follow the reciprocity law, that is, intensity × duration = constant for a given threshold perception (in the absence of any background light). This is known as *Bloch's law*. When a sequence of images is viewed at a rate faster than $1/\tau$, images are perceived as a continuous flow as in a film. The critical frequency that enables a flicker-free perception of image flow is called the *critical flicker frequency*, typically higher than 40 Hz.

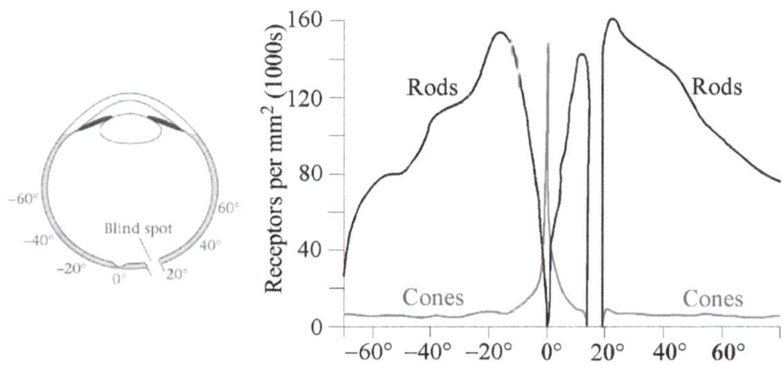

Eye The distribution of rods and cones in the retina about the fovea. (Adapted from B. A. Wandell, *Foundations of Vision*, Sinauer Associates Inc. Publishers, Sunderland, Massachusetts, 1995.)

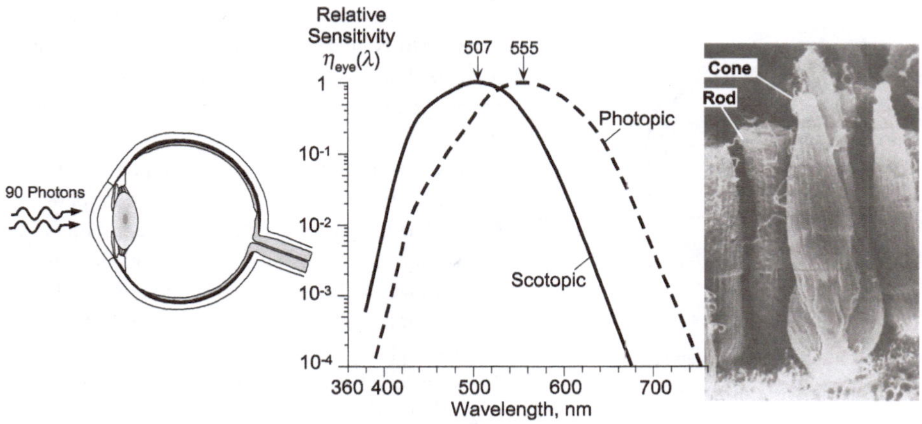

Eye The retina in the eye has photoreceptors that can sense the incident photons on them and hence provide necessary visual perception signal. Minimum visual perception has been estimated to be roughly 90 photons falling on the cornea of the eye. The wavelength dependence of the relative efficiency $\eta_{eye}(\lambda)$ of the eye is different for daylight vision or photopic vision (involves mainly cones) and for vision under dimmed light or for scotopic vision (represents the dark-adapted eye, and involves rods). SEM photo of rods and cones in the retina of the human eye. (Courtest of Dr. Frank Werblin, University of California, Berkeley.)

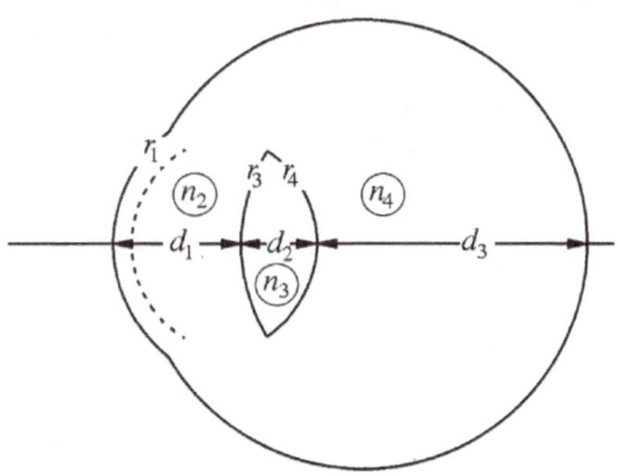

Eye Gullstrand–Emsley simplified schematic of the eye has three curvatures.

The *Gullstrand–Emsley schematic* of the eye is a simplified version of the original Gullstrand model, and has three surfaces, r_1 for the cornea and r_3 and r_4 for the lens. The cornea's refracting power is fixed and roughly 40 diopters. The lens refracting power can be adjusted and brings the total power to 60 diopters. The lens is a GRIN lens, that is the refractive index is greatest at the center ($n = 1.41$) and lowest on the surface ($n = 1.39$). (Gullstrand received the Nobel Prize for Physiology in 1911 for his research on the eye.) The *reduced Emsley schematic* of the eye has only one refracting surface, the cornea with a radius of curvature 5.55 mm; it cannot be used for describing the accommodation of the eye. The original Gullstrand model had four refractive surfaces, but the simplified Gullstrand–Emsley model is probably the most commonly used optical model of the eye.

Eye pattern or **eye diagram** is a very useful way to quickly characterize the performance of an optical system by examining the output of the system on an oscilloscope when many (millions to billions) pulses are sent along the optical system as a nearly random sequence of zeros and ones. The oscilloscope time base sweep is set to a fraction of the bit rate. The displayed waveform therefore represents many waveforms overlapping each other

Eye pattern or eye diagram

Table: Eye (human) The characteristics of the eye show variations not only from person to person within a given age group, but also changes with age. r = radius of curvature (values normally refer to a relaxed eye).

Property	Typical values	Comment
r_1 (Cornea)	+7.80 mm	
r_3 (Lens, anterior)	+10.00 mm	Adjusted by the ciliary muscles
r_4 (Lens, posterior)	−6.00 mm	Adjusted by the ciliary muscles
Lens	5 mm thick; 9 mm in diameter.	Can vary substantially
n_3 (Lens)	Core: 1.406 Surface: 1.386 Equivalent: 1.42	Lens is a GRIN lens with the center having a higher n. n_3 represents the overall equivalent refraction. At the lens surface $n = 1.386$, at center $n = 1.406$.
Cornea index, n_1	1.376	
Aqueous humor index, n_2	1.336	
Vitreous humor index, n_4	1.336	
d_1 (Anterior chamber depth)	3.6 mm	
d_3 (Posterior chamber depth)	16.7 mm	
d_2, "lens thickness"	3.6 mm	
Pupil diameter d in mm	$d = 5–3\ \tanh[(0.4\log(L)]$	in mm, L is mean luminance in cd/cm^2. Le Grand 1969
Rayleigh diffraction limit	28 arcsec	5 mm pupil
Storage time	100–300 ms	
Minimum photons	90	Threshold of perception

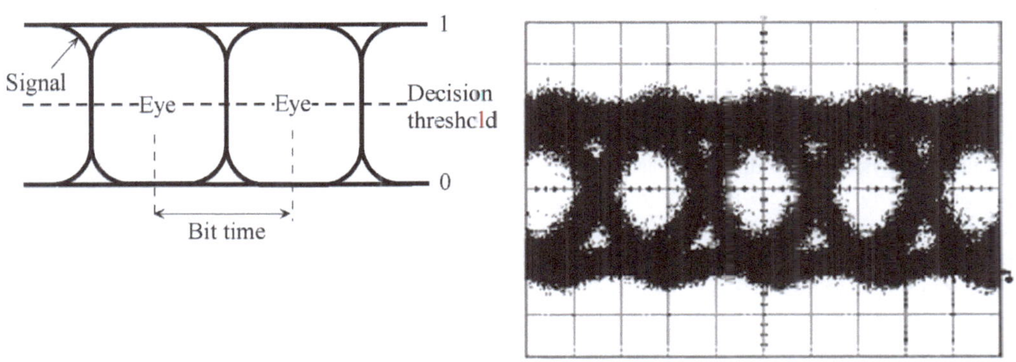

Eye pattern (Left) An example of a nearly ideal eye diagram as observed on the screen of an ideal oscilloscope connected to a receiver at the output of an optical system when many (perhaps millions to billions) of pulses are sent along the optical system. (Right) An example of a practical optical eye diagram obtained on an optical modulation system by using $2^{31}-1$ pseudorandom bit sequence at 43.5 Gb/s. Horizontal time axis is 10 ps/division. (K. Kondo, A. Kondo, K. Aoki, M. Imaeda, T. Mori, Y. Mizuno, S. Takatsuji, Y. Kozuka, O. Mitomi, and M. Minakata, *J. Lightwave Technol.* 20, issue 12. © 2003 IEEE.)

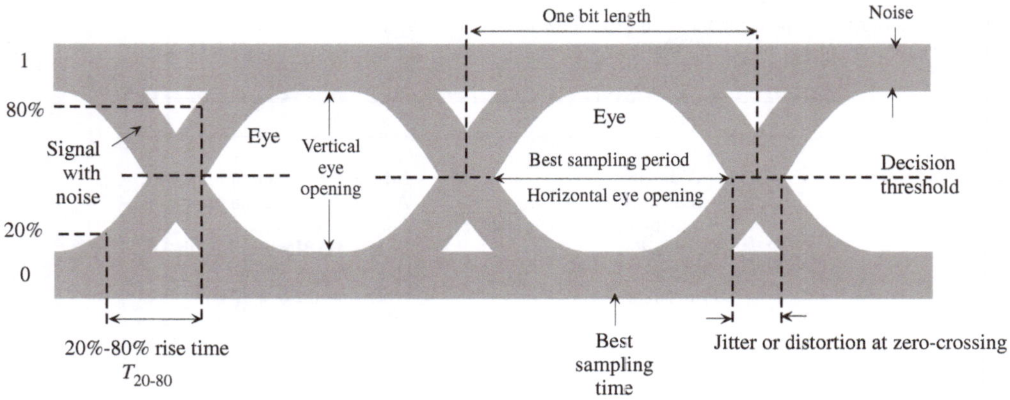

Definitions for an **eye pattern** in the presence of noise and jitter.

and hence the displayed signal is an average over many pulses. In an ideal system, the pulses at the receiver would be perfectly rectangular but, due to the finite bandwidth of the system (including the pulse generator and the receiving system), the waveforms are rounded as shown in the figure. The rounded rectangles fill between the lowest and the highest lines corresponding to 0 and 1. The central open space is called the "eye" and its *openness* is a measure of the goodness of the optical system. The advantage of the eye pattern is that it corresponds to examining the system by sending a very large number of pulses along the system, and then overlaying the output pulses on top of each other to determine the rise and fall times, noise, repeatability and jitter in the system. The eye patterns in practice are far from this ideal picture, have substantial smearing (noise), and have the eye less open as shown for one practical case in the first figure. The eye should be as open as possible to minimize the errors. The eye width is the *horizontal eye opening* and indicates the time over which the received signal is best sampled as shown in the second figure. Best time to sample the received signal is when the vertical eye opening (the height of the eye) is largest. The vertical eye opening also measures the noise margin in the signal. The thickness of the waveform represents the noise in the system. The width of the waveform at the zero-crossing (center line), at the corner of the eye, is the jitter, which is also called the timing jitter and phase distortion. The rise time $T_{20\text{-}80}$ is the time it takes the waveform to rise from 20% to 80% of the final value. In nonlinear communication channels, the eye pattern will not be symmetrical and will also evince overshoot and undershoot at the high (1) and low ends (0).

F

F center, see *color center*.

***f*-number** or **focal ratio of a lens** is an inverse measure of its light gathering and collimating ability and is quantitatively defined as the effective focal length divided by the effective aperture D of a lens,

$$f\text{-number} = f/\# = f/D$$

where the aperture D is the largest aperture the lens sees; this would be its *diameter* if unconstricted. Light waves from a light source placed at the focal point F of a near perfect lens of diameter D will be collected and collimated (neglecting spherical aberrations) by the lens into a beam of diameter D. If a lens has a focal length of 10 cm and a diameter of 2.5 cm, then its *f*-number, 10/2.5 = 4, is written as *f*/4. The image of an extended object receives a flux density that varies as $(D/f)^2$, i.e. it is inversely proportional to $(f\text{-number})^2$. D/f is called the *relative aperture*. The *f*-stop on a camera is an adjustment that changes the aperture (effective lens diameter) by adjusting the iris diaphragm and hence changes the amount of light that is used in forming the image. Since $\sin\alpha_{\max} \approx \tan\alpha_{\max} = (D/2)/f$,

$$f\text{-number} \approx 1/(2\sin\alpha_{\max}) = 1/(2\text{NA})$$

where NA is the numerical aperture of the lens with the object (source) at F and, by definition, $\text{NA} = n_0 \sin\alpha_{\max}$; since $n_0 = 1$, the above follows immediately.

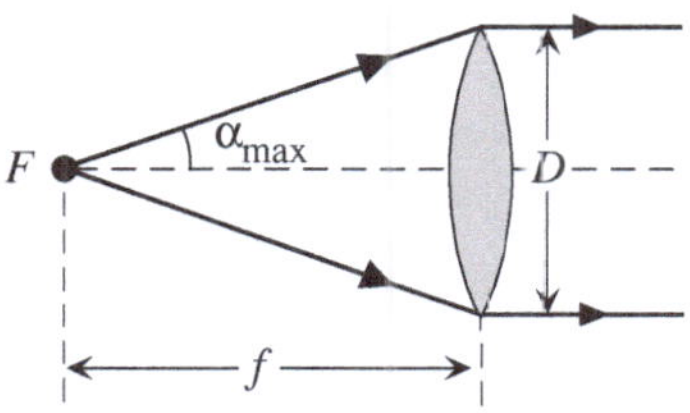

f-number This ideal lens of diameter D provides a collimated beam of diameter D from a source at its focal point F.

***f*-stop** see *f-number*.

Fabry–Perot cavity is made of a cavity between two mirrors, usually (but not necessarily) plane and parallel. Suppose that the reflectors are ideal (lossless). With respect to the inside of the cavity, they exhibit amplitude reflectivities $r_A\exp(j\varphi_A)$ and $r_B\exp(j\varphi_B)$. Let $R_A = r_A^2$, $R_B = r_B^2$, $T_A = (1 - R_A)$, $T_B = (1 - R_B)$. The cavity can be either passive, absorbing or amplifying. In the most general case, the relevant parameters are its length L, its refractive index n, and its optical gain $g\,[\text{m}^{-1}]$. (In the absorbing case, $g < 0$.)

The transfer function $T(\upsilon)$ is the well-known *Airy function*, expressed as a function of the phase change Φ on a round-trip (in normal incidence),

$$T(\upsilon) = T_0/[1 + m\sin^2(\Phi/2)],$$
$$\Phi = (2\pi\upsilon/c)2nL - \varphi_A - \varphi_B,$$

Fabry–Perot cavity

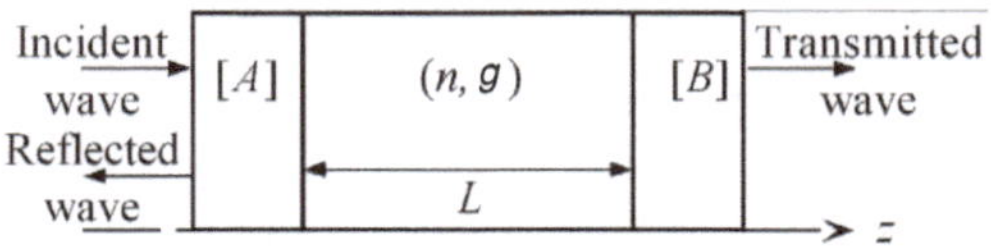

Fabry–Perot cavity of length L, index n, optical gain g between two reflectors [A] and [B].

$$m = \frac{4r_A r_B \exp(gL)}{(1 - r_A r_B \exp[gL])^2},$$

$$T_0 = \frac{T_A T_B \exp[gL]}{(1 - r_A r_B \exp[gL])^2}.$$

In the passive and symmetrical case, when $R_A = R_B = R_m = (1 - T_m)$ and $g = 0$,

$$m = \frac{4R_m}{(1 - R_m)^2},$$

$$T_0 = \frac{T_m^2}{(1 - R_m)^2} = 1.$$

If we neglect the material dispersion, then the Airy function is obviously 2π-periodic with respect to the Φ variable. The frequencies for which the Airy transfer function reaches its maximum T_0 correspond to the *longitudinal modes* of the resonator. $T(\upsilon)$ is also periodic with respect to the frequency υ, and the period is called the *free spectral range*, equal to $\Delta\upsilon = c/(2nL)$. The Airy function is a comb-like function (see figure).

The higher the reflectivities, the sharper the resonance. The full-width at half-maximum (FWHM) of a resonance peak is

$$\delta\upsilon = \frac{2}{\pi\sqrt{m}}\left(\frac{c}{2\,nL}\right),$$

and the *finesse factor* F is defined as

$$F = \Delta\upsilon/\delta\upsilon = \frac{\pi\sqrt{m}}{2} = \frac{\pi\sqrt{R_m}}{(1 - R_m)}.$$

The higher the reflectivity R_m, the greater the finesse factor. A highly resonant Fabry–Perot cavity is a narrow bandpass filter around a given frequency and can be used as a spectrum analyzer, provided the resonance frequency is tuned. Tunability is usually achieved by modulating the optical length nL of the cavity, for instance by controlling the displacement of one mirror with respect to the other, by means of a piezoelectric transducer. In optoelectronic cavity structures, a current injected into the active zone can change (modulate) the refractive index, which varies with the quantity of injected carriers. The optical gain or absorption is also affected by the injected carriers.

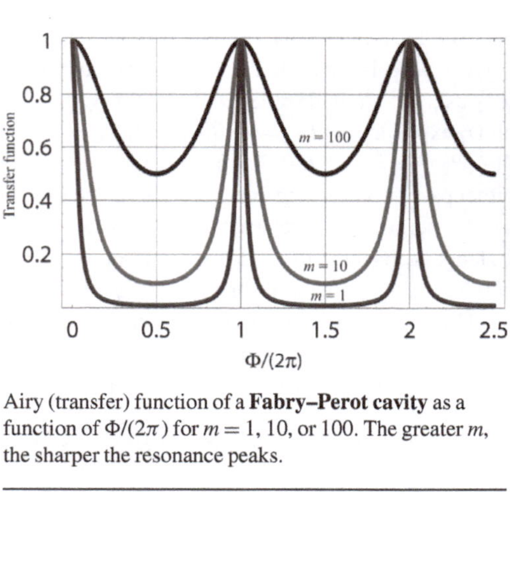

Airy (transfer) function of a **Fabry–Perot cavity** as a function of $\Phi/(2\pi)$ for $m = 1$, 10, or 100. The greater m, the sharper the resonance peaks.

Charles **Fabry** (1867–1945), right, and Alfred **Perot** (1863–1925), left, were the first French physicists to construct an optical cavity for interferometry (Fabry: AIP Emilio Segrè Visual Archives, E. Scott Barr Collection; Perot: *Astrophys. J.*, **64**, 208, 1926.)

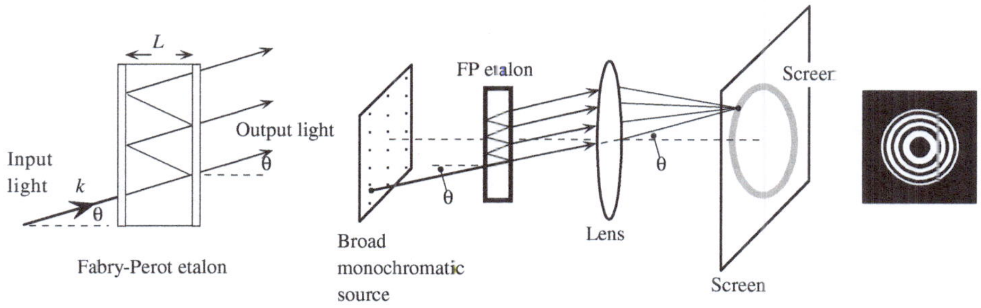

Fabry–Perot optical resonator and the **Fabry-Perot interferometer** (schematic).

Fabry–Perot interferometer is a Fabry–Perot cavity, or etalon, based interferometer that produces an interference ring pattern (bright and dark rings) when illuminated from a broad monochromatic light source. The resonator consists of two parallel flat glass plates facing each other and separated by an adjustable spacer. The inside surface of the glass plates are coated to enhance reflections within the cavity. (Dielectric mirrors can also be used on the inner surfaces.) If L is the separation of the plates, θ the angle at which a particular light wave is incident on the etalon (with respect to the etalon normal), $k = 2\pi/\lambda$ is the wavevector, then the bright rings correspond to the condition $2L\cos(\theta) = m\lambda/n$, where m is an integer (1,2,3 . . .) and n is the refractive index of the cavity medium. The interference ring diameter depends on the wavelength and the optical separation (refractive index × distance) of the plates of the etalon. The interferometer can be used in spectroscopic applications such as in the measurement of the source wavelength. See *Fabry–Perot cavity*.

Fabry–Perot laser amplifier, see *optical semiconductor amplifier*.

Fabry–Perot resonator or **etalon**, see *Fabry–Perot cavity, optical resonator*.

Fall time is the time it takes for a pulse (e.g. an optical pulse) to fall from its 90% peak value to its 10% peak value.

Fano factor in high-energy radiation or particle detectors compares the variation in the statistics of the number of electron and hole pairs created upon the absorption of a high-energy photon (such as from γ-radiation or x-rays) to statistics one would expect if the number of electron and hole pairs created was to follow Poisson statistics. Suppose N is the collected number of carriers upon the absorption of a given amount of high-energy radiation. Then N will have certain statistics over numerous measurements, and the *Fano factor, F*, is defined by

$$F = \text{Fano factor} = \frac{\text{Observed variance in } N}{\text{Poisson predicted variance in } N}.$$

$F = 1$ if N follows Poisson statistics. $F < 1$ represents less spread in N statistics, and hence the detector is "better" if one was interested in relating N to the energy of the absorbed photon. In Si and Ge high-energy radiation detectors, F is observed to be less than 1, and about ~0.15. (U. Fano, *Phys. Rev.*, **70**, 44,1946; **72**, 26, 1947.)

Far-field distribution refers to the field distribution or intensity distribution very far from an aperture or source so that the intensity pattern, by and large, can be examined in terms of its angular dependence (the dependence on the distance from the source can be neglected in examining the intensity distribution).

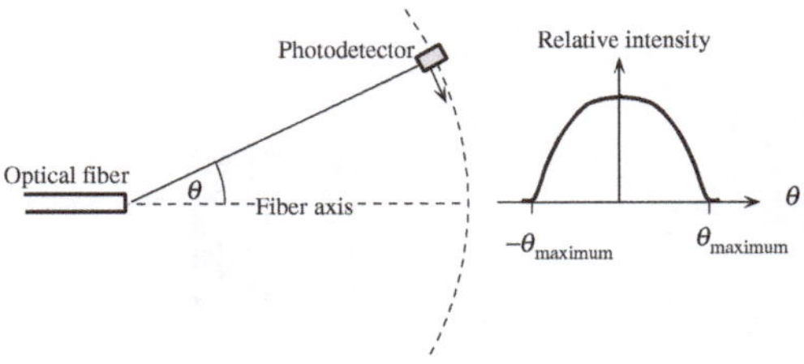

Far-field distribution from an optical fiber.

Michael **Faraday** (English physicist, 1791–1867) (Courtesy of AIP Emilio Segrè Visual Archives, Zeleny Collection.)

Faraday effect, originally observed by Michael Faraday (1845), is the rotation of the plane of polarization of a light wave as it propagates through a medium that has been placed in a magnetic field that is parallel to the light propagation direction. When an optically inactive material such as glass is placed in a strong magnetic field and then plane polarized light is sent along the direction of the magnetic field, it is found that the emerging light's plane of polarization has been rotated. The magnetic field can be applied, for example, by inserting the material into the core of a magnetic coil – a solenoid. The induced specific rotatory power (θ/L) has been found to be proportional to the magnitude of the applied magnetic field, B. The amount of rotation θ is given by

$$\theta = \vartheta B L$$

where B is the magnetic field (flux density), L is the length of the medium, and ϑ is the so-called *Verdet* constant. It depends on the material and the wavelength. The Faraday effect is typically small. A magnetic field of ~0.1 T causes a rotation of about 1° through a glass rod of length 20 mm. It seems to appear that an "optical activity" has been induced by the application of a strong magnetic field to an otherwise optically inactive material. There is, however, an important distinction between the natural optical activity and the Faraday effect. The sense of rotation θ in the Faraday effect, for a given material (Verdet constant), depends only on the direction of the magnetic field **B**. The Verdet constant ϑ is normally positive so that for light propagating along the direction of **B**, the optical field **E** rotates in the same sense as an advancing right-handed screw pointing in the direction of **B**. The direction of light propagation does not change the absolute sense of rotation of θ. If we reflect the wave to pass through the medium again, the rotation increases to 2θ. The Verdet constant depends not only on the wavelength λ but also on the charge-to-mass ratio of the electron and $n(\lambda)$ through

$$\vartheta = -\frac{(e/m_e)}{2c}\lambda\frac{dn}{d\lambda}.$$

Table: Faraday effect Verdet constants of some selected materials.

Material	Quartz 589.3 nm	Flint glass 632 nm	Tb-Ga Garnet 632 nm	Tb-Ga Garnet 1064 nm	ZnS 589.3 nm	Crown glass 589.3 nm	NaCl 589.3 nm
ϑ (rad m^{-1} T^{-1})	4.0	4.0	134	40	82	6.4	9.6

Fermat's principle is a variational principle in geometric optics which states that, between two points A and B, the trajectory (Γ) actually followed by a light ray is the one for which the optical path, $L = \int_{\Gamma} n\, ds$, is a minimum; the curve (Γ) is then said to be *stationary*. Fermat's principle can also be derived as a consequence of the *Eikonal equation*. Stated simply, the actual trajectory of light corresponds to the *least time* of propagation (or the *shortest path*). As a consequence, in a *linear*, *homogeneous* and *isotropic* medium, the light ray propagates as a *straight line*. It should be noted that this principle has a mechanical counterpart, *Maupertuis's principle* of *least action*; this formal equivalence, already noted during the eighteenth century, is one of the arguments which led Louis de Broglie to the unification of two points of view previously thought of as incompatible, the description of light and/or matter as a particle or wave. The so-called *wave mechanics* is the ancestor of our current *quantum physics*.

Pierre de **Fermat** (1601–1665) was a French mathematician who made many significant contributions to modern calculus, number theory, analytical geometry and probability. (AIP Emilio Segre Visual Archives.)

Fermi energy, E_F or Fermi level, may be defined in several equivalent ways. The Fermi level is the energy level corresponding to the energy required to remove an electron from the semiconductor; there need not be any actual electrons at this energy level. The energy needed to remove an electron defines the work function Φ. We can define the Fermi level to be Φ below the vacuum level. E_F can also be defined as that energy value below which all states are full and above which all states are empty at absolute zero of temperature. A difference in the Fermi energy, ΔE_F, in a system is the external electrical work done per electron either on the system or by the system just as the electrical work done when a charge e moves through a potential difference is $e\Delta V$. It can be viewed as a fundamental material property. In more advanced texts it is referred to as the chemical potential of the semiconductor i.e. Gibbs free energy per electron.

Fermi–Dirac statistics determines the probability of occupancy of a state at an energy level E by an electron. It takes into account that when we are considering a collection of electrons, they must obey

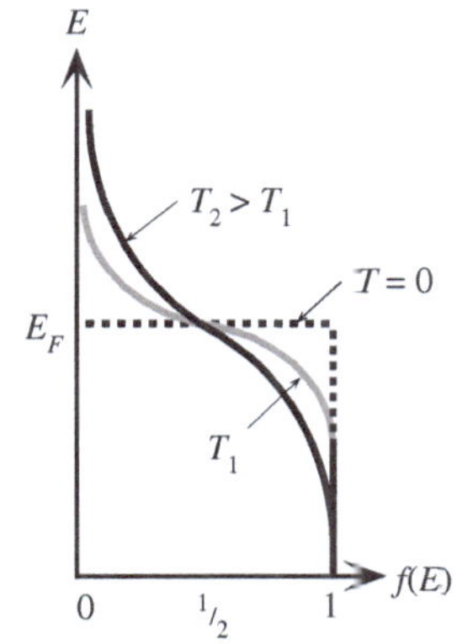

The **Fermi–Dirac** function, *f(E)*, describes the statistics of electrons in a solid. The electrons interact with each other and the environment so that they obey the Pauli exclusion principle.

the Pauli exclusion principle. The Fermi–Dirac function quantifies this probability via

$$f(E) = \frac{1}{1 + \exp\left(\dfrac{E - E_F}{k_B T}\right)}$$

where E_F is the Fermi energy, k_B is the Boltzmann constant and T is the temperature (kelvins).

Fermi's golden rule describes the quantum mechanical transition rate for electrons from an initial state with energy E_i to a final state with energy E_f as a result of an external perturbation to the system applied at some particular time. The external perturbation can, for example, result from the application of an electromagnetic field such as light. The perturbed system states can be found by solving the *time-dependent Schrödinger equation* for this system:

$$H|n\rangle = E_n|n\rangle$$

where H is the Hamiltonian operator, $|n\rangle$ is the eigenvector (wavefunction) and E_n is the eigenvalue, the characteristic energy, of the quantum state $|n\rangle$, and n assumes integer values (quantum number). If the unperturbed Hamiltonian is H^0 and if H' is the perturbation, the perturbed system has a Hamiltonian $H = H^0 + H'$. By using first-order time-dependent perturbation theory, the transition rate W_{if} between two states $|i\rangle$ (initial state) and $|f\rangle$ (final state) under the applied perturbation can be shown to depend on two important factors: the magnitude of the quantum mechanical perturbation integral $\langle i| H' |f\rangle$, and the density of states per unit energy, $N_f(E)$, in the final state; $N_f(E)$ is a function of the electron energy E. The rate of transition is

$$W_{if} = \frac{2\pi}{\hbar} \left|\langle f|H'|i\rangle\right|^2 N_f(E)$$

where $\langle f| H' | i\rangle$ in the Dirac notation is defined by

$$\langle f|H'|i\rangle = \int \psi_f H' \psi_i^* d\tau$$

and $d\tau$ is the unit volume element. From the above, a simplified expression for the transition rate can be derived as

$$W_{if} = \frac{2\pi}{\hbar} \left|\langle f|H'|i\rangle\right|^2 \frac{\sin[(E_f - E_i)t/\hbar]}{\pi(E_f - E_i)}.$$

However, from Heisenberg's uncertainty principle, the transition between the two states $|i\rangle$ and $|f\rangle$ can only occur if $|E_f - E_i|t \gg \hbar$, in which case the last term in the expression for W_{if} reduces to the delta function, $\delta(E_f - E_i)$, giving *Fermi's golden rule* for the first-order transition rate,

$$W_{if} = \frac{2\pi}{\hbar} \left|\langle f|H'|i\rangle\right|^2 \delta(E_f - E_i).$$

As an example, consider the interband transition rate for transitions within an infinite quantum well. The matrix element $\langle f| H' | i\rangle$ will only have finite values for transitions between electron and hole states with the same quantum numbers; that is, for transitions between the first electron state and the first hole state, or between the second electron state and the second hole state. The reason for this is due to the orthogonality of the wavefunctions for the different states. For a finite quantum well, the same selection rules also exist, but they are less rigid; that is, for a symmetric finite quantum well, for example, the wavefunctions will still have a definite parity (being even for odd quantum number states, and being odd for even quantum number states). Fermi's golden rule is so-called because it has been an exceptionally useful equation in calculating quantum mechanical rates of transitions – a "golden equation."

Ferry–Porter law see *critical flicker frequency*.

Fiber see *optical fiber*.

Fiber lasers are lasers in which the gain medium is a long length (typically several km long) of doped optical fiber, disposed between optical mirrors formed by splicing Bragg gratings to its ends. Dopants include rare earth elements such as Er, Yb, Nd and Tl. The fiber laser is normally optically pumped using a high-power diode laser or other fiber lasers. Such lasers can sustain kW level CW output owing to the ease of cooling the very long length of fiber and the efficiency of waveguiding in such high-quality fibers. Moreover, owing to the ability to coil the fiber length, such lasers can be made to be very compact. There are numerous applications of such lasers from laser cutting and welding to medicine and telecommunications that take advantage of the ability to deliver the light that is already coupled into a fiber without a complicated optical system. Fiber lasers are also fabricated using fiber nonlinear optical properties to provide gain – these include stimulated Raman scattering and four-wave mixing.

Fiber optic sensors are sensors that use fiber optics to provide a detectable output caused by an environmental or an external signal input. Generally such sensors are placed in two categories: *intrinsic* and *extrinsic* or hybrid sensors. In the *extrinsic* case, the fiber is merely used as a conduit to bring light to and from a device that itself responds to the external environmental stimuli. For example, this might include two end polished fibers separated by a variable small gap, *d*, with one fiber bringing light in and the other collecting light from the first fiber. The light intensity seen by the collecting fiber will depend on the core/cladding contrast and dimensions of core/cladding in the fibers, as well as *d*. Clearly the system provides for intensity modulation as a function of *d*, and may serve as a sensor of distance, vibration, pressure, and so forth. In the case of *intrinsic* fiber sensors, the environmental effect may directly cause measurable changes to the propagation and confinement of light within the fiber system. In either case, fiber sensors can offer significant advantages over conventional sensors, including immunity to electromagnetic interference and significant weight savings and compactness – this is particularly true for aerospace applications. Intensity-based fiber optic sensors often suffer from limitations resulting from variable losses in a given system unrelated to the environmental effect that is to be measured. Potential sources of error include variable losses due to connectors and splices, microbend and macrobend losses, misalignment errors for light sources and detectors. To circumvent this, dual wavelength sensors are often used, where one wavelength is used to calibrate the system and remove errors by bypassing the sensing region. Alternatively, so-called *spectrally based* sensors may be used which are inherently insensitive to errors introduced by intensity variations. The sensors operate on the principle of sensing based on an environmental effect modulating the wavelength of light within the sensing system – such sensors include black body, absorption and fluorescence sensors. Probably however, the sensor area of most importance is interferometric fiber optic sensors. This includes Sagnac, Mach–Zehnder and Michelson interferometers, as well as ring resonators, dual mode, polarimetric, rating and etalon based interferometers. The Sagnac interferometer is used principally to measure rotation, and is a replacement for ring laser gyroscopes and mechanical gyroscopes. Applications of fiber optic gyroscopes include automotive navigation systems, pointing and tracking of satellite antennas, and inertial measurement systems for aircraft and missiles. Fiber-based Mach–Zehnder and Michelson interferometers have a great advantage of extreme flexibility in their geometry and high sensitivity, enabling planar arrays, omnidirectional elements, line arrays and gradient elements to be realized. Many of the intrinsic and extrinsic types of sensors discussed above may be multiplexed in such a way as to support multiple sensors on a single fiber optic line. Sensing from individual sensors may be accomplished using time, frequency, wavelength, and polarization multiplexing, for example. In the case of time division multiplexing, a pulsed light source is injected into the fiber system and time delay is used to discriminate between sensors. Additional approaches include forming large sensor arrays – for example, $m \times n$ arrays comprising m-input (source) fibers and n-output (detector) fibers. Such arrays can be used to monitor large structures such as bridges and buildings. Some important applications of fiber optic sensors include fiber ring laser gyroscopes for aerospace applications, industrial processes control in applications where ruggedness and safety (i.e. especially in areas where electrical discharges could be hazardous) are at a premium. In the area of medicine and automotive, fiber sensors are becoming more and more pervasive. A strong growth area is that of so-called *smart structures* in which fiber optic sensors can be embedded and provide information on materials evaluation and structure control. See *Mach–Zehnder interferometer, Sagnac effect.*

Fiber-to-fiber attenuation represents the loss of optical power when the optical signal is passed from one fiber to another that have been connected together either through a connector or by *splicing* the fibers. The losses are called connector and splice losses. Connector losses tend to be anywhere from 0.2 to 2 dB, depending on the connectors, and splice losses usually close to 0.1 dB. There are several reasons for the losses: (a) mismatch in fiber cores or mismatch in mode field diameters; (b) lateral offset or misalignment, i.e. the second fiber's axis is laterally shifted; (c) angular offset or misalignment (misalignment of fiber axes); (d) longitudinal offset, i.e. spacing between fiber ends; (e) mismatch in numerical apertures (NAs); (f) reflections at fiber ends; (g) fiber faults such as a noncircular core or an elliptical core, nonconcentric core, etc.; (h) mechanical fiber end faults such as an angled end surface, a hackled surface or a lip at the fiber end generated during the cleaving, etc. All latter mismatch effects lead to some degree of attenuation. In calculating the total fiber-to-fiber loss individual losses are calculated in dB and then simply summed up. Usually two or more of these mismatch losses occur at the same time. The *Nemoto–Makimoro equation* provides a means of calculating the fiber-to-fiber loss when it arises from the two connected fibers having different numerical apertures, and the two fibers

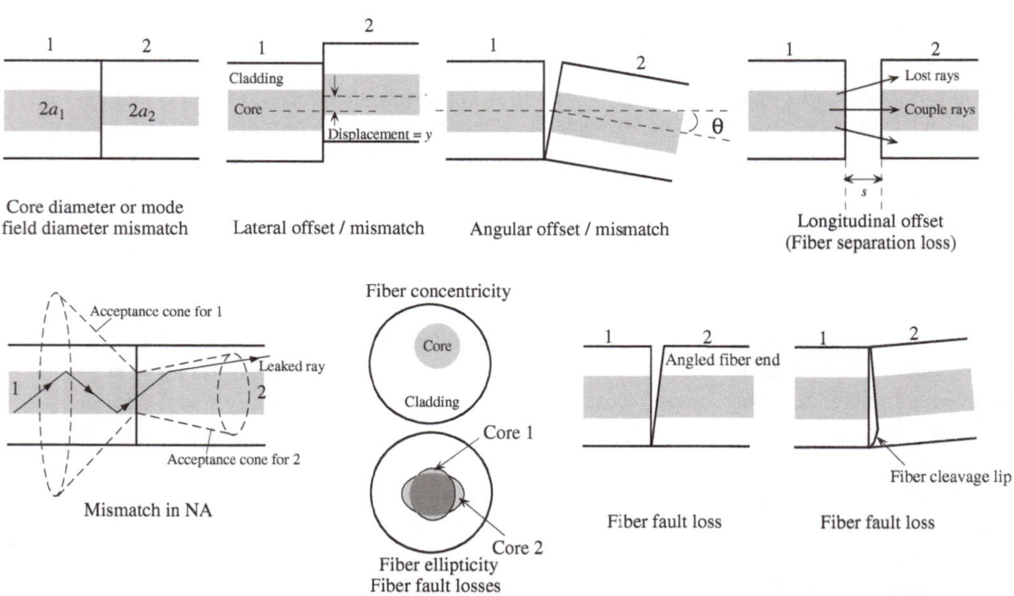

Different types of **fiber-to-fiber attenuation**.

Table: Fiber-to-fiber attenuation Fiber-to-fiber losses in dB. n_1 = core refractive index; n_o = refractive index of the medium between the fibers; a = fiber core radius; NA = numerical aperture; V = V-number; a = core radius; w = mode field radius or spot size, $w = a(0.65 + 1.62V^{-3/2} + 2.88V^{-6})$. Useful equations from various sources including G. Keiser, *Optical Fiber Communications,* Third Edition, McGraw-Hill, Boston, 2000. The factor 4.343 arises from 10/ln(10).

Loss	Loss (dB) =	Comment
Mode field diameter mismatch	$-10\log\left[\frac{4}{(D_1/D_2+D_2/D_1)^2}\right]$	D is the mode field diameter = $2w$ D_1 and D_2 = MFD of fibers 1 and 2
Lateral mismatch, single mode	$4.343\left(\frac{y}{w}\right)^2$	y = lateral displacement between fiber axes $w = a(0.65 + 1.62V^{-3/2} + 2.88V^{-6})$
Lateral mismatch, multimode	$-10\log\left(1-\frac{8y}{3\pi a}\right)$	Approximate for small y/a i.e. $y/a < 0.4$
Angular mismatch, single mode	$4.343\left(\frac{\pi n_2 w\theta}{\lambda}\right)^2$	θ in radians; n_2 = cladding refractive index
Angular mismatch, multimode	$-10\log\left(1-\frac{3n_1\sin\theta}{3\pi \mathrm{NA}}\right)$	θ in radians. $\mathrm{NA} = (n_1^2 - n_2^2)^{1/2}$
Fiber separation, single mode	$10\log\left(\frac{a}{a+s\arcsin(\mathrm{NA}/n_o)}\right)^2$	s = separation n_o = refractive index of the medium between the two fibers.
Numerical aperture mismatch,	$-10\log(\mathrm{NA}_2/\mathrm{NA}_1)^2$	$\mathrm{NA}_2 < \mathrm{NA}_1$, otherwise there is no loss. 1 and 2 are fiber labels.
End reflection	$-\log(1-R)$	R = reflectance $R = [(n_1 - n_o)/(n_1 + n_o)]^2$

being laterally, longitudinally and angularly offset by y, s and θ,

$$L_{\mathrm{dB}} = -10\log\left[\frac{19n_1^2n_0^2}{(n_1+n_o)^4}\frac{4\sigma}{q}\exp\left(-\frac{\rho u}{q}\right)\right]$$

where n_1 = core refractive index of fibers; n_o = refractive index of the medium between the fibers; w_1 = mode field radius (1/e location in the evanescent wave) of fiber 1 (transmitting fiber); w_2 = mode field radius of fiber 2 (receiving fiber); λ = source wavelength; $k = 2\pi n_o/\lambda$; $\sigma = (w_2/w_1)^2$; $\rho = (kw_1)^2$; $q = G^2 + (\sigma+1)^2$; $G = s/(kw_1{}^2)$; $u = (\sigma+1)^2F + 2\sigma FG\sin\theta + \sigma^2(G^2+\sigma+1)\sin^2\theta$; $F = y/(kw_1{}^2)$; y = lateral displacement between fiber axes; s = separation of fiber ends; q angular misalignment in radians. The loss includes reflections.

Fiber Bragg grating refers to an optical fiber with a periodically modulated core index, obtained by photoinscription of an interference pattern through UV irradiation. As in distributed-feedback (DFB) structures, the inside grating is responsible for the spectrally selective behavior; the Bragg fiber acts thus as a *spectral filter,* with a forbidden band centered on the Bragg wavelength $\lambda_B = 2\,n_m\,\Lambda$, where Λ is the period of the corrugation and n_m the average effective refractive index. Its spectral properties are well described within the framework of *coupled-wave equations* (CWE), as a function of a coupling constant κ. The spectral transfer function in intensity is given by

$$T(\delta) = 1/|\cosh(\gamma L) + j(\delta/\gamma)\sinh(\gamma L)|^2$$

with L the total interaction length, $\delta = n_m\omega/c - \pi/\Lambda$ the detuning with respect to the Bragg condition ($\delta = 0$ is associated with $\lambda_0 = \lambda_B$), and $\gamma = [|\kappa|^2 - \delta^2]^{1/2}$. Moreover, any modification of the fiber (due to temperature changes, tension, and so on) leads to a detuning of the spectral response. This is used to realize Bragg sensors, the detuning being

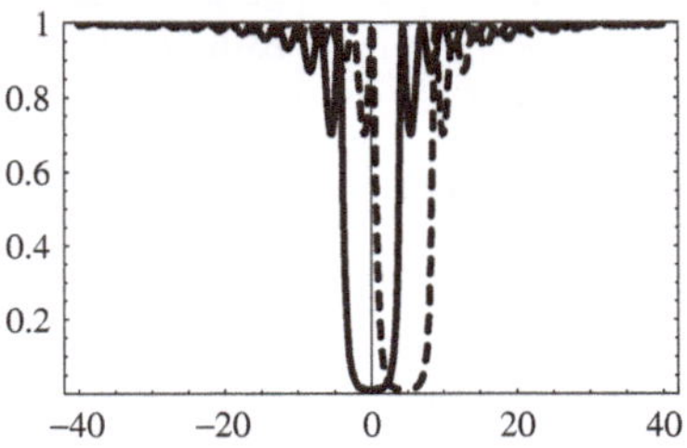

$T(L\delta)$ vs. $L\delta$. Spectral transmission of a **fiber Bragg grating** ($|\kappa L| = 3$) used as a sensor, without constraint (straight line) and under constraint (dashed line). More elaborate index profiles, such as multiperiodic or quasiperiodic corrugations, can also be designed and photoinscripted inside the fiber in order to tailor at will its spectral response.

directly related to the applied constraint, i.e. temperature changes, tension, etc. Fiber Bragg gratings date back to 1978 when Kenneth Hill and coworkers demonstrated the first Bragg reflection gratings. The formation of a Bragg reflection filter relied on the photosensitivity of germanium-doped fibers in the wavelength region from 170 to 400 nm, due to the strong absorption bands originating with the Ge dopant. In their work, the core of a single-mode Ge-doped fiber was exposed to high-power 488 nm laser irradiation causing the formation of a standing wave pattern in the fiber and hence creating an interference filter. The formation of the filter was shown to result from a nonlinear process producing a periodic modulation of the index of refraction within the Ge-doped core region of the fiber. Since Hill's initial work, 193 nm excimer laser irradiation has been used to fabricate *transverse holographic diffraction gratings*. Such an approach permits (a) grating formation at much lower powers, (b) an ability to create highly wavelength selective modal discriminators, (c) the capability to write holographic patterns at practically any longer wavelength than the writing laser, and (d) an inherent facility to write a large number of gratings into a single fiber. These fiber-written gratings have found many applications from sensing to telecommunications.

Field curvature or **curvature of field** is an aberration in which an optical system images a flat object such a plane onto a curved image, that is, the image is a curved surface, instead of a scaled flat image of the object. This type of aberration is associated with extended objects. Even in the absence of astigmatism, the focal plane is not a flat plane as expected by the paraxial approximation but a paraboloidal curved surface, which is known as the *Petzval surface* (after the Hungarian mathematician Josef Max Petzval, 1807–1891). A collimated light beam of rays $A, A', \ldots$ parallel to the lens (optical) axis is focused to F_A whereas parallel rays $C, C', \ldots$ in a beam *inclined* to the lens axis are focused to F_C on the Petzval surface, and not on the paraxial focal plane. Thus, optical systems tend to image better on curved surfaces than on flat surfaces. For a converging lens the Petzval surface bends towards the object whereas for the diverging lens it bends away, so that a suitable combination would eliminate field curvature aberration; the required condition is known as the *Petzval condition* and for m thin lenses, it is given by

$$\frac{1}{n_1 f_1} + \frac{1}{n_2 f_2} + \cdots + \frac{1}{n_m f_m} = 0$$

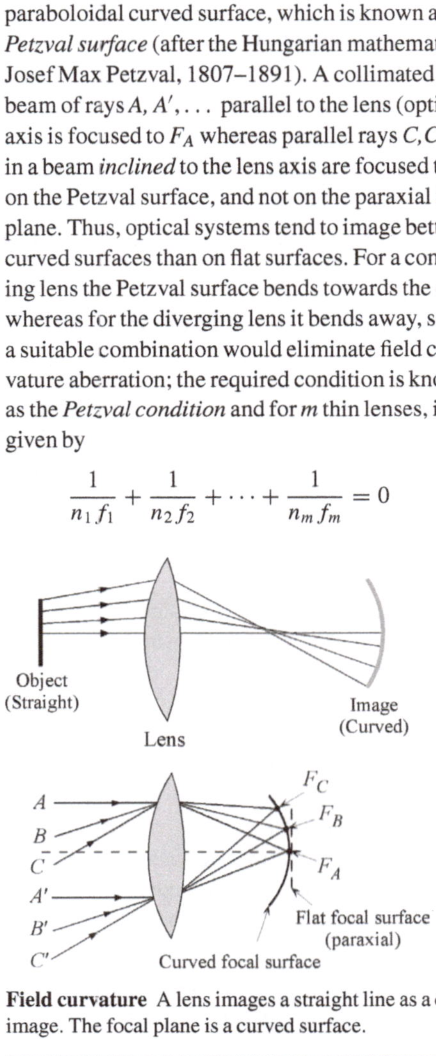

Field curvature A lens images a straight line as a curved image. The focal plane is a curved surface.

where n_1 and f_1 are the refractive index and focal length of lens 1, etc. and diverging lenses have negative focal length. Even if one converging and one diverging lens with $|f_1| = |f_2|$ and $n_1 = n_2$ were used, if they were separated by a distance d, the overall combination would still have a positive focal length (act as a converging lens) with $f = |f_1|^2/d$

Field effect transistor (FET) is a transistor whose normal operation is based on controlling the conductance of a channel between two electrodes by the application of an external field. The effect of the applied field is to control the current flow. The current is due to majority carrier drift from the source to the drain and is controlled by the voltage applied to the gate.

Field flattener lens see *Petzval lens*.

Fill factor (FF) in active matrix image detectors represents the fraction of the area of a pixel in the array that can actually receive the image photons and hence capture the image. The "dead" area that is not able to participate in photon detection usually has some other elements such as a thin film transistor, etc.

Fill factor (FF) in photovoltaics is a figure of merit for a solar cell that gauges what maximum fraction of the power, represented by (short circuit current)×(open circuit voltage), is available as output power deliverable to a load, i.e.

$$\mathrm{FF} = \frac{I_{\mathrm{m}} V_{\mathrm{m}}}{I_{\mathrm{sc}} V_{\mathrm{oc}}}$$

where FF is the fill factor, I_{m} and V_{m} are the current and voltage corresponding to maximum power transfer to load, and I_{sc} and V_{oc} are the short-circuit current and open-circuit voltage. It is clearly advantageous to have FF as close to unity as possible but the exponential *pn* junction *I-V* properties prevent this. Typically FF values are in the range 70–85% and depend on the device material and structure. There is a useful empirical equation for calculating the FF,

$$\mathrm{FF} \approx \frac{\upsilon_{\mathrm{oc}} - \ln(\upsilon_{\mathrm{oc}} + 0.72)}{\upsilon_{\mathrm{oc}} + 2}$$

where $\upsilon_{\mathrm{oc}} = V_{\mathrm{oc}}/(nk_{\mathrm{B}}T/e)$ is the normalized open-circuit voltage (normalized with respect to the thermal voltage $k_{\mathrm{B}}T/e$), T is the temperature (K), k_{B} is the Boltzmann constant, e is the elementary charge and n is the ideality factor. The maximum power output from a solar cell is

$$P_{\max} = \mathrm{FF} I_{\mathrm{sc}} V_{\mathrm{oc}}.$$

See *photovoltaic I-V characteristics*.

(M. Green, Ch 8, "Solar cells" in *Modern Semiconductor Device Physics*, S. M. Sze (Editor), John Wiley, 1997)

Filter is a device that allows only certain frequencies (or wavelengths) to pass and others to be blocked. Filters typically have a certain resonant frequency υ_0 and a bandwidth $\Delta\upsilon$ around υ_0 such that input (incoming) signals within this $\Delta\upsilon$ are passed.

Finesse of a *Fabry–Perot resonator* is a measure of how fine, i.e. narrow, an individual resonant cavity mode is with respect to the separation of two consecutive modes. For a low-loss optical resonator, it is the ratio of mode separation ($\Delta\upsilon_m$) to spectral width ($\delta\upsilon_m$). The narrower the individual mode width and larger the mode separation, the greater is the finesse. Mathematically, the finesse F is defined in terms of the mirror *reflectance* R as

$$F = \frac{\pi R^{1/2}}{1 - R}; \; R > 0.6.$$

Finger electrodes see *photovoltaic devices*.

Finite element method (FEM) is a collection of techniques which can produce near-optimal approximate solutions to initial boundary value partial differential equations common to optoelectronics. The implementation typically involves subdividing the geometry of the problem into a finite number of small regions called *finite elements*. Many convenient shapes are available and include quadrilaterals and triangles in 2D, and tetrahedral, pentahedral and hexahedral in 3D. One is said to use a *weak statement* or weighted residual finite element theory, if one requires that the approximation error be a minimum. All combinations of appropriate boundary conditions can be naturally included in this process. These basic constraints produce a set of algebraic equations written on each finite element, and which are then collected together using a procedure called *assembly* to form the global matrix statement. The approximation-dependent variable distribution is then determined by solving this matrix statement using a chosen linear algebra technique.

Originally, the FEM was conceived of to help structural engineers analyze aircraft structures in the 1950s – as this technique matured the concept of force-balance was replaced by theory founded in variational calculus and classical Rayleigh–Ritz methods. Application of these classical theoretical concepts to unsteady field transport is not directly possible – this is partially due to the fact that velocity terms may be coupled to field gradients, which is not compatible with a variational principle as required by the Rayleigh–Ritz theory. Thus, derivative terms are now typically replaced by algebraic divided difference expressions, or finite differences – a related procedure thus employs integration over cells whereby the gradient terms are replaced with finite volume quotients. Essentially, every finite difference and/or finite volume recursion relation replacement for specific derivative terms can be reconstructed within the weighted residual or weak statement framework intrinsic to the FEM.

As an illustration of this approach, consider the one common problem finding discrete solutions to Maxwell's equations in heterogeneous dielectric and semiconductor domains. Solving these equations in the frequency domain couples fields at all points in space and yields large systems of equations. A more practical approach is to integrate Maxwell's equations in the time domain using finite differences. The equations' fundamental hyperbolicity decouples points separated by Δx if the time step $\Delta t < \Delta x/c$, where c is speed of light. The discrete problem is then solved locally at each time step by summing the nodal contributions from nearest neighbors and integrating each node independently using a leapfrog scheme. "Lumping" permittivity at the nodes can further simplify the calculations. This approach eliminates large systems of frequency-domain equations and yields a simple, fast algorithm well suited to parallel computer architectures. There are two standard methods to spatially discretize Maxwell's equations – using finite differences and finite elements. Finite differences are the most efficient for structurally simple models, while finite elements are better suited to structurally complex models, but increase the demands on computational resources. The latter also more directly yield information on field sampling and include nonlinear material behavior more readily. The dominance of finite elements in the commercial sector is due to geometric adaptability and modeling ease. Computational speed often is not the primary concern in many cases. A critical issue for discrete solvers is grid truncation error caused by spurious reflections at a given model's artificial boundaries. Radiation conditions are necessary to reduce this error. In time-domain analysis these are usually local relations established among nodal boundary values derived from the paraxial wave equation. However, higher order implementations tend to degrade in 3D vector domains and also, the local phase velocity is required.

Flat panel display refers to a video display in which the thickness of the display is very small as compared with conventional cathode ray tube displays. Typically this type of display uses liquid crystal displays, although gas plasma and field-emission displays are also common. The technology offers high-resolution large area displays for a number of applications including portable electronics (e.g. laptops), as well as handheld devices (e.g. personal digital assistants, cellular phones, and electronic games).

Flicker noise see *excess noise*.

Flint glass see *chromatic dispersion.*

Fluence represents the number of photons flowing through a unit area, whose plane is placed perpendicular to their propagation direction. The fluence of x-ray photons can be predicted from the x-ray exposure in roentgens from the conversion equation

$$\frac{X}{\Phi} = \frac{1}{543{,}000} E(\mu_{en}/\rho)_{\text{air}} \equiv \kappa(E)$$

where the exposure X is in roentgens, the fluence Φ is in photons mm^{-2}, E is the photon energy, and $(\mu_{en}/\rho)_{\text{air}}$ is the mass energy absorption coefficient of air. See *roentgen, exposure, dose.*

Fluorescence is the emission of longer wavelength radiation from a material when it absorbs a shorter wavelength light that *causes* the longer wavelength light emission; the persistence of emission after the cessation of excitation is very short (less than 10 ns). Fluorescence occurs, for example, when a substance absorbs UV radiation and emits light in the visible region. See *luminescence, photoluminescence.*

Fluorescent tube (lamp) is a gas discharge lamp in which excited Hg atoms emit UV radiation that is absorbed and remitted as visible light by a phosphor coating on the inside of the tube. The tube contains Hg vapor. The Hg atoms are excited by projectile electrons that are accelerated by a field set up inside the tube by the applied voltage. The electrodes of the tube are filaments that can be heated to initiate electron emission to start the gas discharge. The tube has a mixture of Hg vapor and either argon or xenon gas. When the tube is turned on, the filaments are heated and thermionically emit electrons. Due to the applied voltage, these electrons become accelerated and hence impact ionized gas atoms and thereby form a plasma. This plasma is conducting and initiates an electric discharge i.e. electric conduction. Once the gas is conducting, the voltage across the tube is actually less. With the gas becoming conducting and many projectile electrons in the tube, Hg atoms become excited by impacting projectile electrons. These impact-excited Hg atoms emit UV radiation (mainly at 254 nm) to return to their de-excited state. The UV radiation is absorbed by the phosphor coating on the inside of the tube. The phosphors absorb the UV radiation and emit visible light. The emitted light spectrum depends on the phosphors used in the coating that convert the UV radiation to visible light. Usually several phosphors are used together to obtain the desired overall spectral effect. For example, Eu^{2+} and Eu^{3+} activated phosphors provide emissions in the blue and red regions and the Tb^{3+} activated phosphor

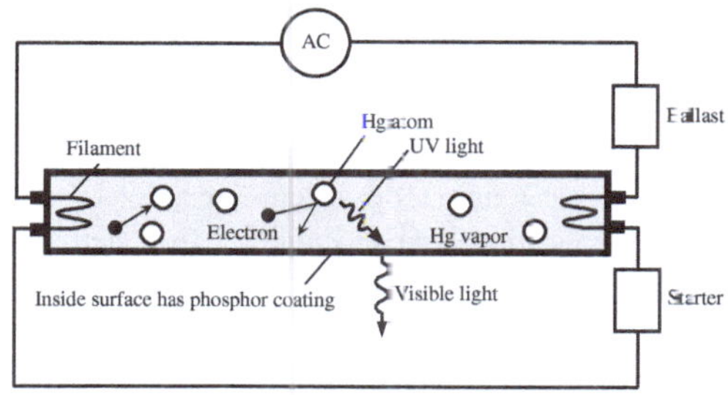

The **fluorescent tube**. The ballast and the starter (the starting switch) are usually combined into a single electronic unit called the *rapid start ballast.*

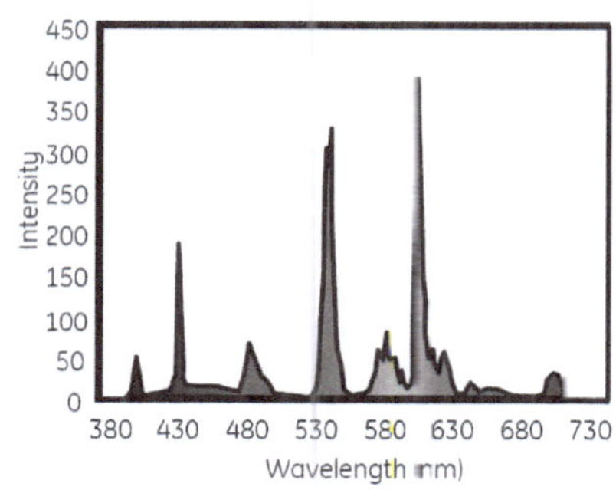

The light output spectrum from a particular triphosphor **fluorescent tube**, GE SPX35 FL, rated as 97 lm/W. The vertical axis is the relative intensity (not in absolute units).

luminesces in the green region so the overall spectrum from using these three rare earth phosphors would appear to be "white." There are fluorescent tubes with various mixtures of phosphors in their coatings to generate smooth spectra for various lighting applications. (An example of a Eu^{2+} activator based phosphor would be $BaMg_2Al_{16}O_{27}$:Eu^{2+}.) The *ballast* is a common accessory device that helps in the initiation of the gas discharge and then controls the flow of current in the gas discharge tube; many modern ballasts are electronic. Compact fluorescent lamps have become quite popular recently toward directly replacing incandescent bulbs, and have the ballast integrated into the bulb. They can be screwed into normal lamp sockets, and are more efficient in converting electrical power to luminous flux. A typical fluorescent tube can have luminous efficacies of 30–100 lm/W whereas an incandescent bulb would have 15–20 lm/W.

Fluorine is a gas that is used to dope silica (and thereby introduce F into the SiO_2 glass network) to reduce its refractive index.

Flux, Φ, in radiometry and photometry has three related definitions, radiant, luminous and photon flux, which correspond to the rate of flow of radiation energy, perceptible visual energy, and photons in "energetic," "visual" and "photonic" systems, respectively. For example, *radiant flux* is the energy flow per unit time in units of watts. When spectral information is required, these quantities are modified into *spectral densities*; for instance, the *spectral density of flux*, or *spectral flux*, Φ_ν or Φ_λ [W Hz^{-1} or W nm^{-1}], are such that

$$\int \Phi_\nu(\nu)d\nu = \int \Phi_\lambda(\lambda)d\lambda = \Phi.$$

The flux in the visual system, the *photometric* system, called the *luminous flux*, fundamental for lighting studies, is relative to a standardized human eye. The photometric flux $\Phi_v(\lambda)$ is related to its energetic counterpart, *radiant flux*, $\Phi_e(\lambda)$ by $\Phi_v(\lambda) = (683\ \text{lm W}^{-1})V(\lambda)\Phi_e(\lambda)$, where $V(\lambda)$ is the relative *luminous efficiency* of the eye, or its "visual spectral efficiency"; it is a maximum at $\lambda_{max} = 555$ nm where $V(\lambda_{max}) = 1$. The photonic system of quantities such as photonic flux is especially useful for monochromatic light. The *photon flux* is expressed as a number of photons per second (instead of W), the photon intensity as a number of photons per second per unit of solid angle, and so on. Flux in optical applications refers to optical power flowing or to radiation energy flow at optical frequencies. See *radiant flux, luminous flux, lumen, luminous efficiency*.

Flux conservation law see *radiation energy conservation*.

Flux density see *irradiation*.

Focal plane of a lens is a plane (σ in the figure) onto which a lens focuses a collimated beam passing through its center but at an arbitrary angle to the lens axis. Stated differently, it is a plane that has all the possible focal points such as F, F', etc. that result when a narrow collimated beam (a bundle of parallel light rays) that is parallel to the lens axis is tilted with the center ray in the beam still passing through the lens center.

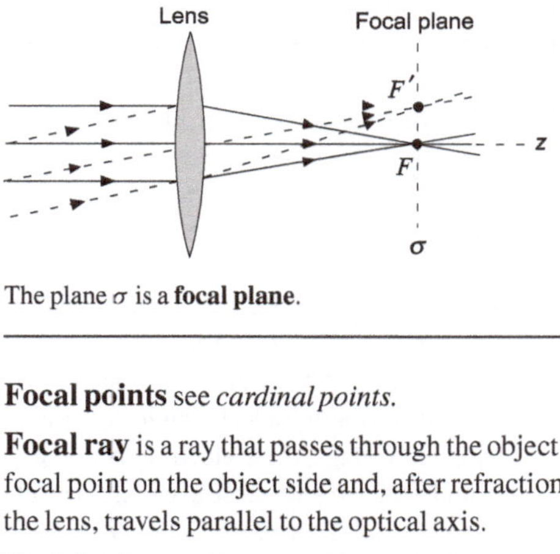

The plane σ is a **focal plane**.

Focal points see *cardinal points*.

Focal ray is a ray that passes through the object focal point on the object side and, after refraction by the lens, travels parallel to the optical axis.

Fock basis provides a complete description of the state of a light field, analogous to that provided by the *Wigner function*. The photon number distribution is given by the diagonal elements of the density matrix in the Fock basis. The diagonal elements of the density matrix contain the information on the intensity distribution of the state, while the phase information is contained within the off-diagonal elements.

Forbidden lines see *parity selection rule*.

Förster–Dexter interaction in luminescence is a nonradiative electromagnetic energy transfer mechanism in which an excited atom becomes de-excited by transferring the de-excitation energy ΔE to a neighboring atom, which becomes excited by an amount ΔE. Thus, the donor atom, called the sensitizer, falls in energy by ΔE and the acceptor atom gains energy by ΔE without the donor emitting and the acceptor absorbing photons. However, the acceptor atom must have energy levels such that it can be excited by an amount ΔE. This is a *resonance energy* transfer from one center to another, and does not involve the donor emitting an actual photon and the acceptor absorbing the same photon, but rather the transfer of energy by resonance coupling between the donor and the acceptor. Two atoms are energetically coupled if an energy difference ΔE between two energy levels in one (the sensitizer) matches an energy difference in the other (the acceptor). As a result of the interactions between the two atoms, one atom (the sensitizer) can be de-excited and the other (the acceptor) can be excited at the same time, leading to an energy transfer. Polyatomic molecules and atoms in solids have numerous energy levels (which contain fine vibrational energy levels as well) and, in practice, it is not unusual for the absorption spectrum of the acceptor to easily overlap the fluorescence spectrum of the sensitizer. For example, the energy from a sensitizer in a phosphor is transferred to the activator through the Förster–Dexter mechanism. Initial observations of this type of energy transfer have been attributed to Cario and Franck in 1922. When a mixture of mercury (Hg) and thallium (Tl) vapor is irradiated with radiation that has the Hg resonance line, the emission from the gas mixture exhibits the emission spectra of both Hg and Tl atoms. Tl atoms do not absorb the exciting ration and cannot reabsorb the emitted photons from the Hg atoms Tl atoms must therefore have been excited by indirect energy transfer, which must be a nonradiative energy transfer. The Hg atom, the donor, acts as the sensitizer that excites the Tl atom, which is the acceptor. The energy transfer can occur over a long range, for example 5–10 nm, and depends on the separation between the donor and acceptor atoms. The resonance energy transfer is strongest when the transitions in the sensitizer and the acceptor are both allowed electric dipole transitions. In the latter case, the interaction effectively corresponds to a *dipole–dipole interaction* and the rate of energy transfer from excited sensitizers to activators can be simplified to

$$\text{Rate} = \frac{1}{\tau_S}\left(\frac{R_0}{R}\right)^6$$

where R is the mean separation between the sensitizer and the activator, τ_S is the mean lifetime of the excited sensitizer, and R_0 is the critical transfer distance for which excitation transfer and spontaneous deactivation of the sensitizer are of equal probability. R_0 is quite complicated and involves the spectral distribution of fluorescence of the sensitizer, the refractive index of the medium, etc. (T. Förster, *Ann. Pjysik,* **2**, 55, 1948; *Naturforsch.*, **44**, 321, 1949; *Z. Electrochem.*, **53**, 93, 1949; D.L. Dexter, *J. Chem Phys.*, **21**, 836, 1953.)

Forouhi–Bloomer dispersion relations are semiempirical dispersion relations for the refractive index n and the extinction coefficient K in terms of the photon energy $\hbar\omega$ for photon energies greater than the bandgap E_g for a wide range of semiconductor materials. Due to various features and singularities in the E–**k** diagrams of crystalline semiconductors, the optical constants n and K for $\hbar\omega > E_g$ are not readily expressible in simple terms. Various authors, for example, Forouhi and Bloomer, have nonetheless provided useful and tractable expressions for modeling n and K in this regime.

Table: Forouchi-Bloomer dispersion relations
Forouhi–Bloomer coefficients for selected semiconductors for four terms $i = 1$ to 4. First entry in the box is for $i = 1$, and the fourth is for $i = 4$.

	A_i	B_i (eV)	C_i (eV^2)	$n(\infty)$	E_g (eV)
Si	0.00405	6.885	11.864	1.950	1.06
	0.01427	7.401	13.754		
	0.06830	8.634	18.812		
	0.17488	10.652	29.841		
Ge	0.08556	4.589	5.382	2.046	0.60
	0.21882	6.505	11.486		
	0.02563	8.712	19.126		
	0.07754	10.982	31.620		
GaP	0.00652	7.469	13.958	2.070	2.17
	0.14427	7.684	15.041		
	0.13969	10.237	26.567		
	0.00548	13.775	47.612		
GaAs	0.00041	5.871	8.619	2.156	1.35
	0.20049	6.154	9.784		
	0.09688	9.679	23.803		
	0.01008	13.232	44.119		
GaSb	0.00268	4.127	4.267	1.914	0.65
	0.34046	4.664	5.930		
	0.08611	8.162	17.031		
	0.02692	11.146	31.691		
InP	0.20242	6.311	10.357	1.766	1.27
	0.02339	9.662	23.472		
	0.03073	10.726	29.360		
	0.04404	13.604	47.602		
InAs	0.18463	5.277	7.504	1.691	0.30
	0.00941	9.130	20.934		
	0.05242	9.865	25.172		
	0.03467	13.956	50.062		
InSb	0.00296	3.741	3.510	1.803	0.12
	0.22174	4.429	5.447		
	0.06076	7.881	15.887		
	0.04537	10.765	30.119		

In particular, Forouhi–Bloomer equations express n and K in terms of the photon energy $\hbar\omega$ in a consistent way that obey the Kramers–Kronig relations.

$$K = \sum_{i=1}^{q} \frac{A_i(\hbar\omega - E_g)^2}{(\hbar\omega)^2 - B_i(\hbar\omega) + C_i} \quad \text{and}$$

$$n = n(\infty) + \sum_{i=1}^{q} \frac{B_{oi}(\hbar\omega) + C_{oi}}{(\hbar\omega)^2 - B_i(\hbar\omega) + C_i}$$

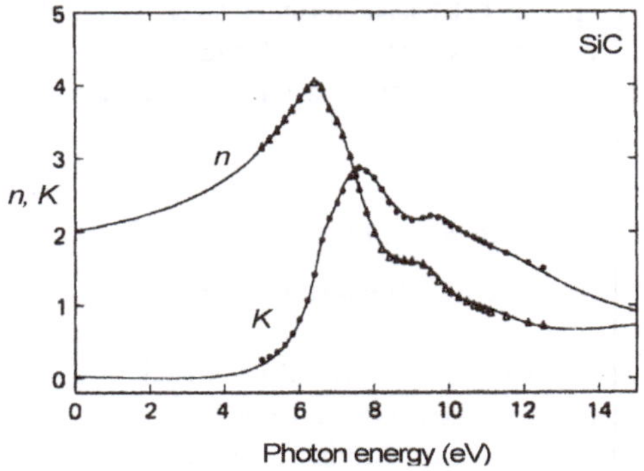

Forouhi–Bloomer dispersion Optical dispersion, n, K, vs. photon energy, for crystalline SiC. The solid line is the Forouhi–Bloomer equations with four terms with appropriate parameters, and points are experimental data. See original reference for details. (Reprinted figure with permission from: A. R. Forouhi, I. Bloomer, *Phys. Rev. B.*, **38**, Issue 3, 1865–1988. © 1988 by the American Physical Society.)

where q is an integer that represents the number of terms needed to suitably model experimental n, K, E_g is the bandgap and A_i, B_i, C_i, B_{oi}, C_{oi} are constants, B_{oi} and C_{oi} depend on A_i, B_i, C_i, and E_g; only the latter four are independent parameters, $B_{oi} = (A_i/Q_i)[-(1/2)B_i^2 + E_gB_i - E_g^2 + C_i]$; $C_{oi} = (A_i/Q_i)[(1/2)(E_g^2 + C_i)B_i^2 - 2E_gC_i]$; $Q_i = (1/2)(4C_i - B_i^2)^{1/2}$. Forouhi and Bloomer provide a table for A_i, B_i, C_i and E_g for four terms in the summations above for a number of semiconductors; an example that shows an excellent agreement between these dispersion relations and the experimental data is shown for SiC.

Four-wave mixing (FWM) occurs in a nonlinear medium when the optical beating of a pump wave with a signal wave with sufficiently close frequencies ω_S and $\omega_P = \omega_S + \Delta\omega$ generates a traveling index and gain modulation. The interaction of the injected waves with these modulations leads to new frequency components at $\omega_S - \Delta\omega$ and especially the up-converted signal at $\omega_C = \omega_P + \Delta\omega$ that can be observed at the output.

Fourier analysis of diffraction

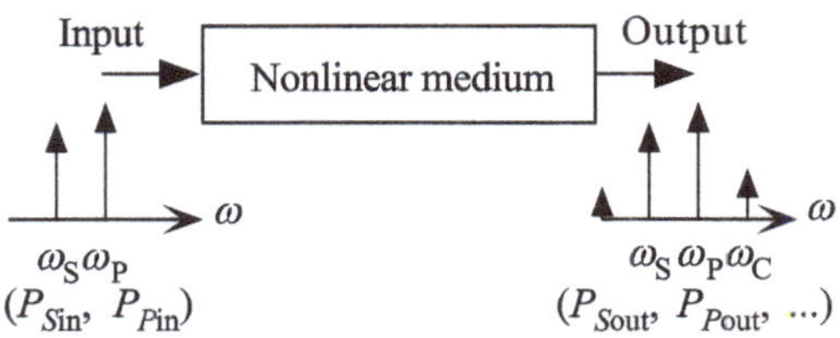

Four-wave mixing $\Delta\omega = \omega_P - \omega_S = \omega_C - \omega_P$. The two optical waves (pump and signal) are injected with input powers P_{Pin}, P_{Sin}, detuning $\Delta\omega$ and the same state of polarization.

The conversion efficiency $\eta_{FWM} = P_{Cout}/P_{Sin}$ is defined as the ratio of the converted output signal at ω_C on the input signal at ω_S. FWM occurs in passive components such as nonlinear crystals or optical fibers as well as in active devices such as semiconductor optical amplifiers (SOAs). In the latter case, FWM can be used for high bit-rate all-optical data processing, frequency conversion of optical data signals, dispersion compensation or time-division demultiplexing. Because of the optical gain, the SOA works with low optical input power ($\approx$1 mW) and short interaction length and is not subject to critical phase matching conditions. When FWM is used as a spectroscopic tool, it enables one to extract its dynamic parameters. The only drawback of SOAs stems from amplified spontaneous emission (ASE) noise added to the output. The theoretical description of FWM involves nonlinear coupled wave equations.

Fourier analysis of diffraction is a mathematical method for describing diffraction in terms of a Fourier transform (FT) of an aperture function A that represents the amplitude and phase of radiation emanating from an illuminated object; the object, for example, can be an opening in an opaque screen which is illuminated by light.

A. *Fraunhofer diffraction from a linear aperture.*

Consider the general linear aperture diffraction problem shown in the figure. The aperture is on the y-axis. Each point on the aperture acts as a source of secondary waves. The relative strength and phase of radiation from each point on the aperture is described by an *aperture function* $A(y)$ that is defined as follows.

$$A(y)dy = \text{Amplitude and phase of the wave leaving region } dy \text{ around the point } Q \text{ at } y \text{ with respect to the origin } O.$$

If we have a single slit of width a and this slit is illuminated with a plane wave, then each point in the slit has the same phase. Further, since the incident wave on the slit is a plane wave, each point in the slit is excited by the same field and hence each has the same strength when emitting radiation

$$A(y) = \begin{cases} 0 & y < 0 \\ \text{constant} & 0 < y < a \\ 0 & y > a. \end{cases}$$

Each point in the aperture emits a spherical wave that has an amplitude proportional to $A(y)dy$. If the screen is far away, we can assume that the wave that

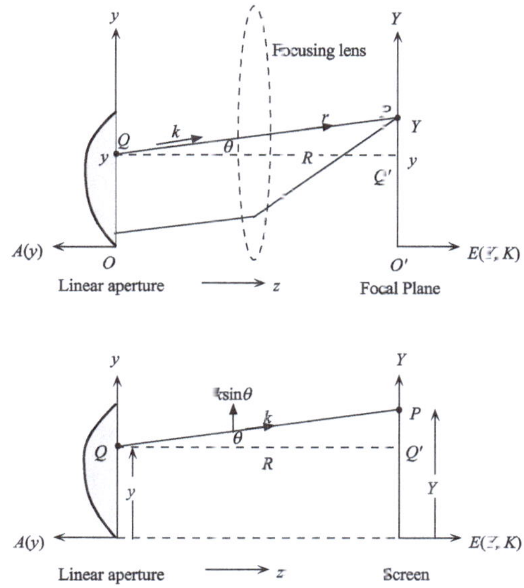

Fourier analysis Diffraction from a linear aperture.

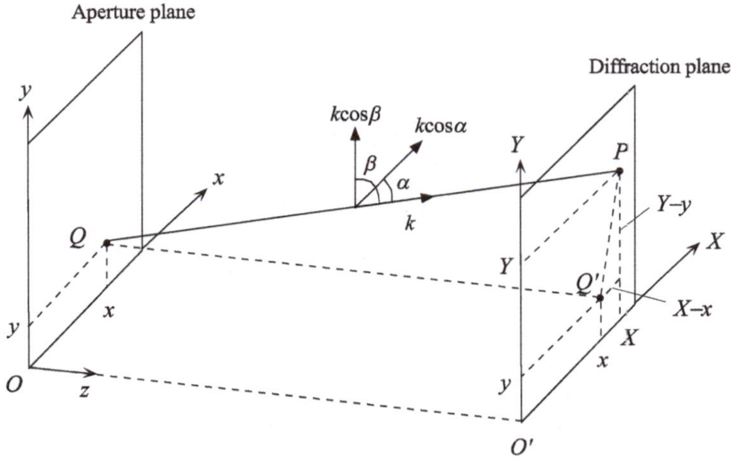

Fourier analysis General diffraction problem.

arrives at P is a plane wave (Fraunhofer approximation). In calculating the field at P coming from Q, we have to consider the total phase change from Q to P. However, we only need to consider the phase change arising from the net propagation along y, that is distance difference $Y-y$ because along z all waves from the aperture have the same phase change at the screen. The wavevector along y is $k \sin \theta$. The field at P due to emission from Q is then

$$dE(r) = CA(y)dy \exp[-jk \sin\theta(Y - y)]$$

where C is a proportionality constant and $\exp(j\omega t)$ has been tacitly assumed. Thus,

$$dE \approx Ce^{-jk_yY} A(y)dy \exp(jk_y y)$$

where we have defined $k_y = k \sin\theta$ as a variable. k_y is called a *spatial frequency*. The total field at P is found by summing all dE from all the points on the y-axis:

$$E(Y, k_y) \approx Ce^{-jk_yY} \int_{-\infty}^{+\infty} A(y) \exp(jk_y y)dy.$$

It is clear that the field E at point P at Y is proportional to the *Fourier transform* of the aperture function and is a function of k_y, the *spatial frequency*.

B. *Two-dimensional aperture*.

The emission from a small elemental area $dxdy$ in the aperture is described by an *aperture function* $A(x,y)$ such that,

$A(x, y)dxdy =$ Relative strength and phase of radiation emanating from a region $dxdy$ at x, y.

Consider the wave arriving at P from Q. We only need to consider the phase change from Q to P along the x- and y-directions because all waves from the aperture travel the same z-distance. Thus the field at P due to Q is

$$dE = CA(x, y)dxdy \exp[-jk \sin\alpha(X - x) - jk \sin\beta(Y - y)]$$

where C is a proportionality constant. Suppose we define $k_x = k \cos\alpha$ and $k_y = k \cos\beta$, then

$$dE = Ce^{-j(k_xX+k_yY)}A(x, y)dxdy \times \exp(jk_x x + jk_y y).$$

The field at P from all the points in the aperture plane is given by summing this field from all Q points in the aperture, that is, by integrating over

 Fourier transform infrared (FTIR) spectroscopy

Joseph von **Fraunhofer** (1787–1826) was a German optician, who produced some of the best optical glasses and lenses at his time. Dark absorption lines in the spectrum of the Sun have been named after him. He has been credited with the invention of the diffraction grating among many other contributions to optics. (AIP Emilio Segre Visual Archives, W. F. Meggers Collection.)

---

all x and y. Thus,

$$E(X, Y, k_x, k_y) = Ce^{-j(k_x X + k_y Y)} \iint_{\text{All } x,y} A(x, y)dxdy \times \exp(jk_x x + jk_y y).$$

It is clear that the field E at point P at X, Y is proportional to the *Fourier transform* of the aperture function and is a function of k_x and k_y, the *spatial angular frequencies.* We can apply the above equation to a plane wave that is incident on a circular aperture of radius a to obtain the diffraction pattern which would be a *Bessel function*. See *Airy disk*.

Fourier transform infrared (FTIR) spectroscopy is a technique for measuring the wavelength-dependent absorption of light passing through a sample with infrared light. Unlike grating or prism spectrometers relying on the dispersion of a broadband light source to provide a wavelength tunable source of radiation to probe the absorption characteristics of a sample, FTIR spectroscopy relies on using an interferometer such as, for example, a so-called Michelson interferometer. This instrument relies on the interference of light from a broadband infrared source traversing two different optical path lengths, one including the sample to be measured, and then combined at a beam splitter. The path length of one beam can be varied by adjusting the location of one of the mirrors with respect to the central beamsplitter. Let $G(k)$ be the intensity spectrum through the sample. At normal incidence to the sample, and for nonmonochromatic incident infrared light with a wavevector k, the detected intensity resulting from the interference of light traversing a given path length difference x, is:

$$\begin{aligned} I(x) &= \int_0^\infty (1 + \cos(kx))G(k)dk \\ &= \int_0^\infty G(k)dk + \int_0^\infty G(k)\frac{[e^{ikx} + e^{-ikx}]}{2}dk \\ &= \frac{1}{2}I(0) + \frac{1}{2}\int_{-\infty}^{\infty} G(k)e^{ikx}dk \end{aligned}$$

giving the more familiar form,

$$W(x) = \frac{2I(x) - I(0)}{\sqrt{2\pi}} = \frac{1}{\sqrt{2\pi}}\int_{-\infty}^{\infty} G(k)e^{ikx}dk$$

which is a Fourier transform – that is, the detected intensity measured as a function of the position of a moving mirror, $I(x)$, can be converted into $G(k)$, the intensity spectrum as a function of frequency by a simple Fourier transform. In practice one cannot measure from $-\infty$ to $+\infty$ and the resolution of an FTIR instrument is simply given by how far in x

one measures – thus, the resolution is proportional to $(2\pi x_{\max})^{-1}$. In practice one therefore normally adopts two approaches to collecting spectral data – rapid scan and step scan measurements. In the case of rapid scan measurements, one sweeps the mirror (and hence x) rapidly and averages many interferograms to provide a quick assessment of the spectra at moderate resolution. Step scan measurements involve stepping to a given x position and taking measurements (with long time averaging or time triggered data) covering very long path lengths. This latter approach is most suitable for high-resolution spectra. Infrared spectroscopy is important in that it is sensitive to local infrared active modes (dipole moments) of given species in molecules (including for example biological matter). Thus, for example, for the case of a molecule such as CO_2, asymmetric stretching or bending modes provide for infrared absorbance at around 2345 cm^{-1} and 550 cm^{-1}, respectively. By using the characteristic fingerprints of a given molecule in an FTIR spectra (e.g. the coincidence of having the two modes discussed above occurring with appropriate oscillator strength) one can unambiguously determine the presence and concentration of such species. This has resulted in such spectroscopy being widely used.

Fournier–Snitzer equation is an equation that describes the nonlinear refractive index n_2, due to the optical Kerr effect, in glasses in which the ions can have three or more energy levels, in terms of the linear refractive index n, photon excitation energies, and the dipole transition probabilities to the higher energy levels and between higher levels. In its simplest form, in CGS units, the equation can be expressed as

$$n_2 \approx \frac{(n^2+2)(n^2-1)}{19n} \times \left(\frac{e^2 x_{2g}^2}{\hbar^2 \omega_g^2}\right)\left(\frac{x_{32}^2}{x_{2g}^2} - 1\right) \qquad \text{[CGS]}$$

where $x_{2g} = \langle 2|x|g\rangle$ is the dipole matrix element that involves the ground state $|g\rangle$ and the next excited state $|2\rangle$, similarly $x_{32} = \langle 3|x|2\rangle$ where $|2\rangle$ and $|3\rangle$ are the first two excited states above the ground state, and $\hbar\omega_g$ is the first excitation energy (or ω_g is a resonance frequency). The expression can be further extended by including a Wemple–DiDomenico dispersion relation in the above expression as done in the original expression. Boling and coworkers (1978) have derived a different equation for n_2 which provides reasonably good values for n_2 close to experimental values. The Boling equation can also be written in terms of n and the Abbe number, and has gained substantial popularity. See *Boling equation.* (J. Fournier and E. Snitzer, *IEEE J. Quantum Electron.* **QE-10**, 473, 1974.)

Franck–Condon principle states that since electronic transitions take place on much shorter time-scales than those required for any significant nuclear (or atomic) motion, electronic transitions can be approximated by so-called vertical transitions on an energy versus configuration coordinate diagram. These considerations are pertinent to analyzing the optical response of molecules as well as so-called deep levels in semiconductors. For the case of a molecule, for example, the Franck–Condon principle states that after a molecule is excited from its ground state, the nuclear conformation must readjust. In general, this leads to an increase in the allowed vibrational transitions, greatly influencing the observed optical spectrum. The Franck–Condon principle for calculating transition probabilities using fixed nuclear positions is a more restrictive approximation than the Born–Oppenheimer approximation which states that nuclear and electronic motions can be separated, but does not demand that the nuclear coordinate be held at a fixed value. An illustration of how the Franck–Condon principle relates to transition energies for an electron in a semiconductor having a deep defect energy level located close to the middle of the energy bandgap is shown in the figure. The Franck–Condon principle assumes that nuclear motion can be neglected on the time-scale of electronic transitions and hence optical transitions can

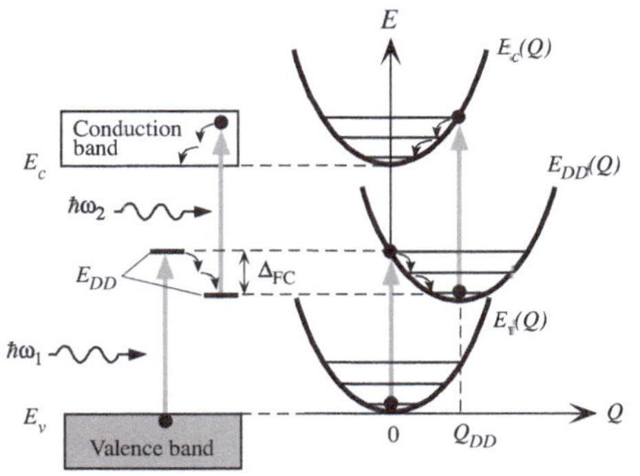

Illustration of the **Franck–Condon principle** for a deep defect (DD) level in a semiconductor.

be denoted by vertical transition on a total energy E (i.e. electronic and lattice or vibrational energy) versus total configurational coordinate Q. Assume that the electronic system has a minimum energy in both its initial and final states at $Q = 0$ and that the defect state has minimum energy at $Q = Q_{DD}$. After the initial absorption of a photon of energy $\hbar\omega_1$, an electron makes a vertical transition to the defect level and vibrationally de-excites itself, losing energy Δ_{FC}, by the emission of an integer number S of optical phonons, each of energy $\hbar\omega_o$ – where S is known as the *Hwang–Rhys factor*. The de-excitation can result in the electron in the defect level occupying a state near the minimum of the $E_{DD}(Q)$ curve in the figure. One can consider two possible cases for the electron in this state – either it radiatively decays back to the ground state $E_v(Q)$ with a vertical transition, or it absorbs a second photon, exciting it by a vertical transition into the state $E_c(Q)$. The former transition results in the emission of a photon of energy less than $\hbar\omega_1$ and the light is said to be *Stokes shifted*. The latter transition involves the absorption of a second photon of energy $\hbar\omega_2$, resulting in the excited electron occupying an empty conduction band state. Subsequent de-excitation of the excited electron can occur to restore the conduction band electron energy to a minimum energy value near the conduction band edge. The Hwang–Rhys factors for various deep levels in III–V semiconductors can be quite large (as much as a tenth of the bandgap energy) owing to strong coupling of defect states to the lattice. (J. Franck, *Trans Faraday Soc* , **21**, 536, 1926; E. U Condon, *Phys. Rev.*, **28**, 858, 1928.)

Franz–Keldysh effect is electric field induced absorption of light with photon energies slightly less than the bandgap in a semiconductor due to the modification of the energy band edge by the applied field. This effect can be easily understood as corresponding to photon-assisted tunneling through the energy gap, where carrier wavefunctions are exponentially decaying functions. In the presence of an applied electric field F, a valence band electron must tunnel through a triangular potential barrier to reach the conduction band. In the absence of photons, the barrier has height E_g and width $t = E_g/eF$. In the presence of photons of energy $\hbar\omega$, the barrier height is reduced to $E_g - \hbar\omega$ and width $(E_g - \hbar\omega)/eF$. Hence the tunneling probability is considerably enhanced with photon absorption. An analysis of this photon-assisted tunneling process for a direct bandgap semiconductor with spherical bands gives the electroabsorption coeffient α:

$$\alpha(\omega) = \alpha_0(\omega)\exp\left[-\frac{4}{3}\sqrt{\frac{2m^*}{\hbar^2}}\frac{(E_g - \hbar\omega)^{3/2}}{eF}\right]$$

where $\alpha_0(\omega)$ is the absorption coefficient in the absence of electric field, m^* is the effective mass, E_g is the optical gap, e is the electron charge and F is the applied electric field. It should be mentioned that the electric field induced change in the absorption coefficient implies a change in the refractive index as well. (W. Franz, *Z. Naturforsch.* **13a**, 484, 1958; L.V. Keldysh, *Zh. Eksperim. i Teor. Fiz.* **34**, 1138, 1958, English translation in *Soviet Phys.—JETP* **7**, 788, 1958; J. Dow and D. Redfield, *Phys Rev B*, **1**, 3358, 1970.)

Fraunhofer diffraction see *diffraction*.

Free carrier absorption (FCA) is the attenuation of an electromagnetic (EM) wave as a result of its interaction with the free carriers (e.g. electrons) in a medium. FCA is therefore intimately connected with the conductivity of the medium. An EM with sufficiently low-frequency oscillations can interact with free carriers in a material and thereby drift the carriers. This interaction results in an energy loss from the EM wave to the lattice vibrations through the carrier scattering processes. Based on the Drude model, the relative permittivity $\varepsilon_r(\omega)$ due to N free electrons per unit volume is given by

$$\varepsilon_r = \varepsilon_r' - j\varepsilon_r'' = 1 - \frac{\omega_p^2}{\omega^2 - j\omega/\tau}; \quad \omega_p^2 = \frac{Ne^2}{\varepsilon_0 m_e}$$

where ω_p is a *plasma frequency* which depends on the electron concentration, and τ is the relaxation time of the scattering process, i.e. the mean scattering time. For metals where the electron concentration is very large, ω_p is of the order of $\sim 10^{16}$ rad s^{-1}, UV frequencies, and for $\omega > \omega_p$ $\varepsilon_r \approx 1$, the reflectance becomes very small. Metals lose their free-electron reflectance in the UV range, thus becoming UV transparent. The reflectance does not fall to zero because there are other absorption processes such as interband electron excitations or excitations from core levels to energy bands. Plasma edge transparency where the reflectance almost vanishes can also be observed in doped semiconductors. For example, the reflectance of doped InSb has a plasma edge wavelength that decreases with increasing free carrier concentration. The above can be written in terms of the conductivity σ_0 at low frequencies (dc) as

$$\varepsilon_r = \varepsilon_r' - j\varepsilon_r'' = 1 - \frac{\tau\sigma_0}{\varepsilon_0[(\omega\tau)^2 + 1]} - j\frac{\sigma_0}{\varepsilon_0\omega[(\omega\tau)^2 + 1]}.$$

In metals, σ_0 is high. At frequencies where $\omega < 1/\tau$, the imaginary part $\varepsilon_r'' = \sigma_0/\varepsilon_0\omega$ is normally much more than 1, and the refractive index $n = K \approx \surd(\varepsilon_r''/2)$, where K is the extinction coefficient, so that the free carrier attenuation coefficient α is then given by

$$\alpha = 2k_0 K \approx \frac{2\omega}{c}\left(\frac{\varepsilon_r''}{2}\right)^{1/2} \approx (2\sigma_0\mu_0)^{1/2}\,\omega^{1/2}.$$

Furthermore, the reflectance R can be calculated also using $n = K \approx \surd(\varepsilon_r''/2)$, which leads to the well-known *Hagen–Rubens relationship*

$$R \approx 1 - 2\left(\frac{2\omega\varepsilon_0}{\sigma_0}\right)^{1/2}.$$

In semiconductors, one encounters typically $\sigma_0/\varepsilon_0\omega < 1$, since the free-electron concentration is small, and we can treat n as constant due to various other polarization mechanisms, e.g. electronic polarization. Since $2nK = \varepsilon_r''$, the absorption coefficient α is

$$\alpha = 2k_0 K \approx \frac{2\omega}{c}\left(\frac{\varepsilon_r''}{2n}\right) = \frac{\sigma_o}{nc\varepsilon_0[(\omega\tau)^2 + 1]}.$$

At low frequencies where $\omega < 1/\tau$, we have $\alpha(\lambda) \sim \sigma_o/n(\lambda)$ so that α should be controlled by the dc conductivity, and hence the amount of doping. Furthermore, α will exhibit the frequency dependence of the refractive index n, i.e. $\alpha(\lambda) \sim 1/n(\lambda)$, thus n would typically be determined by the electronic polarization of the crystal. At high frequencies where $\omega > 1/\tau$,

$$\alpha \sim \sigma_0/\omega^2 \sim N\lambda^2.$$

α is proportional to N, the free carrier concentration, and λ^2. Experimental observations on FCA in doped semiconductors are in general agreement with these predictions. For example, α increases with N, whether N is increased by doping or by carrier injection. However, not all semiconductors show the simple $\alpha \sim \lambda^2$ behavior. A proper account of the field-driven electron motion and scattering must consider the fact that τ will depend on the electron's energy. The correct approach is to use the Boltzmann transport equation with the appropriate scattering mechanism. FCA can be calculated by using a quantum mechanical approach based on second-order time-dependent perturbation theory with Fermi–Dirac statistics.

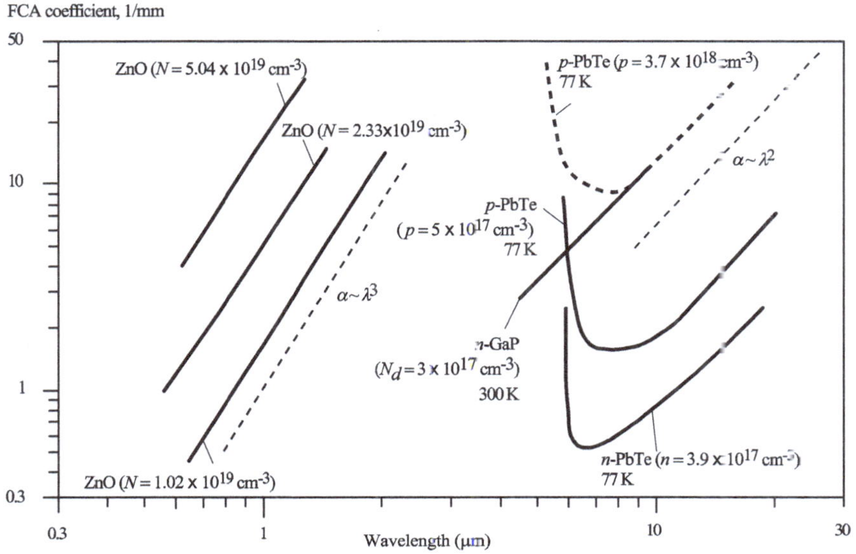

Free carrier absorption in n-GaP at 300 K (J. D. Wiley and M. DiDomenico, *Phys. Rev. B.*, 1, 1655, 1970), p and n type PbTe (H. R. Riedl, *Phys. Rev.*, **127**, 162,1962) at 77 K and In doped n-type ZnO at room temperature (R. L. Weihler, *Phys. Rev.*, **152**, 735, 1966).

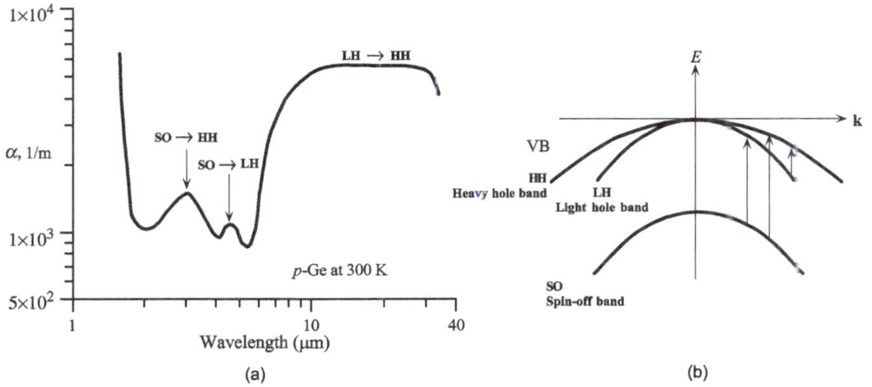

Free carrier absorption (a) Free carrier absorption due to holes in p-Ge (W. Kaiser, E. J. Collins and H. Y. Fan, *Phys. Rev.*, **91**, 1380, 1953).
(b) The valence band of Ge has three bands; heavy hole band, light hole band and a spin-off band. LH, light hole band; HH, heavy hole band; SO, spin-off band.

Absorption due to free carriers is commonly written as $\alpha \sim \lambda^p$, where the index p depends primarily on the scattering mechanism, though it is also influenced by the compensation doping ratio, if the semiconductor has been doped by compensation, and the free carrier concentration. In the case of lattice scattering one has to consider scattering from acoustic and optical phonons. For acoustic phonon scattering $p \approx 1.5$, for optical phonon scattering $p \approx 2.5$, and for impurity scattering, $p \approx 3.5$. The observed free carrier absorption coefficient will then have all three contributions,

$$\alpha = A_{\text{acoustic}}\lambda^{1.5} + A_{\text{optical}}\lambda^{2.5} + A_{\text{impurity}}\lambda^{3.5}.$$

Inasmuch as α for FCA depends on the free carrier concentration N, it is possible to evaluate the latter from the experimentally measured α, given its wavelength dependence and p. One useful empirical relationship for FCA in semiconductors is given by

$$\alpha = K_\alpha \left[\frac{N(\text{cm}^{-3})}{10^{17}}\right]\left[\frac{\lambda(\mu\text{m})}{9}\right]^p.$$

Table: Free carrier absorption Constants K_α and p in the FCA empirical relationship (data from H.Y. Fan, *Semiconductors and Semimetals*, Vol. 3, R. K. Willardson and A. C. Beer (Editors), Academic Press, 1967, Ch. 9)

	GaAs	InP	GaSb	InAs	InSb	AlSb
N (10^{17} cm^{-3})	1–5	0.4–4	0.5	0.3–8	1–3	0.4–4
K_α	3	4	6	4.7	2.3	15
p	3	2.5	3.5	3	2	2

Free carrier absorption in p-type Ge demonstrates how the FCA coefficient α can be dramatically different from what is expected from simple concepts based on the EM field interacting with free carriers. The figure shows the wavelength dependence of the absorption coefficient for p-Ge over the wavelength range from about 2 to 30 μm. The observed absorption is due to excitations of electrons from the spin-off band to the heavy hole band, and from the spin-off band to the light hole band, and from the light hole band to the heavy hole band as marked in the figure. (H. E. Ruda, *J. Appl. Phys.*, **61**, 3035, 1987). See *Drude model.*

Free electron laser (FEL) differs from most conventional lasers by its active medium, a relativistic electron beam generated by a particle accelerator, then injected into a periodic magnetic field (*wiggler*) that leads eventually to a periodic distribution of the electronic density. When accelerated, charged particles emit radiation: the FEL laser wavelength is selected by a phase matching condition on the waves emitted by synchronous electrons.

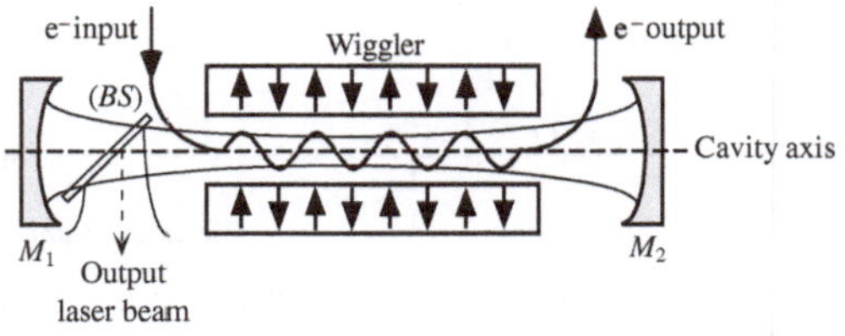

Schematic representation of a **free electron laser**.

As shown in the above figure, to couple out the laser beam, a beam splitter BS is inserted onto the main axis of the optical cavity. As shown in the figure, mirrors M_1 and M_2 define the optical cavity. However, the main selection mechanism comes from the periodicity of the magnetic field. (Protection walls are not represented.)

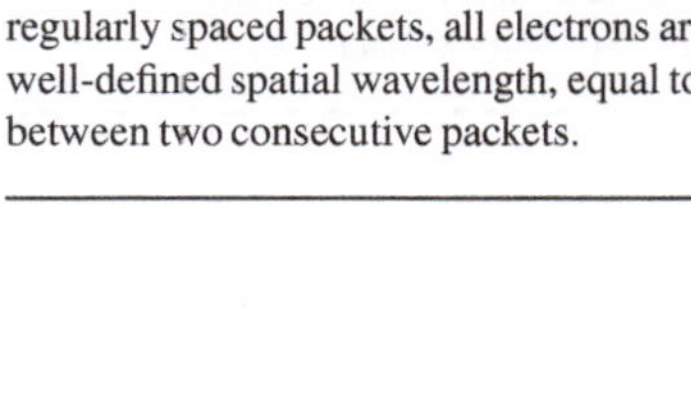

Free electron laser Left: Incoherent radiation emitted by randomly distributed electrons. Right: Distributed in regularly spaced packets, all electrons are "in phase" for a well-defined spatial wavelength, equal to the distance between two consecutive packets.

FELs can achieve coherent, high-power and near diffraction-limited radiation, continuously tunable from the millimetric range up to the near UV, especially valuable for nuclear physics, materials science, biological and biomedical applications. Intense picosecond sources are already available in the mid- to far-IR, enabling the investigation of fast kinetics or time-resolved nonlinear spectroscopy. Far-UV to x-ray sources are currently under research. The FEL wavelength depends on the periodicity λ_w of the wiggler magnetic field, the energy of the electron γ expressed in terms of its rest mass (0.511 MeV),

$$\lambda = \frac{\lambda_w}{2\gamma^2}(1 + K^2/2)$$

where K is the wiggler strength parameter that depends on the beam magnetic field B and λ_w, and given by $K = 0.934B\lambda_w$ (in which B and λ_w are in teslas and cm). Short wavelengths require small wiggler periodicity and a large electron energy. (G. R. Neil and L. Merminga, *Rev. Sci. Intrum.*, **74**, 685, 2002.)

Free spectral range (FSR) is the spectral interval between two consecutive resonance modes in an optical resonator (in a *Fabry–Perot cavity*). For a cavity of length L, of refractive index n, and for normal incidence, the frequency interval $\Delta\upsilon$, and hence FSR, is

$$\Delta\upsilon = c/(2nL).$$

The corresponding value expressed in wavelength (in vacuum) is $\Delta\lambda = \lambda^2/(2nL)$. In a dispersive medium in which $n = n(\upsilon)$, or $n = n(\lambda)$, the FSR becomes

$$\Delta\upsilon = c/(2N_g L),$$

where $N_g = n + \upsilon\,(dn/d\upsilon) = n - \lambda\,(dn/d\lambda)$ is the *group index*. Since $N_g > n$ in normally dispersive media, the FSR is usually reduced with respect to the nondispersive case. Note that it does not depend upon the mirror reflectivities (whereas the *finesse factor* does).

Frenkel exciton see *exciton*.

Frequency chirp describes a light pulse that has a varying instantaneous frequency. A pulse traveling in a linear dispersive medium is subject to spreading because its constituent frequencies travel at different velocities (group-velocity dispersion, GVD). On the other hand, in a nonlinear material like a Kerr medium, self-phase modulation (SPM) can alter the (instantaneous) frequency of the pulse. The phase of the pulse can be written as $\varphi(t) = \omega(t)t + \varphi_0$, and the *chirp* is completely determined by the instantaneous angular frequency $\omega(t)$. The precise relationship depends strongly on the medium (normal or anomalous dispersion, sign of the Kerr effect). The chirp can be used for ultrashort pulse compression, where faster traveling components are forced to follow slightly longer optical paths than slower ones (by the use of well-adjusted prisms or gratings). An exact compensation between linear GVD and nonlinear SPM is also the very mechanism that is required for *soliton* propagation.

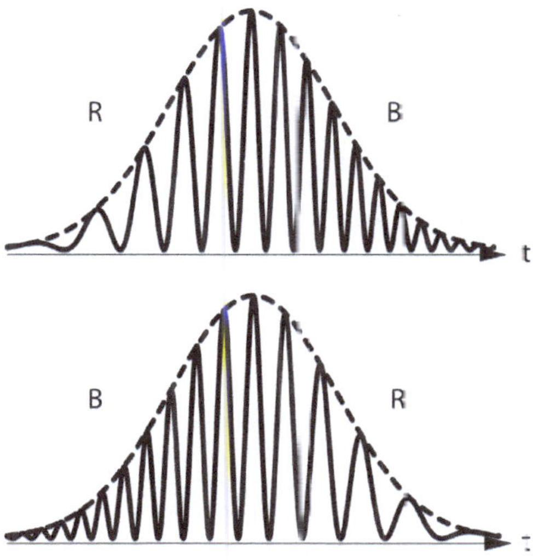

Two classes of **frequency chirp**: top, the red-shifted frequencies (marked R) travel faster (appear sooner) than the blue-shifted frequencies (marked B); bottom, quite the opposite.

Frequency division multiplexing see *wavelength division multiplex.*

Fresnel diffraction see *diffraction.*

Fresnel lens is a special lens, designed by Augustine-Jean Fresnel, that has a large aperture and a relative short focal length; the lens is lighter, and has smaller volume, than an equivalent solid planoconvex lens, and was originally designed for lighthouse applications. The main idea is to break the lens into sections and reduce the thickness of each section while keeping the "bending power" (determined by the curved surface) of each section the same. The original Fresnel lens had the sections as prisms, and looked as if it were constructed by putting together different prisms. Modern Fresnel lenses are continuous, and are fabricated using modern machining techniques. They are used in inexpensive applications where accuracy is not critical, e.g. *traffic lights*, etc.

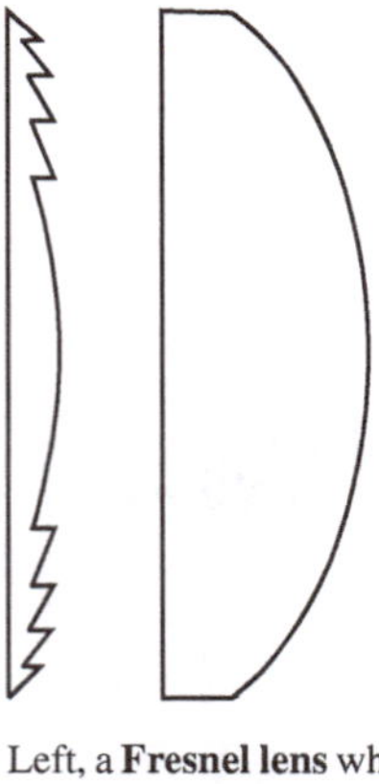

Left, a **Fresnel lens** which has nearly the same diffractive power as the planoconvex lens on the right. The Fresnel lens has sections with curvatues that can bend the light as much as the corresponding curvature on the planoconvex lens. Each section in the Fresnel lens, however, is thinner.

Fresnel prism is a prism that converts incoming unpolarized light into two divergent beams that have opposite circular polarizations. Light that enters the right-handed quartz prism (R-prism), can be thought of as two different circularly polarized waves, right circular polarization (R-polarization) and left circular polarization (L-polarization). These travel with different velocities and experience different refractive indices n_R and n_L, where $n_R < n_L$ so that R-polarization propagates faster. In the left-handed quartz prism (L-prism), however, the indices are reversed $n_R > n_L$ as L-polarization moves faster in L-quartz. Thus R-polarization wave at the interface experiences an increase in n_R and L-polarized wave experiences a decrease in n_L. Therefore they have different angles of refraction and hence enter the L-prism diverging from each other.

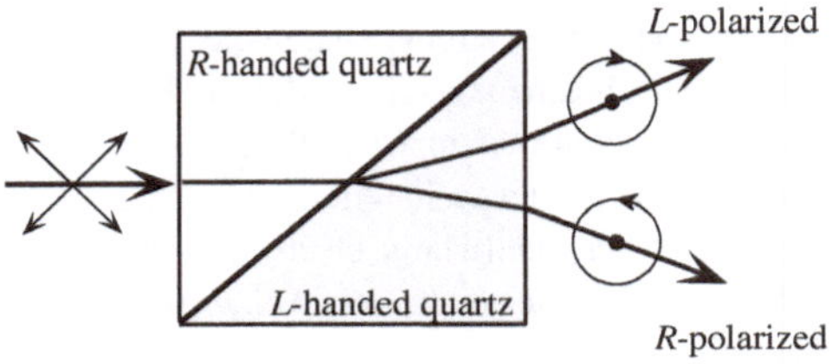

The **Fresnel prism** for separating unpolarized light into two divergent beams with opposite circular polarizations (R = right, L = left; divergence is exaggerated).

Fresnel reflection is partial (or full) reflection of an electromagnetic wave at a dielectric–dielectric boundary. See *Fresnel's equations.*

Fresnel's equations describe the amplitude and phase relationships between the incident, reflected and transmitted waves at a dielectric–dielectric interface in terms of the refractive indices of the two media and the angle of incidence.

Consider an electromagnetic (EM) wave going from medium 1 to 2 as illustrated in the figure. The amplitudes of the reflected and transmitted waves can be readily obtained in terms of n_1, n_2 and the incidence angle θ_i alone. If we define $n = n_2/n_1$, as the relative refractive index of medium 2 to that of 1, then the *reflection and transmission coefficients* for

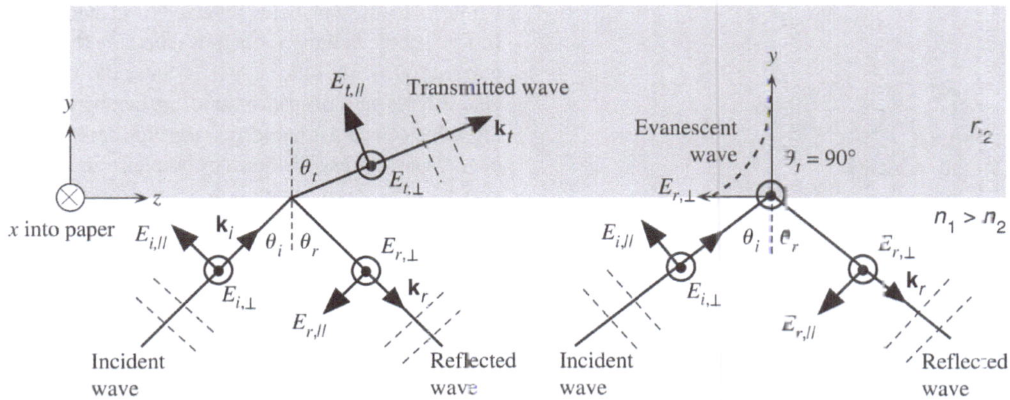

Fresnel's equation A light wave traveling in a more dense medium strikes a less dense medium. The plane of incidence is the plane of the paper and is perpendicular to the flat interface between the two media. The electric field is normal to the direction of propagation. It can be resolved into perpendicular ($\perp$) and parallel (//) components. (a) if $\theta_i < \theta_c$ then some of the wave is transmitted into the less dense medium. Some of the wave is reflected. (b) if $\theta_i > \theta_c$ then the incident wave suffers total internal reflection. There is a decaying evanescent wave into the n_2-medium.

$E_\perp$ are given by

$$r_\perp = \frac{E_{r0,\perp}}{E_{i0,\perp}} = \frac{\cos\theta_i - \left[n^2 - \sin^2\theta_i\right]^{1/2}}{\cos\theta_i + \left[n^2 - \sin^2\theta_i\right]^{1/2}}$$

Reflection coefficient

$$t_\perp = \frac{E_{t0,\perp}}{E_{i0,\perp}} = \frac{2\cos\theta_i}{\cos\theta_i + \left[n^2 - \sin^2\theta_i\right]^{1/2}}$$

Transmission coefficient.

There are corresponding coefficients for the $E_{i/}$ fields with corresponding *reflection and transmission coefficients*, $r_{//}$ and $t_{//}$,

$$r_{//} = \frac{E_{r0,//}}{E_{i0,//}} = \frac{\left[n^2 - \sin^2\theta_i\right]^{1/2} - n^2\cos\theta_i}{\left[n^2 - \sin^2\theta_i\right]^{1/2} + n^2\cos\theta_i}$$

Reflection coefficient

$$t_{//} = \frac{E_{t0,//}}{E_{i0,//}} = \frac{2n\cos\theta_i}{n^2\cos\theta_i + \left[n^2 - \sin^2\theta_i\right]^{1/2}}$$

Transmission coefficient.

Further, the above coefficients are related by

$$r_{//} + nt_{//} = 1 \quad \text{and} \quad r_\perp + 1 = t_\perp$$

Transmission coefficient.

Fresnel–Kirchhoff diffraction integral is the form taken by the *Helmholtz–Kirchhoff integral* when the scalar field $F(r') = A\exp(jkr')/r'$ represents an incident spherical wave centered on a point P' exterior to the volume V, r' denoting the distance to point of interest P. In this expression A is a constant, and the time-dependent part $\exp(j\omega t)$ has been separated out. For any function $f(r)$ with purely spherical symmetry, $\nabla f = (\partial f/\partial r)\mathbf{u}_r$, with $\mathbf{u}_r$ the radial unit vector. As a result,

$$\nabla[\exp(jkr)/r] = [(jkr-1)\exp(jkr)/r^2]\mathbf{u}_r,$$
$$\nabla[\exp(jkr')/r'] = [(jkr'-1)\exp(jkr')/r'^2]\mathbf{u}_{r'}.$$

By neglecting the r^{-2} and r'^{-2} terms and keeping only the greater r^{-1} and r'^{-1}, a straightforward calculation gives eventually the amplitude F_P of F at point P as:

$$F_P = -j\frac{Ak}{4\pi}\int\int_S \frac{\exp[jk(r+r')]}{rr'}(\mathbf{u}_r - \mathbf{u}_{r'})\cdot d\mathbf{S}.$$

Frustrated total internal reflection (FTIR) is total internal reflection in which not all the light is

Augustin Jean **Fresnel** (1788–1827) was a French physicist, and a civil engineer for the French government, who was one of the principal proponents of the wave theory of light. He made a number of distinct contributions to optics including the well-known Fresnel lens that was used in lighthouses in the nineteenth century. He fell out with Napoleon in 1815 and was subsequently put under house-arrest until the end of Napoleon's reign. During his enforced leisure time he formulated his wave ideas of light into a mathematical theory. (Photo: Smithsonian Institution, courtesy of AIP Emilio Segrè Visual Archives.) ("If you cannot saw with a file or file with a saw, then you will be no good as an experimentalist." Augustin Fresnel).

reflected, as would happen in ideal TIR, because the second medium is sufficiently thin to transmit some of the light through it to the other side. Light has tunneled through the second medium, which is called *optical tunneling*. Consider light propagating in a medium (I) of refractive index n_1 falling over a thin slab (of thickness d) of medium (II) of index $n_2 < n_1$ at an incidence angle θ_i greater than the critical angle $\theta_c = \arcsin(n_2/n_1)$. Instead of the whole incident beam being reflected from medium II, some of the radiation penetrates medium II and emerges out on the other side (III). Since the incident wave is partially transmitted, the intensity of the reflected beam is reduced. If the time dependence is written as $e^{j\omega t}$, then the complex transmission coefficient is given by

$$t = \frac{1}{\cosh(\alpha d) - \dfrac{j}{2}\left(\kappa - \dfrac{1}{\kappa}\right)\sinh(\alpha d)}$$

where

$$\alpha = \frac{\omega}{c}\sqrt{n_1^2 \sin^2\theta_i - n_2^2}, \quad \kappa = \eta\frac{\sqrt{n_1^2 \sin^2\theta_i - n_2^2}}{n_1 \cos\theta_i},$$

with $\eta = 1$ for σ-polarization and $\eta = (n_1/n_2)^2$ for π-polarization; σ (or s) and π (or p) polarization refer to the electric field vector orientation perpendicular and parallel to the plane of incidence. The particular phase dispersion of this expression is responsible for the apparent *supraluminal velocity* associated to the tunnel effect. See *supraluminal velocity*.

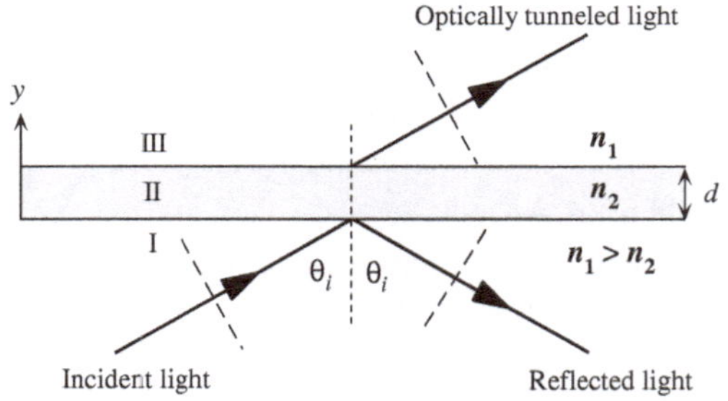

Frustrated total internal reflection When medium II is thin (thickness d is small), the field penetrates to the II–III interface and gives rise to an attenuated wave in medium II. The effect is the tunneling of the incident beam in I through II to III. (The arrows indicate wavevectors)

FSR see *free spectral range.*

Füchtbauer–Ladenburg equations

correspond to the Einstein rate equations for a

system with two energy levels, which are not necessarily discrete levels, cast in such a way to relate the optical gain curve, emission and absorption cross-sections and the radiative lifetime for the excited state at the higher energy to decay to the lower energy by a radiative transition. Suppose the radiative lifetime of the upper level is $\tau_{\text{radiative}}$ (the reciprocal of the Einstein A coefficient), σ_{em} and σ_{ab} are the emission and absorption cross-sections, $g_{\text{em}}(\nu)$ and $g_{\text{ab}}(\nu)$ are the emission and absorption lineshape functions, g_1 and g_2 are the degeneracies of the lower and upper level and ν is the frequency. Then

$$\sigma_{\text{em}}(\nu) = \frac{c^2}{8\pi n^2 \nu^2 \tau_{\text{radiative}}} g_{\text{em}}(\nu)$$

$$\sigma_{\text{ab}}(\nu) = \frac{g_2}{g_1} \frac{c^2}{8\pi n^2 \nu^2 \tau_{\text{radiative}}} g_{\text{ab}}(\nu)$$

and

$$g_2 \int \sigma_{\text{em}}(\nu) d\nu = g_1 \int \sigma_{\text{ab}}(\nu) d\nu.$$

The equation can also be written in different forms. For example, the first equation can be written as

$$\frac{1}{\tau_{\text{radiative}}} = \frac{8\pi n^2}{c^2} \int \nu^2 \sigma_{\text{em}}(\nu) d\nu.$$

(W. L. Barnes, *IEEE J. Quant. Electron.*, **27**, 1004, 1991.)

Full width at half-maximum (FWHM) is the width on the abscissa between half-maximum points of the ordinate. Suppose that y vs. x has a maximum $y = y_{\max}$ at $x = x_0$ and then falls to half the maximum value $^1/_2 y_{\max}$ at $x = x_2$ and $x = x_1$, where $x_2 > x_o > x_1$. Then FWHM is Δx of y vs. x between the half-point $^1/_2 y_{\max}$ so that $\Delta x = x_2 - x_1$. For a Gaussian function $y(x) = [\sigma(2\pi)^{1/2}]^{-1} \exp(-x^2/2\sigma^2)$ with a mean standard deviation σ, FWHM is $\Delta x_{\text{FWHM}} = 2[2\ln(2)]^{1/2}\sigma = 2.355\sigma$.

Fundamental absorption see *band-to-band absorption.*

Fundamental mode is the lowest order mode that can exist in a dielectric guide and it is the traveling electric field pattern that has no nodes in the transverse direction to the waveguide axis (nodes are locations where the electric field is zero).

Fused silica see *silica, optical materials, Sellmeier equation.*

Fusion splicing describes permanently joining two fiber ends together by means of melting the fiber ends. The flat ends of the two fibers are aligned and then locally melted (by an electric arc) and fused together to form a continuous fiber.

FWHM see *full width half maximum*

FWM see *four-wave mixing.*

G

Ga-La-S or **GLS glasses** are glasses made from gallium, lanthanum and sulfur and are a group within the class of chalcogenide glasses; for example, $65Ga_2S_3$:$32La_2S_3$:$3La_2O_3$ is a GLS glass. Their transmission spectrum covers a wide range, 0.5–5 μm. They have high glass transition temperatures (560 °C) and can dissolve Er^{3+} ions (and other rare earth ions) by erbium substitution for lanthanum. Laser operation has been demonstrated in Er^{3+} and Nd^{+} doped GLS glasses and other potential photonics applications have been either demonstrated, such as an erbium doped waveguide amplifier, or have been suggested from fundamental research. See *erbium doped waveguide amplifier*. (T. Schweizer, D.J. Brady and D.W. Hewak, *Optics Express*, **1**, 102, 1997.)

Gain is the output signal per unit input signal. For voltages , $G_V = V_{out}/V_{in}$ and for optical signals with optical power P, usually, $G_{op} = P_{out}/P_{in}$.

Gain coefficient see *gain saturation in an active optical medium.*

Gain cross-section see *gain saturation in an active optical medium.*

Gain equalizer see *equalizer.*

Gain flattening filter *see equalizer.*

Gain guided laser is a laser diode, such as a stripe geometry laser diode, in which the width of the active region, or the optical gain region, is defined by the current density from the smaller electrode to the larger electrode. Optical gain is highest where the current density is greatest. The optical gain width therefore depends on the current density.

Gain saturation in an active optical medium occurs when the signal, amplified by *stimulated emission*, is strong enough to reduce the population inversion noticably. In an idealized (open) two-level system under quasimonochromatic illumination, the gain coefficient [in m^{-1}] is $\alpha = \sigma \Delta N$, in which σ is the *gain cross-section* [in m^2] and $\Delta N = N_2 - N_1$ is the population inversion [in m^{-3}]. Let A and W [in s^{-1}] denote the transition rates of spontaneous and stimulated processes, respectively, Λ_P [in $m^{-3}s^{-1}$] the pumping term into upper level 2, and γ_1 [in s^{-1}] the nonradiative relaxation rate from lower level 1 such that $\gamma_1 > A$. The stimulated term W is directly proportional to the actual optical intensity I inside the medium. The rate equations are related:

$$dN_1/dt = -(W + \gamma_1)N_1 + (A + W)N_2,$$
$$dN_2/dt = +\Lambda_P + WN_1 - (A + W)N_2.$$

From the above equations, the steady state solution for α reads

$$\alpha = \sigma \Delta N = \sigma \frac{\Lambda_P}{\gamma_1}\left(\frac{\gamma_1 - A}{A + W}\right) = \frac{\alpha_0(\Lambda_P)}{1 + S},$$

where α_0 is the *linear* (or unsaturated) gain coefficient, increasing with Λ_P (the pumping) and $S = W/A = I/I_S$ is the dimensionless *saturation parameter*. The *saturation intensity* I_S is a characteristic

Gain saturation in lasers

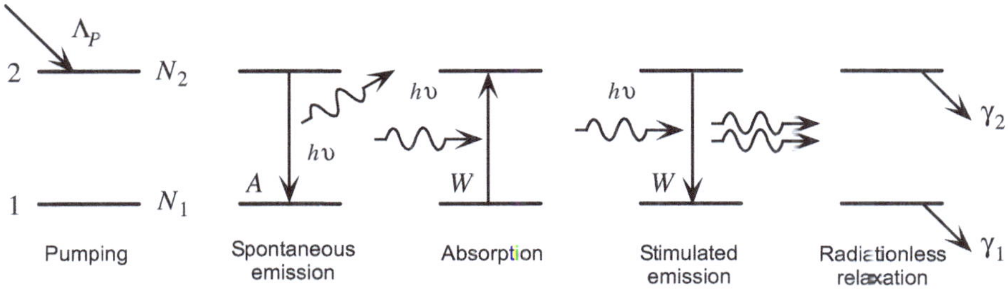

Gain saturation in an active optical medium Schematic illustration of the main processes (with their rates) at work in a simplified active optical medium: pumping, radiative transitions (spontaneous or induced), nonradiative relaxation.

value of a given medium. In a continuous-wave laser oscillator, the steady state saturating intensity is such as to ensure that the saturated gain $\alpha(\Lambda_P, S)$ remains exactly equal to the cavity losses, that is, it is "clamped" to its threshold value g_{th}. At very high levels of incident intensity ($S \gg 1$), $\alpha \to 0$, so the medium becomes transparent (*self-induced transparency*). See also *saturable absorption*.

Gain saturation in lasers occurs when further increase in the pumping power does not proportionally increase the optical gain in the laser cavity. Consider a medium consisting of simplified two-level systems with two eigenstates (1) and (2) of energies E_1 and E_2 in interaction with an electromagnetic wave at frequency υ such as $h\upsilon = E_2 - E_1$. Two stimulated processes should be considered, that is, *induced* by the incident light. Let N_1 and N_2 [m^{-3}] represent the concentration of systems in the lower and upper levels,

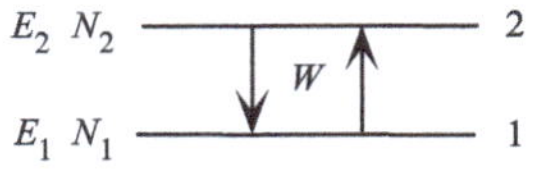

Gain saturation in lasers A simplified two-level system, with only induced transitions considered. Optical gain occurs when stimulated emission prevails over absorption, that is for $N_2 > N_1$.

respectively. Let W denote the induced transition probability, either from (1) to (2) or for (2) to (1), proportional to the flux of the optical field.

The population variation induced by stimulated transitions can be written as

$$(dN_1)_{(1)\to(2)} = -WN_1 dt$$
$$(dN_2)_{(2)\to(1)} = -WN_2 dt.$$

The change of the light wave's intensity (itself proportional to W) induced by such variations is therefore proportional to the population difference $(N_2 - N_1)$. *Population inversion* means that $(N_2 - N_1) > 0$; this is realized through *pumping*. The optical gain α [m^{-1}] is

$$\alpha = \sigma_0(N_2 - N_1)$$

where σ_0 is the gain *cross-section* [m^2]. The higher the intensity I, the more transitions are induced. As a result, downward transitions (2)→(1) tend to diminish N_2 and to augment N_1, thus reducing the population difference $\Delta N = (N_2 - N_1)$. In other words, the stronger the optical field, the lesser is the optical gain, which is called *gain saturation*. Gain saturation is sometimes approximated as

$$\alpha(I) = \frac{\alpha_{lin}}{1 + I/I_S}$$

where α_{lin} is the linear (unsaturated) gain, and I_S the *saturation intensity*, which depends on the material. In the limiting case for $I \gg I_S$, $N_2 \approx N_1$

and $\alpha \to 0$, transparency is reached. A symmetrical phenomenon is observed in absorbers, when a majority of systems, initially in the lower level, have been excited into the upper level, then a balance is reached between N_1 and N_2 and the absorption is said to be saturated. See also *spatial hole burning*.

Gallium arsenide, GaAs, is a direct bandgap III-V semiconductor crystal with the zinc blende structure: $a = 5653$ nm; $\rho = 5.318$ g cm^{-3}; $\alpha_L = 1.6 \times 10^{-6}$ K^{-1}; $c_s = 0.36$ J g^{-1} K^{-1}; $\kappa \approx 0.5$ W cm^{-1} K^{-1}; $T_{\text{Debye}} \approx 350$ K; $\chi = 4.07$ eV; E_g(D) = 1.424 eV (300 K), E_g(D) = 1.519 eV (0 K); Δ_{so} = 0.80 eV; $m_e{}^*/m_e = 0.067$; $m_{hH}{}^*/m_e$ (heavy holes) = 0.48; $m_{hL}{}^*/m_e$ (light holes) = 0.09; $m_h{}^*/m_e = 0.53$; $n_i = 1.6 \times 10^6$ cm^{-3}; $N_c = 4.5 \times 10^{17}$ cm^{-3}; $N_v = 7 \times 10^{18}$ cm^{-3}; $\mu_e = 8500$ cm^2 V^{-1}s^{-1}, $\mu_h = 400$ cm^2 V^{-1}s^{-1}; ε_r(DC) = 13.18, $\varepsilon_r(\infty) = 10.89$; $n \approx 3.255[1 + (4.5 \times 10^{-5})T]$ for T in K. For λ = 1.24 − 4.1 μm, and for λ in μm,

$$n^2 \approx 7.1 + \frac{3.78\lambda^2}{\lambda^2 - 0.5260^2}.$$

For λ = 0.97 to 17 μm, the Sellmeier coefficients in

$$n^2 = A_0 + A_1\lambda^2/(\lambda^2 - \lambda_1^2) + A_2\lambda^2/(\lambda^2 - \lambda_2^2) + A_3\lambda^2/(\lambda^2 - \lambda_3^2)$$

are: λ_1(μm) = $0.4431307 + 0.000050564\Delta T$; λ_2(μm) = $0.8746453 + 0.0001913(\Delta T) - 4.882 \times 10^{-7}(\Delta T^2)$; λ_3(μm) = $36.9166 - 0.011622(\Delta T)$; $A_0 = 5.372514$; $A_1 = 27.83972$; $A_2 = 0.031764 + 4.350 \times 10^{-5}(\Delta T) + 4.664 \times 10^{-7}(\Delta T^2)$; $A_3 = 0.00143636$; $\Delta T = T -$ 295 K; $dn/dT = 2.1 \times 10^{-4}$ K^{-1} at $\lambda = 2$ μm. (T. Skauli, P. S. Kuo, K. L. Vodopyanov, T. J. Pinguet, O. Levi, L. A. Eyres, J. S. Harris, M. M. Fejer, B. Gerard, L. Becouarn, and E. Lallier, *J. Appl. Phys.*, **94**, 6447, 2003.)

Gallium nitride (GaN) is direct bandgap III-V semiconductor crystal with either the wurtzite or zinc blende structure. Properties of the wurtzite form are: $a = 0.3189$ nm; $c = 0.5125$ nm; ρ = 6.15 g cm^{-3}; $T_m = 2500$°C; $\alpha_{L,a} = 5.59 \times 10^{-6}$ K^{-1}; $\alpha_{L,c} = 3.17 \times 10^{-6}$ K^{-1}; $c_s = 0.49$ J g^{-1} K^{-1}; $\kappa \approx 1.3$ W cm^{-1} K^{-1}; $\chi = 4.10$ eV; $T_{\text{Debye}} = 600$ K; E_g(D) = 3.39 eV (300 K), 3.44 eV (0 K); E_g(T) = $3.44 - 5.08 \times 10^{-4}$ T^2/(996−T), (T in K); Δ_{so} = 0.008 eV; $m_e{}^*/m_e = 0.20$; $m_{hH}{}^*/m_e = 1.40$; $m_{hL}{}^*/m_e$ = 0.30; $m_h{}^*/m_e = 1.50$; $N_c = 2.3 \times 10^{18}$ cm^{-3}; N_v = 4.6×10^{19} cm^{-3}; $\mu_e \sim 440$ cm^2 V^{-1}s^{-1}, $\mu_h \sim$ 100 cm^2 V^{-1}s^{-1}; ε_r(DC) = 8.90, $\varepsilon_r(\infty) = 5.35$; $n \approx 2.295$ (infrared); $E_{\text{breakdown}} \approx 5 \times 10^6$ V cm^{-1}; $\mu_e \sim 1000$ cm^{2}V^{-1}s^{-1}; $\mu_h \sim 200$ cm^2 V^{-1}s^{-1}; D_e ~ 25 cm^2 s^{-1}; $D_h \sim 5$ cm^2 s^{-1}; $\hbar\omega_{\text{op}} = 91.2$ meV; $B_{\text{radiative}} = 1.1 \times 10^{-8}$ cm^3/s. Properties of the zinc blende form are: $a = 0.452$ nm; $\rho = 6.15$ g cm^{-3}; $T_m = 2500$ °C; $c_s = 0.49$ J g^{-1} K^{-1}; $\kappa \approx 1.3$ W cm^{-1} K^{-1}; $\chi = 4.10$ eV; $T_{\text{Debye}} = 600$ K; E_g(D) = 3.20 eV (300 K), 3.28(0 K); Δ_{so} = 0.02 eV; $m_e{}^*/m_e = 0.13$; $m_{hH}{}^*/m_e = 1.30$; $m_{hL}{}^*/m_e = 0.19$; $m_h{}^*/m_e = 1.40$; $N_c = 1.2 \times 10^{18}$ cm^{-3}; $N_v = 4.1 \times 10^{19}$ cm^{-3}; $\mu_e \sim$ 500 cm^2 V^{-1}s^{-1}, $\mu_h \sim 400$ cm^2 V^{-1}s^{-1}; ε_r(DC) = 9.70, $\varepsilon_r(\infty) = 5.30$; $E_{\text{breakdown}} \approx 5 \times 10^6$ V cm^{-1}; $\mu_e \sim 1000$ cm^{2}V^{-1}s^{-1}; $\mu_h \sim 350$ cm^2 V^{-1}s^{-1}; D_e ~ 25 cm^2 s^{-1}; $D_h \sim 9$ cm^2 s^{-1}; $n = 2.3$ (infrared); $\hbar\omega_{\text{op}} = 87.3$ meV; $B_{\text{radiative}} = 1.1 \times 10^{-8}$ cm^3 s^{-1}.

Gallium nitride (GaN) optoelectronics is based on wide-bandgap III-V nitride materials that have successfully yielded blue and green light emitting devices, including commercial blue lasers. The latter are of immense technological importance as they are suitable for increasing optical disk storage capacity over current digital video disk (DVD) technology. Another critical development that has been enabled by the presence of blue light emitters is the generation of solid state white light sources. In addition, nitride based photodetectors have enabled *high-efficiency* UV detectors that were previously unavailable. The field of nitride optoelectronics is still however, at an early stage, with applications taking advantage of these materials' unique high-frequency, high-power and high-temperature performance, being developed and refined.

Gamma, or **gamma correction,** is simply a special nonlinear operation that is employed to modify the luminance in an image in such a way that if L_o is the original luminance, then the modified luminance L_m is given by $L_m = L_o{}^\gamma$, where the index gamma γ is chosen according to the image correction, encoding or decoding that is desired. Gamma processing can be used to encode and decode images, and hence γ can be less than 1 for encoding and more than 1 in decoding. It is primarily used in CRTs to make up for nonuniformities in the image brightness, to improve the perception of the image.

Gamma, or γ**-rays,** are highly energetic electromagnetic waves, or photons, that are emitted from the nuclei of atoms, during nuclear transitions, e.g. when an excited nucleus decays to a lower energy state. γ-ray photons are emitted with discrete energies since they involve energy transitions, typically in the range 10 keV to 10 MeV. γ-rays have properties that are similar to x-ray photons.

Gamma detector is capable of detecting more penetrating radiations such as gamma (γ) rays (at least those with energies larger than 0.1 MeV); thus a large detector thickness is required to detect gamma rays. Scintillator, germanium, cadmium-zinc-telluride (CZT) or mercuric iodide (HgI_2) semiconductor detectors are usually used to detect γ-rays. Gamma rays are emitted in nuclear processes during radioactive decay or in nuclear reactions. Detection of these γ rays gives basic information on the nuclear energy levels: their energy and intensity determine the position of the levels; the γ-ray emission times measure the lifetime of levels.

Gamma-ray laser (*GRASER*), if realized, would emit photons in the keV to even MeV range. The concept and plan were introduced as early as 1961 by Soviet scientist Lev A. Rivlin, but experimental work did not begin in earnest until the mid-1980s, and the subject remains up to now at a prospective research stage. It has been proposed to take advantage of electromagnetic transitions between quantized states of the nucleus, the same way as optical transitions occur between quantized states of the atom; hence the term *quantum nucleonics*, coined after *quantum electronics*. In theory, a continuum of x-rays (itself delivered by a particle accelerator) pumping an atomic nucleus could send it toward excited states of higher energy. In rare cases, some of these excited states would relax to so-called isomeric levels, sufficiently long-lived to enable accumulation, and therefore induced gamma-ray emission. According to specialists, hafnium-178 seems a likely candidate.

Gas discharge lamp see *glow discharge lamp, arc lamp.*

Gauss's law is a fundamental law of physics (one of Maxwell's equations) that relates the surface integral of the electric field over a closed (hypothetical) surface to the sum of all the charges enclosed within the surface. If E_n is the electric field normal to a small surface area dA and Q_{total} is the total enclosed charge, then over the whole closed surface

$$\varepsilon_0 \int E_n dA = Q_{total}$$

where the integration is over the whole closed surface. In the presence of a dielectric medium, then we can use

$$\varepsilon_0 \varepsilon_r \int E_n dA = Q_{free}$$

where Q_{free} represents the sum of all *free* charges (excluding polarization charges) within the surface. This equation applies to an isotropic medium in

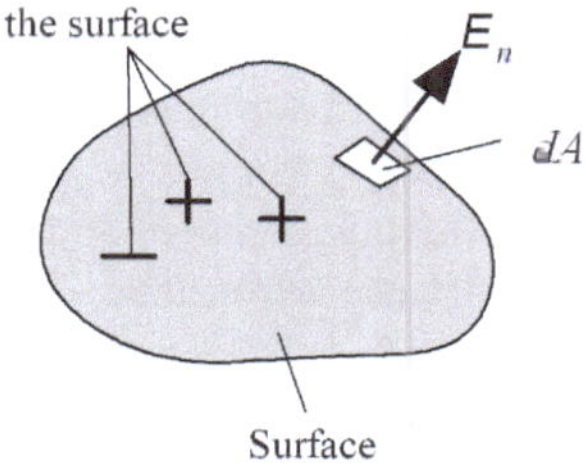

Gauss's law The surface integral of the electric field normal to the surface is the total charge enclosed. The field is positive if it is coming out, negative if it is going into the surface.

which ε_r is independent of the direction of the field. Gauss's law in point form relates the gradient of the electric field at a point to the charge density ρ at that point, i.e. in a one-dimensional problem (as through the space charge region of a *pn* junction) the gradient would be dE/dx, and

$$\varepsilon_0 \varepsilon_r dE/dx = \rho_{net}.$$

Gauss's lens equation for thin lenses is a relationship between the object distance s_o from the lens, image distance s_i from the lens, and object and image focal lengths f_o and f_i in terms of

$$\frac{f_o}{s_o} + \frac{f_i}{s_i} = 1.$$

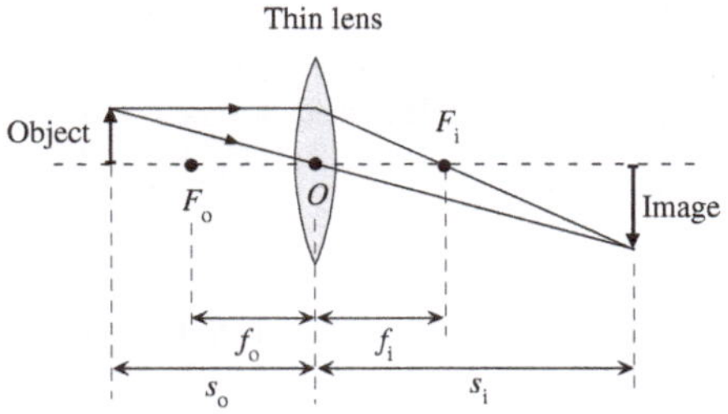

Gauss's lens equation A thin lens of focal lengths $f_o = OF_o$ and $f_i = OF_i$.

Gaussian beam is a propagating electromagnetic beam with an intensity that varies spatially away from the beam axis z, i.e. across the beam cross-section and also *along* the beam axis. The light intensity distribution across the beam cross-section anywhere along z is Gaussian. The *beam diameter* $2w$ at any point z is defined in such a way that the cross-sectional area πw^2 at that point contains 86% of the beam power. At that distance w from the z-axis, the intensity is a factor e^{-2} $(= 0.14)$ of its value at the center of the beam. The beam diameter $2w$ increases as the beam travels along z. A Gaussian beam starts with a finite width $2w_0$ where the wavefronts are parallel and then the beam slowly diverges as the wavefronts curve out during propagation along z. The divergence is due to the self-diffraction

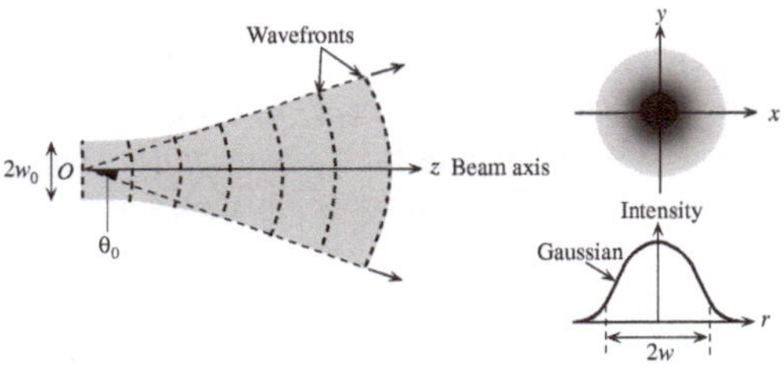

Gaussian beam (a) Wavefronts of a Gaussian light beam. (b) Light intensity across beam cross-section. (c) Light irradiance (intensity) vs. radial distance r from beam axis (z).

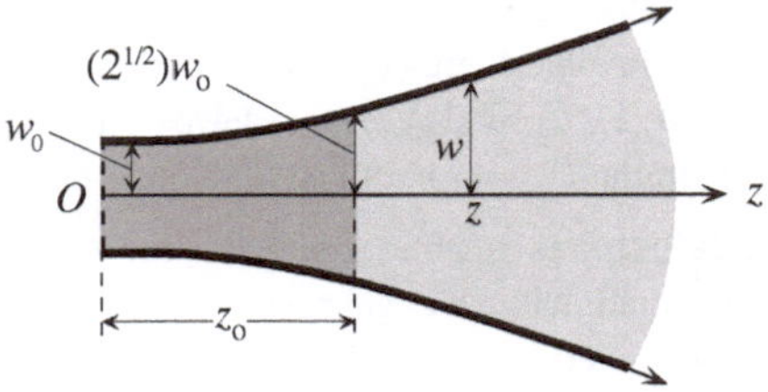

Gaussian beam definitions.

of the beam – the beam is diffraction limited. It can be shown that the Gaussian beam is a solution of Maxwell's equations. The finite width $2w_o$ where the wavefronts are parallel is called the *waist* of the beam; w_o is the *waist radius* and $2w_o$ is the *spot size*. Far away from the source, the beam diameter $2w$ increases linearly with distance z from the center O. The width $2w$ becomes $(2^{1/2})\, 2w_o$ when $z = z_o$, at the so-called *Rayleigh range*, when,

$$z_o = \pi w_o^2/\lambda$$

where λ is the wavelength. The region $-z_o < z < z_o$ of the beam where the waist is less than z_o is called the *depth of focus* or the *confocal parameter*. The width $2w$ at a position along z is given by

$$2w = 2w_o[1 + (z/z_o)^2]^{1/2}.$$

The radius of curvature R of the wavefronts at z is

$$R = z[1 + (z/z_o)^2].$$

If r is the radial distance from the center of the beam (z-axis), the intensity is

$$I(r, z) = I_o(w_o/w)^2 \exp(-2r^2/w^2)$$
$$= [2P/(\pi w^2)] \exp(-2r^2/w^2)$$

where P is the total power in the beam. For large z values, $z \gg z_o$, far away from the source

$$w = \lambda z/(\pi w_o)$$

which shows that the waist expands linearly with the distance, which defines a cone with its vertex at O with half an apex angle θ_o given by

$$\theta_o = w/z = \lambda/(\pi w_o)$$

which is the *far-field angular radius* of the Gaussian beam, and $2\theta_o$ is the *far-field divergence* of the Gaussian beam. For example, a Melles Griot red HeNe laser ($\lambda = 632.8$ nm) has a beam diameter $2w_o$ of 0.8 mm so that $\theta_o = 5.05 \times 10^{-4}$ mrad. The waist at $z =$ 100 m is $2w = 2z\theta_o = 100.8$ mm or 10 cm.

Generally, for many applications, a laser beam can be approximated as a Gaussian beam. A number of gas lasers, such as He-Ne and argon ion lasers, provide a beam with an almost ideal Gaussian cross-section. Other laser beams are not truly Gaussian but can, nonetheless, be represented as a Gaussian beam. A single-mode laser beam, i.e. that with the TEM_{00} mode, is quite close to being a Gaussian beam. The *M^2 factor* measures the deviation of the real laser beam from the Gaussian characteristics, in which $M^2 = 1$ for an ideal (theoretical) Gaussian shape. A useful definition for the *M^2 factor* is

$$M^2 = (w_{or}\theta_{or})/(w_o\theta_o)$$

where $2\theta_{or}$ and $2w_{or}$ are the divergence and waist of the *real* laser beam, and $w_o\theta_o$ is for the ideal Gaussian. M^2 gauges the beam quality of the laser beam. For many lasers, $M^2 > 1$ and can be quite high, e.g. 30, for multimode lasers. We can apply the above Gaussian equations to a real laser beam by replacing w_o with w_{or}/M^2:

$$2w = 2w_o[1 + (z\lambda M^2/\pi w_o)^2]^{1/2}.$$

An *optimized Gaussian beam* has just the right waist $2w_o$ to give a *minimum* waist $2w$ at a particular fixed distance z (e.g. at 100 m):

$$2w_o(\text{optimum}) = 2(\lambda z/\pi)^{1/2}.$$

Gaussian distribution refers to the *normal distribution* found in statistics, that is, it is a particular probability density function, denoted $F(x)$, of a variable x such that $F(x)$ is continuous and has a symmetrical shape about a mean value $\overline{x}$ with the functional form

$$F(x) = \frac{1}{\sigma\sqrt{2\pi}} \exp\left[-\frac{(x-\overline{x})^2}{2\sigma^2}\right]$$

in which σ^2 is the variance, i.e. the mean value of $(x - \overline{x})^2$ found by integrating $(x - \overline{x})^2 F(x)$ over all x (from $-\infty$ to $+\infty$), and σ is the standard deviation. $F(x)$ is called the *Gaussian function*. The integral of $F(x)$ over all x is unity, $\int F(x)dx = 1$. Suppose that we observe a property x of a process or a phenomenon that has a Gaussian probability density function. Then we can find the total probability P_{12} of the occurrence of x in the interval x_1 to x_2 as

$$P = \int_{x_1}^{x_2} F(x)dx.$$

Put differently, the probability dP of observing x with a value in x to $x + dx$ is $F(x)dx$. The Gaussian function is widely used in representing many observable quantities. For example, the output spectrum from gas laser has a Gaussian distribution. The maximum of the Gaussian distribution is $F_{max} = F(\overline{x}) = 0.399/\sigma$. When $x = \sigma$, one standard deviation away from the peak, $F(\sigma) = 0.242/\sigma$, and $F(\sigma)/F_{max} = 0.607$. In moving 2σ from the peak, the Gaussian drops from its maximum by a factor of 0.135. The full width at half maximum, FWHM, is the width $\Delta x = |x_2 - x_1|$ where $F(x_1) = F(x_2) = (1/2)F_{max}$, and FWHM $= \sigma 2\sqrt{2\ln(2)} = 2.355\sigma$.

Gaussian lens equation see *lens*.

Gaussian units see *CGS units*.

General interference law for stationary optical fields see *coherence*.

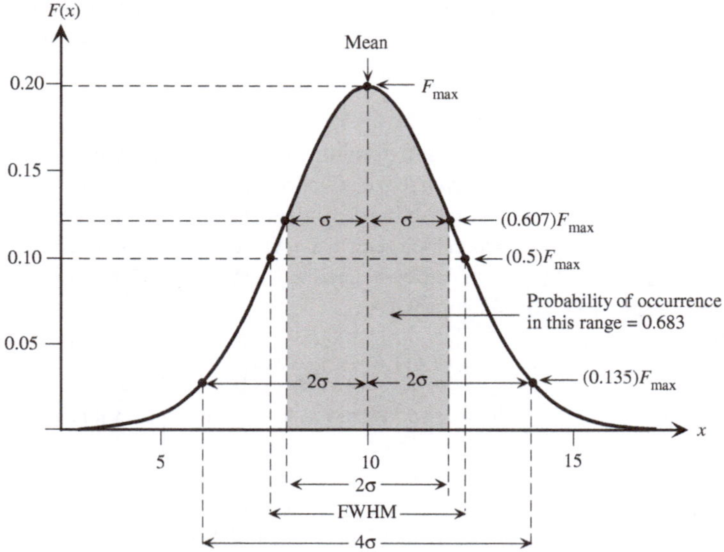

The **Gaussian distribution** that has $\overline{x} = 10$ and $\sigma = 2$. The shaded area is the integration from $x = \overline{x} - \sigma$ to $x = \overline{x} + \sigma$ and has a probability of 0.683.

Table: Gaussian distribution Selected elementary properties of the Gaussian distribution.

Mean	$\overline{x}$	Total area	1
Variance	σ^2	Probability in $\overline{x} \pm \sigma$	0.683
Maximum, F_{max}	$(\sigma\sqrt{2\pi})^{-1}$	Probability in $\overline{x} \pm 2\sigma$	0.955
$F(\sigma)$	$0.607 F_{max}$	FWHM	$\sigma 2\sqrt{2\ln(2)} = 2.355\sigma$

Generation–recombination noise see *excess noise.*

Geometrical extent qualitatively is the product of the cross-sectional area of a light beam and its solid angle. The elementary flux $d^2\Phi$ emitted by an elementary source dS of luminance L at point A, detected by an elementary detector dS' at point A′, is $d^2\Phi = L d^2 G$. The multiplication of luminance by the geometrical extent gives the elementary flux. The quantity d^2G is the elementary geometrical extent of the pencil of light:

$$d^2G = \frac{dS\cos\theta\, dS'\cos\theta'}{r^2}.$$

Note the symmetry between source and receptor. The finite geometrical extent defined by a given distribution of both source and detector is therefore $G = \int\int d^2G$. See also *optical extent.*

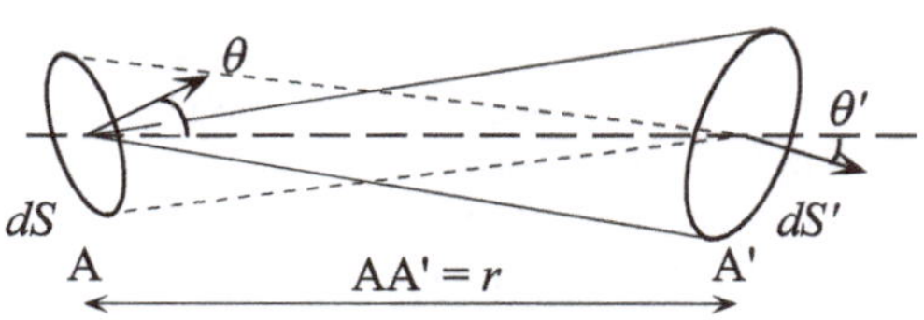

Illustration of **geometrical extent**.

Germania is germanium dioxide, GeO_2, glass.

Gires–Tournois interferometer is a modified Fabry–Perot cavity where the second mirror is totally reflecting ($R_A = r_A{}^2, R_B = 1$, where R is the reflectance and r is the reflectivity and subscripts A and B refer to the two mirros of the cavity). Assuming a lossless cavity, the structure reflects 100% of incident light. Its complex reflectivity r_{GTI} reads:

$$r_{GTI} = \frac{r_A - e^{-j\Phi}}{1 - r_A e^{-j\Phi}},$$

where Φ is the phase change over a round-trip in the cavity. It is obvious that $|r_{GTI}| = 1$. The phase dispersion is all the more pronounced as r_A is enhanced.

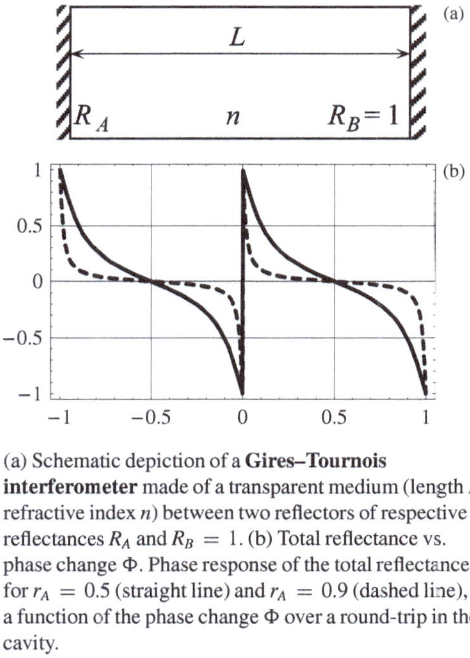

(a) Schematic depiction of a **Gires–Tournois interferometer** made of a transparent medium (length L, refractive index n) between two reflectors of respective reflectances R_A and $R_B = 1$. (b) Total reflectance vs. phase change Φ. Phase response of the total reflectance, for $r_A = 0.5$ (straight line) and $r_A = 0.9$ (dashed line), as a function of the phase change Φ over a round-trip in the cavity.

Gladstone–Dale formula is an empirical equation that allows the average refractive index n of an oxide glass to be calculated from its density ρ and its constituents by

$$\frac{n-1}{\rho} = p_1 k_1 + p_2 k_2 + \cdots$$
$$= \sum_{i=1}^{N} p_i k_i = C_{GD}$$

where the summation is for various oxide components (each a simple oxide), p_i is the weight fraction of the ith oxide in the compound and k_i is the refraction coefficient that represents the polarizability of the ith oxide. The right-hand side of $(n-1)/\rho$ is called the *Gladstone–Dale* coefficient C_{GD} which has been measured and tabulated for various materials. In more general terms, as a mixture rule for the overall refractive index, the Gladstone–Dale formula is frequently written as

$$\frac{n-1}{\rho} = \frac{n_1 - 1}{\rho_1} w_1 + \frac{n_2 - 1}{\rho_2} w_2 + \cdots$$

where n and ρ are the effective refractive index and effective density of the whole mixture, $n_1, n_2, \ldots$ are the refractive indices of the constituents, and ρ_1, $\rho_2, \ldots$ represent the density of each constituent. Gladstone–Dale equations for the polymorphs of SiO_2 and TiO_2 give the average n respectively as

$$n(SiO_2) = 1 - 0.21\rho \quad \text{and}$$
$$n(TiO_2) = 1 - 0.40\rho$$

(D. Dale and F. Gladstone, *Philosophical Transactions,* **148**, 887, 1858 **153**, 317, 1863.)

Glan–Foucault prism is a polarizing prism that produces linearly polarized light. It is made of two right-angle *calcite* prisms with a prism angle of 38.5°. Both have their optic axes parallel to each other and to the block faces as in the figure. The refractive index n_o for the ordinary waves (the o-ray) is larger than the index n_e for extraordinary waves (the e-ray). The critical angle for the o-ray is smaller than 38.5° but larger for the e-ray. Thus, the o-ray suffers total internal reflection at the calcite–air interface whereas the e-ray refracts through the calcite–air–calcite interfaces and comes out from the other side of the cube. It is useful in laser work

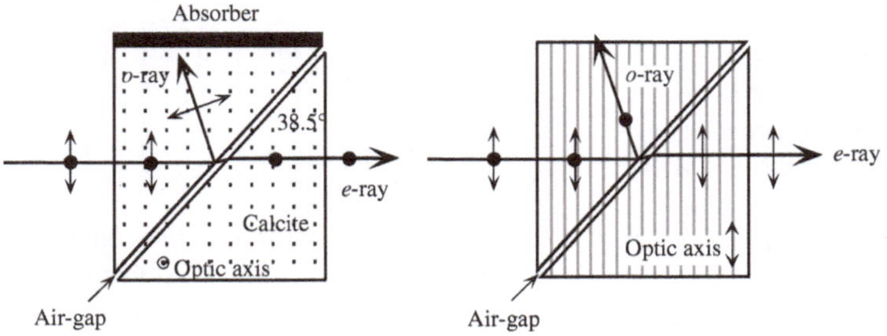

Left: The **Glan–Foucault prism** provides linearly polarized light. Right: The Glan–Taylor prism provides two linearly polarized light beams of nearly orthogonal polarizations. Note the difference in the optic axes in the two prisms. For calcite, at Na wavelength (589 nm), the critical angles are 37° for the *o*-ray, 42° for the *e*-ray.

that has high power densities (e.g. 100 W cm^{-2}). Note that although the transmitted beam is fully polarized (*e*-ray), the reflected beam has all of the *o*-ray and some of the *e*-ray that is reflected (not shown). (Some reflection accompanies a refraction at an interface.) The acceptance angle for providing polarized light output is usually small (7°) but the wavelength range of operations is wide, typically from 230–5000 nm. If a cement is used instead of the air gap, then the prism is called a *Glan–Thompson* polarizer, and has a wider range of acceptance angles (e.g. 30°) but the wavelength range then depends on the properties of the cement, and usually cannot be used in the UV. Unlike the Glan–Foucault prism, it cannot accept high-intensity laser beams. The *Glan–Taylor* prism is similar to the Glan-Foucault but has the optic axes in each prism parallel and in the plane of incident and reflected beams. The prism is cut close to the Brewster angle for the *e*-ray at the calcite–air interface. The angle of incidence at the interface can thus be made close to the Brewster angle, which reduces the reflection component of the *e*-ray. Thus, the reflected beam is mostly *o*-rays and the two emerging beams are close to being orthogonally polarized.

Glan–Taylor prism see *Glan-Foucault prism.*

Glan–Thompson prism or **polarizer** see *Glan–Foucault prism.*

Glare normally refers to the (undesirable) partial reflection of light from the glass or transparent surface of a visual display such as a TV screen. Glare can be reduced by using a glare filter on the surface of the display glass. The filter reduces the transmitted light from the display, which is compensated by increasing the display's brightness.

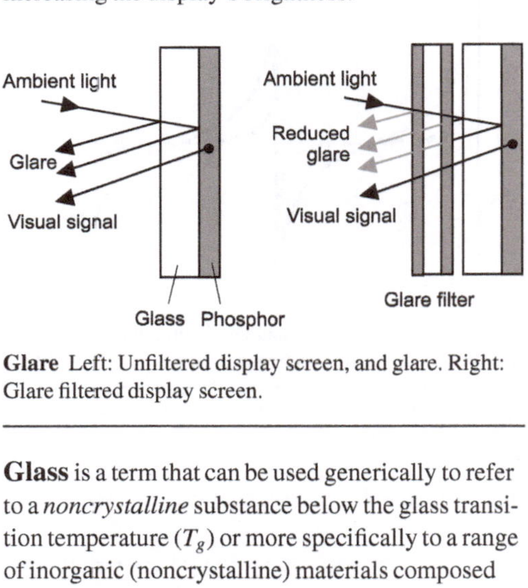

Glare Left: Unfiltered display screen, and glare. Right: Glare filtered display screen.

Glass is a term that can be used generically to refer to a *noncrystalline* substance below the glass transition temperature (T_g) or more specifically to a range of inorganic (noncrystalline) materials composed of such oxides as SiO_2, B_2O_3, Na_2O, P_2O_5, etc. The glass structure has no long-range order of the crystal, but exhibits only *short-range order*; we can identify the nearest neighbors of an atom because

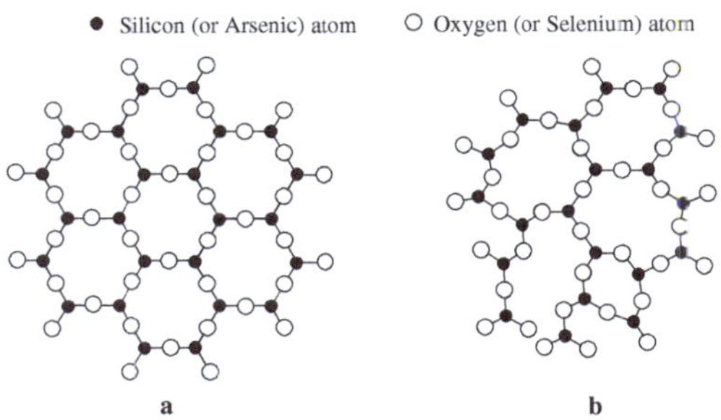

Glass (a) A crystalline solid reminiscent of crystalline SiO_2 (density = 2.6 g cm^{-3}). (b) An amorphous solid reminiscent of vitreous silica (SiO_2) cooled from the melt (density = 2.2 g cm^{-3}). Crystalline and amorphous structures illustrated schematically in two dimensions.

its valency requirements in forming bonds must be satisfied. Glasses are widely used in photonics and optoelectronics in such applications as optical fibers, lenses, prisms, windows, laser active medium, etc. The *glass transition (transformation) temperature,* T_g, is usually a narrow temperature range over which the properties of a material change from solid-like, below T_g, to liquid-like behavior, above T_g. For example, the *heat capacity* of a glass is solid-like (small) below T_g and liquid-like (high) above T_g. The transition is not always very clear and sometimes the distinction between solid-like and liquid-like behavior is made in terms of the viscosity of the material. Below the glass transition temperature, the viscosity of the structure is so great that over the time-scale of observation there is no significant flow. Any shearing stress below T_g results in an elastic response due to the stretching of bonds, and on removal of the stress the glass returns to its original shape. This is a typical solid-like behavior and for all practical purposes we view the structure below T_g as that of a solid. The glass transition temperature is not actually a unique temperature where the liquid on cooling suddenly becomes glass but a narrow temperature range in which the liquid-like behavior gradually changes over to a solid-like behavior as the substance is cooled. It depends not only on the viscosity of the material, which in turn depends on the bonding and structure, but also on the rate of cooling. Clearly, if the cooling is very slow then the atomic arrangements can follow the demands of the cooling rate much more closely and the structure exhibits liquid-like behavior to lower temperatures. If the cooling rate is too slow then there may be sufficient time for atomic diffusions to form the crystal.

When a glass is cooled rapidly, the temperature differences between the exterior and the interior of the sample during cooling develop into permanent internal strains at room temperature. These internal strains may cause the glass sample to break more easily. The strains can be reduced by annealing the glass at an appropriate temperature. At the

Table: Glass Definitions of glassy state based on typical viscosity values.

State of Glass	Viscosity η (Pa s) 1 Pa s = 10 Poise	T °C Alkali borosilicate glass: Corning 7056	T °C Fused silica (SiO_2), °C
Melting point	10^1		
Working point	10^3	1058 °C	
Flow point	10^4		
Softening point	4×10^6	718 °C	1730 °C
Annealing point	10^{12}	512 °C	1130 °C
Glass transition point	10^{12}–4×10^{12}		
Strain point	10^{13} (or 3.2×10^{13})	472 °C	1120 °C (1075 °C)

annealing temperature, e.g. 500 °C for borosilicate glass, the glass viscosity η is 10^{12} Pa s, and the internal strains reduce to 10% of their original value in about 15 minutes. Glass can be worked, i.e. shaped when $\eta = 10^3$ Pa s. The softening point corresponds to $\eta = 10^{6.6}$ Pa s when (roughly) a vertically suspended glass fiber of length 24 cm and diameter 0.6–0.8 mm will extend under its own weight at a rate of 1 mm min^{-1}.

See *optical glasses, optical materials.*

Glass transition (transformation) temperature see *glass*.

Glow or gas discharge lamp has an electric discharge in a glass or quartz tube containing certain selected gases for exciting the gas atoms to emit radiation. While it is similar to an arc lamp in which an electric discharge leads to the generation of light, in the gas discharge lamp, the discharge generates energetic electrons that excite gas atoms which then undergo de-excitation transitions and emit characteristic wavelengths. Glow discharge lamps do not have a bright hot arc. In the arc lamp, both characteristic atomic emissions and blackbody radiation from the "hot plasma" are involved since the arc temperatures are usually quite high and lead to a broad thermal emission over the visible range as in the xenon lamp. For example, neon and fluorescent tube lamps are usually referred to as gas discharge lamps whereas the xenon lamp is an arc lamp. A characteristic feature of gas discharge tubes is that as the discharge current increases, more ions are generated, and the impedance of the tube falls and hence the voltage across it also falls; thus the current must be controlled. Gas discharge lamps have a negative resistance region in which when the current increases, the voltage across it drops. The choice of the gas in the discharge tube

Table: Glow or gas discharge lamp Some characteristics of glow discharge lamps; lm/W is the luminous efficacy and quoted values are typical; T_{color} is the color temperature of the source.

Glow discharge lamp	Spectral range (nm)	Comment
Deutetrium	115–400	Extremely stable. Used in spectrophotometers
Hg		Good efficacy, 20–50 lm/W. $T_{\text{color}} \approx 4000$–6000 K. Roadways, landscapes, residential and commercial security, parking lots, industrial places, etc.
Metal halide		High luminous efficacy, 50–90 lm/W. $T_{\text{color}} \approx 3000$–4000 K. Commercial and retail display lighting, floodlighting, corridors, lobbies, walkways; retail, office, commercial places
Neon	Yellow-red	Color can be modified by mixing other gases such as Ar, Hg vapor, or using a phosphor coating on the inside surface of the tube since Ne also has a weak UV emission (similar to the fluorescent tube)
High-pressure Na (HPS)	White with yellow-orange tint	Very high efficacy, typically 80–120 lm/W output, and low operating cost. $T_{\text{color}} \approx 2000$ K. High pressure broadens the Na emission. Illumination of roads, malls, parks, building floodlighting, security lighting, storage rooms, industrial facilities, gymnasiums, etc. More efficient than Hg lamps
Low-pressure Na (LPS)	Yellow-orange (589 nm)	Very high efficacy lamp, 100–160 lm/W. Not used for color critical lighting. Street lighting. Known as SOX (sodium oxide)
Sulfur lamp	Broad visible spectrum, 400–800 nm, peak at 510 nm	Electrodeless plasma excited by microwave radiation. Very high efficacy of 100 lm/W. Emission from S_2 molecules

determines the predominant wavelength of emission. In many cases, mixtures of gases are used. In addition, some discharge lamps can operate with ac or dc power supplies since once the discharge is initiated, ac or dc current can be passed through the plasma. See *fluorescent tube lamp*. (A. Bogaerts *et al.*, *Spectrochimica Acta*, B., **57**, 609, 2008.)

Goos–Hänchen shift is the lateral shift of the reflected beam upon being reflected from a boundary with an optically less dense medium (Δz in the figure is the Goos-Hänchen shift). The reflection appears as if it had occurred from a virtual plane inside the second medium at a certain distance d from the interface.

Graded bandgap see *bandgap*.

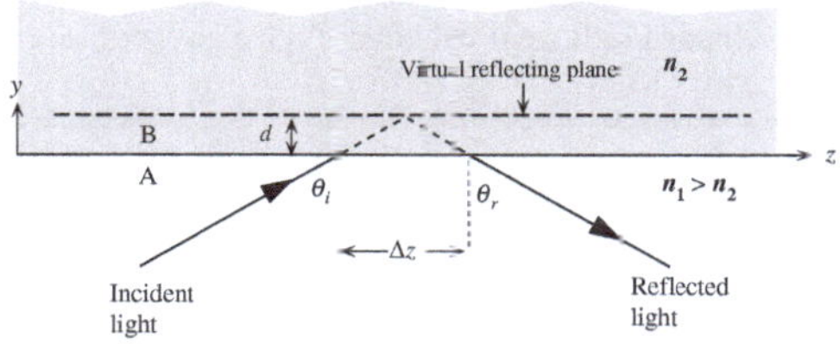

Goos-Hänchen shift The reflected light beam in total internal reflection appears to have been laterally shifted by an amount Δz at the interface.

Graded index optical fiber has a core refractive index that is graded gradually, i.e. changes continuously, towards the cladding. Typically in a graded index fiber, the refractive index profile is approximately parabolic to minimize modal dispersion to

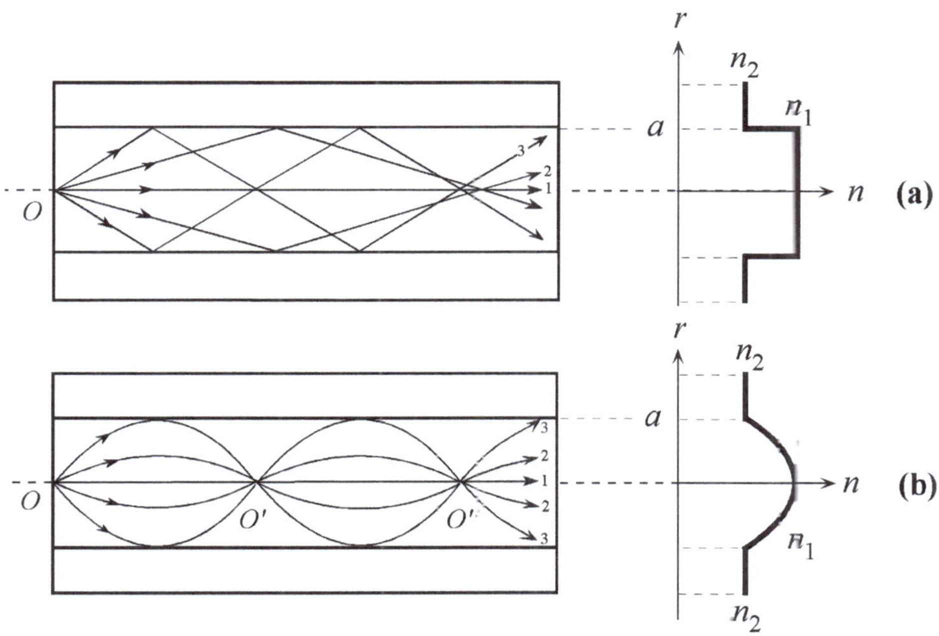

Graded index optical fiber r is the radial distance from the fiber axis. (a) Multimode step index fiber. Ray paths are different so that rays arrive at different times. (b) Graded index fiber. Ray paths are different but so are the velocities along the paths so that all the rays arrive at the same time.

Table: Graded index optical fiber Typical properties of a graded index fiber compared with a multimode step index fiber.

Property	Multimode step-index fiber	Graded index
$\Delta = (n_1 - n_2)/n_1$	0.02	0.015
Core diameter (μm)	100	62.5
Cladding diameter (μm)	140	125
Numerical aperture	0.3	0.26
Bandwidth × distance	20–100 MHz km.	300 MHz km–3 GHz km at 1.3 μm
Attenuation of light	4–6 dB km^{-1} at 850 nm 0.7–1 dB km^{-1} at 1.3 μm	3 dB km^{-1} at 850 nm 0.6–1 dB km^{-1} at 1.3 μm 0.3 dB km^{-1} at 1.55 μm
Typical light source	Light emitting diode (LED)	LED, lasers
Typical applications	Short haul or subscriber local network communications	Local and wide-area networks. Medium haul communications

a virtually innocuous level. All different mode rays in the graded index fiber arrive at the same time. The intuitive reason for this is that the velocity along the ray path, c/n, is not constant and increases as the ray is farther away from the center. A ray such as 2 that has a longer path than ray 1 then experiences a faster velocity during a part of its journey to enable it to catch up with ray 1. Similarly, ray 3 experiences a faster velocity than 2 during part of its propagation to catch up with ray 2, and so on. If $n(r)$ is the index variation from the center of the fiber, and r is the radiant distance from the center, then $n = n_1[1 - 2\Delta(r/a)^{\gamma}]^{1/2}$ for $r < a$ and $n = n_2$ for $r = a$, where γ is a constant called the *profile index*, or the coefficient of index grating. The intermodal dispersion is minimum when $\gamma \approx (4 + 2\Delta)/(2 + 3\Delta)$ where $\Delta = (n_1 - n_2)/n_1$ is the normalized refractive index difference between the center, and the "cladding." The intermodal dispersion $\sigma_{\text{intermode}}$ becomes a minimum and is given by

$$\frac{\sigma_{\text{intermode}}}{L} \approx \frac{n_1}{(20\sqrt{3})c}\Delta^2.$$

The number of modes M in a graded index fiber is given by $M \approx V^2/4$, where V is the usual V-number in terms of n_1, n_2, a and the free-space wavelength λ.

Graded index rod lens see *edge emitting LEDs*.

GRASER see *gamma-ray laser*.

Grassmann's law in color perception, in simple terms, states that if two colors C_1 and C_2 are added to generate a new color C, then the tristimulus values (for example, the red, green and blue fundamental colors) $R_1G_1B_1$ of C_1 and $R_2G_2B_2$ of C_2 can be added to obtain the tristimulus values RGB of C, that is

$$R = R_1 + R_2; \qquad G = G_1 + G_2;$$
$$B = B_1 + B_2.$$

Grassmann's law is based on the nearly linear response of the eye.

Gray see *dose*.

Gray body is a body which cannot be considered as pitch-black, that is a perfect black body radiator. Some of the incident light is either reflected or diffused; only part of it will be converted into thermal radiation. The spectral emissivity of a gray body is smaller than that of a perfect black body by a constant numerical factor. The Stefan–Boltzmann law for the total radiation power P emitted from a gray body is

$$P = \varepsilon\sigma_S T^4$$

where the quantity ε is called the *emissivity*, less than 1, except for a perfect black body. A gray body is sometimes defined by its *color temperature*, that is the temperature of an ideal black body with the same color (same emitted radiation power).

Green's theorem is the very basis of the scalar theory of diffraction. Assume two scalar fields F and G, well-defined and devoid of any singularity over a whole volume V defined by a closed surface S. For each point of S, let $\mathbf{n}$ designate the unit vector normal to S and oriented toward its exterior. Then **Green's theorem** (also called Green's second identity) states that

$$\iiint_V (G\nabla^2 F - F\nabla^2 G)dV = \iint_S (G\nabla F - F\nabla G)\cdot d\mathbf{S}.$$

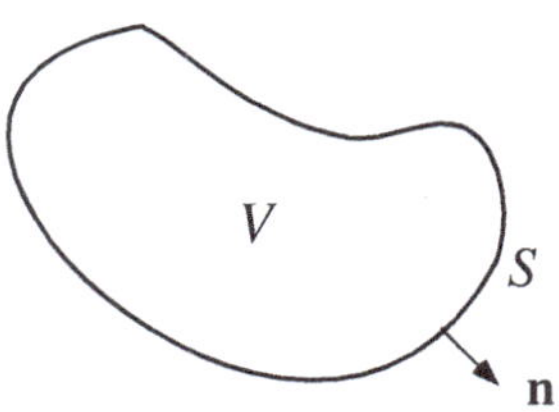

Green's theorem.

GRIN is an acronym for graded index.

GRIN rod lens is a glass rod whose refractive index is highest at the center, on the rod axis, and then decreases with distance from the rod axis. It is

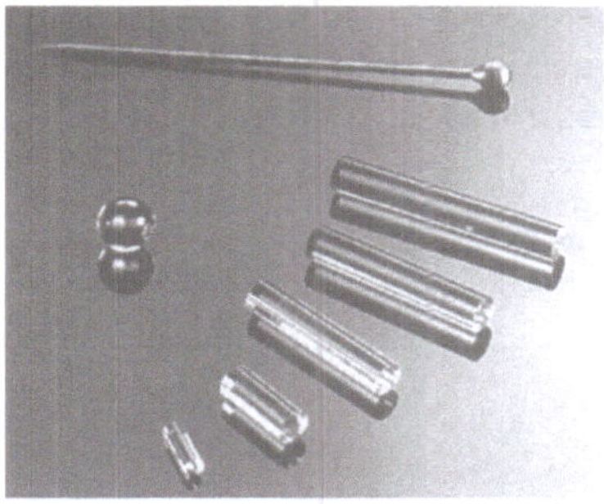

GRIN rod lenses and a spherical lens (a ball lens) used in coupling light into fibers. (Courtesy of Melles Griot.)

typically used to focus or couple light into a fiber. One pitch (P) is a full one-period variation in the ray trajectory along the rod axis

Group delay is the signal delay time per unit distance. The transit time of a light pulse from the input to the output represents a delay time in the information being sent. The group delay τ_g is determined by the *group velocity* v_g as it refers to the transit time of signals (energy). If β_{01} is the propagation constant of the fundamental mode and L is the fiber length then by definition

$$\tau_g = \frac{L}{v_g} = \frac{d\beta_{01}}{d\omega}L.$$

Group index, N_g, represents the factor by which the group velocity of a group of waves in a dielectric medium is reduced with respect to propagation

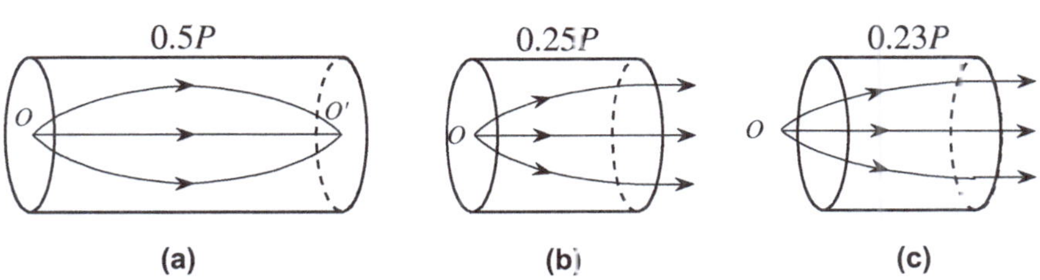

Graded index (GRIN) rod lenses of different pitches. (a) Point O is on the rod face center and the lens focuses the rays onto O' on the center of the opposite face. (b) The rays from O on the rod face center are collimated out. (c) O is slightly away from the rod face and the rays are collimated out.

in free space, i.e. $N_g = v_g/c$ where v_g is the group velocity. The group index can be determined from the ordinary refractive index n through

$$N_g = n - \lambda \frac{dn}{d\lambda}.$$

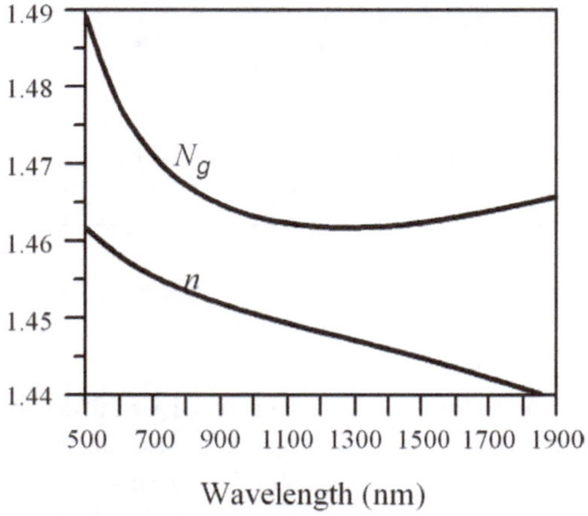

Group index Refractive index n and the group index N_g of pure SiO_2 (silica) glass as a function of wavelength.

Group velocity, v_g, is the velocity with which a group of waves of closely spaced wavelengths (called a wave packet) travel in a medium. It is the velocity with which the energy in the group of waves is propagated in the medium. The group velocity in free space is the same as the velocity of light c in vacuum. Mathematically group velocity is $v_g = d\omega/dk$. Consider the way in which a group of waves differing slightly in wavelength will travel along the z-direction. When two perfectly harmonic waves of frequencies $\omega - \delta\omega$ and $\omega + \delta\omega$ interfere, they generate a *wave packet* which contains an oscillating electric field at the mean frequency ω that is amplitude modulated by a slowly varying field of frequency $\delta\omega$. The maximum amplitude moves with a *group velocity* that is given by

$$v_g = \frac{d\omega}{dk}.$$

The group velocity therefore defines the speed with which energy or information is propagated since it defines the speed of the envelope of the amplitude variation. The maximum electric field advances with a velocity v_g whereas the phase variations in the electric field are propagating at the *phase velocity v*. See *group velocity in optical waveguides*.

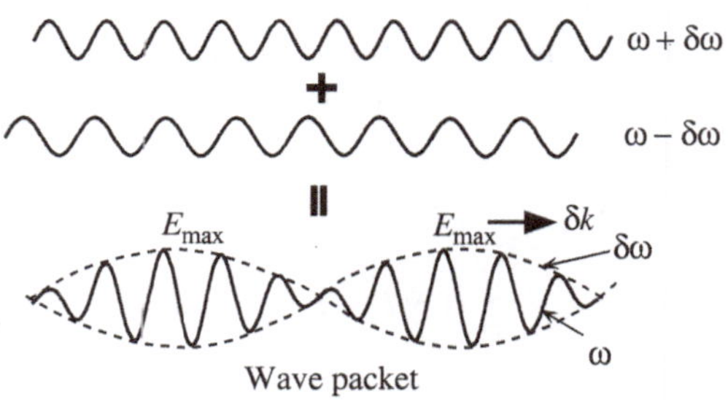

Group velocity Two slightly different wavelength waves traveling in the same direction result in a wave packet that has an amplitude variation which travels at the group velocity.

Group velocity in optical waveguides is the slope of the angular frequency vs. propagation constant of dispersion curve, ω vs. β, and represents the velocity of energy transport in a particular mode propagating in the optical waveguide along its axis. First, one has to solve the Maxwell wave equation for the waveguide with the appropriate boundary conditions and refractive index variation and obtain ω vs. β for each of the allowed propagating modes and then differentiate this dispersion relation to find the group velocity for each mode, since $v_g = d\omega/d\beta$. There are no simple expressions and in most cases numerical solutions are necessary. For a *planar dielectric waveguide*, the propagation constant β for the mth mode ($m = 0, 1, 2, \ldots, M$) is

$$\beta_m \approx \frac{\tan\theta}{a}\left[\arctan\left(\sec\theta\sqrt{\sin^2\theta - (n_2/n_1)^2}\right) + m(\pi/2)\right] = F_m(\theta)$$

where θ is the angle of incidence at the core-cladding boundary for the mode of interest, $2a$ is

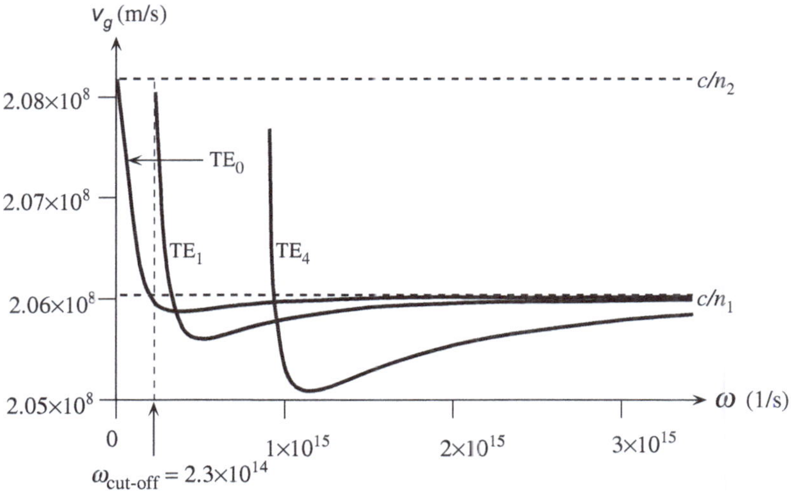

Group velocity in optical waveguides Group velocity vs. angular frequency for three modes for a planar dielectric waveguide which has $n_1 = 1.455$, $n_2 = 1.440$, $a = 10$ μm. TE_0 is for $m = 0$, etc.

the core thickness, and $F_m(\theta)$ is defined as above to be the second term. The corresponding group velocity v_m for the mth mode is

$$v_m \approx \frac{c}{n_1 \sin\theta}\left[1 - \cot\theta \frac{F_m(\theta)}{dF_m(\theta)/d\theta}\right]$$

in which c is the velocity of light and n_1 is the core refractive index. An example is shown in the figure to illustrate that group velocities for the modes are not simply between c/n_1 (core velocity) and c/n_2 (cladding velocity). In a multimode step index fiber in which the number of modes $M \gg 1$,

$$v_{lm} \approx \frac{c}{n_1}\left[1 - \frac{(l+2m)^2}{M}\Delta\right]$$

where l and m are the mode integers for the mode LP_{lm} in the fiber, and Δ is the normalized refractive index difference, $\Delta = (n_1 - n_2)/n_1$.

Guard ring is an annular region around a *pn* junction to reduce or eliminate undesirable *avalanche breakdown* in the junction periphery; this limits avalanche multiplication to the illuminated region.

Guiding layer see *distributed feedback laser.*

Gullstrand's equation see *thick lens.*

Gullstrand's eye model see *eye.*

Gullstrand–Emsley schematic of the eye see *eye.*

Gyromagnetic ratio, γ, is the ratio of the magnetic moment of an electron to its angular momentum. It depends whether the angular momentum is due to orbital motion, spin or both. The term can also apply to a nucleus in which case it is the nuclear gyromagnetic ratio.

H

Hagen–Rubens relationship see *free carrier absorption*.

Half-wave plate is a retarder plate (made from a birefringent refracting crystal) that results in a relative phase change of a half-wave (π) between the *ordinary* and *extraordinary* waves propagating inside the crystal from entry to exit. Linearly polarized light incident with its **E**-field at an angle α to the optic axis leaves the crystal with its field rotated by 2α. See *retarding plates*.

Halogen lamp, quartz halogen or **tungsten–halogen lamp** is a tungsten (W) filament incandescent lamp in which the bulb has some halogen gas (such as bromine), which prevents the preferential loss of tungsten from parts of the filament and allows the filament to operate at a higher temperature. The bulb is a *quartz* tube and close to the filament. In a normal incandescent lamp, the filament temperature is not uniform at every point along the filament since thin regions will be hotter. There is some inevitable evaporation of W from the hot filament, and W is evaporated more from the hot regions, which become even thinner and hotter, leading to an early failure of the filament. The halogen atoms combine with evaporated W atoms to form tungsten oxyhalide molecules. These tungsten-halide molecules disassociate on the hottest regions of the filament, and thereby redeposit the tungsten back onto the filament; the halogen is now available to repeat the cycle. This is the so-called the "halogen regeneration cycle." Since the filament evaporation is suppressed, a higher filament temperature of operation can be achieved. Halogen lamps have higher efficacies, typically 20–24 lm/W, than tungsten lamps; they appear brighter and have higher color temperatures.

Hamiltonian operator see *Fermi's golden rule*.

Harpicon or **HARP video tube** is an ultrasensitive TV camera or a video tube that uses a stabilized a-Se (amorphous selenium) as a photoconductor at very high fields to obtain avalanche multiplication and hence achieves internal gain for very high signal-to-noise ratio and ultrasensitivity. The HARP is an acronym for "high-gain avalanche rushing amorphous photoconductor." A vidicon (a TV pickup tube) using the HARP target is called the Harpicon or the HARP video tube. The HARP was pioneered by Tanioka and coworkers at NHK and it is presently used at NHK in high-definition TV. It can even capture star light images. The basic structure of the HARP target and the principle of operation are shown in the first figure. The entire target is several microns thick. The transparent signal electrode (SnO_2) is biased positively with respect to the cathode. The CeO_2 and SbS_3 layers act as blocking contacts for hole and electron injection, respectively. The incident light from the object is absorbed mainly in the a-Se layer (the thin a-Se:Te layer is inserted to increases the red response). The electron–hole pairs (EHPs) photogenerated in the

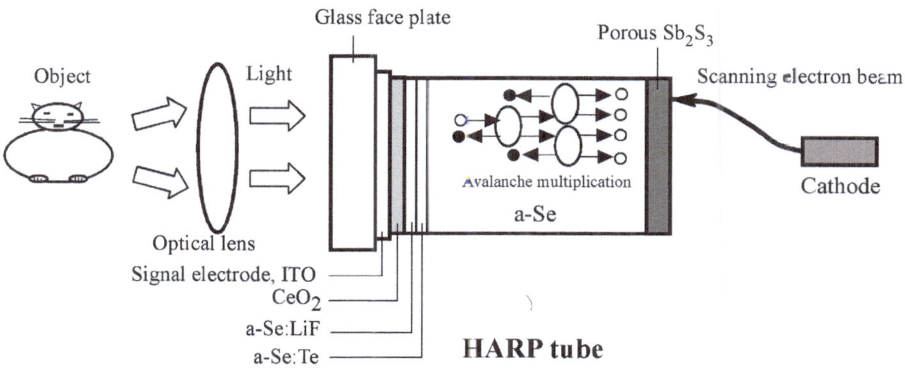

Harpicon The principle of operation of the HARP video camera (or Harpicon).

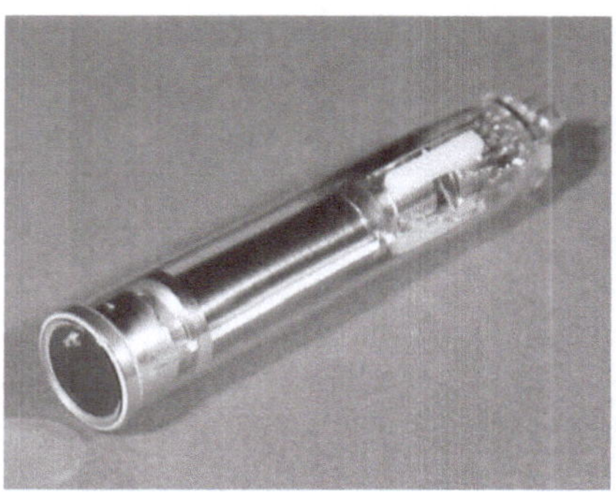

The HARP video tube, or the Harpicon.

Kenkichi Tanioka, the inventor of the Harpicon. He started at NHK as a technician, working through their research and development laboratories to become a vice-president (director general) at the company.

a-Se layer are then drifted by the applied electric field and constitute the signal current. As photogenerated holes drift through the a-Se layer towards the back electrode, as a result of the large applied electric field (greater than 8×10^5 V/cm or 80 V/μm), they experience avalanche multiplication and hence yield an effective quantum gain greater than unity (quantum efficiency is the number of EHPs generated per absorbed photon and the quantum gain is the number of EHPs collected per absorbed photon). The effective quantum gain resulting from avalanche multiplication depends on the field as well as the photoconductor thickness. For example, in a 25-micron thick HARP target, the quantum gain is about 10^3 at a field of 100 V/μm (10^6 V cm^{-1}). TV pickup tubes using such HARP targets have far

Harpicon One frame from a real time image of a ship captured by the Harpicon at night. (Courtesy of Dr. Kenkichi Tanioka, NHK)

superior sensitivity than conventional TV pickup tubes and solid state imaging sensors, and accordingly constitute ultrahigh-sensitive video tubes. The need for high gain vidicons in HDTV is due to the use of small aperture lenses with much greater depth of field than in conventional TV, which allows realistic large screen effects. The origin of avalanche multiplication in a-Se is believed to be due to a lucky drift impact ionization mechanism. (K. Tanioka *et al.*, *Adv., Electron. Elec. Phys.*, **74**, 389, 1988 (Edited by P. W. Hawkes, Academic Press, London).) See *impact ionization.*

Hartmann's equation or **formula** is an empirical dispersion relation with three constants for the refractive index n of glasses in terms of the wavelength of light λ of the form

$$n = n_o + \frac{c}{(\lambda - \lambda_o)^a}$$

where c (not to be confused with the speed of light), a and λ_o are constants. It shows a better representation of dispersion for glass in the visible region than the *Cauchy relation*. Typically $a \approx 1.2$. (H. W. Lee, *Trans. Opt. Soc.*, **28**, 161–167, 1927.)

Hecht collection efficiency describes the efficiency η of charge collection in a detector in terms of what fraction of photogenerated carriers that have been generated inside the detector material have been collected by the external circuit; it excludes the quantum efficiency (conversion efficiency from photons to carriers). The Hecht collection efficiency includes only the efficiency of transporting and collecting the photogenerated carriers. Suppose that we inject a total charge of Q_0 electrons *all* at one end of a detector (radiation receiving side) at $x = 0$, and that the electrons have to drift by an applied electric field E across the detector thickness L to $x = L$ before they can be collected. If μ_e is the electron drift mobility and τ_e is the electron lifetime (mean capture time), then the amount of collected charge Q_c is given by the *Hecht collection efficiency*,

$$\eta_c = \frac{Q_c}{Q_o} = \frac{\mu_e \tau_e E}{L}\left[1 - \exp\left(-\frac{L}{\mu_e \tau_e E}\right)\right]$$

where $\mu_e \tau_e E$ is the electron *schubweg*, the mean distance an electron drifts before becoming captured. Suppose that incident ration is absorbed exponentially in the detector with distance x from the radiation receiving electrode, biased negatively, and α is the absorption coefficient. Both electrons and holes drift and contribute to the total collected charge by virtue of *Ramo's theorem*, and

$$\eta_c = \frac{Q_c}{Q_o} = \frac{\mu_e \tau_e E}{L}\left[(1 - e^{-\alpha L}) + \frac{(e^{-L/\mu_e\tau_e E} - e^{-\alpha L})}{(\alpha\mu_e\tau_e E)^{-1} - 1}\right] + \frac{\mu_h \tau_h E}{L}\left[(1 - e^{-\alpha L}) - \frac{(1 - e^{-\alpha L - L/\mu_h\tau_h E})}{(\alpha\mu_h\tau_h E)^{-1} + 1}\right]$$

where μ_h and τ_h are the hole drift mobility and capture time, and $\mu_h \tau_h E$ is the hole schubweg. (K. Hecht, *Z. Phys.*, **77**, 235, 1932.)

Helium–neon (HeNe) laser is a gas laser in which the Ne atoms are made to lase by *pumping* them through atomic collisions with He atoms that have been excited by an electrical discharge. The emissions may be in the infrared (1523 nm), red (632.8 nm), orange (612 nm), yellow (594.1 nm) or green (543.5 nm). Most popular is the red-emitting He-Ne laser. Ne is an inert gas whose ground state can be represented in terms of the outer subshell as ($2p^6$). If one of the electrons from the $2p$ orbital is excited to a $5s$-orbital then the excited configuration ($2p^5 5s^1$) is a state of the Ne atom that has higher energy. Similarly He is also an inert gas which has the ground state configuration of ($1s^2$). The state of He when one electron is excited to a $2s$-orbital can be represented as ($1s^1 2s^1$) and has higher energy. The He-Ne laser consists of a gaseous mixture of He and Ne atoms in a gas discharge tube. The ends of the tube are mirrored to reflect the stimulated radiation and build-up intensity within the cavity. In other words, an optical cavity is formed by the end-mirrors so that reflection of photons back into the lasing medium builds up the photon concentration in the cavity, a requirement of an efficient *stimulated emission* process as discussed above. By using a dc or RF high voltage, electrical discharge is obtained within the tube which causes the He atoms to become excited by collisions with the drifting electrons. The excitation of the He atom by an electron collision puts the second electron in He into a $2s$ state and changes its spin so that the excited He atom, He*, has the configuration ($1s^1 2s^1$) with parallel spins which is *metastable* (long lasting) with respect to the ($1s^2$) state. He* cannot spontaneously emit a photon and decay down to the ($1s^2$) ground state because the *orbital quantum number l* of the electron must change by ± 1, i.e. Δl must be ± 1 for any photon emission or absorption process. Further, the emission or absorption of electromagnetic radiation cannot change the electron's spin. If m_s is the *magnetic spin quantum number*, then a radiation emitting transition must have $\Delta m_s = 0$. Thus a large number of He* atoms build up during the electrical discharge because they are not allowed to simply decay back to ground state. When an excited He atom collides with a Ne atom, it transfers its energy to the Ne atom by resonance energy exchange because Ne happens to have an empty energy level, corresponding to the ($2p^5 5s^1$) configuration, matching that of ($1s^1 2s^1$) of He*. Thus the collision process excites the Ne atom and de-excites He* down to its ground energy. The excited Ne atom, Ne*, has the ($2p^5 5s^1$)

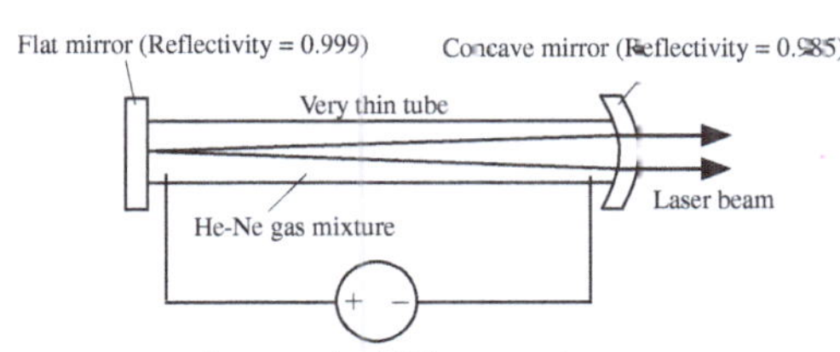

A schematic illustration of the principle of the **He-Ne laser**.

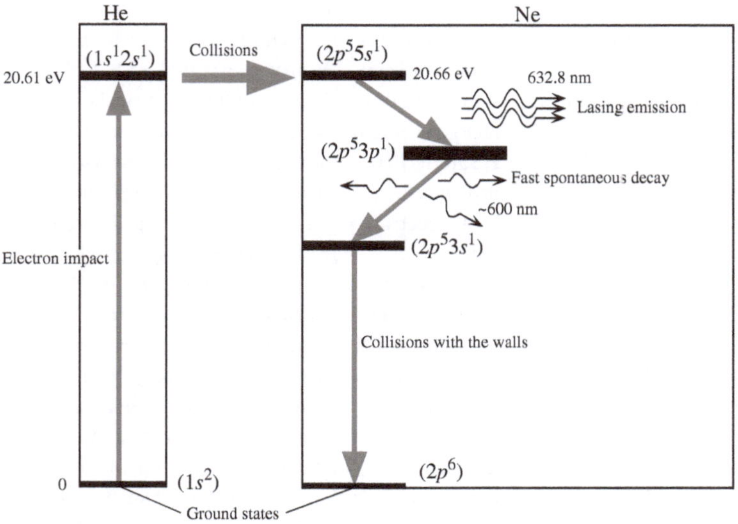

The principle of operation of the **He-Ne laser**. He-Ne laser energy levels (for 632.8 nm emission).

configuration. With many He*-Ne collisions in the gaseous discharge we end up with a large number of Ne* atoms and a *population inversion* between $(2p^5 5s^1)$ and $(2p^5 3p^1)$ states of the Ne atom. A *spontaneous emission* of a photon from one Ne* atom falling from $5s$ to $3p$ gives rise to an avalanche of stimulated emission processes which leads to a lasing emission with a wavelength 632.8 nm in the red. From the $(2p^5 3p^1)$ energy levels, the Ne atoms decay rapidly to the $(2p^5 3s^1)$ energy levels by spontaneous emission. Most of the Ne atoms with the $(2p^5 3s^1)$ configuration, however, cannot simply return to the ground state $2p^6$ by photon emission because the return of the electron in $3s$ requires that its spin is flipped to close the $2p$ subshell. Electromagnetic radiation cannot change the electron spin. Thus the Ne$(2p^5 3s^1)$ energy levels are *metastable* states. The only possible return to the ground state (and for the next repumping act) is by collisions with the walls of the laser tube. We cannot therefore increase the power obtainable from a He-Ne laser by simply increasing the laser tube diameter because that will accumulate more Ne atoms at the metastable $(2p^5 3s^1)$ states. A typical He-Ne laser consists of a narrow glass tube which contains the He and Ne gas mixture. The lasing emission intensity (optical gain) increases with the tube length since then more Ne atoms are used in stimulated emission. The intensity decreases with increasing tube diameter since Ne atoms in the $(2p^5 3s^1)$ states can only return to the ground state by collisions with the walls of the tube. The ends of the tube are generally sealed with a flat mirror (99.9% reflecting) at one end and, for easy alignment, a concave mirror (99% reflecting) at the other end to obtain an *optical cavity* within the tube. The outer surface of the concave mirror is ground to behave like a convergent lens to compensate for the divergence in the beam arising from reflections from the concave mirror. The output radiation from the tube is typically a beam of diameter 0.5–1 mm and a divergence of 1 milliradians at a power of few milliwatts. In high-power He-Ne lasers, the mirrors are external to the tube. In addition, *Brewster windows* are typically

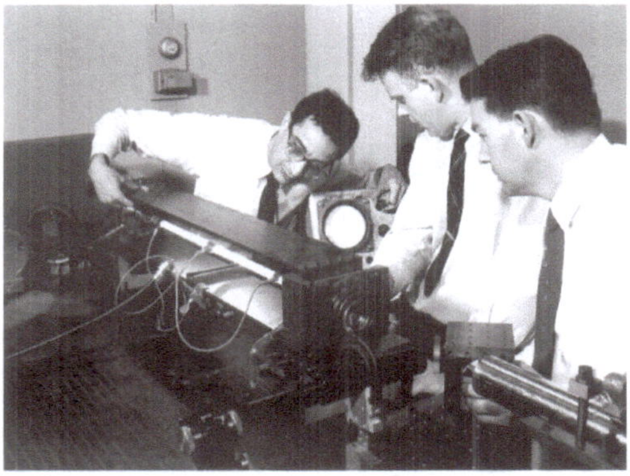

He-Ne laser Ali Javan and his associates William Bennett Jr. and Donald Herriott at Bell Labs were the first to successfully demonstrate continuous wave (cw) helium–neon laser operation (1960–1962). (Courtesy of Alcatel-Lucent Bell Labs.)

used at the ends of the laser tube to allow only polarized light to be transmitted and amplified within the cavity so that the output radiation is polarized (has electric field oscillations in one plane). Even though we can try to get as parallel a beam as possible by lining up the mirrors perfectly, we will still be faced with diffraction effects at the output. When the output laser beam hits the end of the laser tube it becomes diffracted so that the emerging beam is necessarily divergent. Simple diffraction theory can readily predict the divergence angle. Further, typically one or both of the reflecting mirrors in many gas lasers are made concave for a more efficient containment of the stimulated photons within the *active medium* and for easier alignment. The beam within the cavity and hence the emerging radiation is approximately a *Gaussian beam*. Due to their relatively simple construction, He-Ne lasers are widely used in numerous applications such interferometry, for example, accurately measuring distances or flatness of an object, laser printing, holography, and various pointing and alignment applications (as in civil engineering).

Typical He-Ne laser parameters for 632.8 nm emission are (from W. T. Silfvast, *Laser Fundamentals, Second Edition*, Cambridge University Press, 2004): Laser wavelength = 632.8 nm; upper laser level lifetime = 3×10^{-8} s; stimulated emission cross-section = 3×10^{-7} m^2; spontaneous emission linewidth and gain bandwidth = 1.5×10^9 Hz (Doppler broadened); population inversion density = 5×10^{15} m^{-3}; small-signal gain coefficient = 0.15 m^{-1}; typical laser gain medium length (L) = 0.1–1.0 m; single-pass gain = 1.015 – 1.16; gas pressure ≈ 2.5 torr; gas temperature ≈ 400 K; mode of operation is cw; typical output power = 0.5–100 mW, TEM_{00} mode.

Helmholtz equation see *electromagnetic wave equation*.

Helmholtz–Kirchhoff integral links the (scalar) complex amplitude at a given point P to the distribution of amplitude across a closed surface S surrounding P. Consider two scalar fields F and G, both solutions to the Helmholtz equation,

$$\nabla^2 \Psi + k^2 \Psi = 0$$

where k is the wavevector and G represents spherical waves centered on P, of amplitude $G(r) = \exp(jkr)/r$. We apply *Green's theorem* to F and G over the volume V', that is V minus a small sphere of infinitesimal radius ε surrounding the singular point P. The total surface delimiting V' is $S + S'$.

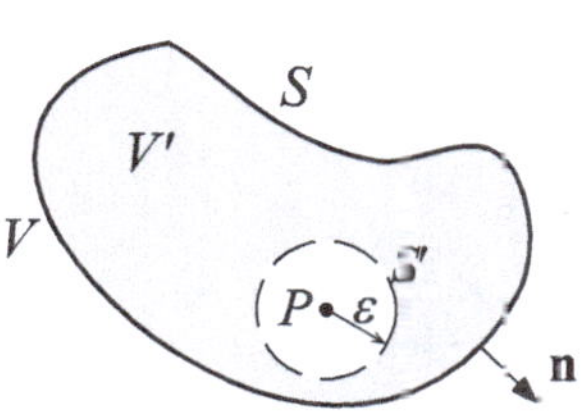

Illustration of the **Helmholtz–Kirchhoff integral**.

We have $\nabla^2 F = -k^2 F$ and $\nabla^2 G = -k^2 G$, so that

$$\int\int\int_V (G \cdot \nabla^2 F - F \cdot \nabla^2 G) dV$$
$$= \int\int_S (G\nabla F - F\nabla G) \cdot d\mathbf{S}$$
$$+ \int\int_{S'} (G\nabla F - F\nabla G) \cdot d\mathbf{S} = 0,$$

with $\nabla G = (r^{-2} - j\,kr^{-1})\exp(j\,kr)\,\mathbf{n}$ and $\nabla F \cdot d\mathbf{S} = (-\partial F/\partial r) r^2 d\Omega$, if $d\Omega$ denotes the elementary solid angle. Therefore, over S',

$$\int\int_S (F\nabla G - G\nabla F) \cdot d\mathbf{S}$$
$$= \int\int_{S'} [F - jkFr + r(\partial F/\partial r)]$$
$$\times \exp(j\,kr) d\Omega$$

which tends toward $4\pi F_P$ for $\varepsilon \to 0$. As a result, over S,

$$\int\int_S (F\nabla G - G\nabla F) \cdot d\mathbf{S} = -4\pi F_p.$$

The Helmholtz–Kirchhoff integral consists in the following expression,

$$F_P = \frac{1}{4\pi} \int\int \left\{ \left(\frac{\exp(jkr)}{r} \right) \nabla F - F\nabla \left(\frac{\exp(jkr)}{r} \right) \right\} \cdot d\mathbf{S}.$$

See also *Fresnel–Kirchhoff diffraction integral.*

Hermite–Gaussian beam is another solution to Maxwell's paraxial equation for an electromagnetic beam with intensity varying with distance, not only along the Oz propagation axis, but also in the transverse plane Oxy. It generalizes the Gaussian beam (*see this term*). If the electric field is written as $E(x,y,z)\exp(-jkz)$, then it can be shown that:

$$E(x, y, z) = A(z) G_x \left(\frac{x\sqrt{2}}{w(z)} \right) G_y \left(\frac{y\sqrt{2}}{w(z)} \right) \times \exp \left(-j \frac{k\rho^2}{2Q(z)} \right)$$

with $A(z)$ an axially slowly varying amplitude, $w(z)$ the beam radius, $\rho = (x^2 + y^2)^{1/2}$ and $Q(z)$ the complex curvature such as $1/Q(z) = 1/R(z) - 2j/[kw^2(z)]$.

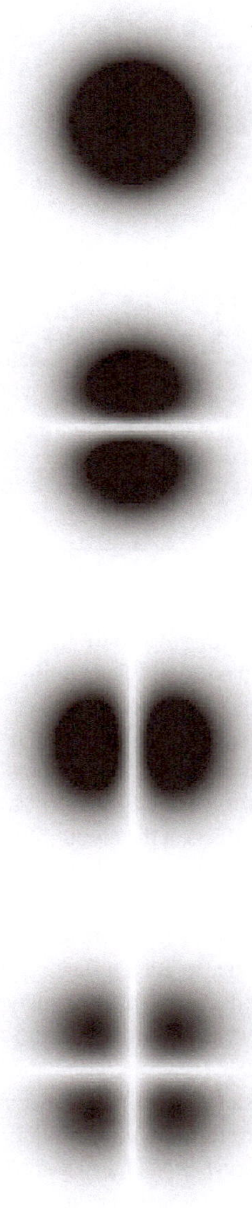

Intensity profile in the transverse plane (x,y) for **Hermite-Gaussian** modes TEM_{00} (fundamental Gaussian beam), TEM_{10}, TEM_{01} and TEM_{11}. The darker the zone, the higher the intensity.

The function G_x is solution to the following differential equation; $G_x''-2xG_x'+2mG_x=0$, with m a positive integer. G_y obeys the same type of equation. The solutions are Hermite polynomials H_m, such as: $H_0(x)=1, H_1(x)=2x, H_2(x)=4x^2-2$, and any subsequent order n is defined by the recurrence: $H_n(x)=2x\,H_{n-1}(x)-2\,(n-1)\,H_{n-2}(x)$. The field becomes:

$$E_{mn}(x,y,z)=\frac{w_0}{w(z)}H_m\left(\frac{x\sqrt{2}}{w(z)}\right)H_n\left(\frac{y\sqrt{2}}{w(z)}\right)\times\exp\left(-\frac{\rho^2}{w^2(z)}-j\Phi_{mn}(\rho,z)\right)$$

with w_0 the beam waist and Φ_{mn} the total phase of the field (z_R the Rayleigh range):

$$\Phi_{mn}(\rho,z)=kz-(m+n+1)\arctan(z/z_R)+k\rho^2/[2R(z)].$$

Any couple of integers (m,n) defines what is called the transverse electromagnetic mode TEM_{mn}, expressed as the product of a radial Gaussian distribution by polynomials. They form a complete set of modes on which any freely propagating beam can be projected and fully described. The lowest mode TEM_{00} is the fundamental Gaussian mode. The transverse field of higher mode TEM_{mn} exhibit m zeros along the x-axis and n zeros along the y-axis. See also *Laguerre-Gaussian beam*.

Hervé–Vandamme relationship see *refractive index*.

Heteroepitaxy refers to epitaxial growth where the layer of material grown on top of the substrate and the substrate itself are different materials. An example of this would be a single-crystal layer of InGaAsP grown on top of a single-crystal GaAs substrate.

Heterojunction is a junction between different semiconductor materials, for example between GaAs and AlGaAs ternary alloy. The terms *heterojunction* and *heterostructure* are frequently used interchangeably. There may or may not be a change in the doping. The doping in the wider bandgap semiconductor is usually denoted with a capital letter *N* or *P*, and that in the narrower bandgap semiconductor with lower case *n* or *p*. Doped heterojunctions are normally distinguished by type – when both materials are of the same (different) doping type they are referred to as *isotype* (*anisotype*) heterojunctions. At the junction between the two materials there will be an energy discontinuity in the conduction and valence band energies ΔE_c and ΔE_v respectively. These are *band offsets* and are extremely important in heterojunction devices. Which material offers the lowest energy location for electrons and holes, respectively, is determined by the bandgaps for each material and the band edge discontinuities. There are three key types of heterojunction determined by these parameters, known as Type I, Type II and Type III heterojunctions. In a Type I *straddled alignment* heterojunction, the smaller bandgap material offers the lowest energy for both electrons and holes as in $GaAs/Al_xGa_{1-x}As$ heterostructures. Type I is the most common heterostructure in optoelectronic devices, e.g. $Ga_xIn_{1-x}As/InP$, $Ga_xIn_{1-x}P/GaAs$. In a Type II *staggered lineup* heterojunction, minimum energies for holes and electrons are in the different materials as shown in the first figure. $Ga_xIn_{1-x}As/GaAs_ySb_{1-y}$ over wide compositions follow Type II behavior. In Type III, the bandgap is "broken up" as in InAs/GaSb. Type III is also called "Type II broken gap" alignment. For example double heterostructures are formed by having a given semiconductor sandwiched between another semiconductor resulting in a pair of heterojunctions. For identical materials in each case, this is called a *double heterostructure* and forms the basis of high-power laser diode structures. It should be noted that given the dependence of refractive index on alloy composition for many alloy materials, one can engineer structures for the confinement of both carriers and light using such heterostructures. Indeed, when the distance between the heterojunctions is of the order of the de Broglie wavelength of the carriers, quantized states may be formed, or a so-called *quantum well*. If the double heterostructure is formed from Type I junctions, then the quantum well acts as a region to quantum confine both electrons and holes and can be used to increase radiative efficiency of optical devices. Similarly, for photodetector structures, it may be advantageous to spatially separate photogenerated electrons and holes, and Type II heterojunctions offer a unique advantage. The band edge profiles for a given heterostructure are determined by the doping levels and carrier transport and recombination, as they are for a simple *pn* homojunction. Examples of such band diagrams are given in the figure below. Of critical importance in the practical implementation of heterostructures is control over the formation of interface states between the component materials. That is, when the interfacial strain due to lattice mismatch between the component materials becomes too large, the lattices of both materials can no longer adjust coherently, but result in the formation of dislocations at the interface. The critical strain level will be determined by both the lattice constant mismatch, and the thickness of the mismatched material being elastically accommodated. Once dislocations form, these act as defect states, typically located within the energy bandgap, which can hold charge when occupied, and hence influence the band edge profiles, as well as act as traps for carriers being injected into the heterojunction depletion region. Not only does this influence the steady state characteristics of the heterojunction, but also the dynamical behavior as one must contend with the additional capture and emission of charges from the interfacial trap states. Finally, it should be noted that superlattices can also be viewed as generated from double heterostructures in which the double heterostructure is repeated a number of times. If these are very thin, quantum well structures, then the repeated structures are referred to as multi-quantum well structures. Although to this point, the heterostructures discussed were all abrupt in their composition, there are numerous examples of so-called band

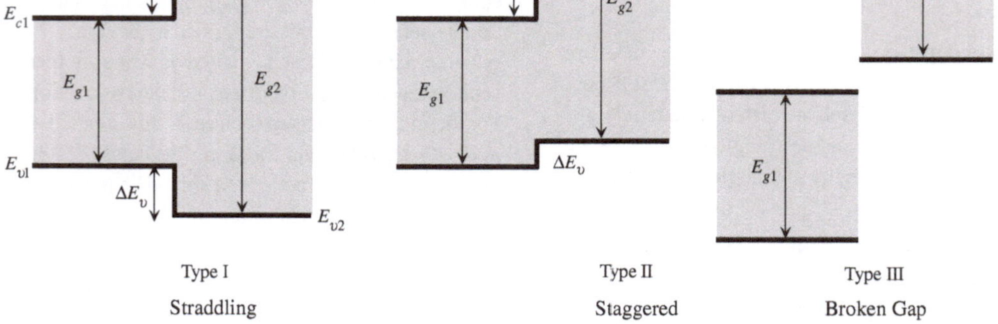

Heterojunction Three types of heterostructures.

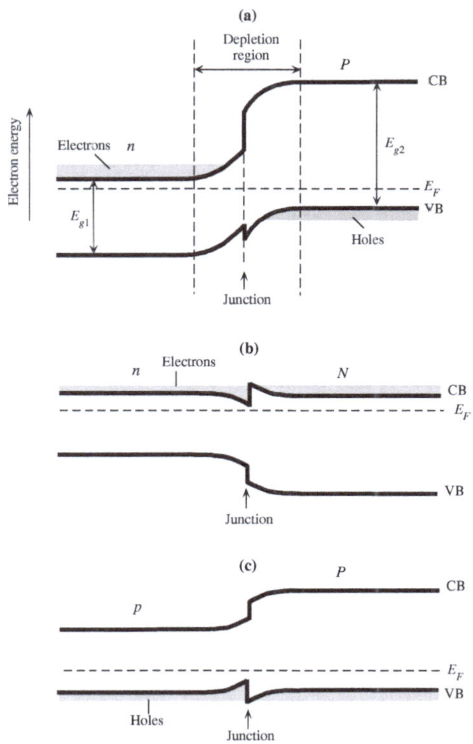

n-P, *n-N* and *p-P* **heterojunctions** and their band diagrams (schematic only to illustrate general features). Under open circuit and equilibrium conditions, the Fermi level E_F must be uniform, i.e. continuous throughout the device. If E_F is close to the conduction band (CB) edge, E_c, it results in an *n*-type, and if it is close to the valence band (VB), E_v, it results in a *p*-type.

structure engineered heterostructures involving both composition and doping that may be continuously varied to create almost arbitrarily designed structures.

Electrical properties of heterojunction diodes such as *pN* or *Pn* are similar to the ordinary *pn* junction. In both cases the *J–V* characteristic can be shown to be governed by the same type of exponential dependence as for the *pn* homojunction:

$$J = J_0[\exp(eV/k_BT) - 1]$$

where V is the applied voltage, k_B is Boltzmann's constant, T the temperature, and J_o is the reverse saturation current. J_o is given by the following expressions:

$$J_o = q[(D_n n_{po}/L_n) + (D_P p_{No}/L_P)];$$
for a pN heterojunction

$$J_o = q[(D_p p_{no}/L_p) + (D_N n_{Po}/L_N)];$$
for an nP heterojunction

where q is the elementary charge, D_n (electrons) and D_P (holes) are the diffusion coefficients of minority carriers in the *p*-side and *N*-side respectively, L_n (electrons) and L_P (holes) are the diffusion lengths of minority carriers in the *p*-side and *N*-side, respectively, and n_{po} and p_{No} are the equilibrium minority carrier concentrations in the *p* and *N* sides. Similarly, D_N (electrons) and D_p (holes) are the diffusion coefficients of minority carriers in the *P*-side and *n*-side respectively, L_N (electrons) and L_p (holes) are the diffusion lengths of minority carriers in the *P*-side and *n*-side respectively, and n_{Po} and p_{no} are the equilibrium minority carrier concentrations in the *P* and *n* sides. Most modern optoelectronic devices such as super-intensity LEDs, laser diodes, photodetectors, etc. rely on heterostructures. Herbet Kroemer and Zhores Alferov shared the 2000 Nobel Prize "for developing semiconductor heterostructures used in high-speed and optoelectronics." (H. Kroemer, *Rev. Mod. Phys.*, **73**, 783, 2001.) See *heterostructure laser diodes, quantum well device*.

Heterostructure laser diodes have a thin narrower-bandgap semiconductor layer sandwiched between two wider-bandgap semiconductors so that there are two heterojunctions. The wider-bandgap (E_g) layers provide potential energy steps and hence carrier confinement within the narrower E_g layer and also provide a step change in the refractive index and hence optical.

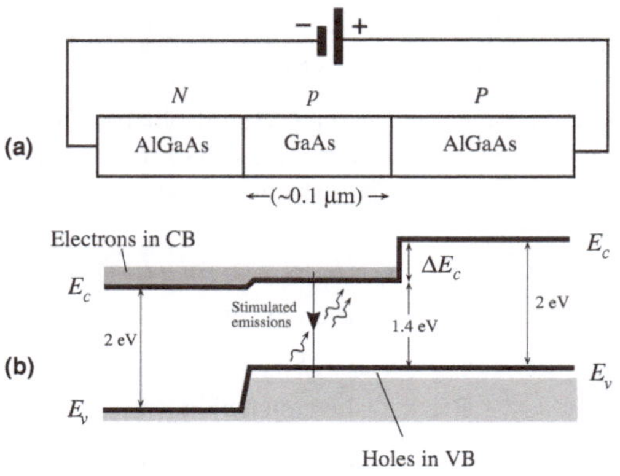

A simplified example of a **heterostructure laser diode** that has a *pN* junction between a wider-bandgap AlGaAs(*N*) and narrower-bandgap GaAs(*p*). The letters in parantheses represent the doping in which the upper case letter implies a wider-bandgap material; *P* is a *p*-type wider-bandgap material with respect to the neighboring semiconductor. The heterostructure between the GaAs(*p*) and AlGaAs(*P*) has a band offset of ΔE_c which prevents the injected electrons from spreading out, i.e. it enables carrier confinement. (The small offset ΔE_v not shown.)

The sandwiched narrower E_g layer acts as the *active region* in which the electron and hole concentrations are sufficiently high to lead to stimulated emission. The figure shows a forward biased AlGaAs(*N*)/GaAs(*p*)/AlGaAs(*N*) diode laser in which electrons are injected from AlGaAs(*N*) into the conduction band of GaAs(*p*), and these electrons become confined by the band offset ΔE_c between the conduction bands of GaAs(*p*) and AlGaAs(*P*); this increases the electron concentration and hence the rate of stimulated emissions since the GaAs(*p*) layer is heavily *p*-doped. See *laser diodes, laser diode equation.*

Hg_{1-x} Cd_xTe (MCT) photoconductive IR detectors are typically used to detect IR radiation 1.5–20 μm. MCT is a solid ternary alloy obtained by the alloying of HgTe and CdTe, i.e. $(HgTe)_{1-x}$ + $(CdTe)_x$ to obtain $Hg_{1-x}Cd_x$Te. The bandgap of MCT, $E_g(x,T)$ can be modified from 0 to 1.6 eV for $x = 0.14$ to 1. MCT is a versatile semiconductor whose properties, e.g. E_g, can be tailored for

Heterostructure laser diode Izuo Hayashi and Morton Panish (1971) at Bell Labs were able to design the first semiconductor laser that operated continuously at room temperature. (Courtesy of Alcatel-Lucent Bell Labs.)

various wavelength applications. A useful empirical relationship for E_g in eV is

$$E_g = -0.302 + 1.93x + 5.35(10^{-4})T(1 - 2x) - 0.810x^2 + 0.832x^3$$

which has been proposed to hold over the full compositional range (from HgTe to CdTe) and from 4.2 to 400 K. MCT detectors are usually cooled to achieve the required detectivity. (G. L. Hansen, J. L. Schmit, and T. N. Casselman, *J. Appl. Phys.*, **53**, 7099, 1982.)

See *detectivity, infrared detectors*.

HID lamp is a high-intensity discharge lamp, see *arc lamp*.

High-intensity discharge lamp see *arc lamp*.

High-purity (HP) and ultrahigh-purity (UHP) are terms that are used to refer to solids, liquids and gases that contain only trace amounts of impurities. The quantity of these impurities is usually measured in units of ppm (parts-per-million), ppb (parts-per-billion) and ppt (parts-per-trillion).

Hole (h^+) is a missing electron in an electronic state that is in the valence band. Intuitively, it is a missing electron in a bond between two neighboring atoms in the semiconductor crystal. The region around this "ruptured" bond is a net positive charge of $+e$. It can drift in an applied field because an electron in a neighboring bond can tunnel into this vacant site and thereby cause the positively charge bond-vacancy to become displaced, shifted. Thus holes contribute to electrical conduction in semiconductors as well. In a full valence band there is no net contribution to the current. There are equal numbers of electrons (e.g. at b and b') with opposite momenta. If there is an empty state, *hole*, at b at the top of the valence band (VB) then the electron at b' contributes to the current. The reason that the presence of a hole makes conduction possible is the fact

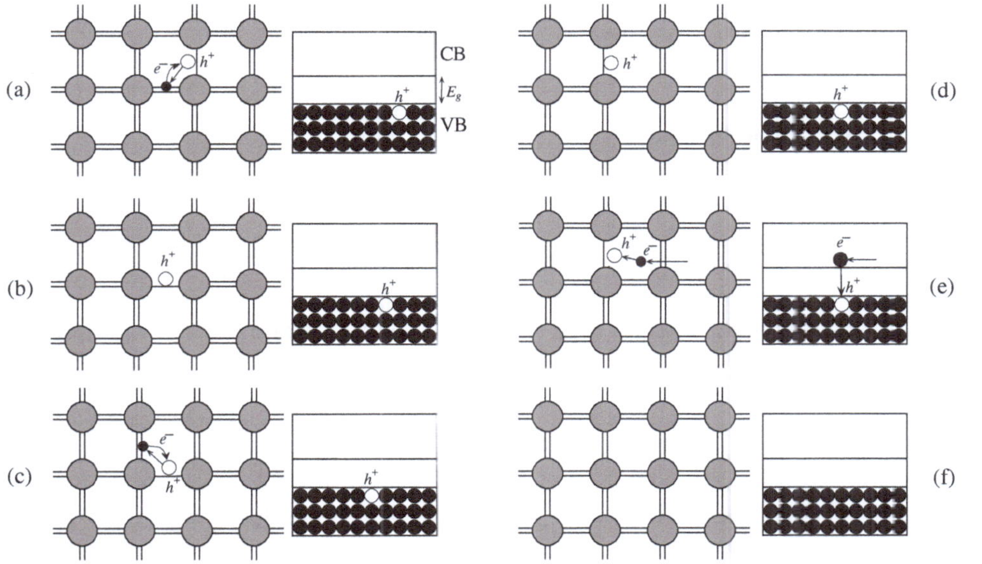

A pictorial illustration of a **hole** in the valence band wandering around the crystal due to the tunneling of electrons from neighboring bonds.

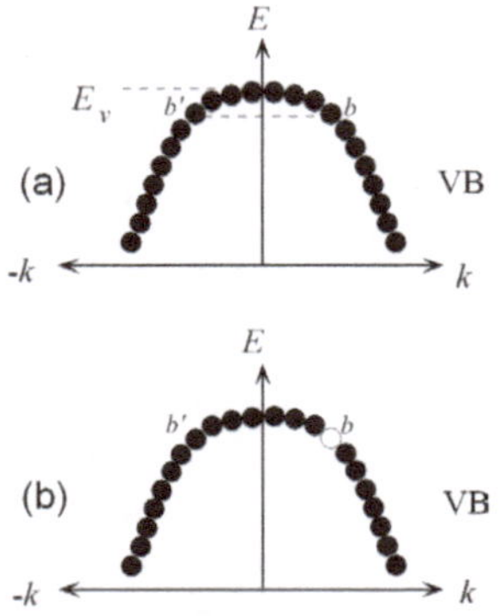

Hole In a full valence band there is no net contribution to the current. There are equal numbers of electrons (e.g. at b and b') with opposite momenta. (b) If there is an empty state (hole) at b at the top of the band then the electron at b' contributes to the current.

that the momenta of all the VB electrons are cancelled except that at b'. Thus, we can consider the net result of the motions of *all* the electrons in the VB just by examining the behavior of the missing electron at b' and assigning to it a positive charge $+e$ and an effective mass $m_h{}^*$.

Hole burning see *spectral hole burning*.

Holey fibers are optical fibers incorporating arrays of air holes in their core forming a two-dimensional *photonic crystal* structure along the fiber length. The result is an extraordinary performance that is unattainable with standard fibers, including an ability to tailor dispersion, and to achieve single-mode behavior over many wavelengths, as well as engineered nonlinear properties. Such fibers are also termed as *photonic crystal fibers*, and are expected to have important applications in spectroscopy, metrology, biomedicine, imaging, and telecommunications. Initially, interest in these materials focused on the occurrence of a *photonic bandgap* (see *photonic crystal*) as a means for guiding light through the fibers. One approach has been to take a pure silica core region and to surround it by a series of air holes – these air holes effectively lower the refractive index and create a step-index profile mimicking standard fibers that typically would have a high-refractive-index Ge doped core region surrounded by a lower-refractive-index pure silica region. However, the patterning of the holes in the outer sheath region into a photonic crystal structure presents a number of advantages. Since the core is made of undoped silica, this results in much lower losses than for a Ge doped core, and permits the fiber to sustain high power and temperature levels, as well as withstand nuclear radiation. Moreover, the air in the cladding region yields a fiber with a huge index step because of the large difference between the refractive index of air ($n = 1$) and silica ($n = 1.45$), translating into fibers with very high numerical apertures (as high as 0.9). Such fibers may support single-mode operation over wavelengths from about 300 to 2000 nm. Also, since the mode field area can exceed 300 μm^2 (i.e. several times larger than about 80 μm^2 – about the most provided by standard fibers), holey fibers can transmit higher powers without running into nonlinear effects or optically induced damage. Alternatively, highly nonlinear fibers can be made as single-mode fibers with extremely small mode field areas (i.e. $\sim$3 μm^2) and efficiently confine light to the core region. Designers can manipulate the dispersion characteristics to create fibers with zero, low, or anomalous dispersion at visible wavelengths. The dispersion can also be flattened. Combining these features with a small mode field area, results in outstanding nonlinear fibers. By altering the pattern of air holes or the materials used, it is possible to manipulate other characteristics including the single-mode cut-off wavelength, the numerical aperture and the nonlinear coefficient. A triangular arrangement of circular air holes in the cladding is typically used to create single-mode fibers, while increasing the air-filling fraction in the cladding typically leads to multimode behavior. An elliptical core can be used to create a highly birefringent fiber that is polarization maintaining.

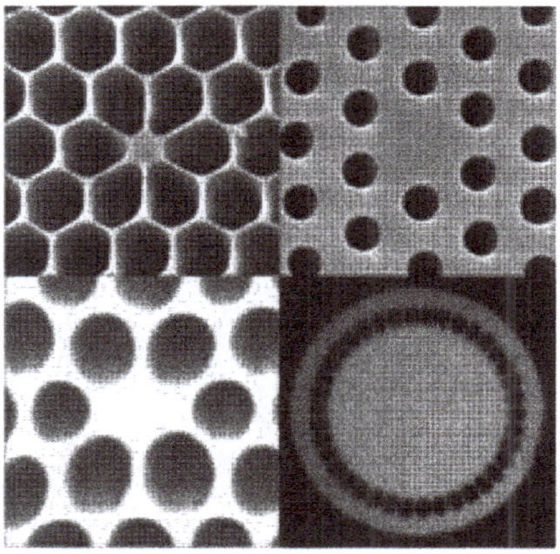

Holey fiber structures vary according to the application (clockwise from the top left): (a) highly nonlinear fiber; (b) single-mode fiber; (c) polarization maintaining fiber; and (d) high numerical aperture fiber. (Courtesy *Crystal Fibre AS, Blokken 84, DK-3460 Birkerød, Denmark.*)

Silica is the material of choice for most applications in the wavelength region between 200 and 2500 nm, but other materials can be used to enhance specific parameters such as nonlinearity or waveguiding outside of this spectral region. Furthermore, one can combine the silica with many dopant species, yielding fiber lasers and amplifiers, for example. Manufacturing of these fibers typically starts with an array of hollow capillary silica tubes bundled around a pure silica rod replacing the center capillary. A sleeve tube is then used to surround the assembly and forms the so-called *preform*. In a fiber draw tower, the manufacturer heats the preform to about 2000 °C and carefully pulls the preform, to form a fiber typically 125 μm in diameter. A polymer film is then applied to the outside of the fiber to protect it. It should be noted that the drawn fiber maintains the structure of the preform. One of the most remarkable properties of these new fibers is spectral broadening – the so-called *supercontinuum* process exploits the high peak powers available from mode-locked femtosecond or picosecond pulsed lasers. Since the fibers have a small core and low or zero dispersion close to the pumping wavelength, they can broaden the spectral width of pulses to previously unknown levels. Such supercontinua can cover wavelengths from 500 to 1300 nm with intense coherent light and have applications in areas such as metrology, spectroscopy, imaging, and microscopy and provide submicron resolution for *optical coherence tomography*. Since the phase coherent spectrum spans more than one optical octave, that makes it useful for a number of metrology and spectroscopy applications. Telecommunications applications of these fibers include low-nonlinearity large-core transmission fibers, dispersion-compensating fibers (including slope compensation), nonlinear fibers for wavelength conversion, switching, amplification and signal regeneration, and doped large mode-area fibers for high-power amplification. Large mode-area fibers can also provide high-power delivery for applications in astronomy, lithography, materials processing, imaging, and femtosecond pulse guidance.

Hollow cathode lamp is a metal vapor arc lamp or a discharge lamp which has been filled with a metal vapor to obtain certain wavelengths of emission. They are often used in spectral work such as atomic absorption analysis. For example, the Na-vapor lamp has emissions at 589.00 and 589.59 nm. See *arc lamp*.

Hollow optical fiber (HOF) is an optical fiber that has been fabricated with a hole at its center to be hollow. It has similarities to microwave waveguides. The hollow core is filled with air or an inert gas. The cladding is a metal or a dielectric, or both. For example, one type of simple hollow fiber has a thin metal (e.g. silver) coated on the inside of a hollow tube such as a capillary glass tube. In most cases the inside of a hollow tube is coated with a thin metal film, or a metal and dielectric film. Metal, polymer and glass tubes have been used. HOFs are very useful for delivering high optical powers, and obviously do not suffer from core material nonlinearities. They are excellent for delivering high laser powers in industrial applications (e.g. laser cutting) and IR laser surgery. They also have low insertion losses, and do not suffer from end reflections. The electromagnetic wave guiding principles are different for hollow fibers than those for the usual glass optical fiber waveguides – the core has a lower refractive index. There is no total internal reflection, and the light is transmitted along the core as a leaky mode and experiences attenuation. The attenuation coefficient depends not only on the reflectance R but also on the roughness of the core-cladding interface. The attenuation can be minimized by maximizing R. Various techniques have been proposed to increase R such as using silver cladding with a dielectric coating over it (used with IR lasers), Al cladding for use at UV wavelengths, and cladding made from dielectric multilayers. The attenuation also depends on the bore radius a, and fiber bend radius R. Very roughly, for a given hollow waveguide, α increases with decreasing a as $1/a^3$, and increases with decreasing fiber bending radius as $1/R$. The power attenuation coefficient $2\alpha_{nm}$ for mode HE_{nm} is given by the *Miyagi–Kawakami equation* which, for single layer on the inside of the tube, in simplified form is

$$\alpha_{nm} \approx \left(\frac{u_{nm}}{2\pi}\right)^2 \left(\frac{\lambda^2}{a^3}\right) \mathrm{Re}\left[\frac{\frac{1}{2}(\tilde{n}^2+1)}{(\tilde{n}^2-1)^{1/2}}\right]$$

where u_{nm} is a modal parameter, 2.405 for the lowest order HE_{11} mode, $\tilde{n}$ is the complex refractive index of the inner wall material, $2a$ is the inner diameter of the hollow core, and λ is the free-space wavelength. In the case of metal/dielectric reflector,

$$\alpha_{nm} \approx \frac{\lambda^2}{a^3}\left(\frac{u_{nm}}{2\pi}\right)^2\left(\frac{n}{n^2+k^2}\right) \times \frac{1}{2}\left[1+\frac{n_1^2}{(n_1^2+1)^{1/2}}\right]^2,$$

where n and k are the optical constants of the metal film (real and imaginary parts of the complex refractive index) and n_1 is the refractive index of the dielectric film. (T. Abel, J. Hirsch, and J. A. Harrington, *Opt. Lett.*, **19**, 1034–1036, 1994; J. A. Harrington, *Fiber Integrat. Optics*, **19**, 211, 2000; M. Miyagaki and S. Kawakami, *J. Lightwave Technol.*, **LT2**, 116, 1984.)

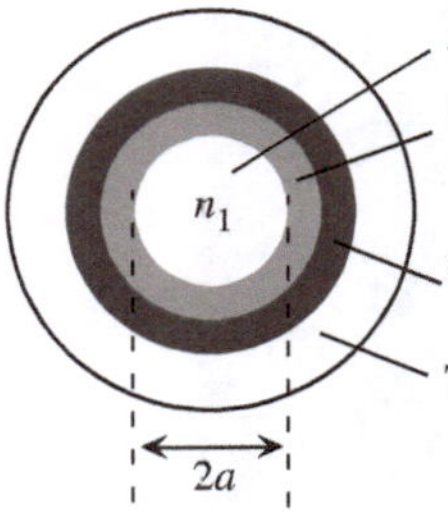

A typical **hollow optical fiber** which has a thin metal and dielectric film coated on the inside of a tube. Usually there is a protective plastic coating on the outside, which is not shown.

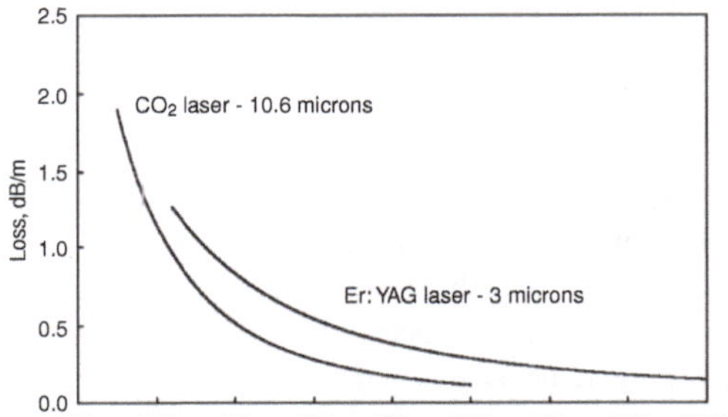

Hollow optical fiber Loss vs. bore diameter for straight losses measured in hollow glass waveguides with Ag/AgI films. The guide labeled CO_2 laser was designed for optimal transmission at 10.6 μm while that labeled Er:YAG laser was designed for optimal transmission at 3 μm. Note that the loss varies approximately as $1/a^3$. From the IR fibers website at Rutgers University.

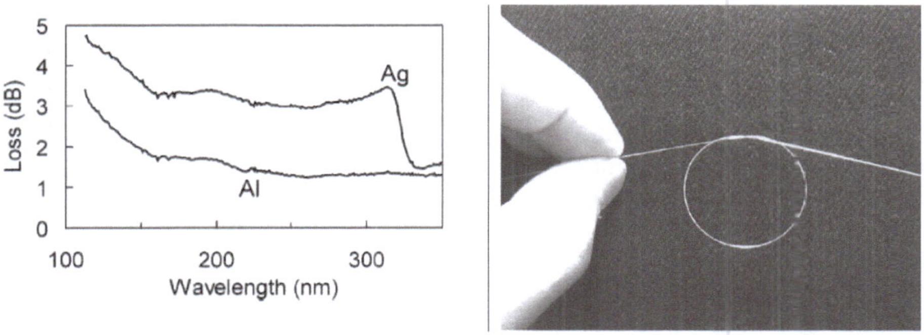

Hollow optical fiber Loss spectra of Al and Ag coated hollow fibers (thin glass capillary tubes) for use in the UV region. The fiber diameter is 1 mm and length 17 cm. The photograph shows that HOF can be flexible. (Inner bore diameter 0.32 mm.) (From Y. Matsura and M. Miyagi, *J. Selected Topics Quant. Electron.*, **10**, 1430, 2004).

Holography is a technique of reproducing three-dimensional optical images of an object by using a highly *coherent radiation* from a laser source. The object, e.g. a cat, is illuminated by a highly coherent beam as would normally be available from a laser. The laser beam has both *spatial* and *temporal coherence*. Part of the coherent wave is reflected from a mirror to form a *reference beam* E_{ref} and travels towards a fine-grained photographic plate. The waves that are reflected from the cat, E_{cat}, will have both amplitude and phase variations that represent the cat's surface (topology). If we were looking at the cat, our eyes would register the wavefront of the reflected waves E_{cat}. Moving our head around we would capture different portions of the reflected wave and we would see the cat as a three-dimensional object. The reflected wave from the cat, E_{cat}, is made to interfere with the reference

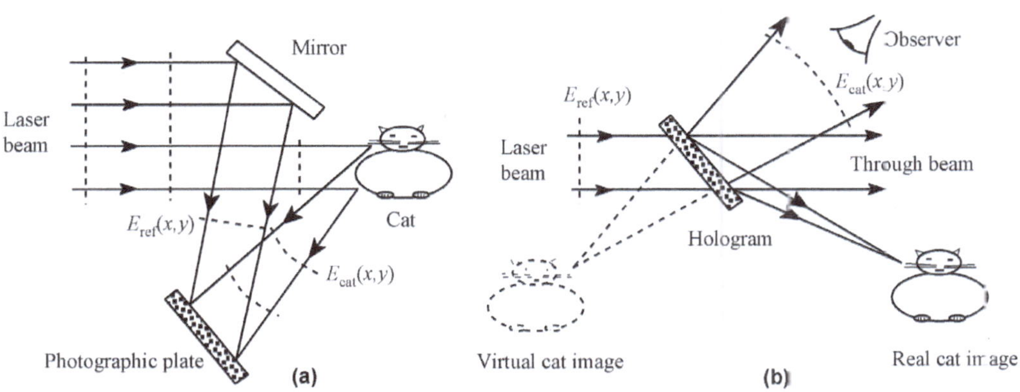

A highly simplified illustration of **holography**. (a) A laser beam is made to interfere with the diffracted beam from the subject to produce a hologram. (b) Shining the laser beam through the hologram generates a real and a virtual image.

wave, E_{ref}, at the photographic plate and give rise to a complicated *interference pattern* that depends on the magnitude and phase variation in E_{cat}. The recorded interference pattern in the photographic film (after processing) is called a *hologram*. It contains all the information necessary to reconstruct the wavefront, E_{cat}, reflected from the cat and hence produce a three-dimensional image. To obtain the three-dimensional image, we simply have to illuminate the hologram with the reference beam, E_{ref}. Most of the beam goes right through but some of it becomes *diffracted* by the interference pattern in the hologram. One diffracted beam is an exact replica of the original wavefront, E_{cat}, from the cat. The observer sees this wavefront as if the waves were reflected from the original cat and registers a three-dimensional image of the cat. This is the *virtual image*. We know that first-order diffraction from a grating has to satisfy the Bragg condition, $d\sin\theta = m\lambda$ ($m = \pm 1$) where λ is the wavelength and d is the separation of the slits. We can qualitatively think of a diffracted beam from one locality in the hologram as being determined by the local separation d between interference fringes in this region. Since d changes in the hologram depending on the interference pattern produced by E_{cat}, the whole diffracted beam depends on E_{cat} and the diffracted beam wavefront is an exact scaled replica of E_{cat}. Just as normally there would be another diffracted beam on the other side of the zero-order (through) beam, there is a second image, called the *real image*, which is of lower quality. (It may help to imagine what happens if the object consists of black and white stripes. We would then obtain periodic dark and bright interference fringes in the hologram. This periodic variation is just like a diffraction grating; the exact analysis is more complicated.) Holography is therefore a method of *wavefront reconstruction*.

Homoepitaxy refers to epitaxial growth where the layer of material grown on top of the substrate and the substrate itself are of the same material. An example of this would be a layer of GaAs single crystal grown on a single crystal GaAs substrate.

Dennis **Gabor** (1900–1979), inventor of **holography**, is standing next to his holographic portrait. Professor Gabor was a Hungarian born British physicist who published his holography invention in *Nature* in 1948 while he was at Thomson-Houston Co. Ltd, at a time when coherent light from lasers was not yet available. He was subsequently a professor of applied electron physics at Imperial College, University of London. (From M.D.E.C. Photo Lab, Courtesy AIP Emilio Segrè Visual Archives, AIP.)

Homogeneous or **uniform medium** in photonics implies a medium in which the refractive index n (or the relative permittivity ε_r) is the same everywhere in the medium. Put differently, the polarization **P** induced by a given field **E** is the same everywhere in the medium.

Homogeneous broadening of an atomic spectral line occurs when all identical atoms in a collection are subject to exactly the same energy-decay and dephasing mechanisms, for instance, in defect-free crystals where the local surroundings remain uniform. All the dipoles react therefore in the same way, so that the response of each individual oscillator is equally broadened. The overall spectral response is well described by a *Lorentzian function*

$L(\omega)$ centered at an angular frequency ω_0, with half-width at half-maximum γ,

$$L(\omega) = \{\gamma^2/[(\omega - \omega_0)^2 + \gamma^2]\}.$$

The natural linewidth of such a transition can be related to Heisenberg's uncertainty principle, $\tau\ \Delta E \approx \hbar$: since $\Delta E = \hbar\gamma$, we get $\gamma = 1/\tau$. The finite lifetime $\tau = 1/\gamma$ is attributable to decay into other levels. In an individual quantum dot, often thought of as an artificial atom-like system with its discrete states, ideal delta function-like emission and absorption spectra can be thermally broadened due to temperature fluctuations in the dot. Analytically, this line broadening can be related to the heat capacity per unit volume of the quantum dot material, its thermal conductivity and temperature dependence of the bandgap, as well as the dot size. At room temperature, for example, in an InAs quantum dot of 2 nm diameter, the broadening corresponds to about 1 meV, and decreases to about 0.1 meV for a 12 nm dot.

Homojunction is a junction between differently doped regions of the same semiconducting material, for example a *pn* junction in the same silicon crystal; there is no change in the bandgap energy, E_g.

Hooge's law see *excess noise*.

Hue see *color*.

Human eye see *eye*.

Hund's rule states that electrons in a given subshell, *nl*, try to occupy separate orbitals (different m_ℓ) and keep their spins parallel (same m_s). In doing so they achieve a lower energy than pairing their spins (different m_s) and occupying the same orbital (same m_ℓ). Hund's rule must be consistent with the Pauli exclusion principle.

Huygens–Fresnel principle states that each point of a wavefront can be regarded as the center of a secondary vibration giving rise to a secondary spherical wavelet, and that the position of the new wavefront at any later time is given as the envelope of all such wavelets. (Put differently, the amplitude of the optical field at any point is the superposition of all these wavelets, considering their amplitudes and relative phases.) The mutual interference between secondary wavelets, as postulated by Fresnel, explains the propagation of plane or spherical waves in free space. Moreover, it describes effectively the diffraction of light through a small aperture on an opaque screen (pinhole or slit). This principle has played a decisive role in establishing the wave theory of light, in the first half of the nineteenth century. See *diffraction*.

Hydrogenated amorphous silicon (a-Si:H) see *amorphous semiconductor*.

Hyperpolarizability refers to terms higher than the linear term in the dependence of the induced polarizability of a molecule on the local electric field acting on it. In a linear medium the induced dipole moment p_i would depend on the local field E_L as $p_i = \alpha E_L$, where α is the polarizability of the molecule. In general p_i will not depend on the local field in a simple linear fashion (except at sufficiently low fields), and the induced dipole is then expressed as

$$p_i = \alpha E_L + \beta E_L^2 + \gamma E_L^3$$

where E_L is the local field acting on the molecule, and the coefficients β and γ are called the *hyperpolarizabilities* of the molecule; β being the second-order and γ the third-order hyperpolarizability. There are various models in the literature that allow β and γ to be calculated or estimated, some models being purely empirical; for example, γ can be estimated from α. In some cases, upper quantum limits on β and γ values have been calculated as in the *Kuzyk limit of hyperpolarizabilities*. Some authors use a factor (1/2) before β and (1/6) before γ. See *electric susceptibility, Kuzyk susceptibility or hyperpolarizability limit*.

Hwang–Rhys factor, see *Frank–Condon principle*.

I

Illuminance, E_v [in lux, or lumens/m^2], is the incident luminous flux, from all directions, per unit of area of a surface; its units are in lux or lm m^{-2}. It is useful to define illuminance in terms of small or elemental quantities. If an elementary surface area dS is irradiated by an element of luminous flux $d\Phi_v$, arriving on the surface from all directions, then the illuminance E_v of the surface is defined by $E_v = d\Phi_v/dS$. Most light meters measure illuminance in lux.

Table: Illuminance

Illumination source	Illuminance (Lux), typical
Bright sunny day	50,000–100,000
Skylight	20,000
Cloudy day	2000–25,000
Sunset	1–100
Good street lighting	20
Moonlight	0.1
Starlight	0.001
Recommended for reading	500
Recommended for manufacture inspection	800
Recommended for schoolrooms	240

Image is what is observed (that is, visualized) by an eye when an object is viewed through an optical instrument as in looking through a simple lens, a telescope, looking at an image on a screen from a projector, etc. A *real image*, as observed by looking through an optical instrument (e.g. a simple lens) is formed by the intersections of light rays emanated from the object; the rays actually pass through the image. A real image can be viewed on a screen placed at the location of the image. A *virtual image* is formed when the light rays from the object do *not* pass through the image, but, instead, give the appearance of coming from it. The eye's focusing mechanism coverts a virtual image to a real image on the retina which is then detected by the photoreceptors. A virtual image cannot be displayed on a screen placed at the location of the "image." A virtual image is an apparent image.

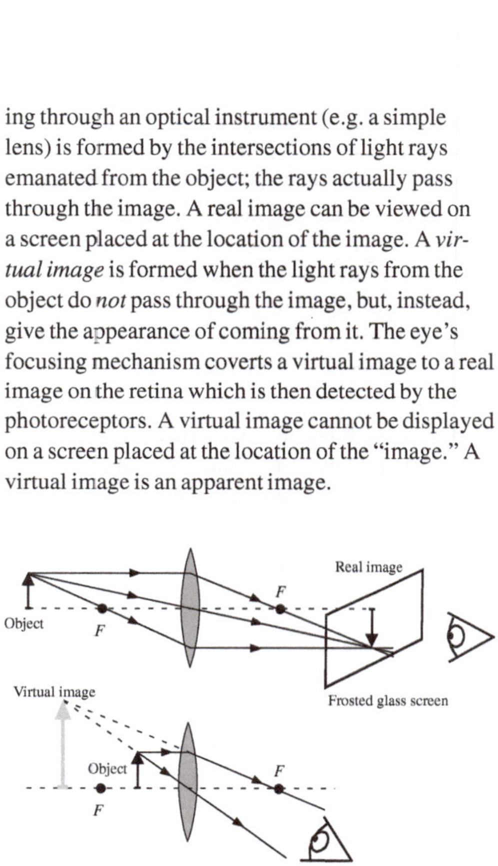

Image Upper: real image. Lower: virtual image.

Image illuminance see *illuminance, cosine fourth law.*

Image sensor in general terms is a device that is able to capture an image and provide an output in the form of an electrical signal such as current, charge or voltage. The output is usually put through a multiplexer and an analog-to-digital converter to obtain a digital form of the image. Early image sensors were "electron tubes" (cathode ray tube based imaging devices such as the TV pickup tubes) but today nearly all imaging sensors are solid state devices. As a general description, the sensor usually consists of an array of elements, called pixels (*pixel* from "picture element"), in N rows and M columns, where each element or pixel has a detector and is able to provide an electrical signal that is proportional to the light intensity received at this pixel. A lens forms an image of the object on the image sensor, and the light intensity $I(X,Y)$ at each point of this image excites the corresponding pixel at location (X, Y); the image point (X, Y) excites the pixel at that location that is in row i and column j, and the pixel (i,j) has an electrical signal (e.g. charge) proportional to $I(X, Y)$. Thus, each pixel carries a piece of the image information, a pixel of it. The array is addressed from a master timer to read out the signals (charges) at the pixels and hence provide an electronically recorded image. There are two major solid state technologies: CCD (charge coupled device) and CMOS (complementary metal oxide semiconductor) sensors, each with its own advantages and drawbacks. CMOS refers to the technology that is used to fabricate the sensor. In the CCD technology, the incident light at a pixel generates charge that is stored in the CCD well at this location. This charge (signal) is read out by a sequential clocking mechanism (a characteristic of the CCD operation) to shift the charges to an eventual shift register. The CMOS sensor is more like the image sensor chip shown in the figure. Each pixel has MOS transistors and a photodiode. The electrical signals at pixels are read out line by line. A row i is addressed by the controller, and all the pixels in this row provide their signals to the columns (data lines) $j = 1$ to M. The "parallel" data are multiplexed and digitized to provide a digital image. The image sensor quantizes the image because it breaks it into $N \times M$ number of pixels. The pixel size imposes a resolution limit because the image is sampled by the pixels, which are of finite size. Color rendering can

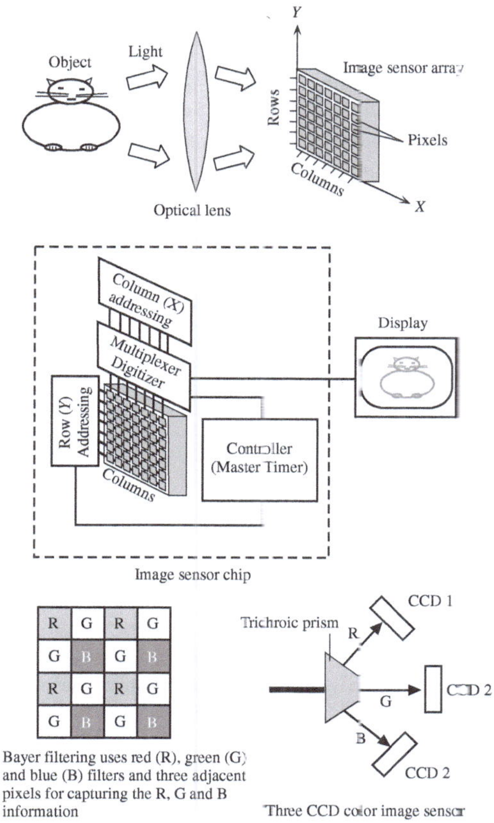

Image sensor Top: The basic image sensing operation using an array of photosensitive pixels. Middle: The image sensor chip that incorporates the auxiliary electronics that run the sensor array (CMOS technology). Lower: Color imaging by using two different techniques. Bayer filtering uses red, green and blue filters on three pixels. 3CCD uses a trichroic prism to separate the image colors into red, green and blue and uses three CCD chips for each.

be done in three ways. The most common is to use red, green and blue filters to separate out the colors at a point onto three different adjacent pixels as in the *Bayer color image sensor*. In *three CCD* imaging, a trichroic prism assembly is used to separate the light from the object into its red, green and blue components, and three different CCDs are used to detect each component. The red, green and blue light have different penetration depths into silicon and one can use photodiodes at different depths to distinguish between the colors. See *active matrix array, convolution, charged coupled devices.*

Imaging equations enable the calculation of the position and dimension of the image of a small object through a centered (aligned) optical system, within the paraxial approximation. If (A,A') denotes a couple of conjugated points, then

$$\overline{FA}.\overline{F'A'} = \overline{HF}.\overline{H'F'},$$

and the lateral magnification g_y is

$$g_y = \frac{\overline{A'B'}}{\overline{AB}} = -\frac{\overline{F'A'}}{\overline{H'F'}} = -\frac{\overline{HF}}{\overline{FA}}.$$

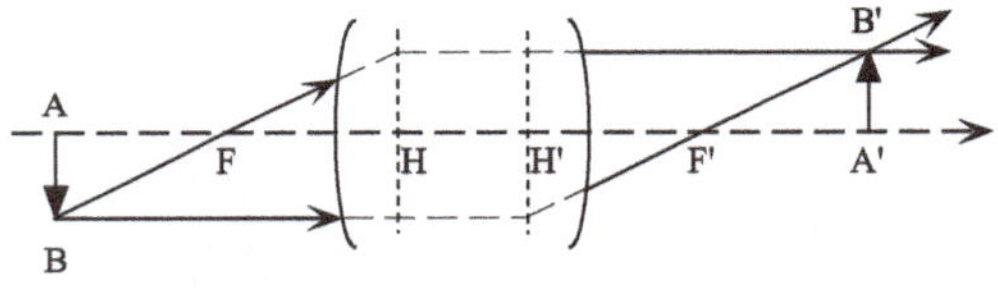

Imaging equations Within the paraxial approximation, any incident ray passing by object point B gives rise to an emergent ray passing by image point B′. Object AB produces image A′B′.

For a thin lens of optical center O between media with refractive indices n and n',

$$\frac{n'}{\overline{OA'}} - \frac{n}{\overline{OA}} = \frac{n'}{\overline{OF'}} = \frac{-n}{\overline{OF}} = P$$

where P, the *power* of the lens, is expressed in *diopters*.

For a spherical diopter of center C, summit S, separating media with indices n and n',

$$\frac{n'}{\overline{SA'}} - \frac{n}{\overline{SA}} = \frac{n'-n}{\overline{SC}} = P$$

or, with the center C as the origin,

$$\frac{n}{\overline{CA'}} - \frac{n'}{\overline{CA}} = \frac{n-n'}{\overline{CS}}.$$

For a spherical mirror of center C, summit S,

$$\frac{1}{\overline{SA'}} + \frac{1}{\overline{SA}} = \frac{2}{\overline{SC}} = \frac{1}{\overline{SF'}} = \frac{1}{\overline{SF}} = P.$$

Impact ionization is an *avalanche multiplication process* in which hot carriers gain sufficient energy from the applied field so that when they collide with the lattice they are able to excite electrons from the valence band, VB (E_v) to the conduction band, CB (E_c), which generates electron–hole pairs (EHPs), hence avalanche multiplication. A hot carrier means that an electron in the conduction band or a hole in the valence band has a kinetic energy greater than $(3/2)k_BT$. Intuitively, impact ionization is the process by which a high electric field accelerates a free charge carrier which then impacts with a covalent bond in the semiconductor to rupture this bond and thereby release a free electron from the bond, and leave behind a free hole. The impact ionization (II) process leads to an avalanche multiplication of charge carriers and hence it is fruitfully utilized in avalanche photodiodes that have internal (intrinsic) gain; photogenerated carriers are multiplied at high fields. If n is the number of electrons in the conduction band, then the electron *impact ionization coefficient* (IIC), α_e (i.e. II caused by electrons), is defined as the fractional increase in electron concentration per unit distance:

$$\alpha_e = (1/n)(dn/dx). \quad (1)$$

There is a similar IIC α_h for holes, $\alpha_h = (1/p)(dp/dx)$, where p is the hole concentration.

The field dependence of the IIC α, at least over the limited fields where avalanche is observed, is usually modeled by experimentalists by using

$$\alpha = A\exp\left[-\left(\frac{B}{F}\right)^n\right] \quad (2)$$

where F is the electric field, A, B, n are constants that depend on the semiconductor material properties such as the E-**k** electronic structure, phonon

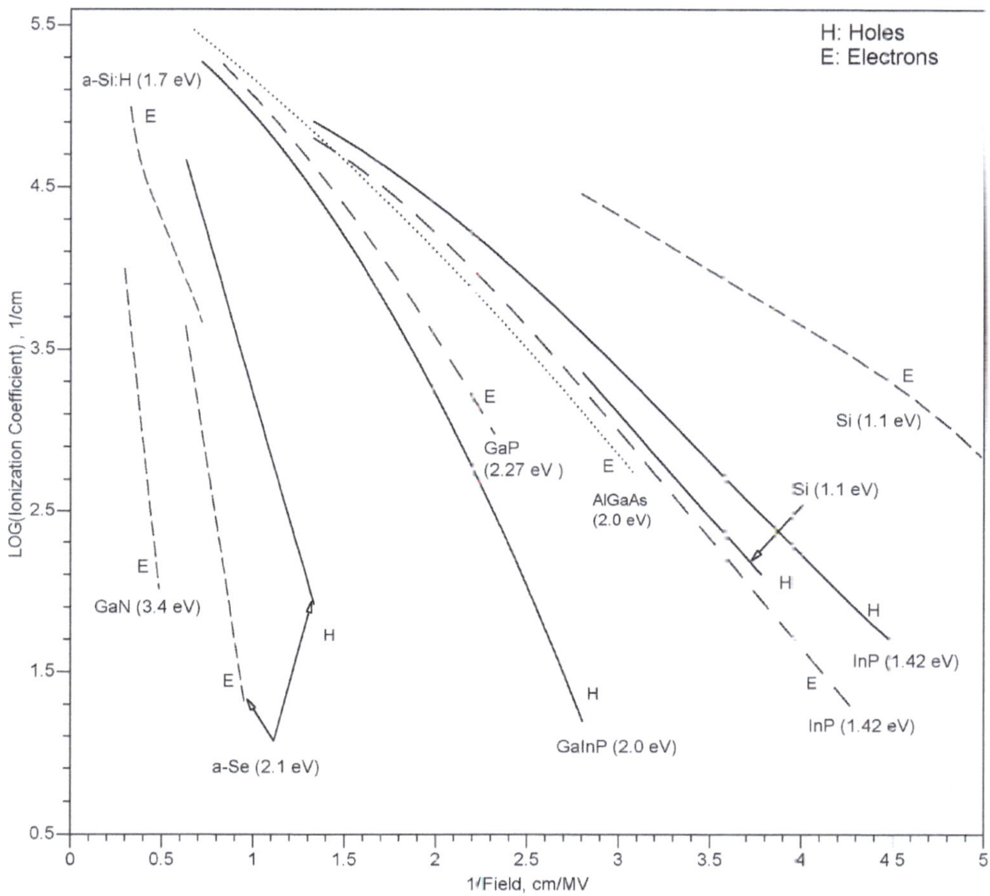

Impact ionization Plot of IIC vs. reciprocal field for various semiconductors. The numbers in parentheses are the bandgaps in eV. (From S. Kasap, J. A. Rowlands, S. Baranovski, K. Tanioka, *et al.*, *J. Appl. Phys.*, **96**, 2037, 2004.)

energies and spectra, scattering mechanisms, and so on. Typically, it has been found that $n \approx 1$ at low fields and $n \approx 2$ for high fields. Equation (2) with $n = 1$ is sometimes called *Chynoweth's law*. The origin of Equation (2) in the simple $n = 1$ form lies in Shockley's *lucky electron* model that dates back to 1961. When a carrier moves a distance z downstream (along the field) without being scattered, it gains an energy eFz. An unlucky carrier is scattered so frequently that its eFz never reaches the threshold ionization energy E_I for impact ionization. On the other hand, a *lucky electron* is a ballistic electron that avoids scattering for a substantial distance, and hence is able to build its eFz to reach E_I and thereby cause impact ionization. If λ is the mean free path of collisions, then Shockley's model gives

$$\alpha = \frac{eF}{E_I} \exp\left(-\frac{E_I}{e\lambda F}\right). \qquad (3)$$

The main problem with the Shockley model is that there are just not enough ballistic electrons to cause sufficient impact ionizations to explain the experiments. A better model was developed by Baraff in 1962 who numerically solved the Boltzmann transport equation for a simple parabolic band and energy independent mean free path to provide a relationship between α and the F in terms of four parameters, viz., threshold energy for impact ionization, mean free path associated with ionizations, optical phonon energy, and mean free path for optical phonon scattering. The model is not intuitive and is limited in terms of its assumptions and applicability to real semiconductors. More recent major advancement in the II theory appeared as the *lucky drift* (LD) *model*, and it was based on the realization that at high fields, hot electrons do not relax momentum and energy at the same rates (as formulated by Ridley, and Burt, and Mackenzie and Burt). The momentum relaxation rate is much faster than the energy relaxation rate. An electron can drift, being scattered by phonons, and have its momentum relaxed, which controls the drift velocity, but it can still gain energy during this drift. Stated differently, the mean free path λ_E for energy relaxation is much longer than the mean free path λ for momentum relaxation. In the Mackenzie–Burt version of the LD model, the probability $P(E)$ that a carrier attains an energy E is given by,

$$P(E) = \exp\left[-\int_0^E \frac{dE'}{eF\lambda(E')}\right] + \int_0^E \frac{dE_1}{eF\lambda(E_1)} \exp\left[-\int_0^{E_1} \frac{dE'}{eF\lambda(E')}\right] \times \exp\left[-\int_{E_1}^E \frac{dE'}{eF\lambda_E(E')}\right] \quad (4)$$

where λ is the mean free path associated with momentum relaxing collisions and λ_E is the mean energy relaxation length associated with the energy relaxing collisions. The first term is the Shockley lucky electron probability, i.e. the electron moves ballistically to energy E. The second term is the lucky drift probability term which is composed of the following: the electron first moves ballistically to some intermediate energy E_1 ($0 < E_1 < E$) from where it begins its lucky drift to energy E; hence the integration over all possible E_1. The IIC can then readily be evaluated from

$$\alpha = \frac{eFP(E_I)}{\int_0^{E_I} P(E)dE}. \quad (5)$$

The model above is based on a hard threshold ionization energy E_I; that is, when a carrier attains the energy E_I, ionization ensues. The model has been further refined by the inclusion of soft threshold energies which represent the fact that the ionization does not occur immediately when the carrier attains the energy E_I, and the carrier drifts further to gain more energy than E_I before impact ionization. Assuming λ and λ_E are energy independent, which would be the case for a single parabolic band in the crystalline state, Equations (4) and (5) can be solved analytically to obtain

$$\alpha = \frac{1}{\lambda} \times \frac{(\lambda/\lambda_E)\exp(-E_I/eF\lambda_E) + (\lambda/\lambda_E)^2\exp(-E_I/eF\lambda)}{1 - \exp(-E_I/eF\lambda_E) - (\lambda/\lambda_E)^2[1 - \exp(-E_I/eF\lambda)]}. \quad (6)$$

For $\lambda_E > \lambda$, and in the "low field region" where typically $(\alpha\lambda) < 10^{-1}$, or $x = E_I/eF\lambda > 10$, we have

$$\alpha = \left(\frac{1}{\lambda_E}\right)\exp\left(-\frac{E_I}{eF\lambda_E}\right). \quad (7)$$

For crystalline semiconductors, one typically also assumes that λ_E depends on the field F, λ and the optical phonon energy $\hbar\omega$ as

$$\lambda_E = \frac{eF\lambda^2}{2\hbar\omega}\coth\left(\frac{\hbar\omega}{2kT}\right).$$

As the field increases λ_E, it eventually exceeds λ, and allows lucky drift to operate and the LD carriers to reach the ionization energy. (W. Shockley, *Solid St. Elec.*, **2**, 35, 1961; G. A. Baraff, *Phys. Rev.*, **128**, 2507, 1962; B. K. Ridley,

J. Phys. C, **16**, 3373 and 4733, 1983; *Semicond. Sci. Technol.* **2**, 116, 1987; S. McKenzie and M. G. Burt, *J. Phys. C*, **19**, 1959, 1986; J. S. Marsland, *Solid St. Elec.* **30**, 125, 1987; S. O. Kasap *et al.*, *J. Appl. Phys.* **96**, 2037, 2004.)

See *avalanche, avalanche photodiode.*

Impactron is a CCD (charge coupled device) based image sensor that has internal gain through the impact ionization of electrons. There is a large field between two neighboring gates of a CCD, and this large field enables the avalanche multiplication of injected electrons. As charge is transferred from one well to the next as in the shift operation of the CCD (gate to gate shift of charge), the electrons go through these high-field regions and become multiplied. Gains up to 100 have been demonstrated. (J. Hynecek, *IEEE Trans Elec Dev.*, **48**, 2238, 1991.)

See *charge coupled device, image sensor.*

Impedance of a dielectric medium is the ratio of the magnitude of the electric field **E** to the magnitude of the magnetic field intensity **H** at the same location and time in a traveling plane wave that is propagating in the medium. For a homogeneous, isotropic and linear medium, the impedance Z for an EM wave is given by

$$Z = \frac{E}{H} = \left(\frac{\mu}{\varepsilon}\right)^{1/2} = \left(\frac{\mu_0}{\varepsilon_0\varepsilon_r}\right)^{1/2} = \frac{Z_0}{n}.$$

where n is the refractive index. The impedance of free space ($n = 1$) is 377 Ω.

Impulse response function or **point spread function**, is the response, IRF(t), of any linear system H to a Dirac pulse $\delta(t)$. The response $y(t)$ to

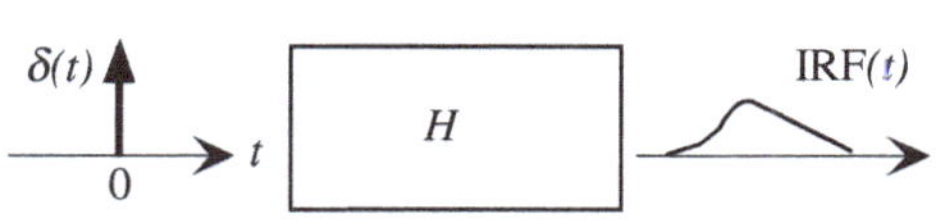

Impulse response function Response of a system to an impulse at its input.

any complex signal $x(t)$ is given by the convolution product,

$$y(t) = \int_0^t x(\tau)\mathrm{IRF}(t-\tau)d\tau.$$

For example, the response to a step-like input function of a linear system with a Lorentzian-like IRF is a smoother step; the width of the IRF is directly related to the "rise time" of the overall response. Such an effect can also be interpreted as low-pass spectral filtering: the "response time" of the system limits its resolution.

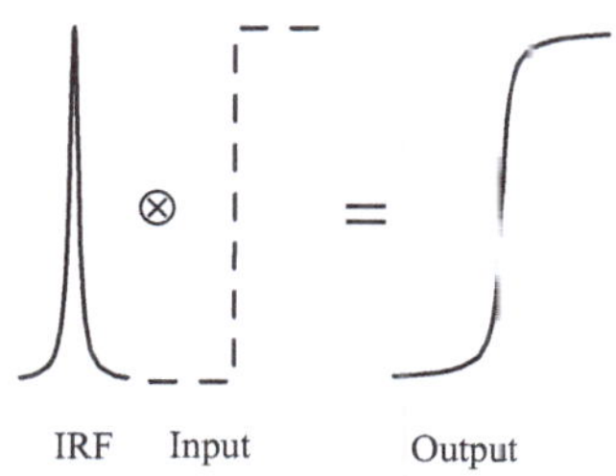

Impulse response function Response of a system to a step-like excitation.

Another example is the temporal broadening of a short pulse propagating in an optical fiber, which sets a limit to the permissible modulation rate. But the variable t doesn't necessarily represent time. The same effect occurs in any linear detection system, whether a spectrograph, an imaging system or a photodiode. In two dimensions, an out-of-focus

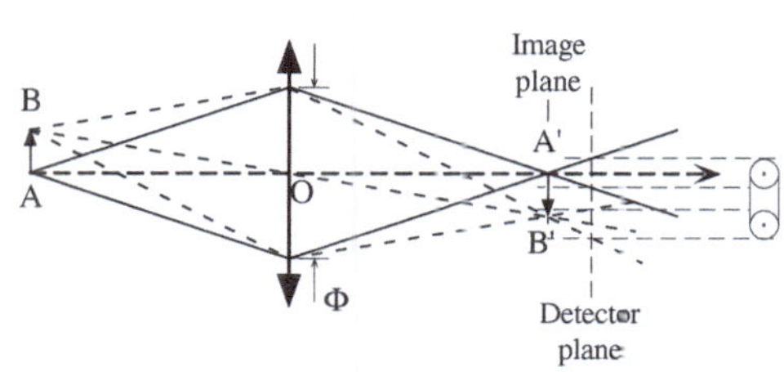

Impulse response function Illustration of the convolution of the correct image by the out-of-focus IRF: each point of segment A′B′ is to be replaced, in the detector plane, by a circular spot. Consequently, the whole image is hopelessly blurred.

blurred photograph can be seen as resulting from the convolution of the correct ideal image by the disk-like IRF that stems from the distance between the quasistigmatic image plane and the actual detector plane. Another classical example is provided in astronomy by the *Airy disk*, which explains the *finite* resolution of even the best diffraction-limited telescopes: see *Rayleigh criterion*.

Incandescence is the emission of radiation from a heated object. See *black body radiation, luminescence*.

Index guided laser diode is a double heterostructure semiconductor laser device that has its *active region* defined by a refractive index variation to guide the lasing radiation. For example, a *buried double heterostructure diode* has its active region surrounded by high-refractive-index semiconductor materials so that the photons are confined to the active region which acts as an optical waveguide or confines the radiation to this region.

Index matching see *second harmonic generation*.

Indirect bandgap see *bandgap*.

Indirect transitions see *band-to-band absorption*.

Indium antimonide (InSb) is a direct bandgap III–V semiconductor crystal with zinc blende structure: $a = 0.6479$ nm; $\rho = 5.770$ g cm^{-3}; $T_m = 527$ °C; $\alpha_L = 5.3\times10^{-6}$ K^{-1}; $c_s = 0.2$ J g^{-1} K^{-1}; $\kappa \approx 0.18$ W cm^{-1} K^{-1}; $\chi = 4.59$ eV; $T_{\text{Debye}} = 160$ K; E_g(D) $= 0.17$ eV (300 K), E_g(D) $= 0.24$ eV (0 K); E_g(D) $= 0.24 - 6.0\times10^{-4}T^2/(T+500)$ (eV); $\Delta_{\text{so}} = 0.80$ eV; $m_e{}^*/m_e = 0.014$; $m_{hh}{}^*/m_e = 0.43$ $m_{lh}{}^*/m_e = 0.015$; $n_i = 1.9\times10^{16}$ cm^{-3}; $N_c = 4.2\times10^{16}$ cm^{-3}; $N_v = 7.3\times10^{18}$ cm^{-3}; ε_r(DC) $=$ 16.8, $\varepsilon_r(\infty) = 15.7$; $E_{\text{breakdown}} \approx 1\times10^3$ V cm^{-1}; $\mu_e \sim 7.7\times10^4$ cm^{2}V^{-1}s^{-1}; $\mu_h \sim 850$ cm^2 V^{-1}s^{-1}; $D_e \sim 2\times10^3$ cm^2 s^{-1}; $D_h \sim 22$ cm^2 s^{-1}; $n = 4$; $\hbar\omega_{\text{op}} = 25$ meV; $B_{\text{radiative}} = 5\times10^{-11}$ cm^{3}s^{-1}; $B_{\text{Auger}} \approx 5\times10^{-26}$ cm^6/s.

Indium arsenide (InAs) is a direct bandgap III–V semiconductor crystal with zinc blende structure: $a = 0.6058$ nm; $\rho = 5.667$ g cm^{-3}; $T_m = 942$ °C; $\alpha_L = 4.52\times10^{-6}$ K^{-1}; $c_s = 0.352$ J g^{-1} K^{-1}; $\kappa \approx$ 0.43 W cm^{-1} K^{-1}; $\chi = 5.06$ eV; $T_{\text{Debye}} = 280$ K; E_g(D) $= 0.36$ eV (300 K), E_g(D) $= 0.415$ eV (0 K); E_g(D) $= 0.415 - 2.7610^{-4}T^2/(T+83)$ (eV); $\Delta_{\text{so}} =$ 0.38 eV; $m_e{}^*/m_e = 0.024$; $m_{hh}{}^*/m_e = 0.36$; $m_{lh}{}^*/m_e = 0.026$; $n_i = 1.3\times10^{15}$ cm^{-3}; $N_c = 9\times10^{16}$ cm^{-3}; $N_v = 1.51\times10^{19}$ cm^{-3}; ε_r(DC) $= 15.15$, $\varepsilon_r(\infty) =$ 12.25; $E_{\text{breakdown}} \approx 40\times10^3$ V cm^{-1}; $\mu_e \sim 4\times10^4$ cm^{2}V^{-1}s^{-1}; $\mu_h \sim 500$ cm^2 V^{-1}s^{-1}; $D_e \sim 10^3$ cm^2 s^{-1}; $D_h \sim 13$ cm^2 s^{-1}; $\hbar\omega_{\text{op}} = 27$ meV; $B_{\text{radiative}} =$ 1.1×10^{-10} cm^{3}s^{-1}; $B_{\text{Auger}} \approx 2.2\times10^{-27}$ cm^6/s. For $\lambda = 0.517$ to 25 μm,

$$n = 3.4899 + \frac{0.0345}{\lambda^2} + \frac{0.0671}{\lambda^4} - 0.0004\lambda^2.$$

Indium nitride (InN) is a direct bandgap III–V semiconductor crystal with wurtzite crystal structure: $a = 0.3533$ nm, $c = 0.5693$ nm; $\rho =$ 6.81 g cm^{-3}; $T_m = 1100$ °C; α_L *(a-direction)* $=$ 3.80×10^{-6} K^{-1}; α_L *(c-direction)* $= 2.90\times10^{-6}$ K^{-1};$c_s = 0.320$ J g^{-1} K^{-1} ; $\kappa \approx 0.45$ W cm^{-1} K^{-1}; $\chi = 5.80$ eV; $T_{\text{Debye}} = 660$ K; E_g(D) $=$ 1.97 eV (300 K), E_g(D) $= 1.994$ eV (0 K); E_g(D) $= 1.994 - 2.45\times10^{-4}T^2/(T+624)$ (eV); $\Delta_{\text{so}} = 0.003$ eV; $m_e{}^*/m_e = 0.11$; $m_{hh}{}^*/m_e =$ 1.63; $m_{lh}{}^*/m_e = 0.27$; $n_i \sim 243$ cm^{-3}; $N_c = 9\times10^{17}$cm^{-3}; $N_v = 5.30\times10^{19}$ cm^{-3}; ε_r(DC) $= 15.3$, $\varepsilon_r(\infty) = 8.4$; $\mu_e \sim 250$ cm^{2}V^{-1}s^{-1}; $D_e \sim <80$ cm^2 s^{-1}; $\hbar\omega_{\text{op}} = 73$ meV; and $B_{\text{radiative}} = 2\times10^{-10}$ cm^{3}s^{-1}.

Indium phosphide (InP) is a direct bandgap III–V semiconductor crystal with zinc blende structure: $a = 0.5867$ nm; $\rho = 4.81$ g.cm^{-3}; $T_m =$ 1060 °C; $\alpha_L = 4.6\times10^{-6}$ K^{-1}; $c_s = 0.31$ J g^{-1} K^{-1}; $\kappa \approx 0.68$ W cm^{-1} K^{-1}; $\chi = 4.38$ eV; $T_{\text{Debye}} = 425$ K; E_g(D) $= 1.344$ eV (300 K), E_g(D) $= 1.421 -$ $4.9\times10^{-4}[T^2/(T+327)]$ (eV), for T in K ($0 < T <$ 800 K); $\Delta_{\text{so}} = 0.11$eV; $m_e{}^*/m_e = 0.08$; $m_{hh}{}^*/m_e =$ 0.6; $m_{lh}{}^*/m_e = 0.089$; $n_i = 1.3\times10^7$ cm^{-3}; $N_c =$ 5.7×10^{17} cm^{-3}; $N_v = 1.1\times10^{19}$ cm^{-3}; ε_r(DC) $=$ 12.5, $\varepsilon_r(\infty) = 9.61$; $E_{\text{breakdown}} \approx 5\times10^5$ V cm^{-1}; $\mu_e \sim 5400$ cm^{2}V^{-1}s^{-1}; $\mu_h \sim 200$ cm^2 V^{-1}s^{-1};

$D_e \sim 130\ \text{cm}^2\ \text{s}^{-1}$; $D_h \sim 5\ \text{cm}^2\ \text{s}^{-1}$; $n = 3.075\cdot(1 + 2.7\times10^{-5}T)$; $\hbar\omega_{\text{To}} = 38.1$ meV; $\hbar\omega_{\text{Lo}} = 42.6$ meV; $B_{\text{radiative}} = 1.2\times10^{-10}\ \text{cm}^3\text{s}^{-1}$; $B_{\text{Auger}} \approx 9\times10^{-31}\ \text{cm}^6\ \text{s}^{-1}$. For $\lambda = 0.399$ to 15 μm

$$n = 3.0704 + \frac{0.2748}{\lambda^2} - \frac{0.0177}{\lambda^4} - 0.0002\lambda^2$$

Other dispersion relations for undoped and doped InP are available in the literature. (P. Martin, E. Skouri, L. Chusseau, C. Alibert, *J. Appl. Phys.*, **67**, 881, 1995.)

Induced emission see *stimulated emission*.

Infrared (IR) is radiation with wavelengths that are longer than 700 nm but shorter than 1 mm, i.e. from just outside the visible, to wavelengths below microwaves. Infrared radiation was discovered by *William Herschel* (1738 – 1822) in 1800 in the spectrum of the Sun. Ordinary glass absorbs infrared wavelengths greater than ~2 μm. Various schemes have been used to classify the IR radiation into different bands, for example, dividing the spectrum into regions based on the application or the detector response. In astronomy, the IR radiation has typically been classified into three regions which are roughly: *near-IR*: 770 nm – 5 nm; *middle-IR*: 1.5 μm – 30 μm; *far-IR*: 30 – 200 μm (the values are approximate). A different classification is used in telecommunications where the interest is over 1260–1675 nm.

Infrared absorption see *Restrahlen absorption*.

Infrared (IR) imagers or **cameras** are comprised of arrays of optical detectors that respond in the infrared spectral region. Typically, such cameras will operate in one of three wavelength regions – namely, the near-infrared (i.e. typically for wavelengths between about 1 and 2 μm), the mid-wave infrared [MWIR] (i.e. about 3–5 μm) and the long-wave infrared [LWIR] (i.e. about 8–12 μm). These wavelength bands have historically been selected because they correspond to regions of weak atmospheric absorption and hence permit remote imaging in these regions. Black-body spectra of objects

Infrared imagers A thermographic image of a cat taken by an infrared camera. You can see the cat's warm glowing eyes and the heat from the cat's ears and mouth area. The cat's fur keeps her warm by trapping her body heat. You can also see the cat's warm infrared pawprints. (Courtesy of IPAC, Image Processing and Analysis Center, NASA. Caption adapted from the IPAC's description.) The color bar represents the temperature in °F.

with a surface temperature around room temperature emit in the LWIR and also hotter bodies emit in the MWIR and even near-IR. The MWIR and LWIR are also called the first and second atmospheric windows. HgCdTe, InSb InGaAs, PbS, PbTe, and even quantum well infrared sensors, have all been used for fabricating photodiodes for fabricating arrays of detectors. Since the light absorption corresponds to relatively low energy differences (or bandgaps) in these materials, significant background carrier generation can occur even at room temperature, leading to so-called *thermal noise*. Cooling to cryogenic temperatures, which increases the bandgap and reduces the thermal generation rate, can significantly reduce the background noise and enhance the performance of such cameras. There are also uncooled IR imagers that use arrays of pyroelectric detectors. The photodetector arrays generate an image that represents the temperature distribution on the body. A warmer region generates a higher signal at the corresponding detector in the camera Usually a color bar is placed next to the image to indicate the correspondence between the color and the temperature. Applications of such cameras are widespread, including military (i.e. for detection of tanks, personnel, rockets, etc.), for surveillance (e.g. personnel perimeter intrusion), for analysis (e.g. building heat loss, roof inspection, medical imaging including mammography), for atmospheric and earth condition monitoring (including pollutant and other species in the air, earth and sea surface mapping), and many others. IR cameras are also called *thermographic cameras*. See *image sensor, pyroelectric detector*. (J. F. Ralph and M. Bernhard, *Contem. Phys.*, **43**, 259, 2002; S. Hargreaves and R. A. Lewis, *J. Mater Sci.: Mater Electron*, **18**, S299, 2007.)

Infrared thermometers directly or indirectly measure the heat emitted from a hot object and thereby assign a temperature to the body. Optical detectors permit the temperature to be measured in a noncontact fashion and are often termed radiation thermometers. These instruments make use of the known relationship between the emitted radiation spectrum and temperature of a target material for a *black body*. The emitted light when focused onto a suitable optical detector provides a response related to the target temperature. Such thermometers are often classified into broadband, bandpass, ratio thermometers and fiber optic thermometers. In the case of fiber optic thermometers, the emitted radiation is collected and guided using an optical fiber having an appropriate spectral response, onto a detector – the detector system being equivalent, in principle, to other types of IR detectors. Broadband detectors, also termed total radiation thermometers, respond typically over a band from about 0.3 μm to as high as 20 μm, measuring a significant fraction of all of the radiation emitted by the hot target. However, such devices tend to have lower sensitivity as compared with the other types of thermometer, and are susceptible to absorption errors. Bandpass thermometers operate by analyzing emitted light over a narrow selected window of wavelengths. These devices can provide good temperature accuracy, and can be used to look through hostile environments to collect data (e.g. viewing through flames) – however, the devices tend to only operate over narrow temperature ranges. Ratio thermometers, often called two-color thermometers, measure the ratio of the detector signal for two wavelength bands. If the emissivity of the target is the same in both bands, the ratio only depends on the temperature of the target and is absolute (requiring no calibration). Moreover, the readings will be independent of the size of the target. The spectral luminance $L(\lambda T)$ of a *black body* with temperature T and for a wavelength interval $d\lambda$, about center wavelength λ, may be written as,

$$L(\lambda T)d\lambda = \frac{Ad\lambda}{\lambda^5[\exp(B/\lambda T) - 1]}$$

which is also known as the Wien–Planck law, where $A = 3.742 \times 10^{-16}$ W/m^2 and $B = 14388$ μm K when λ is expressed in units of μm and T in kelvin. If $\lambda T < B$, as is typical for most radiation thermometers, then the above equation may be simplified to

$$L(\lambda T)d\lambda \approx \frac{Ad\lambda}{\lambda^5 \exp(B/\lambda T)}.$$

The detector response $R(T)$ over its operating wavelength range λ_1 to λ_2 can thus be expressed as

$$R(\lambda T) \approx AC \int_{\lambda_1}^{\lambda_2} \frac{d\lambda}{\lambda^5 \exp(B/\lambda T)} \approx KT^{\frac{B}{\lambda_e T}}.$$

where C and K are constants. An equivalent wavelength λ_e of the detector is defined using the above equation – an effective wavelength at which the integrated response over the detectors spectral bandwidth is assigned to that at one particular wavelength λ_e, given by

$$\lambda_e^5 \exp(B/\lambda_e T) \approx AC \int_{\lambda_1}^{\lambda_2} \frac{d\lambda}{\lambda^5 \exp(B/\lambda T)}.$$

λ_e may be used to establish how to design a thermometer with the highest sensitivity. Using the third equation, one can relate the relative change in response of the instrument $\Delta R/R$ to the target's relative temperature change, $\Delta T/T$, as approximately linear in λ_e, which means that when designing an optical thermometer, the highest resolution will be achieved when selecting the smallest equivalent wavelength. In a ratio thermometer, from the third

Table: Infrared detectors

Detector type	Wavelengths (μm) peak (μm)	D^* ($cm^2\,Hz^{1/2}\,W^{-1}$) Typical	Response time	Type or effect
PbS	1–3.2 Peak: 2.2–2.5	10^{11}	0.01–1 ms	Photoconductive, ($E_g = 0.25$ eV)
PbSe	1.5–5.2 Peak: 4.0–4.2	10^9–10^{10}	1–10 μs	Photoconductive
InSb	1–6.7 Peak: 5.5	10^{10}	0.1–1 μs	Photoconductive, photovoltaic
InAs	1–3.8 Peak: 3–3.5	10^{11}–10^{12}	< 1 μs	Photovoltaic
HgCdTe (MCD)	2–25 Peak: 3.5–17	10^9–10^{11}	1 ns–1 μs	Photoconductive
PZT or $LiTaO_3$	1–100*	10^8	0.01–10 ms	Pyroelectric
Thermopile	1–300*	10^8	1–100 ms	Thermoelectric
Bolometer	1–100*	10^8	0.01–1 ms	Thermistor

* Choice of detector window critical to allow the required wavelengths to pass

equation, it is clear that

$$\frac{R(\lambda_{e1}T)}{R(\lambda_{e2}T)} = T^{B/[T(\lambda_{e1}-\lambda_{e2})]}.$$

This ratio is independent of everything except temperature and hence does not require calibration, assuming the emissivity of the target at the two effective wavelengths is the same.

Infrared detectors are designed to respond over the wavelengths typically covering the infrared (IR) from about 1 μm to about 25 μm. Most solid state (semiconductors) detectors are photoconductive, i.e. the change in the conductivity due to incident radiation is used to sense the radiation. An infrared photoconductive detector uses a narrow bandgap II–VI type semiconductor such as PbS, PbSe or MCD (HgCdTe) as a photoconductor for typically sensing in the infrared region; for PbSe, the infrared range is 1.5–5 μm. (PbSe has a bandgap of 0.25 eV.) Many *pyroelectric detectors* are highly efficient in detecting infrared radiation over a very wide wavelength range, up to ~1000 μm. *Thermal detectors*, such as *thermopiles*, have a wide and reasonably flat spectral range (an advantage) but suffer from a slow response and low sensitivity. Modern bolometers utilize the change in the resistance of a semiconductor (termed a *thermistor*) when heated by the absorption of IR radiation. They are faster than thermopiles but slower than photoconductive detectors. The choice of window for the detector is important as well to allow the desired radiation to pass to the detecting element.

See *pyroelectric detectors, photoconductor, windows*.

Inhomogeneous broadening of an atomic spectral line occurs when, for various reasons, otherwise identical constituent atoms of a collection exhibit slightly different resonant frequencies, "randomly" distributed around some central value ν_0. The overall response of the atomic collection can be thought of as the convolution of the "natural" (Lorentzian) line by the inhomogeneous distribution of resonant frequencies. If the spread $\Delta\nu_{inhom}$ is much greater than the Lorentzian width, then the form is that of the inhomogeneous distribution only. For example, inhomogeneous broadening occurs not only in Doppler-broadened spectral widths observed from gas discharges (a different shift in the emission frequency for each velocity), but also in amorphous solids like glass, or crystals with lattice impurities, due to variations of the local surroundings from site to site. In the case of emission from gases, if one assumes that the distribution of velocities for the atoms (or molecules) that emit the radiation is Maxwellian, then the Doppler broadened full width at half maximum (FWHM) in the output intensity vs. frequency spectrum is given by,

$$\Delta\nu_{1/2} = 2\nu_o \left[\frac{2k_B T \ln(2)}{Mc^2}\right]^{1/2}$$

where M is the mass of the atom or molecule emitting the radiation, ν_o is the actual emission frequency of the atom, c is the velocity of light, k_B is the Boltzmann constant, and T is the absolute temperature. In the case of quantum dots, for example, dots can vary in their size as well as in their local environment (e.g. strain), corresponding to different transition energies and hence inhomogeneous broadening. Indeed, inhomogeneous broadening is used to advantage with quantum dot lasers to provide a broad gain spectrum permitting a wide range of laser wavelength tuning. Depending on the material and the quantum dot size distribution, the inhomogeneous broadening can vary from a few meV to 50 meV with tens of meV being typical for molecular beam epitaxially grown InAs quantum dots.

Injection (electrode) limited current through a solid occurs when the current flow is limited by the rate at which the electrical contacts can inject carriers into the sample. As the applied field increases, the potential energy barrier against the injection of carriers from the electrode into the sample is lowered by the Schottky effect.

Injection electroluminescence see *light emitting diode*.

Injection laser diode see *laser diode*.

Injection pumping is the achievement of *population inversion* in a region of a semiconductor laser device by injecting a sufficiently large number of electrons (in the conduction band) and holes (in the valence band) by virtue of passing a sufficiently large current.

Inokuti–Hirayama photoluminescence decay model represents the decay of luminescence in phosphors (or sensitized phosphors) in which the luminescence centers are able to exchange energy with each other or other molecules through dipole–dipole or dipole–quadrupole interactions. The time evolution of the luminescence intensity $I(t)$ is

$$I(t) = I(0)\exp\left[-\frac{t}{\tau_r} - \Gamma\left(1-\frac{3}{s}\right)\frac{N}{C_o}\left(\frac{t}{\tau_r}\right)^{3/s}\right]$$

where τ_r is the radiative lifetime, N is the concentration of the luminescence centers, $C_o = 3/4\pi R_o{}^3$ is the critical concentration, R_o is the critical transfer distance (radius), that is, the distance at which an isolated donor–acceptor pair has the same transfer rate as the spontaneous decay of the donor, and $s =$ 6, 8, 10 representing electric dipole–dipole, dipole–quadrupole, and quadrupole–quadrupole interactions, respectively. (M. Inokuti and F. Hirayama, *J. Chem. Phys.*, **43**, 1978, 1965.)

Insertion loss is an attenuation caused by the insertion of a component into an optical transmission system. See *fiber-to-fiber losses*.

Insertion loss technique is a two-step measurement technique that allows the *total attenuation* in a fiber (in a cable) to be measured. (The cut-back method is not suitable for cables.) First, in the reference measurement, the fiber cable of interest is not used, and a launch cable and a receive cable are connected through their connectors 1 and 2. The received power P_1 is measured, and represents the so-called launch power for the next step. In the second measurement, the fiber cable of interest with its connectors A and B at its ends is inserted between the launch and receive cables, and the power P_2 is measured again. Since P_1 is without the cable and P_2 with the cable, the *total attenuation* A in dB is given by

$$A = -10\log(P_1/P_2).$$

If the powers P_1 and P_2 are measured in dB, then the total cable attenuation is $P_1 - P_2$. The measured attenuation A includes not only the attenuation in the fiber of the cable, but also the attenuation in the connectors A and B and any splice, etc. Usually a wavelength filter needs to be used to measure A at a given wavelength. For multimode fibers, a mode scrambler is inserted before the cable under test. See *cut-back method*.

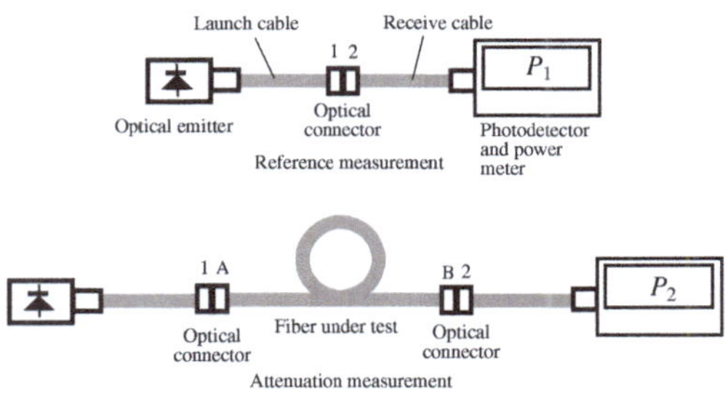

The **insertion loss technique** involves two measurements. First without the actual fiber and then with the fiber inserted and everything else remaining the same.

Instantaneous irradiance is the instantaneous flow of energy per unit time per unit area and is given by the instantaneous value of the *Poynting vector*, $\mathbf{S} = v^2\varepsilon_0\varepsilon_r\mathbf{E}\times\mathbf{B}$, where v is the velocity of the electromagnetic wave in the medium ($v^2 = c^2/\varepsilon_r$), ε_r is the relative permittivity, $\mathbf{E}$ is the electric field and $\mathbf{B}$ is the magnetic field.

Integrated optics refers to the integration of two or more optical or optoelectronic functionalities, that is various optical devices and components, onto a single common crystal substrate, for example, lithium niobate. Just as in integrated electronics all the necessary devices for a given overall function are integrated in the same semiconductor crystal substrate (the silicon chip). There is a distinct advantage to implementing various optically communicated devices, e.g. laser diodes, waveguides, splitters, modulators, photodetectors, and so on, on the same substrate as it leads to miniaturization and also to an overall enhancement in performance and usability (typically). In principle, with the right choice of the substrate material, one should be able to achieve sophisticated integrated optic devices which can carry out quite complicated operations, analogous to ICs in the 1960s. To date, however, commercial integrated optoelectronic devices have integrated only a small number of devices. It should be apparent that since integrated optical circuits handle photons these photons have to be coupled into the integrated optic device, guided and manipulated, from one device to another within the monolithic assembly. Thus the most important

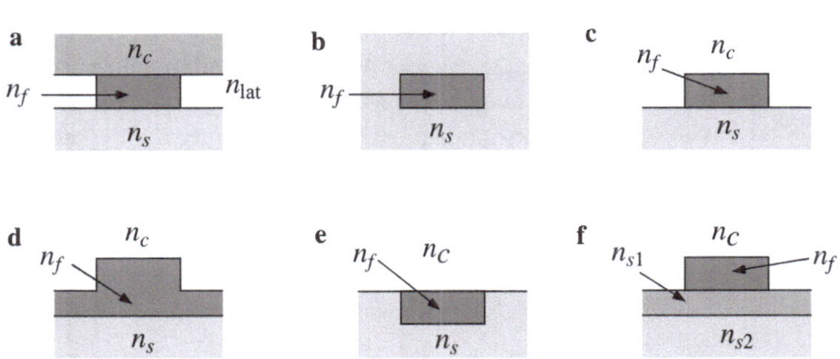

Integrated optics Cross-sections of six channel guide structures: (a) general channel guide; (b) buried channel; (c) raised strip; (d) rib guide; (e) embedded strip; (f) ridge guide. Refractive indices are: n_f (film material); n_s (substrate); n_c (cover or cap); n_{lat} (lateral layer).

---

consideration in integrated photonics is the guiding of light beams along dielectric layers, often realized by depositing semiconductor films on wafer-like substrates. Technological advances have made possible the development of passive or active components such as emitters, receivers, modulators, switches, couplers, multiplexers, and so on, with numerous applications to telecommunications. The confinement can occur in one dimension only (see *slab waveguide*) or in two dimensions (channel guide). According to the geometry, various denominations are used.

The channel waveguide may be fabricated in different configurations, either along a straight line or with bends, or branches; a Y branch may play the role of a beamsplitter or a combiner. The energy exchange along a directional coupler (between two waveguides in close proximity) is well described within the frame of *coupled-wave equations* (CWE). Integrated interferometric devices are also possible.

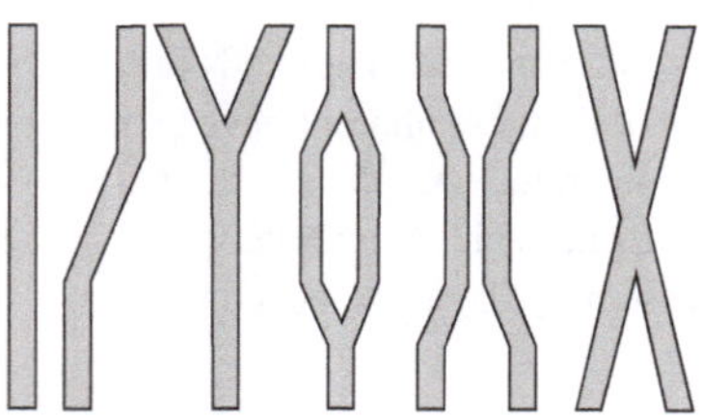

Integrated optics Schematic top view of various configurations for waveguides: (a) straight; (b) S bend; (c) Y branch; (d) Mach–Zehnder; (e) directional coupler; (f) X branch (intersection).

At present there are essentially two important classes of substrates that are commonly used in integrated optoelectronics. The $LiNbO_3$ substrates are used for various electrooptic based functionalities. Optoelectronic integrated circuits that use lasers and photodetectors are based on III–V semiconductors such as GaAs or InP as substrates. III–V compounds are the materials of choice for fabricating LEDs and semiconductor laser diodes as well as detectors for optical communications. Integrated optoelectronic circuits have a number of distinct advantages over those circuits that are based on connecting discrete optoelectronic devices. The reduction in *RC* time constants in integrated optoelectronic circuits leads to an increase in the electrical bandwidth, and thus to a higher modulation frequency. The small size of the integrated optoelectronic circuit implies smaller lengths and widths which translate to smaller external voltages for applying the necessary electric field for electrooptic effects. When properly designed and fabricated, the final integrated optoelectronic device is free from alignment problems. As in the case of ICs, integrated optical and optoelectronic circuits offer cost-effective mass production of reliable devices.

Integrating sphere, also known as the *Ulbricht sphere*, is a light collector that can be used for measuring either the optical flux from a light source (laser, LED, halogen, etc.), or scattering losses from a rough surface. It is made of a hollow sphere with an input port for incident light and an output port for detection; the internal coating (white diffuse reflector coating) is chosen to scatter the light uniformly, so that the radiation arriving on the detector is more or less independent of the spatial properties such as the polarization state of the incident light. In the ideal case, the detected power depends almost only on the total (integrated) input power – hence the name. Alternatively, an integrating sphere can be used as a source producing a highly uniform distribution of radiation. Consider a source S at the center of an integrating sphere, and a very small detector D on the wall. The illuminance at an arbitrary point D is

$$E_{\mathrm{D}} = E_{\mathrm{SD}} + \frac{\Phi}{4\pi^2 r^2} \cdot \frac{R}{1-R}$$

where E_{SD} is the "direct" illuminance at D produced by light falling onto D directly from S, and the second term is due to multiple reflections, that is, due to scattered light from the inside wall reaching

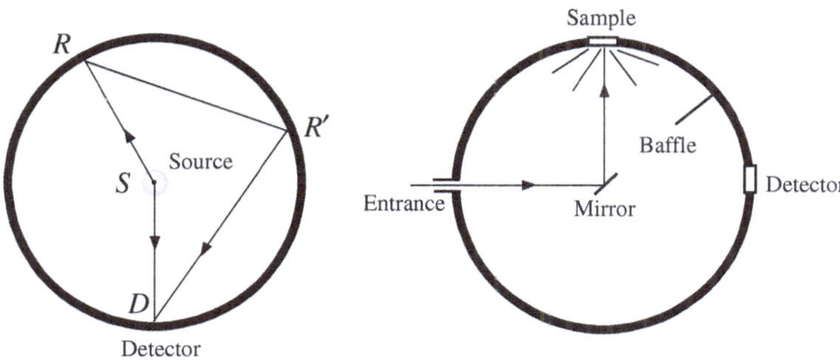

Integrating sphere Left: A point D (representing a detector) on the inside surface of an integrating sphere receives radiation not only directly from the source S at the center but also from multiple reflections (e.g. R and R') from the inside walls. Right: An example in which the diffuse reflectance of a sample is measured using an integrating sphere.

the detector; r is the radius of the sphere, R is the reflectance of the inside coating of the sphere, and Φ is the luminous flux emitted from S. If the direct illumination of D is baffled (prevented to reach D), then E_D is given by the second term and is proportional to the luminous flux.

Intensity in photometry represents the optical power (flux) emitted by a point source per unit solid angle [W sr^{-1}] (see *candela*), whereas in optoelectronics, intensity is the quantity of optical power crossing a unit of area [W m^{-2}]. The latter intensity and the electric field $\boldsymbol{E}$ are related by

$$I = (1/2)\varepsilon_0 cn\langle|E|^2\rangle,$$

where $\langle f(t)\rangle$ denotes the time average of any time-varying function $f(t)$ over the integrating time T_D of the detector.

Interband absorption *in direct bandgap* semiconductors such as III–V semiconductors (e.g. GaAs, InAs, InP, GaP) and in many of their alloys (e.g. InGaAs, GaAsSb) is a photon absorption process in which there is no assistance from lattice vibrations. The photon is absorbed and the electron is excited directly from the valence band to the conduction band without a change in its k-vector (or its crystal momentum $\hbar k$) inasmuch as the photon momentum is very small. The change in the electron momentum from the valence band (VB) to the conduction band (CB) is

$$\hbar k_{CB} - \hbar k_{VB} = \text{photon momentum} \approx 0.$$

This process corresponds to a vertical transition on the E-k diagram, that is electron energy (E) vs. electron momentum ($\hbar k$) in the crystal. The absorption coefficient of these semiconductors rises sharply with decreasing wavelength from λ_g (the bandgap wavelength) as for GaAs and InP. *Interband absorption in indirect bandgap* semiconductors such as Si and Ge, is a photon absorption process for photon energies near and above the bandgap (E_g) energy and requires the absorption and emission of lattice vibrations, that is phonons, during the absorption process. If K is the wavevector of a lattice wave (lattice vibrations travel in the crystal), then $\hbar K$ represents the momentum associated with such a lattice vibration, that is $\hbar K$ is a *phonon momentum*. When an electron in the valence band is excited to the conduction band there is a change in its momentum in the crystal and this change in the momentum cannot be supplied by the momentum of the incident photon which is very small. Thus, the momentum difference must be balanced by a phonon momentum:

$$\hbar k_{CB} - \hbar k_{VB} = \text{phonon momentum} = \hbar K.$$

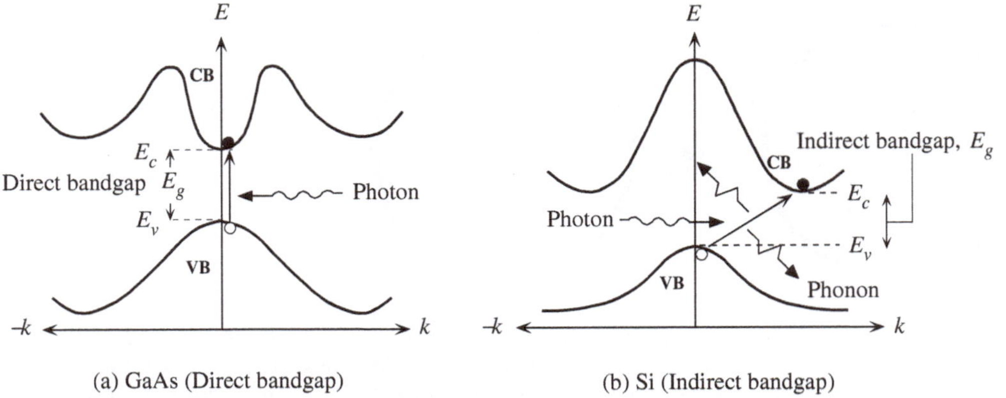

(a) GaAs (Direct bandgap) (b) Si (Indirect bandgap)

Interband absorption (a) Photon absorption in a direct bandgap semiconductor. (b) Photon absorption in an indirect bandgap semiconductor (VB, valence band; CB, conduction band).

The absorption process is said to be *indirect* as it depends on lattice vibrations which in turn depend on the temperature. Since the interaction of a photon with a valence electron needs a third body, a lattice vibration, the probability of photon absorption is not as high as in a direct transition. Furthermore, the cut-off wavelength is not as sharp as for direct bandgap semiconductors. During the absorption process, a phonon may be absorbed or emitted. If ϑ is the frequency of the lattice vibrations then the phonon energy is $h\vartheta$. The photon energy is $h\upsilon$ where υ is the photon frequency. Conservation of energy requires that

$$h\upsilon = E_g \pm h\vartheta.$$

Thus, the onset of absorption does not exactly coincide with E_g, but typically it is very close to E_g inasmuch as $h\vartheta$ is small (< 0.1 eV). The absorption coefficient initially rises slowly with decreasing wavelength from about λ_g.

Interface scattering in optical waveguides arises as waves propagating along the guide become scattered at the core-cladding interface due to the roughness of the interface surface. The fluctuations in the uniformity of the interface between the core and cladding, due to the surface roughness, leads to the scattering of guided waves just as waves become scattered from small particles. The attenuation due to interface scattering scales with the square of the "roughness" σ^2 (mean square value of the fluctuations about the average thickness), the square of the fractional field at the interface. Other factors also control the attenuation. See *Tien's equation*.

Interference is the interaction of two or more waves of the same type (e.g. all waves are oscillations of the electric field) that are at the same location. Consider the interference of two electromagnetic waves in terms of their electric field oscillations. When the two waves are at the same location, they will interact and the result of the interaction is an instantaneous net electric field $\mathbf{E}$ that is the vector sum of the individual electric fields $\mathbf{E}_1$ and $\mathbf{E}_2$: $\mathbf{E} = \mathbf{E}_1 + \mathbf{E}_2$. *Constructive interference* occurs when the two individual waves reinforce each other and the resultant is greater than the individual amplitudes. *Destructive interference* occurs when the two waves have opposite instantaneous vector fields (i.e. oppositely directed fields) so that the net resultant is zero if the individual waves have the same amplitude.

Interference order see *Young's interference fringes.*

Interferometer is an optical instrument that uses wave-interference phenomena to produce interference fringes (e.g. bands, rings) which can be used to measure the wavelength of light, surface flatness or small distances, etc. A nearly monochromatic light wave is split into two coherent waves traveling along two different paths, and then the two waves are brought together and made to interfere on a screen (or a detector array); the result is an interference pattern. The pattern's bright regions correspond to two waves, traveling along different paths, arriving in phase (or having a phase difference that is a multiple of 2π) and the dark regions correspond to the two waves arriving 180° out of phase (having a phase difference that is an odd multiple π). A phase change $\Delta\phi = \Delta(kd)$, where k is the wavevector $(2\pi n/\lambda)$ and d is the path length, in one of the paths will shift the interference pattern, and if $\Delta\phi$ is π the bright patterns become dark. Even a small change in the optical path, distance × refractive index, *nd,* can cause a measurable shift in the diffraction pattern or a displacement in the fringes, which can be used to infer *nd*, or $(2\pi n/\lambda)d$. There are many types of interferometers, see for example, *Fabry–Perot*, *Gires–Tournois*, *Mach–Zehnder*, *Michelson*, *Sagnac Twyman–Green Interferometers.*

Internal quantum efficiency (η_{int}) **of a light-emitting diode** gauges what fraction of electron–hole recombinations in the forward biased *pn* junction are radiative and therefore lead to photon emission. Nonradiative transitions are those in which an electron and a hole recombine through a recombination center such as a crystal defect or an impurity and emit phonons (lattice vibrations). By definition,

$$\eta_{\text{int}} = \frac{\text{Rate of radiative recombination}}{\text{Total rate of recombination (radiative and nonradiative)}}$$

or

$$\eta_{\text{int}} = \frac{\dfrac{1}{\tau_r}}{\dfrac{1}{\tau_r} + \dfrac{1}{\tau_{nr}}}$$

where τ_r is the mean lifetime of a minority carrier before it recombines radiatively and τ_{nr} is the mean lifetime before it recombines via a recombination center without emitting a photon. The total current I is determined by the total rate of recombinations whereas the number of photons emitted per second (Φ_{ph}) is determined by the rate of radiative recombinations. Thus,

$$\eta_{\text{int}} = \frac{\text{Photons emitted per second}}{\text{Total carriers lost per second}} = \frac{\Phi_{ph}}{I/e} = \frac{P_{op(int)}/h\upsilon}{I/e}$$

where $P_{op(int)}$ is the *optical power* generated internally (not yet extracted), e is the electronic charge, and $h\upsilon$ is the photon energy.

Internal quantum efficiency of a photodetector is the number of *free* electron–hole pairs photogenerated per *absorbed photon*; this is not per incident photon on the device. Inasmuch as internal quantum efficiency is defined in terms of per absorbed photon, it is greater than the external quantum efficiency which is defined in terms of per incident photon; not all incident photons are absorbed.

Internal reflection is the reflection of an electromagnetic wave initially traveling in a medium of high refractive index n_1 at the boundary with a medium of lower refractive index n_2 $(< n_1)$; a light wave traveling in n_1 is reflected at the $n_1 - n_2$ boundary (some of the energy may however be transmitted). For example, internal reflection occurs when a light wave traveling in water is incident on the water–air surface and becomes reflected.

Intraband absorption refers to absorption of electromagnetic radiation through electronic transitions within a given band or indeed subband, rather than between allowed bands or subbands. Typically, such transitions involve much smaller differences than interband transitions, and thus tend to occur for light of much longer wavelengths than for interband transitions. Devices that operate based on intrasubband transitions include so-called *quantum well infrared photodetectors* (QWIPs) as well as *quantum dot infrared photodetectors* (QDIPs). In the case of both QWIPs and QDIPs, intrasubband electron transitions from the bound state in the quantum well/dot to continuum states lying above the barriers between them are used. Often for QWIPs, for example, multiple quantum wells are used to provide a high density of states in the subbands, and device operation can involve a number of mechanisms after photoexcited carriers are generated from intrasubband absorption – these include the transport of photoexcited carriers, carrier capture and gain. Also of importance has been the influence of the applied field to change the spectral response as well as multiplication gain.

Intramodal or intramode dispersion describes the temporal spread of a propagating optical pulse in a fiber when the propagation is through a single mode (and hence intramode) and the dispersion arises because the guide properties force different frequencies present in the optical pulse to travel with different group velocities. Put differently, the input signal into an optical guide has a range of frequencies or wavelengths and these propagate with different group velocities and arrive at different times at the destination. The cause is the nonlinear ω vs. β *dispersion diagram* of the guide. Intramode dispersion can be broken into *material* dispersion, *waveguide* dispersion and *profile* dispersion.

Intrinsic carrier concentration, n_i, is the electron concentration in the conduction band of an intrinsic semiconductor (an ideal perfect crystal that has not been doped). The hole concentration in the valence band is equal to the electron concentration. The intrinsic concentration is a material property that depends on the energy bandgap E_g and the temperature and can be written as

$$n_i^2 = N_c N_v \exp(-E_g/k_B T)$$

where N_c and N_v are the effective densities of states at the conduction and valence band edges, E_g is the bandgap, k_B is the Boltzmann constant and T is the absolute temperature.

Intrinsic semiconductor has equal number of electrons and holes due to thermal generation across the bandgap, E_g. It corresponds to a pure semiconductor crystal in which there are no impurities and crystal defects to cause an effective doping of the crystal.

Inverse dispersion slope fiber see *compensating fiber, dispersion compensation.*

Inversion layer is a region in the semiconductor laser device in which there is a *population inversion* (more *stimulated* emission than *absorption*) due to the presence of sufficiently large number of injected electrons and holes brought in by the diode current. Holes represent empty electron states in the valence band and electrons occupy states in the conduction band.

Ion traps are electromagnetically defined cavities for confining ions – typically these traps involve the use of electric and magnetic fields. Shown in the figure is a linear ion trap; – ac voltages of about 1 kV are applied between the rods at a frequency of a few MHz to define the trap region confining ions (shown as black dots) along the central axis of the trap. The end-rings are electrically charged to prevent the ions from escaping from the two ends of the trap. In the case of quantum computing studies, finely focused lasers are used to prepare and study the individual ions. The outer electron of each ion is manipulated to be in one of two different orbits about the nucleus. Each ion therefore represents a *qubit*. As an ion scatters photons from the laser it recoils. Its recoil motion is felt by the other ions. This recoil motion is equivalent to the data bus of a classical computer.

A schematic of a linear **ion trap** consisting of four rod electrodes fed from a high-frequency high-voltage source, and two annular ring electrodes at the ends held at an appropriate dc potential to confine ions in the trap.

Ionic polarization see *polarization of a dielectric.*

Ionization is a process that either removes an electron or adds an electron to an atom or a molecule which results in the creation of an ion (an atom or a molecule with a net charge). Ionization of an atom or a molecule can occur by a collision with a projectile particle (e.g. a projectile electron) with a sufficient energy, by the absorption of a photon with the right energy, or by an application of a very large electric field. The term ionization is most commonly use to describe the removal of one or more electrons from an initially neutral atom or molecule, which results in positively charged ions.

Ionization energy in radiation detectors see *electron–hole pair creation energy.*

Irradiance (average) is the average flow of energy per unit time per unit area where averaging is typically carried out by the light detector (over many oscillation periods). Average irradiance can also be defined mathematically by the average value of the Poynting vector $\mathbf{S} = \mathbf{E} \times \mathbf{H}$ where $\mathbf{E}$ is the electric and $\mathbf{H}$ is the magnetic field intensity. For a linear isotropic medium, $\mathbf{S} = v^2 \varepsilon_0 \varepsilon_r \mathbf{E} \times \mathbf{B}$ where $\mathbf{B}$ is the magnetic field, v is the velocity of the electromagnetic field in the medium, and ε_r is the relative permittivity. To find the average flow of energy per unit area per unit time, we need to average the latter over one period. The result is

$$S_{\text{average}} = (1/2)c\varepsilon_0 n E_0^2 = c\varepsilon_0 n E_{\text{rms}}^2$$

where E_0 is the amplitude of the electric field, E_{rms} is the rms (root mean square) value of the electric field, $E_{\text{rms}} = E_0/2^{1/2}$, and c is the speed of light. The instantaneous irradiance can only be measured if the power meter can respond more quickly than the oscillations of the electric field, and since this is in the optical frequency range, all practical measurements invariably yield the average irradiance because all detectors have a response rate much slower than the frequency of the wave. See *illuminance, Poynting vector.*

Irradiance or **radiant incidence,** E_e in W/m^2, of a surface is the radiant flux incident on the surface per unit area from all directions. If an elementary surface dS is irradiated by an element of radiant flux $d\Phi_e$ from all directions, the irradiance is $E_e = d\Phi_e/dS$.

Irradiation usually means *total hemispherical irradiation*, and represents the rate at which a surface area receives radiation energy per unit time from all directions per unit surface area; a surface can only receive radiation from the hemisphere above it. If a surface receives an amount of radiation ΔQ_e in time Δt over a surface area ΔA then $G = \Delta Q/\Delta A \, \Delta t$, in units of W m^{-2}. *Spectral irradiation* G_λ is defined as $G_\lambda = dG/d\lambda$; that is over a small wavelength range $d\lambda$ centered at λ, a surface receives an irradiation of $G_\lambda d\lambda$. See *emissivity.*

Isotropic medium or **optically isotropic medium** in optics is a medium in which all directions are equivalent in terms of light propagation. The refractive index (or the relative permittivity ε_r) is independent of the direction of propagation of the light beam (or the electromagnetic wave). Liquids, glasses, amorphous or noncrystalline solids and cubic crystals are optically isotropic. In an isotropic medium ε_r is a scalar quantity not a tensor and the induced polarization **P** and the field **E** are in the same direction. Maxwell's equations for light propagation become highly simplified in an isotropic medium; in particular, one can use the electric field **E** with the appropriate relative permittivity without needing the displacement and polarization vectors **D** and **P**. See *anisotropy, electromagnetic wave propagation, Maxwell's equations*.

Isotropic source has a radiant intensity that is the same in all directions. See *radiant intensity*.

J

Jitter noise is undesirable fluctuations in signal properties due to inevitable fluctuations in the timing or control functions in electronic and optoelectronic circuits. For example, due to jitter, the pulses in a circuit may not be generated at exactly the desired times but at some offsets, which introduces fluctuations in the positions of the pulses around their mean values. Similarly, the pulse widths may fluctuate about some mean value. Random jitter can affect both the amplitude and phase of a signal, and hence generates *jitter noise*. In many cases, random jitter follows a Gaussian distribution. See *phase noise, eye pattern*.

Joint density of states, $N_{\text{joint}}(\hbar\omega)$, is the combined *density of states* available in both conduction and valence band for a photon of energy $\hbar\omega$. It may be found by using a similar derivation to that for the *density of states* in a given band. Since an optical transition must obey the *Franck–Condon principle*, that is, the transition is vertical on the E vs. $\mathbf{k}$ (energy vs. electron wavevector) curves, the emission or absorption of a photon of energy $\hbar\omega$ will satisfy the following equation:

$$\begin{aligned}\hbar\omega &= E_c(k) - E_v(k)\\ &= E_c(0) - E_v(0) + h^2k^2/2m_r\\ &= E_g + \hbar^2k^2/2m_r\end{aligned}$$

where $m_r^{-1} = m_e^{-1} + m_h^{-1}$ is a reduced or joint effective mass, and E_g is the semiconductor direct bandgap. In 3D a spherical shell of thickness dk and of radius k, in k-space, has volume $4\pi k^2 dk$. Since the volume of k-space per state is $(2\pi)^3/V$, where V is the physical volume of the crystal, the shell of radius k will comprise $k^2 dk V/(2\pi^2)$ states. The number of states per unit photon energy, and per unit volume of crystal, is then found using this expression and the equation above as:

$$N_{\text{joint}}(\hbar\omega) = [(2m_r/\hbar^2)^{3/2}/4\pi^2]\sqrt{(\hbar\omega - E_g)}.$$

Based on this expression, it is apparent why the absorption coefficient $\alpha(\omega)$ in a direct bandgap semiconductor depends on the photon energy above the bandgap with a square root dependence. This is apparent since $\alpha(\omega)$ is proportional to the transition rate for valence to conduction band transitions, W_{vc}, and may be written as

$$\alpha(\omega) = \frac{2\hbar W_{vc}}{c\varepsilon_0\omega n(\omega)|A_0|^2\Omega}$$

where the electric and magnetic fields are expressed in terms of a vector potential $\boldsymbol{A} = A_0\cos(\boldsymbol{q}\cdot\boldsymbol{r}-\omega t)$, where $\boldsymbol{q}$ is the photon wavevector, and V is the sample volume. W_{vc} may be found by using *Fermi's golden rule* to give

$$W_{vc} = \frac{e^2V|A_0|^2P^2}{8\pi^2\hbar m^2}\int \delta\left(E_g + \frac{\hbar^2k^2}{2m_r} - \hbar\omega\right) \times f_v(1 - f_c)d^3k$$

where P is the momentum expectation value for the crystal based on the cell periodic components of the crystal *Bloch functions*, and f_v and f_c are *Fermi–Dirac* distribution functions for the valence and conduction bands. At zero temperature (where $f_v = 1$ and $f_c = 0$), and assuming spherical, parabolic energy bands, neglecting spin–orbit interactions and the degeneracy of the valence band, the integral in the expression for W_{vc} may be simplified:

$$\int \delta\left(E_g + \frac{\hbar^2 k^2}{2m_r} - \hbar\omega\right) f_v(1 - f_c) d^3k$$
$$= 4\pi \int_0^\infty k^2 \delta\left(E_g + \frac{\hbar^2 k^2}{2m_r} - \hbar\omega\right) dk$$
$$= \begin{cases} 8\pi^3 N_{\text{joint}}(\hbar\omega) : & \hbar\omega \geq E_g \\ 0 : & \hbar\omega < E_g \end{cases}$$

giving an expression for the absorption coefficient,

$$\alpha(\omega) = \frac{2\pi e^2 N_{\text{joint}} P^2}{cm^2 \varepsilon_0 \omega n(\omega)} \text{ for } \hbar\omega \geq E_g \text{ and 0 otherwise}$$

with the expected square root dependence on photon energy, for energies in the vicinity of the direct bandgap, coming from N_{joint} as discussed above.

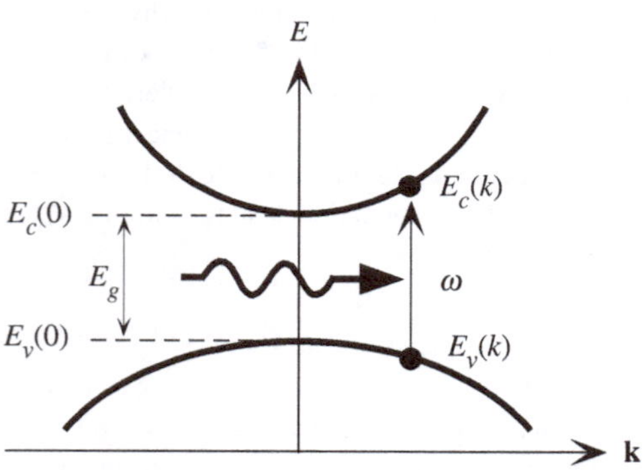

Joint density of states Semiconductor E vs. **k** diagram, showing an optical transition with corresponding photon energy $\hbar\omega$, coupling valence and conduction band states $E_v(k)$ and $E_c(k)$, respectively.

Jones as a unit is a specific detectivity (D^*) unit usually in cm $\text{Hz}^{-1/2}\ \text{W}^{-1}$. See *detectivity*.

Jones matrices represent the state of polarization of a light wave. Various operations on the polarization state correspond to multiplying this matrix with another matrix that represents the optical operation. Consider a light wave traveling along z with field components E_x and E_y along x and y. These components are orthogonal and, in general, would be of different magnitude and have a phase difference ϕ between them. If we use exponential notation then

$$E_x = E_{xo} \exp[j(\omega t - kz + \phi_x)] \quad \text{and}$$
$$E_y = E_{yo} \exp[j(\omega t - kz + \phi_y)].$$

A *Jones matrix* is a column matrix whose elements are E_x and E_y without the common $\exp j(\omega t - kz)$ factor:

$$\mathbf{E} = \begin{bmatrix} E_x \\ E_y \end{bmatrix} = \begin{bmatrix} E_{xo}\exp(j\phi_x) \\ E_{yo}\exp(j\phi_y) \end{bmatrix}.$$

Usually this equation is normalized by dividing by the total amplitude $E_o = (E_{xo}^2 + E_{yo}^2)^{1/2}$. We can further factor out $\exp(j\phi_x)$ to further simplify to obtain the Jones matrix

$$\mathbf{J} = \frac{1}{E_o}\begin{bmatrix} E_{xo} \\ E_{yo}\exp(j\phi) \end{bmatrix}$$

where $\phi = \phi_y - \phi_x$. For example linearly polarized light that is 45° to the x-axis is

$$\frac{1}{\sqrt{2}}\begin{bmatrix} 1 \\ 1 \end{bmatrix}$$

Passing a wave of given Jones vector $\mathbf{J}_{\text{in}}$ through an optical device is represented by multiplying $\mathbf{J}_{\text{in}}$ by the *transmission matrix* $\mathbf{T}$ of the device. If $\mathbf{J}_{\text{out}}$ is the Jones vector for the output light through the device, then $\mathbf{J}_{\text{out}} = \mathbf{T}\,\mathbf{J}_{\text{in}}$.

Judd–Ofelt analysis or **theorem** allows the calculation of the radiative transition rate and radiative lifetime of an electric dipole transition between two states of a trivalent rare-earth ion embedded in different host lattices. Rare earth (RE) is the common name for the elements from lanthanum (La) to lutetium (Lu). A specific feature of these elements is the incompletely filled 4*f* shell. REs may be embedded in different host materials in the form of divalent or trivalent ions. As divalent ions, REs exhibit broad absorption–emission lines related to allowed $4f - 5d$ transitions. In trivalent form, the RE atom loses two $6s$ electrons and one $4f$ or $5d$ electron. As a result of the Coulomb interaction of the $4f$ electron with the positively charged core, the $4f$ level becomes split into a complicated set of manifolds whose positions, in a first approximation, are virtually independent of the host matrix because the $4f$ level is well screened by $5s$ and $5p$ shells from outer influence. The optical transitions between $4f$ manifold levels are forbidden by the *parity selection rule* which states that for a permitted atomic (ionic) transition, the wavefunctions of initial and final states must have different parities. However, the parity selection rule is relaxed by the admixture of a $5d$ state to $4f$, and by the disturbance of RE ion symmetry due to the influence of the host, which increases with the growth of covalency. The higher covalency implies stronger sharing of electrons between the RE ion and ligands. This effect is known as the *nephelauxetic effect*. The resulting absorption–emission lines are characteristic for individual RE ions and quite narrow because they are related to forbidden inner-shell $4f$ transitions.

The possible states of a RE ion are often referred to as $^{2S+1}L_J$, where $L = 0, 1, 2, 3, 4, \ldots$ determines the electron's total angular momentum and is conventionally represented by letters S, P, D, F, G. The term $2S + 1$ is called *spin multiplicity* and represents the number of spin configurations, while J is the *total angular momentum*, which is the vector sum of the overall (total) angular momentum and overall spin. The value $2J + 1$ is called the term's multiplicity and is the number of possible combinations of overall angular momentum and overall spin, which yield the same J. Thus, the notation $^4I_{15/2}$ for the ground state of Er^{3+} corresponds to the term $(J, L, S) = (15/2, 6, 3/2)$, which has a multiplicity $2J + 1 = 16$ and a spin multiplicity $2S + 1 = 4$. If $(^{2S'+1}L_{J'})$ is the initial and $^{2S+1}L_J$ is the final state of a RE ion, then the probability of spontaneous emission per unit time can be expressed as

$$A(J'; J) = \frac{64\pi^4 e^2}{3h(2J' + 1)\lambda^3} \times \left[\chi_{\mathrm{ed}} S_{\mathrm{ed}}(J; J') + \chi_{\mathrm{md}} S_{\mathrm{md}}(J; J')\right]$$

where $S_{\mathrm{ed}}(J'; J)$ and $S_{\mathrm{md}}(J'; J)$ are the electric and magnetic line strengths of this transition, respectively. The factors $\chi_{\mathrm{ed}} = n(n^2 + 2)^2/9$ and $\chi_{\mathrm{md}} = n^3$ are so-called *local field correction factors* in the medium with refractive index n for an electric dipole moment transition and for a magnetic dipole moment transition, respectively. The radiative lifetime τ_r is the reciprocal of A; $\tau_r = 1/A$. S_{ed} and S_{md} are given by

$$S_{\mathrm{ed}}(J; J') = \sum_{k=2,4,6} \Omega_k \left|\langle SLJ \left| U^{(k)} \right| S'L'J'\rangle\right|^2$$

and

$$S_{\mathrm{md}}(J; J') = \left(\frac{\hbar}{2mc}\right)^2 \sum_{k=2,4,6} \Omega_k \left|\langle SLJ \,||\mathbf{L} + 2\mathbf{S}||\, S'L'J'\rangle\right|^2$$

where Ω_k are the coefficients, called *intensity parameters*, representing the influence of the host material and $\left|\langle SLJ \left| U^{(k)} \right| S'L'J'\rangle\right|^2$ are reduced matrix operator components, which are virtually independent of the host material and their values are calculated using the so-called intermediate coupling approximation and are tabulated (e.g. M. J. Weber, *Phys. Rev.* **157**, 262, 1967). For example, for $^4I_{13/2}$ to $^4I_{15/2}$ transition at 1.55 μm:

$$S_{\mathrm{ed}} = 0.0188\Omega_2 + 0.1176\Omega_4 + 1.4617\Omega_6$$

where the coefficients of Ω_2, Ω_4 and Ω_6 are reduced matrix components. If $\alpha(\nu)$ is the experimentally observed absorption spectrum for a particular absorption band, the experimental oscillator strength is given by

$$f_{\mathrm{exp}} = \frac{mc^2}{\pi e^2 N} \int_{\mathrm{band}} \alpha(\nu) d\nu$$

where N is the concentration of rare earth ions in the host, and ν is the frequency. The theoretical or

Table: Judd-Ofelt analysis Selected examples of intensity parameters for Er^{3+} in various hosts

	Host	$\Omega_2 \times 10^{-20}$ cm^2	$\Omega_4 \times 10^{-20}$ cm^2	$\Omega_6 \times 10^{-20}$ cm^2
Er^{3+}	Phosphate	6.28	1.03	1.39
Er^{3+}	$(TeO_2)_{75}(ZnO)_{0.25}$ glass	5.60	1.87	1.03
Er^{3+}	GeO_2-PbO-BaO-ZnO-K_2O	5.14	1.24	0.68
Er^{3+}	$LiNbO_3$	7.29	2.24	1.27

Table: Judd-Ofelt analysis The reduced matrix elements of $U^{(k)}$ and typical wavelengths $\overline{\lambda}$ for the Er^{3+} transitions are taken from M. J. Weber, *Phys. Rev. B*, **157**, 262, 1967; W. T. Carnall, P. R. Field and B. G. Wybourne, *J. Chem. Phys.*, **49**, 4424, 1968; S. Tanabe, *Phys. Rev. B*, **46**, 3305, 1992.

$(S, L)J$	$(S', L')J'$	$(U^{(2)})^2$	$(U^{(4)})^2$	$(U^{(6)})^2$	Band	$\overline{\lambda}$ (nm)
$^4I_{15/2}$	$^4I_{13/2}$	0.0188	0.11176	1.4617		1520
	$^4F_{9/2}$	0	0.5655	0.4651	1	652
	$^4S_{3/2}$	0	0	0.2285	2	521
	$^2H_{11/2}$	0.7056	0.4109	0.0870	2	
	$^4F_{7/2}$	0	0.1467	0.6273	3	487
	$^4F_{5/2}$	0	0	0.2237	4	450
	$^4F_{3/2}$	0	0	0.1204		
	$^2H_{9/2}$	0	0.078	0.17	5	407
	$^2G_{11/2}$	0.9178	0.5271	0.1197	6	378
	$^2G_{9/1}$	0	0.2416	0.1235	6	
	$^2K_{15/2}$	0.0219	0.0041	0.0758	6	
	$^2G_{7/2}$	0	0.0174	0.1163	6	

calculated oscillator strength is related to S_{ed}:

$$f_{\text{cal}} = \frac{8\pi^2 mc}{3h\lambda(2J+1)n^2}\chi_{\text{ed}} S_{\text{ed}} = \frac{8\pi^2 mc}{3h\lambda(2J+1)n^2}\chi_{\text{ed}}(0.0188\Omega_2 + 0.1176\Omega_4 + 1.4617\Omega_6).$$

By minimizing the error $|(f_{\text{exp}}{}^2 - f_{\text{cal}}{}^2)/f_{\text{exp}}{}^2|$ we can find the best fit intensities Ω_2, Ω_4 and Ω_6. The key idea of JO analysis is to minimize the discrepancy between experimental and calculated values of oscillator strengths, f_{exp} and f_{cal}, by appropriate choice of coefficients Ω_k, which are used to characterize and compare different materials. The Ω_2 value is of prime importance because it is the most sensitive to the local structure and material composition and is correlated with the degree of *covalency*.

The values Ω_k are used to calculate the probabilities of radiative transitions and appropriate radiative lifetimes of excited states, which are very useful for numerous optical applications. (B. R. Judd, *Phys. Rev.* **127**, 750, 1962; G. S. Ofelt, *J. Chem. Phys.* **37**, 511, 1962.)

Junction in a semiconductor device such as a *pn* junction is a plane in a crystal which has differently doped materials, such as *p*-type and *n*-type semiconductors, on opposite sides.

K

K-absorption edge refers to the sudden increase in the attenuation coefficient for x-rays (or high-energy radiation) when the x-ray photon energy is able excite an electron from the K-shell of the material with which it interacts.

KDP (KH_2PO_4), potassium dihydrogen phosphate, is an important nonlinear and birefringent crystal that is used in electrooptic modulation and frequency conversion. It is a tetragonal ($\bar{4}2m$) crystal; $\rho = 2.338$ g cm^{-3}; its refractive indices over the 200–2000 nm wavelength range are given in terms of Cauchy's equation with λ in μm as:

$$n_o = 1.504527 + \frac{2.78038 \times 10^{-3}}{\lambda^2} + \frac{7.84718 \times 10^{-5}}{\lambda^4} - 1.12959 \times 10^{-2}\lambda^2$$

$$n_e = 1.460924 + \frac{2.85724 \times 10^{-3}}{\lambda^2} + \frac{5.06736 \times 10^{-5}}{\lambda^4} - 2.83348 \times 10^{-3}\lambda^2.$$

(n vs. λ data from F. Zernicke Jr., *J. Op Soc Am.* **54**, 1215, 1964, and then fitted.)

Kerma is the total kinetic energy released per unit mass of medium when incident high-energy radiation (uncharged) interacts with the medium and releases charged particles with kinetic energies; Kerma stands for "kinetic energy released in media." If $\Phi(E)$ is the photon fluence (photons cm^{-2}), E is the photon energy and we consider only a monochromatic beam, then kerma K is defined by

$$K = k\Phi(E)E(\mu_{\text{tr}}/\rho)$$

where (μ_{tr}/ρ) is the mass energy transfer coefficient of the medium (cm^2/g), $\Phi(E)$ is the photon flux (photons cm^{-2}), k is a constant, 1.6021×10^{-11} rad/keV, and K is in rad. If k is in 1.6021×10^{-13} Gy/keV, then K is in Gy (Gray). Except at very high energies, for x-rays, K is numerically close to the absorbed DOS. Sometimes K is quoted in grays per roentgen, and, expressed in this way, Kerma in air is typically 8.8 mGy/R for x-ray radiation from 30 to 140 kVp and it is 9.4 mGy/R in water, with very little energy dependence.

Kerr effect is a second-order electrooptic effect in which an applied strong electric field modifies the refractive index in such a way that the refractive index change is proportional to the square of the applied field. For example, when a strong field is applied to an otherwise optically isotropic material such as glass (or liquid), the change in the refractive index will be due only to the Kerr effect, since such materials exhibit no Pockels effect. If E_a is the applied field, then the change in the refractive index for light polarization parallel to the applied field is given by

$$\Delta n = \lambda K E_a^2$$

where K is the Kerr coefficient and λ is the wavelength. Suppose that we arbitrarily set the z-axis

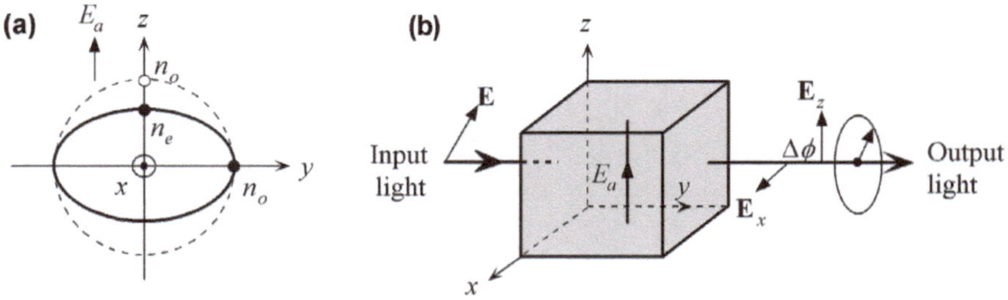

Kerr effect (a) An applied electric field, via the Kerr effect, induces birefringences in an otherwise optically istropic material. n_o and n_e are the ordinary and extraordinary refractive indices. (b) A Kerr cell phase modulator.

of a Cartesian coordinate system along the applied field. The applied field distorts the electron motions (orbits) in the constituent atoms and molecules, including those valence electrons in covalent bonds, in such a way that it becomes "more" difficult for the electric field in the light wave to displace electrons parallel to the applied field direction. Thus a light wave with a polarization parallel to the z-axis will experience a smaller refractive index, reduced from its original value n_0 to n_e. A light wave with polarizations orthogonal to the z-axis will experience the same refractive index n_o. The applied field thus induces birefringence with an optic axis parallel to the applied field direction. The material becomes birefringent for waves traveling off the z-axis. The polarization modulator and intensity modulator concepts based on the Pockels cell can be extended to the Kerr effect. In the Kerr case, the applied field again induces birefringence. The Kerr phase modulator uses the fact that the applied field along z induces a refractive index n_e parallel to the z-axis whereas that along the x-axis will still be n_o. The light components $\mathbf{E}_z$ and $\mathbf{E}_x$ then travel along the material with different velocities and emerge with a phase difference $\Delta\phi$ resulting in elliptically polarized light. However, the Kerr effect is small as it is a second-order effect, and therefore only accessible for modulation use at high fields. The advantage, however, is that all materials, including glasses and liquids, exhibit the Kerr effect and the response time in solids is very short, much less than nanoseconds, leading to high modulation frequencies (greater than GHz). See also *electrooptic effects*.

Ketteler–Helmholtz formula describes the refractive index vs. photon energy dispersion behavior of semiconductors in terms of a simple relationship of the form

$$n^2 = A + \frac{B}{1-(\nu/\nu_o)^2} = A + \frac{B'\lambda^2}{\lambda^2 - \lambda_o^2}$$

where ν is the light frequency, λ is the wavelength, and A, B, B', ν_o, λ_o are constants specific to the material. This expression is equivalent to the Sellmeier expression. For example for GaAs, over $\lambda = 0.89$ to 4.1 μm, $A = 7.9$; $B' = 3.78$ μm^{-2}; $\lambda_o = 0.16634$ μm.

KH_2PO_4 see *KDP*.

Kirchhoff's law of emissivity states that the emissivity of a surface at a particular temperature is the same as its absorptivity; $\varepsilon = \alpha$. The law also applies to monochromatic (spectral) emissivity and absorptivity. A good absorber of radiation is also a good emitter of radiation. *Kirchhoff's spectral laws of black-body radiation* state that (i) a hot body emits radiation with a continuous spectrum; (ii) a hot gaseous body at low pressure emits radiation that has discrete emission lines; (iii) a gaseous medium at low pressure placed in front of a continuum of radiation (such as that from a hot body) will produce missing discrete lines in the spectrum, that is, the spectrum will have absorption lines.

See *black-body radiation, emissivity, radiation energy conservation law.*

Gustav **Kirchhoff** (1824–1887), a German physicist, is known for his contributions to electricity, black-body radiation and spectroscopy. Kirchhoff originated the term "black-body radiation" in 1862. He received the Rumford Medal (1862) for his work on spectral lines in the solar spectrum. (Photo by Meisenbach Riffarth und Co., Leipzig, courtesy AIP Emilio Segre Visual Archives, gift of Emil Wolf, University of Rochester from their gallery at Optical Scientists, Institute of Optics.)

Klein's rule see *electron–hole pair (EHP) creation energy.*

Klein–Nishina equation see *Compton effect.*

Kleinman's conjecture relates to the *second-order harmonic generation coefficients* $d_{ijk}^{(2)}$ discussed in *nonlinear optical properties*. In general, these coefficients are the elements of a nonlinear 3×6 tensor. According to the *Kleinman conjecture*, $d_{ijk}^{(2)} = d_{kij}^{(2)} = d_{jki}^{(2)}$, that is, the coefficients remain invariant under a permutation of indices, provided that the frequencies are also permuted. This conjecture applies to lossless media and is a good approximation for most relevant nonlinear materials. See *nonlinear optical properties.*

Kramer's spectral model is a simple model for the x-ray spectrum emitted from a metal target bombarded with an energetic beam of electrons, or the x-ray absorption through a material, in which the fluence (number of photons flowing through a unit area per unit second) decreases monotonically with the x-ray photon energy up to the maximum photon energy given by kVp (peak voltage on the x-ray tube). (H. A. Kramer, *Phil Mag.* **46**, 836, 1923.)

Kramers–Kronig relations relate the frequency dependence of the real part of the relative permittivity $\varepsilon_r'(\omega)$ to the frequency dependence of the imaginary part, $\varepsilon_r''(\omega)$. If we know the frequency dependence of the real part ε_r' of the relative permittivity of a material, we can also determine the frequency dependence of the imaginary part ε_r'', and vice versa. This may seem remarkable but it is true provided that we know the frequency dependence of either the real or imaginary part over as wide a range of frequencies as practically possible (ideally from dc to infinity) and the material is *linear*, i.e. it has a relative permittivity that is independent of the applied field; the polarization response must be linearly proportional to the applied field. Similarly one can relate the real and imaginary parts of the polarization, $\alpha'(\omega)$ and $\alpha''(\omega)$, and the real and imaginary parts of the complex refractive index, $n(\omega)$ and $K(\omega)$. For example Kramers–Kronig (KK) relations for the relative permittivity $\varepsilon_r(\omega) = \varepsilon_r''(\omega) + j\varepsilon_r''(\omega)$ are

$$\varepsilon_r'(\omega) = 1 + \frac{2}{\pi} P \int_0^\infty \frac{\omega' \varepsilon_r''(\omega')}{\omega'^2 - \omega^2} d\omega'$$

and

$$\varepsilon_r''(\omega) = -\frac{2\omega}{\pi} P \int_0^\infty \frac{\varepsilon_r'(\omega')}{\omega'^2 - \omega^2} d\omega'$$

where P represents the Cauchy principal value of the integral and the singularity at $\omega = \omega'$ is avoided. For $\alpha = \alpha'(\omega) + j\alpha''(\omega)$,

$$\alpha'(\omega) = \frac{2}{\pi} P \int_0^\infty \frac{\omega' \alpha''(\omega')}{\omega'^2 - \omega^2} d\omega' \quad \text{and}$$

$$\alpha''(\omega') = -\frac{2\omega}{\pi} P \int_0^\infty \frac{\alpha'(\omega')}{\omega'^2 - \omega^2} d\omega'$$

and for the real and imaginary parts n and K in the complex refractive index $\tilde{n} = n + jK$,

$$n(\omega) = 1 + \frac{2}{\pi} P \int_0^\infty \frac{K(\omega')}{\omega' - \omega} d\omega' \quad \text{and}$$

$$K(\omega) = -\frac{2}{\pi} P \int_0^\infty \frac{n(\omega')}{\omega' - \omega} d\omega'.$$

Note that if the harmonic dependence on time is taken as $\exp(-j\omega t)$ then it is usual to take $\varepsilon_r(\omega) = \varepsilon_r''(\omega) + j\varepsilon_r''(\omega)$ whereas for $\exp(-j\omega t)$, $\varepsilon_r(\omega) = \varepsilon_r''(\omega) - j\varepsilon_r''(\omega)$. The above relationships remain the same.

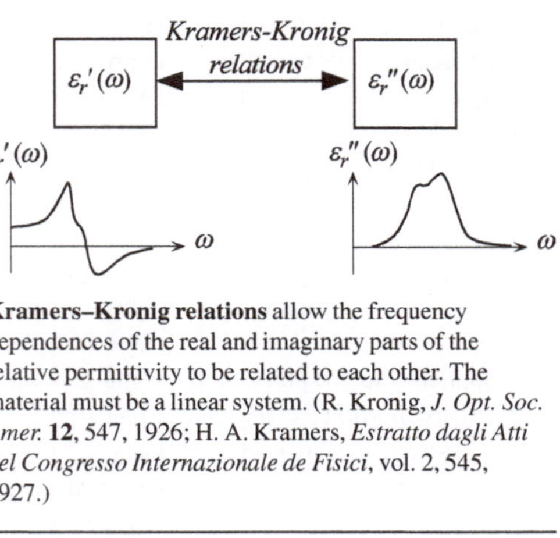

Kramers–Kronig relations allow the frequency dependences of the real and imaginary parts of the relative permittivity to be related to each other. The material must be a linear system. (R. Kronig, *J. Opt. Soc. Amer.* **12**, 547, 1926; H. A. Kramers, *Estratto dagli Atti del Congresso Internazionale de Fisici*, vol. 2, 545, 1927.)

***kTC* noise** in active matrix array based imaging devices (sensors) refers to the "switching noise" that is associated with the channel conductance of the thin film transistor (TFT) or the field effect transistor (FET) that is used to read out the signal at a pixel of the sensor. Each pixel has a sensor that converts the photons it receives to an electrical signal, e.g. converts light to a photocurrent. This electrical signal, the photocurrent, is integrated and converted to a charge signal at the pixel capacitance C_{px}. An FET then switches on (when activated by an external control signal applied to the gate) to transfer the signal charge to an output amplifier and electronics. If there is no input signal, the charge read out by the FET switch should be zero. However, due to thermal fluctuations, every time the FET is switched, the root mean square charge that is read out is not zero, and constitutes what is called the "*kTC*" noise. Even if the collected charge across C_{px} is zero, there will nonetheless be charge fluctuations across C_{px} as a result of the equipartition of energy in this system. This noise is called *reset* or *kTC* noise and has been extensively studied. Suppose that ΔQ is the instantaneous charge across C. Then, from the equipartition theorem, the average energy stored on C_{px}, $\langle \Delta Q^2 \rangle/(2C)$, must be $k_B T/2$ where k_B is the Boltzmann constant and T is the temperature. Even if $\langle \Delta Q \rangle$ is zero, the mean squared fluctuation in ΔQ, $\langle \Delta Q^2 \rangle$, would not be, which constitutes the *kTC* noise. In terms of rms fluctuations N_{reset} in the number of electrons, the above *kTC* noise is approximately equivalent to

$$N_{reset} = (k_B T C_{px})^{1/2}/e.$$

The actual treatment of reset noise is much more complicated than the intuitive simplistic description given above. (J. Hynecek, *IEEE Trans. Electron. Devices*, **37**, 640, 1990.)

KTP ($KTiOPO_4$), potassium titanyl phosphate, is a nonlinear optical crystal that has various optical applications such as frequency doubling, frequency mixing, parametric oscillators, electrooptic (EO) modulators, etc. It has high EO coefficients and a low dielectric constant. It has an orthorhombic crystal structure.

Kubelka–Munk theory is a theory that relates the diffuse reflectance $R(\lambda)$ of a substance to multiple scattering and absorption within the substance. The model assumes diffuse reflectance, and

Kuzyk susceptibility or hyperpolarizability limit

represents the absorption and scattering in the medium by two phenomenological coefficients labeled K and S, respectively; K and S do not simply correspond to the intrinsic optical constants α (absorption coefficient) and s (scattering coefficient) of the medium, though they are related. K and S are used as coefficients that gauge the amount of absorption and diffusion per unit thickness in the medium. For a homogenous slab of thickness d, one form of the equation for the diffuse reflectance R at a given wavelength is given by

$$R = \frac{(1-\beta^2)\sinh(\kappa d)}{(1-\beta^2)\sinh(\kappa d) + 2\beta\cosh(\kappa d)}$$

in which $\beta = [K/(K+2S)]^{1/2}$ and $\kappa = [K(K+2S)]^{1/2}$. Consider diffuse reflection from a layer of thickness d on a substrate with a reflectance R_g. Then

$$R = \frac{1 - R_g[a - b\coth(bSd)]}{a + b\coth(bSd) - R_g}$$

in which $a = 1 + K/S$ and $b = [a^2 - 1]^{1/2}$. It is assumed that the light arrives from a medium that has the same refractive index as the layer, and scattering is isotropic in the medium. There are various corrections to the Kubelka–Munk equations such as corrections due to refractive index mismatches, interface or lateral scattering, etc. The usefulness of the Kubelka–Munk theory arises from the fact that the scattering and the absorption constants K and S scale linearly with the concentration of colorants or pigments; and overall contributions of colorants, inks, pigments can be found by applying an additive rule, that is, adding the individual Kd together for absorption and individual Sd for scattering. A useful simple form of the equation for a thick medium takes the form

$$R = (1 + K/S) - [(1 + K/S)^2 - 1]^{1/2}$$

or equivalently,

$$\frac{K}{S} = \frac{(1-R)^2}{2R}.$$

The above equation provides a useful way of using the diffusive reflectance spectrum to extract the absorption spectrum.

(P. Kubelka and F. Munk, *Z. Tech. Phys.* (Leipzig), **12**, 593, 1931; P. Kubelka, *J. Opt. Soc.*, **38**, 448, 1948; L. Yang and B. Kruse, *J. Opt. Soc. Am.*, **21**, 1933, 2004.)

Kuzyk susceptibility or hyperpolarizability limit is the maximum possible value for the second- or third-order molecular nonlinear susceptibility that is allowed from quantum mechanical and sum rule predictions; most maximum experimental values actually measured are far below this limit, and the difference has been called the *Kuzyk quantum gap*. The original calculations were actually for the theoretical upper values of the hyperpolarizabilities β and γ of molecules given the quantum mechanical sum rule. The maximum value for the hyperpolarizability turns out to depend on two factors: the number of electrons in the molecule (N) and the energy (E_{10}) to the first excited state. For example, the limit or maximum value for positive γ is (in Gaussian units)

$$\gamma_{\text{limit}} = 4\frac{e^4\hbar^4}{m_e^2}\left(\frac{N^2}{E_{10}^6}\right)$$

where m_e is the mass of the electron.

(M. G. Kuzyk, *Phys. Rev. Letts.*, **85**, 1218, 2000; *Opt. Lett.*, **25**, 1183, 2000.)

L

Laguerre-Gaussian beams form, along with *Hermite-Gaussian beams*, another complete set of solutions to Maxwell's paraxial equation. If (ρ,θ) denote the polar coordinates, then the field can be expressed as:

$$E_{pli}(\rho,\theta,z) = \frac{w_0}{w(z)}\left(\frac{\rho\sqrt{2}}{w(z)}\right)^l L_p^l\left(\frac{2\rho^2}{w^2(z)}\right) \times \exp\left(\frac{-\rho^2}{w^2(z)} - j\Phi_{pl}(\rho,z)\right) \times \begin{cases}\cos l\theta, i=1\\ \sin l\theta, i=2.\end{cases}$$

The phase is given by:

$$\Phi_{pl}(\rho,z) = kz - (2p+l+1)\arctan(z/z_R) + k\rho^2/[2R(z)],$$

where the integer p is the radial index, and the integer l the angular index (for $l=0$, we note $E_{p0i} \equiv E_{p0}$). As for a Gaussian beam, $2w_0$ is the waist, $w(z)$ the radius at a distance z, z_R the Rayleigh range, and $R(z)$ the curvature. They remain strictly identical to that of Hermite-Gaussian beams.

Generalized Laguerre polynomials $L^l{}_p$ obey the differential equation

$$x\frac{d^2L_p^l}{dx^2} + (l+1-x)\frac{dL_p^l}{dx} + pL_p^l = 0,$$

or can be derived alternatively from

$$L_p^l(x) = \frac{x^{-l}}{p!}e^x\frac{\partial^p}{\partial x^p}(x^{p+l}e^{-x}).$$

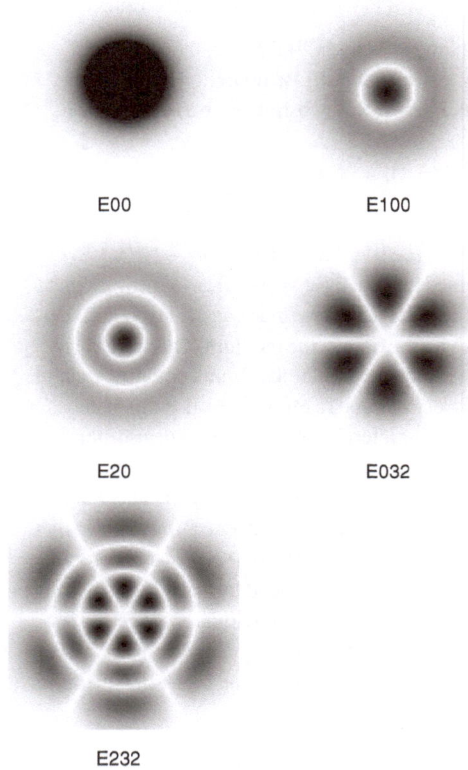

Intensity profile in the transverse plane (x,y) for **Laguerre-Gaussian** modes E_{00} (fundamental Gaussian beam), E_{100}, E_{20}, E_{032} and E_{232}. The darker the zone, the higher the intensity.

The first three polynomials read:

$$L_0^l(x) = 1, \quad L_1^l(x) = l + 1 - x,$$
$$L_2^l(x) = \frac{(l+1)(l+2)}{2} - (l+2)x - \frac{x^2}{2}.$$

The fundamental mode is purely Gaussian. For $l = 0$, the solutions E_{p0} present cylindrical symmetry and the intensity exhibits a transverse structure with p rings around a central spot. For $l \neq 0$, the intensity structure is made of $p + 1$ rings (no central spot) with $2l$ angularly distributed maxima.

Lamb shift refers to the small energy difference between the states $2s_{1/2}$ and $2p_{1/2}$ in hydrogen. These two states have the same principal quantum number n, and the same total angular momentum quantum number j which is 1/2 for both. These two states have the same energy in the Dirac quantum mechanical theory that only treats the Coulombic potential. These two states, however, have slightly different energies, which can be explained by a rigorous treatment of the interaction between the electron, the proton and the *quantized* electromagnetic field. The Lamb shift therefore represents a radiative correction to the Dirac approximation. This effect was first discovered by Willis Eugene Lamb (1913–2008).

Lambert's absorption law describes the exponential attenuation of a light beam as it propagates along a particular direction, arbitrarily set to be the z-direction, inside a lossy medium with an attenuation coefficient α [in m^{-1}],

$$I(z) = I_0 \exp(-\alpha z).$$

where I is the intensity at z and I_0 is the intensity at $z = 0$. Frequently this law is simply called Lambert's absorption law and is often stated as the light transmittance T of a medium of thickness d is given by $T = \exp(-\alpha d)$. If scattering losses in the medium can be neglected, then α is simply the absorption coefficient.

Lambert's cosine law governs the intensity I of light leaving an isotropic plane light source of luminous intensity in a particular direction with respect to the normal to the surface. If the intensity is I_0 for light emitted normal to the surface and if I is the intensity for light emitted at an angle θ to the normal, then

$$I = I_0 \cos\theta.$$

Lambertian diffusion refers to an "ideal" surface which, either by reflection or transmission, reemits any incident light according to a luminance law that is independent of the direction of the incident radiation and of the direction of observation. Such surfaces, with no memory of the incident flux, are called *orthotropic* or *Lambertian.* Their luminance is directly proportional to their illumination.

Lambertian source emits radiation uniformly over all angles. It is a surface emission source which exhibits an angularly uniform luminance, independent of the direction of emission. Consequently, its intensity in the angular direction θ with respect to the normal is $I(\theta) = LA_S\cos\theta = I_0\cos\theta$, where L is the luminance, A_S is the surface area of the source, and I_0 is the maximum intensity in the direction normal to the source surface. It is a good approximation for most thermal sources. A Lambertian source has a flux $\Phi = \pi I_0$, which corresponds to integrating the intensity $I(\theta)$ over a hemisphere above the emitting surface.

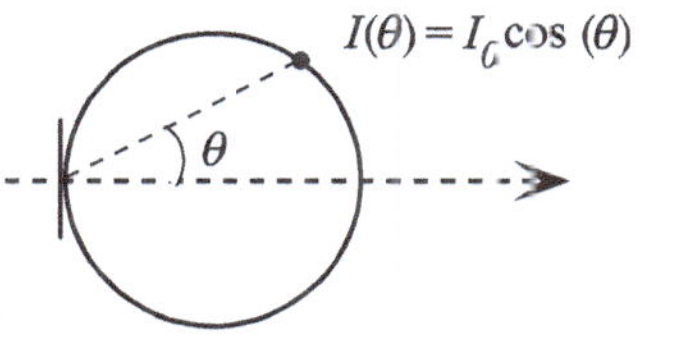

Lambertian source The radiation pattern takes the form of a perfect sphere: the source appears equally bright in all directions.

Lamé's constants see *elastic coefficients.*

Langmuir–Blogett (LB) films are typically formed by a layer-by-layer transfer of monomolecular films of lipids from a water surface to a substrate. This process enables complex multilayered materials to be designed in a layer-by-layer fashion with tailored optical response. Ultrathin films and multilayered film structures are important for many applications, including x-ray optics, nonlinear optics, microelectronics and for biosensors.

Laser, an acronym for *light amplification by stimulated emission of radiation*, is a device that produces an output beam of radiation that is nearly monochromatic, coherent and has a very low divergence, by the use of two phenomena: stimulated emission in a special medium and optical feedback by the use of an optical cavity. In contrast, a tungsten light bulb emits incoherent light, that is, an incoherent stream of photons over a very wide range of frequencies. The light beam from an LED is a stream of photons with random phases. The LED beam has a much larger divergence than a laser beam and the spectral width of the LED light is also much wider than that from a laser. The principle of the laser can be easily explained by considering the ruby laser, which is one of the first lasers built in the 1960s. Laser operation is first based on achieving *population inversion* between two energy levels in the medium which populates an excited energy level at E_2 with a large number of atoms or ions compared with a lower energy level E_1 as explained below. The process that establishes population inversion is called *pumping*. In an optically pumped solid state laser, the laser medium is usually a crystal or a glass rod that has suitable ions, e.g. the ruby laser has Cr^{3+} ions in a Al_2O_3 crystal (sapphire) rod, that can be excited by the absorption of photons. The ions in the crystal must have at least *three* suitable energy levels. In the ruby laser, the Cr^{3+} ions in the ground state at E_1 are optically pumped, by photons from a xenon flash lamp, to a band of energies around E_3. (E_3 is not a single level in this case but a narrow band.) The excited Cr^{3+} ions decay quickly to an energy level E_2 (above E_1) by emitting phonons (lattice vibrations). The E_2-state, on the other hand, does not rapidly and spontaneously decay to lower-energy state. In other words, the state at E_2 is a *long-lived state*; it has a lifetime of ~ 3 ms, which is very long on the atomic scale. Since the Cr^{3+} ions cannot decay rapidly from E_2 to E_1 they accumulate at this energy level causing a population inversion between E_2 and E_1 as pumping takes more and more ions to E_3 and hence E_2, which eventually becomes more populated than the E_1-level. (Remember that in thermal equilibrium we expect more ions at E_1 than at E_2; hence the name *population inversion* when the opposite is achieved.) When one ion at E_2 decays spontaneously, it emits a photon (a "random photon") which can go on to a neighboring ion and cause that to execute stimulated emission. The photons from the latter can go on to the next ion at E_2 and cause that to emit by stimulated emission, and so on. The result is an avalanche effect of stimulated emission processes with all the photons in phase so that the light output is a large collection of coherent photons. As the ions at E_2 are stimulated to emit their photons, they return to E_1 and can be pumped again to repeat the stimulated emission cycle. The emission from E_2 to E_1 is called *lasing emission*. The pumped ruby crystal with excited Cr^{3+} ions (at E_2) acts as an *active medium* with an *optical gain*, that is, a medium in which stimulated emissions exceed absorption and spontaneous emission so that radiation traveling along the ruby rod at the frequency $\nu_{21} = (E_2 - E_1)/h$ can be amplified. The ends of the ruby crystal are silvered to reflect back and forward the stimulated radiation so that its intensity builds up in the rod in much the same way we build up voltage oscillations in an electrical oscillator circuit by positive feedback. Thus, an *optical cavity* is used to build up stimulated emissions and the intensity of the radiation inside the ruby at ν_{21}. One of the mirrors is partially silvered to allow some of this radiation to be tapped out. In fact, the mirrors can be external to the active medium as in many modern lasers. The ruby laser operates as

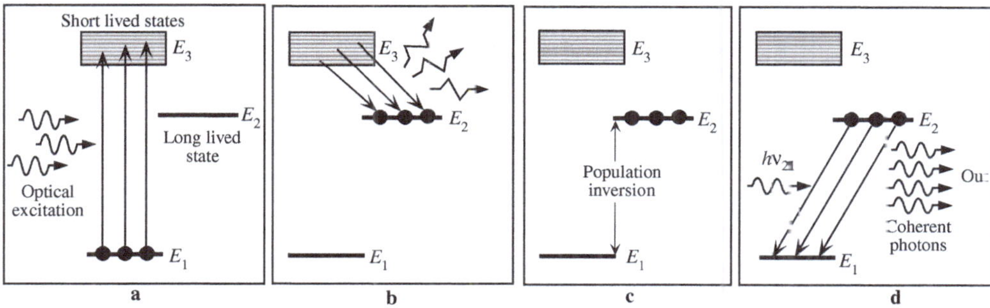

Laser The principle of the laser. (a) Ions in the ground state are pumped up to the energy band E_3 by photons from an optical excitation source. (b) Ions at E_3 rapidly decay to the long-lived state at energy level E_2 by emitting lattice vibrations (phonons). (c) As the states at E_2 are long-lived, they quickly become populated and there is a population inversion between E_2 and E_1. (d) A random photon of energy $h\nu_{21} = E_2 - E_1$ can initiate stimulated emission. Photons from this stimulated emission can themselves further stimulate emissions leading to an avalanche of stimulated emissions and coherent photons being emitted.

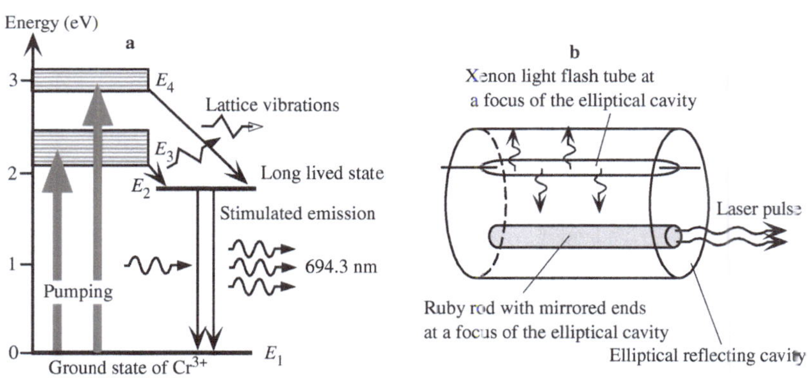

Laser (a) The ruby laser Cr^{3+} energy diagram and the principle of the ruby laser. Cr^{3+} ions in the ground state are pumped up to energy bands E_3 and E_4 by photons from a xenon flash. The Cr^{3+} ions decay rapidly to the long-lived state at energy level E_2 by emitting phonons. As the states at E_2 are long-lived, they quickly become populated and there is a population inversion between E_2 and E_1, which leads to optical amplification by stimulated emission at photon energy $(E_2 - E_1)$. (b) The ruby laser needs an *optical cavity*, which is obtained by polishing and silvering the ends of the ruby rod or by placing external mirrors at the ends of the rod. Efficient optical pumping can be obtained by using an elliptical reflector.

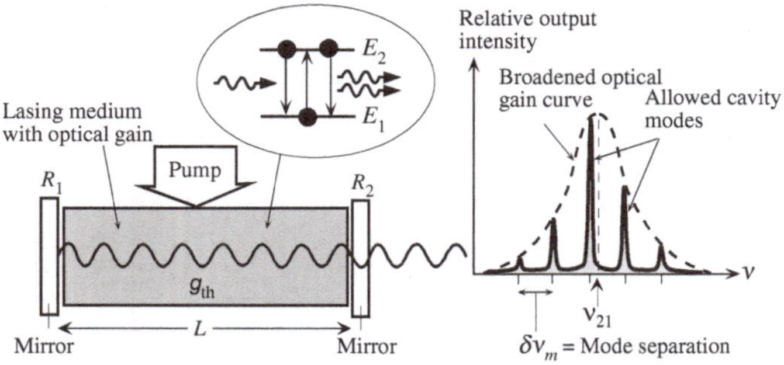

Laser The basic principle of a CW laser. The active medium is pumped continuously, for example by using an electric discharge (not shown) in a gas laser. This achieves the necessary threshold population inversion that allows the medium to have a threshold gain for self-sustained electromagnetic oscillations in the cavity. The gain in the active medium just balances all the losses as in a self-sustaining oscillator. There are two mirrors to enclose the lasing medium in an optical cavity, and the radiation is allowed to escape from one of the mirrors that is not fully reflecting. The optical gain in the active medium is not at ν_{21} but has a broadened shape. The output spectrum (relative intensity vs. wavelength) consists of one or more modes, determined by those modes that are allowed in the cavity and, at the same time, occur inside the optical gain curve of the medium.

a pulsed laser with the lasing emission at 694.3 nm. Ruby lasers are used mainly in pulsed holographic work. Many lasers are designed to operate under steady state and provide a continuous wave (CW) output when switched on. CW lasers need steady state maintenance of population inversion by using a continuous pumping process. Under steady state operation, the CW laser has self-sustained electromagnetic oscillations in the optical cavity, and these oscillations are maintained because the optical gain in the medium just balances all the losses, including losses at the mirrors and losses when some of the radiation is tapped out. The threshold optical gain g_{th} needed for a CW laser can be found from the fact that when the radiation travels in one round-trip from one mirror to the other and back, the overall power gain must be unity as in all self-sustaining oscillators. If γ is the loss coefficient that represents all the losses in the medium, including scattering, and if R_1 and R_2 are the reflectances of the mirrors, then we need g_{th} to satisfy

$$\text{Round-trip power gain} = R_1 R_2 \exp(2g_{th}L) \times \exp(-2\gamma L) = 1$$

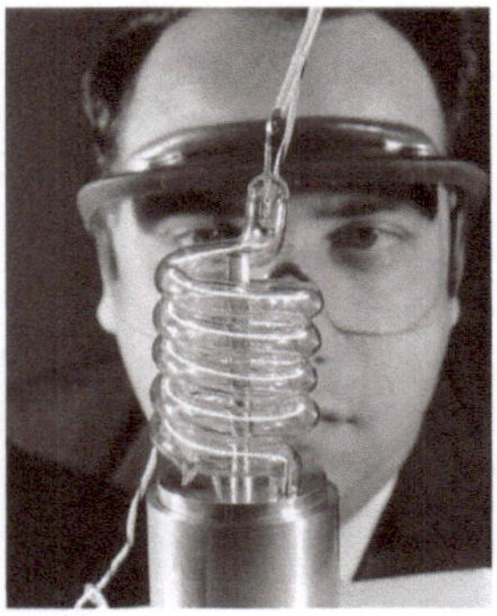

Laser Theodore Harold Maiman was born in 1927 in Los Angeles, son of an electrical engineer. He studied engineering physics at Colorado University, while repairing electrical appliances to pay for college, and then obtained a Ph.D. from Stanford. Theodore Maiman constructed this first laser in 1960 while working at Hughes Research Laboratories (T. H. Maiman, *Nature*, **187**, 493, 1960). There is a vertical chromium ion doped ruby rod in the center of a helical xenon flash tube. The ruby rod has mirrored ends. The xenon flash provides optical pumping of the chromium ions in the ruby rod. The output is a pulse of red laser light. (Courtesy of HRL Laboratories, LLC, Malibu, California.)

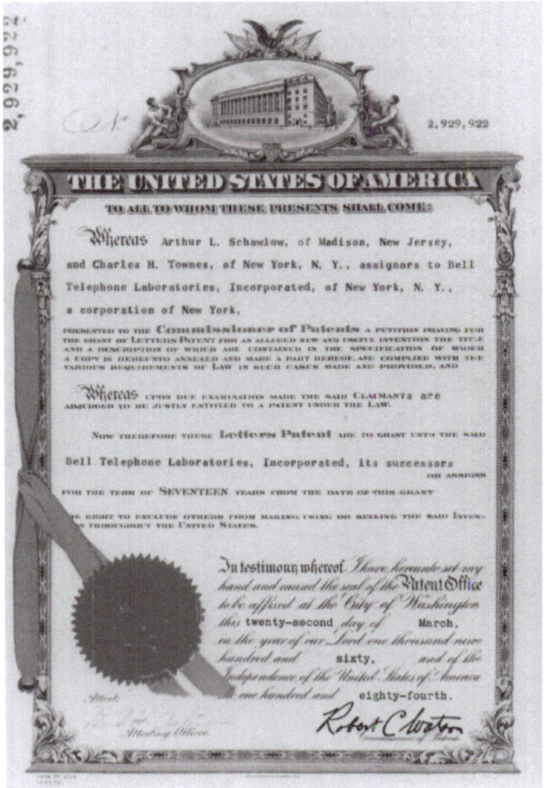

2,929,922

2,929,922

THE UNITED STATES OF AMERICA

TO ALL TO WHOM THESE PRESENTS SHALL COME:

Whereas Arthur L. Schawlow, of Madison, New Jersey, and Charles H. Townes, of New York, N. Y., assignors to Bell Telephone Laboratories, Incorporated, of New York, N. Y., a corporation of New York,

presented to the Commissioner of Patents a petition praying for the grant of Letters Patent for an alleged new and useful invention the title and a description of which are contained in the specification of which a copy is hereunto annexed and made a part hereof, and complied with the various requirements of Law in such cases made and provided, and

Whereas upon due examination made the said Claimants are adjudged to be justly entitled to a patent under the Law.

Now therefore these Letters Patent are to grant unto the said Bell Telephone Laboratories, Incorporated, its successors or assigns for the term of SEVENTEEN years from the date of this grant the right to exclude others from making, using or selling the said Invention throughout the United States.

In testimony whereof I have hereunto set my hand and caused the seal of the Patent Office to be affixed at the City of Washington this twenty-second day of March, in the year of our Lord one thousand nine hundred and sixty, and of the Independence of the United States of America the one hundred and eighty-fourth.

Attest:

Attesting Officer.

Robert C. Watson

Commissioner of Patents.

Laser The patent for the invention of the laser by Charles H. Townes and Arthur L. Schawlow in 1960 (Courtesy of Bell Laboratories). The laser patent was later bitterly disputed for almost three decades in "the patent wars" by Gordon Gould, an American physicist, and his designated agents. Gordon Gould eventually received the US patent for optical pumping of the laser in 1977 since the original laser patent did not detail such a pumping procedure. In 1987 he also received a patent for the gas discharge laser, thereby winning his 30 year patent war. His original notebook even contained the word "laser." (See Jeff Hecht, "Winning the laser-patent war", *Laser Focus World*, December 1994, pp. 49–51).

which makes the required threshold gain in the pumped active medium to be

$$g_{\text{th}} = \gamma + \frac{1}{2L} \ln\left(\frac{1}{R_1 R_2}\right).$$

This equation is the key to designing lasers because it allows the threshold gain g_{th} and the device characteristics (γ, L, R_1 and R_2) to be linked. For gas lasers g_{th} tends to be small, typically 0.1–5 m^{-1}, and hence gas lasers need a good active medium length ($L > 10$ cm) and excellent reflectances at the mirrors (R_1 and $R_2 > 99\%$) to work. The laser is the optical analog of an electrical oscillator in which positive feedback of radiation back into the optical amplifier (pumped medium) leads to powerful standing waves inside the optical cavity some of which are

 Laser cooling

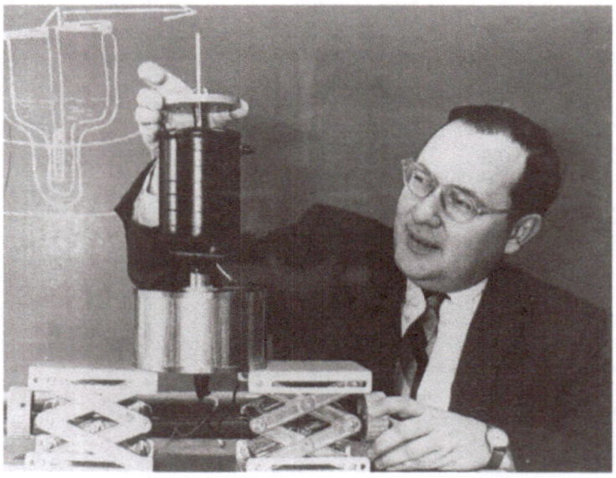

Laser Arthur Schawlow (1921–1999), one of the co-inventors of the laser, was well-known for his humor and has, apparently, said that "Anything will lase if you hit it hard enough." In 1971, Schawlow and Ted Hänsch were able to develop the first edible laser made from "Jell-O," which was actually gelatin with fluorescent dye. He won the Nobel prize in physics with Nicolaas Bloembergen and Kai Siegbahn in 1981 for their contributions to the development of laser spectroscopy. (Copyright Stanford University, courtesy AIP Emilio Segre Visual Archives)

allowed to escape as laser output by having one of the mirrors partially transmitting. What comes out from a laser is highly coherent radiation which has high intensity. The coherence and the well-defined wavelength of this radiation is what makes it distinctly different from a random stream of different wavelength photons emitted from a tungsten bulb, or randomly phased photons from an LED. The output spectrum from a laser has a finite width due to various broadening mechanisms (see *laser output spectrum*). Further, the wavelengths in the output spectrum must be those that are permitted by the optical cavity; they must be modes of the optical cavity. The pumping mechanism that achieves population inversion depends on the type of laser that is being constructed, for example, optical pumping at a wavelength different from the lasing wavelength, electrical discharge as in gas laser, an electric current as in semiconductor laser diodes, or a chemical reaction as in chemical lasers. See *argon ion laser, diode pumped laser, dye laser helium–neon laser, free electron laser, semiconductor laser, laser output spectrum, threshold gain, threshold population inversion.*

Laser cooling is an experiment in which atoms are cooled by irradiating the selected atoms (e.g. sodium, rubidium and cesium) using pairs of oppositely pointing laser beams with frequency slightly detuned (i.e. lower in energy) from the electronic resonance condition for the specific atom's transition energy. Atoms moving toward a given laser beam will have their transition energy lowered as a result of the Doppler effect, resulting in photon absorption – correspondingly, momentum

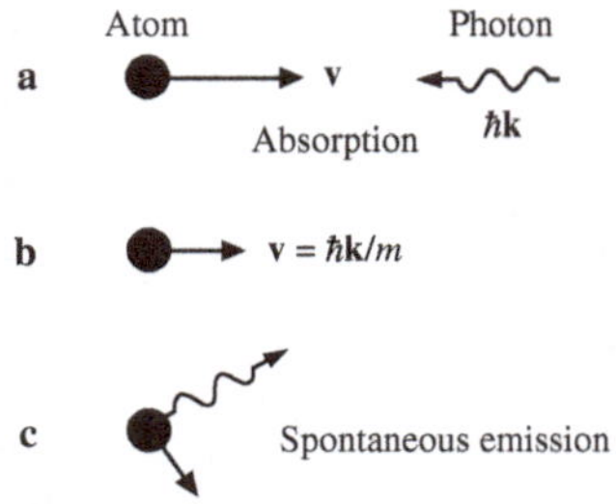

Laser cooling occurs primarily through the Doppler effect. (a) An atom moving with velocity **v** in the $+x$-direction, is retarded owing to momentum transfer occurring when the atom absorbs a photon with momentum $\hbar\mathbf{k}$ moving in the $(-x)$-direction, from the incident laser beam. (b) The absorption produces a so-called braking force that slows the atom's velocity in the $(+x)$-direction to $\mathbf{v} = \hbar\mathbf{k}/m$, where m is the mass of the atom. (c) The atom returns to its ground state by emitting a photon in a process known as *spontaneous emission*. The spontaneous emission occurs in a random direction. Thus on average, spontaneous emission results in no net contribution to the atom's momentum, and the atom's velocity remains lower than it was previously. The atom may also undergo another so-called stimulated emission. If this occurs, the radiated photon rejoins the optical field in the same mode as the original photon and there is no net momentum transfer, with the atom retaining its original velocity. The sum of these two processes is a slowing of the atom, dependent on the relative rates of spontaneous versus stimulated emission.

absorption will cause the atom to reduce its momentum equal to that of the photon. Random reradiation of a photon by the atom will then occur, resulting in a small net momentum for the whole process. Repeating this process many times will reduce the momentum of the atom and its associated kinetic energy. As a result, the mean kinetic energy E_{KE} of the ensemble of such atoms is reduced, and hence the average temperature of the ensemble, $T = E_{KE}/k_B$, is also reduced, or said to be *laser cooled*.

Laser diode is a semiconductor diode which emits *coherent radiation* in contrast to an LED (light-emitting diode) which emits incoherent radiation. Laser diodes operate on the principle of *stimulated emission* resulting from the direct recombination of injected electrons and holes in the active region of the device. Sufficient carrier injection and hence stimulated recombination in the active region is achieved by passing a current through the laser diode that exceeds some threshold current. For example, in the AlGaAs(n)/GaAs(p)/AlGaAs(n) heterostructure diode, the GaAs is heavily p-type doped and acts as the active region. The current under forward bias injects electrons from the n-AlGaAs layer into the p-GaAs layer. There is a step increase in the conduction band edge from p-GaAs to p-AlGaAs and this potential energy step acts as a barrier that prevents the electrons from spreading out. Due to the latter *confinement effect*, the electron concentration and hence the rate of stimulated emissions are large in the active layer even under a small forward current. When the current exceeds the threshold current, the optical gain in the active region is sufficient to allow stimulated emissions to exceed absorption, spontaneous emission, and cavity losses so that the net effect is the generation of coherent radiation by stimulated emissions. See *heterostructure laser diode*.

Laser diode bandwidth is the frequency range over which a laser diode output, a coherent beam, can be modulated by the modulation of the input current into the diode. Typically, the relative radiation output vs. modulation frequency characteristics is relatively flat until high frequencies, and then peaks at a certain frequency, called the *relaxation oscillation frequency* f_r, as shown in the figure, which is given by

$$f_r = \frac{1}{2\pi(\tau_r\tau_{ph})^{1/2}}\left(\frac{I_0}{I_{th}} - 1\right)^{1/2}$$

where τ_r is the effective carrier recombination time (radiative and nonradiative processes acting in parallel), sometime simply called the *spontaneous lifetime*, τ_{ph} is the *photon lifetime*, the average time it takes a photon to be lost from the laser diode cavity (by absorption or emission through the cavity facets), I_0 is the operating diode bias current and I_{th} is the threshold current. τ_r and τ_{ph} are typically in the nano- and picosecond range so that for I_0 twice I_{th}, f_r is a few GHz. The useful bandwidth, such as the 3 dB point, would be at $1.5f_r$. When the laser diode is excited with a step current pulse of magnitude $I_0 > I_{th}$, the lasing radiation output begins only

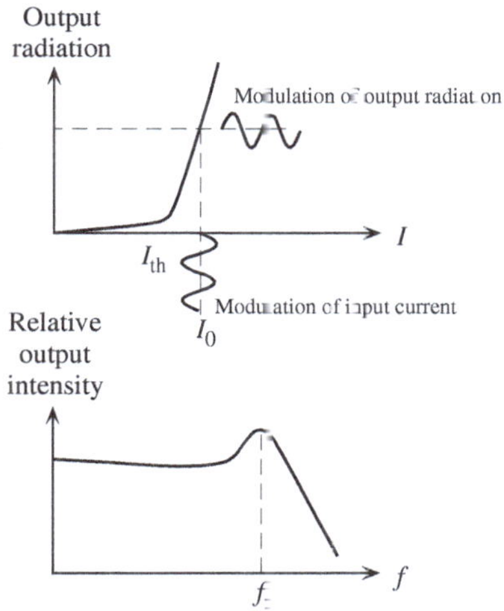

Laser diode bandwidth The modulation of a laser diode and the resulting frequency response which has a relaxation oscillation peak.

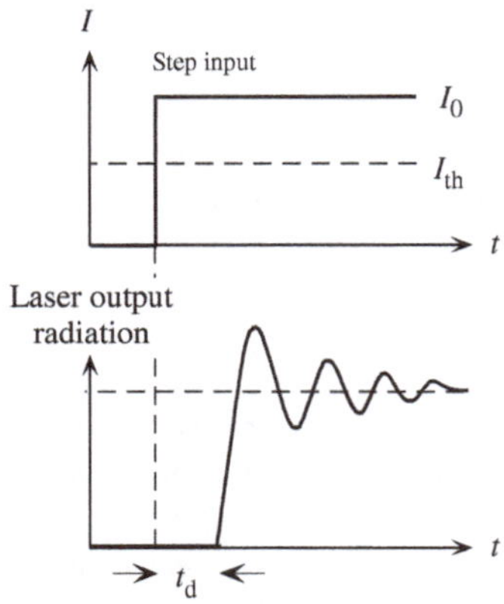

Laser diode bandwidth The step excitation of a laser diode by the application of a step current above the threshold value, and the delayed and oscillatory response of the output radiation.

after a delay time t_d, and reaches a steady state value after a few damped oscillations, where

$$t_d = \tau \ln[I_0/(I_0 - I_{th})]$$

where τ is the overall effective recombination time in the recombination region around the threshold operating conditions.

Laser diode beam collimation is an optical collimation set up not only to change the beam profile from an elliptical cross-section to a circular cross-section but to provide a collimated beam with the required beam diameter. See *laser diode output beam.*

Laser diode equation relates the output optical power P_o to the diode current density J (or current I) above the threshold current density J_{th}. In simple terms,

$$P_o = \left[\frac{hc^2\tau_{ph}W(1 - R)}{2en\lambda}\right](J - J_{th})$$

where R is the reflectance of the optical cavity reflectors, n is the refractive index, λ is the wavelength, W is the width of the active region and τ_{ph} is the photon decay time in the cavity; the power in a light wave, in the absence of amplification, decreases as $\exp(-\alpha_t x)$ which is equivalent to a decay in time as $\exp(-t/\tau_{ph})$ where $\tau_{ph} = n/(c\alpha_t)$ and α_t is the *total attenuation coefficient* representing all these loss mechanisms.

Laser diode external differential quantum efficiency, η_{EDQE}, of a laser diode is defined as

$$\eta_{EDQE} = \frac{\text{Increase in number of output photons from diode per second}}{\text{Increase in number of injected electrons into diode per second}}$$

$$= \left(\frac{e}{E_g}\right)\frac{dP_o}{dI}$$

where P_o is the emitted optical power, E_g is the bandgap of the active region semiconductor and I is the diode current.

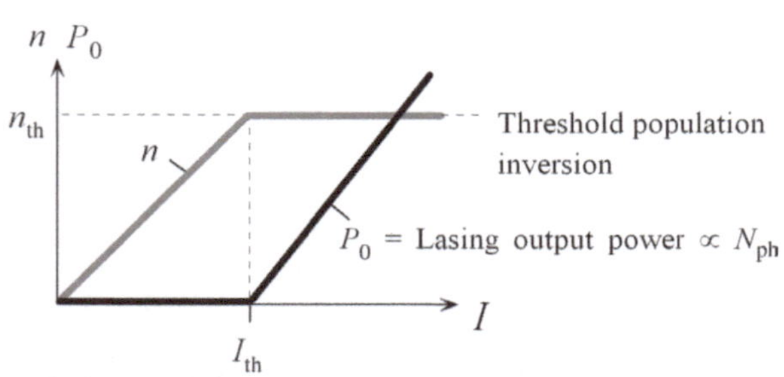

Laser diode equation Simplified and idealized description of a semiconductor laser diode based on rate equations. Injected electron concentration n and coherent radiation output power P_o vs. diode current I.

Laser diode external power efficiency, η_{EPE}, of the laser diode is defined by

$$\eta_{\text{EPE}} = \frac{\text{Optical output power}}{\text{Electical input power}} = \eta_{\text{EQE}}\left(\frac{E_g}{eV}\right)$$

where η_{EQE} is the laser diode external quantum efficiency, E_g is the bandgap of the active region semiconductor and V is the diode voltage.

Laser diode external quantum efficiency, η_{EQE}, is defined as

$$\eta_{\text{EQE}} = \frac{\begin{array}{l}\text{Number of output photons from}\\ \text{the diode per second}\end{array}}{\begin{array}{l}\text{Number of injected electrons into}\\ \text{the diode per second}\end{array}} = \frac{eP_o}{E_g I}$$

where P_o is the emitted optical power, E_g is the bandgap of the active region semiconductor and I is the diode current.

Laser diode output beam is an astigmatically diverging nearly-coherent beam that emerges from the laser diode's semiconductor crystal emission facet. The angular spread in the vertical (y) and horizontal (x) directions (perpendicular or parallel to the plane of the diode junction) are not the same which leads to a nearly elliptical beam cross-section. Normally the active cavity height is smaller than the cavity width. Since the angular spread of the emerging laser beam is diffraction controlled by the emission aperture, that is, the far field intensity distribution is a Fourier transform of the intensity distribution at the cavity aperture, the vertical angular spread $\theta_\perp$ is more than the horizontal spread $\theta_{//}$. The far-field intensity distributions in the y- and x-directions are (nearly) Gaussian in the angular displacements θ_y and θ_x. The spreads $\theta_\perp$ and $\theta_{//}$ refer to the angular separation of the e^{-2} points on the Gaussian curve as indicated in the figure. In some specifications, $\theta_\perp$ and $\theta_{//}$ refer to the angular separation of half-intensity points (full width at half maximum, FWHM); the two are related by $\theta(e^{-2}) = 1.7\theta(\text{FWHM})$. The output divergence is quoted as $\theta_\perp \times \theta_{//}$. The output beam from the laser diode (LD) in most applications is either put through a

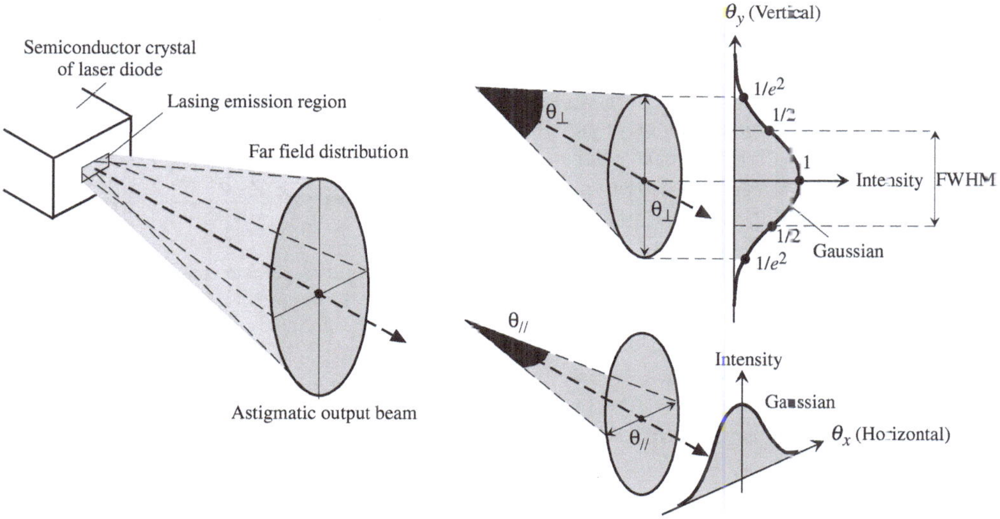

Laser diode output beam The output beam characteristics from a laser diode.

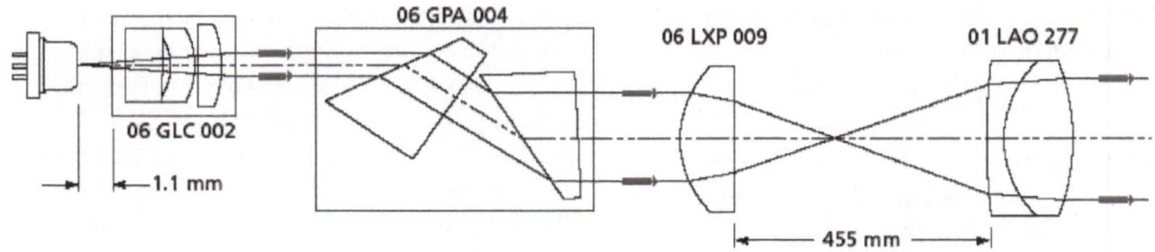

Laser diode output beam A collimated beam of 25 mm diameter from a 780 nm laser diode that has 60° × 20° divergence. The collimation lens (06 GLC 002) has a NA of 0.5, and provides 8 × 2.7 mm collimated elliptical beam. This is circularized by a 3:1 anamorphic prism pair (06 GPA 004) to obtain an 8 mm diameter beam. The beam is expanded by using two lenses, a singlet and a doublet, to a 25 mm diameter collimated beam. (Courtesy of Melles Griot.)

lens–prism system to collimate the beam, or it is coupled, using an anamorphic lens, into a fiber. The manipulation of the LD output beam through optics into a well-defined collimated beam with a circular cross-section ($\theta_\perp = \theta_{//}$) is simply referred to as the *collimation* and *circularization*. (Circularization should not be confused with circularly polarized light.) The circularization and collimation can be quite complicated depending on the exact requirements and space available for the optics. Usually a collimating lens (a multielement lens system) is used to capture and collimate the diverging beam. Because $\theta_\perp$ can be quite large, e.g. 80° (e^{-2} points), and the lens needs to capture most of the radiation, its NA = sin($\theta_\perp$/2) has to be 0.64. Suppose that we want the collimated beam to be 1 mm in diameter. Since NA = $D/2f$, where D is the diameter of the lens (and the beam), we need f = 0.78 mm. Such a focal length and the required NA can only be obtained by a multilens collimator. Once the beam is collimated, it is circularized by passing it through an *anamorphic lens* pair. After the latter optics, one can expand the beam further by using more optics (beam expansion is relatively straightforward).

Laser diode output vs. current characteristics typically refer to the lasing output power vs. diode current characteristics at various temperatures. Above the *threshold current* (I_{th}), the optical power in the lasing emission increases sharply with the diode current and, to first order, the optical power is proportional to the diode current above the threshold value ($I - I_{th}$).

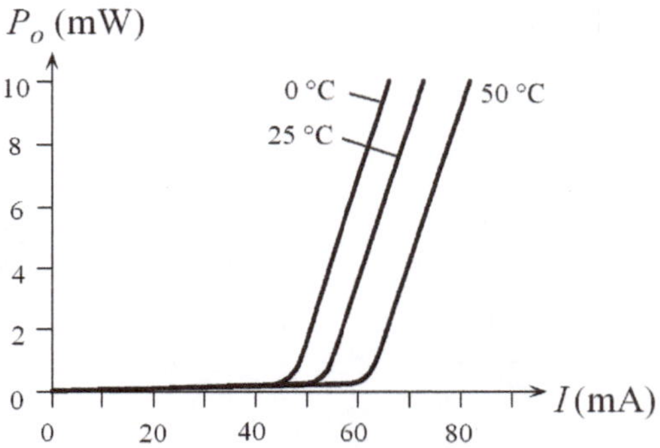

Laser diode output vs. current characteristics Output optical power vs. diode current at three different temperatures. The threshold current shifts to higher temperatures.

Laser modes are distinct electric field patterns that are allowed by the active optical resonator structure of a laser diode. The output spectrum from a laser diode (LD) depends on two factors: the nature of the optical resonator used to build the laser oscillations and the optical gain curve (lineshape) of the active medium. The optical resonator is essentially a Fabry–Perot cavity which can be assigned a length (L), width (W) and height (H). The length L determines the *longitudinal mode* separation whereas the width W and height H determine the transverse modes, or *lateral modes*, in LD nomenclature. If the transverse dimensions (W and H) are sufficiently small, only the lowest transverse mode, TEM_{00} mode, will exit. This TEM_{00} mode, however, will have longitudinal modes whose separation depends on L.

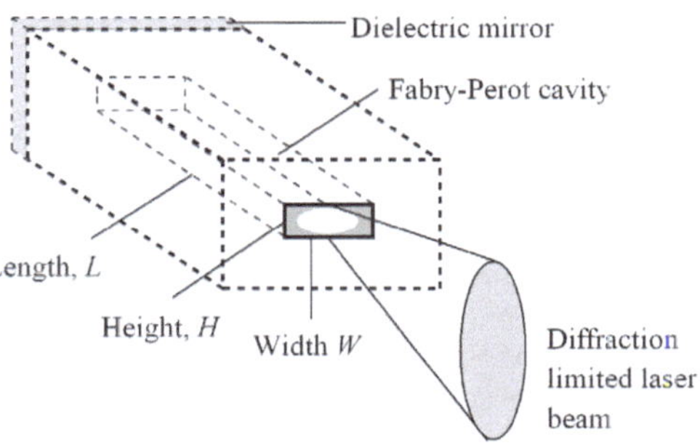

Laser modes The laser cavity definitions and the output laser beam characteristics.

Laser oscillations occur in a laser device when it has achieved lasing emission, that is, there are sustainable electromagnetic oscillations in the laser optical cavity as a result of optical gain (from stimulated emissions) which overcomes the cavity losses.

Laser output spectrum refers to the spectral intensity vs. wavelength characteristics of the laser beam. The output spectrum is the result of the broadening of the optical gain curve of the lasing medium and the requirement that the optical cavity can only sustain oscillations with certain wavelengths. For simplicity, consider a gas laser. The frequency of the emitted radiation from the gas atoms is broadened around the stimulated emission transition frequency υ_o due to the Doppler effect. Consequently, the optical gain of the medium vs. frequency characteristic is also broadened. We recall from kinetic molecular theory that gas atoms are in random motion with an average kinetic energy of $^3/_2 k_B T$, when k_B is the Boltzmann constant and T is temperature. Suppose that these gas atoms emit radiation of frequency υ_o which we label as the source frequency. Then, due to the Doppler effect, when a gas atom is moving *away* from or *towards* an observer, the latter detects a lower or higher frequency υ_1 or υ_2 respectively given by

$$\upsilon_1 = \upsilon_o\left(1 - \frac{\upsilon_x}{c}\right) \quad \text{and} \quad \upsilon_2 = \upsilon_o\left(1 + \frac{\upsilon_x}{c}\right)$$

where υ_x is the relative velocity of the atom along the laser tube (x-axis) with respect to the observer and c is the speed of light. Since the atoms are in random motion the observer will detect a range of frequencies due to this Doppler effect. As a result, the frequency or wavelength of the output radiation from a gas laser will have a "linewidth" $\Delta\upsilon = \upsilon_2 - \upsilon_1$. This is what we mean by a *Doppler broadened linewidth* of a laser radiation. There are other mechanisms which also broaden the output spectrum but we will ignore these in the present case of gas lasers. From kinetic molecular theory we know that the velocities of gas atoms obey the Maxwell distribution. Consequently, the stimulated emission wavelengths in the lasing medium must exhibit a distribution about a central wavelength $\lambda_o = c/\upsilon_o$ Stated differently, the lasing medium therefore has an optical gain (or a photon gain) that has a distribution around $\lambda_o = c/\upsilon_o$. The variation in the optical gain with the wavelength is called the *optical gain lineshape*. For the Doppler broadening case, this lineshape turns out to be a Gaussian function. For many gas lasers, this spread in the frequencies from υ_1 to υ_2 is 2–5 GHz (for the He-Ne laser the corresponding wavelength spread is ~0.02 Å). When we consider the Maxwell velocity distribution of the gas atoms in the laser tube, we find that the linewidth $\Delta\upsilon_{1/2}$ between the half-intensity points (full width at half maximum FWHM) in the output intensity vs. frequency spectrum is given by

$$\Delta \upsilon_{1/2} = 2\upsilon_{o}\sqrt{\frac{2k_{B}T\ln(2)}{Mc^{2}}}$$

where M is the mass of the lasing atom or molecule. This width $\Delta \upsilon_{1/2}$ can be taken to be the FWHM of the optical gain curve of nearly all gas lasers. It does not apply to solid state lasers in which other broadening mechanisms operate. Suppose that for simplicity we consider an optical cavity of length L with parallel end mirrors. Such an optical cavity is called a *Fabry–Perot optical resonator* or *etalon*. The reflections from the end mirrors of a laser give rise to traveling waves in opposite directions within the cavity. These oppositely traveling waves interfere constructively to set up a standing wave, that is stationary electromagnetic (EM) oscillations. Some of the energy in these oscillations is tapped out by the 99% reflecting mirror to get an output just like the way we tap out the energy from an oscillating field in an LC circuit by attaching an antenna to it. Only standing waves with certain wavelengths, however, can be maintained within the optical cavity just as only certain acoustic wavelengths can be obtained from musical instruments. Any standing wave in the cavity must have an integer number of half-wavelengths $\lambda/2$ that fit into the cavity length L,

$$m\left(\frac{\lambda}{2}\right) = L$$

where m is an integer that is called the *mode number* of the standing wave. The wavelength λ in this equation is that within the cavity medium but for gas lasers the refractive index is nearly unity and λ is the same as the free space wavelength. Each possible standing wave within the laser tube (cavity) satisfying this standing wave equation is called a *cavity mode*. Further, each allowed wavelength is a mode wavelength and is labeled λ_m. Modes that exist along the cavity axis are called *axial* (or *longitudinal*) *modes*. Other types of modes, that is stationary EM oscillations, are possible when the end mirrors are not flat. The laser output thus has a

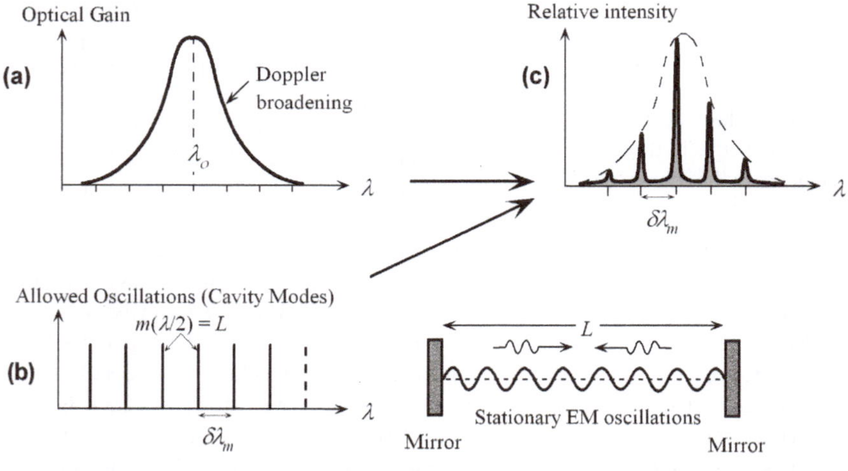

Laser output spectrum (a) Optical gain vs. wavelength characteristics (called the optical gain curve) of the lasing medium; λ_o is the source wavelength corresponding to υ_o. (b) Allowed modes and their wavelengths due to stationary EM waves within the optical cavity. (c) The output spectrum (relative intensity vs. wavelength) is determined by satisfying (a) and (b) simultaneously, assuming no cavity losses.

broad spectrum with peaks at certain wavelengths corresponding to various cavity modes existing within the Doppler broadened optical gain curve. At wavelengths satisfying the mode wavelength λ_m, that is representing certain cavity modes, we have spikes of intensity in the output. The net envelope of the output radiation is a Gaussian distribution which is essentially due to the Doppler broadened linewidth. Notice that there is a finite width to the individual intensity spikes within the spectrum which is primarily due to nonidealities of the optical cavity such as acoustic and thermal fluctuations of the cavity length L and nonideal end mirrors (less than 100% reflection). Typically, the frequency width of an individual spike in a He-Ne gas laser is ~1 MHz, though in highly stabilized gas lasers widths as low as ~1 kHz have been reported. It is important to realize that even if the laser medium has an optical gain, the optical cavity will always have some losses inasmuch as some radiation will be transmitted through the mirrors, and there will be various losses such as scattering within the cavity. Only those modes that have an optical gain that can make up for the radiation losses from the cavity can exist.

Lasher–Stern model, formulated in 1964, describes the spectra obtained when excess carriers in semiconductors undergo band-to-band spontaneous and stimulated recombination processes, and thereby emit radiation. The momentum matrix element is assumed to be the same for the initial and final states in the transitions. The peak of the stimulated emission spectrum falls at a lower photon energy than the peak for the spontaneous emission, except at $T = 0$ K. (G. Lasher and F. Stern, *Phys. Rev.* **133**, A553, 1964.)

Lasing conditions are the conditions for obtaining continuous wave lasing emissions from a laser device consisting of a pumped medium with an optical gain within an optical resonator structure that is within an optical cavity with end reflectors. The optical gain of the medium must just overcome the losses in the medium and also the losses from the end reflectors (i.e. radiation escaping from the cavity). The net round-trip gain must be unity and the phase change must be a multiple of 2π; otherwise the electromagnetic wave cannot replicate itself. The corresponding optical gain of the medium the threshold optical gain g_{th}. See *threshold gain.*

Lateral displacement of light, or **beam displacement**, occurs when a beam of light passes obliquely through a plate of transparent material, such as a glass plate. When a light beam is incident on a plate of transparent material of refractive index n, it emerges from the other side traveling parallel to the incident light but displaced from it by a distance d, called the *lateral displacement.* If the angle of incidence is θ_i, and the angle of refraction is θ_r, then

$$d = L\left[\frac{\sin(\theta_i - \theta_r)}{\cos\theta_r}\right]$$

where L is the thickness of the plate, and $n\sin\theta_r = n_0\sin\theta_i$, assuming that the plate is in an air surrounding. If the plate is in a medium of refractive index n_0, then the above equation in terms of the incidence angle and n/n_0 is given by

$$\frac{d}{L} = \sin\theta_i\left[1 - \frac{\cos\theta_i}{\sqrt{\left(\frac{n}{n_0}\right)^2 - \sin^2\theta_i}}\right]$$

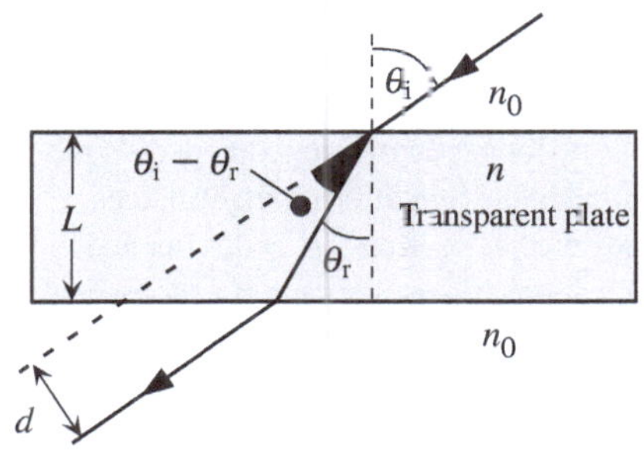

Lateral displacement of light passing obliquely through a transparent plate.

which is maximum with $\delta \to L$ when $\theta_i \to 90°$, glazing incidence.

Lateral laser diode mode is a distinct electric field pattern within the semiconductor laser cavity that is determined by the lateral dimension of the cavity. Each lateral mode has many longitudinal modes.

Lattice is a regular array of points in space with a periodicity. There are 14 distinct lattices in three-dimensional space. When an atom or molecule is placed at each lattice point the resulting regular structure is the crystal structure.

Lattice parameter mismatch describes the difference in the unit cell dimensions between layers in an epitaxial layer structure or between an epitaxial layer and the single-crystal substrate on which it is being prepared. It is assumed that all materials have the same crystallographic space group. Epitaxial growth requires that the lattice parameters for both substrate and epilayer must be closely matched. Otherwise, any difference between the lattice parameters will require the lattice planes on either side of the interface to bend to accommodate the differences. The amount of lattice plane bending and how much of it can be tolerated will depend on the elastic stiffnesses of both the substrate and the layer, as well as the actual layer thickness grown. Thus the difference in the lattice parameters results in a strain at the interface between the layer and substrate, commonly referred to as the *mismatch strain*. Beyond a critical mismatch strain, the planes at either side of the interface can no longer deform together or *coherently*, resulting in the formation of *misfit* dislocations. These defects are often highly detrimental in device applications, since they can act as nonradiative recombination centers or deep traps. The usual way to define the lattice mismatch, ε, is in terms of the average lattice parameter of layer and substrate a_{average},

$$a_{\text{average}} \equiv \tfrac{1}{2}(a_{\text{epilayer}} + a_{\text{substrate}}),$$

and

$$\varepsilon = \frac{a_{\text{epilayer}} + a_{\text{substrate}}}{a_{\text{averarge}}}.$$

As a guide, for $\varepsilon \leq 10^{-3}$, epitaxial growth can occur coherently, whereas for larger mismatch strains, there is an increased tendency to generate misfit dislocations, and nucleation and epilayer growth become increasingly difficult. The issue of lattice matching is further complicated by the fact that although materials may be lattice matched at one temperature (e.g. the growth temperature), they may be mismatched at another (e.g. on cooling, due to thermal expansion coefficient mismatch).

Lattice parameters are the lengths of the sides of the unit cell and the angles between the sides.

Lattice scattering limited mobility is the mobility of the electrons when their motion is limited by scattering from thermal vibrations of the lattice atoms. Generally lattice scattering limited drift mobility decreases with increasing temperature.

Law of the junction relates the injected minority carrier concentration just outside the depletion layer to the applied voltage. For holes in the n-side, that is for minority carriers, it is, $p_n(0) = p_{no}\exp(eV/k_BT)$ where $p_n(0)$ is the hole concentration just outside the depletion layer, p_{no} is the equilibrium hole concentration (both in the n-side), V is the external applied voltage, k_B is the Boltzmann constant and T is the temperature. The law of the junction states that the injected minority carrier concentration depends exponentially on the voltage across the pn junction device.

Laydown is the first stage in the outside vapor deposition process by which soot glass products (from the chemical burning of silicon tetrachloride and germanium tetrachloride vapors with a fuel) are deposited onto a target rod (a fused silica glass rod, or a ceramic rod such as alumina). This target rod acts as a mandrel and is rotated during the deposition. The required glass material for the preform with the right composition is grown on the outside surface of this target rod by depositing glass soot particles.

Leaky modes in an optical fiber are modes in which the field monotonically decays as the mode propagates along the fiber, transferring energy out of the core as the mode propagates. Modes will only remain within the core of an optical fiber if the propagation constant $\beta = 2\pi n/\lambda$ for core and cladding satisfy the condition, $\beta_{\text{core}} > \beta_{\text{cladding}}$, where n is the refractive index and λ the wavelength. When $\beta < \beta_{\text{cladding}}$, power leaks out of the core into the cladding, and the associated leaky mode(s) are generally lost in a few centimeters along the fiber.

Least time principle, see *Fermat's principle*.

LED, see *light-emitting diode*.

Left-handed material is an artificial medium in which, over a range of frequencies, the refractive index is negative; the result is that the medium has a number of unusual properties. If the relative dielectric permittivity ε_r and magnetic permeability μ_r are simultaneously negative in a definite frequency range, then the equivalent refractive index $n = [\varepsilon_r]^{1/2}[\mu_r]^{1/2}$ is also negative. All the usual laws of refraction are inverted, the refracted ray is transmitted without crossing the normal to the diopter, the sign of the Doppler effect is inverted, and so on. This nonintuitive behavior was theoretically demonstrated by the Soviet scientist Veselago in 1967–1968, but the first experimental realizations (initiated by John Pendy, David Smith and collaborators) in the microwave range are only a few years old and confirm the reality of negative refraction. Veselago called them "left-hand materials," because the wavevector is in the opposite direction to the $\mathbf{E} \times \mathbf{H}$ vector product and hence follows the left-hand rule rather than the right-hand rule for generating $\mathbf{k}$ from $\mathbf{E} \times \mathbf{H}$. (V. Veselago, *Soviet Physics USPEKHI*, **10**, 509, 1968; D. R. Smith *et al.*, *Phys. Rev. Lett.* **84**, 4184; D. R. Smith, J. B. Pendry, M. C. K. Wiltshire, *Science*, **305**, 788, 2004; J. Pendry, *Contemp. Phys.* **45**, 191, 2004.) See also *negative refractive index medium*.

Lens is a transparent optical component bound by two curved surfaces to refract light; most lenses are optical glasses that have been shaped to have the right surface curvatures to either converge or diverge a collimated beam of light. A converging lens brings the rays of a collimated beam to a "point," a focus, whereas a divergent lens diverges the collimated beam's rays in such a way that the rays appear as if originating from a single point. Lenses are classified into two types. *Convex, converging* or *positive* lenses have a thicker central region and they bring, or converge, incident parallel light rays into a focus that defines the focal point (F) of the convex lens. *Concave, diverging*, or *negative* lenses are thinner in the central region and when parallel light rays impinge on a concave lens, they diverge out in such a way that they appear as if they are spreading out from a point F, which is the focal point (F) of a concave lens. The distance of F from the lens center is the *focal length* f. In general, two focal points are defined. The rays that pass through the *object focal point* (F_o) are refracted to become parallel rays to the optical axis, whereas those rays that are parallel to the optical axis become refracted to pass through the *image focal point* (F_i).

Table: Lens Definitions of lenses and sign conventions for R_1, R_2 and f.

Biconvex	Planoconvex	Meniscus convex	Biconcave	Planoconcave	Meniscus concave
$R_1 > 0$	$R_1 = \infty$	$R_1 < 0$	$R_1 < 0$	$R_1 = \infty$	$R_1 > 0$
$R_2 < 0$	$R_2 < 0$	$R_2 < 0$	$R_2 > 0$	$R_2 > 0$	$R_2 > 0$
$f > 0$	$f > 0$	$f > 0$	$f < 0$	$f < 0$	$f < 0$

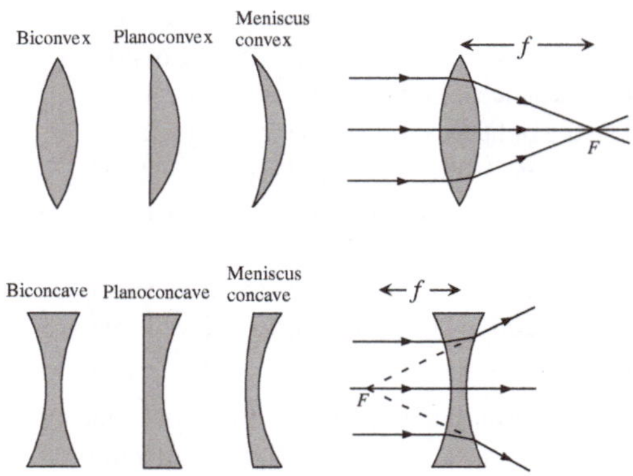

Lens Types of lenses, focal points (F) and focal lengths (f).

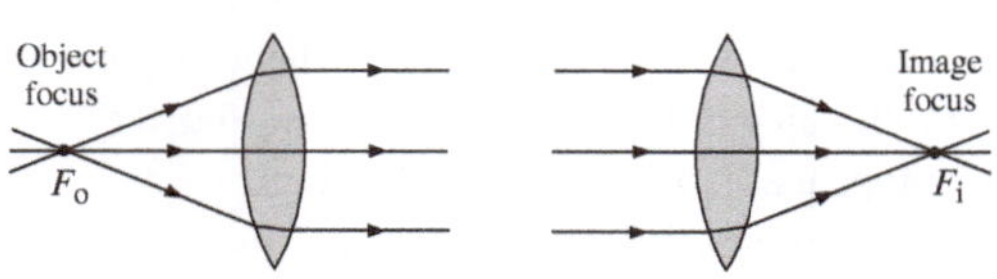

Lens Definitions of object and image focuses.

A *thin lens* is such that its thickness t is much less than the radii of curvature R_1 and R_2 of its surfaces. Many lenses (but not all) used in optical instruments are thin lenses. A thin lens can be characterized with two well-defined focal lengths, an object focal length f_o and an image focal length f_i, both measured from a single plane at the location of the lens. If s_o is the distance of the object from the lens and s_i is the distance of the image from the lens, then the *Gaussian lens equation* is

$$\frac{1}{s_o} + \frac{1}{s_i} = \frac{1}{f}$$

where the *focal length* f is given (or defined) by the *lens-maker's equation*

$$\frac{1}{f} = (n-1)\left(\frac{1}{R_1} - \frac{1}{R_2}\right).$$

Decarstes's lens equation is the above equation with $1/f$ replaced by $1/s_o + 1/s_i$. Some books use u or p for s_o, and v or q for s_i. When $s_o \to \infty$, $s_i \to f_i$ and $s_i \to \infty$, $s_o \to f_o$. When the lens thickness t is not totally negligible, the Gaussian equation can still be used provided that f is calculated with an additional correction term, i.e

$$\frac{1}{f} = (n-1)\left(\frac{1}{R_1} - \frac{1}{R_2}\right) + \left[\frac{(n-1)^2}{n}\left(\frac{t}{R_2 R_2}\right)\right].$$

The *power of a lens,* or *lens power*, is the reciprocal focal length, $P = 1/f$, and is measured in diopters ($D = m^{-1}$).

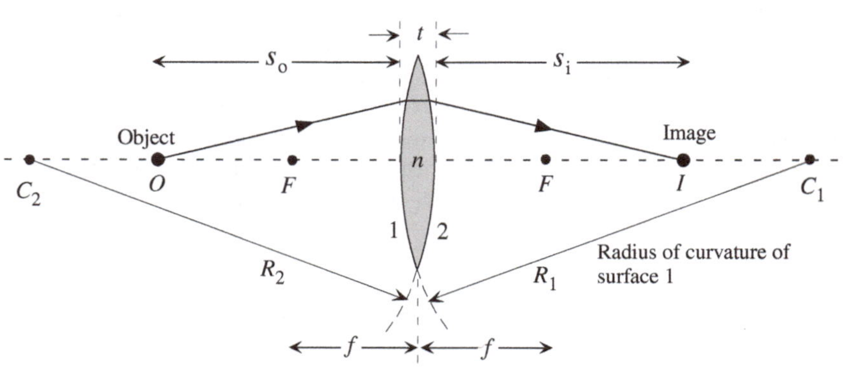

Lens A thin lens: definitions.

Table: Lens Usual sign convention in lenses. Sometimes the object and image distances are represented by s and s'; u and v or p and q respectively.

Quantity	Positive	Negative
s_o (s or u or p)	Real object	Virtual object
s_i (s' or v or q)	Real image	Virtual image
f	Converging lens	Diverging lens

When two thin lenses L_1 and L_2, with focal lengths f_1 and f_2, and separated by a distance a, are used in an aligned combination, then the combination behaves as if it has an effective focal length F given by

$$\frac{1}{F} = \frac{1}{f_1} + \frac{1}{f_2} - \frac{a}{f_1 f_2}.$$

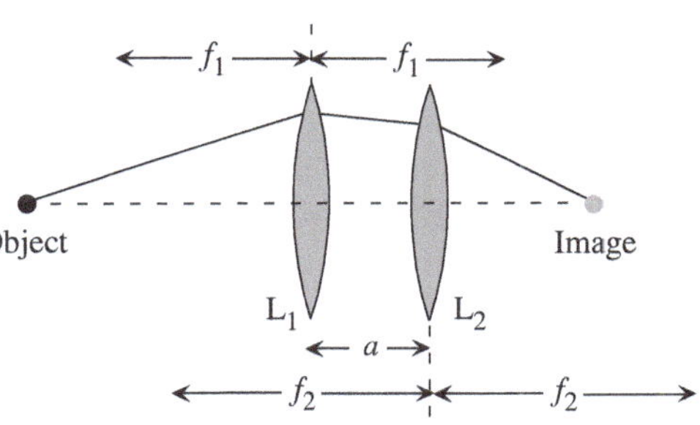

Lens Two lenses in an aligned combination have an effective focal length of F.

See *lens-maker's equation, cardinal points, Newton's lens equation, refraction at a spherical surface, thick lenses.*

Lens-maker's formula relates the algebraic positions of a point object A and its conjugated image A' through a thin lens of image focal length f (of negligible thickness, so that the two summits S_1 and S_2 of subsequent facets are merged into a single point O), in the paraxial approximation. With $s = \overline{OA}$ (object distance), $s' = \overline{OA}'$ (image distance), $R_1 = \overline{S_1 C_1}$, $R_2 = \overline{S_2 C_2}$ and n being the refractive index of the lens surrounded by air, the lens-maker's formula is

$$P = \frac{1}{f} = \frac{1}{s} + \frac{1}{s'} = (n-1)\left(\frac{1}{R_1} - \frac{1}{R_2}\right)$$

where $P = 1/f$ is the *power* of the lens (positive if convergent, negative otherwise), and f is the focal length. This equation is also known as Decrastes's thin lens equation. See *lens, thick lens.*

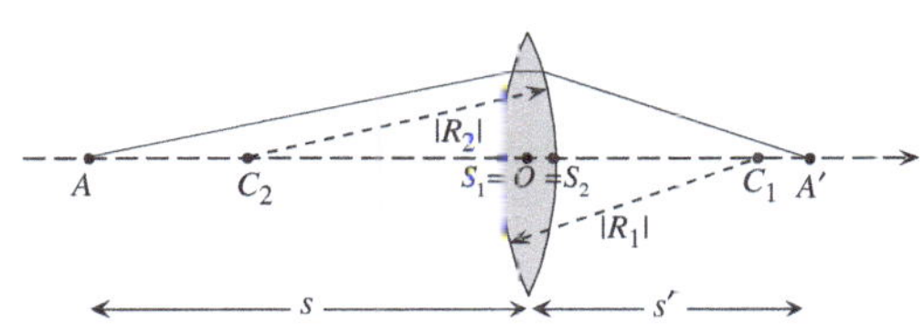

Definitions of symbols for the **lens-maker's formula**: C_1 (C_2) is the center, S_1 (S_2) is the summit of the first (second) facet. The lens is considered thin so that $S_1 = S_2 = \text{O}$. A and A' are conjugate points.

Table: Lens Images formed from a real object. (Adapted from E. Hecht, *Optics, Fourth Edition*, Addison Wesley, 2002.)

Object	Image			
		Convex lens		
Object location	*Type*	*Location*	*Orientation*	*Relative size*
$\infty > s_o > 2f$	Real	$f > s_i > 2f$	Inverted	Reduced
$s_o = 2f$	Real	$s_i = 2f$	Inverted	Same
$2f > s_o > f$	Real	$\infty > s_i > 2f$	Inverted	Magnified
$s_o < f$	Virtual	$\|s_i\| > s_o; s_i < 0$	Upright	Magnified
		Concave lens		
Anywhere	Virtual	$\|s_i\| < \|f\|$; $\|s_i\| < s_o$	Upright	Reduced

Levorotatory see *optical activity*.

Ligand see *crystal field, rare earth*.

Ligand field splitting parameter, $10Dq$, for an octahedral symmetry metal ion, gives the magnitude of the energy splitting between allowed d-orbital states. The splitting arises owing to the fact that as surrounding ligands approach the metal ion, the energy of the orbitals changes, with some being stabilized and others destabilized. In particular, the splitting occurs between e_g and t_{2g} irreducible representations, providing a ligand field splitting $\Delta = 10Dq$. This parameter is a function of the particular set of ligands, the particular metal ion and the metal–ligand separation. Theoretically, using a point charge model for the crystal field provides the following expression:

$$10Dq = \frac{5}{3}\frac{q\langle r^4\rangle}{\overline{R}^5}$$

where q is the effective charge of the ligands, $\overline{R}$ is the mean ion–ligand separation and $\langle r^4\rangle$ is the mean value of the fourth power of the radial separation of the d-electrons from the nucleus. Although this equation is widely applied, it is only valid when (a) one type of ligand is present, (b) the ligands can be assumed to act as point charges, and (c) the ligands octahedrally coordinate the central metal ion with perfect O_h symmetry. See *crystal field*.

Light-emitting diode (LED) is a semiconductor diode which emits incoherent radiation when forward biased. LEDs operate on the principle of spontaneous emission resulting from electron–hole pair (EHP) injection and *direct recombination* under forward bias. The emitted photon energy $h\upsilon$ is approximately equal to the bandgap energy E_g. The figure shows the energy band diagram of an unbiased pn^+ junction device in which the n-side is more heavily doped than the p-side. (The superscript + on n or p indicates heavy doping.) The Fermi level E_F is uniform through the device which is a requirement of equilibrium with no applied bias. The depletion region extends mainly into the p-side. There is a potential energy (PE) barrier eV_0 from E_c on the n-side to E_c on the p-side where V_0 is the

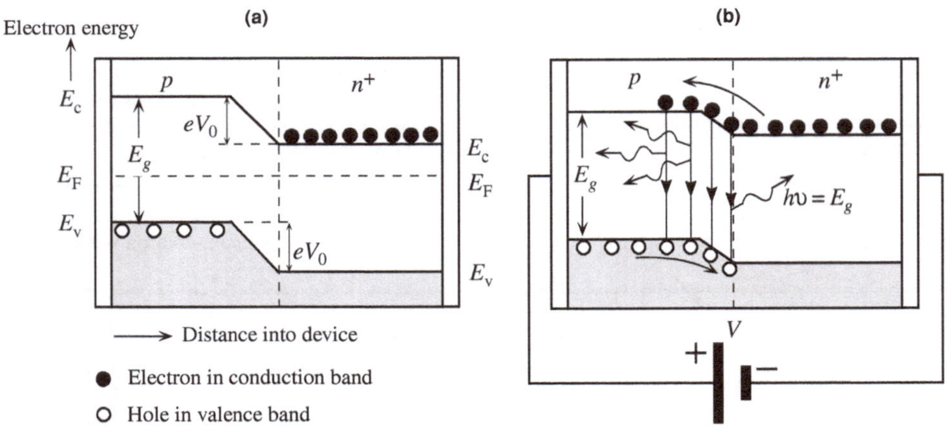

Light-emitting diode (a) The energy band diagram of a p-n^+ (heavily n-type doped) junction without any bias. The built-in potential V_0 prevents electrons from diffusing from n^+ to p side. (b) The applied bias reduces V_0 and thereby allows electrons to diffuse, be injected, into the p-side. Recombination around the junction and within the diffusion length of the electrons in the p-side leads to photon emission.

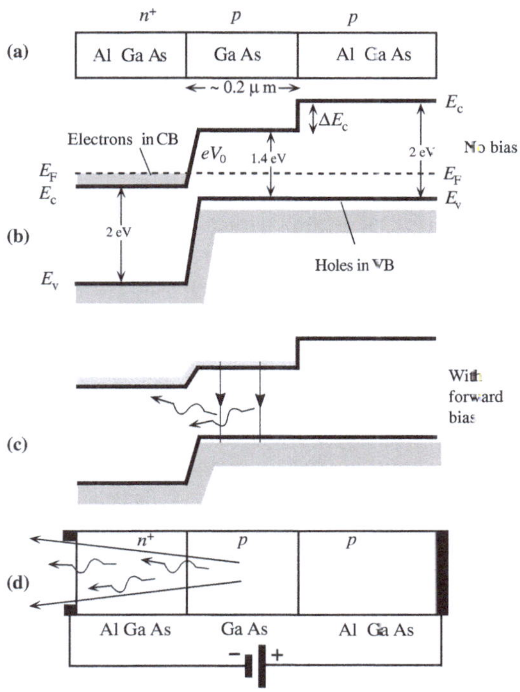

Light-emitting diode A double heterostructure LED. (a) A double heterostructure diode has two junctions that are between two different bandgap semiconductors (GaAs and AlGaAs). (b) A simplified energy band diagram with exaggerated features. E_F must be uniform. (c) Forward biased simplified energy band diagram. (d) Forward biased LED. Schematic illustration of photons escaping reabsorption in the AlGaAs layer and being emitted from the device.

built-in voltage. The PE barrier eV_0 prevents the diffusion of electrons from the n-side to p-side. When a forward bias V is applied, the built-in potential V_o is reduced to $V_0 - V$ which then allows the electrons from the n^+ side to diffuse, that is, become injected, into the p-side. The hole injection component from p into the n^+ side is much smaller than the electron injection component from the n^+ to p-side. The recombination of injected electrons in the depletion region and within a volume extending over the electron diffusion length in the p-side leads to photon emission. The phenomenon of light emission from EHP recombination as a result of minority carrier injection is called *injection electroluminescence*. Due to the statistical nature of the recombination process between electrons and holes, the emitted photons are in random directions; they result from spontaneous emission processes. The LED structure has to be such that the emitted photons can escape the device without being reabsorbed by the semiconductor material. This means the p-side has to be sufficiently narrow or we have to use *heterostructure* devices.

LED constructions for increasing the intensity of the output light make use of the double heterostructure. The figure shows a double-heterostructure (DH) device based on two junctions between different semiconductor materials with different bandgaps. In this case the semiconductors are AlGaAs with $E_g \approx 2$ eV and GaAs with $E_g \approx 1.4$ eV. The double heterostructure has an n^+p heterojunction between n^+-AlGaAs and p-GaAs. There is another heterojunction between p-GaAs and p-AlGaAs. The p-GaAs region is a thin layer, typically a fraction of a micron and it is lightly doped. In the absence of an applied voltage, the Fermi level E_F is continuous through the whole structure. There is a potential energy barrier eV_0 for electrons in the CB of n^+-AlGaAs against diffusion into p-GaAs. There is a bandgap change at the junction between p-GaAs and p-AlGaAs which results in a step change ΔE_c in E_c between the two conduction bands of p-GaAs and p-AlGaAs. This ΔE_c is effectively a *potential energy barrier* that prevents any electrons in the CB in p-GaAs passing to the CB of p-AlGaAs. (There is also a step change ΔE_v in E_v but this is small and is not shown.) When a forward bias is applied, most of this voltage drops between the n^+-AlGaAs and p-GaAs and reduces the potential energy barrier eV_0, just as in a normal pn junction. This allows electrons in the CB of n^+-AlGaAs to be injected into the CB of p-GaAs.

These electrons, however, are *confined* to the CB of *p*-GaAs since there is a barrier ΔE_c between *p*-GaAs and *p*-AlGaAs. The wide bandgap *p*-AlGaAs layers therefore act as confining layers that restrict injected electrons to the *p*-GaAs layer. The recombination of injected electrons and the holes already present in this *p*-GaAs layer results in spontaneous photon emission. The *p*-GaAs layer is called the *active layer* inasmuch as this is where light is generated. Since the bandgap E_g of AlGaAs is greater than GaAs, the emitted photons do not get reabsorbed as they escape the active region and can reach the surface of the device. Since light is also not absorbed in *p*-AlGaAs, it can be reflected to increase the light output. The external efficiency η_{external} of an LED quantifies the efficiency of conversion of electrical energy into an emitted external optical energy. It incorporates the internal efficiency of the radiative recombination process and the subsequent efficiency of photon extraction from the device. (N. Holonyak, *Am. J. Phys.* **69**, 864, 2000.)

Light-emitting diode (LED) characteristics typically refer to the emitted light spectral characteristics in terms of spectral intensity vs. wavelength, output light intensity vs. diode current, current–voltage characteristics and the angular variation of the emitted light intensity. There may be other additional associated characteristics as well such as changes in the light output characteristics with temperature, etc. The width of the output light spectrum depends not only on the temperature but also on the LED material, whether a direct or indirect bandgap semiconductor is used. Typically the output spectral width corresponds to a few $k_B T$ in the energy distribution of the emitted photons. Typical current–voltage characteristics exhibit a *turn-on* or the *cut-in voltage* from which point the current increases sharply with voltage. The turn-on voltage depends on the semiconductor and generally increases with the energy bandgap E_g. For example, typically, for a blue LED it is about 3.5–4.5 V, for a yellow LED, it is about 2 V, and for a GaAs infrared LED it is around 1 V. The LED's *luminous intensity* I_v is the luminous flux emitted by an LED per unit solid angle in a particular direction; I_v is a function of the solid angle Ω and hence depends on θ and ϕ where θ and ϕ are the spherical coordinates centered at the LED's chip. The luminous flux Φ_v can be found by integrating $I_v(\Omega)$ with respect to the solid angle Ω, that is with respect to θ and ϕ, over the hemisphere above the LED:

$$\Phi_v = \int_0^{2\pi} d\phi \int_0^{\pi/2} I_v(\theta) \sin\theta d\theta$$

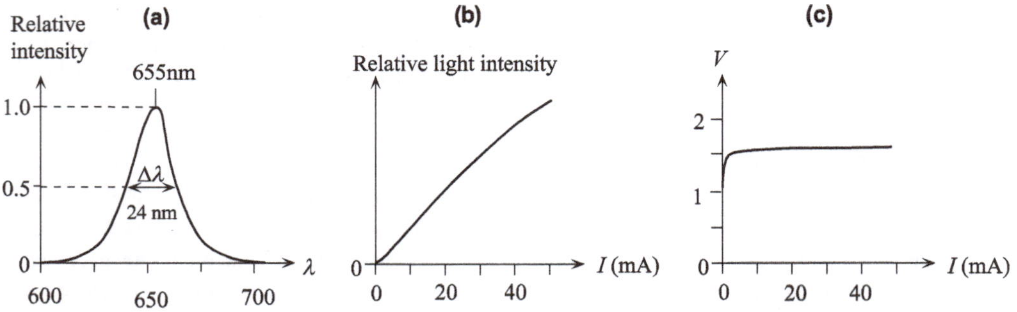

LED characteristics (a) A typical output spectrum (relative intensity vs. wavelength) from a red GaAsP LED. (b) Typical output light power vs. forward current. (c) Typical *I–V* characteristics of a red LED. The turn-on voltage is around 1.5V.

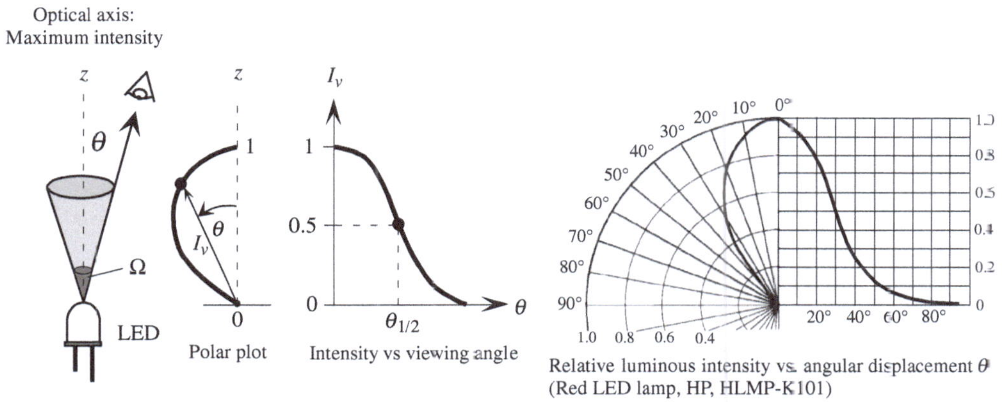

LED characteristics Angular dependence of emitted luminous intensity: I_v vs. angle θ from the optical axis z (maximum intensity axis).

If I_{v0} is the maximum intensity along the LED's optical axis, defined as the maximum intensity axis, and $\theta_{1/2}$ is the "half-angle" where the intensity is half I_{v0}, a first-order estimate for Φ_v would be

$$\Phi_v \approx I_{v0}\pi(\theta_{1/2})^2$$

where $\theta_{1/2}$ is in radians, and the emission is symmetric about the optical axis. $I_v(\theta)$ vs. θ for an LED can be approximated by $I_v(\theta) \approx I_v(0)(\cos\theta)^n$ where n is an index that depends on $\theta_{1/2}$ through $n = \ln(0.5)/\ln[\cos(\theta_{1/2})]$.

Light-emitting diode (LED) materials are typically various direct bandgap semiconductor materials that can be readily doped to make commercial *pn* junction LEDs which emit radiation in the visible range of wavelengths.

An important class of commercial semiconductor materials which cover the visible spectrum is the *III–V ternary alloys* based on alloying GaAs and GaP which are denoted as $GaAs_{1-y}P_y$. In this compound, As and P atoms from group V are distributed randomly at normal As sites in the GaAs crystal structure. When $y < 0.45$, the alloy $GaAs_{1-y}P_y$ is a direct bandgap semiconductor and hence the electron–hole pair (EHP) recombination process is direct. The rate of recombination is directly proportional to the product of electron and hole concentrations. The emitted wavelengths range from about 630 nm, red, for $y = 0.45$ ($GaAs_{0.55}P_{0.45}$) to 870 nm for $y = 0$, GaAs.

$GaAs_{1-y}P_y$ alloys (which includes GaP) with $y > 0.45$ are indirect bandgap semiconductors. The EHP recombination processes occur through recombination centers and involve lattice vibrations rather than photon emission. However, if we add *isoelectronic impurities* such as nitrogen (in the same group V as P) into the semiconductor crystal then some of these N atoms substitute for P atoms. Since N and P have the same valency, N atoms substituting for P atoms form the same number of bonds and do not act as donors or acceptors. The electronic cores of N and P are, however, different. The positive nucleus of N is less shielded by electrons compared with that of the P atom. This means that a conduction electron in the neighborhood of a N atom will be attracted and may become trapped at this site. N atoms therefore introduce localized energy levels, or electron traps, E_N, near the conduction band edge. When a conduction electron is captured

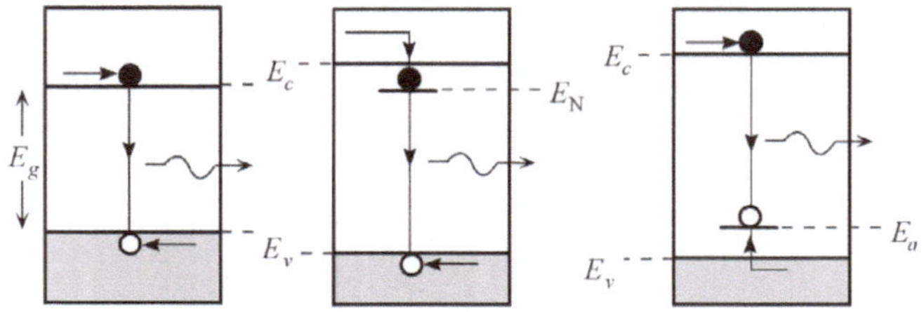

LED materials (a) Photon emission in a direct bandgap semiconductor. (b) GaP is an indirect bandgap semiconductor. When doped with nitrogen there is an electron trap at E_N. Direct recombination between a trapped electron at E_N and a hole emits a photon. (c) In Al doped SiC, EHP recombination is through an acceptor level like E_a.

at E_N, it can attract a hole (in the valence band) in its vicinity by Coulombic attraction and eventually recombine with it directly and emit a photon. The emitted photon energy is only slightly less than E_g as E_N is typically close to E_c. As the recombination process depends on N doping, it is not as efficient as direct recombination. Thus, the efficiency of LEDs from N doped indirect bandgap $GaAs_{1-y}P_y$ semiconductors is less than those from direct bandgap semiconductors. Nitrogen doped indirect bandgap $GaAs_{1-y}P_y$ alloys are widely used in inexpensive green, yellow and orange LEDs. There are two types of blue LED materials. GaN is a direct bandgap semiconductor with an E_g of 3.4 eV. The blue GaN LEDs actually use the GaN alloy InGaN which has a bandgap of about 2.7 eV which corresponds to blue emission. The less efficient type is the Al doped *silicon carbide* (SiC), which is an indirect bandgap semiconductor. The acceptor type localized energy level captures a hole from the valence band and a conduction electron then recombines with this hole to emit a photon. As the recombination process is not direct and therefore not as efficient, the brightness of blue SiC LEDs is limited. Recently there has been considerable progress made towards more efficient blue LEDs using direct bandgap compound semiconductors such as II–VI semiconductors, for example ZnSe (Zn and Se are in groups II and VI in the periodic table). The main problem in using II–VI compounds is the current technological difficulty in appropriately doping these semiconductors to fabricate efficient *pn* junctions.

There are various commercially important direct bandgap semiconductor materials that emit in the red and infrared wavelengths which are typically ternary (containing three elements) and quarternary (four elements) alloys based on III and V elements, so called III–V alloys. For example, GaAs with a bandgap of about 1.43 eV emits radiation at around 870 nm in the infrared. But ternary alloys based on $Al_{1-x}Ga_xAs$ where $x < 0.43$ are direct bandgap semiconductors. The composition can be varied to adjust the bandgap and hence the emitted radiation from about 640 to 870 nm, from deep red light to infrared. In-Ga-Al-P is a quarternary III–V alloy (In, Ga, Al from III and P from V) that has a direct bandgap variation with composition over the visible range. It can be lattice matched to GaAs substrates when in the composition range $In_{0.49}Al_{0.17}Ga_{0.34}P$ to $In_{0.49}Al_{0.058}Ga_{0.452}P$. Recent high-intensity LEDs have been based on this material which is likely to eventually dominate the high-intensity visible LED range. The bandgap of quarternary alloys $In_{1-x}Ga_xAs_{1-y}P_y$ can be varied with

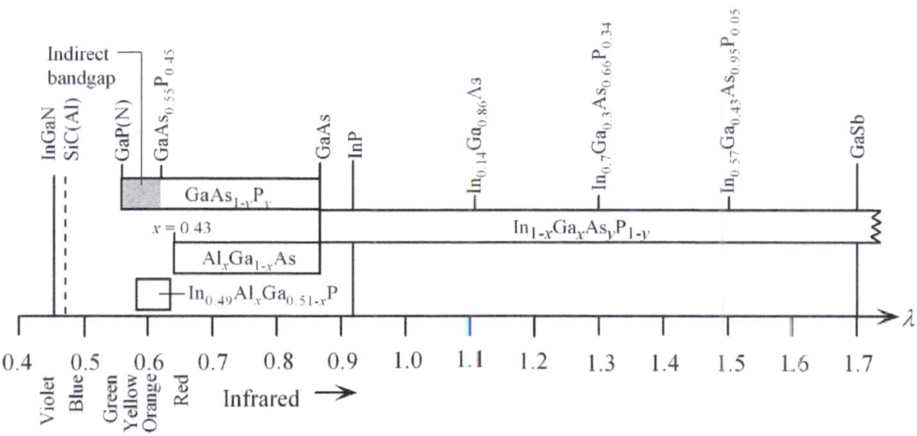

LED materials Free-space wavelength coverage by different LED materials from the visible spectrum to the infrared including wavelengths used in optical communications. Hatched region and dashed lines are indirect E_g materials.

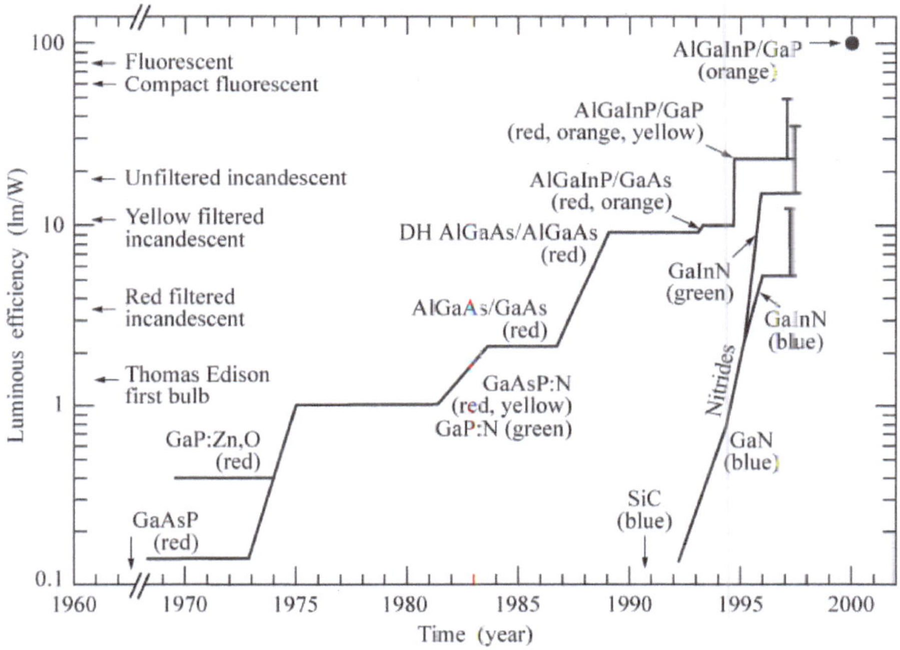

LED materials Luminous efficiency of different visible wavelength light emitting diodes. (From E.F. Schubert, *Light Emitting Diodes*, Cambridge University Press, 2003.)

Table: LED materials Selected LED semiconductor materials, wavelengths of emission, and typical external quantum efficiencies in commercial LEDs. Optical communication channels are at 850 nm (local network) and at 1.3 and 1.55 μm (long distance). D: direct, I: Indirect bandgap. $\eta_{external}$ is typical and may vary substantially depending on the device structure. DH: double heterostructure, HJ: homojunction

Semiconductor active layer	Structure	D or I	λ (nm)	$\eta_{external}$ (%)	Comment
GaAs	DH	D	870–900	10	Infrared (IR)
$Al_xGa_{1-x}As$ $(0 < x < 0.4)$	DH	D	640–870	3–20	Red to IR
$In_{1-x}Ga_xAs_yP_{1-y}$ $(y \approx 2.20x, 0 < x < 0.47)$	DH	D	1–1.6 μm	> 10	LEDs in communications
$In_{0.49}Al_xGa_{0.51-x}P$	DH	D	590–630	> 10	Amber, green, red. High luminous intensity.
InGaN/GaN Quantum Well	QW	D	450–530	5–20	Blue – green
$GaAs_{1-y}P_y$ $(y < 0.45)$	HJ	D	630–870	< 1	Red – IR
$GaAs_{1-y}P_y$ $(y > 0.45)$ (N or Zn, O doping)	HJ	I	560–700	< 1	Red, orange, yellow
SiC	HJ	I	460–470	0.02	Blue. Low efficiency
GaP (Zn-O)	HJ	I	700	2–3	Red
GaP (N)	HJ	I	565	< 1	Green

Table: LED materials Wavelength ranges and colors as usually specified for LEDs. (Color perception is a difficult topic and the table only indicates typical values as quoted by various manufacturers of LEDs and as perceived by an average person.)

Color	Blue	Emerald green	Green	Yellow	Amber
λ (nm)	λ < 500	530–564	565–579	580–587	588–594
Color	Orange	Red-Orange	Red	Deep red	Infrared
λ (nm)	595–606	607–615	616–632	633–700	λ > 700

composition (x and y) to span wavelengths from 870 nm (GaAs) to 3.5 μm (InAs) which includes the optical communication wavelengths of 1.3 and 1.55 μm.

The external efficiency $\eta_{external}$ measures the optical power emitted P_{out} by the device per unit electrical power input into the device: $\eta_{external} = P_{out}/(IV)$ where IV = LED current × LED voltage. For indirect bandgap semiconductors $\eta_{external}$ is generally less than 1% whereas for direct bandgap semiconductors with the right device structure, $\eta_{external}$ can be substantial (> 10%).

Light-emitting diodes (LEDs) for optical fiber communications are used as emitters in short-haul applications, e.g. local networks because they are simpler to drive, more economic, have a longer lifetime and provide the necessary output power even though their output spectrum is much wider than that of a laser diode. LEDs are frequently

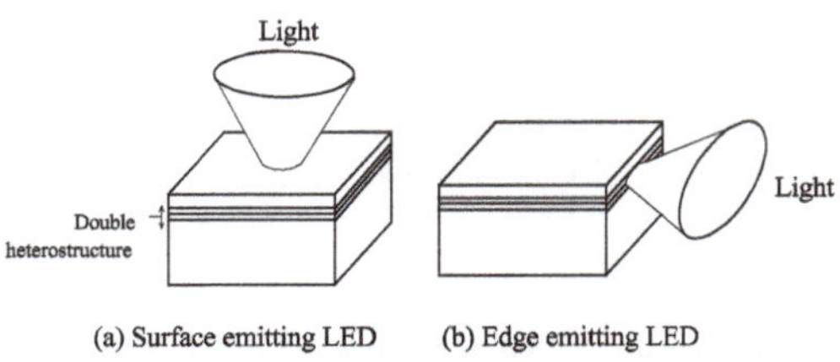

Light-emitting diodes for optical communications.

used with graded index fibers inasmuch as typically dispersion in a graded index fiber is primarily due to *modal dispersion* rather than *intramodal dispersion*. For long-haul and wide-bandwidth communications invariably laser diodes are used because of their narrow linewidth, high output power and higher signal bandwidth capability. There are essentially two types of LED devices. If the emitted radiation emerges from an area in the plane of the recombination layer then the device is a *surface emitting* LED (SLED). If the emitted radiation emerges from an area on an edge of the crystal, i.e. from an area on a crystal face perpendicular to the active layer, then the LED is an *edge-emitting* LED (ELED).

Linear attenuation coefficient see *mass attenuation coefficient.*

Linear electrooptic effect see *electrooptic effects.*

Linear medium in photonics implies a medium in which the relative permittivity ε_r is independent of the electric field, and the refractive index n is independent of the light intensity. Put differently, the induced polarization **P** is linearly proportional to the field **E**. The latter may not hold when the medium is subjected to an intense laser beam in which the field can be very large.

Linear stability analysis see *reduced rate (Statz–De Mars) equations.*

Linear Stark effect see *Stark effect.*

Linearly polarized light has its electric field oscillations contained within a well-defined plane. See *polarization.*

Linearly polarized (LP) modes are allowed nearly plane polarized, transverse electric and magnetic (TEM) propagating modes in a weakly guiding step optical fiber, that is a fiber with a small refractive index difference ($\Delta \ll 1$). The field magnitudes in an LP mode, however, are not constant in a plane. A guided LP mode along the fiber can be represented by the propagation of an electric field distribution $E(r,\varphi)$ along z where r, φ and z represent cylindrical coordinates in which z is along the fiber axis. This field distribution, or pattern, is in the plane normal to the fiber axis and hence depends on

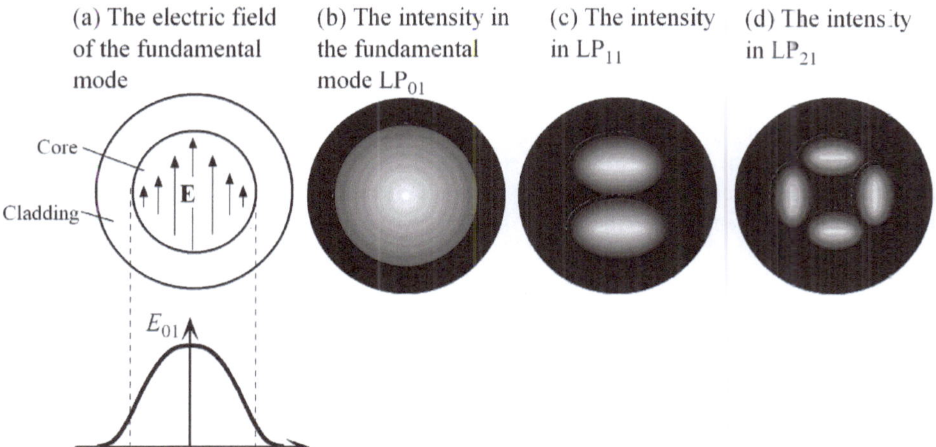

Linear polarized modes The electric field distribution of the fundamental mode in the transverse plane to the fiber z-axis. The light intensity is greatest at the center of the fiber. Intensity patterns in LP_{01}, LP_{11} and LP_{21} modes.

r and φ but not on z. A guided LP mode along the fiber can be represented by the propagation of an electric field distribution $E(r,\varphi)$ along z. This field distribution, or pattern, is in the plane normal to the fiber axis and hence depends on r and φ but not on z. Further, because of the presence of two boundaries it is characterized by two integers, l and m. The propagating field distribution in an LP mode is therefore given by $E_{lm}(r,\varphi)$ and we represent the mode as LP_{lm}. Thus an LP_{lm} mode can be described by a traveling wave along z of the form,

$$E_{LP} = E_{lm}(r, \varphi) \exp j(\omega t - \beta_{lm} z)$$

where E_{LP} is the field of the LP mode and β_{lm} is its *propagation constant* (wavenumber) along z. It is apparent that for a given l and m, $E_{lm}(r, \phi)$ represents a particular field pattern at a position z that is propagated along the fiber with an effective wavevector β_{lm}.

The electric field pattern (E_{01}) in the fundamental mode, which corresponds to $l = 0$ and $m = 1$, the LP_{01} mode, has maximum intensity at the center of the core (or fiber axis) and penetrates somewhat into the cladding due to the accompanying evanescent wave. The extent of penetration depends on the V-number of the fiber (and hence on the wavelength). The light intensity in a mode is proportional to E^2 which means that the intensity distribution in the LP_{01} mode has a maximum along the fiber axis. The integers l and m are related to the intensity pattern in an LP_{lm} mode. There are m number of maxima along r starting from the core center and $2l$ number of maxima around a circle around the fiber axis. l is called the *azimuthal mode number*.

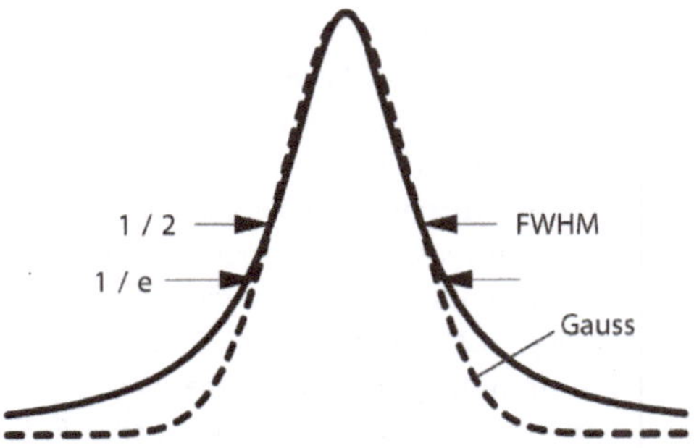

Comparison between Gaussian (dashed curve) and Lorentzian (solid) **lineshapes** with the same peak value and the same full width at half maximum ($\Delta\nu = 2\gamma = 2\Gamma[\ln(2)]^{1/2}$). The Gaussian decreases much more quickly, whereas the far-reaching "tails" characterize the Lorentzian.

Lineshape, $F(\nu)$ or $g(\nu)$, is a probability density function that describes the spectral behavior of atomic resonance, optical gain, emission or absorption curves. $g(\nu)$ is usually "normalized" in order to ensure that $\int g(\nu)d\nu = 1$. For example, for emission, $g(\nu)d\nu$ is the relative probability that the emission occurs in the range ν to $\nu + d\nu$; and if $I(\nu)$ is the intensity per unit frequency in the range ν to $\nu + d\nu$ then $g(\nu)d\nu = I(\nu) / \int I(\nu)d\nu$. Two normalized distributions are frequently met, the Lorentzian $g(\nu) = L(\nu)$ and the Gaussian $g(\nu) = G(\nu)$, both of which are summarized in the table. If $g(\nu_o)$ is the

Table: Lineshape $g(\nu)$ for homogeneous and inhomogeneous broadening of radiation in photon emission. $\Delta\nu$ = FWHM, full width half maximum, γ = HWHM, half width at half maximum, Γ the half-width at (1/e) of the Gaussian.

Lorentzian homogeneous broadening	Doppler inhomogeneous broadening
$L(\nu) = \dfrac{\Delta\nu}{2\pi[(\nu - \nu_o)^2 + (\Delta\nu/2)^2]}$	$G(\nu) = \dfrac{2}{\Delta\nu}\left(\dfrac{\ln 2}{\pi}\right)^{1/2} \exp\{-[2(\nu - \nu_o)/\Delta\nu]^2 \ln 2\}$
$L(\nu) = \dfrac{\gamma/\pi}{(\nu - \nu_o)^2 + \gamma^2}$	$G(\nu) = \dfrac{1}{\pi^{1/2}\Gamma} \exp[-(\nu - \nu_o)^2/\Gamma^2]$

maximum at $\nu = \nu_0$, and $\Delta\nu$ is FWHM, full width half maximum, the frequency width between the points where $g(\nu)$ has fallen to the half value of the peak, then very roughly we can take $\int g(\nu)d\nu \approx g(\nu_0)\Delta\nu$ so that $g(\nu_0) \approx 1/\Delta\nu$.

Liquid crystals (LCs) are materials that have a rod-like molecular structure, with molecules or *mesogens* having strong dipoles or easily polarizable substituents. LCs have properties that lie between a liquid phase and crystalline solid phase; e.g. they can flow like a liquid but at the same time possess crystalline domains that lead to anisotropic optical properties. A distinguishing characteristic of the liquid crystal state is the tendency of the mesogens to point along a common axis called the *director* (a preferred common axis in the liquid crystal) imparting an orientationally ordered state. This behavior is distinct from that of molecules in a normal liquid phase, where there is no intrinsic order. The orientational order in the liquid crystal state lies between that of a normal crystalline solid (i.e. fully ordered) and that of a normal liquid (i.e. nearly fully disordered), and hence is given the name *mesogenic state*. As a result of thermally induced motion of the mesogens, the number of mesogens pointing along the director, or the degree of anisotropy, will depend on temperature. The degree of alignment will be a maximum at low temperature and decreases with increasing temperature, until at some critical temperature thermal agitation will result in a state of no net ordering. There are a number of possible liquid crystal phases, or ordered arrangements including *nematic*, *smectic*, *cholesteric* and *columnar* phases. *Nematic phases* are characterized by mesogens that have no positional order, but tend to point along the same direction (i.e. along the director). Physical properties of these materials tend to depend sensitively on the degree of alignment, and can be highly *anisotropic* for well-aligned materials. A distinct advantage is that the molecular orientation and consequently the optical properties can be controlled by an applied field. The molecules have rod-like shapes with lengths typically in the 20–30 nm range. *Smectic phases* have mesogens that exhibit a degree of translational ordering, while maintaining the general orientational ordering of nematics – the mesogens tend to align themselves in layers. In *cholesteric phases*, the liquid crystal is typically composed of nematic mesogens that are aligned within a given plane, but with individual planes rotated in their

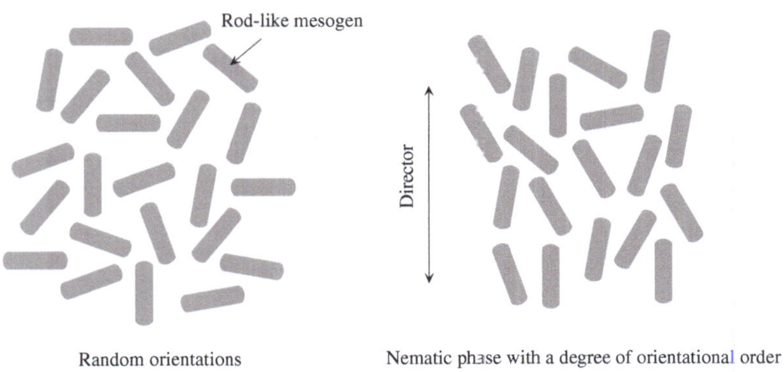

Liquid crystal Schematic illustration of orientational disorder in a liquid with rod-like mesogens. Left, no order, and rods are randomly oriented. Right, there is a tendency for the rods to align with the *director*, the vertical axis, in this example.

alignment to each other, leading to a *chiral* structure. Traversing an appropriate distance along the chiral structure (i.e. the pitch), the ordering recurs. This is the length along the helix required for the director to rotate one full turn. Columnar liquid crystal phases are distinct from the previous phases owing to the morphology of the mesogens being disk shaped as opposed to rod shaped for the above phases. The columnar phase is characterized by stacked columns of the mesogens packed, with individual columns closely packed into arrays. An important parameter that characterizes LCs is the *liquid crystal ordering parameter* S defined by

$$S = \langle P_2(\cos\theta)\rangle = \left\langle \frac{3\cos^2\theta - 1}{2} \right\rangle$$

in which P_2 is the second-order Legendre polynomial whose argument is $\cos\theta$ with the averaging done over all molecules, and θ is the angle between the axis of mesogen molecule and the local director. For a completely random arrangement obviously $S = 0$ while for a perfectly ordered arrangement of molecules ($\theta = 0$) $S = 1$. For many LCs S is around 0.3–0.8. S decreases with temperature and obviously can be controlled by the application of a field. (T. J. Slucklin, *Contemp. Phys.*, **41**, 37 2000.)

Liquid crystal display (LCD) is a display that is based on utilizing a liquid crystal medium whose optical properties can be controlled by an applied field. It can also be considered as a LC *light*

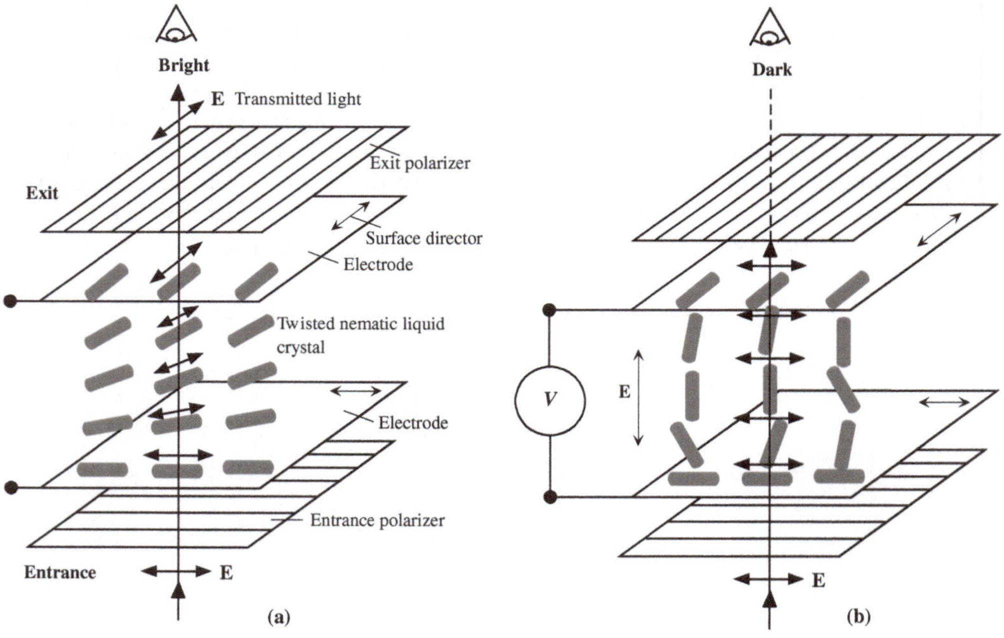

Liquid crystal display Transmission-based LCD. (a) In the absence of a field, the liquid crystal has the twisted nematic phase and the light passing through it has its polarization rotated by 90°. The light is transmitted through both polarizers. (b) When a voltage is applied, the molecules in the liquid crystal align with the field and are unable to rotate the polarization of the light passing through it; light therefore cannot pass through the exit polarizer.

modulator or a *light valve.* It is a display comprised of a thin film of *liquid crystal* (e.g. a few microns thick) interposed between two semitransparent electrically conducting electrodes to form a cell. Most LCDs are based on the *twisted nematic effect* discovered by Schadt and Helfrich (Switzerland) in 1970–1971 and Fergason (USA) in 1971. In a twisted nematic liquid crystal cell, the two electrodes have surfaces that have been treated (have an orientational layer) to acts as "directors" for the molecules and the "directors" are at right angles to each other. Molecules next to the surfaces are forced to align along these surface directors and hence the molecules homogeneously twist through 90° from one electrode to the other. The *twisted nematic medium* has molecules arranged in a helical structure, and is able to "twist," or rotate the field in light that passes through it. Two polarizers lie at the entrance and the exit ends of the cell. Polarized light thus enters the cell and has its polarization rotated by 90° as it arrives at the exit polarizer; since this light has its polarization aligned with the optical axis of the exit polarizer it passes through. Thus, without an applied field, the light is transmitted; the display is bright. If an electric field is applied by applying an ac voltage (usually a few volts) to the two electrodes comprising the faces of the cell, the molecular alignment of the nematic liquid crystals is disturbed by the field. The field acts as an imposed director and the molecules align with the field, which results in the twisted molecular arrangement being destroyed. The "helix becomes unwound." The polarization in the light is unaffected and therefore it cannot pass through the exit polarizer. The LCD cell appears dark. Under an appropriate applied field, light transmission is completely extinguished. If a mirror is placed behind the second polarizer, the display can be operated under reflection instead of transmission. A reverse switching behavior can be achieved by using parallel polarizers. Gray-scale modulation can be achieved by varying the voltage between the threshold for reorientation and the saturation field for unwinding the twisted nematic structure. By patterning the electrodes (indium-tin-oxide) by lithographic techniques into appropriate shapes various display configurations can be implemented. LCDs are widely used in various display applications.

Liquid phase epitaxy (LPE) is an epitaxial deposition technique for preparing crystalline layers of material, suitable for subsequent processing to form electronic devices. To produce an epitaxial layer by LPE, a substrate is chosen that is closely lattice-matched to the layer material that is to be grown. The substrate is immersed in a solution containing the required solute to form the epitaxial layer. Since growth occurs at a relatively low temperature compared with the melting point of the substrate and epitaxial layer, the equilibrium amount of solute that can be dissolved in the solution to saturate it, is relatively low. Reducing the temperature of the solution results in an excess of solute at the lower temperature (due to the reduction in solubility of solvent with decreasing temperature), known as a *supersaturation*. The supersaturation can be relieved by precipitation of the excess solute onto a substrate, leading to epitaxial growth. Doping of epilayers can easily be achieved by dissolving selected species into the solution being used for growth. These impurities are then incorporated into the growing epitaxial layer, doping it A series of epitaxial layers can be grown sequentially on top of one another in a given process, by using a so-called slider system. In this case, a series of solutions with different dopants in each, or different alloy compositions (to prepare alloy layers such as, for example, AlGaAs), can be placed over the substrate or another epitaxial layer, and growth from that solution initiated. Thus complex multilayer structures (both *homoepitaxial* and *heteroepitaxial*) can be prepared. Since the growth process operates at relatively low temperatures, the solubility of extraneous impurities is low and high purity materials can be prepared. Moreover, for many device applications, control over layer thickness is important. Since at the low growth temperatures, the growth rate (controlled by diffusion of species to the growing interface) is slow, it is easy to control layer thickness to the level of fractions of a micron. The technique has been successfully used to prepare high-power laser diodes, solar cells and photodiodes. However, for many applications, difficulties with scaling up this technique to simultaneously deposit a plurality of epitaxial layers, has led to a decline in the use of this approach as compared with vapor phase based epitaxial growth techniques.

Lithium niobate ($LiNbO_3$) is a birefringent crystal that has a wide range of uses in electrooptic and acoustooptic devices. Its ordinary and extraordinary refractive indices can be approximated by

$$n_0^2 = 4.9130 + \frac{0.1173 + 1.65 \times 10^{-8}T^2}{\lambda^2 - (0.212 + 2.7 \times 10^{-8}T^2)^2} - 0.0278\lambda^2$$

$$n_e^2 = 4.5567 + 2.605 \times 10^{-7}T^2 + \frac{0.0970 + 2.70 \times 10^{-8}T^2}{\lambda^2 - (0.201 + 5.4 \times 10^{-8}T^2)^2} - 0.0224\lambda^2.$$

(E. L. Wooten, K. M. Kissa, Yi-Yan, E. J. Murphy, D. A. Lafaw, P. F. Hallemeier, D. Maack, D. V. Attanasio, D. J. Fritz, G. J. McBrien, D. E. Bossi, *IEEE J. Selected Topics Quantum Electronics*, **6**, 69–82, 2000.)

Lithium niobate shows excellent low optical absorption or transparency over the ultraviolet to the first atmospheric window in the infrared, making it ideal for developing planar optical circuits for telecommunications applications, for example. Owing to its noncentrosymmetric crystallographic structure, lithium niobate exhibits a strong Pockels effect. Generally for device applications, pure lithium niobate is limited in its utility owing to a strong photorefractive effect – or strong variation in refractive index induced by optical illumination. This effect is also referred to as optical damage in lithium niobate and appears to be related to the presence of impurity ions such as Fe^{2+}. The problem is usually addressed by MgO-doping which considerably increases the photoconductivity and eases charge relaxation processes; this is especially important for high optical power density device applications.

Another important property of lithium niobate relates to the relationship between its elastic properties and its dielectric properties. Namely, by setting up elastic deformations in the material, one can tune the optical properties. This so-called acoustooptical effect is used to advantage in guided wave devices, based on index gratings, with characteristics controlled through the frequency and power of a surface acoustic wave. Interaction of a guided optical mode with such a grating can be applied to fabricate tunable wavelength switches and filters.

Owing to lithium niobate's excellent electrooptic and acoustooptic properties, combined with its good transparency at telecommunications based wavelengths, it was soon recognized as an excellent material for the fabrication of optical waveguides in planar optical circuits. Key aspects of realizing this technology include fabricating regions with the appropriate refractive index variation for guiding regions, including waveguides that support both TE and TM modes, single-mode waveguides, and waveguides with low propagation loss and high coupling efficiency to a single-mode optical fiber. Currently, waveguide fabrication is primarily through both proton exchange and by titanium diffusion. In the case of titanium diffusion, the largest reported index contrast is $\sim 10^{-2}$ to 10^{-3}, corresponding to optical mode sizes for single-mode 1.5 μm devices or ~6–8 μm. These characteristics are quite comparable to those of single-mode optical fibers and explain why lithium niobate devices can be made with very low coupling losses to optical fibers (e.g. ~ 0.5 dB/facet). In addition, such waveguides can also exhibit low propagation losses (e.g. typically ~0.2 dB/cm).

Proton exchange (PE) involves the exchange of lithium ions and hydrogen ions, causing an increase in the extraordinary refractive index and a decrease in the ordinary index – PE waveguides can be made for only TM modes and this has been exploited as a simple process for fabricating polarization-selective devices, with much lower sensitivity to optical damage than their titanium diffused counterparts. Problems are encountered when the H^+ concentration is too high and so in practice devices are made with low H^+ concentrations – such devices show propagation losses <0.5 dB/cm with extraordinary index contrasts of $<10^{-2}$. Recently, prospects of combining the two processes – titanium diffusion (TI) and PE – have been explored, with the combined process referred to as TIPE. This process enables waveguides supporting both TE and TM modes to be fabricated, while still retaining low sensitivity to optical damage. In addition, the TE-TM birefringence of such waveguides is controllable.

Recently, doping of lithium niobate with rare earth materials has been established as an efficient means of fabricating optical amplifiers and lasers for telecommunications applications. This approach enables the full advantages of lithium niobate to be exploited in planar optical circuits – that is, signal processing devices such as modulators, optical space switches, wavelength tunable filters, can then be combined with diode pumped optical amplifiers or lasers.

Another important property that lithium niobate can offer is a large second-orders susceptibility, suitable for second harmonic generation, frequency difference mixing, spontaneous parametric fluorescence, stimulated Raman and Brillouin scattering. The ability to use low-loss planar optical waveguides provides an excellent capability for achieving the high power densities required to obtain these nonlinear optical effects, together with long interaction lengths owing to the two-dimensional confinement provided by these waveguide structures. In addition, phase matching can be used in such structures. Currently, this makes use of so-called periodically poled lithium niobate (PPLN).

PPLN can provide significantly increased nonlinearity compared with bulk lithium niobate, while affording phase matching over any frequency in its wide transmission bandwidth. In addition, PPLN can be made with special structures, such as aperiodic gratings which allow effects to occur that have no counterparts in bulk nonlinear media. These properties offer novel devices and major enhancements in many existing devices – examples include application to parametric devices such as optical parametric oscillators, amplifiers and generators. The high parametric gain achievable at modest pump powers has underlined the convergence between pump power requirements for PPLN devices and the optical power available from fiber laser devices.

Lithium niobate polarization modulator is an integrated optic device in which the phase of a light wave propagating through a waveguide channel is modulated by an external applied voltage. The modulator has an embedded waveguide fabricated by implanting a $LiNbO_3$ substrate with Ti atoms which increase the refractive index. Two coplanar strip electrodes run along the waveguide and enable the application of a transverse field E_a to the light propagation direction z as shown in the figure. The external modulating voltage $V(t)$ is applied between the coplanar drive electrodes and, by virtue of the Pockels effect, induces a change Δn in the refractive index and hence a voltage-dependent phase shift through the device. We can represent light propagation along the guide in terms of two orthogonal modes, E_x along x and E_y along y. These two modes experience symmetrically opposite phase changes. The applied field is not uniform between the electrodes and, further, not all applied field lines lie inside the waveguide. The *electrooptic effect* takes place over the spatial overlap region between the applied field and the optical fields. This spatial

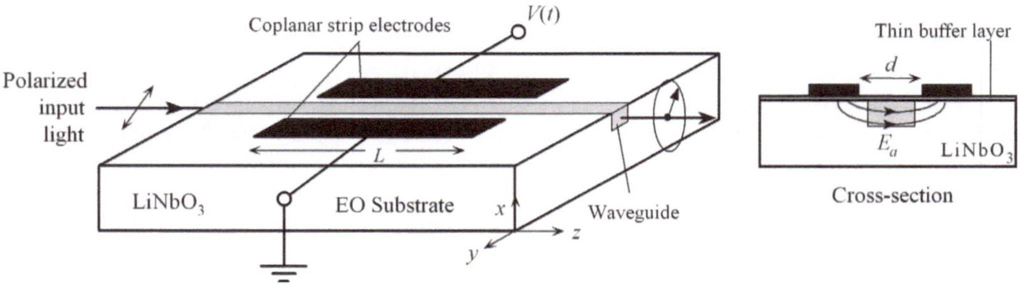

Lithium niobate polarization modulator Integrated transverse Pockels cell phase modulator in which a waveguide is diffused into an electrooptic (EO) substrate. Coplanar strip electrodes apply a transverse field E_a through the waveguide. The substrate is an x-cut $LiNbO_3$ and typically there is a thin dielectric buffer layer (e.g. ~200 nm thick SiO_2) between the surface electrodes and the substrate to separate the electrodes away from the waveguide.

overlap efficiency is lumped into a coefficient Γ and the phase shift $\Delta\phi$ due to an applied voltage V is written as

$$\Delta\phi = \Gamma\frac{2\pi}{\lambda}n_o^3 r_{22}\frac{L}{d}V$$

where typically $\Gamma \approx 0.5$–0.7 for various integrated polarization modulators of this type, λ is the free space wavelength, n_o is the ordinary wave refractive index, L and d are defined in the figure and r_{22} is the appropriate Pockel's coefficient. Since the phase shift depends on the product of V and L, a comparative device parameter would be the $V \times L$ product for a phase shift of π (half-wavelength), i.e. $V_{\lambda/2}L$. At $\lambda = 1.5$ µm for an x-cut $LiNbO_3$ modulator, with $d \approx 10$ µm, $V_{\lambda/2}L \approx 35$ V·cm. For example, a modulator with $L = 2$ cm has a half-wave voltage $V_{\lambda/2} = 17.5$ V. By comparison, for a z-cut $LiNbO_3$ plate, that is for light propagation along the y-direction and E_a along z, the relevant Pockels coefficients (r_{13} and r_{33}) are much greater than r_{22} which leads to $V_{\lambda/2}L \approx 5$ V·cm.

Lithium tantalate ($LiTaO_3$) is a birefringent electrooptic crystal that has useful piezoelectric and pyroelectric properties; it is widely used in pyroelectric detectors as in the measurement of laser power, especially in the measurement of IR radiation intensity. Trigonal crystal; $3m$ point group, $R3c$ space group; $a = 5.154$, $c = 13.784$; $\rho = 7.45$ g cm^{-3}; transparent from 0.4–5 µm; $n_o = 2.183$, $n_e = 2.188$ at 0.6 µm; $n_o = 2.176$, $n_e = 2.180$ at 633 nm; $n_o = 2.131$, $n_e = 2.134$ at 1.2 µm; $d_{22} = 2$; $d_{12} = -1$, $d_{33} = -21$ pm V^{-1} at 1.06 µm; $r_{13} = 8$, $r_{22} = -0.2$, $r_{33} = 30$ pm V^{-1} at 633 nm; $\varepsilon_{11}/\varepsilon_o = 52$, $\varepsilon_{33}/\varepsilon_o = 44$; $p = 180 - 230$ µC m^{-2} K^{-1}. (I. Tomeno and S. Matsumura, *Phys. Rev. B*, **38**, 606, 1988.) See *pyroelectricity*.

Lithography is a process by which a pattern is defined on the surface of the wafer which identifies the crystal regions to be doped.

Load line see *photovoltaic or solar cell I–V characteristics*.

Local area network (LAN) see *optical network*.

Local or Lorentz field, E_{local}, is the true field experienced by a molecule in a dielectric that arises from the free charges on the plates and all the induced dipoles surrounding the molecule. The true field at a molecule is not simply the applied field (applied voltage/sample thickness) because the field inside a polarized dielectric at the atomic scale is not uniform. Each molecule also experiences the field of the neighboring induced dipoles. The local field is the actual field that acts on a molecule. It can be calculated by removing that molecule and evaluating the field at that point due to the charges on the plates and the dipoles surrounding the point. There are various models for the local field; the most popular is the Lorentz field. If **P** is the polarization in the medium by an applied field **E**, for noncrystalline materials (liquids and glasses) and cubic crystals, then, according to Lorentz, $\mathbf{E}_{\text{local}} = (1/3\varepsilon_0)\mathbf{P}$, which means that $\mathbf{E}_{\text{local}} = [(\varepsilon_r + 2)/3]\mathbf{E}$.

Longitudinal chromatic aberration see *chromatic aberration.*

Longitudinal Pockels cell has the applied field along the direction of light propagation, both parallel to the z-axis (optic axis). The light is allowed to enter the crystal along the applied field direction. Suppose that a KDP crystal is used. KDP is uniaxial and $n_1 = n_2 = n_o$ (polarizations parallel to x and y) and $n_3 = n_e$ (polarization parallel to z). The principal axes x and y are rotated by 45° to become x' and y' and

$$n_1' \approx n_o - \tfrac{1}{2}n_o^3 r_{63} E_a \quad n_2' \approx n_o + \tfrac{1}{2}n_o^3 r_{63} E_a$$
$$\text{and} \quad n_3' = n_3 = n_e.$$

We can calculate the half-wave voltage required to induce a retardation of π between the emerging components of the electric field, for a free-space wavelength of 633 nm since for KDP at 633 nm, $n_o \approx 1.51$, $r_{63} \approx 10.5 \times 10^{-12}$ m/V. Light propagates along z with polarizations parallel to the x'- and y'-axes and these orthogonal waves propagate with

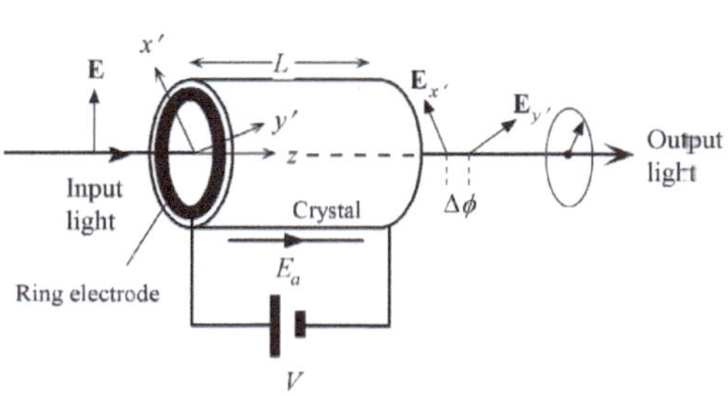

A **longitudinal Pockels** cell phase modulator.

different refractive indices n'_1 and n'_2. The applied field $E_a = V/L$ is along z so that the relative phase change $\Delta\phi$ between E_x' and E_y' is

$$\Delta\phi = \frac{2\pi}{\lambda} L(n_2' - n_1') = \frac{2\pi}{\lambda} n_o^3 r_{63} V$$

and for $\Delta\phi = \pi$ we have

$$V_\pi = \frac{\lambda}{2n_o^3 r_{63}} = \frac{(633 \times 10^{-9})}{2(1.51)^3(10.5 \times 10^{-12})}$$
$$= 8.75\,\text{kV}$$

Longitudinal spherical aberration see *aberrations.*

Lorentz distribution in physics is a particular probability distribution function that is also known as the *Cauchy–Lorentz distribution* (see *Cauchy distribution*). In photonics, the term is used to describe the spectral lineshape of emission or the optical gain curve that results from homogeneous broadening of the emission spectrum. See *lineshape, homogeneous broadening.*

Lorentz oscillator is one of the simplest models that describes the polarization of an atom as if the electrons in the atom were attached by springs to the nucleus; trying to displace an electron either stretches or compresses the spring. In the Lorentz model, an electron experiences a driving force $eE_o e^{j\omega t}$ (charge × field) from the light (electromagnetic wave), and a restoring force $F_r = -\beta x$ from the spring (which represents the "pull" of the positive nucleus), where β is the "spring constant" and x is its displacement from its center of motion induced by the driving field. Its equation of motion is similar to a damped forced mechanical oscillation

$$m_e \frac{d^2x}{dt^2} = eE_o e^{j\omega t} - \beta x - \gamma' \frac{dx}{dt}$$

where the last term $\gamma'(dx/dt)$ represent energy losses by the oscillating electron (e.g. emission of electromagnetic waves). The induced dipole moment p is ex and the polarizability α is the dipole moment per unit field. Defining $\omega_o = (\beta/m_e)^{1/2}$ and γ as γ'/m_e,

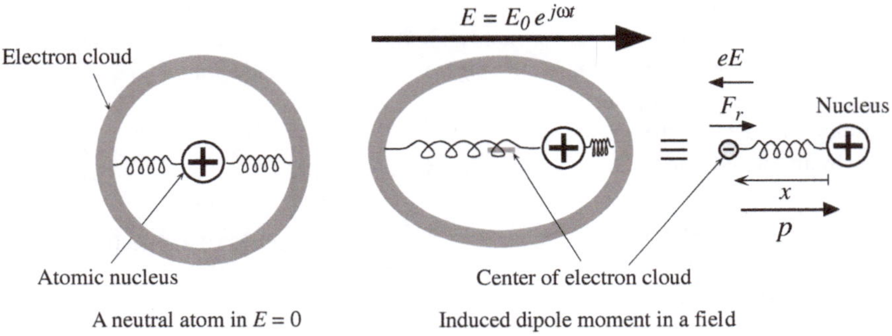

A highly simplified **Lorentz oscillator** model for the electron bound to a nucleus and driven by an applied field. The force between the nucleus and the electron cloud is "spring" like, and acts to restore the electron cloud to its equilibrium position in the presence of an external field.

gives α as

$$\alpha = \frac{e^2}{m_e[(\omega_0^2 - \omega^2) + j\gamma\omega]} = \frac{\alpha(0)}{1 - (\omega/\omega_o)^2 + j(\gamma/\omega_o)(\omega/\omega_o)}$$

where $\alpha(0) = e^2/(m_e\omega_o{}^2)$ is the low-frequency (dc) polarizability. This is the classical Lorentz oscillator polarizability, and is only for one electron. There will be similar contributions from all the Z electrons. The total polarizability must be obtained by suitably summing terms as above for the Z electrons, which involves using oscillator strengths, a kind of weighting factors f_i, for each contribution from each electron i in the atom ($i = 1, \ldots, Z$). α has real and imaginary parts and exhibits resonance as ω around ω_0 as shown. The real part α' of α represents oscillations of the induced dipole p in phase with the field oscillations (and determines the refractive index n), and the imaginary α'' part represents the out-of-phase (quadrature) oscillations of p. The losses are embedded in the imaginary part α''. Explicitly,

$$\alpha' = \alpha(0)\frac{1 - (\omega/\omega_o)^2}{[1 - (\omega/\omega_o)^2]^2 + (\gamma/\omega_o)^2(\omega/\omega_o)^2},$$

and

$$\alpha' = \alpha(0)\frac{(\gamma/\omega_o)(\omega/\omega_o)}{[1 - (\omega/\omega_o)^2]^2 + (\gamma/\omega_o)^2(\omega/\omega_o)^2}.$$

The α vs. ω behavior is called the *polarizability dispersion relation*. The Lorentz oscillator model can also be applied to ionic crystals in which an anion is attached with a "spring" to a cation. Since we can calculate the atomic polarizability from the Lorentz oscillator model, we can use the Lorentz–Lorenz (or Clausius–Mossotti) equation to obtain the refractive index N from the

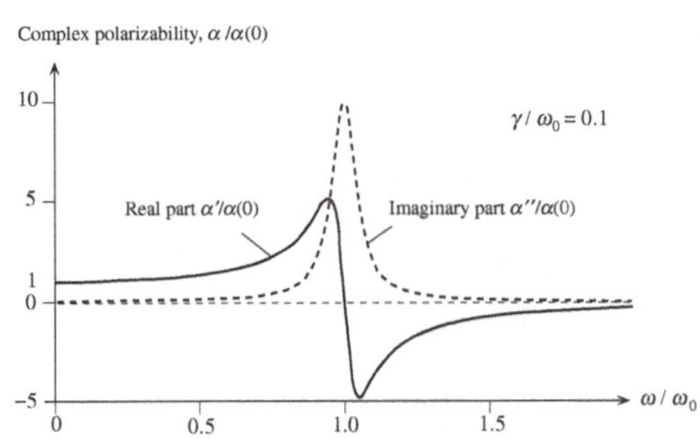

Dependence of the real and imaginary parts of α on the frequency ω for the **Lorentz oscillator** model.

polarizability α, and the concentration of atoms N_{at} in the medium,

$$\frac{N^2-1}{N^2+2}=\frac{N_{at}}{3\varepsilon_0}(\alpha'-j\alpha'')$$

which leads to a frequency-dependent complex refractive index N whose real part n behaves as the usual real refractive index (determines the propagation constant or the phase velocity) and the imaginary part, called the extinction coefficient K, represents the loss in the medium. If one knows the frequency dependence of α' (or α'') then the frequency dependence of α'' (or α') can be found from Kramers–Kronig relations. See *Lorentz–Lorenz equation, polarizability, oscillator strength.*

Lorentz–Lorenz refractive index equation as used in the literature represents the Clausius–Mossotti relative permittivity equation in which $\varepsilon_r{}^2$ is set to the refractive index n, and the polarizability contributions are summed up for all polarization mechanisms that operate in the frequency range of interest, provided that the Lorentz local field concept is valid. For example, in a crystal with i types of ions, the Lorentz–Lorenz refractive index n is

$$\frac{n^2-1}{n^2+2}=\frac{\rho/M}{3\varepsilon_0}\sum_i f_i\alpha_i$$

where ρ is the density, M is the molecular weight, f_i is the mole fraction of the ion type i in the structure, and α_i is its polarizability. In principle, there is no major fundamental difference between the Lorentz–Lorenz and the Clausius–Mossotti equations. The Lorentz–Lorenz equation for a gas expresses n approximately as

$$n^2\approx 1+\frac{3R_mP}{R_{gas}T}$$

where R_m is the molar refractivity, P is the gas pressure, R_{gas} is the universal gas constant and T is the temperature in K. The equation is named after Hendrik Lorentz (1853–1928), a Dutch physicist and a Nobel laureate (1902), and Ludvig Lorenz (1829–1891), a Danish mathematician and physicist.

Lorentz–Lorenz refractive index mixture rule is a rule that allows the effective refractive index n of a mixture to be determined from its constituents via

$$\frac{n^2-1}{n^2+2}=\phi_1\frac{n_1^2-1}{n_1^2+2}+\phi_2\frac{n_2^2-1}{n_2^2+2}+\cdots$$

where n_1 and ϕ_1 are the refractive index and the volume fraction of component 1, and n_2 and ϕ_2 are the refractive index and the volume fraction of component 2 and so on, and $\phi_1+\phi_2+\cdots=1$. The volume fraction can be easily obtained from the mole fraction. See *effective medium approximation.*

Low-water peak fiber see *attenuation in optical fibers.*

Lubberts effect describes the reduction in detective quantum efficiency (DQE) with spatial frequency caused by different incident x-ray photons producing signals with different amounts of signal spreading (point spread function, PSF) in image sensors. The expression for DQE(f) is given by

$$\text{DQE}(f)=\frac{\text{DQE}(0)\text{MTF}^2(f)}{\text{NPS}_0(f)}$$

where MTF(f) is the overall system modulation transfer function and $\text{NPS}_0(f)$ is the noise power spectrum of the system normalized to unity as the spatial frequency f approaches 0. X-rays are absorbed at different depths in the sensor layer. As a result, lights or electron and hole pairs generated at different depths travel different distances through the sensor layer before they reach the surface of the detector, and are scattered differently. X-rays deposited at various depths within the sensor layer produce signals with different point spread functions. This may affect the propagation of signal and noise differently, and as a result, the ratio of $\text{MTF}^2(f)$ to $\text{NPS}_0(f)$ decreases with increasing spatial frequency. Thus, the DQE decreases with increasing spatial frequency. If every absorbed x-ray photon produced a signal with the same point spread function, $\text{MTF}^2(f)$ would have the same frequency dependence as $\text{NPS}_0(f)$ if the effect of

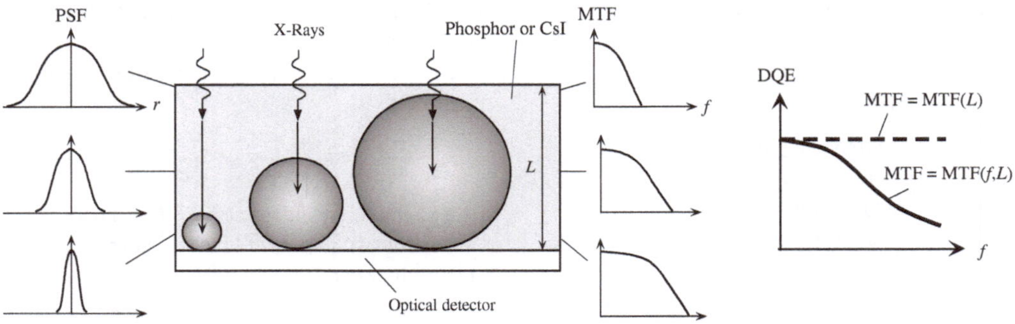

Lubberts effect X-rays deposited at various depths within the sensor layer (phosphor) produce signals with different point spread functions, as shown on the left. Thus both PSF and MTF depend on the x-ray interaction depth, and the overall MTF depends on the thickness L. The DQE(f) decreases with increasing f when the depth dependence is large but if there is no depth dependence of MTF and no aliasing, then DQE is independent of f (adapted from, Rowlands and Yorkston, in *Handbook of Medical Imaging*, edited by J. Beutel, H. L. Kundel, and R. L. Van Metter, SPIE Press, Washington, DC, 2000, chapter 4).

aliasing is ignored, and the DQE is independent of f as shown in the figure. This effect is more important when the active layer (conversion layer) of an image detector is thicker.

Lumen (lm) is the basic visual photometry unit for measuring the *luminous power*, or *luminous flux* of a light source. One lumen is also the luminous flux emitted within a unit solid angle by a uniform isotropic point source of luminous intensity 1 candela; 1 lm = 1 cd sr. At a reference frequency $\nu = 540 \times 10^{12}$Hz, or $\lambda \approx 555$ nm, 1 lm = (1/683)W. The wavelength of 555 nm corresponds to the response of the daylight adapted eye at which the luminous efficiency is maximum. At any other wavelength in the visible range, the conversion factor varies according to the *luminosity* or *the visibility function*. A 60 W tungsten light bulb emits a luminous flux of roughly 800 lumens. See *luminous flux*.

Luminance or **brightness sterance** (L_v in lm m^{-2} sr^{-1} or cd m^{-2}) is a measure of the "brightness" of an emitter (a source or a reflecting surface) and it is defined as the luminous flux emitted per unit surface area of the emitter per unit solid angle. Quantitatively, if the elementary luminous flux $d^2\Phi_v$ is

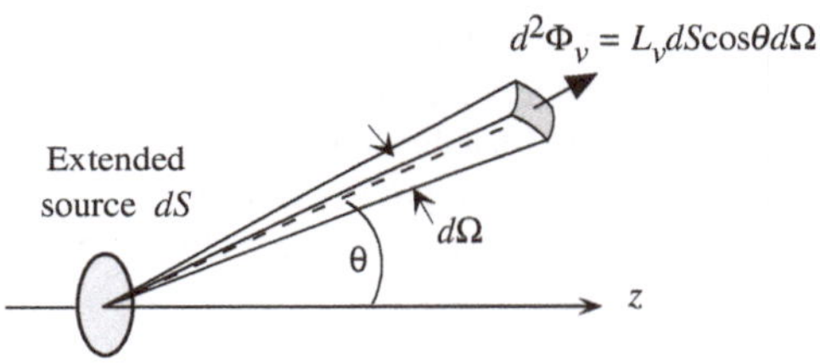

Definition of **luminance**.

Table: Luminance

Emitter	Luminance (cd m^{-2}, nits), approximate
Sun	1.6×10^9
Carbon arc lamp	1×10^9
Incandescent filament (100 W)	1×10^6
Snow in sunlight	10,000
Candle flame	5000
Moon	1000–3000
Laptop computer screen	100–250
CRT	50–150
Overcast sky	1000
Clear night sky, no moon	10^{-3}

emitted by an elementary surface dS in an elementary solid angle $d\Omega$, in the direction θ with respect to its normal, then the *luminance* L_v is defined by

$$L_v = \frac{d^2\Phi_v}{dS\cos\theta d\Omega}$$

where the quantity $L_v = L_v(\theta)$ is the *luminance* of the source and has the units of candela per m^2, nit, or lumens per steradian per m^2 [cd m^{-2}, nit, or lm sr^{-1} m^{-2}], and θ is the angle between the normal to the surface and "direction" of interest or observation. If $L_v(\theta) = \text{constant}$, the source or the reflecting surface is said to be *Lambertian*; emission is angularly uniform. A Lambertian surface is a flat radiating surface whose luminance does not depend on direction (but its luminous intensity does depend on direction). The total luminance would be $L_v(\theta)$ summed over all θ angles.

Luminescence in general terms, is the emission of light by a luminescent material, also called a *phosphor*, due to the absorption and conversion of energy into electromagnetic radiation over and above thermal radiation. The luminescent radiation emitted by the phosphor material is considered to be quite separate from thermal radiation emitted by virtue of its temperature. Luminescence is light emitted by a nonthermal source *when it is excited,* in contrast to the emission of radiation from a heated object such as the tungsten filament of a light bulb; the latter is called *incandescence*. Typically the emission of light occurs from certain dopants, impurities or even defects, called *luminescent* or *luminescence centers*, purposefully introduced into a *host matrix*, which may be a crystal or glass as shown in the figure The luminescent center is also called an *activator*. There are many examples of phosphors. For example, in ruby, the Cr^{3+} ions are the luminescent centers in the sapphire (Al_2O_3) crystal host. Cr^{3+} ions can absorb UV or violet light and then emit red light. This phosphor system is written as Al_2O_3:Cr^{3+}. The excitation and emission involves only the Cr^{3+} ion. In other cases, the activator excitation may also involve the host.

Luminescence is normally categorized according to the source of excitation energy. *Photoluminescence* involves excitation by photons (light). *X-ray luminescence* involves incident x-rays exciting a phosphor to emit light. *Cathodoluminescence* is light emission when the excitation is the bombardment of the phosphor with energetic electrons as in TV cathode ray tubes. *Electroluminescence* is light emission due to the passage of an electric current. Electroluminescence in semiconductive materials appears as a result of an excited electron transiting down to the ground energy level, which would correspond to the recombination of an electron and a hole; the excited electron is the conduction band (CB), and its ground state corresponds to a hole in the valence band (VB). The direct electron–hole recombination mechanism generally occurs very quickly. For

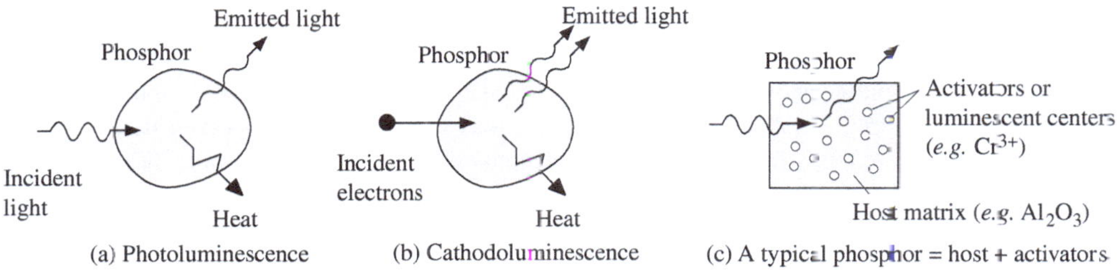

Luminescence Photoluminescence, cathodoluminescence and a typical phosphor.

example, typical minority carrier lifetimes are in the range of nanoseconds so that light emission from a semiconductor stops within nanoseconds after the removal of excitation. Such quick luminescence processes occurring over a nanosecond time scale or shorter are normally identified as *fluorescence*. The emission of light from a fluorescent tube is actually a fluorescence process. The tube contains a gas mixture of argon and mercury. The Ar and Hg gas atoms become excited by the electric discharge process and emit light mainly in the ultraviolet region. This UV light is absorbed by the fluorescent coating on the inside of the tube. The excited activators in the phosphor coating then emit radiation in the visible region. A number of phosphors are used to obtain "white" light from the tube.

There are also phosphors from which light emission may continue for milliseconds to hours after the cessation of excitation. These slow luminescence processes are normally referred to as *phosphorescence* also known as *afterglow*.

Many phosphors are based on activators doped into a host matrix; for example, Eu^{3+} (europium ion) in a Y_2O_3 (yttrium oxide) matrix is a widely used modern phosphor. When excited by UV radiation, it provides an efficient luminescence emission in the red (around 613 nm). It is used as the red-emitting phosphor in color TV tubes and in modern tricolor fluorescent lamps. Consider an activator with a ground state at E_1. Upon excitation by an incident radiation of suitable energy $h\upsilon_{ex}$ the activator becomes excited to E_2. From this energy level, it decays, or *relaxes*, down relatively quickly (on a time-scale of the order of picoseconds) to an energy level E_2' by emitting phonons or lattice vibrations. This type of decay is called *radiationless* or *nonradiative decay*. From E_2', the activator decays down to a lower level E_1' by emitting a photon (spontaneous emission), which is the emitted luminescent radiation. The emitted photon energy is $h\upsilon_{em}$, which is less than the excitation photon energy $h\upsilon_{ex}$. The return from E_1' to the ground state E_1 involves phonon emissions. Further, for some activators, E_1' is either very close to E_1, or it is E_1. The energy levels such as E_2, E_2', E_1', *etc.* are not well-defined single levels but involve finely spaced multilevels. Higher levels may form multilevel narrow energy "bands." In this example, the activator absorbed the incident radiation, and was directly excited, which is known as *activator excitation*. The Cr^{3+} ions in Al_2O_3:Cr^{3+} can be excited directly by blue light, and would then emit in the red. There are many phosphors in which the excitation involves the host. In host excitation, the host matrix absorbs the incident radiation, and transfers the energy to the activator, which then becomes excited to E_2, and so on. In x-ray phosphors, for example, the x-rays are absorbed by the host, which subsequently transfers the energy to the activators. It is apparent that the emitted radiation ($h\upsilon_{em}$) has a *longer* wavelength than the exciting radiation ($h\upsilon_{ex}$), that is, $h\upsilon_{em} < h\upsilon_{ex}$. The downshift in the light frequency from absorbed to emitted radiation is called the *Stoke's shift*. It should be emphasized that the energy levels of the activator also depend on the host. The internal electric fields within the host crystal act on the activator, which results in the splitting and/or shifting of these levels up and down. The emission characteristics depend firstly on the activator, and secondly on the host.

There are a number of host excitation mechanisms. In one possible process, which involves a semiconductor host, an incident photon initially excites a valence band (VB) electron to the conduction band (CB). The electron then *thermalizes*, *i.e.* loses the excess energy as it collides with lattice vibrations, and falls close to E_c, and wanders around in the crystal. In one process, shown as *a*, the electron can be captured into an excited state D of a luminescent center or an activator. The electron then falls down in energy to the ground state A of the activator releasing a photon, which is the luminescent emission. The electron at state A then recombines with a hole in the VB. Thus the activator acts as a *radiative recombination center*. In some

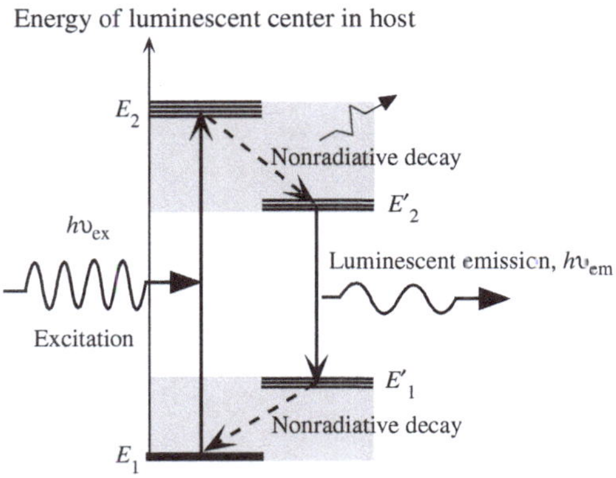

Luminescence Photoluminescence: light absorption, excitation, nonradiative decay and light emission, and return to the ground state E_1. (The energy levels have been displaced horizontally for clarity.)

cases D and A may be separate centers representing *donor*- and *acceptor*-like centers, hence the labels D and A. In other cases, the radiative recombination center may simply be a single energy level in the bandgap, which is shown as R. The electron can emit a photon as it is captured into R, shown as process *b* in the figure, or emit the photon after it is captured by R, as it recombines with a hole, shown as process *c* in the figure. Processes *a* and *b* occur in various ZnS based phosphors. For example, in ZnS:Cu^+ phosphors the activator is Cu^+, which has an energy level at A. The luminescent emission is enhanced by using a coactivator, such as Al in ZnS:Cu^+. Aluminum acts as a shallow donor D, and the luminescence is due to process *a* in the figure.

There may also be traps in the semiconductor because of various crystal defects or they may be added impurities. The electron can become captured by a trap at a localized energy level E_t in the bandgap, but close to E_c. These electron traps temporarily capture an electron from the conduction band and thereby immobilize it. The time the electron spends trapped at E_t depends on the energy depth of the trap from the conduction band, $E_c - E_t$. After a while a strong lattice vibration returns the electron back into the conduction band (by thermal excitation). The time interval between photogeneration and recombination can be relatively long if the electron remains captured at E_t for a

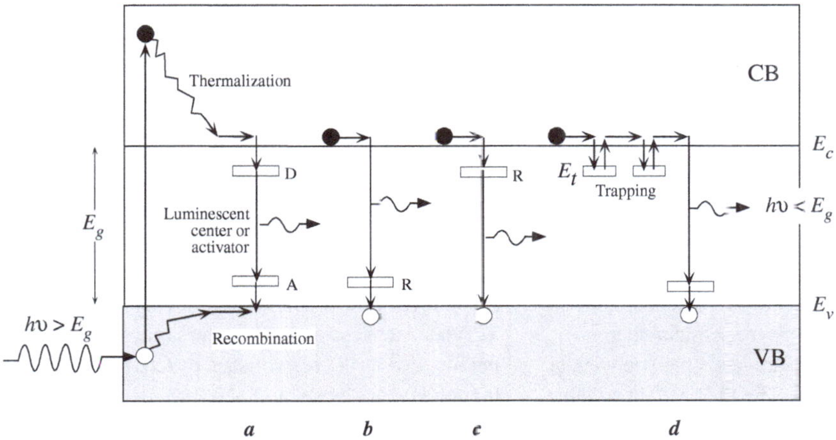

Luminescence Optical absorption generates an electron–hole pair (EHP). Both carriers thermalize. There are a number of recombination processes via a dopant or an impurity that can result in a luminescent emission.

Table: Luminescence Selected example of phosphors.

Phosphor	Activator	Useful emission	Example excitation	Comment or application
Y_2O_3:Eu^{3+}	Eu^{3+}	Red	UV	Fluorescent lamp, color TV
$BaMgAl_{10}O_{17}$:Eu^{2+}	Eu^{2+}	Blue	UV	Fluorescent lamp
$CeMgAl_{11}O_{19}$:Tb^{3+}	Tb^{3+}	Green	UV	Fluorescent lamp
$Y_3Al_5O_{12}$:Ce^{3+}	Ce^{3+}	Yellow	Blue, Violet	White LED
Sr_2SiO_4:Eu^{3+}	Eu^{3+}	Yellow	Violet	White LED (experimental)
ZnS:Ag^+	Ag^+	Blue	Electron beam	Color TV blue phosphor
$Zn_{0.68}Cd_{0.32}S$:Ag^+	Ag^+	Green	Electron Beam	Color TV green phosphor
ZnS:Cu^-	Cu^-	Green	Electron Beam	Color TV green phosphor

considerable length of time. In fact, the electron may become trapped and detrapped many times before it finally recombines so that the emission of light can persist for a relatively long time after the cessation of excitation (e.g. milliseconds or longer) as indicated by process *d* in the figure.

It is also possible to excite electrons into the CB by bombarding the material with a high-energy electron beam, which leads to cathodoluminescence. Color CRT displays are typically coated uniformly with three sets of phosphor dots which exhibit cathodoluminescence in the blue, red and green wavelengths. In electroluminescence, an electric current, either ac or dc, is used to inject electrons into the CB which then recombine with holes and emit light. For example, passing a current through certain semiconducting phosphors such as ZnS doped with Mn causes light emission by electroluminescence. The emission of light from a light-emitting diode (LED) is an example of *injection electroluminescence* in which the applied voltage causes charge carrier injection and recombination in a device (diode) that has a junction between a *p*-type and an *n*-type semiconductor. Zinc sulfide with various activators has been one of the traditional phosphors. The ZnS:Ag^+ in which Ag^+ is the activator, is still used as a blue emitting phosphor, though in some cases Cd is substituted for some of the Zn. ZnS:Cu^+ emits in the green, which is also a useful phosphor. Most modern phosphors, on the other hand, have been based on using rare earth activators in various hosts. For example, Y_2O_3:Eu^{3+} absorbs UV radiation and emits in the red. $Y_3Al_5O_{12}$:Ce^{3+} absorbs blue light and emits yellow light. Some of the most popular activators are Eu^{3+} for red, Eu^{2+} for blue, and Tb^{3+} for green.

Recent inexpensive white LEDs (light-emitting diodes) that have appeared in the market seem to emit white light by emitting a mixture of blue and yellow light which are registered visually by the eye as appearing white. See *photoluminescence, white LED.*

Luminosity function or the **luminous efficiency** is a dimensionless and a normalized function, usually denoted $V(\lambda)$, that describes the *efficiency* of an average (or the standard) human eye to sensing different wavelengths, i.e. it represents the spectral efficiency of the eye. Its functional form and the position of the maximum depends on whether the eye is light (*photopic*) or dark (*scotopic*) adapted. $V(\lambda)$, also known as the *visibility function*, was originally established by CIE in 1924 after the spectral visibility studies of K. S. Gibson and E. P. T. Tyndall and corresponds to the same color matching function as $\bar{y}(\lambda)$ established by CIE in 1931, though $V(\lambda)$ has been modified since then to better match the eye's spectral efficiency. Judd-Vos (1978) modified and Sharpe, Stockman, Jagla and Jägle (2005) modified luminosity curves are some recent well-known examples, which make up for the

Luminous efficacy of a light source

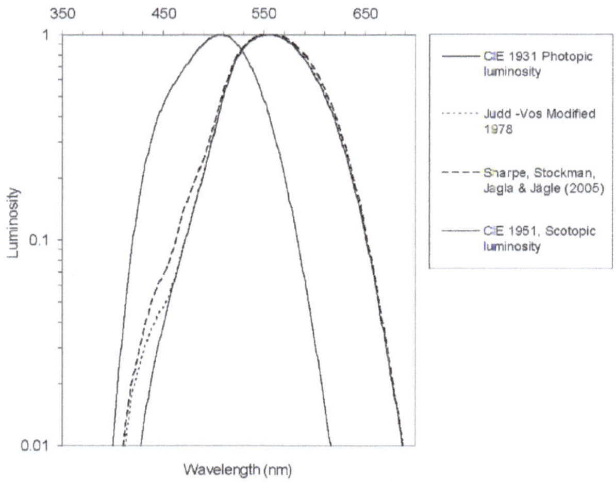

Luminosity function Luminosity vs. wavelength for the photopic and scotopic eye. Tree functions are shown for the photopic luminosity behavior: CIE 1931 (solid curve), Judd-Vos modified (1978) (dots) and Sharpe, Stockman, Jagla and Jägle modified (2005) (dashes) curves. Data from Dan Plummer (1966–2006)'s color and vision data base at http://www.cvrl.org/.

deficiency of the original luminosity curve in representing the response of the eye to wavelengths in the blue region. The *scotopic luminosity* curve has been established by CIE in 1951 based on the data from Wald (1945) and Crawford (1949). Various luminosity curves are shown in the figure.

The total luminous (visible) flux Φ_v in lumens (lm) is obtained by integrating the radiant spectral density (radiant flux per unit wavelength) corrected by the visibility or luminosity function $V(\lambda)$

$$\Phi_v = (683\text{lm/W}) \int_0^\infty P(\lambda)V(\lambda)d\lambda$$

in which $P(\lambda)$ is the radiant spectral density (W m^{-1}). (K. S. Gibson and E. P. T. Tyndall, *Scientific Papers of the Bureau of Standards*, **19**, 131, 1923; *Journal of the Optical Society of America*, **9** (4), 403, 1924; J. J. Vos, *Color Research and Application*, **3**, 125, 1978.) See *color matching functions, eye, luminous flux*.

Luminous efficacy of a light source (such as a lamp) as commonly used in *lighting applications* is the efficiency with which an electrical light source converts the input electric power (watts) into an emitted luminous flux (lumens). A 100 W light bulb produces 1700 lumens and has an efficacy of 17 lumens/watt (lm/ W). While at present LED efficacies are below those of fluorescent tubes, rapid advances in LED technologies are bringing the

Table: Luminous efficacy of a light source Typical approximate efficacies. Maximum possible efficacy is 683 lm/W (*In research labs).

Light source	Incandescent	Fluorescent	Compact arc and gas discharge	Yellow high intensity LED	White LED
Efficacy (lm/W)	15–20	60–80	60–100	20–30	20–40; 60–75*

Table: Luminous efficiency Typical and approximate efficiencies. Maximum possible efficiency is 100%.

Light source	Incandescent	Fluorescent	Compact arc	Gas discharge	White LED
Efficiency	2%	10%	5–20%	10–20%	10–30%

expected efficacies to around 50 lm/W or higher. LEDs as solid state lamps have much longer lifetimes and much higher reliability, and hence are expected to be more economical than incandescent and fluorescent lamps. Note: The term *efficacy* is preferred to efficiency when the ouput units are different from the input units as in a lamp. The output, the luminous flux, is in lumens whereas the input is in watts. Efficacy has units whereas efficiency does not. (B. F. Jones, *J. Phys. D: Appl. Phys.* **6,** L80, 1973; S. T. Henderson and A. M. Marsden, *Lamps and Lighting* Edward Arnold, London, 1972.)

Luminous efficacy of radiation (not that of a lamp) is the ratio K of the luminous flux Φ_v to the radiant flux Φ_e, a measure of the brightness of radiation for a given amount of radiant energy flow; $K = \Phi_v/\Phi_e$, with the maximum value of 683 lm/W if all the radiation is monochromatic at the maximum luminous efficiency of the daylight-adapted eye at 555 nm. Suppose that the spectral density of radiant flux of a radiation is $\Phi_{e,\lambda}(\lambda)$ so that $\Phi_{e,\lambda}(\lambda)d\lambda$ is the radiant flux $d\Phi_e$ in an interval $d\lambda$ at λ, and $V(\lambda)$ is the luminosity (visibility) function of the daylight-adapted eye. Then the luminous efficacy of radiation is defined by

$$\text{Luminous efficacy} = \frac{(683\text{lm/W})\int_0^\infty V(\lambda)\Phi_{e,\lambda}d\lambda}{\int_0^\infty \Phi_{e,\lambda}d\lambda}(\text{lm/W}).$$

Luminous efficiency refers to the fraction of radiant energy that falls into the eye's spectral response characteristics, that is, its luminous efficiency characteristic. The luminous efficiency term can be applied to both light sources such as lamps or to radiation. Suppose that the spectral density of radiant flux is $\Phi_{e,\lambda}(\lambda)$ so that $\Phi_{e,\lambda}(\lambda)d\lambda$ is the radiant flux $d\Phi_e$ in an interval $d\lambda$ at λ, and $V(\lambda)$ is the luminosity (visibility) function of the daylight-adapted eye. Then the luminous efficiency is defined by

$$\text{Luminous efficiency} = \frac{\int_0^\infty V(\lambda)\Phi_{e,\lambda}d\lambda}{\int_0^\infty \Phi_{e,\lambda}d\lambda}.$$

See *luminous efficacy.*

Luminous flux or **photometric flux**, F_v, or Φ_v, is the visual "brightness" of a source as observed by an average daylight-adapted eye, and is proportional to the radiant flux (radiation power emitted) of the source and the efficiency of the eye to detect the spectrum of the emitted radiation. While the eye can see a red color source it cannot see an infrared source and the luminous flux of the infrared source would be zero. Similarly, the eye is less efficient in the violet than in the green region, and less radiant flux is needed to have a green source at the same luminous flux as the blue source. Luminous flux Φ_v is measured in *lumens* (lm) and at a particular wavelength it is given by

$$\Phi_v = \Phi_e \times (683\ \text{lumens/Watt}) \times V$$

where Φ_e is the radiant flux (in watts) and V is the *relative luminosity* (or simply *the luminosity*) of the daylight-adapted eye or its *luminous efficiency*, which is unity at 555 nm. A typical 60 W incandescent lamp provides roughly 900 lumens. Fluorescent tubes provide more luminous flux output than incandescent lamps for the same electric power input as they have more spectral emission in the

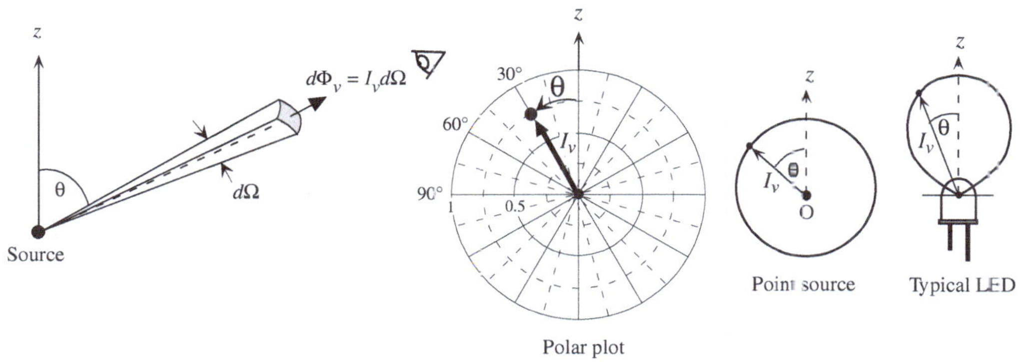

Definition of **luminous intensity** and the luminous intensity of a typical LED.

visible region and make better use of the eye's spectral sensitivity. Examples are: 100 W incandescent lamps: 1300–1800 lm, depending on the filament operating temperature (hence bulb design); 25 W compact fluorescent lamps: 1500–1750 lm. The spectral distribution of luminous efficiency is different for daylight-adapted eye, *photopic* response, and dark adapted eye, *scotopic* response. See *luminosity, eye, lumen.*

Luminous intensity is the luminous flux per unit solid angle along a particular direction, and measures the visual brightness of radiation as observed by a daylight-adapted eye within a unit solid angle. If $d\Phi_v$ is the luminous flux within a solid angle $d\Omega$ for a direction θ, ϕ, then

$$I_v = \frac{d\Phi_v}{d\Omega} = r^2\frac{d\Phi_v}{dA}.$$

The units are lumen per steradian or *candela*. If the source is isotropic, then the luminous intensity is the same in all directions. The source emits over 4π steradians and if the point source has a luminous flux of 1 lm, then its luminous intensity is $1/(4\pi)$ cd. A 1 candela point source emits 4π lumens into space in all directions. For many practical sources, luminous intensity is directional as in the case of LEDs. Conventionally, the intensity is plotted on a polar plot in which a point on the plot has an angle θ with respect to z (usually taken along the maximum intensity or direction of interest) and a distance from the origin that is either I_v or proportional to I_v. Such polar plots represent the angular distribution of intensity from a source. For a *point source*, $I_v(\theta)$ is a circle, that is $I_v(\theta) = \text{constant}$. For a *Lambertian source*, $I_v(\theta) = I_v(0)\cos\theta$. For an exponential intensity source, I_v falls exponentially (very rapidly) with θ, and can be approximated by using $I_v(\theta) = I_v(0)(\cos\theta)^n$ where n is usually greater than unity; e.g. if $I_v(\theta_{1/2}) = I_v(0)(1/2)$ when $\theta = \theta_{1/2}$, then $n = \ln(0.5)/\ln[\cos(\theta_{1/2})]$.

See *radiant intensity, candela, illuminance, lumen.*

Luttinger band parameters are dimensionless parameters γ_1, γ_2 and γ_3 that are frequently used to describe the shape of the *E–k* surface of the valence band in zinc blende semiconductors. The energies of light (+) and heavy (−) holes are thus given by the following expression:

$$E_{h,\pm} = \frac{\hbar^2}{2m}\left[\gamma_1 k^2 \pm [4\gamma_2^2 k^4 + 12(\gamma_3^2 - \gamma_2^2)\right.$$
$$\left.\times\sqrt{(k_x^2k_y^2 + k_y^2k_z^2 + k_z^2k_x^2)}\right].$$

Table: Luttinger band parameters Dimensionless Luttinger parameters for valence band energy dispersion of selected compound semiconductors.

Semiconductor	γ_1	γ_2	γ_3
GaSb	13.30	4.40	5.70
GaAs	6.95	2.25	2.86
GaP	4.05	0.49	1.25
InSb	33.50	14.50	15.60
InAs	20.40	8.30	9.10
InP	5.15	0.94	1.62
ZnSe	3.77	1.24	1.67
CdTe	5.29	1.89	2.46

Note that for spherical bands, $\gamma_2 = \gamma_3$. The table gives values for Luttinger parameters for selected semiconductors.

Lux [lx] is the basic visual photometry unit for measuring the *illuminance* of an illuminated surface. 1 lx = 1 lm/m^2. The illuminance of the Earth due to the Sun is close to 70,000 lx in bright summer, less than 6000 lx in winter. The limit of color perception is close to 3 lx (full Moon is only 0.25 lx).

Lyddane–Sachs–Teller relation. see *Reststrahlen absorption.*

Lyman ghost see *diffraction grating*.

M

***M*² factor** see *Gaussian beam.*

Mach–Zehnder (MZ) interferometer is an optical instrument that splits a collimated input beam into two beams through a beam splitter and, using two mirrors, recombines these two beams at another beam splitter. Suppose an input wave E_0 is split into two output waves E_1 and E_2, each of intensity I_0, at the input beam splitter. The intensity of the combined output beam is $I = |E_1 + E_2|^2$ and will depend on the phase of the two waves, $\varphi = 2\pi nd/\lambda$, where λ is the free space wavelength, d is the difference in the path length traveled by the two beams, and n is the refractive index. Thus, writing the two waves as:

$$E_1 = I_0^{1/2} e^{-ikz}$$
$$E_2 = I_0^{1/2} e^{-ik(z-d)}$$

then, $I = 2I_0[1 + \cos\ kd)] = 2I_0[1 + \cos(\varphi)]$, and thus the MZ interferometer can be used to measure

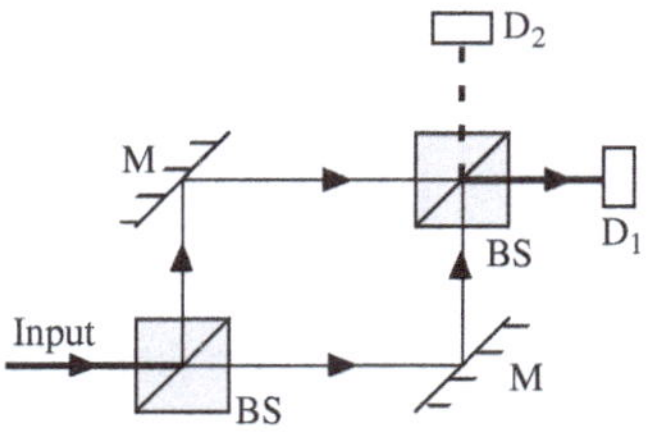

Mach–Zehnder interferometer At the input beam splitter (BS), an input wave E_0 is equally split into two waves E_1 and E_2 that recombine at the output beam splitter, assumed identical to the first one. Both mirrors (*M*) are also identical. The output intensity is measured at detectors D_1 and D_2. If the system is perfectly balanced, the superposed waves interfere constructively on D_1, destructively on D_2. If the two internal paths differ by a phase factor, both outputs are complementary.

Ernst Mach (1838–1916) was a Bohemian-Austrian physicist (born in the Czech Republic) and obtained his PhD from University of Vienna. He became a professor of experimental physics in 1867 at Charles-Ferdinand University, Prague. The mach number is named after him. The Mach-Zehnder interferometer was actually named after his son Ludwig Mach (AIP Emilio Segre Visual Archives, Physics Today Collection.)

Ludwig Louis Albert Zehnder (1854–1949) was a Swiss physicist who held professorships at the Universities of Freiburg and Basle and the inventor of the Mach–Zehnder interferometer in 1891. (AIP Emilio Segre Visual Archives, Sawyer Collection.)

small variations in distance, changes in refractive index or wavelength. When used to measure the wavelength of polychromatic light, one usually refers to the instrument as a spectrometer.

Mach–Zehnder modulator is a lithium niobate based electrooptic device whose light transmittance is controlled by an applied external voltage. It uses the Pockels effect of $LiNbO_3$ and the interference of two waves which have a relative phase difference, induced by an applied external voltage. It converts the induced phase shift by an applied voltage to an amplitude variation by using an interferometer, a device that interferes two waves of the same frequency but different phase. As shown in the figure, the device has implanted single-mode waveguides in a $LiNbO_3$ (or other electrooptic) substrate in the geometry shown. The waveguide at the input branches out at C to two arms A and B and these arms are later combined at D to constitute the output. The splitting at C and combining at D involve simple Y-junction waveguides. In the ideal case, the power is equally split at C so that the field is scaled by a factor $\sqrt{2}$ going into each arm. The structure acts as an interferometer because the two waves traveling through the arms A and B interfere at the output port D and the output amplitude depends on the phase difference (optical path difference) between the A and B branches. Two back-to-back identical phase modulators enable the phase changes in A and B to be modulated. Notice that the applied field in branch A is in the opposite direction to that in branch B. The refractive index changes are therefore opposite which means the phase changes in arms A and B are also opposite. For example, if the applied voltage induces a phase change of $\pi/2$ in arm A, this will be $-\pi/2$ in arm B so that A and B would be out of phase by π. These two waves will then interfere destructively and cancel each other at D. The output intensity would then be zero. Since the applied voltage controls the phase difference between the two interfering waves A and B at the output, this voltage also controls the output light intensity, though the relationship is not linear. It is apparent that the relative phase difference between the two waves A and B is therefore doubled with respect to a phase change ϕ in a single arm. We can predict the output intensity by adding waves A and B at D:

$$\frac{P_{out}(\phi)}{P_{out}(0)} = \cos^2\phi.$$

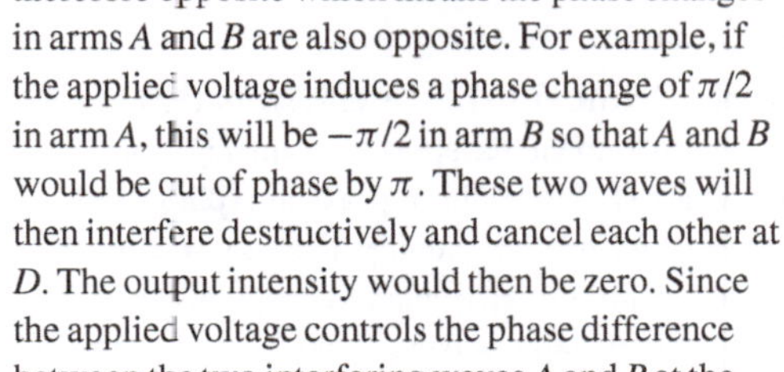

An integrated **Mach-Zehnder optical intensity modulator**. The input light is split into two coherent waves A and B, which are phase shifted by the applied voltage, and then the two are combined again at the output.

The power transfer is zero when $\phi = \pi/2$ as expected. In practice, the Y-junction losses and uneven splitting results in less than ideal performance; A and B do not totally cancel out when $\phi = \pi/2$.

Macrobending loss is due to small changes in the refractive index of the fiber due to induced strains when it is bent during its use, e.g. when it is cabled and laid. Induced strains change n_1 and n_2, and hence affect the mode field diameter, that is the field penetration into the cladding. Some of this increased cladding field will reach the cladding boundary to become lost into the outer medium (radiation, absorption, etc.). Typically macrobending loss crosses over into *microbending loss* when the radius of curvature becomes less than a few centimeters.

Magnesium fluoride (MgF_2) is an optical material that has good transmission from 0.15 to 7 μm, which includes the UV down to 150 nm. It is used in various optics applications, including antireflection coatings. The crystal structure is tetragonal. Single crystals are grown by the vacuum Stockbarger technique. It is a positive birefringent crystal with the optic axis along the crystal c-axis. MgF_2 has been reported to have good resistance against thermal and mechanical shock. $\rho = 3.177$ g cm^{-3}; $\nu_d = 106$; $E_g \approx 11$ eV; $T_m = 1255\,°$C; $\alpha_L = 14 \times 10^{-6}$ // to c-axis $8.5 \times 10^{-6} \perp$ to c-axis K^{-1}; $\kappa = 30$ // to c-axis W cm^{-1} K^{-1}; $c_s = 0.92$ J g^{-1} K^{-1}; $E = 138.5$ GPa; $G = 54.66$ GPa; $K = 101.32$ GPa; $\upsilon = 0.271$; $H_{\text{Knoop}} = 415$ kg mm^{-2}; fracture strength = 110–179 MPa, depending on the surface; crystalline MgF_2 is birefringent, $dn_o/dT \approx 0.19 \times 10^{-5}$ K^{-1}, $dn_e/dT \approx 0.10 \times 10^{-5}$ K^{-1} at 0.707 μm;

$$n_o = 1.36957 + \frac{3.5821 \times 10^{-3}}{\lambda - 0.14925};$$

$$n_e = 1.38100 + \frac{3.7415 \times 10^{-3}}{\lambda - 0.14947}.$$

where λ is the wavelength in μm (A. Duncanson and R. W. H. Stevenson, *Proc. Phys. Soc.* **72**, 1001, 1958). See *optical materials*.

Magnetic field of propagating electromagnetic (EM) radiation is related to its electric field E. In an isotropic medium of refractive index n (n = (relative permittivity)$^{1/2}$), the magnetic field is $B = E/v$ where $v = c/n = c/\sqrt{\varepsilon_r}$ is the phase velocity in the medium which is assumed to be nonmagnetic and ε_r is the relative permittivity. B and E are at right angles to each other. In general terms, in an isotropic medium, in vector notation, we have

$$\omega \mathbf{B} = \mathbf{k} \times \mathbf{E}$$

where ω is the angular frequency of the EM wave and $\mathbf{k}$ is the wavevector whose magnitude is $2\pi/\lambda$ in which λ is the wavelength.

Magnetic photonic crystal is a *photonic crystal* constructed from a magnetic material, or in which the application of an external magnetic field causes a change in the optical response of such a crystal. One important consequence of the action of an external field on such crystals would be the tuning of the optical response, including the polarization state of light propagating through the photonic crystal. Changing the temperature of such crystals also has important consequences on the optical response. This is because above the Curie (or Neél) temperature, the magnetically ordered medium undergoes a transition to a paramagnetic state, in turn changing the magnetooptical properties. An attractive aspect of these structures is the ability to exploit magnetic fields, magnetic ordering and magnetic domains as a means for controlling the optical properties of a photonic crystal. These properties can include both linear and nonlinear optical response. Moreover, such materials can have applicability in both visible/near-visible wavelengths, as well as in the microwave regime. Materials of interest can include both magnetic solids and thin films, as well as magnetic liquids and colloids.

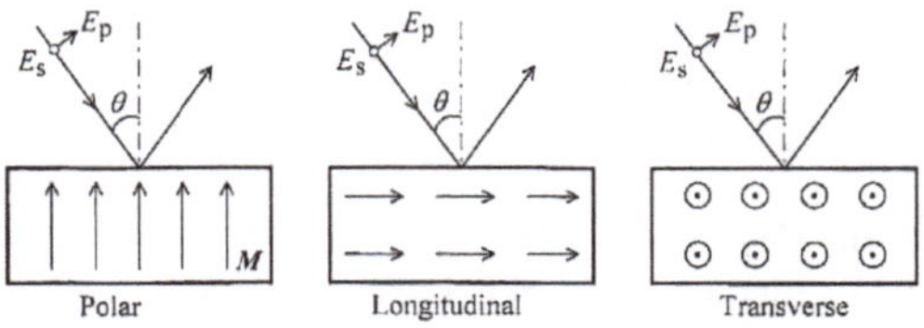

Magnetooptic effects A schematic of the geometry of the magnetooptic Kerr effect in polar, longitudinal and transverse geometries. Three orientations of the magnetization ***M*** in the sample are shown relative to the sample surface and to the plane of the incident light. The incident light is characterized as *p*- or *s*-polarized depending on whether the incident electric field lies in the plane of incidence (i.e. E_p) or perpendicular to it (i.e. E_s). (From M. Mansuripur, *Classical Optics*, Cambridge University Press, 2002; Figue 11.1.)

Magnetooptic effects are changes in the optical properties due to magnetization ***M*** of a material. The magnetization is typically induced by the presence of an external magnetic field. Magnetooptic effects can be observed over a broad range of electromagnetic radiation energy from the microwave region (e.g. so-called spin resonance) to the optical and x-ray (e.g. magnetic x-ray diffraction). In the case of the optical region two important effects are the *Faraday effect* and *magnetooptical Kerr effect* (MOKE) – both of these effects are linear in ***M***. Effects that are quadratic in ***M*** include the so-called Voigt effect. In the optical regime, the Faraday and MOKE effects are the most commonly studied effects. The Faraday effect is concerned with the change in the polarization (both rotation and ellipticity) of linear polarized light upon transmission through a sample with ***M*** pointing perpendicular to the sample surface. In the case of MOKE, one is concerned with the change in the polarization of linearly polarized light reflected from a sample surface in which three geometries are usually distinguished: (a) polar – where ***M*** lies perpendicular to the sample surface, and (b) and (c) in-plane magnetization. These latter two cases are as follows: (b) longitudinal – where ***M*** is parallel to the plane of incidence,

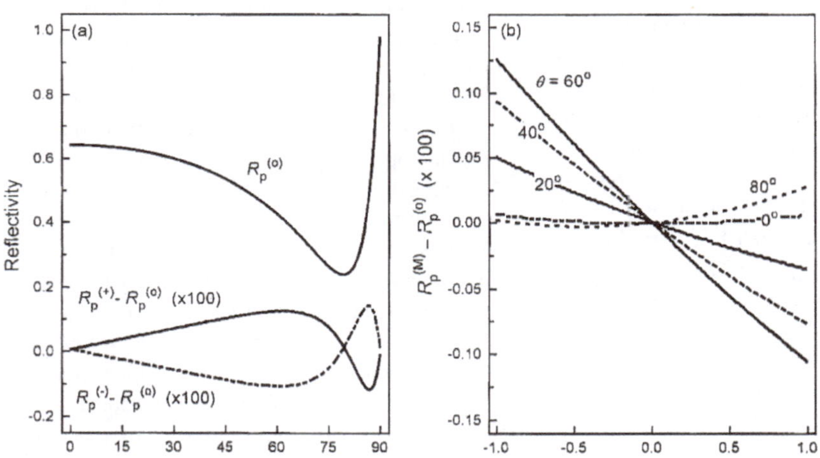

Magnetooptic effects Dependence of reflectivity R_p on the magnitude and/or direction of ***M*** in the transverse geometry. The incident beam is *p*-polarized in all cases as there are no transverse effects for *s*-polarized light. (a) Dependence of R_p on angle of incidence θ, where the superscript (o) refers to $M = 0$. When the medium is fully magnetized in the $\pm Y$ direction the reflectivity is denoted by $R_p^{\pm}$. (b) The variation of R_p with M for different angles of incidence. At $\theta = 0°$, the dependence on M is quadratic, while for 20°, 40° and 60°, it is nearly linear. (From M. Mansuripur, *Classical Optics*, Cambridge University Press, 2002; Figue 11.4.)

and (c) transverse – where $\boldsymbol{M}$ is perpendicular to the plane of incidence. These effects can be described by reference to the dielectric tensor for a magnetized medium:

$$\varepsilon = n^2 \begin{bmatrix} 1 & -iQm_z & iQm_y \\ iQm_z & 1 & -iQm_x \\ -iQm_y & iQm_x & 1 \end{bmatrix}$$

where m_x, m_y, and m_z are the direction cosines of the magnetization vector $\mathbf{M}_s$, and Q is the magnetooptical (or Voigt) constant such that $Q = i(\varepsilon_{xy}/\varepsilon_{xx})$, and $i = \sqrt{-1}$. The effects of the off-diagonal terms typically are not dominant because the usual values of Qm_x, Qm_y, and Qm_z are very small ($\sim 10^{-2}$). Based on this tensor, expressions can simply be derived for the change in the polarization of linearly polarized light reflected from a sample surface in the polar, longitudinal and transverse geometries.

Magnetooptical (MO) recording is a method of storing and retrieving data using a laser and magnetic media. The laser induces the local temperature of the magnetic medium to rise near or above its *Curie temperature* (typically about 200 °C) causing it to become almost demagnetized, i.e. have a very small magnetization (M_s) and a coercivity (H_c) without changing the physical structure of the material (and hence allowing for a high level of rewritability). A stationary magnet (or an electromagnet) below a disc of the MO material creates a magnetic field through the disk. Once the material is heated to its Curie temperature, the spinning disk moves the heated spot away from the laser. As this heated material cools, the magnetic field from the stationary magnet magnetizes this volume of material; the magnetization is retained by quenching in the written magnetic moment as the heated material rapidly cools, resulting in a written bit (a definite magnetization orientation) on the disk. Because both heat and magnetization are utilized, the recoding is called *thermomagnetic recording.*

Conventional MO recording requires at least two rotations of the disk of MO material. In the first or erase pass, all locations on the disk are scanned

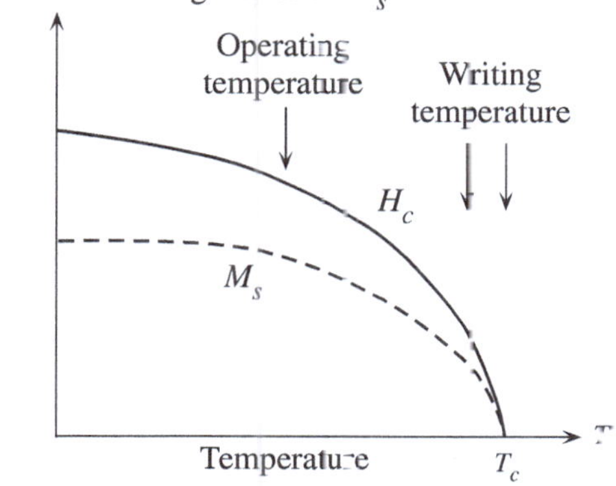

Magnetooptical recording Magnetic properties vs. temperature behavior of the magnetic medium in which the information is written. T_c is the Curie temperature at which magnetization would be lost.

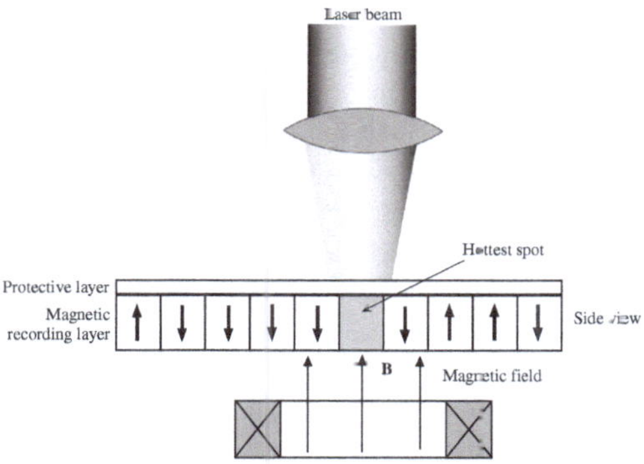

Magnetooptical recording principles.

with an intense laser beam sufficient to cause all bits to rise above the Curie temperature and be magnetized by a stationary magnetic field to a common uniform magnetization. During the second or write pass, the magnetic field of the stationary magnet is reversed. Data written in the write pass creates magnetic bits in the opposite direction to the erase pass. When the laser is pulsed at appropriate intervals, a pattern of recorded data is written to the disk at selected locations. With current MO technology, recorded data marks are typically 0.4 μm or less. A third pass, called a verify pass, is often performed to reread the data recorded on the disk and verify that it is correct. Data reading is accomplished by making use of the *magnetooptic Kerr effect* – when a low intensity beam of polarized laser light is reflected off a magnetic material, the polarity of the reflected beam is tilted. Similarly, if a laser is shone onto the recorded data pattern on the disk surface, the polarity of the reflected beam is rotated, either positively or negatively, depending on the magnetization of the recorded magnetic bit. MO diskettes used in current MO drives typically have a capacity from 100 MB to several gigabytes, and are capable of being read from, and written to, several times. The media of choice for current MO recording technology are amorphous rare earth transition metal (RE-TM) alloys such as TbFeCo or GdTbFe thin films. In order to increase the area recording density (quantity of bits that can be stored per unit area of storage material) for future high-density storage media, the wavelengths of the read and write laser are being decreased from the current wavelengths of 650–830 nm to obtain a smaller focused spot size. However, the current media exhibit a significant reduction in the Kerr effect at shorter wavelengths. Future high-density MO recording technology is focusing on novel materials with high Kerr activity at short wavelengths. Attractive candidates include magnetic multilayers such as Co/Pt multilayers. These have an easy direction of magnetization perpendicular to the film plane and 100% remanence when the thickness of the individual Co layers exceed about 2.5–5Å and the thickness of the individual Pt layers are about 10Å. As compared to RE-TM films, Co/Pt multilayers exhibit a higher corrosion resistance and a larger Kerr rotation angle at short wavelengths (300–400 nm), which promotes a higher recording density using a blue laser diode in the near future. Also CoNi/Pt multilayers can be applied for MO recording media, because its Curie temperature T_c is lower than that of Co/Pt multilayers. A lower T_c is required to satisfy low-power recording (high write sensitivity) and a high number of write/erase cycles. (S. Tsunashima, *J. Phys. D: Appl. Phys.*, **34**, R87, 2001.) See *magnetooptic effects*.

Magnification of an optical instrument is the ratio of the size of the observed image through the optical instrument to the actual object size. There are three types of magnification: angular, transverse, and longitudinal. If an object subtends an angle θ_o at an unaided eye while it subtends an angle θ_i when viewed through an optical instrument, then *the angular magnification* is simply θ_i/θ_o. Consider the magnification of a thin lens. If y_o and Δx_o are the height (transverse size) and longitudinal length of a "thin" object, and y_i and Δx_i are the height and longitudinal length of the image, then the *transverse* (or *lateral*) and *longitudinal magnifications*, M_T and M_L, may be defined generally as

$$M_T = \frac{y_i}{y_o} \quad \text{and} \quad M_L = \frac{\Delta x_i}{\Delta x_o}$$

and are not normally equal; thus the image of a three-dimensional object will have distortion due to unequal M_T and M_L. Further, M_L is usually defined for infinitesimally short lengths (Δx_o is very small) because M_L depends on the location of the object (x_o). $M_L = -M_T{}^2$.

The magnifications M_T and M_L for the thin lens system shown can also be written in terms of the object and image distances x_o and x_i from the object and image focal points F_o and F_i, respectively, as

$$M_T = -\frac{x_i}{f} = -\frac{f}{x_o} \quad \text{and} \quad M_L = -\frac{f^2}{x_o^2}$$

Table: Magnification of an optical instrument Sign convention in magnification by lenses.

Quantity	Positive sign	Negative sign
f	Converging lens	Diverging lens
Object height, y_o	Upright	Inverted
Image height, y_i	Upright	Inverted
Magnification, M_T	Upright image	Inverted image

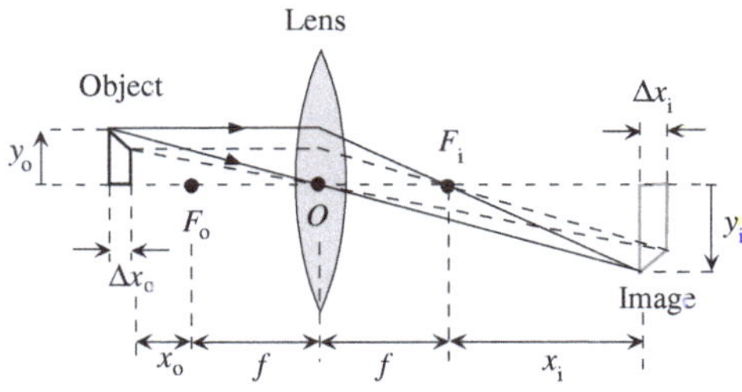

Transverse and longitudinal **magnification** of a thin lens. F_o and F_i are the object and image focal points.

where *f* is the focal length $f = OF_o = OF_i$; the above equation reflects the dependence of the magnification on the position of the object.

Magnifying power (MP) of an optical instrument is the *angular magnification* M_θ registered by the eye in viewing an object through the instrument. Put differently, it is the ratio of the image size on the retina produced by the optical instrument to the image size on the retina in the absence of the instrument (in the unaided eye). Suppose that an object (at a distance d_o) viewed by the unaided eye subtends an angle θ_u at the eye's pupil, angle subtended by

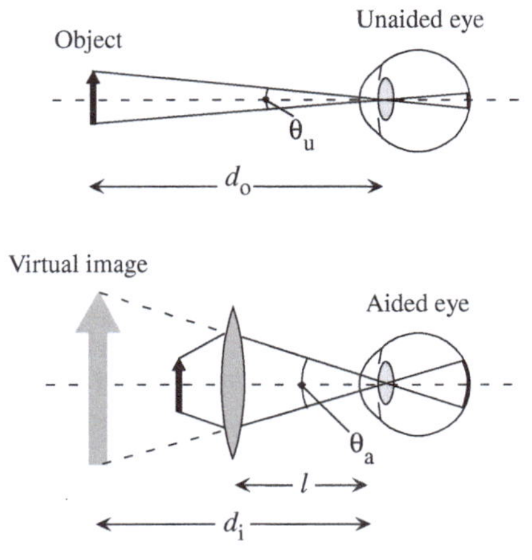

Magnifying power for a simple lens (magnifying glass) that produces a virtual image.

the chief rays from the top and bottom of the object. When the object is viewed through a magnifying glass, its image (virtual) at a distance d_i will subtend an angle θ_a (aided eye). The angular magnification and hence magnifying power is

$$\mathrm{MP} = M_\theta = \theta_a/\theta_u.$$

In the paraxial approximation (small angles), we obtain

$$\mathrm{MP} = (d_o/d_i)[1 + (d_i - l)/f]$$

where *f* is the focal length and *l* is the distance of the lens from the eye. The closest distance for an object to be clearly visible is the *near point*, for which $d_o = 0.25$ m. The clearest image will also occur at $d_i = 0.25$ m. Thus, $\mathrm{MP} \approx 0.25/f + 1 = 0.25D + 1$, where $D = 1/f$ is the *dioptric power* measured in diopters.

See *magnification of an optical instrument*.

Malus's law relates the output light intensity to the input light intensity when a linearly polarized light is passing through two polarizers that have their transmission axes aligned at different orientations. There are various optical devices that operate on the polarization state of a wave passing through it and thereby modify the polarization state.

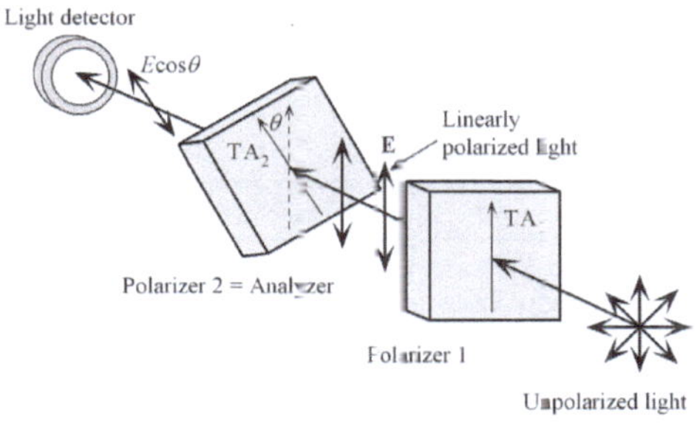

Malus's law Randomly polarized light is incident on a Polarizer 1 with a transmission axis TA_1. Light emerging from Polarizer 1 is linearly polarized with **E** along TA_1 and becomes incident on Polarizer 2 (called the "analyzer") with a transmission axis TA_2 at an angle θ to TA_1. A detector measures the intensity of the incident light. TA_1 and TA_2 are normal to the light direction.

Consider light passing through two polarizers with their transmission axes making an angle θ. The field in the radiation passing through the first polarizer will have the E-field aligned with the transmission axis of this first polarizer. However, only the component $E\cos\theta$ of the field will be allowed to pass through the second polarizer. The irradiance (intensity) of light passing through the second polarizer is proportional to the square of the electric field which means that the detected intensity varies as $(E\cos\theta)^2$. Since all the electric field will pass when $\theta = 0$, this is the maximum irradiance condition. The irradiance I at any other angle θ is then given by *Malus's law*, $I(\theta) = I(0)\cos^2\theta$.

Manca's rule allows the bandgap of various binary semiconductors to be predicted from the single bond energies using the equation $E_g = b(E_S - a)$ in which E_g is the bandgap, E_S is the single bond energy that can be calculated by using Pauling's equation and the electronegativities of A and B in the binary compound AB, that is, $E_S = D(AB) = [D(AA)D(BB)]^{1/2} + (X_A - X_B)^2$, where $D(AB)$ refers to the bond energy between A and B, $D(AA)$ is the A-A bond energy, $D(BB)$ is the B-B bond energy, $X_A - X_B$ is the electronegativity difference between A and B. (P. Manca, *J. Phys. Chem. Solids*, **20**, 268, 1961.)

Table: Manca's rule, $E_g = b(E_S - a)$.

Group	Examples	Manca's original relation
$A^{IV}A^{IV}$	Ge, Si	$E_g = 2.17(E_S - 1.34)$; eV
$A^{III}B^{V}$	GaAs, InAs	$E_g = 1.88(E_S - 1.43)$; eV
$A^{II}B^{VI}$	CdSe, CdTe	$E_g = 1.65(E_S - 2.23)$; eV

Marcuse's equation see *mode field diameter*.

Marginal rays in an optical system, e.g. a lens, are those rays that enter and travel near the periphery of the optical aperture of the system, that is, rays that pass through near the edges of the lens. If the lens is large, marginal rays do not obey the paraxial approximation, and hence cannot be focused to the same point as paraxial rays passing through the center of the lens. An aperture stop may be used to limit marginal rays to achieve paraxial imaging.

Mass action law in semiconductor science refers to the law $np = n_i^2$ (where n and p are electron and hole concentrations and n_i is the intrinsic concentration) which is valid under thermal equilibrium conditions and in the absence of external biases and illumination.

Mass attenuation coefficient is the *linear attenuation coefficient*, μ of a medium for high-energy radiation divided by the density, ρ, of the medium, i.e. μ/ρ. The most important property is the linear attenuation coefficient μ which is defined as the fraction of photons lost from the radiation beam per unit distance of propagation in the medium; the loss may be due to absorption and/or scattering. If $N(E)$ is the number of photons each with energy E traveling in the direction of a collimated beam of radiation along z in a medium, then $\mu(E)$ at the photon energy E is defined by

$$\mu(E) = -\frac{1}{N(E)} \cdot \frac{dN(E)}{dz}$$

so that the photon concentration decays exponentially $N(E) = N_o(E)\exp[-\mu(E)z]$, where N_o is the number of photons at $z = 0$. The attenuation depth δ is the reciprocal attenuation coefficient, $\delta(E) = 1/\mu(E)$. The coefficient $\mu(E)$ depends on the photon energy E as well as the density ρ and the atomic number Z of the medium. For high-energy radiation such as x-rays, it has been found that μ depends linearly on the density ρ. The *mass attenuation coefficient* is the linear attenuation coefficient of a medium normalized by its density, i.e. it is μ/ρ with units cm^2g^{-1}. Usually standard tables list the mass attenuation coefficient μ/ρ rather than μ. Thus the number of photons N at z becomes $N = N_o\exp[-(\mu/\rho)\rho z]$. Attenuation can be due to a number of reasons, such as the photoelectric effect (absorption), Compton scattering, Rayleigh scattering, and triplet production. The *total mass*

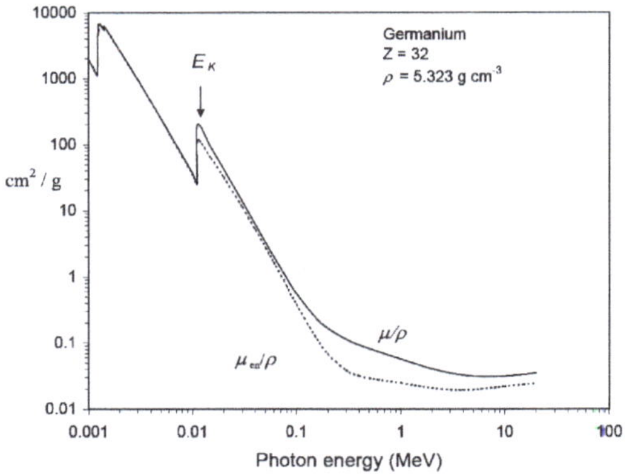

Mass attenuation coefficient μ/ρ and μ_{en}/ρ vs. photon energy E for Ge. At high energies μ_{en}/ρ is smaller than μ/ρ. (Data from NIST website. http://physics.nist.gov.)

attenuation coefficient can be written as

$$\frac{\mu}{\rho} = \frac{\tau}{\rho} + \frac{\sigma}{\rho} + \frac{\sigma_r}{\rho} + \frac{\pi}{\rho} + \frac{\gamma}{\rho}$$

where $\tau, \sigma, \sigma_r, \pi$ and γ are attenuation coefficients due to the photoelectric effect (τ), Compton scattering (σ), Rayleigh scattering (σ_r), pair production (π) and triplet production (γ). Over the photon energy (E) range from 10–100 keV, τ depends on atomic number Z as Z^3 and on E as E^{-3}, whereas σ and σ_r have no Z dependence but σ depends on E as $E^{-1.2}$. The relative contributions of various terms depend on the material; an example for carbon is shown in the figure. It should be emphasized that μ, or μ/ρ, does not specify how much energy has been deposited into the material; it simply characterizes the attenuation. μ/ρ for mixtures and compounds can be easily calculated from the simple summation (additivity) rule

$$\frac{\mu}{\rho} = \sum_i w_i(\mu/\rho)_i$$

where w_i is the weight fraction of the ith constituent in the compound.

The *mass energy transfer coefficient* μ_{tr}/ρ is the fraction of the mass attenuation coefficient μ/ρ

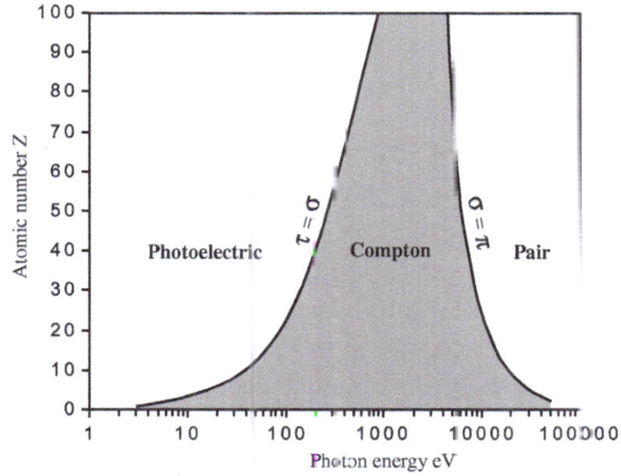

Mass attenuation coefficient The region where each x-ray interaction process is most likely is shown as a function of atomic number and x-ray photon energy. The transition zones between regions correspond to the two equal cross-sections; $\tau = \sigma$ and $\sigma = \tau$. (Data extracted and replotted from J. M. Boone, in *Handbook of Medical Imaging* Vol I. Physics and Pyschophysics, J. Beutel, H. L. Kundel, R. L. van Metter, Editors, SPIE, Belingham, WA, 2000, Ch 1, p.30.)

which goes to generate kinetic energy in charged particles. Usual contributions to μ_{tr}/ρ come from the photoelectric effect, and Compton scattering. For example, in the photoelectric effect, the absorbed x-ray photon knocks out an electron from an atom and imparts to it kinetic energy. The ionized atom can undergo a transition emitting a characteristic fluorescent x-ray photon. The latter can be absorbed as well. Alternatively the ionized atom can undergo a radiationless Auger transition, passing its energy to other electrons. In Compton scattering, the photon passes some of its energy to an electron. All these processes are encompassed in the μ_{tr}/ρ coefficient which can be written as the sum of τ_{tr}/ρ (fraction of τ_{tr} involved in energy transfer) and σ_{tr} which represents the energy transfer coefficients due to the Compton effect (fraction of σ that is involved in energy transfer). If the Compton scattering component is small, above the K-edge absorption,

$$\frac{\mu_{tr}}{\rho} = \frac{\tau}{\rho}\left[\frac{E - P_K Y_K \bar{E}_K}{E}\right];$$
$$E \geq E_K,\ K\text{-edge energy}$$

where Y is the probability of producing characteristic x-rays, Y_K (Y_L) is the K-shell fluorescent yield, P_K is the fraction of photoelectric attenuation coefficient that results in the K-shell interaction, and $\bar{E}_K$ is the average K-shell fluorescent energy.

The actual energy E_{absorbed} deposited into a medium by high-energy radiation per photon of energy E is characterized by the *mass energy absorption coefficient* μ_{en}/ρ, which can be defined by

$$E_{\text{absorbed}} = E\frac{(\mu_{en}/\rho)}{(\mu/\rho)}.$$

μ_{en}/ρ is not exactly the same as μ_{tr}/ρ because the latter characterizes how much kinetic energy has been given to charged particles and not necessarily deposited into the material. As these charges decelerate in the medium they lose energy by radiation (Bremsstrahlung) that escapes the material so that μ_{en}/ρ is less than μ_{tr}/ρ. Thus

$$\frac{\mu_{en}}{\rho} = \frac{\mu_{tr}}{\rho}(1 - g)$$

where g is the fraction of the kinetic energy of all charged particles produced by the absorbed photons that is lost as radiation as the charges decelerate. g is usually negligible for small Z materials.

If $\Phi(E)$ is the incident photon fluence of ionizing radiation onto a material of thickness L, the total energy deposited can be found by

Total energy absorbed

$$= \int_0^{E_{\max}} \Phi(E)E\frac{\mu_{en}}{\mu}[1 - \exp(-\mu(E)L)]dE.$$

(J. M. Boone, in *Handbook of Medical Imaging* Vol I. Physics and Pyschophysics, J. Beutel, H. L. Kundel, R. L. van Metter, Editors, SPIE, Belingha, WA, 2000 Ch 1, p. 30); NIST website on mass attenuation coefficient; http://physics.nist.gov. See *kerma*.

Mass energy absorption coefficient see *mass attenuation coefficient*.

Mass energy transfer coefficient see *mass attenuation coefficient*.

Material dispersion is the spread in the velocities of different wavelength waves in a dielectric medium due to the wavelength dependence of the refractive index of the medium. As long as the propagating electromagnetic (EM) wave is not monochromatic (single wavelength), and the medium has a group index that changes with the wavelength, there will always be material dispersion.

Material dispersion coefficient, D_m, is the time spread of a propagating optical pulse in an optical guide per unit length and per unit spectral wavelength width due to the wavelength dependence of the refractive index of the medium. If $\delta\lambda$ is the spread of the excitation wavelengths coupled into the fiber, $\delta\tau$ is the spread in propagation times of these different wavelengths due to material dispersion, and L is the fiber length, then,

$$D_m = \frac{1}{L}\frac{\delta\tau}{\delta\lambda}.$$

D_m can be determined from the n vs. λ relationship via

$$D_m \approx -\frac{\lambda}{c}\left(\frac{d^2n}{d\lambda^2}\right).$$

Matrix emitter see *vertical cavity surface emitting lasers*.

Maxwell's equations are a set of equations that describe the space and time dependence of the electric and magnetic fields in a medium through partial derivatives. In an isotropic linear dielectric material with a relative permittivity ε_r, Maxwell's equations lead to a wave equation for the electric field E,

$$\frac{\partial^2 E}{\partial x^2} + \frac{\partial^2 E}{\partial y^2} + \frac{\partial^2 E}{\partial z^2} = \varepsilon_o\varepsilon_r\mu_o\frac{\partial^2 E}{\partial t^2}$$

where ε_o and μ_o are the permittivity and permeability of free space. It is assumed that the dielectric

James Clerk Maxwell (1831–1879) was a Scottish physicist born in Edinburgh. He held professorships at Aberdeen (1856–1860), London (1860–1865) and Cambridge (from 1871) Universities; at Cambridge he set up the Cavendish Laboratory (1856). He is renowned for his mathematical formulation for describing light in terms of the behavior of time-varying electric and magnetic fields. He also made distinct contributions to the kinetic molecular theory. (Source: *Physics World*, Institute of Physics, March 1999, p. 43). See *electromagnetic wave*.

material is nonmagnetic, i.e relative permeability $\mu_r = 1$. The solution of this equation is an electromagnetic wave that is traveling with a velocity v given by

$$v = \frac{1}{\sqrt{\varepsilon_o \varepsilon_r \mu_o}}.$$

Maxwell–Garnett equation, see *effective medium approximation*.

McCumber theory relates the absorption and emission cross-sections $\sigma_a(\nu)$ and $\sigma_e(\nu)$, each a function of the photon frequency ν, for transitions in impurities or dopants embedded in a dielectric medium (such as Er^{3+} ions in a glass medium) through a relatively simple expression derived for a two-level system through detailed balance and the applications of Einstein relations as

$$\sigma_a(\nu) = \sigma_e(\nu) \exp\left(\frac{h\nu - \varepsilon_T}{k_B T}\right)$$

in which $h\nu$ is the photon energy, k_B is the Boltzmann constant, T is the absolute temperature and ε_T is a temperature-dependent effective excitation energy that is obtained by summing transition energies between the different levels between the two manifolds of energies:

$$\varepsilon_T = \Delta E_{21} - kT \ln \left[\frac{1 + \sum_{j=2}^{g_2} \exp(-\delta E_{2j}/kT)}{1 + \sum_{i=2}^{g_1} \exp(-\delta E_{1i}/kT)} \right]$$

where g_1 and g_2 are the degeneracies of the levels E_1 and E_2, ΔE_{21} is the separation between the lowest levels of the two manifolds, δE_{1i} are the energy differences between the sublevels with respect to the lowest level in the $1i$ manifold, and δE_{2j} in the $2j$ manifold (Stark energy levels). It is assumed that the system is in thermal equilibrium, and it is a dilute system, that is impurity–impurity interactions are negligible. The rate of spontaneous emission of photons per unit solid angle $f(\nu)$ is given by (in CGS)

$$f(\nu) = \sigma_e(\nu) \left(\frac{\nu n}{c}\right)^2$$

where n is the refractive index of the host material. (D. E. McCumber, *Phys Rev*. **136**, 954, 1964; B. S. Neporent, Dok. Akad. Nauk. SSR, **112** 682, 1958 (Sov. Phys. Dokl. **3**, 337, 1858).)

Mean free path is the mean distance traversed by an electron between scattering events. If τ is the mean free time between scattering events, and u is the mean speed of the electron, then the mean free path is $\lambda = u\tau$.

Mean free time is the average time it takes to scatter a conduction electron. If t_i is the free time between collisions (between scattering events) for an electron labelled as i, then $\tau = \overline{t_i}$ averaged over all the electrons. The drift mobility μ_d is related to the mean free time by $\mu_d = e\tau/m_e$. The reciprocal of the mean free time is the mean probability per unit time that a conduction electron will be scattered, or, put differently, the mean frequency of scattering events.

Meinel–Meinel equation is a useful formula that allows the calculation of the integrated solar intensity (total solar intensity) arriving at a point on Earth where the *solar latitude* is α,

$$I = 1.353(0.7)^{(\mathrm{cosec}\,\alpha)^{0.678}}\,\mathrm{kWm^{-2}}$$

where $\mathrm{cosec}\alpha = 1/\sin\alpha$. The solar latitude α is the angle between the Sun's rays and the horizon. This intensity is maximum at 0.95 kW m^{-2} when $\alpha = 90°$. If ϕ is the geographic latitude, then $\alpha = 90° - \phi$ in degrees. Around September 23rd and March 22nd, the Sun's rays arrive parallel to the plane of the equator.

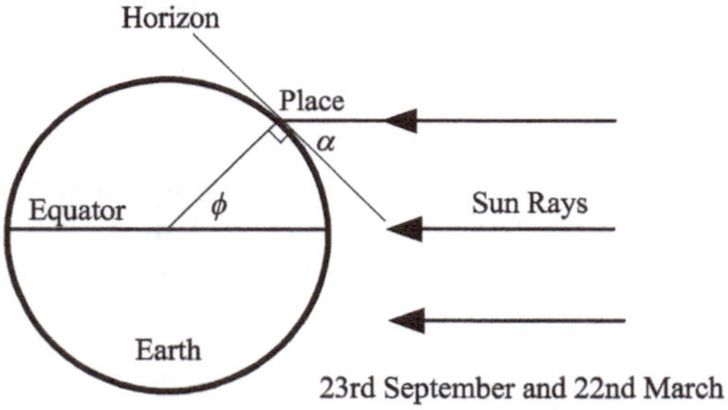

Meinel–Meinel equation Noon insolation at a location of latitude ϕ during 23rd September and 22nd March. The angle α is the solar latitude and ϕ is the geographic latitude.

Melt growth refers to a class of techniques for the crystallization of a material from a liquid phase (a melt) at the material's melting temperature. A large variety of technologically important electronic materials can be prepared using melt growth including, for example, Si, Ge, GaAs, InP, InSb, GaP and CdTe. Different melt growth techniques are suitable, depending on the specific requirements for the material to be prepared: these include *Czochralski*, *liquid encapsulated Czochralski (LEC)*, *magnetic LEC*, *horizontal Bridgman*, *vertical Bridgman* and the *floating-zone* growth techniques.

MEMS systems technology, or *microelectromechanical systems*, results from the integration of mechanical elements, sensors, actuators, and electronics on a common substrate. The key to MEMS is the use of microfabrication technology. Thus, the well-known processes used to fabricate integrated circuits (IC) for computers are applied to provide precision manufacturing of integrated systems at great cost advantage. The electronics which provide for on-board intelligence are typically fabricated using traditional CMOS, Bipolar, or BICMOS processes, while the micromechanical components are fabricated using compatible "micromachining" processes that selectively etch away parts of the silicon wafer, also known as *bulk micromachining*, or add new structural layers, also known as *surface micromachining*, to form the mechanical and electromechanical devices. Microsystems use sensors and actuators. Sensing functions can include sensing of mechanical, thermal, biological, chemical, optical, and magnetic phenomena. Actuation can include moving, positioning, regulating, pumping, and filtering, thereby controlling the environment for some desired outcome or purpose. Because MEMS devices are manufactured using batch fabrication techniques similar to those used for ICs, unprecedented levels of functionality, reliability, and sophistication can be placed on a small chip at a relatively low cost. An example of a current MEMS device is the micromirror array used in digital video projector displays, and also for optical cross-connects in telecommunications systems (see the figure).

Another example is the so-called lab-on-a-chip technology in which very small volumes (e.g. picoliters) of fluid are pumped around a 'circuit' and some property of the fluid examined. These so-called microfluidic chips can involve optical monitoring through integration of optical waveguides on

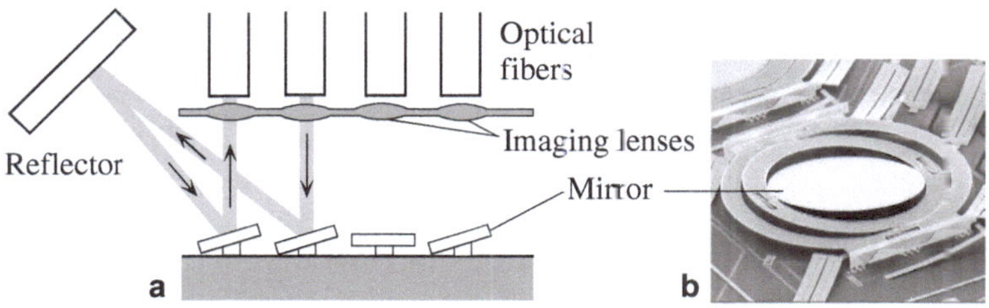

(a) An example of **MEMS** optical switch in which very tiny mirrors can be made to redirect light from one incoming fiber to another, rerouting the optical data. (b) Each mirror measures about several hundred microns in diameter.

the same chip, as well as light sources and detectors. The latter systems are also referred to as *micro-electrooptomechanical systems* or MEOMS.

Mercury cadmium telluride see $Hg_{1-x}Cd_xTe$.

Meridional or tangential plane in an optical system is the plane that contains the optical axis and the principal (or chief) ray from the object to the image. All rays in this plane are *tangential or meridional rays*. Rays that do not lie in this plane are *skew rays*. See *astigmatism* for an illustration.

Meridional ray is a light ray that enters an optical fiber through the fiber axis and hence also crosses the fiber axis on each reflection as it zigzags down the fiber. It travels in a plane that contains the fiber axis. Guided modes in optical fibers resulting from meridional rays are either TE or TM type.

Mesogen see *liquid crystal*.

Metal–insulator transition was proposed by Sir Nevil Mott in 1949, and is usually termed a Mott transition. It was proposed to occur for a crystal with atoms having an odd number of electrons per primitive unit cell, and at a temperature of 0 K, having a critical lattice constant $a = a_c$ of about $4.5a_B$, where a_B is the Bohr radius. For $a < a_c$, the crystal will behave as a metal and for $a > a_c$ it will behave as an insulator. The origin of this transition lies in the fact that as the separation between atoms increases, the conduction electron(s) of the atom can form a bound and therefore insulating state with the ion, while for small separations, the electron(s) will be screened from the attractive potential of the positive ion potential, and the bound state cannot form, resulting in a metallic state. Naturally, the screening depends on the electron concentration, n. This

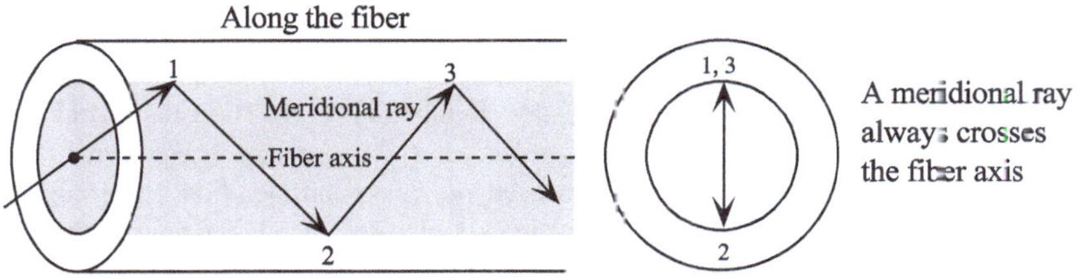

A **meridional ray** always crosses the fiber axis.

is more quantitatively expressed in terms of the screening length – for example, in the case of the so-called Thomas–Fermi approximation, the screening length λ for a screened Coulomb potential $\varphi(r) = (q/r)\exp(-r/\lambda)$ is:

$$\lambda^2 = 2E_F\varepsilon_r\varepsilon_0/3q^2n$$

where $E_F = (h^2/8m)(3n/\pi)^{2/3}$ is the Fermi energy. Screening thus weakens the binding of the electron–ion pair, and at a critical electron concentration, the screening constant becomes so small that the electron and ion can no longer occupy the same bound state. This happens when λ exceeds the electron–ion separation or Bohr radius, $a_B = \varepsilon_r\varepsilon_0 h^2/(\pi mq^2)$. By substitution, $\lambda^2 = (a_B/4n^{1/3}).(\pi/3)^{1/3} \sim a_B/4n^{1/3}$ and thus using the known condition for a bound state to form as $\lambda > 0.84\, a_B$ gives $n^{1/3} < (0.354/a_B)$ as the critical concentration required to have an insulator, or the Mott density. If the crystal is a simple cubic lattice of lattice constant a, then $n = a^{-3}$ and the condition for a Mott insulator is that $a > 2.8a_B$. An interesting example of a Mott transition is that for a metallic electron–hole (*e-h*) plasma and an *exciton* gas in a semiconductor. In semiconductors, when photons of energy close to the fundamental bandgap E_g are absorbed, they generate *e-h* pairs. At energies significantly above E_g, the electrons and holes move largely independent of each other and form a conducting gas of unbound *e-h* pairs. However, electrons and holes always interact via the Coulomb force and – due to their opposite charge – can form bound states called excitons. Due to the attractive binding, excitons exhibit a narrow resonance at energies slightly below the band gap. Since excitons are analogous to neutral hydrogen atoms, they are insulating, and their formation from unbound *e-h* pairs corresponds to a metal–insulator transition (and vice versa). Although the properties of excitons have been extensively studied for over 50 years many aspects of the excitonic Mott transition remain elusive.

Metalization refers to the use of sputtering or *chemical vapor deposition* (CVD) processes to create conductors by applying a thin layer of metal to a device. This may be a single metal such as aluminum, gold or platinum, or may be an alloy such as gold-germanium, or may involve a multilayer structure of different metals.

Metallurgical junction in a *pn* junction is where there is an effective junction between the *p*-type and *n*-type doped regions in the crystal. It is where the donor and acceptor concentrations are equal or where there is a transition from *n*- to *p*-type doping.

Metalorganic chemical vapor deposition (MOCVD) is a type of CVD process in which metalorganic species are used as precursor source gases for the deposition process. This is currently one of the techniques of choice for fabricating epitaxial layers of compound semiconductors, the other technique being MBE. For example, to grow the III-V alloy compounds AlGaAs, the usual choice of feedstock gas is metalorganics of the group III elements (Ga and Al) and hydrides of the group V elements. These chemicals must satisfy several requirements to be suitable for growth: (a) they must decompose at the growth temperature (typically in the range between 500 and 800 °C) to permit incorporation into a growing layer to occur; (b) they must be gaseous or have high vapor pressures at room temperature, such that they can easily be transported through tubes from the sources to the reaction vessel; and (c) they must be available with reasonable purity. Metalorganic sources typically used in this case include the two gallium sources, trimethyl-gallium (TMGa) and triethyl-gallium (TEGa) and the aluminum sources trimethyl-aluminum (TMAl) and triethyl-aluminum (TEAl). The usual hydride source of arsenic is AsH_3 (arsine). These source gases are fed into a gas manifold prior to entering the reaction vessel to provide a homogeneous feedstock composition which, in turn, assists in producing epitaxial layers with a more homogeneous composition. A key feature of MOCVD systems is the ability to rapidly switch the gas phase composition feeding the reaction chamber. These well-controlled compositional changes have made superlattice growth and abrupt doping profiles attainable and have been instrumental in the success of MOCVD for advanced optoelectronic device development. Key to this capability is the use of mass flow controllers that can precisely control the flow of each gas supplied to the system manifold, coupled with fast acting control valves and a low volume reaction chamber. The chemical reaction that occurs on a hot substrate using TMGa as a gallium source at about 700 °C is $(CH_3)_3Ga + AsH_3 \Rightarrow GaAs + 3CH_4$ while a similar reaction could have been written for TMAl. Thus, simple mixing of TMGa and TMAl can yield the whole range of AlGaAs alloys: $(1 - x)\text{TMGa} + x\text{TMAl} + AsH_3 \Rightarrow Al_xGa_{1-x}As + 3CH_4$. Dopants are added to the gas feedstock in the same manner as other components. Typically, either metalorganic group II sources, or group IV and group VI hydrides are used as the *p*- and *n*-type doping sources for III-V growth. This approach permits growth and doping on a number of substrates in a single growth run, and therefore offers higher throughput. One problem with MOCVD is that many of the growth parameters are interdependent, leading to problems in controlling the process. Another issue of concern with MOCVD is the handling of toxic and flammable gases. In spite of its limitations, MOCVD can produce layers of almost any III-V compound as well as high-quality *superlattices* and *heterostructures* with quality matching that of both LPE and MBE. The total cost of an MOCVD system (including safety equipment) is about an order of magnitude higher than for an LPE system, but about half that of an MBE system.

Metamers or **metamer colors** are two (or more) colors that are perceived to be the same but have different physical origins. A yellow color can be generated by a yellow LED and can also be generated by mixing red and green. These two yellows are perceived to be the same and are metamers.

Metropolitan area networks (MAN) see *optical networks*.

Michelson contrast see *contrast*.

Microbending loss is the attenuation of light in a fiber due to the local (sharp) bending of the fiber, typically for fiber bending with a radius of curvature less than ~10 mm. Microbending loss is due to a "sharp" local bending of the fiber which changes the guide geometry and refractive index profile locally which leads to some of the light energy being radiated away from the guiding direction. A sharp bend will change the local waveguide geometry in such a way that a zigzagging ray suddenly finds itself with an incidence angle θ', narrower than its normal angle θ ($\theta' < \theta$) which gives rise to either a transmitted wave (a refracted wave into the cladding), or to a greater cladding penetration. If $\theta' < \theta_c$, the *critical angle*, then there will be no *total internal reflection* and substantial light power will be radiated into the cladding and eventually to the outside medium (polymer coating, etc.). Greater penetration can lead to the optical field reaching the cladding-outer boundary and hence to some of the light being lost away into the outer coating. Attenuation increases sharply with the extent of bending; as θ' gets narrow and TIR is lost, substantially more energy is transferred into the cladding. Further, highest modes propagate with incidence angles θ close to θ_c which means that these modes are most severely affected. Multimode fibers therefore suffer more from bending losses then single mode fibers. Microbending loss α_B increases rapidly with increasing bend "sharpness," i.e. with decreasing radius of bend curvature, *R*. The microbending loss α_B increases exponentially with decreasing *R* which depends on the wavelength and the fiber characteristics (e.g. *V*-number). Typically, bend radii less than ~10 mm can lead to appreciable microbending loss. Microbending loss α_B depends on the fiber characteristics and wavelength and can be calculated approximately given various fiber parameters using the single-mode fiber microbending loss equation

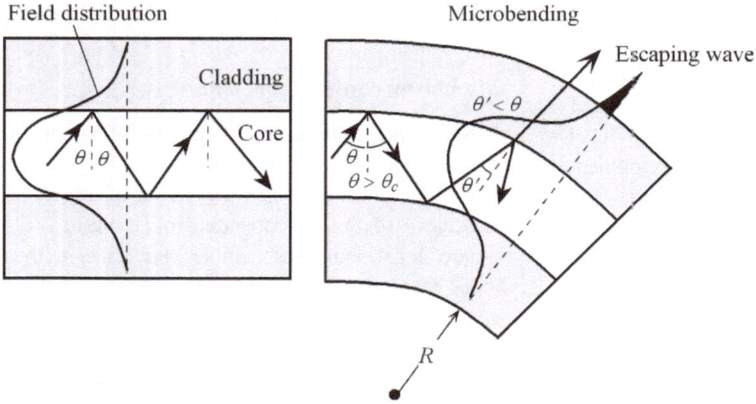

Microbending loss Sharp bends change the local waveguide geometry that can lead to waves escaping. The zigzagging ray suddenly finds itself with an incidence angle θ' that gives rise to either a transmitted wave, or to a greater cladding penetration; the field reaches the outside medium and some light energy is lost.

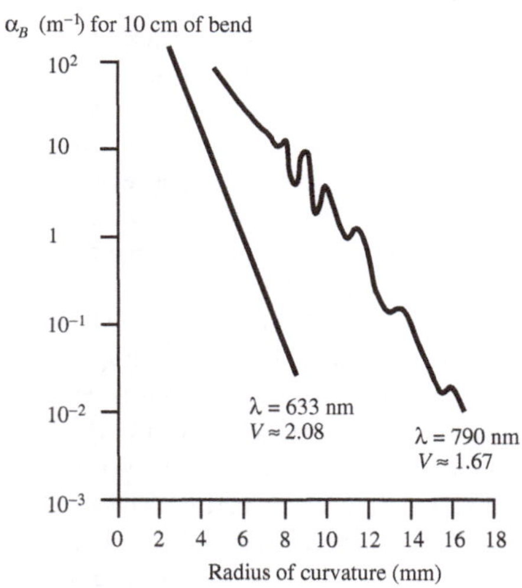

Microbending loss Measured microbending loss for a 10 cm fiber bent by different amounts of radius of curvature R. A single-mode fiber with a core diameter of 3.9 μm, cladding radius 48 μm, $\Delta = 0.00275$, NA ≈ 0.10, $V \approx 1.67$ and 2.08. (Data extracted and replotted from A. J. Harris and P. F. Castle, *IEEE J. Light Wave Technology*, **LT**14, 34, 1986; see original article for a discussion of peaks in α_B vs. R at 790 nm.)

(D. Marcuse, *J. Op. Soc. Am.*, **66**, 216, 1976):

$$\alpha_B \approx \frac{\pi^{1/2}\kappa^2}{2\gamma^{3/2}V^2[K_1(\gamma a)]^2} R^{-1/2} \exp\left(-\frac{2\gamma^3}{3\beta^2}R\right)$$

where R is the bend radius of curvature, $a =$ fiber radius, β is the propagation constant, determined by b, *normalized propagation constant*, which is related to V, $\beta = n_2k[1 + b\Delta]$; $k = 2\pi/\lambda$ is the free-space wavevector; $\gamma = \sqrt{[\beta^2 - n_2{}^2k^2]}$; $\kappa = \sqrt{[n_1{}^2k^2 - \beta]}$, and $K_1(x)$ is a first-order modified Bessel function, available in mathematical software packages. Note that $b \approx (1.1428 - 0.996V^{-1})^2$ for a single mode fiber with $1.5 < V < 2.5$. (A. J. Harris and P. F. Castle, *IEEE J. Light Wave Technol.*, **LT**4, 34, 1986.)

Microelectrooptomechanical systems or **MEOMS** see *MEMS*.

Microlaser see *vertical cavity surface emitting lasers*.

Mie scattering refers to the scattering of light from scatterers with size about the same as the wavelength of the incident light. The scatterers are assumed to have a refractive index significantly different from that of the surrounding medium. The scattering from molecules and very tiny particles (i.e. with size $< \lambda/10$) is predominantly governed by *Rayleigh scattering*. For particle sizes exceeding a wavelength, Mie scattering predominates. Mie scattering is weakly dependent on the incident light wavelength. In the case of sunlight, Mie scattering produces the almost white glare around the Sun when a lot of particulates are present in the air. Such scattering is also responsible for the white light seen with mist and fog. Mie scattering is described by Mie theory and produces a pattern like an antenna lobe, with a sharper and more intense forward lobe for larger particles. *Mie theory* represents a complete physical and mathematical theory for light scattering from spherical particles and was developed by Gustav Mie in 1908. As distinct from *Rayleigh scattering*, Mie theory treats all possible ratios of particle diameter to wavelength. Mie theory assumes a homogeneous, isotropic and optically linear material irradiated by a plane wave. Mie theory is very important for meteorological optics, where diameter-to-wavelength ratios on the order of unity and larger are characteristic of many problems such as haze and cloud light scattering. Also the scattering of radar waves by raindrops has been another significant application of Mie theory. Mie theory is a rigorous treatment of the diffraction of a plane monochromatic wave by a homogeneous sphere of any diameter and of any composition situated in a homogeneous medium. The solution, derived for diffraction by a single sphere, also applies to diffraction by any number of spheres, provided that they are all the same diameter and composition, and provided that they are randomly distributed and separated from each other by distances larger than the wavelength of the incident radiation. Mie scattering normally increases with the wavelength. (C. L. Adlera and J. A. Lock, *Am. J. Phys.*, **70**, 01, 2002.)

Mie theory see *Mie scattering*.

Miller's rule is an empirical rule that is based on Miller's observation that the ratio of the second-order susceptibility $\chi^{(2)}$ to the cube of the first-order susceptibility $\chi^{(1)}$ in the expression relating the polarization and the electric field, $P = \chi^{(1)}E + \chi^{(2)}E^2$, expressed in esu units, in many noncentrosymmetric crystals, is very approximately constant. The rule is not rigorous. (R. C. Miller, *Appl. Phys. Letts.*, **5**, 17, 1964; R. W. Boyd, *Nonlinear Optics, Second Edition*, Academic Press, San Diego, 2003.)

Minority carrier diffusion length, L, is defined by the equation $L = (D\tau)^{1/2}$ where D is the diffusion coefficient and τ is the lifetime of the minority carriers. Suppose that the diffusion occurs because of an exponential concentration gradient, for example, the minority carrier concentration profile $p(x)$ in an n-type semiconductor follows a simple exponential $\exp(-x/L)$ form. Then, the mean distance a minority carrier diffuses before recombination is L because the average distance diffused along x is $\int x \exp(-x/L)/\int \exp(-x/L)$ which gives L.

Minority carrier injection is the flow of electrons from the n-side into the p-side and holes from the p-side into the n-side of a pn junction when a voltage is applied to reduce the built-in voltage across the junction. It can also refer to an increase in the minority carrier concentration due to photogeneration.

Minority carrier lifetime, τ, is the mean existence time of a minority carrier before it disappears by recombination. $1/\tau$ is the mean probability per unit time that a minority carrier recombines with a majority carrier. Suppose that excess electrons and holes have been injected, as would be in a pn-junction under forward bias, and that Δn_p is the excess electron, minority carrier, concentration and Δp_p is the excess hole, majority carrier, concentration in the *neutral* p-side of a GaAs pn junction. Injected electron and hole concentrations would be the same to maintain charge neutrality, that is, $\Delta n_p = \Delta p_p$. In many instances the rate of change $\partial \Delta n_p/\partial t$ is proportional to Δn_p and an *excess minority carrier recombination time (lifetime)* τ_e is defined by

$$\frac{\partial \Delta n_p}{\partial t} = -\frac{\Delta n_p}{\tau_e}.$$

In practice, the injected excess minority carrier concentration Δn_p is much greater than the actual equilibrium minority carrier concentration n_{po}. There are two conditions on Δn_p corresponding to *weak* and *strong* injection based on Δn_p compared with the majority carrier concentration p_{po}. In *weak injection*, $\Delta n_p \ll p_{po}$, where p_{po} is the majority carrier concentration. Then $n_p \approx \Delta n_p$ and $p_p \approx p_{po} + \Delta p_p \approx p_{po} \approx N_a$ = acceptor concentration. The rate of recombination in this case is given by

$$\partial \Delta n_p/\partial t = -B(n_p p_p - n_{po} p_{po}) \approx -B_o N_a \Delta n_p$$

where B is the direct recombination capture coefficient. Thus, under weak injection, the lifetime is constant and given by

$$\tau_e = 1/BN_a.$$

In *strong injection*, $\Delta n_p \gg p_{po}$. Then the rate of recombination becomes

$$\partial \Delta n_p/\partial t = -B_o \Delta p_p \Delta n_p = B(\Delta n_p)^2$$

so that $\tau_e = 1/B\Delta n_p$ which implies that under high-level injection conditions the lifetime τ_e is inversely proportional to the injected carrier concentration. When a light-emitting diode (LED) is modulated under high injection levels, for example, the lifetime of the minority carriers is therefore not constant, which in turn leads to distortion of the modulated light output. See *direct recombination capture coefficient*.

Minority carriers are electrons in a p-type and holes in an n-type semiconductor.

Mismatch strain see *lattice parameter mismatch*.

Miyagi–Kawakami equation see *hollow optical fiber*.

Modal noise, or *speckle noise*, in an optical system is random fluctuations in the transmitted power due to the optical power carried by various modes experiencing random phase changes which result in random "mode mixing" *and* the fact that attenuation losses are mode dependent; a single-mode fiber has no modal noise.

Modal or intermodal dispersion is due to the spread in time of an infinitesimally thin light pulse of single wavelength as it propagates along a fiber through various modes. Each mode has a different propagation constant so that at the end of the fiber the modes arrive at different times and constitute an output pulse that is broadened in time. Modal dispersion is zero in a single-mode fiber that allows

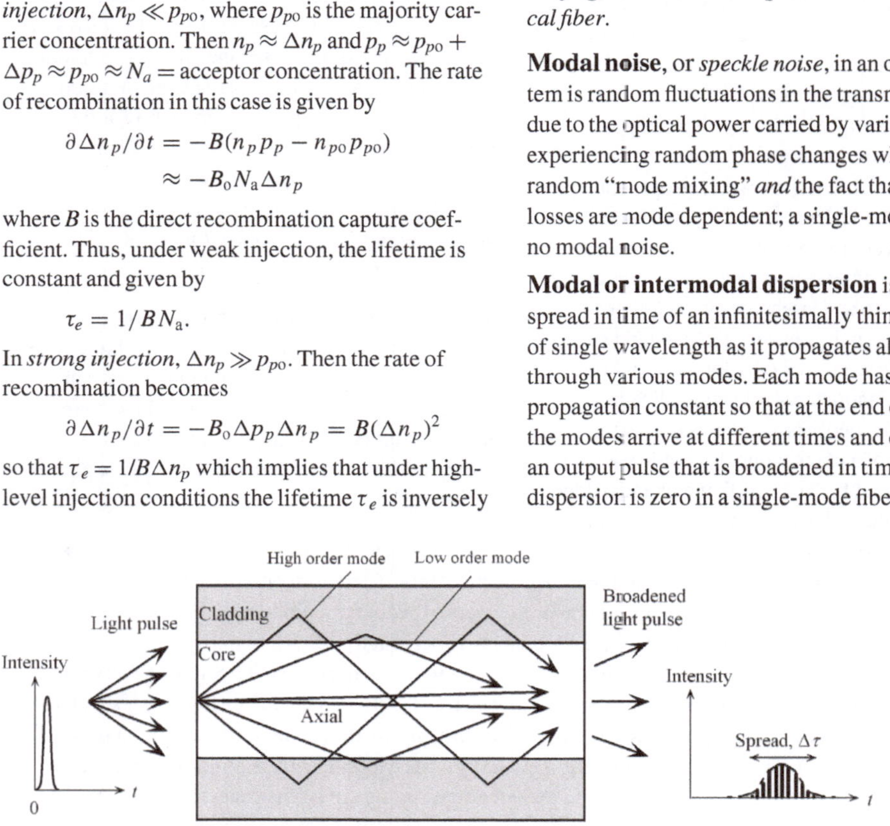

Modal dispersion Schematic illustration of light propagation in a slab dielectric waveguide. A light pulse entering the waveguide breaks up into various modes which then propagate at different group velocities down the guide. At the end of the guide, the modes combine to constitute the output light pulse which is broader than the input light pulse.

only one mode, the fundamental mode, to propagate. In a multimode fiber, modal dispersion in terms of the full spread $\Delta\tau$ is approximately given by

$$\frac{\Delta\tau}{L} \approx \frac{n_1 - n_2}{c}$$

where L is the guide length, n_1 and n_2 are the refractive indices of the *core* and the *cladding* and c is the velocity of light in free space.

Mode field diameter (MFD), $2w$, is the extent of the field distribution across the optical guide in the *fundamental mode.* The field penetrates the *cladding* so that the mode field diameter is not simply the diameter of the core. Usually the optical power or field squared, $E(r)^2$, distribution across the fiber as a function of the radial distance r from the fiber axis can be written as a *Gaussian*, which can then be used to define a MFD by

$$E(r)^2 = E(0)^2 \exp[-(r/w)^2]$$

where $E(0)$ is the field at the center. At $r = w$, the power has dropped by $1/e$. One approximation for w is

$$w = a(0.65 + 1.62V^{-3/2} + 2.88V^{-6})$$

where V is the *V-number.* In many single-mode fibers, V is typically 2, and the MFD is about 1.27 times the core diameter. In the range $1.6 < V < 2.4$,

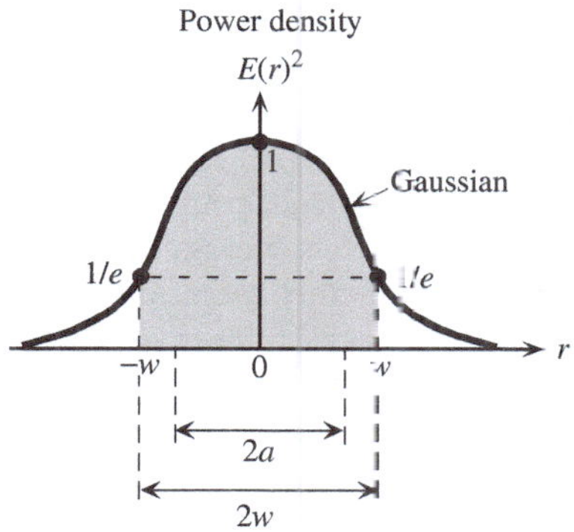

Mode field diameter Definition of MFD, mode field diameter $2w$.

the MFD can be calculated by an even simpler equation, $w = 2.6a/V$. (D. Marcuse, *Bell Syst. Tech. J.* **56**, 703, 1977). See *Petermann mode radius.*

Mode hop is a sudden change in the mode appearing in the lasing output spectrum from a laser as a result of temperature change or an increase in the pumping (e.g. an increase in the diode current).

Mode locked laser is a laser that has been constructed to have one transverse mode and many longitudinal modes that have the same phase so that the longitudinal modes reinforce each other to

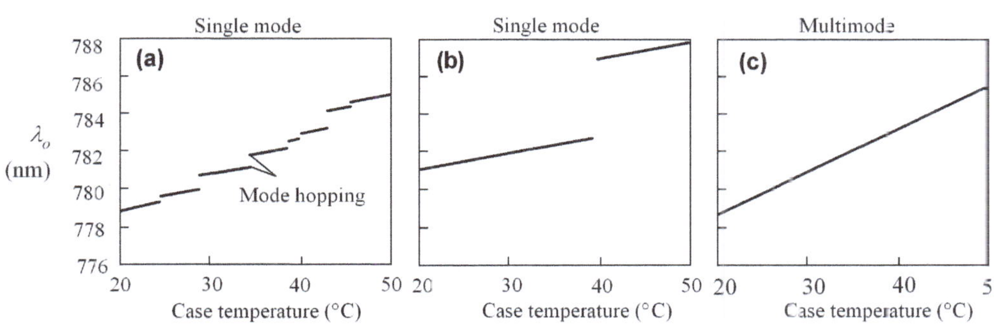

Mode hop Peak wavelength vs. case temperature characteristics. (a) Mode hops in the output spectrum of a single-mode laser diode. (b) Restricted mode hops and none over the temperature range of interest (20 – 40 °C). (c) Output spectrum from a multimode laser diode.

generate an intense pulse of lasing emission at fixed time intervals which depends on the laser cavity length and the speed of light in the lasing medium. Normally the longitudinal modes of a laser tube would be "independent" with random relative phases. In such a case, the output intensity from the laser would simply be the sum of these many individual modes. If, on the other hand, the modes have the same phase then they would reinforce each other in such a way that they generate an intense optical pulse that repeats itself every $T_{\text{round-trip}} = 2L/c$ seconds where L is the optical resonator length, c is the velocity of light in the medium and $T_{\text{round-trip}} = 2L/c$ is the round-trip time of a light pulse in the optical resonator. The situation is analogous to taking a fixed number of sine waves with multiple frequencies, adjusting their phases so that one can obtain the maximum amplitude from their summation. The achievement of such an intense light pulse at every $T_{\text{round-trip}}$ seconds requires that the "modes are locked," that is their relative phases have been correctly adjusted and fixed, to yield the required maximum output intensity. When modes have been locked, there must be an optical pulse in the resonator that is traveling between the mirrors with a round-trip time of $T_{\text{round-trip}} = 2L/c$. The width Δt of an individual pulse depends on the frequency width $\Delta\upsilon$ of the optical gain curve between points where the gain just exceeds cavity losses (threshold condition). Of course, $\Delta\upsilon$ also determines how many modes M there are in the output. Thus, $\Delta t \sim 1/\Delta\upsilon \sim T_{\text{round-trip}}/M$. In practice, it is not very difficult to obtain mode locking since the required output must correspond to an intense pulse train of repetition rate $T_{\text{round-trip}}$. This output itself corresponds to a single optical pulse in the resonator bouncing back and forward between the mirrors with a round trip time $T_{\text{round-trip}}$. Each time this pulse impinges on a partially reflecting mirror, a portion of it is transmitted as an output pulse. Suppose that we insert an electrooptic switch that is switched and made transparent at every $T_{\text{round-trip}}$ seconds. The switch is on to be transparent only when the pulse is there and only for the duration of the pulse. Then the only possible situation is where a single optical pulse can bounce back and forward in the resonator and this corresponds to locking the modes. This is an example of *active* mode locking. Another possibility is *passive* mode locking where a saturable absorber is used in the optical cavity. A saturable absorber's absorption *decreases* with *increasing* light intensity so that it becomes transparent only at high intensities. This means the optical cavity is lossy for low light intensities. Such an absorber would only allow a high-intensity light pulse to exist in the cavity which would correspond to various modes having the right

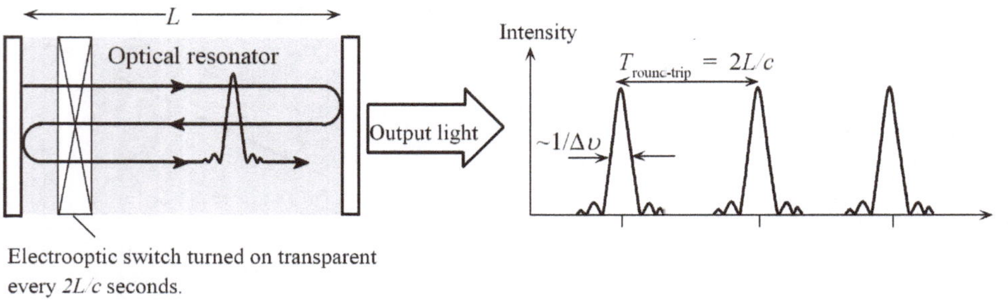

Output light intensity from a **mode locked laser** is a periodic series of short intense optical pulses that are separated in time by $2L/c$, the round-trip time for the pulse in the resonator. Each time the pulse in the resonator impinges on the right mirror, every $2L/c$ seconds, a portion of it is transmitted which constitutes the output.

phases; mode locking. The output pulses are spatially separated by $2L$ because this is the effective distance the pulse in the resonator travels at every $T_{\text{round-trip}}$ seconds to come back to the output mirror and generate another output pulse. A mode-locked He-Ne laser typically has a pulse width of about 600 ps whereas a mode locked Nd:glass laser pulse width is around 400 fs.

Mode mixing is the exchange of radiation energy between various modes as light propagates along an optical fiber. Due to small variations in the sample geometry and the refractive index profile along the fiber, there is a small but finite energy exchange between the propagating modes. One propagating mode gradually excites and transfers energy to another mode, and so on, leading to *mode mixing*. Consequently, the bandwidth of a long multimode fiber is generally wider than the expected bandwidth obtained by simple extrapolation from a short fiber. Thus, the bandwidth–distance (BL) product is strictly not a constant and is expressed as $BL^{-\gamma} =$ constant where γ is an index between 0.5 and 1 depending on the extent of mode mixing. The index γ is 1 for a short fiber but approaches 0.5 for a long fiber.

Mode number is an integer quantity, analogous to a quantum number, that identifies and distinguishes various allowed modes, that is, allowed electromagnetic field (**E** and **B**) distributions, in an optical waveguide.

Mode of a laser is a distinct field pattern at a reflector that can propagate to the other reflector and back again and return the same field pattern. If a particular field pattern can travel from one reflector to the other and back again, experience the right phase change and experience the right amplitude changes to replicate itself, then it is an allowed field pattern and hence a mode of the laser device (the active optical resonator). All such laser modes can be represented by fields (**E** and **B**) that are nearly normal to the cavity axis; they are referred to as *transverse modes* or *transverse electric and magnetic (TEM) modes*. Each allowed mode corresponds to a distinct spatial field distribution at a reflector. These modal field patterns at a reflector can be described by three integers p, q, m and designated by TEM_{pqm}. The integers p, q represent the number of nodes in the field distribution along the transverse directions y and z to the cavity axis x (put differently across the beam cross-section). The integer m is the number of

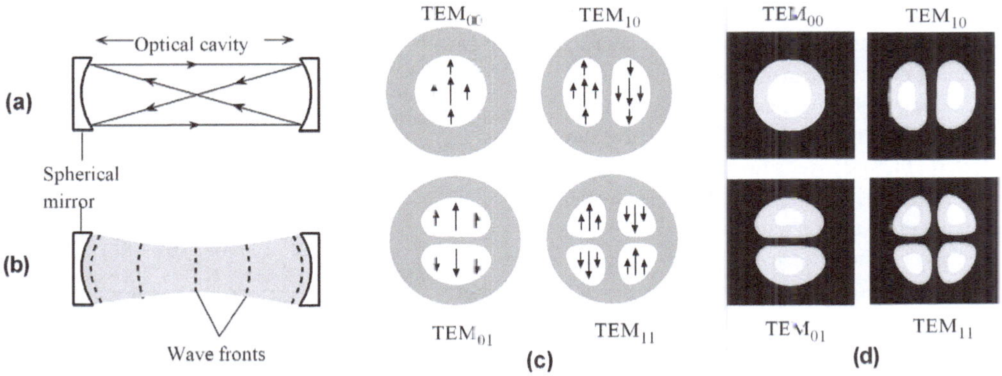

Mode of a laser (a) An off-axis transverse mode is able to self-replicate after one round-trip. (b) Wavefronts in a self-replicating wave. (c) Four low-order transverse cavity modes and their fields. (d) Intensity patterns in the modes of (c).

nodes along the cavity axis x and is the usual longitudinal mode number. Each transverse mode with a given p, q has a set of longitudinal modes (m values) but usually m is very large ($\sim 10^6$ in gas lasers) and is not written, though understood. Thus, transverse modes are written as TEM_{pq} and each has a set of longitudinal modes ($m = 1, 2,$). Moreover, two different transverse modes may not necessarily have the same longitudinal frequencies. (For example, the refractive index may not be spatially uniform so that different TEM modes have different spatial field distributions.)

Mode of an optical fiber is a distinct transverse (to the waveguide axis) electric field pattern that can propagate and hence be guided along the waveguide. These are allowed modes, field patterns that can ideally propagate unattenuated in a perfect optical fiber that has no losses.

Mode spot size is taken as the radius of the mode field diameter or the radial distance from the fiber axis where the optical power has fallen by a factor of $1/e$ with respect to its value at the fiber axis. See *Petermann mode radius*, *mode field diameter*.

Modulation index see *Young's interference fringes*.

Modulation transfer function (MTF) measures the efficiency of an imaging system such as a lens, or an image detector or an imager, to resolve (transfer) different spatial frequencies of information in an image. In other words, MTF is the *relative* signal response of the system as a function of *spatial frequency f*. Consider three objects A, B and C, each of which is made up of a sinusoidal variation in its brightness or its luminance along the x-axis with a certain spatial frequency, A with 2, B with 4 and C with 8 cycles/mm. Suppose that the detector is able to 100% resolve A, to a good extent B, but the information in C is totally lost, which represents 0% resolving ability. MTF is a convenient way to describe the spatial response of an imaging system because an imaging system can be described as a cascade of several stages and the overall MTF is the product of the MTF of all the individual stages. The MTF can be quantified by using the *contrast* in the object, which is defined in terms of the maximum and minimum luminance appearing in the object, $C_{object} = [(L_{max} - L_{min})/L_{max} + L_{min})]_{object}$, which is the input into the detector. A similar definition of

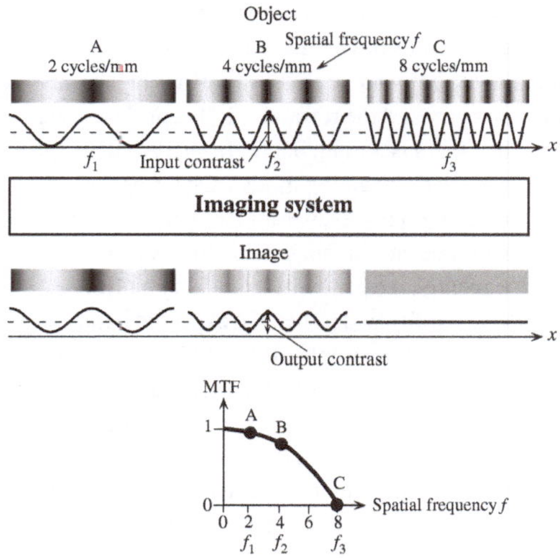

Modulation transfer function (MTF) measures the efficiency of a detector to resolve (transfer) different spatial frequencies of information. It compares the contrast in the image with that in the object at a given spatial frequency. The detector is able to almost fully resolve A, to a good extent B, but all of the information in C has been lost, which represents 0% resolving ability. Maximum MTF is 1 or 100%.

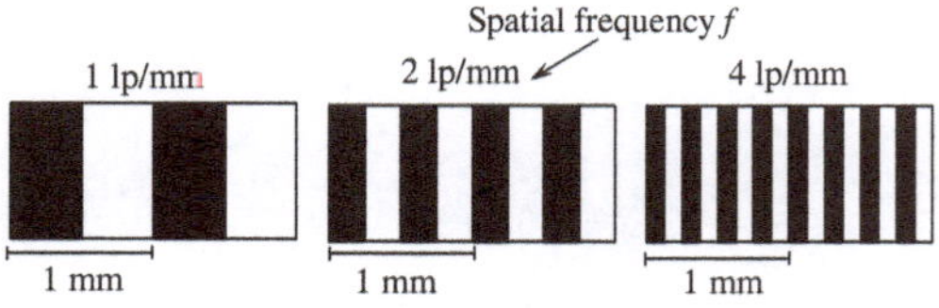

Modulation transfer function Bar patterns of different spatial frequencies for resolution testing.

contrast C_{image} in the image (at the detector output) is in terms of the maximum and minimum luminance in the image, $C_{image} = [(L_{max} - L_{min})/L_{max} + L_{min})]_{image}$. The MTF at a given spatial frequency f is defined as

$$\text{MTF} = \frac{\text{Contrast at the output}}{\text{Contrast at the input}} = \frac{C_{output}}{C_{input}}.$$

Its maximum value is 1 (or 100%). In the testing of many imaging systems, especially x-ray imaging systems, it is more convenient to use a *bar pattern* which is a periodic array of black lines separated by white spaces as shown in the second figure. Then, each cycle becomes a line pair – the bright stripe and its neighboring dark stripe. The distance of a dark stripe and its neighboring bright stripe in A is 1 mm. The spatial frequency of this periodic object (or image) is therefore 1 line pair per millimeter, 1 lp/mm. In such cases, the units of spatial frequency are usually expressed as line pairs/mm (lp/mm).The spatial frequencies of B and C are 2 lp/mm and 4 lp/mm. The *resolving power of the imaging system* is the spatial frequency when its MTF falls to some threshold value, which for many systems, for practical reasons, is taken to be 20%. See *resolving power, contrast, spatial frequency*.

Modulation transfer function of a "perfect" lens, in which the image formation is limited by diffraction effects, depends on the lens's f-number $F = f/D$, where f is the focal length and D is the lens diameter, and for a circular, spherical lens given by

$$\text{MTF} = (2/\pi)[\arccos(\nu/\nu_c) - (\nu/\nu_c)\sqrt{1 - (\nu/\nu_c)^2}];$$
$$\nu_c = 1/fF$$

where ν is the spatial frequency, and ν_c is the cut-off frequency given by $1/fF$. Clearly the MTF depends on the f-number. An ideal point source at infinity will be imaged by this lens as an Airy diffraction pattern at the focal plane whose central bright spot (the Airy disk) has a radius $r_d = 1.22\lambda F$. Thus, $\nu_c = 1.22/r_d$. The *resolving power of the lens* corresponds to a spatial frequency $\nu_r = 1/r_d$, and hence it corresponds to the case when MTF = 0.09 (or 9%).

Molar refractivity, R_m, is an optical property that is a measure of the molar polarizability of a substance. Suppose that n is the refractive index, M is the molecular weight, and ρ is the density of a simple substance in which each molecule (this can be a simple atom as well) has a polarizability α. Its molar refractivity R_m is defined by

$$R_m = \frac{M}{\rho} \cdot \frac{n^2 - 1}{n^2 + 2} = \frac{N_A\alpha}{3\varepsilon_0}$$

where M/ρ is the molar volume V_m so R_m is $(n^2-1)/(n^2+2)$ per unit molar volume. The first term provides a macroscopic *definition* for R_m, whereas the second term relates it to the microscopic processes through the polarizability α. Since R_m is proportional to $N_A\alpha$, it represents "a molar polarizability." If each distinct species of units (atoms, molecules and ions) in the structure with a polarizability α_i is identified with an index i, then the right-hand side of the above equation must have a suitable average for the polarizability α such as $\alpha = \sigma n_i \alpha_i$, where n_i is the molar fraction of the species i, and M represents the mean molecular weight. In a dilute system such as gases, n is close to unity and $R_m = (n^2 - 1)M/\rho$ and directly proportional to the electric susceptibility. Molar refractivity is an additive property so that in a glass with various constituents each of which has a molar refractivity $R_{m,i}$ and a molar fraction X_i, the overall molar refractivity is given by

$$R_m = \sum_i X_i R_{m,i}.$$

Molecular beam epitaxy (MBE) is one of the most advanced epitaxial techniques available today, offering arguably the most flexibility during growth and highest quality material. MBE is a refined form of vacuum evaporation. The refinement comes from the nature of the sources used and the vacuum system itself. The vacuum levels typical for this technique are on the order of 10^{-11} torr, permitting molecular flow (i.e. molecules from the source arrive at the substrate without suffering

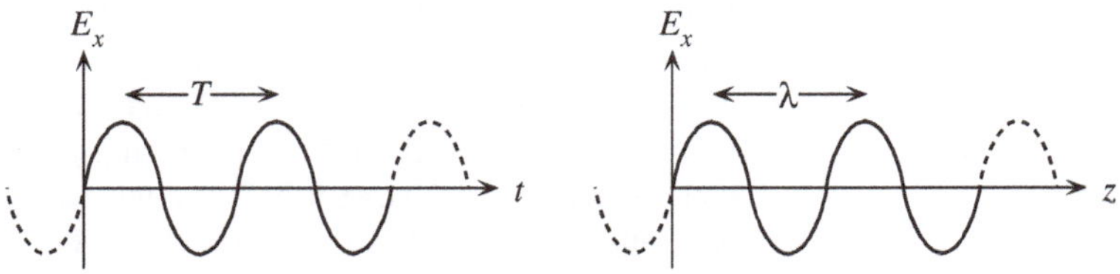

A plane **monochromatic wave** is a wave in which a property or a quantity such as the electric field (E_x, field along the x-direction) is varying with time and space in a sinusoidal fashion over an infinite time-scale and for all space.

collisions with other molecules). These conditions are commonly referred to as *ultrahigh vacuum* (UHV) conditions. MBE sources provide beams of material used for deposition. These beams result in very slow growth rates (about 1 to 3 Å/second for GaAs). Source beams originate from *Knudsen* or *effusion cells*, which are essentially heated crucibles with a defined aperture at their open end. The cells are filled with high-purity charges of appropriate elements or compounds and are heated to provide a flux of pure vapor of the appropriate constituent. By blocking the aperture with a shutter the beams can be turned on and off. Given the extremely slow growth rates, this enables atomically engineered device structures to be fabricated (so-called *nanostructures*). Growth is usually performed at relatively low temperatures (from about 580 to 630 °C for GaAs). Doping is achieved by adding appropriate additional Knudsen cells. In general, complex materials structures can be implemented; this is done in practice by adding appropriate Knudsen cells, and by adjusting the flux of atoms from the given cell by adjusting its temperature, while applying appropriate shuttering. Because of the extremely slow growth rates, film quality is excellent, as is the capability to vary structures (e.g., thickness, composition and doping) on the nanoscale. Moreover, layer purity is also excellent as source purity is always strictly controlled and the environment itself is conducive to strict purity standards. Another key feature of the MBE UHV environment is that it is ideally suited for incorporating *in situ* characterization tools including, for example, techniques for real-time layer-by-layer thickness monitoring.

Monochromatic wave is an untruncated sinusoidal wave of the form $\sin(\omega t - kz + \phi_0)$ that has a well-defined frequency ω and wavevector k (or wavelength λ); it is of infinite space and time extent and hence an idealization. The quantity ϕ_0 is a phase constant (independent of z and t) that simply accounts for the fact that the wave may not be zero when $t = 0$ and $z = 0$. The period T of the wave represents the periodicity on the time domain and is $2\pi/\omega$. The wavelength λ of the wave represents the periodicity of the spatial variations and is $2\pi/k$.

Monochromator is a spectrometer-type optical instrument that produces nearly monochromatic output light of wavelength λ' from an input source light beam that has a broad spectrum range over a range λ_1 to λ_2 that encompasses λ'. Usually the input light comes from a broadband source which may be an incandescent lamp, a hollow cathode lamp, a plasma in an atomic absorption spectrometer, an ultraviolet source (such as deuterium lamp) in a spectrometer with a UV range, or emission from a fluorescing sample, etc. A diffraction grating (or a prism) generates a spatially separated spectrum and a slit allows a narrow range of wavelength $\Delta\lambda$ around λ' to pass through to the output slit; the intensity and $\Delta\lambda$ depend on the slit width. By rotating the grating (or the prism), the output light can be scanned from λ_1 to λ_2, covering the wavelength content in the source light beam. Hence

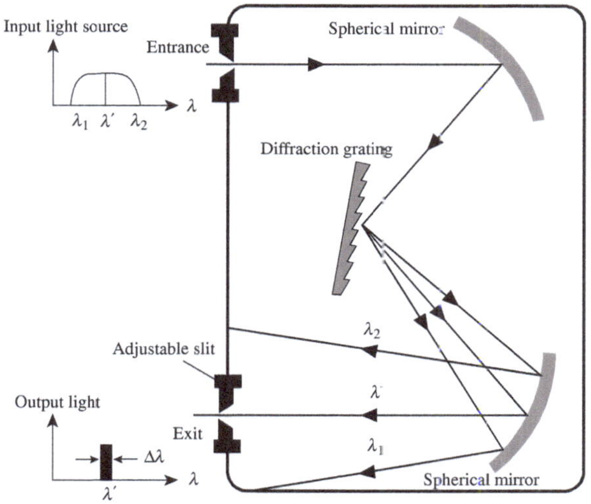

A typical **monochromator** design (Czerny–Turner monochromator).

the input's light spectrum can be determined or a scanned wavelength output light is generated from a suitable input light source. A typical monochromator design (the *Czerny–Turner monochromator*) is shown in the figure. It consists of a diffraction grating (the dispersing element), slits, and spherical mirrors. The entrance slit forms a slightly diverging polychromatic entrance beam towards the first spherical mirror. The latter transforms the diverging beam into nearly parallel beam and redirects it toward the grating. The surfaces of the mirrors must be reflective in the wavelength region of the light involved. This can be polished aluminum, silver or gold. The grating diffracts the beam and thereby disperses it into different wavelengths of light in different directions; the rays are no longer parallel but instead leave the grating at slightly different angles depending on the wavelength. These diverging waves are made to reflect from a second spherical mirror to further build-up the spatial separation between the different wavelength waves. When these waves arrive at the exit slit they are so substantially separated in space that they fall like a rainbow across the plane of the exit slit. The fine adjustment of the "rainbow" with respect to the exit slit is accomplished by rotating the grating. In modern instruments, this is controlled by a servomotor controlled by a computer. The final choice of desirable wavelength λ' is done by the exit slit whose width defines the narrow range of wavelength $\Delta\lambda$ around λ' to pass through. The edges of the slit are very carefully machined so that the distance between the two edges through which the light passes are equidistant all along the slit opening. The monochromator's *bandpass* is the wavelength range that a monochromator transmits. *Dispersion* is the wavelength dispersing power of the monochromator, usually given as the spectral range/slit width (nm/mm). Dispersion depends on the focal lengths of the spherical mirrors, grating resolving power, and the grating order. *Resolution* is the minimum bandpass of the spectrometer, usually determined by the aberrations in the optical system. The *acceptance angle* is a measure of light collecting ability of the monochromator, normally defined as focal length / mirror diameter. The *blaze wavelength* is the wavelength of maximum intensity in the first-order diffracted beam.

Monomode optical fiber is a single-mode optical fiber. See *single-mode fiber.*

Moss–Burstein effect is the apparently large bandgap energy deduced from the onset of band-to-band absorption in optical measurements on degenerate semiconductors in which the Fermi level is within either the conduction band or the valence band. Suppose that the Fermi level E_F is inside the conduction band. The least photon energy needed for excitation would be $E_F - E_v > E_c - E_v$, where E_v and E_c are the valence and conduction band edges. See *band-to-band absorption.*

Moss's rule see *refractive index.*

Multimode fiber allows the propagation of light via many guided electromagnetic waves, called guided modes, along the fiber. Typically it has a core diameter that is much larger than the wavelength of light which leads to a V-number greater than 2.405 and hence to many guided modes.

Multiple quantum well device is a semiconductor structure that has more than one quantum well and the quantum wells form a periodic structure.

Multiple quantum well (MQW) lasers have the structure of alternating ultrathin layers of wide and narrow bandgap semiconductors. The smaller bandgap layers are the active layers where electron confinement and lasing transition take place

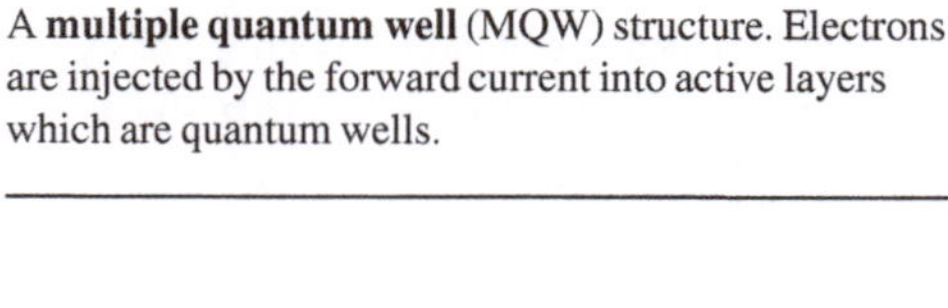

A **multiple quantum well** (MQW) structure. Electrons are injected by the forward current into active layers which are quantum wells.

whereas the wider bandgap layers are the barrier layers.

Although the optical gain curve is narrower than the corresponding bulk device, the output spectrum from a quantum well device is not necessarily a single mode. The number of modes depends on the individual widths of the quantum wells. It is, of course, possible to combine a MQW design with a distributed feedback structure to obtain single-mode operation. Many commercially available laser diodes are currently MQW devices.

Multiplexer (MUX) is a device where two or more signals (channels) are combined into a single output.

Multiplied photocurrent see *avalanche multiplication.*

Mutual coherence function see *coherence.*

Mutual temporal coherence between two waves is the time duration over which there is a correlation between two waves. Suppose that two finite length (truncated) identical frequency sine waves coincide only over the time interval Δt. Thus, they can only give rise to interference phenomena over this time interval. They therefore have *mutual temporal coherence* over the time interval Δt.

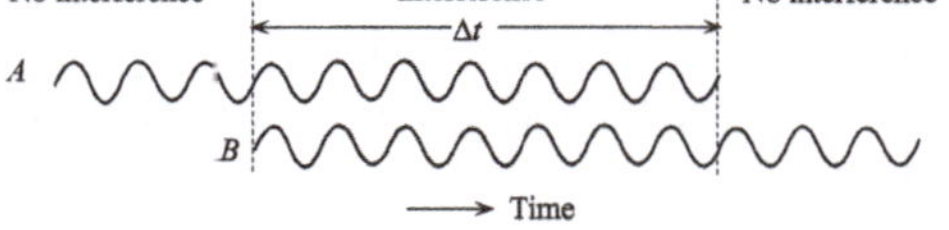

Mutual temporal coherence Two waves can only interfere over the time interval Δt which defines their *mutual temporal coherence.*

Myburg relation is an empirical relationship between the density ρ and the refractive index n for hydrogenated amorphous silicon films (a:Si:H) given as

$$n^2 = 1 + 19.57(\rho - 1.36)$$

in which ρ is in g cm^{-3}. (G. Myburg and R. Swanepoel, *J. Non-Cryst. Solids*, **89**, 13, 1987.)

N

Nanocrystals are single crystals of materials with size on the nanometer scale. An attractive feature of such nanocrystals is that their characteristics can be strongly size-dependent permitting the tuning of their properties through size control. As an example, semiconducting nanocrystals (e.g. CdSe, CdS, CuCl, InP, InAs, GaAs, GaN and PbS) have size-dependent electronic and optical properties (e.g. absorption and luminescence through the dependence of bandgap on size, as well as the enhancement in exciton binding energy in low dimensional systems), making them attractive candidates for use in a wide range of optoelectronic and photonic devices where the absorption or emission may be tuned through control of the nanocrystal size – two such applications are for the development of lasers, and for incorporation into polymeric systems for plastic light emitting devices. In addition, the nanocrystal emission spectrum can be further modified, not only by control of the particle size, but also by changing the nanocrystals' optical environment. An example of this is to locate nanocrystals within a microcavity resulting in a so-called *giant Purcell effect* where the emission is strongly coupled into the cavity modes of a very small-volume high-quality-factor cavity. Such efforts are ongoing and overlap with developments in *photonic crystals*. It should be noted that a given nanocrystal's luminescence is also sensitive to its chemical environment and can be used as a means for nanoscale detection and imaging in biological systems – for example, to detect specific pathogens, especially when combined with appropriate chemical functionalization. Other types of nanocrystals include those formed from magnetic materials. An important development in this field has been the development of techniques for making monodispersed (i.e. in size) nanocrystals, each of which is a hybrid structure. The hybrid structures include purely inorganic structures such as semiconductor heterostructures, on the one hand, and composite inorganic/organic structures, on the other. In the former case, these structures enable control over surface recombination as well as refractive index profiling, while in the latter case, exterior organic ligands coordinating an inorganic core have been used to form lateral ordered superlattices. One very attractive feature of nanocrystals is that they may be synthesized from solution using relatively inexpensive technology, promising a host of exciting future applications.

Nanoelectronics refers to the technology for fabricating electronic devices with feature sizes ranging from a few nanometers to the submicron range. The general motivation for implementing this technology is to achieve enhanced functionality of electronic devices; this includes (i) operating speed, (ii) memory capacity, and (iii) novel devices based on quantum-confinement effects. Ultimately, this can lead to a "superchip," which would consist of a three-dimensional network of trillions of switching elements operating at terahertz switching speeds. This will be made possible by reducing feature sizes to the level of several nanometers. Since, on this scale, the quantum-mechanical nature of electrons determines their behavior, novel structures, devices and circuits are possible. In such cases, electron tunneling and energy quantization are of paramount importance and novel types of nanoelectronic devices have been developed, including, for example, *quantum dot and quantum wire based devices, resonant tunneling devices* and *single electron devices*. Nanofabrication refers to the nanometer-scale fabrication of semiconductor structures and devices and nanometer-scale modification of surfaces. This field is rapidly developing to realize the benefits of nanoelectronic devices. We can distinguish between several types of nanofabrication methods: (i) *conventional lithography* and patterning using electron beams, x-rays, and ions; (ii) methods employing self-assembly, and (iii) nanometer-scale fabrication of semiconductor structures and modification of surfaces using *scanning probe microscopy* (SPM) techniques, such as *scanning tunneling microscopy* (STM) and *atomic force microscopy* (AFM).

Nanometer (nm) is a unit of linear measurement equal to 1/1000,000,000 (one billionth) of a meter. It is equivalent to 10 angstroms or typically to several atomic diameters. It is also used in the measurement of the wavelength of light.

Nanotechnology refers to a multidisciplinary field (or set of technologies) for designing, fabricating, and applying nanometer-scale materials, structures and devices.

Natural broadening see *homogeneous broadening*.

Nd:YAG (neodymium:yttrium aluminum garnet) laser see *diode pumped laser*.

Nd:YVO_4 (neodymium:yttrium orthovanadate) lasers see *diode pumped laser*.

Near-field distribution refers to the field distribution or intensity distribution very near an aperture or source so that the intensity pattern differs from the far-field pattern. The near-field distribution normally has distance dependence. In the case of an optical fiber, the near-field distribution normally refers to the intensity distribution over the cross-sectional area of the fiber.

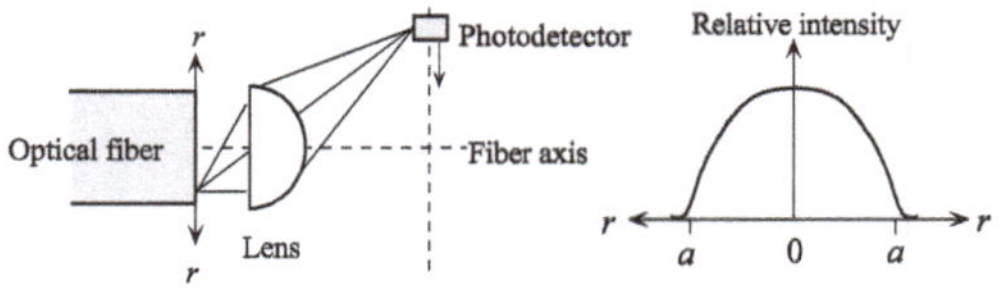

Near-field distribution from an optical fiber.

Negative electron affinity (NEA) refers to materials in which the conduction band edge E_c lies above the vacuum level – this occurs in diamond and some wide-bandgap III-V nitrides such as AlN. In the case of (111) diamond surfaces, it has been found that different surface treatments can change the electron affinity from a positive value to a NEA. For example, hydrogen termination causes a NEA. Materials with a NEA can emit electrons from their surface into vacuum without any barrier, providing for so-called cold cathode emitters. NEA arises from the relatively low electron affinity χ typical of wide-bandgap materials, and the presence of a surface dipole which acts to further lower the vacuum level, E_{vac}. Often a distinction is made between *true* NEA and *effective* NEA materials. In the former case, E_c lies above E_{vac} at the surface – this requires a small initial χ and significant lowering of E_{vac} as a result of the surface dipole, and is independent of any surface band bending. More commonly, so-called effective NEA occurs, where there is a small positive χ for a p-type semiconductor. A depleted surface space charge region bends the bands to lower energy toward the surface from the bulk, and results in the conduction band edge lying above E_{vac} at the edge of the depletion region

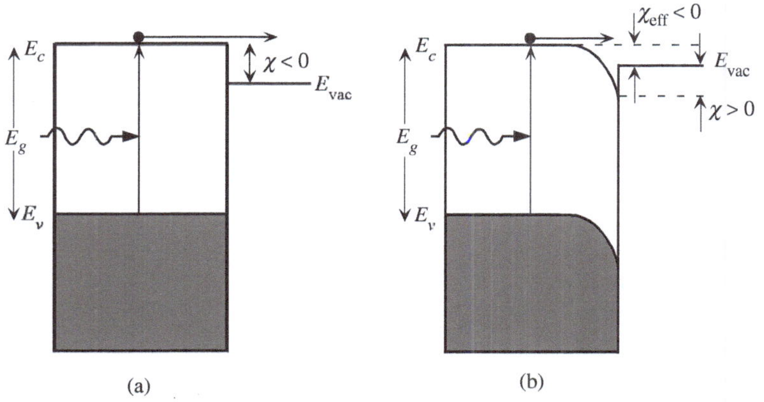

Negative electron affinity (a) true NEA (b) effective NEA.

within the semiconductor (i.e. in the bulk), fulfilling the condition of NEA with respect to the conduction band edge there. As examples, it has been found that deposition of cesium on AlN surfaces provides true NEA, while on p-GaN it provides effective NEA.

Negative refractive index medium is an engineered medium in which both the relative permittivity ε_r and relative permeability μ_r are negative, over a particular frequency, so that the overall refractive index of the medium is negative; such media exhibit various peculiar properties. When ε_r and μ_r are both positive, the refractive index $n = (\varepsilon_r \mu_r)^{1/2}$ is positive, and the electromagnetic (EM) wave obeys the familiar laws for EM waves; for example it is refracted at an interface between two positive refractive index materials, and the transmitted wave makes a positive angle with the normal as shown in the figure. It either moves closer or away from the normal. On the other hand, when ε_r and μ_r are both

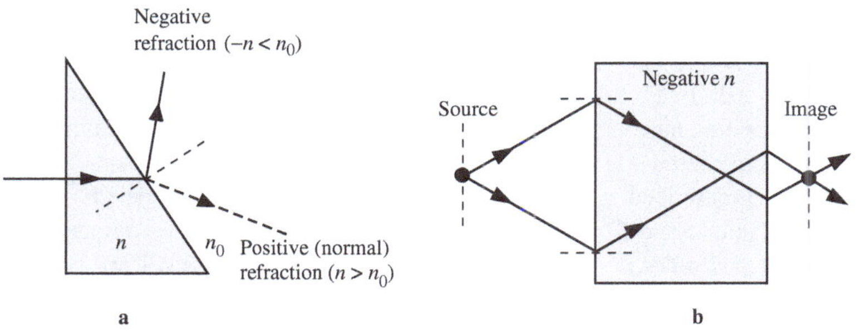

Negative refractive index (a) A prism made from a medium of refractive index n and n_0 is the outside refractive index, assumed positive. If n is positive, then the light entering the prism suffers normal (positive refraction). If n is negative, then the refracted light makes a negative angle to the normal. (b) A slab of a negative index material can focus light, that is, it act as a lens, and is capable of removing the wavelength limitations of lenses made from positive index materials provided that the losses in the negative n medium can be kept very low.

negative (a so-called doubly negative medium), the product $\varepsilon_r\mu_r$ is real but the refractive index $n = (\varepsilon_r\mu_r)^{1/2}$ is actually *negative*, and the medium is called a *negative index medium*. The transmitted wave in refraction makes a *negative* angle with respect to the normal as shown in the first figure. Negative refractive medium was initially proposed in 1968 by Victor Veselago, and demonstrated in 2000 at microwave frequencies by a specially engineered medium by Smith *et al.* Negative refractive index media have various unusual peculiar properties: the refracted light makes a negative angle to the normal; the group velocity (energy flow direction) and phase velocity (direction of wavefront motion); the Doppler shift is reversed, i.e. the detected frequency from a source moving towards the observer is lower; an obtuse angle for Cerenkov radiation, and so on. The striking demonstration by John Pendry at Imperial College that such media can be used to make perfect lenses with resolution capabilities not limited by the conventional diffraction limit has also contributed to the strong interest in such negative *n* media. (J. B. Pendry and D. R. Sminth, *Physics Today*, pp. 37–43, June 2004.) See *left-handed materials, photonic crystals, superlens.*

Negative uniaxial see *principal refractive indices.*

Nematic phase see *liquid crystals.*

Nemota–Makimoto equation represents fiber-to-fiber connection losses arising from the two fibers having different numerical apertures, lateral and longitudinal offsets, angular offset (mismatch) and reflections. The equation is described under *fiber-to-fiber attenuation.* (S. Nemota and T. Makimoto, *Opt. Quant. Electron,* **11**, 447, 1979.)

Neon lamp is a neon gas based arc discharge lamp. See *arc lamp.*

Nephelauxetic effect see *rare earth, Judd–Ofelt analysis.*

Newton's lens equation for thin lenses relates the object distance x_o from the object location to the object focal point F_o to the image distance x_i of the image from the image focal point F_i through the object and image focal lengths f_o and f_i in a very simple form that resembles the "mass action law" first published by Newton in *Optiks* (1704):

$$x_o x_i = f^2.$$

where $f = f_o = f_i$

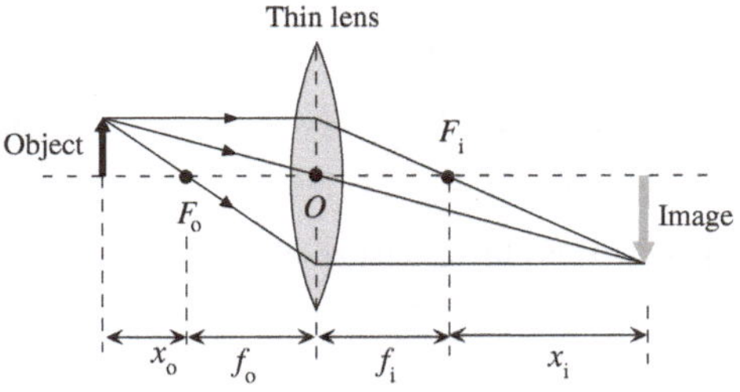

Newton's lens equation A thin lens of focal lengths $f_o = OF_o$ and $f_i = OF_i$.

Nicol prism, due to William Nicol, is a polarizing prism that breaks the incident light into two light beams that emerge in different directions and have perpendicular polarizations; it uses the birefringence of the crystal (calcite) that is used to make the prism. The prism is actually two long calcite rhombohedron crystals cut, ground and polished in such a way that each is a prism that has angles of approximately 68°, 32° and 90°. The prisms are joined on their large face with a thin Canada balsam cement. The incident light is refracted into two beams, an ordinary beam (*o*-ray) and an extraordinary beam (*e*-ray) because the crystal has different refractive indices, n_o and n_e for these two waves. The *o*-ray is incident at the thin Canada balsam interface at an angle greater than the critical angle for the calcite-balsam interface and hence suffers total internal reflection. The *e*-ray, on the other hand, experiences only refraction through the balsam layer and emerges from the other side of the crystal. The two beams that emerge are in different directions and

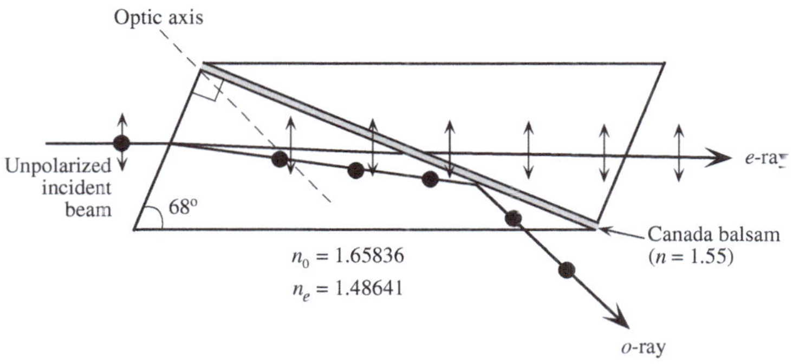

The principle of the Nicol prism. Indices are for Na light (589 nm).

have perpendicular polarizations. The Nicol prism has been superseded by other prisms such as the Glan–Foucault and Wollaston prisms. See *Glan–Foucault prism, Wollaston prism.*

Nit is a unit for luminance in terms of candela per square meter; [nit] = [cd m^{-2}] or [lm sr^{-1} m^{-2}]. See *luminance*.

NLFP see *nonlinear Fabry–Perot (NLFP) resonator*.

Nodal points see *cardinal points*.

Node see *optical network*.

Noise in optoelectronics and photonics can refer to a number of phenomena, each of which, one way or another, involves fluctuations in the quantity (or parameter) of interest about a mean value. The quantity of interest may depend on time, for example, a current or a voltage in a circuit, or the frequency or phase of a light wave emitted from a laser, or some other quantity that changes with time. The quantity may also involve measurements over a time period such as the number of photons collected or emitted over a specified time interval. In nearly all cases we are interested in the extent of fluctuations about some mean value. Suppose that the quantity of interest is x (strictly it should be written as a capital X since this would be a random variable) and its probability density function is $p(x)$, that is $p(x)dx$ is the probability that x has a value that lies in the interval x to $x + dx$. The average value of x is its *expected value*, and is given by

$$\bar{x} = \int_{-\infty}^{+\infty} x p(x) dx.$$

The variance σ^2 of the quantity x measures the average square fluctuations about the mean value, i.e.

$$\sigma^2 = \int_{-\infty}^{+\infty} (x - \bar{x})^2 p(x) dx$$

and σ is the standard deviation, the root mean square value of the fluctuations about the mean. If we know the probability density function $p(x)$, we can calculate the noise in the quantity of interest. The noise in a variable x is frequently characterized by examining its *power spectral density* $S(f)$, where f is the frequency, which is defined as average power in the fluctuations per unit frequency. $S(f)$ represents how the energy or power associated with x is distributed over frequencies. As an example, consider the voltage $v(t)$ across a 1 Ω conductor. $v(t)$ will fluctuate about zero due to the random motions of the conduction electrons inside the conductor. The instantaneous power would be v^2. Suppose that we measure the average power δP_{av} in a small frequency interval δf around the frequency f. Then the power spectral density $S(f)$ for the voltage fluctuations across a resistor is defined as $\delta P_{av} = S(f)\delta f$

or, using $\omega = 2\pi f$, $\delta P_{av} = (1/2\pi)S(\omega)\delta\omega$ so that the average power is

$$P_{av} = \int_{-\infty}^{+\infty} S(f)df = \frac{1}{2\pi}\int_{-\infty}^{+\infty} S(\omega)d\omega.$$

If $F(\omega)$ is the Fourier transform of $v(t)$, then $S(\omega)$ is defined as the ensemble averaged $|V(\omega)|^2$ over a very long time, that is,

$$S(\omega) = \text{Limit}\frac{\overline{|F(\omega)|^2}}{T} \text{ as } T \to \infty.$$

We can obtain $S(\omega)$ by Fourier transforming the signal $v(t)$ over a sufficiently long time, taking its magnitude, squaring it, and then repeating this many times and then averaging it. If the signal $v(t)$ has an autocorrelation function $R(\tau)$ then $S(\omega)$ and $R(\tau)$ are related through the *Wiener–Khintchine* theorem, that is, $S(\omega)$ and $R(\tau)$ form a Fourier transform pair,

$$R(\tau) = \int_{-\infty}^{+\infty} S(f)\exp(-j2\pi f\tau)df$$

and

$$S(f) = \int_{-\infty}^{+\infty} R(\tau)\exp(-j2\pi f\tau)d\tau.$$

Voltage fluctuations across a conductor constitute thermal noise and have $S(f) = 4k_B TR$ where k_B is the Boltzmann constant, T is the temperature and R is the resistance of the conductor and the $S(f)$ spectrum is flat over frequencies of interest, which corresponds to *white noise*. There are, however, limits at very low and at very high frequencies. *Shot noise* also has a white spectrum with $S(f) = 2eI$ where e is the electronic charge and I is the dc current value (average current). *Flicker noise*, on the other hand, has a $1/f$ type of spectrum; there is more power at low-frequency fluctuations. See *autocorrelation function, shot noise, quantum noise, flicker noise, burst noise*.

Noise equivalent power (NEP) is an important property of a photodetector that is defined as the optical signal power required to generate a photocurrent signal that is equal to the total noise current in the photodetector at a given wavelength and within a bandwidth of 1 Hz. It is numerically equal to the optical power necessary to give a photocurrent equal to the noise current in the detector when the detector bandwidth is 1 Hz. NEP represents the required optical power to achieve a signal-to-noise (SNR) of 1 within a bandwidth of 1 Hz. The detectivity D is the reciprocal of NEP, $D = 1/\text{NEP}$. Let R be the responsivity, I_{ph} and I_d be the photocurrent and the dark current in the detector. Suppose that the incident optical power P_1 is such that the photocurrent I_{ph} is equal to the total noise current in the detector. NEP is given by

$$\text{NEP} = \frac{P_1}{B^{1/2}} = \frac{1}{R}\left[2e(I_d + I_{ph})\right]^{1/2}.$$

The quantity $P_1/B^{1/2}$ represents the optical power necessary, per square root of frequency bandwidth, that generates a photocurrent equal to the noise current which is the quantitative definition of NEP. The units for NEP are W $\text{Hz}^{-1/2}$. NEP comparisons between detectors are only useful over the same bandwidth, B. The smaller NEP implies a better detector. I_{ph} and I_d are related because SNR must be unity,

$$I_{ph} = eB\left[1 + \left(1 + \frac{2I_d}{eB}\right)^{1/2}\right].$$

Typical NEP for a Si *pin* is $10^{-14} - 10^{-13}$ W $\text{Hz}^{-1/2}$. In the above, we assumed only shot noise in the detector. See *detectivity*.

Noise figure is an important metric of an amplifer in terms of specifying the signal-to-noise ratio (SNR) at its output with respect to that at its input, $\text{NF} = \text{SNR}_{\text{output}}/\text{SNR}_{\text{input}}$.

Noise power spectrum (NPS) is the Fourier transform of the covariance function of a random process. The autocovariance describes the correlation of a random variable $g(x')$ with itself at a location displaced by x about the expected values. If the process is wide-sense-stationary (WSS), then the NPS is commonly called the *Wiener spectrum*, which is the Fourier transform of the autocovariance function. The NPS(f) describes the spectral decomposition of the noise variance, i.e. the noise variance between spatial frequencies f and $f + df$. If a one-dimensional random variable $g(x)$ is real, $NPS_g(f)$ is also real. X-rays transmitted through a patient and incident on an imaging detector form an x-ray quantum image. The quantum image must be interpreted as a distribution in the mathematical sense, having the dimensions of area^{-1} for a two-dimensional image. If the two-dimensional image quanta are uncorrelated, the NPS is flat and simply equal to the expected number of quanta per unit area. For example, a uniform distribution of x-rays coming from a medical x-ray tube is uncorrelated and the NPS $= \bar{\Phi}_0$, where $\bar{\Phi}_0$ is the mean incident x-ray fluence (photons per unit area) on a detector. But the distribution of secondary quanta (x-rays to charge carriers or light) always has an uncorrelated component; they may also have a correlated component. Therefore, the NPS of a distribution of secondary quanta may have a frequency-dependent component extending to essentially infinite frequencies.

Noncentrosymmetric see *centrosymmetric*.

Nonlinear Fabry–Perot (NLFP) resonator has a cavity made of a nonlinear medium. Two main types can be distinguished, according to the type of nonlinearity (refractive or absorptive).

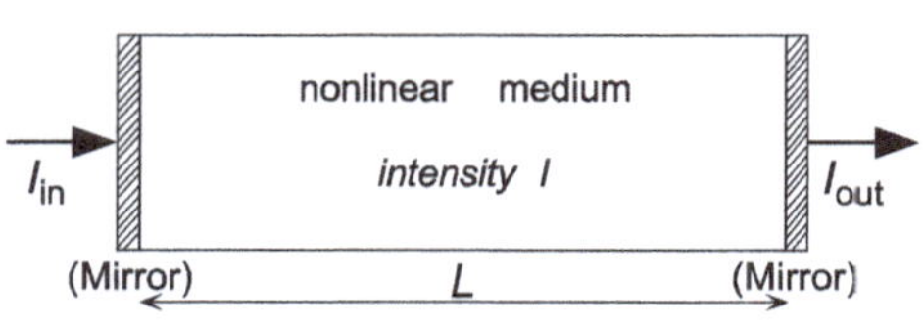

NLFP A cavity made of a nonlinear material of length L, whose optical properties (refractive index and/or absorption) depend on internal intensity I.

In a dispersive medium, the refractive index can be written as $n(I) = n_0 + n_2 I$, so that the round-trip phase change Φ depends on the internal intensity I. If R and T denote the reflectance and transmittance of the mirrors, it can easily be shown that the internal intensity I is expressed as a function of incident intensity I_{in} as:

$$\frac{I}{I_{in}} = \frac{(1/T)}{1 + (4R/T^2)\sin^2(\Phi/2)},$$

$$\text{with } \Phi = \Phi_0 + 4n_2 L(\omega/c)I.$$

For a given value of I_{in}, I can be determined graphically as the intersection of a straight line (slope $1/I_{in}$) with the Airy function. For a definite range of incident intensity, the resonator exhibits *bistability*. See *bistability*.

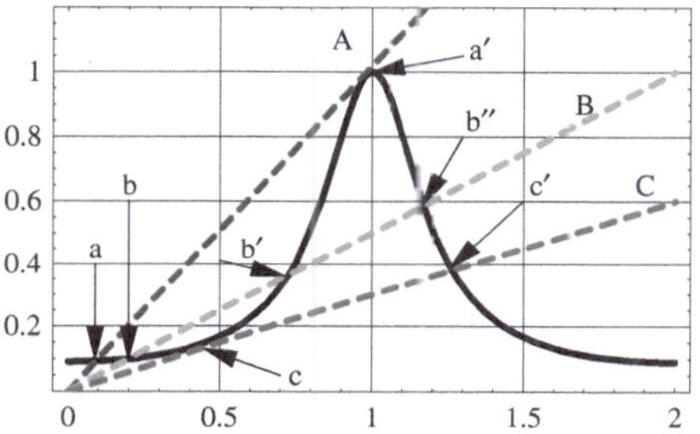

NLFP Functions (TI/I_{in}) and $[1 + (4R/T^2)\sin^2(\Phi/2)]^{-1}$ vs. (Φ/π). Dotted lines (A, B, C) correspond to increasing values of I_{in}.

Intersections (arrows) with the normalized Airy function show that in some cases, various internal states, thus various outputs, can correspond to the same input. The result is schematically reported in the second figure:

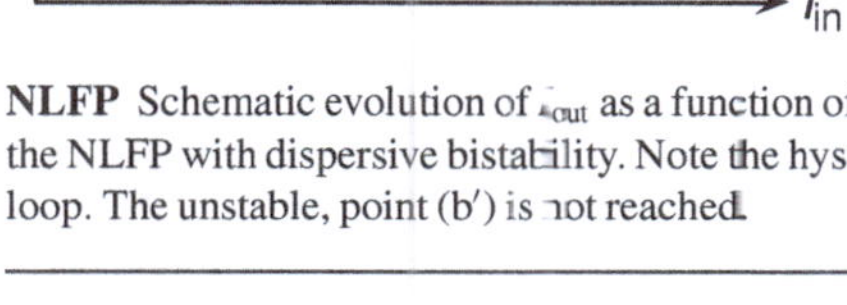

NLFP Schematic evolution of I_{out} as a function of I_{in} in the NLFP with dispersive bistability. Note the hysteresis loop. The unstable, point (b′) is not reached.

With an absorptive medium, the operating wavelength is tuned at a cavity resonance ($\Phi = 2q\pi$, with q an integer). Assume that the absorption can be written as $\alpha = \alpha_0/(1+S)$, with α_0 a constant [m^{-1}] and $S = I/I_S$ the dimensionless saturation parameter (I the internal intensity, I_S the typical saturation intensity for the medium). Suppose also that $\alpha_0 L \ll 1$. One can show that it is possible to express I_{in} as an explicit function of I_{out}, as:

$$I_{in} = I_{out}\left(1 + \frac{C_0}{1 + 2I_{out}/(TI_S)}\right)^2,$$

$$\text{with } C_0 = \alpha_0 L\left(\frac{R}{1-R}\right).$$

For values of the constant C_0 greater than 8, the function is not univocally invertible, so that in a definite range, I_{out} becomes a multivalued function of I_{in} (see third figure.)

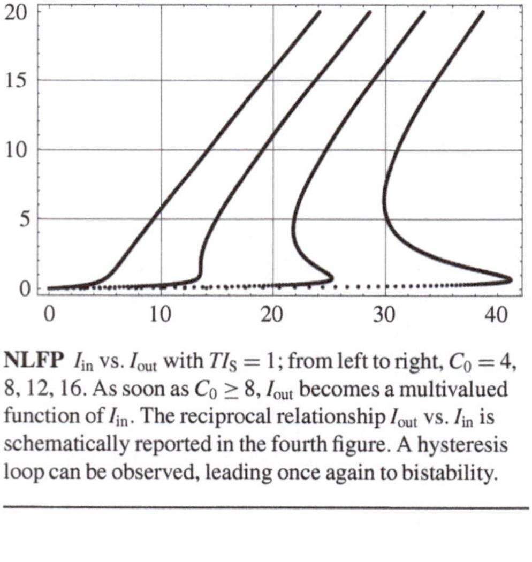

NLFP I_{in} vs. I_{out} with $TI_S = 1$; from left to right, $C_0 = 4$, 8, 12, 16. As soon as $C_0 \geq 8$, I_{out} becomes a multivalued function of I_{in}. The reciprocal relationship I_{out} vs. I_{in} is schematically reported in the fourth figure. A hysteresis loop can be observed, leading once again to bistability.

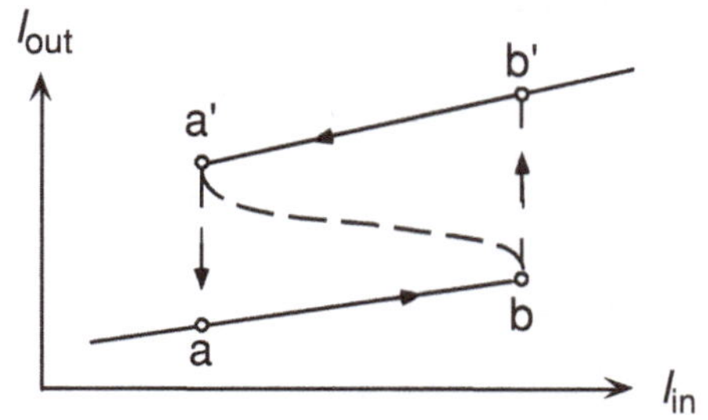

NLFP I_{out} vs. I_{in} for $C_0 > 8$. In the range between (a) and (b), I_{out} is a multivalued function of I_{in}. A hysteresis loop is observed. For increasing input, the path followed is (a – b – b′), whereas for decreasing output, it is (b′ – a′ – a).

Nonlinear optical properties represent the nonlinear response of an optical medium or system subject to a large electric field, for example, when the medium has a very intense laser beam passing through it. In a linear dielectric medium, the induced polarization P is proportional to the electric field E at that point and the two are simply related by $P = \varepsilon_0 \chi E$ where χ is the *electric susceptibility*. However, at high fields, the behavior deviates from this linear relationship. P becomes a function of E and can be expressed as a power law expansion in E as

$$P = \varepsilon_0\big[\chi^{(1)}E + \chi^{(2)}E^2 + \chi^{(3)}E^3 + \cdots + \chi^{(n)}E^n\big]$$

in which the terms $\chi^{(n)}$ are constants known as *susceptibilities* that depend on the material being illuminated. Thus, $\chi^{(1)}$ in the first term corresponds to *linear susceptibility* – being the polarization response, linear in the applied optical field. At higher optical fields, the first two nonlinear terms in the field become significant with corresponding second- and third-order susceptibilities, $\chi^{(2)}$ and $\chi^{(3)}$, respectively. The coefficients decrease rapidly for higher order terms. The importance of the second- and third-order terms (i.e. the nonlinear effects), depend on the field strength E. Nonlinear effects begin to become observable when the fields are very large (e.g. $\sim 10^7$ V m^{-1}). To attain such high fields requires light intensities (e.g. $\sim$1000 kW cm^{-2}) that invariably require lasers to achieve them. All materials, whether crystalline or noncrystalline,

Table: Nonlinear optical properties Table of selected second-order nonlinear coefficients d m V^{-1} or C V^{-2}. Combined table from data from C. C. Davis (*Lasers and Electro-Optics: Fundamentals and Engineering*, Cambridge University Press, Cambridge, 1996) and A. Yariv (*Optical Electronics in Modern Communications*, Oxford University Press, Oxford, 1997).

Crystal	$d_{ij}(10^{-12}$ m $V^{-1})$
Ag_3AsS_2 (proustite)	$d_{15} = 15.3; d_{22} = 24 \pm 1.5; d_{36} = 13.5$
$AgGaS_2$	$d_{36} = 18 \pm 3$
$AgGaSe_2$	$d_{36} = 27 \pm 3$
$AgSbS_3$	$d_{36} = 9.5$
$AlPO_4$	$d_{11} = 0.38 \pm 0.03$
$Ba_2NaNb_5O_{15}$(banana)	$d_{15} = 7.4 \pm 2.3; d_{24} = 6.5 \pm 0.7; d_{32} = 7.4 \pm 0.7; d_{33} = 9.7 \pm 1$
$BaTiO_3$	$d_{15} = 17.35 \pm 2.5; d_{31} = 18 \pm 2; d_{31} = -18.8 \pm 1.5;$ $d_{33} = -7.1 \pm 0.5; d_{33} = 6.4 \pm 0.5$
$CdGeAs_2$	$d_{36} = 877 \pm 175; d_{36} = 363 \pm 70$
CdS	$d_{15} = 29 \pm 7; d_{31} = 26.4 \pm 6.3; d_{33} = 44 \pm 13; d_{36} = 33$
CdSe	$d_{31} = 22.5 \pm 3$
CdTe	$d_{14} = 168 \pm 67$
GaAs	$d_{14} = 72; d_{36} = 135.5 \pm 53$
GaP	$d_{14} = 100 \pm 18; d_{14} = 80 \pm 14$
HgS	$d_{11} = 77 \pm 26$
InAs	$d_{14} = 419 \pm 126$
KD_2PO_4 (KD*P)	$d_{14} = 0.42 \pm 0.02; d_{36} = 0.44 \pm 0.03$
KH_2ASO_4	$d_{14} = 0.51 \pm 0.03; d_{36} = 0.48 \pm 0.03$
KH_2PO_4 (KDP)	$d_{14} = 0.35; d_{36} = 0.48 \pm 0.04$
$LiIO_3$	$d_{15} = 4.4; d_{31} = 6.03 \pm 0.51; d_{33} = 6.28 \pm 0.51$
$LiNbO_3$	$d_{15} = 4.4; d_{22} = 2.45 \pm 1.4; d_{31} = -5.4 \pm 0.5; d_{33} = -49 \pm 9$
$NH_4D_2PO_4$ (AD*P)	$d_{36} = 0.56$
$NH_4H_2PO_4$ (ADP)	$d_{14} = 0.63 \pm 0.03; d_{36} = 0.57 \pm 0.07$
$NH_4H_2PO_4$ (ADP)	$d_{14} = 0.50 \pm 0.02; d_{36} = 0.45$
Se	$d_{11} = 130 \pm 30$
SiO_2	$d_{11} = 0.345 \pm 0.03; d_{14} = 0.003$
Te	$d_{11} = 922 \pm 293$
Te	$d_{11} = 517$
ZnO	$d_{15} = 2.1 \pm 0.2; d_{31} = 1.95 \pm 0.2; d_{33} = 6.5 \pm 0.2$
ZnS	$d_{14} = 30.6 \pm 8.4; d_{36} = 13$

possess a finite $\chi^{(3)}$ coefficient. However, only certain classes of crystals have a finite $\chi^{(2)}$, and the reason is the same as for the observation of the Pockels effect. Only those crystals, such as quartz, that have no center of symmetry have a nonzero $\chi^{(2)}$ coefficient; these crystals are also piezoelectric. The above expression is a highly simplified statement of the nonlinear nature of polarization at high fields. A proper representation of polarization involves *tensors*. In general, an optical field along a certain direction with respect to crystal axes can generate a response along a different direction. For example, the ith component $P_i^{(2)}$ of the *second-order nonlinear polarization* can be written as

$$P_i^{(2)} = \varepsilon_0 \sum_{j,k=x,y,z} \chi_{ijk}^{(2)} E_j E_k$$
$$= \sum_{j,k=x,y,z} d_{ijk}^{(2)} E_j E_k$$

where $\chi_{ijk}^{(2)}$ is the second-order *nonlinear susceptibility tensor*, i, j, k correspond to the Cartesian coordinates x, y, z, usually taken along suitable crystal axes to simplify the above equation. $d_{ijk}^{(2)} = \varepsilon_0 \chi_{ijk}^{(2)}$, and $d_{ijk}^{(2)} = d_{ikj}^{(2)}$; switching j and k does not change anything. $d_{ijk}^{(2)}$ are the *second-order harmonic generation coefficients* (or second-order nonlinear polarization coefficients) and constitute a *tensor*. In matrix notation, subscripts 1–6 are used to correspond to $1 = xx$; $2 = yy$; $3 = zz$; $4 = yz = zy$; $5 = zx = xz$; $6 = xy = yx$ so that

$$\begin{bmatrix} P_1 \\ P_2 \\ P_3 \end{bmatrix} = \begin{bmatrix} d_{11} & d_{12} & d_{13} & d_{14} & d_{15} & d_{16} \\ d_{21} & d_{22} & d_{23} & d_{24} & d_{25} & d_{26} \\ d_{31} & d_{32} & d_{33} & d_{34} & d_{35} & d_{36} \end{bmatrix} \begin{bmatrix} E_x^2 \\ E_y^2 \\ E_z^2 \\ 2E_zE_y \\ 2E_zE_x \\ 2E_xE_y \end{bmatrix}.$$

According to the *Kleinman conjecture*, $d_{ijk}{}^{(2)} = d_{kij}{}^{(2)} = d_{jki}{}^{(2)}$, which means there are 10 independent d_{ijk} values. For example, $d_{26} = d_{yyx} = d_{xyy} = d_{12}$. Crystal symmetry introduces further simplifications by requiring certain d_{ijk} to be zero. For example, potassium dihydrogen phosphate (KDP) is a tetragonal crystal with $\bar{4}2m$ symmetry which has

$$\mathbf{d}(\text{KDP}) = \begin{bmatrix} 0 & 0 & 0 & d_{14} & 0 & 0 \\ 0 & 0 & 0 & 0 & d_{14} & 0 \\ 0 & 0 & 0 & 0 & 0 & d_{36} \end{bmatrix}$$

in which there are only two independent d-values since $d_{25} = d_{14}$. Intense laser light impinging on KDP that has a field in the xy plane (i.e. with components, E_x and E_y), such as light traveling along the optical axis (i.e. z-axis), generates a polarization along the optical axis given by $P_z = d_{36}(2E_xE_y)$. See *optical Kerr effect, Pockels effect, Kerr effect*.

Nonlinear optics is concerned with effects induced by the propagation of light through a medium – in particular, the focus is on using the high fields associated with lasers to generate nonlinear optical effects. Such effects include *second-harmonic generation (SHG)*, *third-harmonic generation (THG)*, *stimulated Raman scattering* and *self-focusing*. It should be noted, however, that although laser-induced nonlinear optical effects became a focus of studies following the availability of lasers after the 1960s, nonlinear optical effects such as the *Pockels* and *Kerr* effects had been observed in the nineteenth century. To appreciate the nonlinear optical response of a medium, one can consider the dependence of the polarization of such a medium on the laser field which may be expressed as a power law expansion in the field with coefficients relating to the linear and nonlinear susceptibility of the medium. Thus, for example, an input electric field (from the optical excitation), $E_\omega \cos(\omega t)$, can result in a number of second-order nonlinear responses $\chi^{(2)}$: these include $\chi^{(2)} E_o E_\omega \cos(\omega t)$, where E_o is an applied field, corresponding to the linear electrooptic effect discovered by Pockels in 1883, and the modification of the refractive index due to the dc applied electric field, and can be used for optical switching and for phase modulation of light. Another second-order response, $\chi^{(2)} E_\omega{}^2 \cos(2\omega t)$, for example, corresponding to polarization generated at a frequency of 2ω, and an optical field generated at 2ω, can be used for second-harmonic generation. Similar nonlinear phenomena and their use can be discussed for third-order nonlinearities: for example, in the case of $\chi^{(3)}(E_\omega)^3 \cos(\omega t)$, this can give rise to self-focusing and self-phase modulation.

Nonlinear polarization occurs when the induced polarization is not linearly proportional to the electric field. The application of an electric field E to a dielectric material causes the constituent atoms and molecules to become polarized. The medium responds to the field E by developing a polarization P which represents the net induced dipole moment per unit volume. In a linear dielectric medium the induced polarization P is proportional to the electric field E at that point and the two are related by $P = \varepsilon_o \chi E$ where χ is the *electric susceptibility*. However, the linearity breaks down at high fields and the P vs. E behavior deviates from the linear relationship. P becomes a function of E which means we can expand it in terms of increasing powers of E.

At high levels of illumination, the polarization induced in a nonlinear medium by the exciting field $\mathbf{E}$ deviates from the expected simple $\mathbf{P} = \varepsilon_o \chi \mathbf{E}$

type linear relationship. The total polarization **P** can be written as the sum of two terms,

$$\mathbf{P} = \varepsilon_0 \chi_{\text{lin}} \mathbf{E} + \mathbf{P}_{\text{NL}},$$

with $\mathbf{P}_{\text{NL}} = \varepsilon_0 \chi_{\text{NL}} \mathbf{E}$, where χ_{NL} is the *nonlinear susceptibility* of the medium. $\mathbf{P}_{\text{NL}}$ depends on the tensor symmetry of χ_{NL} and on the state of polarization of the incident field(s). See for instance *Kerr effect* and *second harmonic generation*.

Nonlinear refractive index (n_2) see *optical Kerr effect*.

Nonradiative lifetime or nonradiative recombination lifetime, τ_{nr}, is the average time it takes for a carrier (electron) to recombine with an oppositely charged carrier (hole) through an indirect recombination process, that is, a recombination process that involves the participation of a recombination center (such as an impurity or a crystal defect) and the emission of phonons (lattice vibrations). If Δn is the concentration of excess carriers, then the nonradiative lifetime is defined by

$$\partial \Delta n / \partial t = -\Delta n / \tau_{\text{nr}}.$$

A recombination center near the middle of the bandgap is the most effective in encouraging indirect (radiationless) recombination, and in such cases, a useful equation for estimating τ_{nr} is

$$\tau_{\text{nr}} \approx 10^8 / N_{\text{r}}$$

where N_{r} is the concentration of recombination centers in cm^{-3}, and τ_{nr} is in seconds. Under weak injection conditions, that is, the injected carrier concentration is much less than the majority carrier concentration, the nonradiative minority recombination time is given by

$$\tau_{\text{nr}} = \frac{1}{S_{\text{r}} v_{\text{th}} N_{\text{r}}}$$

where S_{r} is the recombination capture cross-section, v_{th} is the mean speed of the carriers (thermal velocity) and N_{r} is the concentration of recombination centers; $S_{\text{r}} v_{\text{th}}$ is typically of the order of $10^{-8}\ \text{cm}^3\ \text{s}^{-1}$. A more rigorous analysis for τ_{nr} involves the *Shockley–Reed–Hall statistics*. See *radiationless recombination*.

Nonradiative transitions, see *radiationless recombination, nonradiative lifetime*.

Nonreturn-to-zero, see *return-to-zero*.

Nonzero dispersion shifted fiber see *dispersion shifted fiber*.

Normal dispersion represents a dispersion behavior, refractive index, n vs. wavelength, λ, in which the refractive index decreases with the frequency of light or n decreases with increasing λ. See *refractive index, chromatic dispersion*.

Normal distribution see *Gaussian distribution*.

Normalized frequency is another name for the V-number.

Normalized index difference, Δ, for a step index optical fiber is defined by $\Delta = (n_1 - n_2)/n_1$ where n_1 and n_2 are the core and cladding refractive indices respectively.

Normalized propagation constant, b, is a dimensionless propagation constant that is used for mathematical convenience to represent a waveguide

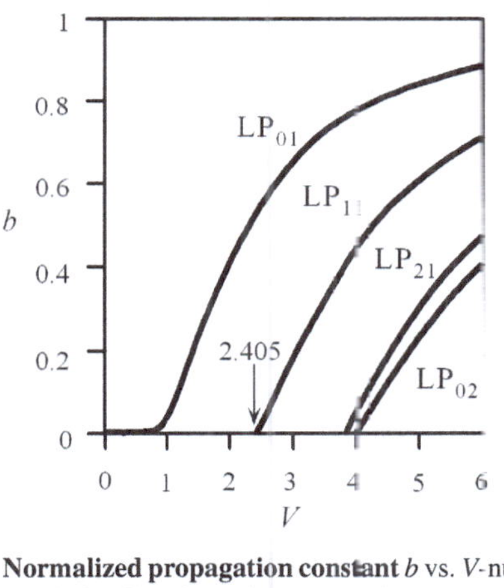

Normalized propagation constant b vs. V-number for a step index fiber for various LP modes.

mode with a propagation constant β through the definition

$$b = \frac{(\beta/k) - n_2}{n_1 - n_2} \qquad \text{i.e. } \beta = n_2 k[1 + b\Delta]$$

where n_1 and n_2 are the core and cladding refractive indices, k is the free-space wavevector $2\pi/\lambda$ and Δ is the normalized refractive index difference, $(n_1 - n_2)/n_1$. The b vs. V-number characteristics for various linearly polarized modes are then universal for weakly guiding step index fibers as shown in the figure.

Normalized thickness is another name for the V-number.

Nuclear magnetic resonance is the absorption of electromagnetic (EM) radiation energy at a suitable frequency by a nucleus with a nonzero magnetic moment in the presence of a sufficiently strong applied magnetic field; the EM radiation has to have an appropriate frequency to be able to be absorbed by the nucleus which then becomes excited. In such a case, the nucleus behaves as a magnet with associated magnetic moment that precesses about the direction of the applied magnetic field. Quantum mechanics only allows certain spin orientations, each having an associated energy. Nuclear magnetic resonance occurs when the incident photon energy is exactly equal to the difference in energy between two of these allowed states.

Nuclear magnetic resonance in computing is a technique that is used to manipulate and control the nuclear magnetic moments as *qubits* for *quantum computing*. The nucleus of an atom can possess a magnetic moment behaving as a tiny permanent magnet. When such a nucleus is placed in a strong magnetic field, owing to its quantum nature, it can have its nuclear magnetic moment aligned in either the same direction or in the opposite direction to the surrounding magnetic field. The orientation of the nuclear magnetic moment is therefore confined to two distinct values and so can be used to represent a qubit. Adjacent nuclei also influence each other analogous to the interaction of magnetic moments in traditional magnetic materials. This interaction allows one nuclear qubit to control an adjacent nuclear qubit. Quantum computation is normally carried out by using radio-frequency fields to flip the nuclear moments. Output data is obtained by measuring the radio-frequency fields generated by the oscillating moments.The largest quantum computers built to date are less than ten qubit devices based on NMR. Unfortunately, technological difficulties limit the size of these devices to about ten qubits. Nevertheless, NMR has proven to be a very useful approach for demonstrating the operating principles of a quantum computer.

Numerical aperture of a lens is a measure of its light gathering and resolving ability. The higher the numerical aperture NA, the greater is the light captured by the lens, and the greater the resolving power. Higher NA means a shorter working distance. If the maximum angle $2\alpha_{\max}$ is the total maximum cone angle defined by marginal rays from the object that are incident on the edges of a lens (see figure), and if D is the lens diameter, then its objective NA (NA on the object side) is

$$\text{NA} = n_o \sin(\alpha_{\max})$$

where n_o is the refractive index of the medium. For small D, NA $\approx n_o D/2s_o$, where s_o is the object distance to the lens. If light is launched from air, $n_o = 1$ and since $\alpha_{\max}$ is 90°, the maximum NA would be 1. NA can be increased by having a medium between the object and the lens.

It is important to identify the "diameter" D (and hence $\alpha_{\max}$) that captures as much of the light as possible in forming the image. For an isolated lens this is either the lens diameter (as in the figure), or the diameter of an aperture placed immediately in front of the lens to limit imaging to paraxial rays, etc. For an optical system, on the other hand, D may be determined by an aperture or other imaging apertures (or NAs) inside the system; in the latter case D is the equivalent maximum diameter on the lens for taking into account the internal apertures, and the NA refers to the optical system (and not the

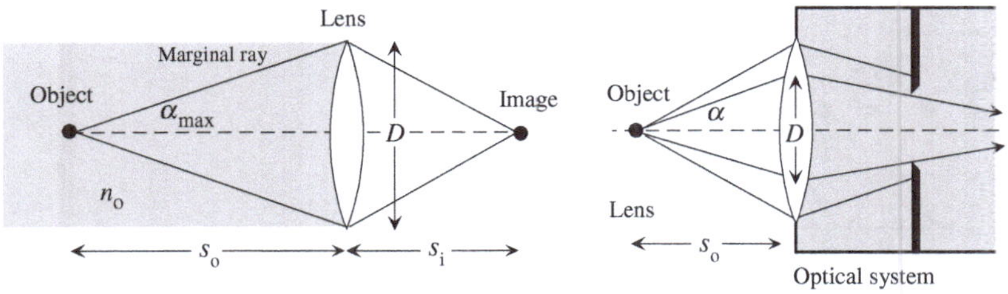

Numerical aperture of an object–lens–image system (left) and an optical system (right).

lens). For an object at the focal point F of a lens, $\mathrm{NA} = n_0 D/2f' = 1/2(f/\#)$, where $f/\#$ is the f-number of the lens, and f' (focal length) of a lens in a medium is n_0 times longer than the focal length f in air (hence n_0 cancels).

Numerical aperture of an optical fiber is a measure of its light collecting ability. Fibers with higher numerical aperture can capture more light. If the maximum acceptance angle into the fiber, as measured from the fiber axis, is $\alpha_{\max}$, and if the light is launched from a medium with refractive index n_0, then

$$\mathrm{NA} = n_0 \sin \alpha_{\max} = \left(n_1^2 - n_2^2\right)^{1/2}$$

where n_1 and n_2 are the refractive indices of the core and cladding of the fiber. For a single-mode fiber, NA would be less than 10%, whereas for multimode fibers, it can be as high as 30%. See *acceptance angle*.

Nyquist criterion see *Nyquist–Shannon theorem*.

Nyquist–Shannon theorem states that a signal $v(t)$ whose highest frequency component is B (the signal has a Fourier transform $V(f)$ with components over a bandwidth B) must be sampled at a rate f_s greater than $2B$, otherwise the information cannot be unambiguously represented; $f_s > 2B$ is called the *Nyquist criterion*. The critical frequency $f_c = 2B$ is called the *Nyquist rate* or the *critical frequency*. The same concept also applies to imaging in which an image $y(x)$ along the x-axis must be sampled, for example, by an array of photodiodes along x. Suppose that the photodiodes are separated by a. Then the image is being sampled at a rate $1/a$. If K is the highest spatial frequency present in the image (that is, the bandwidth of the Fourier transform of the image) then we need $1/a > 2K$ to be able to reconstruct the image. Suppose that $V(f)$ is the Fourier transform of the signal $v(t)$ and is a band that contains frequencies up to B. When $v(t)$ is sampled at a rate f_s, the sampled signal has a Fourier transform that has the original $V(f)$ repeating itself at every f_s. The bands are repeated at every f_s. We can then use a low-pass filter to allow the lowest band to pass, and then use

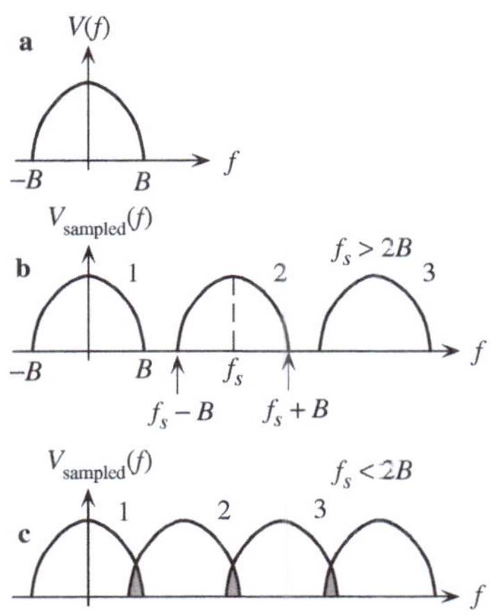

Nyquist–Shannon theorem (a) The spectrum of a signal either in time or space. (b) The spectrum of a sampled signal when sampled at a rate $f_s > 2B$. (c) The spectrum of an undersampled image which has overlapping spectra, which makes it impossible to reconstruct the original image accurately.

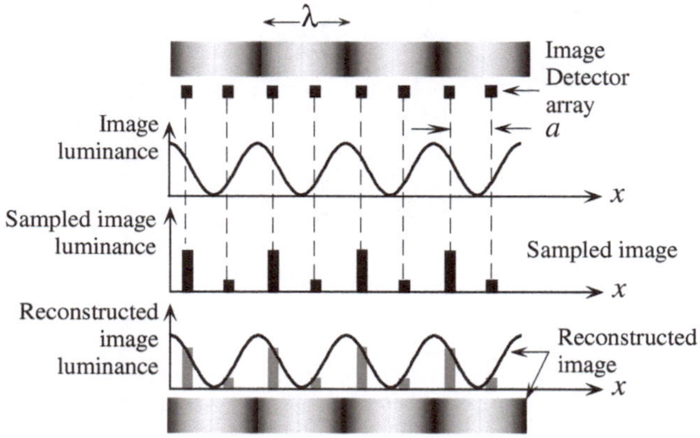

Nyquist–Shannon theorem An example of a sinusoidal image pattern that is sampled and then reconstructed.

this band to reconstruct the original signal $v(t)$. However, we can only do this if the bands remain distinguishable and do not overlap. The bands overlap if the Nyquist criterion is not met; f_s is less than $2B$ in which the signal is undersampled. When we use the overlapping spectra to attempt to reconstruct the signal, we obtain a signal $v_{alias}(t)$ that does not represent the original signal $v(t)$ but can correspond to a number of different signals, that is to aliases; and the effect is called *aliasing*. To avoid aliasing an *antialiasing filter* must be used. The latter band-limits the spectrum $V(f)$, so that the high frequencies are lost and hence there is a loss of fine resolution. See *aliasing, aliasing noise*.

O

Objective lens is normally the closest lens in an optical system to the object under study and forms the first image of the object; this first image may then act as an "object" to other lenses behind the objective lens.

Ohmic conduction refers to electrical conduction that is governed by $J = \sigma E$ where σ is the conductivity of the material. The current flow is limited by the rate of transport of intrinsic charge carriers (e.g. electrons and holes) in the material and their drift mobility and not by the rate of carrier injection from the electrical contacts. The contacts ideally simply replenish the electrons and holes in the material that reach the electrodes and exit the sample.

Ohmic contact is a contact that can supply charge carriers to a semiconductor at a rate determined by charge transport through the semiconductor and not by the contact properties itself. Thus the current is limited by the conductivity of the semiconductor and not by the contact.

Okuto–Crowell equation is an equation for the impact ionization of charge carriers in semiconductors which has the impact ionization coefficient α given as

$$\alpha = \left(\frac{eF}{E_I}\right) \exp\left[0.217(E_I/E_r)^{1.14} - [0.217(E_I/E_r)^{1.14}]^2 + \left(\frac{E_I}{e\lambda F}\right)^2\right]^{1/2}$$

where e is the elementary charge, E_I is the ionization energy, E_r is the *optical phonon* energy, and λ is the mean free path due to scattering from optical phonons. The equation provides a good fit over a wide range of fields for a number of semiconductors. For example, for Si, for electron impact ionization, the equation requires $E_I = 1.1$ eV, $E_r =$ 51 meV, $\lambda = 480$ nm. The model has been largely superseded by the lucky drift model of impact ionization. See *impact ionization.* (Y. Okuta and C. R. Crowell, *Phys. Rev. B*, **6**, 3076, 1992.)

OLED see *organic light-emitting devices (OLED).*

Olivero–Longbothum approximation see *Voigt lineshape.*

Omni directional reflector is a light reflector, usually made from thin film multilayers of different dielectric materials, that is able to reflect light for all polarization states and for nearly all angles of incidence over a specified range of wavelengths, i.e. over an *omnidirectional reflectance band.* Typically, the reflector is a stack of alternating thin films made from high (n_H) and low (n_L) refractive index materials. Suppose that n_H, d_H, and n_L, d_L are the refractive index and thickness of the high and low index layers, respectively, $\Lambda = d_L + d_H$ is the period of the stack, and n_0 is the refractive index of the incident medium (often air). The existence of an *omnidirectional band* requires that both $\delta n_0 = n_L/n_0$ and

$\delta n_1 = n_H/n_L$ are relatively large. If the multilayer stack design is such that the parameters δn_0 and δn_1 exceed some minimum threshold, the Brewster line lies outside the acceptance angle for light rays incident on the mirror from the external medium. Under these conditions there are finite ranges of wavelengths for which no propagating modes can exist inside the dielectric stack. One needs a large difference between n_H and n_L and between n_L and n_0. For example, one particular design (Y. Fink *et al.*, *Science*, **27**, 1679, 1998) has used a simple stack of nine alternating layers of polystyrene ($n_L \approx 1.6$ and $d_L = 1.65$ μm) and tellurium ($n_H \approx 4, d_L = 0.8$ μm) which has resulted in an omnidirectional reflection of nearly 95% over incidence angles 0–80°, in the 10–15 μm band. Another example is an omnidirectional reflector covering the 1475–1635 nm range for incidence angles up to 76° made from a stack of polymer and Ag-diffused amorphous chalcogenide thin films. (T. J. Clement *et al.*, *Optics Express*, **14**, 1789, 2006.)

Operating point see *photovoltaic or solar cell I–V characteristics*.

Optic axis of an anisotropic crystal is a direction in the crystal along which all waves have the same phase velocity whatever their polarization. For example the *ordinary* and the *extraordinary* waves traveling along the optic axis have the same phase velocity and experience the same refractive index.

Optical absorption by charge transfer see *charge transfer absorption*.

Optical absorption in amorphous semiconductors near the absorption edge is usually characterized by three types of optical transitions corresponding to transitions between defects and extended states, between localized and extended states, and between extended states. The absorption coefficient vs. photon energy, α vs. $\hbar\omega$, dependence has three different regions A, B and C, respectively, which correspond to these three characteristic optical transitions, as shown in the figure.

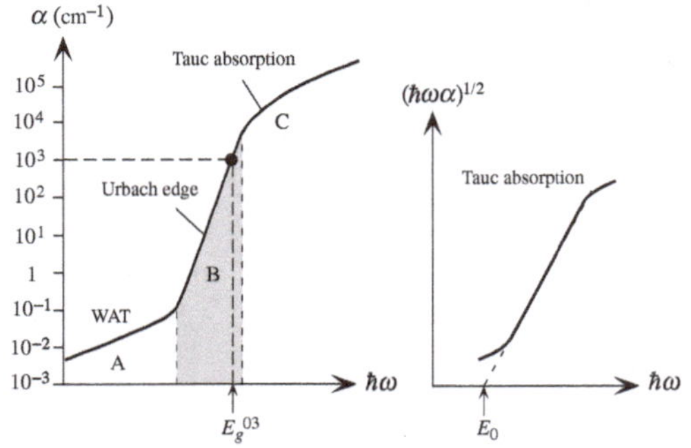

Typical spectral dependence of **optical absorption coefficient in amorphous semiconductors**. In A and B, the absorption is controlled by the optical transitions between localized and extended states. In spectral domain A, "defect" absorption prevails. In domain B the absorption coefficient follows the Urbach rule. In C the absorption is controlled by the optical transitions between extended states. The plot of $(\hbar\omega\alpha)^{1/2}$ vs. $\hbar\omega$ in region C, shown on the right, is called the Tauc plot.

In the small absorption coefficient range A (also called the weak absorption tail, WAT), where $\alpha <$ (10^{-1}–1 cm^{-1}), the absorption is controlled by optical transitions from defect-related localized states to extended states and vice versa. The defects in amorphous semiconductors are considered to be all cases of departure from the normal nearest-neighbor coordination (or normal valence requirement). The examples of defects are broken and dangling bonds (typical for amorphous silicon), over- and under-coordinated atoms (such as "valence alternation pairs" in chalcogenide glasses), voids, pores, cracks and other macroscopic defects. To some extent, the absolute value of absorption in region A may be used to estimate the density of defects in the material. In region B, where *typically* $10^{-1} < \alpha < 10^4$ cm^{-1}, the absorption is related to transitions from localized tail states above the valence band edge to extended states in the conduction band and/or from extended states in the valence band to localized tail states below the conduction band. (The optical transition probability from valence band

Table: Optical absorption in amorphous semiconductors Optical bandgap of a-$Si_{1-x}C_x$:H films (O. Kodolbas, *Mat Sci Eng.*, **B98**, 161, 2003.)

	E_g at $\alpha = 10^3$ cm^{-1}	E_g at $\alpha = 10^4$ cm^{-1}	$E_g = E_o$ (Tauc)	$E_g = E_o$ (Cody)	$E_g = E_o$ (Sokolov)	ΔE meV
a-Si:H	1.76	1.96	1.73	1.68	1.60	46
a-$Si_{0.88}C_{0.18}$:H	2.02	2.27	2.07	2.03	1.85	89

tail states to conduction band tail states is insignificant.) Which of the latter processes is dominant depends mostly on the density of states in the corresponding band tail. The conduction band tail states are shallow localized states which lie in the vicinity of the conduction band mobility edge; similarly valence band localized tail states lie in the vicinity of the valence band edge. They appear due to the distortions of the amorphous matrix with respect to the perfect crystalline lattice, that is, as the result of the deviation of atomic separations and bond angles from their ideal values in the perfect crystal. Usually, the spectral dependence of α follows the so-called *Urbach rule*, which is can be written as

$$\alpha = \alpha_{03} \exp[\hbar\omega - E_{g,03}]/\Delta E$$

where $\hbar\omega$ is the photon energy, $\alpha_{03} = 10^3$ cm^{-1}, $E_{g,03}$ is the photon energy when $\alpha = \alpha_{03}$ ($E_{g,03}$ is sometimes called the optical bandgap at $\alpha = 10^3$ cm^{-1}) and ΔE is the *Urbach width*, which is also written as $1/\Delta E = \Gamma$, the Urbach parameter. For many amorphous semiconductors, ΔE has been related to the width of the valence (or conduction) band tail states, and may be used to compare the "widths" of such localized tail states in different materials; ΔE is typically 0.05–0.2 eV. The onset of the Urbach edge depends on the material; for a-Si:H it is roughly 5×10^4 cm^{-1}. In region C, the absorption is controlled by transitions from extended to extended states. Unlike in the case of crystalline semiconductors, in amorphous semiconductors the momentum conservation rules are not applicable and consequently there are no sharp characteristics peaks (as in the crystal) in α vs. $\hbar\omega$ behavior. For many amorphous semiconductors, the α vs. $\hbar\omega$ behavior follows the so-called *Tauc relation,*

$$(\hbar\omega)\alpha = B_{\mathrm{Tauc}}(\hbar\omega - E_o)^2$$

where B_{Tauc} is a constant, independent of the photon energy, and E_o is a constant called the optical bandgap or Tauc bandgap. E_o in amorphous semiconductors should not be confused with the bandgap in crystalline materials nor is it directly (or simply) related to the mobility gap. While a frequency-independent optical matrix element M combined with a parabolic energy dependence for the density of states, $N_{\mathrm{band}}(E)$, for the bands would result in the Tauc equation for α, in reality, M is frequency dependent, and $N_{\mathrm{band}}(E)$ are not parabolic, but the combination of the two result in an approximate Tauc relation for amorphous semiconductors. Other expressions for α vs. $\hbar\omega$ in region C have also been suggested and used; these are the Cody and Sokolov relationships, each with its own characteristic optical gap E_o, and given by

$$(\hbar\omega)^{-1}\alpha = B_{\mathrm{Cody}}(\hbar\omega - E_o)^2$$
$$(\hbar\omega)\alpha = B_{\mathrm{Sokolov}}(\hbar\omega - E_o)^3.$$

The so-called optical bandgap E_g determined for a given material from α vs. $\hbar\omega$ using definitions can vary as shown in the table. (J. Tauc *et al.*, *Phys. Stat Sol.*, **15**, 627; 1966; G. D. Cody *et al.*, *Phys. Rev. Lett.*, **47**, 1480, 1981; A. P. Sokolov *et al.*, *J. Phys. Cond. Matt.*, **3**, 9887, 1991; R. A. Street, *Hydrogenated Amorphous Silicon*, Cambridge University Press 1991)

Optical activity refers to the rotation of the electric field **E**-vector of a linearly polarized light wave as it passes through certain crystals, such as quartz, along the optic axis. This rotation increases continuously with the distance traveled through the crystal (about 22° per mm of quartz at a wavelength of 600 nm). The rotation of the plane of polarization by a substance is called *optical activity*. In very simply intuitive terms, optical activity occurs in materials in which the electron motions induced by the external electromagnetic field follows spiraling or helical paths (orbits). Electrons flowing in helical paths resemble a current flowing in a coil and thus possess a magnetic moment. The optical field in light therefore induces oscillating magnetic moments which can be either parallel or antiparallel to the induced oscillating electric dipoles. Wavelets emitted from these oscillating induced magnetic and electric dipoles interfere to constitute a forward wave that has its optical field rotated either clockwise or counterclockwise. If θ is the angle of rotation, then θ is proportional to the distance L propagated in the optically active medium. For an observer receiving the wave through quartz, the rotation of the plane of polarization may be *clockwise* (to the right) or *counterclock*wise (to the left) which are called the *dextrorotatory* and *levorotatory* forms of optical activity. The structure of quartz is such that atomic arrangements spiral around the optic axis either in clockwise or counterclockwise sense. Quartz thus occurs in two distinct crystalline forms, right-handed and left-handed, which exhibit dextrorotatory and levorotatory types of optical activity, respectively. Although we used quartz as an example, there are many substances that are optically active, including various biological substances and even some liquid solutions (e.g. corn syrup) that contain various organic molecules with a rotatory power. The *specific rotatory power* is defined as the extent of rotation per unit length, θ/L, of distance traveled in the optically active substance. Specific rotatory power depends on the wavelength. For example, for a 1 mm quartz crystal, along the optic axis, this is 49° at 400 nm but 17° at 650 nm.

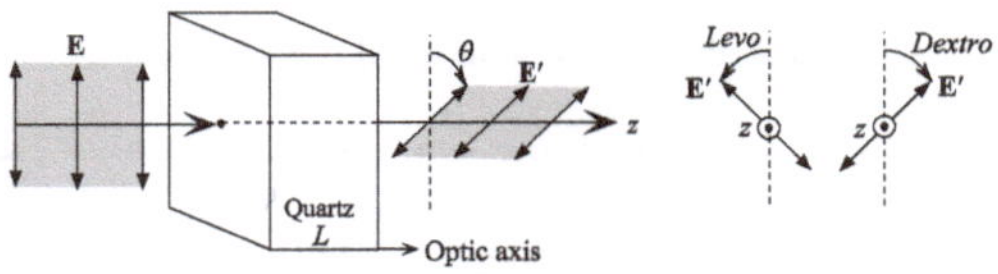

Optical activity An optically active material such as quartz rotates the plane of polarization of the incident wave: The optical field **E** is rotated to **E**′. If we reflect the wave back into the material, **E**′ rotates back to **E**.

Optical amplifier is an amplifier that amplifies an optical signal, that is light. A common type of optical amplifier is an *erbium ion doped optical fiber amplifier* (EDFA) which is a silica fiber that has been doped with Er^{3+} ions. The erbium ions are *pumped* to an excited state by using a laser diode (or an LED) at 980 nm. This *pumping* leads to a *population inversion* of Er^{3+}-ions and hence to an optical gain at wavelengths in a small range around 1550 nm, which corresponds to long-haul communications wavelength. There are other types of optical amplifiers, the most important of which are rare-earth doped fluoride and telluride fibers, *semiconductor laser amplifiers*, *Raman fiber amplifiers*.

Optical axis for a simple lens is a line that joins the centers of curvature of the lens surfaces, and for an optical lens system, with two or more lenses, it is normally the central line passing through the centers of the lenses. The optical axis of an optical system is distinct from the optic axis of a birefringent crystal.

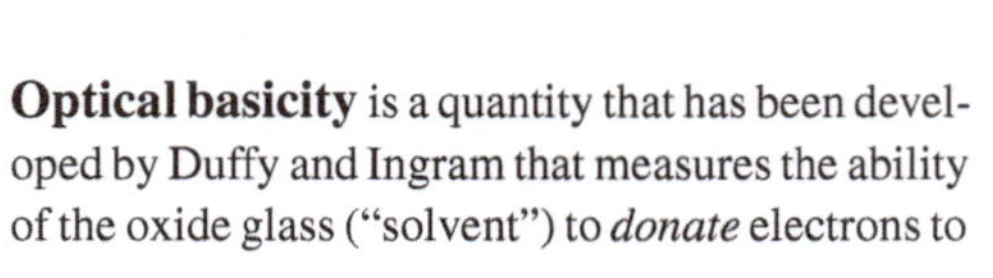

Optical axis of a lens and a simple two-lens system.

Optical basicity is a quantity that has been developed by Duffy and Ingram that measures the ability of the oxide glass ("solvent") to *donate* electrons to a probe ion ("solute" ion) that is added to the oxide, and hence the changes in the optical transitions of these probe ions. For example, there is a strong UV absorption band (an absorption peak) in oxide glasses with Pb^{2+} ions, which is due to the internal transitions of the Pb^{2+} ions from the $^1S_0$ to the $^3P_1$ state. The position of this UV absorption peak depends strongly on the composition of the oxide glass. The shift is due to action of the oxygen atoms in the structure on Pb^{2+} and depends on the "basicity" of the oxide. The free Pb^{2+} ion has an absorption at $\nu_{\text{free-ion}} = 60{,}700\ \text{cm}^{-1}$. When Pb^{2+} is inside an oxide glass, there is a large shift in the absorption frequency to some value ν_{glass}. The shift $\Delta\nu = \nu_{\text{glass}} - \nu_{\text{free-ion}}$ depends on the basicity of the glass – its ability to donate electrons to Pb^{2+}. The shift $(\nu_{\text{glass}} - \nu_{\text{free-ion}})$ can therefore be used to define an "optical basicity" Λ based on using a Pb^{2+} ion as a probe ion. To allow Λ to have values between 0 to 1 (or slightly higher values), one can normalize the shift $(\nu_{\text{glass}} - \nu_{\text{free-ion}})$ to a large shift in the UV peak when Pb^{2+} is added to CaO, a strong ionic oxide. (The choice of CaO is somewhat arbitrary and may not give the largest shift.) If $\nu_{O^{2-}}$ is the frequency of the peak Pb^{2+} absorption in CaO, then the *optical basicity* of an oxide glass for the Pb^{2+} ion is defined as

$$\Lambda = \frac{\nu_{\text{free-ion}} - \nu_{\text{glass}}}{\nu_{\text{free-ion}} - \nu_{O^{2-}}}.$$

The overall basicity Λ can be calculated from the constituent basicities Λ_i ($i = i$th type of oxide) and their molar fractions X_i,

$$\Lambda = \sum_i X_i \Lambda_i.$$

One can calculate Λ for any oxide glass from its chemical constituents and hence predict the UV absorption band using

$$\Lambda = 1 - \tfrac{1}{2} z_A r_A (1 - \gamma_A^{-1}) + \tfrac{1}{2} z_B r_B (1 - \gamma_B^{-1}) + \cdots$$

where $z_A, z_B \ldots$ are the oxidation numbers of the cations in the glass, $r_A, r_B, \ldots$ are their ionic ratios with respect to the total number of ions, γ_A, $\gamma_B, \ldots$ are the basicity moderating parameters of the ions; these values are tabulated. Conversely, one can use Λ to predict the UV absorption peak in various Pb^{2+} oxide glasses. (J. A. Duffy and M.D. Ingram, *J. Am. Ceram. Soc.*, **93**, 6448, 1971; *J. Non-Cryst. Solids*, **21**, 373, 1976; J. A. Duffy, *J. Sol. State Chem.*, **62**, 145, 1986; *Phys Chem. Glass*, **42**, 151, 2001.) See *optical electronegativity*.

Optical beating between two quasimonochromatic waves of slightly different frequencies ω_1 and $\omega_2 > \omega_1$ represents the low-frequency modulation of the resulting superposition. Consider two plane waves $\mathbf{E}_1$ and $\mathbf{E}_2$ sharing the same direction of propagation, for instance, along the z-axis, and the same state of polarization, for instance rectilinear parallel to the x-axis

$$\mathbf{E}_1(r,t) = A_1 \exp[j(\omega_1 t - k_1 z + \varphi_1(t))]\mathbf{u}_x,$$
$$\mathbf{E}_2(r,t) = A_2 \exp[j(\omega_2 t - k_2 z + \varphi_2(t))]\mathbf{u}_x,$$

where $\mathbf{u}_x$ is the unit vector along the x-axis. The *intensity* of the total field $\mathbf{E}(\mathbf{r},t) = \mathbf{E}_1(\mathbf{r},t) + \mathbf{E}_2(\mathbf{r},t)$ is

$$I = (1/2)\varepsilon_0 cn\langle|\mathbf{E}(\mathbf{r},t)|^2\rangle,$$

where $\langle f(t)\rangle$ denotes the time average of any time-varying function $f(t)$ over the integrating time T_D of the detector (its bandwidth $F_D \approx 1/T_D$).

With $\Delta\varphi(t) = \varphi_2(t) - \varphi\ (t)$ and $\Delta\omega = \omega_2 - \omega_1$, the signal detected at $z = 0$ is

$$I = (1/2)\varepsilon_0 cn\{A_1^2 + A_2^2 + 2A_1A_2\langle\cos[\Delta\omega t + \Delta\varphi(t)]\rangle\},$$

$$I(t) = I_1 + I_2 + 2(I_1 I_2)^{1/2}\langle\cos[\Delta\omega t + \Delta\varphi(t)]\rangle.$$

If the two fields are *mutually coherent* ($\Delta\varphi(t) = \Delta\varphi_0 =$ constant, at least on the scale of T_D), then the sine modulation can be detected as long as $\Delta\omega < 2\pi F_D$.

Optical bistability refers to optical materials and devices in which, for a given incident light intensity, the transmitted light intensity can be at one of two possible values depending on the history of the illumination process, that is, if a current output level was achieved by lowering the input signal intensity from a higher level or by raising its intensity from a lower signal intensity level. This phenomenon is also often referred to as optical hysteresis. Note that, as shown in the figure, there are two stable states in the so-called hysteresis loop – if the incident beam intensity is maintained at such a level as to provide an output determined by the domain of the hysteresis loop, the transmitted intensity will remain at the same level (i.e. high or low). Such optically bistable systems can therefore be used for making optical memories, optically activated switches, optical amplifiers and optical logic gates.

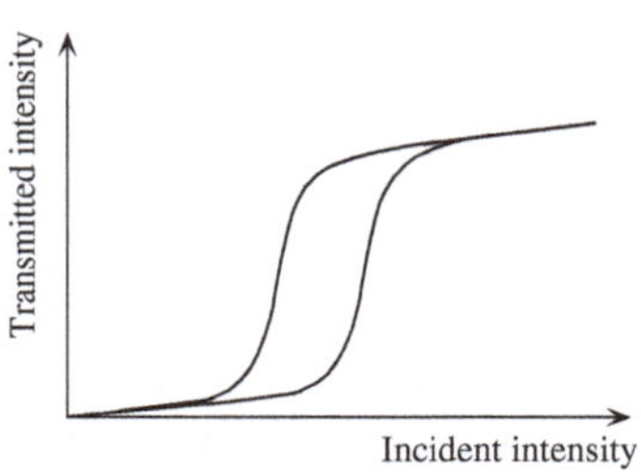

Optical bistability A central so-called hysteresis loop is shown for the dependence of transmitted intensity on incident light intensity. This results in two stable states for the output intensity for a given input intensity and is referred to as optical bistability.

Optical communications is an optoelectronic system in which the information signal is transmitted in terms of light pulses. Usually the information is encoded into a digital signal so that the presence or absence of light represents a "bit." The figure shows an example in which a laser diode generates a continuous wave light at 1550 nm with a narrow spectrum. The light is externally modulated by an electrooptic modulator. The modulator modulates the light passing through it according to the encoded information applied to its electrical input terminal. The optical signal in the form of *light pulses* is fed into an optical fiber for transmission towards its destination. 1550 nm is preferred for long-haul communications because the attenuation along the optical fiber is minimum at this wavelength. Optical switches enable the optical path to be switched at various locations as desired by the user; for example, if there is a break in the fiber, the optical switch in the figure will switch the optical path to the back-up link. Optical amplification is typically used to amplify an optical signal that has been attenuated as a result of being guided along a transmission medium such as an optical fiber. An *optical isolator* allows the optical signal to travel only in one direction and is useful in preventing the amplified

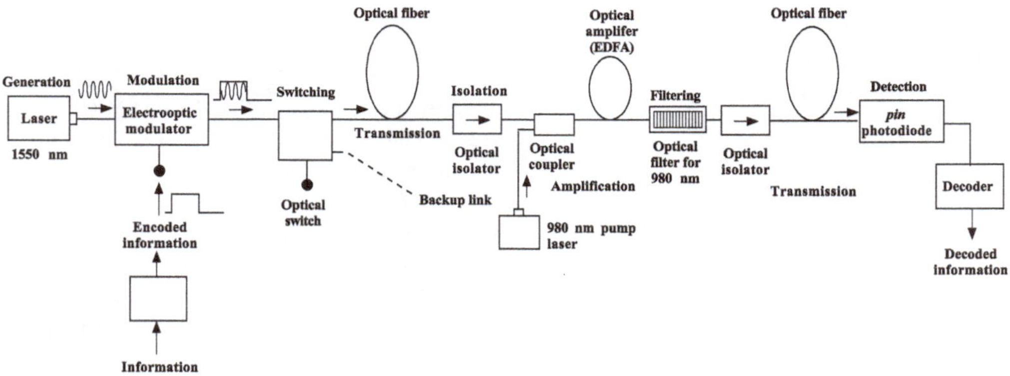

Optical communications Fiber optic communications.

signal traveling back to the transmitter. An optical coupler allows light in one fiber to be coupled into another fiber. An optical filter allows either a given wavelength to pass through (such as the 1550 nm), or it blocks the passage of a certain wavelength (such as the 980 nm pump wavelength of an Er-doped fiber amplifier) by reflecting it. Various optical connectors are obviously needed to connect the devices and components to each other. *Optical connectors*, while simple in functionality, nonetheless play an undeniably critical role in the overall optoelectronic system design since various devices and components have to be connected together efficiently and economically, and without failure. For example, an inefficient connection system may lead to an additional use of optical amplifiers to correct the mistake. In optical communications, typically, various devices are connected through optical fibers and cables using connectors and splices. The connector and splice technology have been by and large standardized to facilitate the rapid and easy connection of fibers and cables to each other.

Optical cross-connect (optical crossbar) is an optical switch that can switch the signal from one optical path to another path. The figure shows how a very simple 4 × 4 *optical cross-connect* (optical crossbar) *switch* can switch the light from any input fiber to any output fiber depending on the electrical control signal. (Practical cross-connect switches have more input–output connections.) The output of each input fiber is converted to an electrical signal (via photodetectors) and then a programmed logic circuit drives LEDs (or laser diodes) that emit light into the appropriate output fibers. This type of switch is limited by the electronic speed with which the detector can respond, and the light emitters can be driven; depending on the semiconductors used, typical speeds are less than a few Gb s^{-1}. See *MEMS*

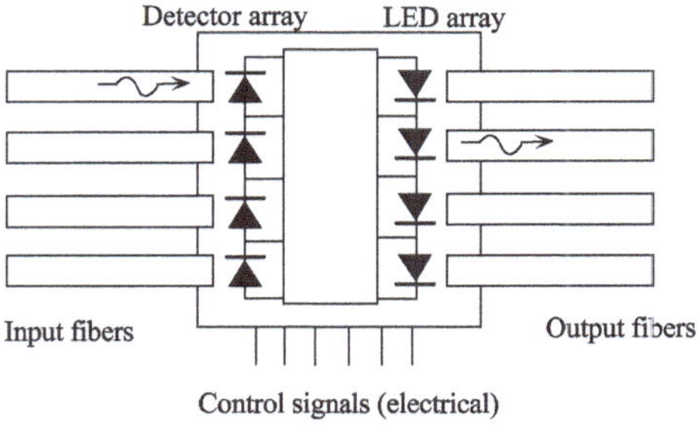

One simple **optical cross-connect**.

Optical density (OD) is a logarithmic quantity (to base 10) that characterizes the fractional transmittance of an optical material. If I_o is the incident light irradiance and I_t is the transmitted light irradiance then

$$\mathrm{OD} = \log_{10}(I_o/I_t).$$

The optical density of a perfectly transmitting optical material is 0, whereas that which transmits only 10% has an OD = 1. If α is the linear attenuation coefficient of the optical medium, $I_t = I_o\exp(-\alpha L)$, where L is the thickness of the optical material through which light passes, then

$$\mathrm{OD} = (0.4343)\alpha L.$$

Optical divergence refers to the angular separation of wavevectors on a given wavefront. Put differently, it describes how a propagating wave spreads out as it travels. A spherical wave has a 360° of divergence in any plane that contains the source; it diverges in all directions. On the other hand, a perfect plane wave has no divergence. Plane and spherical waves represent two extremes of wave propagation behavior from perfectly parallel to fully diverging wavevectors. They are produced by two extremes of sizes of electromagnetic (EM) wave source; infinitely large for the plane wave, and a point source for the spherical wave. In reality, an EM source is neither of infinite extent nor in point form, but would have a finite size and finite power. See *diverging waves*.

Optical electronegativity is an electronegativity-like quantity that can be used to predict the optical properties of certain classes of oxide glasses. Suppose that χ_{op}(anion) and χ_{op}(cation) are the optical electronegativities of the anion and cation in the structure of a simple oxide glass such as SiO_2 (silica). Then the maximum frequency ν_{max} of the absorption band is related to χ_{op}(anion) − χ_{op}(cation) by

$$\nu_{max} = 3000[\chi_{op}(\text{anion}) - \chi_{op}(\text{cation})].$$

This type of empirical approach is useful because one is usually interested in changes in the absorption characteristics with glass composition rather than exact frequencies. Further, in many cases, modified optical electronegativity is used depending on the optical property of interest; for example the bandgap optical electronegativity $\chi_{op}{}^*$ allows the calculation of the bandgap for cation–anion oxide compounds and glasses through Duffy's relationship

$$E_g = 3.72[\chi^*_{op}(\text{anion}) - \chi_{op}{}^*(\text{cation})] = 3.72\Delta\chi^*_{op}$$

where E_g is in eV. Qualitatively, electronegativity itself is a relative measure of the ability of an atom to attract the electrons in a bond it forms with another atom. The Pauling scale of electronegativity assigns an electronegativity value χ, a pure number, to various elements, the highest being 4 for F, and the lowest values being for the alkali metal atoms, for which χ are less than 1. In this scheme, the difference $\chi_A - \chi_B$ in the electronegativities of two atoms A and B is a measure of the polar or ionic character of the bond A–B between A and B. What is important is the electronegativity difference. For example, the Pauling electronegativities χ_A and χ_B for the elements A and B are defined in terms of their differences as in the original definition

$$\Delta\chi = \chi_A - \chi_B = \sqrt{Q/n}$$

where Q is the heat of formation of AB in eV and n is the average number of bonds per atom. There are various semiempirical equations that have been proposed to relate optical properties to the optical electronegativity difference $\Delta\chi_{op}{}^*$. The *Reddy equation* relates $\Delta\chi_{op}{}^*$ to n for silicates through

$$\Delta\chi^*_{op} = 9.8e^{-n}.$$

As for $\Delta\chi$, $\Delta\chi_{op}{}^*$ depends on the nature of the bonding, being high for ionic bonding and low for covalent bonding. See *Moss's rule.* (J. A. Duffy, *Phys. Chem. Glass,* **42**, 151, 2001; R. R. Reddy *et al.*, *Optical Materials*, **10**, 95, 1998.)

Table: Optical electronegativity Electronegativity values for metal ions and anions. Values are to the nearest 0·05 for χ_{opt} and $\chi_{opt}{}^*$, and 0·1 for χ (χ is for the element). Values from J. A. Duffy, *Phys. Chem. Glass,* **42**, 151, 2001.

Metal ion/anion	χ_{opt}	χ^*_{opt}	Pauling χ
Na^+		0·95	0·9
Mg^{2+}		1·1	1·2
Ca^{2+}		1·0	1·0
Sr^{2+}		1·0	1·0
Ba^{2+}		1·0	0·9
Zn^{2+}	< 2·1	1·1	1·6
Cd^{2+}	1·35	1·45	1·7
Hg^{2+}	1·5	1·55	1·9
Al^{3+}		0·95	1·5
Ga^{3+}		1·15	1·6
In^{3+}	<1·4	1·45	1·7
Sn^{4+}	1·5		1·8
P^{3-}		1·75	2·1
As^{3-}		1·55	2·0
Sb^{3-}		1·35	1·8
S^{2-}	2·2	2·15	2·5
Se^{2-}	2·0	1·9	2·4
Te^{2-}		1·8	2·1
Cl^-	3·0	3·0	3·0
Br^-	2·8	2·8	2·8
I^-	2·5	2·5	2·5

Optical extent of a beam is the product n^2G, where n is the refractive index of the medium and G is the beam's *geometrical extent.* In an aplanatic centered system, the optical extent is conservative.

Optical fiber is a thin cylindrical glass fiber, perhaps 100 microns or so in total thickness (as thin as hair), that acts as a dielectric waveguide, that is guiding light (electromagnetic waves) from one end to the other. An optical fiber has a core region of higher refractive index and a surrounding region, cladding, of lower refractive index which "clads" the core. In the case of a step index fiber, the change in the refractive index from the core to the cladding occurs as a step and typically this change is very small (less than 1%). The core region must have as low optical attenuation as possible to allow the transmission of light through the optical fiber. Light entering the fiber can only propagate along the fiber as certain distinct electric field patterns across the fiber, each of which is a distinct *mode* of the fiber. If the wavelength of the light is *longer* than the cut-off wavelength of the waveguide, then only the fundamental mode can propagate and there is no intermode dispersion.

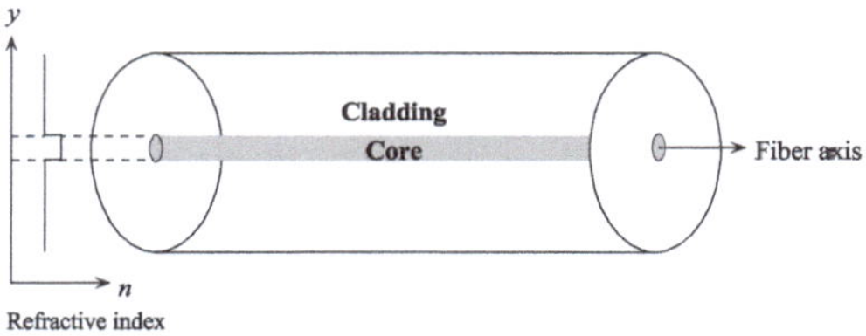

A step index optical fiber

A step index **optical** fiber.

Optical fiber Professor Charles Kao, who has been recognized as one of the pioneers of fiber optics, is receiving an IEE prize from Professor John Midwinter (1998 at IEE Savoy Place, London, UK; courtesy of IEE, now IET).

Optical fiber amplifier is an *optical amplifier* consisting of a glass fiber doped with certain rare-earth ions such as Er^{3+} ions, operating on the principle of stimulated emission. Typically it is a glass optical fiber whose *core* is doped with rare earth ions which are optically excited from an external light source to achieve *population inversion* and hence *optical amplification*. See *Raman fiber amplifier, erbium (rare-earth) doped fiber amplifier*.

Optical field is the electric field in a light wave.

Optical frequencies generally refer to electromagnetic waves in the ultraviolet, visible and infrared regions of the spectrum; or vibrations of the electromagnetic field from 3×10^{11} to 3×10^{16} Hz, which corresponds to a wavelength range of 10 nm to 1 mm. (They do not necessarily correspond to light in the visible range.)

Optical gain occurs whenever the irradiance of electromagnetic radiation is amplified as the radiation propagates through a medium. The amplification is usually described by an *optical gain coefficient g* which is the fractional increase in the optical power per unit distance along the direction of propagation. The amplification generally involves the rate of *stimulated emissions* exceeding *absorption* and *spontaneous emission*.

Optical gain coefficient see *optical gain*.

Optical gain lineshape describes the optical gain g vs. wavelength λ (or frequency υ) behavior of a medium. Normally the medium is pumped to achieve an optical gain. The lineshape function is the function $g = g(\upsilon)$ that represents the optical gain g vs. frequency (υ) behavior. See *laser output spectrum*.

Optical glasses are a range of noncrystalline transparent solids used in various optical components such as lenses, prisms, light pipes, windows, etc. Most (but not all) optical glasses are either

Table: Optical glasses Refractive index n_d at $\lambda_d = 587.6$ nm, the Abbe number υ_d and Sellmeier coefficients for some glasses when λ is in μm. (From Schott AG, Germany.)

Glass	n_d	υ_d	B_1	B_2	B_3	C_1	C_2	C_3
FK51	1.48656	84.47	0.971247817	0.216901417	0.904651666	0.00472301995	0.0153575612	168.68133
K7	1.5112	60.41	1.1273555	0.124412303	0.827100531	0.00720341707	0.0269835916	100.384588
BK7	1.51680	64.17	1.03961212	0.231792344	1.01046945	0.00600069867	0.0200179144	103.560653
F5	1.60342	38.03	1.3104463	0.19603426	0.96612977	0.00958633048	0.0457627627	115.011883
SF15	1.69892	30.20	1.57055634	0.218987094	1.50824017	0.0116507014	0.0597856897	132.709339
LASF31	1.88067	41.01	1.71317198	0.718575109	1.7233247	0.00819172228	0.0297801704	138.461313

Table: Optical glasses Experimental *dn/dT* and various coefficients for calculating *dn/dT*. (From Schott AG, Germany).

Glass	*dn/dT* 20 °C, 0.546 μm	D_0	D_1	D_2	E_0	E_1	λ_{TK}
FK51	-7×10^{-6} K^{-1}	-1.83×10^{-9}	-7.89×10^{-9}	-1.63×10^{-12}	3.74×10^{-7}	3.46×10^{-10}	0.150
BK7	2×10^{-6} K^{-1}	1.86×10^{-6}	1.31×10^{-11}	-1.37×10^{-11}	4.34×10^{-7}	6.27×10^{-10}	0.170

crown (K) or flint (F) types. K-glasses are usually soda-lime-silica glasses whereas flint glasses contain substantial lead oxide and hence F-glasses are denser, have higher refractive power and dispersion. Barium glasses contain barium oxide instead of lead oxide and, like lead glasses, have a high refractive index but lower dispersion. There are other high refractive index glasses such as lanthanum and rare earth containing glasses. Optical glasses can also be made from various other glass formers such as boron oxide, phosphorus oxide, germanium oxide glasses, etc. The Schott *glass code or number* is a special number designation (511604.253 for Schott glass K7) in which the first three numbers (511) represent the three decimal places in the refractive index ($n_d = 1.511$), the next three numbers (604) represent the Abbe number ($\nu_d = 60.4$), and the three numbers after the decimal (253) represent the density ($\rho = 2.53$ g cm^{-3}). In a different numbering system a colon is used to separate n_d and ν_d, e.g. 517:645 for a particular borosilicate crown means $n_d = 1.517$, $\nu_d = 64.5$.

In the Schott glass coding system, optical glasses are represented by letters in which the last letter K is for crown, and F is for flint. The first letters usually represent the important glass component in the glass, e.g. P for *phosphate*. The letters Kz ("Kurtz"), L ("leight") and S ("schwer") before K or F represent *short*, *light* and *dense* (*heavy*), respectively (from the German). S after K or F means *special*. Examples: BK, borosilicate crown; FK, fluor crown; PK, phosphate crown; PSK, dense phosphate crown; BaLK, light barium crown; BaK, barium crown; BaSK, dense barium crown; SSK, extra dense barium crown; ZnK, zinc crown; LaK, lanthanum crown; LaSK, dense lanthanum crown; KF, crown flint; SF, dense flint; SFS, special dense flint; BaF, barium flint; BaLF, light barium flint; BaSF, dense barium flint; LLF, extra light flint; LaF, lanthanum flint.

The refractive indices of optical glasses are normally given in terms of the *Sellmeier* equation

$$n^2 = 1 + \frac{B_1\lambda^2}{\lambda^2 - C_1} + \frac{B_2\lambda^2}{\lambda^2 - C_2} + \frac{B_3\lambda^2}{\lambda^2 - C_3}$$

in which *B*s and *C*s depend on the material; some values are listed in the table.

For many glasses, *n* increases with increasing temperature, i.e. *dn/dT* is positive, but there are also glasses, e.g. FK and PK glasses, in which *dn/dT* is

Table: Optical glasses Typical compositions of various optical glasses. The size of the circle represents a very rough relative fraction of the compound in the structure.

Name	SiO_2	Na_2O and K_2O	CaO and BaO	B_2O_3	Al_2O_3	PbO	P_2O_5	F	Typical n
Fused silica	●								1.46
Crown, K	●	•	•						1.50–1.52
Fluor crown, FK	●	•		•				•	1.49
Borosilicate crown, BK	●	•	•	•					1.50–1.51
Phosphate crown, PK	•	•	•	•	•		●		1.52–1.53
Light flint, LF	●	•				•			1.58–1.60
Flint, F	●	•				●			1.61 – 1.65
Dense flint, SF	•	•				●			1.65 – 1.80

negative. Schott recommends the following expression for *dn/dT* at a wavelength λ and a temperature T from −40 to 80 °C

$$\frac{dn}{dT} = \frac{n_0^2 - 1}{2n_0}\left[D_0 + 2D_1(\Delta T) + 3D_2(\Delta T)^2 + \frac{E_0 + 2E_1(\Delta T)}{\lambda^2 - \lambda_{TK}^2}\right]$$

where *dn/dT* is at T and at λ, n_0 is n at the same wavelength λ, but at a reference temperature T_0, $\Delta T = T - T_o$ is the temperature difference, and $D_0, D_1, D_2, E_0, E_1, \lambda_{TK}$ are constants.

See *Abbe number (Abbe diagram)*, *optical materials*.

Optical gyroscope see *Sagnac effect*.

Optical indicatrix (Fresnel's ellipsoid) is a refractive index surface placed in the center of the principal axes x, y and z of a crystal; the axis intercepts are n_1, n_2, and n_3. We can represent the optical properties of a crystal in terms of three refractive indices along three orthogonal axes, the principal axes of the crystal, x, y and z. These are special axes along which the polarization vector and the electric field are parallel. (Put differently, the electric displacement **D** and the electric field **E** vectors are parallel.) The refractive indices along these x, y and z axes are the principal indices n_1, n_2 and n_3, respectively, for electric field oscillations along these directions (not to be confused with the wave propagation direction). For example, for a wave with a polarization parallel to the x-axes, the refractive index is n_1. The refractive index associated with a particular electromagnetic (EM) wave in a crystal can be determined by using this optical indicatrix, which is Fresnel's *refractive index ellipsoid*. If all three indices were the same, $n_1 = n_2 = n_3 = n_o$ we would have a spherical surface and all electric field polarization directions would experience the same refractive index, n_o. Such a spherical surface would represent an optically isotropic crystal. For positive uniaxial crystals such as quartz, $n_1 = n_2 < n_3$. Suppose that we wish to find the refractive indices experienced by a wave traveling with an arbitrary wavevector **k**, which represents the direction of phase propagation. Suppose that this phase propagation direction is OP and is at an angle θ to the z-axis. We place a plane perpendicular to OP and passing through the center O of the indicatrix. This plane intersects the ellipsoid surface in a curve $ABA'B'$ which is an *ellipse*. The major (BOB') and minor (AOA') axes of this ellipse determine the field oscillation directions and the refractive indices associated with this wave. Put differently, the original wave is now represented by two orthogonally polarized EM waves called *ordinary* (o-) and *extraordinary* (e-) waves. The line AOA', the *minor axis*, corresponds to the polarization of the ordinary wave and its semiaxis AA' is the refractive index $n_o = n_2$ of this o-wave. The electric displacement and the electric field are in the same direction and parallel to AOA'. If we were to change the direction of OP we would always find the same minor axis.

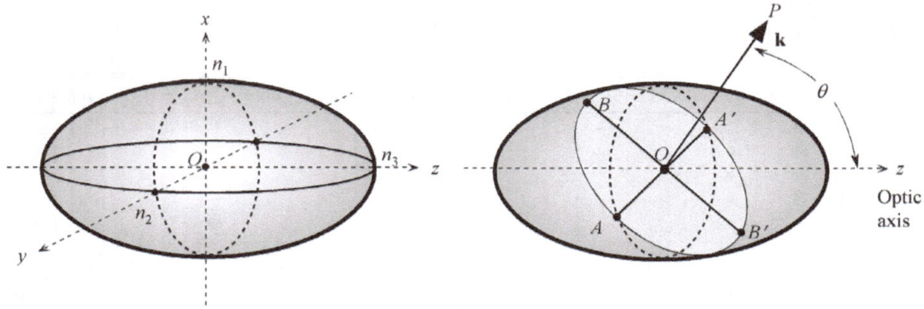

Optical indicatrix (a) Fresnel's ellipsoid (b) An EM wave propagating along OP at an angle θ to optic axis.

i.e. n_o is either n_1 or n_2 whatever the orientation of OP (try orientating OP to be along y and along x). This means that the o-wave always experiences the same refractive index in all directions. The o-wave behaves just like an ordinary wave (hence the name). The line BOB', the *major axis*, corresponds to the electric displacement field (**D**) oscillations in the extraordinary wave and its semiaxis OB is the refractive index $n_e(\theta)$ of this e-wave. This refractive index is smaller than n_3 but greater than n_2 ($= n_o$). The e-wave therefore travels more slowly than the o-wave in this particular direction and in this crystal. If we change the direction of OP, we find that the length of the major axis changes with the OP direction. Thus, $n_e(\theta)$ depends on the wave direction, θ. As apparent, $n_e = n_o$ when OP is along the z-axis, that is, when the wave is traveling along z. This direction is the *optic axis* and all waves traveling along the optic axis have the same phase velocity whatever their polarization. When the e-wave is traveling along the y-axis, or along the x-axis, $n_e(\theta) = n_3 = n_e$ and the e-wave has its slowest phase velocity. Along any OP direction that is at an angle θ to the optic axis, the e-wave has a refractive index $n_e(\theta)$ given by

$$\frac{1}{n_e(\theta)^2} = \frac{\cos^2\theta}{n_o^2} + \frac{\sin^2\theta}{n_e^2}$$

Clearly, for $\theta = 0°$, $n_e(0°) = n_o$ and for $\theta = 90°$, $n_e(90°) = n_e$. The major axis BOB' determines the e-wave polarization by defining the direction of the displacement vector **D** and not **E**. Although **D** is perpendicular to **k**, this is not true for **E**. The electric field of the e-wave is orthogonal to that of the o-wave, and it is in the plane determined by **k** and the optic axis. **E** is orthogonal to **k** when the e-wave propagates along one of the principal axes.

Optical information processing refers to systems where information is processed – involving logical operations and/or memory – using light. Consider a material with nonlinear refraction providing a nonlinear dependence of transmitted light intensity as a function of incident light intensity. A so-called optical transistor can be made from such a material by placing it within a Fabry–Perot resonator. Assume that the front face of the resonator is illuminated with two laser beams – a strong constant intensity I "pump" beam and a weak variable intensity "probe" beam – when I lies below the threshold of the device, the transmitted light intensity is in its "low" state, while adding a ΔI using the probe beam switches the device into its "high" transmitted intensity state. Such a device thus acts as an all-optical switch suitable for performing logical functions. In a conventional electronic computer, logical functions may be accomplished using 16 boolean functions. However, in practice only two of them (e.g. AND or OR) are sufficient; these can be combined to generate the other 14. The optical switch discussed above may easily be used to represent either an AND or an OR gate. The AND gate may be formed from the two incident beams acting as the two inputs for the gate. Each beam is held in its "high" state at an intensity below the switching knee of the device, but when their energy is combined in the cavity, the total intensity lies above the threshold and drives the device to its "high" state – thus, only if *both* beams incident on the front surface of the cavity are in this "high" state, will there be a "high" output from the device. Otherwise, the output of the device will be "low." This is the truth table of an AND gate. To represent an OR gate, we set the "high" intensities of each of the two incident beams to be equal and having a value above the switching threshold of the device. In this case, only if one or both of the beams is "high" will the output be "high," otherwise the output will be "low." Thus, this configuration acts as an OR gate. A NOT gate can be constructed by using the reflected beam as the output, since this beam acts as the inverse of the transmitted beam.

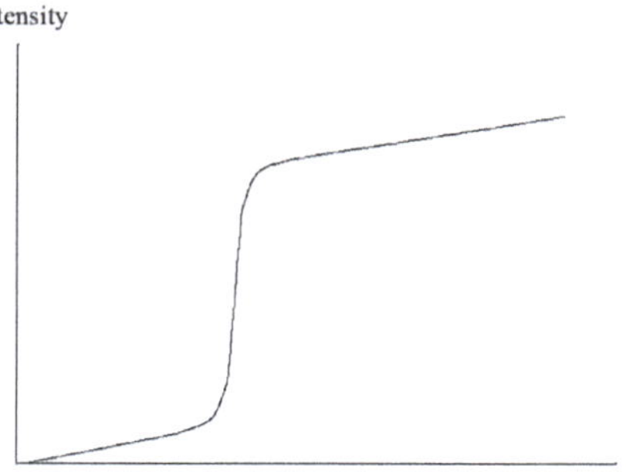

Optical information processing Nonlinear refractive index material provides a nonlinear characteristic for the dependence of transmitted signal intensity on incident light intensity.

Although the schemes described above are suitable for implementing logical operations, a full optical information processing system requires a means for storing information. In a binary computing system one requires storage elements able to represent the two stable states of the system. *Optical bistability* can be used to provide such a capability. The memory functions as follows. Assume that one uses a beam with intensity lying within the domain of the hysteresis loop (i.e. comprising two stable states) If the transmitted beam *was* of high-level intensity, then it *will remain* high-level. If it *was* of low-level intensity, it *will remain* at the low-level. Assuming that the high-level intensity represents a "1" and the low-level intensity represents a "0," putting a "1" in the device may be done by just using a short pulse of another beam, such that the cumulative intensity is sufficient to provide a high-level transmitted intensity. Putting a "0" in the device may be accomplished by blocking the beam for a short duration. Clearly, although optical information processing may be accomplished with the individual components discussed above, an information processing system relies on the interconnection or integration of a plurality of such devices Such optical circuits require careful synchronization of the signals using a reference clock to form a pipeline processor. Operations can be combined using functional logical units interconnected to form the processor.

Optical integrator sometimes refers to an optical device that detects the total integrated power (over its spectrum, spatial distribution, states of polarization) of incoming radiation as in the *integrating sphere*. On the other hand, when used in the context of *optical computing*, it refers to a two-dimensional *analog optical processor* that performs the operation of integration (continuous summation) of a given function over a definite spatial range. See *integrating sphere*.

Optical isolator allows light to pass in one direction and not in the opposite direction. For example, the light source can be isolated from various reflections by placing a polarizer and a Faraday rotator that rotates the field by 45°. The reflected light will have a $2\theta = 90°$ rotation and will not

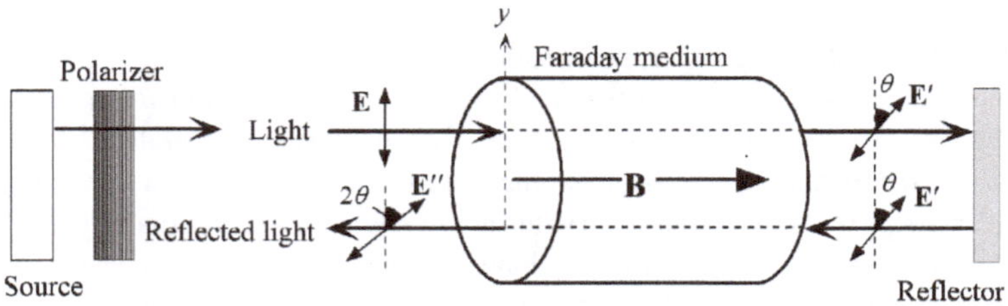

Optical isolator The sense of rotation of the optical field **E** depends only on the direction of the magnetic field for a given medium (given Verdet constant). If light is reflected back into the Faraday medium, the field rotates a further θ in the same sense to come out as **E**″ with a 2θ rotation with respect to **E**.

Optical isolators Faraday effect optical isolators: Air-path isolators (left) and laser diode-to-fiber isolator (right). The cylindrical case contains a rare-earth magnet. (Courtesy of OFR.)

pass through the polarizer back to the source. The magnetic field is typically applied by enclosing the Faraday medium in a rare-earth magnet ring.

Optical Kerr effect is the change in the refractive index of an optically isotropic medium when the intensity of a propagating light wave is sufficiently high. Consider a material in which the polarization P does not have the second-order term

$$P = \varepsilon_0 \chi^{(1)} E + \varepsilon_0 \chi^{(3)} E^3$$

$$\text{or} \qquad P/(\varepsilon_o E) = \chi^{(1)} + \chi^{(3)} E^2$$

where the constants $\chi^{(1)}$ and $\chi^{(3)}$ are called the linear and third-order susceptibility, respectively, and the superscript is only a label to indicate the order of the susceptibility. The first term with the *electric susceptibility* $\chi^{(1)}$ corresponds to the relative permittivity ε_r and hence to the refractive index n_o of the medium in the absence of the third-order term, i.e. under low fields. The E^2 term represents the irradiance I of the beam. Thus, the refractive index depends on the intensity of the light beam, a phenomenon called the *optical Kerr effect*,

$$n = n_o + n_2 I \qquad \text{where}$$

$$n_2 = \frac{1}{n_o^2 c\varepsilon_o}\text{Re}[\chi^{(3)}] \approx \frac{1}{n_o^2 c\varepsilon_o}\chi^{(3)} \text{ [SI units]}$$

where n_2 is a constant often called the *nonlinear refractive index*, and c and ε_o are the speed of light and absolute permittivity of free space, respectively. The optical Kerr effect occurs very fast, typically within picoseconds. The phase ϕ of an electromagnetic wave at a point z is given then by

$$\phi = \omega_o t - \frac{2\pi n}{\lambda} z = \omega_o t - \frac{2\pi[n_o + n_2 I]}{\lambda} z.$$

It is clear that the phase depends on the light intensity and the change in the phase along Δz due to the light intensity alone is

$$\Delta\phi = \frac{2\pi n_2 I}{\lambda}\Delta z.$$

As the light intensity modulates the phase, this is called *self-phase modulation*. Obviously light is controlling light. Consider the phase at time t and point z:

$$\phi = \omega_o t - \frac{2\pi n}{\lambda} z = \omega_o t - \frac{2\pi[n_o + n_2 I]}{\lambda} z.$$

When the light intensity is small $n_2 I \ll n_o$, obviously $\omega = \partial\phi/\partial t = \omega_o$. However, suppose we

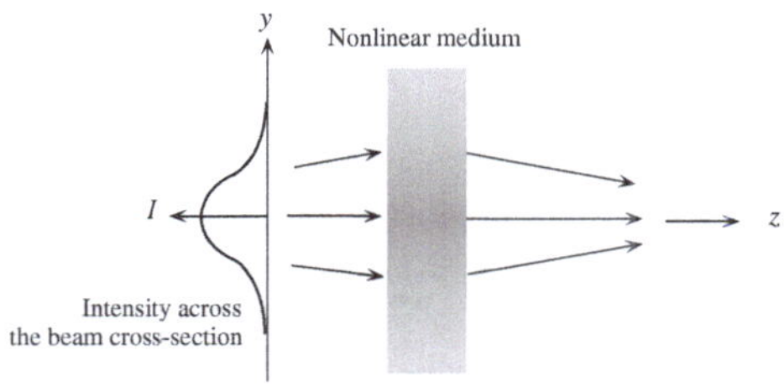

Optical Kerr effect The intensity variation across the beam cross-section leads to a similar refractive index in the nonlinear medium. Thus, the medium resembles a graded index guide or a GRIN rod and can focus the beam.

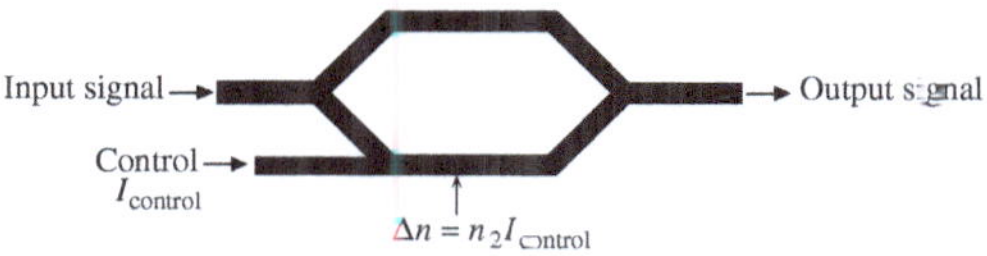

Optical Kerr effect Schematic illustration of an optical switch using the optical Kerr effect. $I_{control}$ is the intensity of the control light (actuating signal) in the lower arm of the Mach–Zehnder interferometer.

have an intense beam and the intensity I is time dependent $I = I(t)$. Consider a pulse of light traveling along the z-direction and the light intensity vs. *time* shape is a "Gaussian" (this is approximately so when a light pulse propagates in an optical fiber, for example). Then the instantaneous frequency is

$$\omega = \frac{\partial \phi}{\partial t} = \omega_0 - \frac{2\pi n_2 z}{\lambda} \frac{\partial I}{\partial t}$$

so that the frequency changes with "time," or across the light pulse. Since $\partial I/\partial t$ is rising at the leading edge and falling in the trailing edge, the two ends of the pulse contain different frequencies. The change in the frequency over the pulse is called *chirping*. Self-phase modulation therefore changes the frequency spectrum of the light pulse during propagation. The second term above results in the broadening of the frequency spectrum and hence leads to more dispersion for a Gaussian pulse propagating in an optical fiber. A Gaussian beam passing through a plate of nonlinear medium can become *self-focused* by the optical Kerr effect. The optical Kerr effect has a number of potential photonics applications including optical switches. For example, consider the Mach–Zehnder optical circuit shown in the second figure. If we pump high-intensity light into the lower arm, the refractive index in this path can be varied, and if the variation

Table: Optical Kerr effect Typical nonlinear properties (approximate values) from various sources including Boyd, *Nonlinear Optics*, Academic Press, 2001.

Material	Silica	As_2S_3	Fluoride phosphate glasses	CS_2 liquid	Azobenzenes	PPV: Poly (p-phenylenevinylene)
n_o (Typical)	1.5	2.40	1.44–1.51	1.63		
n_2 cm^2 W^{-1}	2.7×10^{-16} (1.95 μm)	2×10^{-14}	$(2\text{–}3) \times 10^{-16}$	3.2×10^{-14}	1×10^{-14}–1×10^{-8} May be + or −	2×10^{-12} (0.8 μm)
Material	CdS crystal	GaAs crystal	Ge crystal	Si crystal	ZnSe crystal	Au nanoparticles in glass
n_o (Typical)	2.34	3.47	4	3.4	2.7	1.5
n_2 cm^2 W^{-1}	5.1×10^{14} (1.06 μm)	3.3×10^{-13} (1.06 μm)	9.9×10^{-14}	6×10^{-14} (1.54 μm)	3×10^{-14} (1.06 μm)	2.6×10^{-10}

is sufficient, the phase of the signal in this arm will experience more retardation than that in the upper arm, leading eventually to destructive interference at the output – light being switched off.

The n_2 coefficient in the CGS system appears in $n = n_o + n_2\langle E^2\rangle$ where E is the amplitude of the optical field and has different dimensions and is in esu. Many works report n_2 in esu so that the conversion requires

$$n_2\left[\text{SI, m}^2\text{W}^{-1}\right] = \left(\frac{40\pi}{cn_o}\right) n_2\,[\text{esu}]$$

in which c, the speed of light, is in m/s. Further, $\chi^{(3)}$ is often measured in esu, so that n_2 in cm^2/W is

$$n_2\left[\text{cm}^2\text{W}^{-1}\right] = \left(\frac{12\pi^2}{n_o^2 c}\right) 10^7 \chi^{(3)}\,[\text{esu}] = \left(\frac{0.0395}{n_o^2}\right) \chi^{(3)}\,[\text{esu}]\,.$$

(K. Tanaka, in *The Handbook of Electronic and Photonic Properties*, S. Kasap and P. Capper (Editors) Springer, Heidelberg, 2007 Chaper 44; L. Brzozowski and E. H. Sargent, *J. Mater. Sci. Mater. Elec.*, **12**, 483, 2001.)

Optical Kerr shutter see *optical Kerr effect.*

Optical materials are those crystalline solids and glasses that are commonly used in the construction of optical components, such as lenses, prisms, windows, mirrors, polarizers, etc. In addition to the required optical properties, e.g. a well-defined refractive index n with a known dependence on the wavelength, $n = n(\lambda)$, and temperature, such optical materials should also have various other desirable material properties, such as good homogeneity (e.g. negligible macroscopic variations in the refractive index), negligible thermal expansion (small α_L, thermal expansion coefficient), resistance to mechanical damage (e.g. resistance to scratching), resistance to chemical degradation (e.g. chemical corrosion, staining), etc. Further, for optical glasses, it is desirable to have negligible incorporation of air (or gas) bubbles in the glass structure during fabrication (e.g. shaping), negligible stress induced birefringence, negligible nonlinear properties (unless used specifically for its nonlinearity), negligible fluorescence, etc. In all above cases, negligible implies less than a tolerable quantity within the context of the particular optical application. For example, n at a given wavelength should not vary for a given glass material by more than 10^{-5} for many optical components, and 10^{-6} for certain optics applications in astronomy. In addition, they should have good or reliable manufacturability at an affordable cost. There are various useful optical materials that encompass not only single crystals, such as CaF_2, CdTe, Ge, MgF_2, Si, ZnSe, quartz, sapphire, etc., but also a vast range of glasses (which are supercooled liquids with high viscosity), such as flint and crown glasses, fused silica, etc. Higher refractive index materials have more refractive power and allow lens designs that need less curvature for focusing light, and hence generally less aberrations. Flint glasses have a larger refractive index than crown glasses. On the other hand, crown glasses are chemically more stable, and can be produced more to specification. While most optical materials are used for their optical properties (e.g. in optical transmission), certain "optical" materials, (auxiliary materials) are used in optical applications such as mirror substrates, optical spacers, etc. for their nonoptical properties such as their negligible thermal expansion coefficient. See various individual optical material entries.

Optical mode confinement refers to how well the electric field in a propagating light mode is contained within the guiding region of a waveguide, and depends on the relative difference between the refractive index of the film in which the light actually propagates, and that of the adjacent layers. The higher the index difference, the greater the confinement. The (transverse) confinement factor Γ is a function of the guide geometry, and can be

Table: Optical materials Selected optical materials (Compiled from Oriel, Newport and Melles-Griot websites.) n_d at $\lambda_d = 587.6$ nm.

Glasses	Transmission Typical μm	n_d	υ_d	Typical applications	Comment
Fused silica	0.175–2	1.45846	67.8	Lenses, windows, prisms, interferometric FT-IR components. UV lithography	Sythetic. Has UV transmittance and excellent thermal properties; low α_L. Resistant to scratching
SF 11, flint	0.380–2.35	1.78472	25.76	Lenses, prisms	Flint glass have $\upsilon_d < 50$
LaSFN9, flint	0.420–2.30	1.85025	32.17	Lenses, prisms	High refractive index. More lens power for less curvature
BK7, borosilicate crown	0.380–2.1	1.51680	64.17	Visible and near IR optics. Lenses, windows, prisms, interferometric components	All around excellent optical lens material. Not recommended for temperature sensitive applications.
BaK1, barium crown	0.380–2.1	1.57250	57.55	Visible and near IR optics. Lenses, windows, prisms, interferometric components	All around excellent optical lens material. Not recommended for temperature sensitive applications.
Optical crown	0.380–2.1	1.52288	58.5	Lenses, windows, prisms, interferometric components	Lower quality than BK7
Pyrex, borosilicate glass		1.43385	66	Mirrors	Low thermal expansion
Crystals					
BaF_2 crystal	0.150–13	1.4748	83	UV to Far IR lenses, windows and prisms. Antireflection coatings. Astronomical applications	Sensitive to mechanical and thermal shock. Less hard than CaF_2
CaF_2 crystal	0.170–7	1.43385	95	Lenses, windows for UV optics, especially for excimer laser optics, cryogenically cooled thermal imaging	Sensitive to thermal shock

(*cont.*)

Table: Optical materials (*cont.*)

Glasses	Transmission Typical μm	n_d	υ_d	Typical applications	Comment
CdTe crystal	1–25			Lenses, wedges, prisms, beam splitters, antireflection coatings, windows, thermal imaging, etc. over 10–25 μm IR region. Low power CO_2 laser applications	Very wide IR transmission
Ge	2–16	4.0		IR windows; CO_2 laser windows	50–100 W throughput power
KCl	0.3–16	1.472 at 4 μm		FTIR spectroscopy, laser systems, imaging systems, analytical instruments, lenses, windows, wedges, prisms, etc.	Water soluble; must be protected from humidity and moisture
LiF	0.12–8.5	1.349 at 4 μm		Lenses, windows, wedges, prisms, excimer laser, astronomy, thermal imaging	Very low refractive index among IR transmitting materials
MgF_2 crystal	0.150–7	$n_o = 1.37774$ $n_e = 1.38956$		Lenses, windows, polarizers, filters; UV transmittance	Positive birefringent crystal. Resistant to thermal and mechanical shock
Quartz, SiO_2 crystal	0.150–2.5	$n_o = 1.54431$ $n_e = 1.55343$		UV optics. Wave plates. Polarizers	Positive uniaxial birefringent crystal
Sapphire, Al_2O_3 crystal	0.150–6.0	1.765	72	UV–IR high power laser windows, prisms, lenses	Good resistance to scratching, abrasion
Si (pure, transmission grade)	1.2–7 25–300	$n = 3.4289$ at 4 μm		IR applications, thermal imaging. Lenses, windows, beam splitters, wedges, prisms,	Low density, light and hard
ZnSe	0.5–19	2.625	9.6	Visible – Far IR windows, mirrors, lenses	Low absorption in IR. Popular for use with CO_2 laser at 10.6 μm
Auxiliary optical materials					
ULE™ SiO_2-TiO_2 glass				Optical spacers	Very small thermal expansion
Zerodur™, glass ceramic composite		1.5424	56–66	Mirror substrates. Not suitable for transmission optics due to internal scattering	Ultra low α_L. Fine mixture of glass and ceramic crystals (very small size)

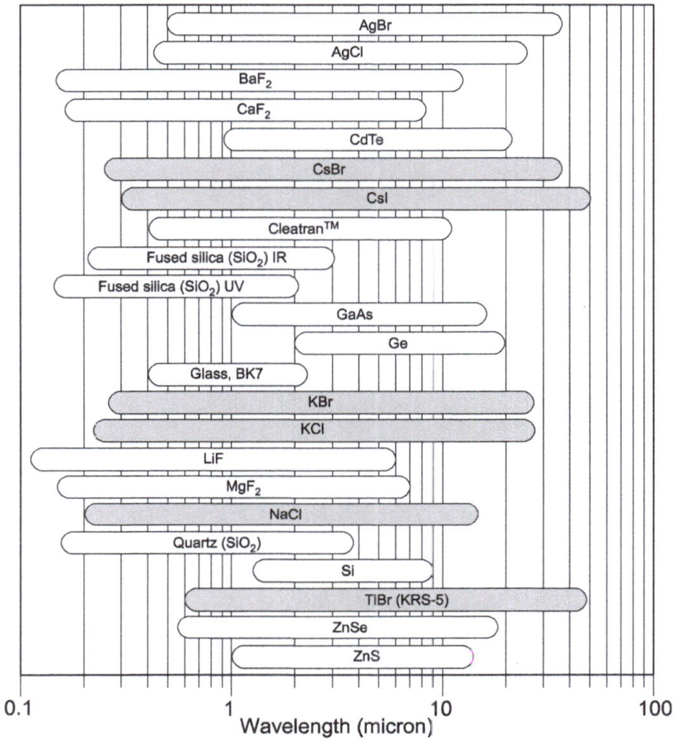

Transmission range of various useful **optical materials**. Gray shade represents water soluble materials. (From various sources.)

expressed as the fraction of the field intensity lying within the guide:

$$\Gamma = \frac{\int\int_{\text{guide}} |E(x, y)|^2 dx dy}{\int\int_{(xy)\text{plane}} |E(x, y)|^2 dx dy}.$$

In an active structure, with optical material gain α_{mat} effectively present inside the guide, only the fraction represented by Γ is actually affected by amplification, so that the *modal* gain is usually written as $\alpha_{\text{mod}} = \Gamma\alpha_{\text{mat}} - \alpha_{\text{d}}$, where a_{d} represents the overall losses, not confined to the guide.

Optical network is a group of stations that have been connected to each other by a number of transmission lines or optical links to allow user stations

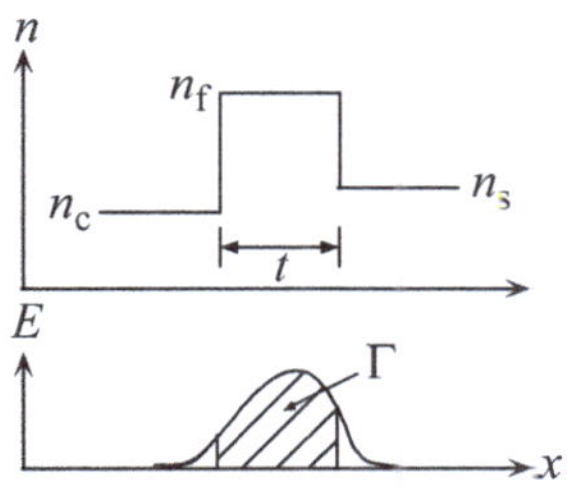

Optical mode confinement Mode confinement factor Γ in an asymmetric slab guide: cross-section of the guided field in an asymmetric slab waveguide of thickness t, index n_f. If the total area under the curve is normalized to unity, the confinement factor corresponds to the hatched area.

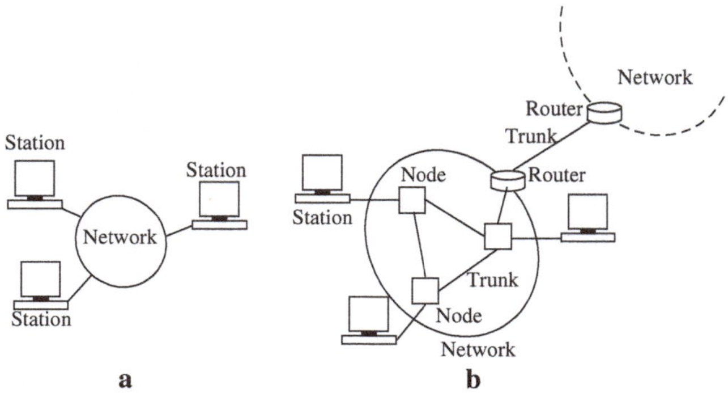

Optical network (a) A simple optical network, (b) A more complicated network that has nodes, trunks connecting nodes, and routers to route the signal to another network

to communicate with each other. Two or more optical networks can be connected to each other by using *routers* to direct (route) data from one network to another. A *station* or a *data terminal equipment* is a device, typically a computer or a phone, the user needs to connect to the network to be able to receive and send data. A node is a connection point in the network that receives communication lines; the latter terminate at nodes. *Trunks* are transmission lines that run between nodes, and carry a large amount of data traffic. They also connect routers of different networks. There are four basic types network. A *local area network* (LAN) normally involves one or more buildings in close proximity that form a group of interconnected buildings, for example, a factory, or a university campus. Metropolitan area networks (MAN) are those networks involved in neighborhoods or even a city that has several LANs. *A wide area network* covers a wide geographic area, which may involve trunks between cities or long-haul cross-country or country-to-country links. *Topology* represents

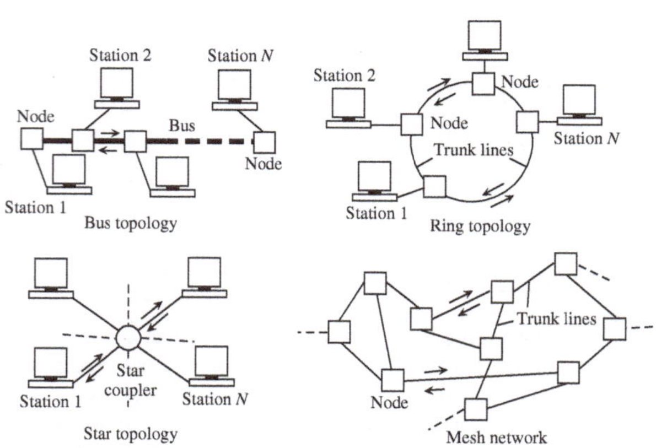

Optical network Four main optical network topologies.

a way of linking nodes by transmission channels to achieve a reliable network. There are four main topologies: (i) bus; (ii) ring; (iii) star; (iv) mesh, as shown in the second figure. In the mesh topology, each node has at least three connections. A switch is used to connect nodes or change links in a network as desired or specified by data. The router directs the data to another router by examining the information package header. This information package contains an address that tells the sending router where it should be sent, and the receiving router examines the same package (the address) upon arrival to see if it is supposed to receive that package. While a switch creates a temporary link, a router simply redirects the data package by examining the information it contains, and redirects packages one at a time. It is an efficient way to process data and send it to the right destination. Data transfer in networks occurs primarily by multiplexing data packages, known as packet switching, and as long as the data stream has a header that identifies the packet, and tells routers where it should be sent, many data packets can be sent one after another. There are various standards such as *asynchronous transfer mode* (ATM) or *internet protocol* (IP) to specify the rules that should be used for packet switching, e.g. the data length and structure of a packet. The fiber transmission system itself has standards as well, one of which is *synchronous optical networking* (SONET).

Optical path, L, along a given light trajectory Γ is

$$L = \int_{\Gamma} n ds$$

where n is the refractive index and s the curvilinear coordinate along the curve. It represents the length of the trajectory that light would travel in vacuum during the same time. It can be calculated along a trajectory actually followed by light, or along a purely virtual path. Fermat's least time principle stipulates which path is actually traveled: the shortest. See *Fermat's principle*.

Optical path difference (OPD) is a critical parameter for investigating two-wave interference phenomena. If two coherent rays, initially in phase, travel different paths Γ_1 and Γ_2 before recombining at point P, the optical path difference

$$\delta = L_2 - L_1 = \int_{\Gamma 2} n ds - \int_{\Gamma 1} n ds$$

is responsible for introducing a phase change between the waves, which usually depends on P. For $\delta = 0$, the two vibrations are in phase; for $\delta \neq 0$ (positive or negative), the relative dephasing at wavelength λ (in vacuum) is $(2\pi\delta/\lambda) = 2\pi p$, where p is the order of interference. Interference is constructive for integer p, destructive for half-integer p. See *Young's interference fringes*.

Optical power is the electromagnetic wave (light) energy flowing per unit time through an area of interest. It is also called the *radiant flux*. Light intensity, also known as irradiance or radiant flux density, is the optical power flowing per unit area See *flux*.

Optical properties of semiconductors typically consist of their refractive index n and extinction coefficient K or absorption coefficient α (or equivalently the real and imaginary parts of the relative permittivity) and their dispersion relations, that is their dependence on the electromagnetic radiation wavelength λ, or photon energy $h\nu$, and the changes in the dispersion relations with temperature, pressure, alloying, impurities, etc. The most important features in the α vs. $h\nu$ behavior as the photon energy increases can be summarized as follows: (a) free carrier absorption due to the presence of free electrons and holes, an effect that decreases with increasing photon energy; (b) an *impurity absorption* band (usually narrow) due to the various dopants; (c) *Reststrahlen* or *lattice absorption* in which the radiation is absorbed by vibrations of the crystal ions; (d) *exciton absorption* peaks that are usually observed at low temperature

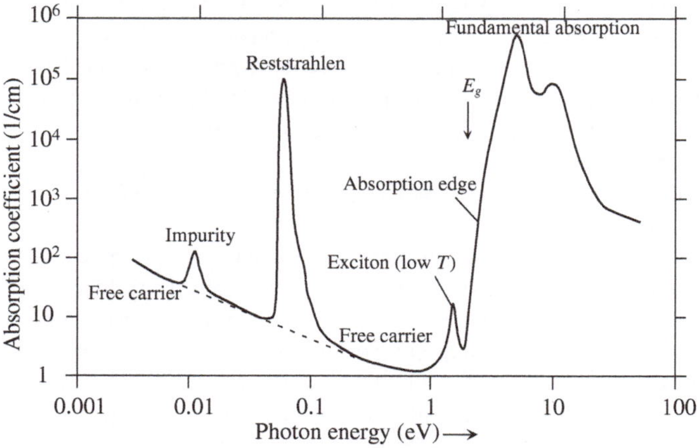

Optical properties of semiconductor Absorption coefficient of a hypothetical semiconductor to illustrate various possible absorption processes.

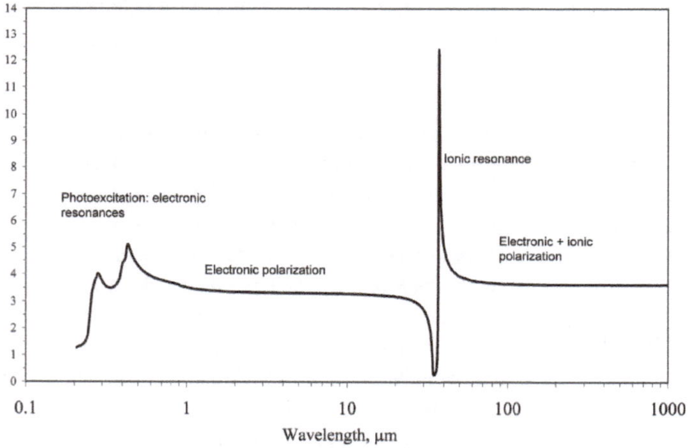

Optical properties of semiconductor Refractive index of GaAs vs. wavelength over a wide range.

and are close to the fundamental absorption edge; (e) *band-to-band* or *fundamental* absorption which photoexcites an electron from the valence to the conduction band. The latter type of absorption has a large absorption coefficient and occurs when the photon energy reaches and exceeds the bandgap energy E_g. It is probably the most important absorption effect; its value for $h\upsilon > E_g$ can be predicted by using band-to-band absorption physics. The value of E_g, and its temperature shift, dE_g/dT, are

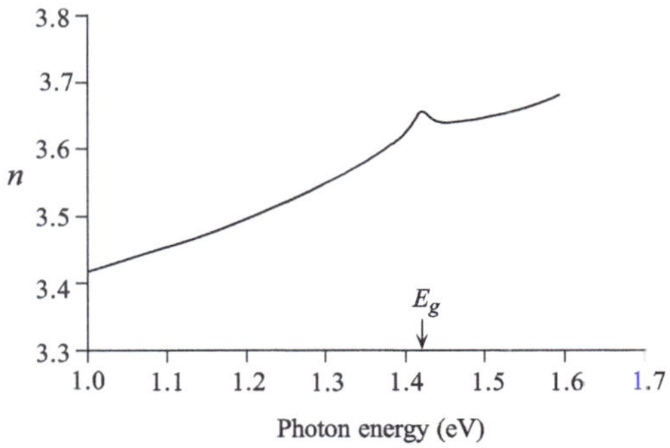

Optical properties of semiconductor Refractive index of GaAs vs. wavelength below and above the bandgap energy.

therefore important factors in semiconductor-based optoelectronic devices. In nearly all semiconductors E_g decreases with temperature, hence shifting the fundamental absorption to longer wavelengths. The refractive index n also changes with temperature. dn/dT depends on the wavelength, but for many semiconductors $(dn/dT)/n$ is of the order of 10^{-4} K^{-1}; e.g. for GaAs, $(dn/dT)/n = 4\times10^{-4}$ K^{-1} at $\lambda = 2$ μm. There is a good correlation between the refractive index and the bandgap of semiconductors in which, typically, n decreases as E_g increases; wider bandgap semiconductors have lower refractive indices.

The refractive index n and the extinction coefficient K (or α) are related by virtue of the *Kramers–Kronig relation*. Thus, large increases in the absorption coefficient for $h\upsilon$ near and above the bandgap energy E_g also result in increases in the refractive index n vs. $h\upsilon$ in this region. See *absorption coefficient*, *Moss's* and *Harve–Vandamme rule*.

Optical rectification is the generation of a dc voltage, or polarization, P, in a nonlinear medium from an electromagnetic wave (light). The electromagnetic oscillations at the light frequency are "rectified" by the medium to result in a constant (dc) polarization.

Optical resonator is an *optical cavity* defined by two reflectors such that only certain light frequencies are allowed to exist within the cavity just as only certain frequencies are allowed in an electrical *LC* circuit. The cavity may contain an optical medium with a refractive index. If the reflectors are flat and parallel mirrors then the resonator is called a *Fabry–Perot resonator* or an *etalon*. Only certain frequencies υ_m are allowed within the cavity, or to pass through the resonator. For example, if υ_m is an allowed frequency, a *mode of the resonator*, then $\upsilon_m = mc/(2L)$ where L is the cavity length and m is an integer 1,2, . . . The width of each mode depends on the cavity losses including mirror reflectances.

Optical semiconductor amplifier see *semiconductor laser amplifier*.

Optical storage see *phase change recording*.

Optical switches enable optical signals in various optical channels to be redirected, cross-connected, and managed in dense wavelength division multiplexed (DWDM) telecommunications systems. The keys to fulfilling the information demands of DWDM systems are scalability, flexibility, and dynamic delivery of communication services. The two types of optical switches are intelligent optical-electrical-optical (O-E-O) switches, or all-optical optical-optical-optical (O-O-O) switches. The latter are particularly attractive longer term solutions for networks where improved speed and performance can potentially be achieved in *all-optical networks* (AON). Arguably, the leading technology for developing an economically viable, scalable all-optical, O-O-O switch is the 3D *microelectromechanical system* (MEMS). 3D MEMS uses control mechanisms to tilt mirrors in multiple directions (three-dimensional). Other competing optical cross-connect switching technologies include liquid crystal, thermooptical bubble and thermooptical/electrooptic waveguide technologies. Only the MEMS technology is a viable scale-up technology suitable for channel counts that could grow to

Table: Optical properties of semiconductors Crystal structure, lattice parameter a, bandgap energy E_g at 300 K, type of bandgap (D = Direct and I = Indirect), change in E_g per unit temperature change (dE_g/dT) at 300 K, bandgap wavelength λ_g and refractive index n close to λ_g. (A = amorphous, D = diamond, W = wurtzite, ZB = zinc blende.)

Semiconductors	Crystal	a nm	E_g eV E_g eV	Type	dE_g/dT meV K^{-1}	$\lambda_g(\mu m)$	$n(\lambda_g)$	$dn/dT \times 10^{-5}\ K^{-1}$
Group IV								
Diamond	D	0.3567	5.48	I	−0.05	0.226	2.74	1.1
Ge	D	0.5658	0.66	I	−0.37	1.87	4	27.6 42.4 (4 μm)
Si	D	0.5431	1.12	I	−0.25	1.11	3.45	13.8 16 (5 μm)
a-Si:H	A		1.7–1.8			0.73		
SiCα	W	0.3081a 1.5120c	2.9	I	−0.33	0.42	2.7	9
III–V Compounds								
AlAs	ZB	0.5661	2.16	I	−0.50	0.57	3.2	15
AlP	ZB	0.5451	2.45	I	−0.35	0.52	3	11
AlSb	ZB	0.6135	1.58	I	−0.3	0.75	3.7	
GaAs	ZB	0.5653	1.42	D	−0.45	0.87	3.6	15
$GaAs_{0.88}Sb_{0.12}$	ZB		1.15	D		1.08		
GaN	W	0.3190 a 0.5190 c	3.44	D	−0.45	0.36	2.6	6.8
GaP	ZB	0.5451	2.26	I	−0.54	0.40	3.4	
GaSb	ZB	0.6096	0.73	D	−0.35	1.7	4	33
$In_{0.53}Ga_{0.47}As$ on InP	ZB	0.5869	0.75	D		1.65		
$In_{0.58}Ga_{0.9}As_{0.1}P_{0.1}$ on InP	ZB	0.5870	0.80	D		1.55		
$In_{0.72}Ga_{0.28}As_{0.62}P_{0.38}$ on InP	ZB	0.5870	0.95	D		1.3		
InP	ZB	0.5869	1.35	D	−0.40	0.91	3.4–3.5	9.5
InAs	ZB	0.6058	0.36	D	−0.28	3.5	3.8	2.7
InSb	ZB	0.6479	0.18	D	−0.3	7	4.2	29
II–VI Compounds								
ZnSe	ZB	0.5668	2.7	D	−0.50	0.46	2.3	6.3
ZnTe	ZB	0.6101	2.3	D	−0.45	0.55	2.7	

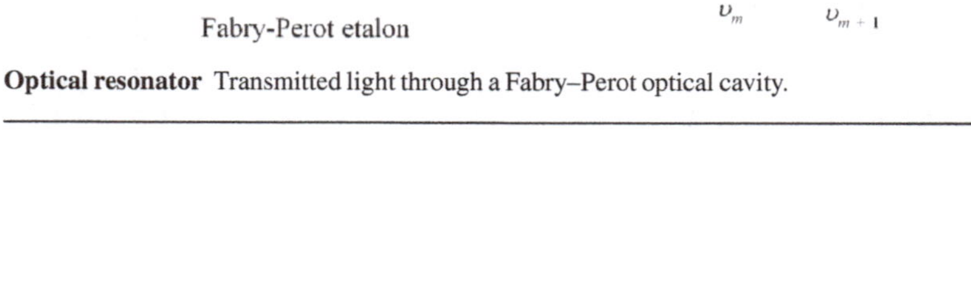

Optical resonator Transmitted light through a Fabry–Perot optical cavity.

hundreds as envisaged in future DWDM systems. Other advantages include low loss, fast switching time, low cross-talk and polarization sensitivity, as well as wavelength and bit rate independence, and low power consumption. At this time, none of the other competing technologies are able to satisfy all of these criteria. An O-O-O switch holds the promise of managing light signals without converting the signals to electrical form and then back again. This is especially attractive to those carriers operating large offices where up to 80% of the traffic is expected to pass through the office on its way to locations around the globe. MEMS currently affords the best chance of providing an all-optical switch matrix that can scale to the size needed to support a global communications network node with multiple fibers, each carrying hundreds of wavelengths. Network-management functions, which are an important part of operating a network, are available today using an optical switch having an electronic-based switching matrix. Available today with proven technology, these intelligent O-E-O switches address the need for high-bandwidth management while continuing to provide easy fault location and performance-monitoring information necessary to monitor and report on the health of a network. By using an electronic-based fabric, the intelligent O-E-O switch can overcome network impairments that currently limit the use of an all-optical switch in dynamic mesh architectures. The intelligent O-E-O switch encourages the use of a mesh, which is more bandwidth efficient and supports a flexible set of bandwidth-intensive service offerings. The electronics used in an intelligent optical switch also allows it to make use of the well-accepted synchronous optical network (SONET) standards. This not only helps with network management, but it also encourages the use of best-of-breed network elements by furthering interoperability among devices from multiple vendors. Not only does the intelligent O-E-O switch offer advantages from the reuse of SONET standards, but it also includes an evolution path to maximize the use of a set of data standards to improve data-centric communications and make the network more dynamic while greatly reducing provisioning times. See *optical networks*, *optomechanical switch*.

Optical thickness of a plate with refractive index n and length L is the quantity $\Delta = nL$, that is the distance which light would travel in vacuum during the same time it takes to travel through the plate.

Optical time domain reflectometry (OTDR) is a technique in which light attenuation along an optical fiber is determined as a function of distance along the fiber by measuring the backscattered light power. A very short light pulse is launched at the front end of a fiber. As this pulse propagates, portions of it become scattered (Rayleigh scattering). A further portion of these scattered waves in the backward direction will enter a guided mode of the fiber and become guided back; this is the *backscattered wave*. When the backscattered waves reach the front end, they are directed towards a detector where the backscattered power P_{BS} is measured as a function of time t from the launch instant. The time t of observation represents a "round trip" time from the fiber front end to the scattering point x and back to the front end. Thus, time t corresponds to a distance $2x = t \times$ (velocity in core) which means that $x = ct/2n_1$, where n_1 is the core refractive index. The detected backscattered signal is analyzed and displayed as backscattered power (usually in dB) vs. distance, x. As x (i.e. time) increases, representing further propagation along the fiber, P_{BS} decreases as a result of attenuation along the fiber. A splice typically leads to a drop in P_{BS} due to a degree of attenuation at the splice. An end-to-end fiber connection may exhibit a spike due to a Fresnel reflection (dielectric–dielectric reflection). When the pulse reaches the far end, it experiences a Fresnel reflection which occurs as a final spike in the OTDR signal. The power in the backscattered wave, P_{BS}, is a fraction of the forward traveling wave but attenuated by $\exp(-\alpha x)$, where

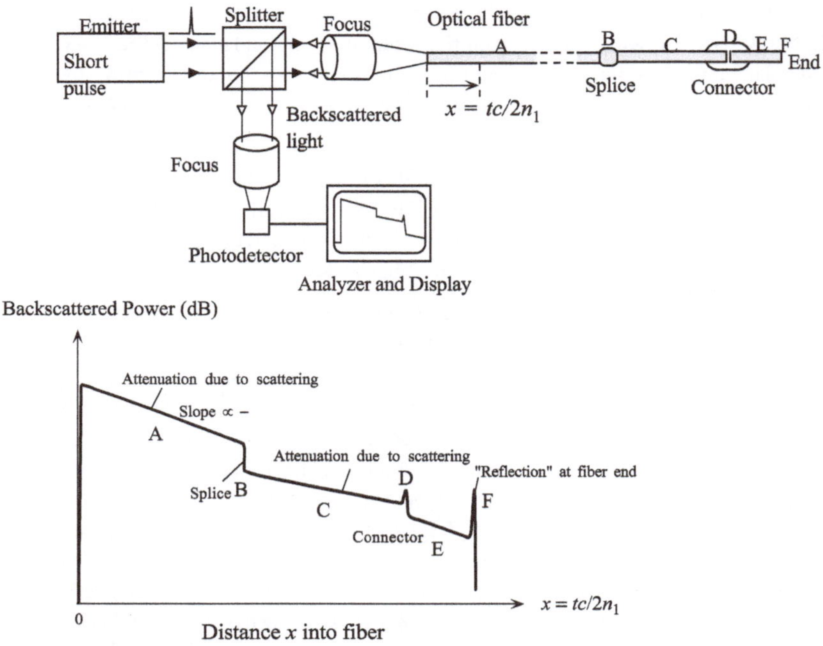

Optical time domain reflectometer (schematic) and a typical result on a fiber that has a splice and a connector (the curve is idealized).

α is the attenuation coefficient. This is because the backscattered wave travels a distance x back to the front end. Since the forward wave at x is $\exp(-\alpha x)$ of the launch power,

$$P_{BS} \propto \exp(-2\alpha x),$$

and

$$d\ln(P_{BS})/dx = -2\alpha.$$

Clearly, the slope of P_{PB} (dB) vs. x provides an indication of the local attenuation α at x.

Optical transfer function (OTF) represents the ability of an optical imaging system to transfer the magnitude and the phase information in the wavefronts it receives from the object onto an imaging plane as a function of different spatial frequencies, v_x and v_y. OTF can be broken into two factors: (i) the modulation transfer function, MTF, and (ii) the phase transfer function, PTF so that $\text{OTF}(v_x, v_y) = \text{MTF}(v_x, v_y) \times \text{PTF}(v_x, v_y)$, and PTF has a magnitude of 1 and only carries the phase information, i.e. $\text{PTF} = \exp[-j\phi(v_x, v_y)]$, where $\phi(v_x, v_y)$ is the change in phase at v_x, v_y. See *modulation transfer function*.

Optical tunneling is a phenomenon in which light is able to tunnel through a thin optical barrier and appear on the other side. The intensity of the transmitted light through the barrier depends on the thickness of the barrier; and decreases exponentially with this thickness. See *frustrated total internal reflection*.

Optical waveguide is a dielectric waveguide which has a *core* region of higher refractive index surrounded by a region of lower refractive index called the *cladding*. Electromagnetic waves with wavelengths in the optical region propagate in the core region of the guide by total internal reflection. See *planar waveguide, rib waveguide, optical fiber; interface scattering in optical waveguides.*

Optical wavelength add/drop multiplexer see *wavelength add/drop multiplexer (WADM).*

Optically denser medium has a higher refractive index.

Optically isotropic see *anisotropy.*

Optimal profile index, γ_{optimal}, is the value of the profile index that results in a minimum of intermodal (intermode) dispersion; typically $\gamma \approx 2$.

Optocoupler (photocoupler) or an **optoisolator** is an optoelectronic device that couples an electrical signal in one electronic circuit to another by using light in a process that coverts the electrical signal to an optical signal and then back to an electrical signal; the light emitter and the photodetector are all packed in a light-tight package. The signal to be coupled, the input signal, is made to drive an LED (light-emitting diode), whose light output is detected by a neighboring photodiode or a phototransistor in an independent electronic circuit. The output from the phototransistor is the coupled output signal and it is *electrically isolated* from the input, which is the main advantage of the optocoupler. Low-voltage circuits, for example, can be used to couple signals into high-voltage circuits, while the two are not electrically connected. Optocouples are used in a variety of applications where *isolation* is important. There must be not only good electrical isolation between the input and the output, but also low (ideally negligible) capacitive coupling. There should also be a high signal (e.g. current) transfer ratio from the input to the output circuit with negligible (or tolerable) noise addition. Typically, GaAs LEDs are used and the detector is either a Si photodiode or a Si phototransistor. Sometimes the phototransistor is part of a Darlington pair to provide a high current gain at the expense of speed. Many optocouplers are available with various logic elements already incorporated into the package. Optocouplers that are used for solid state relay applications (e.g. switching large currents in power circuits) have a phototriggerable silicon controlled rectifier (photo-SCR) or a phototriac that switches when the input LED is turned on. Critical parameters for an optocoupler are the isolation voltage, isolation resistance and capacitance, turn-on time t_{on}, rise time t_{r} (from 10% to

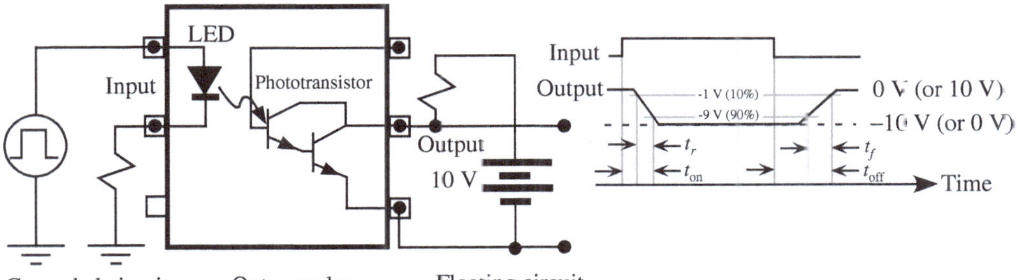

An **optocoupler** that has an LED optically coupled to a phototransistor pair. A Darlington pair connection provides a greater current gain than a single transistor. An input pulse turns on the LED which photogenerates a photocurrent in the phototransistor, which in turn generates sufficient collector current to take the output from 10 V to close to zero. The optocoupler acts as an inverter with the input and the output circuits electrically isolated. (Note: The turn-on and turn-off times have been defined in relation to whether the transistor is "on" or "off" respectively.)

90% of the output pulse change), turn-off time t_{off}, and the fall time t_f (from 90% to 10% of the full output change). For a phototransistor with a Darlington pair circuit, the turn-on and turn-off times are typically in the microsecond (1–10 μs) range. Most optocoupler applications are in digital and switching electronics where typically they fulfill isolation needs. Their speed in digital applications are typically in the 1–5 Mb/s range. Nonetheless, they are also used in analog (or linear) applications. Linear optocouplers are usually designed to have either a simple or a *pin* photodiode optically coupled to a high efficiency (AlGaAs) LED to increase the speed and obtain good linearity which typically results in a linearity in the 0.01–0.25% range depending on the input signal magnitude, and a bandwidth from dc to cut-off frequencies above 200 kHz. Optocouplers are widely and indispensably used in all types of applications in industrial, consumer and medical electronics, instrumentation, data communication systems, etc.; e.g. in isolated process controls, solid state relays, medical sensor isolation, microprocessor system interfaces, computer peripheral interfaces (isolating various signals in ubiquitous modems), digital telephone line isolation, etc.

Optoisolator see *optocoupler*.

Optomechanical switch is usually a mechanical device that relies on mechanical motion that moves one optical guide (e.g. an optical fiber) to line up with another guide and thereby change the optical path; these switches are limited by the speed of the mechanical motion. The mechanical motion may be implemented by an electromagnetic relay, for example, in which case the speed would be of the order of milliseconds. By having tiny electromechanical structures it is, however, possible to construct fast and reliable switches. The tiny electromechanical structures are called *microelectromechanical systems* (MEMS) that are fabricated by various semiconductor fabrication techniques that use special lithographic processes. For example, a MEMS optical switch can tilt a mirror and thereby deflect an optical signal from one guide to another. Such MEMS based switches are fast (microseconds) and reliable.

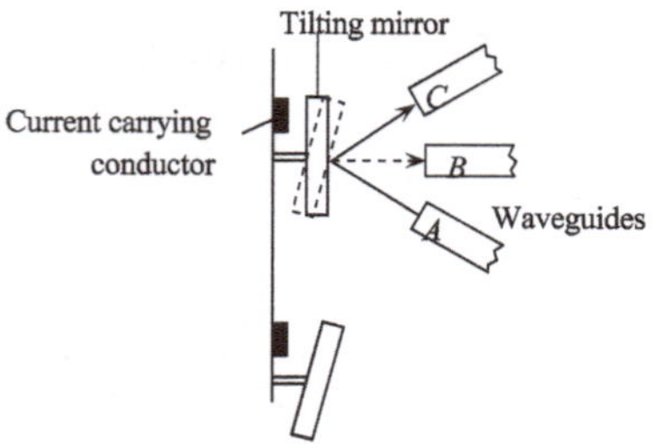

A MEMS **optomechanical switch**.

Order of grating see *diffraction grating*.

Organic light-emitting devices (OLEDs) are devices formed from thin films of organic materials sandwiched between two electrodes, and from which light is emitted when a potential is applied. This originates from the efficient recombination of *excitons* formed from carriers injected from the electrodes, generating light. The organic material is typically either composed of small molecular weight organic materials, or a polymer – in either case, these materials act as semiconductors – in which the organic component acts as the electroluminescent component. OLEDs are particularly attractive as they offer low-cost, high-energy efficiency (i.e. low power consumption, and operating at low voltages), large area, bright multicolor displays. In terms of manufacturing, these devices can literally be printed using inkjetting or screen printing over large areas and even on flexible substrates – the ease of manufacturing (no need for any high-temperature processing) and cheap materials are key reasons for the low cost. Because of the simplicity of their structure, they can be thin and lightweight, and even transparent when off. Unlike *LCDs*, which require backlights and lossy polarizers making them bulky and energy inefficient, OLEDs emit light directly, resulting in much higher light output for a given electrical input. Moreover, OLEDs respond much faster than LCDs (i.e. less than 1 ms compared with ~10 ms for LCDs). They may also be viewed over a wide range of viewing angles, as distinct from LCDs. Currently, research is still being focused on improving OLED display lifetime which is typically less than 10^4 hours. A key problem is permeation and reaction of the organic material with water, which is addressed by improving the sealing technology. Some of the different types of OLED include flexible OLED (FOLED), transparent OLED (TOLED), and stacked OLED (SOLED). FOLED offer the possibility of displays that can be rolled up like a newspaper, while TOLED use transparent top and bottom contacts to make transparent displays that might provide a display on the windshield of a car or eyeglasses, and SOLEDs offer compact high-resolution color displays by stacking the blue, green and red pixels of the displays on top of each other.

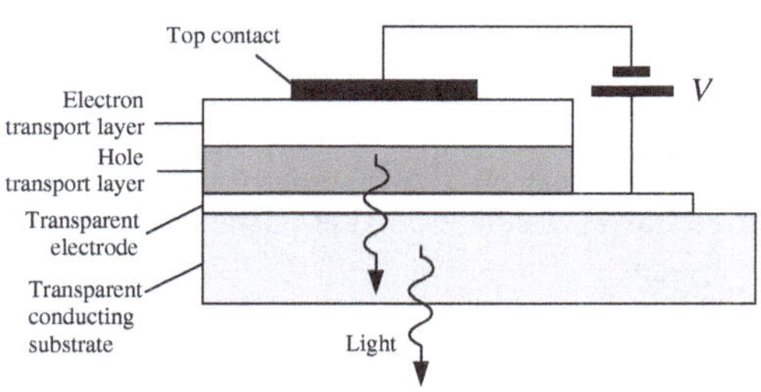

Schematic of an **OLED** structure showing the different organic and contact layers that comprise a standard OLED display pixel.

Orthotropic surface see *Lambertian diffusion.*

Oscillator strength in optical sciences refers to the fractional contribution f of a given "oscillator" to the polarizability of an atom or a molecule and hence to the optical properties of the medium. The simplest case is a single atom that has N electrons. If the first electron was the only electron involved in the polarization process in an oscillating field from a passing light wave, then the polarizability of the atom would be $\alpha_1 = (e^2/m_e)[1/(\omega_{01}^2 - \omega^2)]$ where ω is the angular frequency of light, ω_{01} is the resonance frequency associated with the oscillations of the first electron. This is the simplest Lorentz oscillator model. However, we must also include the other electrons, and this is done by assigning an oscillator strength f_i to each oscillator (in this case each electron) each of which has a resonance frequency ω_{oi}, so that the polarizability of the atom becomes

$$\alpha = \left(\frac{e^2}{m_e}\right)\sum_{i=1}^{N}\frac{f}{\omega_{oi}^2 - \omega^2}.$$

Oscillator strengths can be calculated quantum mechanically using first-order perturbation theory and by using the initial ϕ_i and final ϕ_j wavefunctions and the electric potential of an approximately homogeneous electric field. Note that the oscillator strengths in this case will obey the oscillator sum rule, $\sum_i f_{ij} = 1$. It can be shown that the oscillator strength for the transition from state i to j, is given by:

$$f_{ij} = \frac{2m_e\omega_{ij}|x_{ij}|^2}{\hbar};\ x_{ij}$$
$$= |\langle j|x|i\rangle|^2 = \int \phi_i^*\, x\phi_j\, dV$$

where x_{ij} is defined by the integration over the volume V, and is the so-called *electric dipole matrix element,* and ω_{ij} is the angular frequency for the excitation $E_j - E_i\ (=\hbar\omega_{ij})$. In the case of interband transitions in a semiconductor at energy $E_{ij} = E_j - E_i$,

$$f_{ij} \approx \frac{2m^* E_{ij}|x_{ij}|^2}{3\hbar^2}.$$

An extension of the oscillator sum rule was developed by Bardeen, providing an order of magnitude estimate for f_{cv} by summing over all transitions from $E_j = E_v$ as a function of k in all possible bands to give,

$$f_{vc} = \sum_{j\neq v} f_{vj}(\vec{k}) = 1 - \frac{m_e}{\hbar^2}\frac{\partial^2 E_{\vec{k}v}}{\partial k^2} \approx 1 + \frac{m_o}{m_h^*}.$$

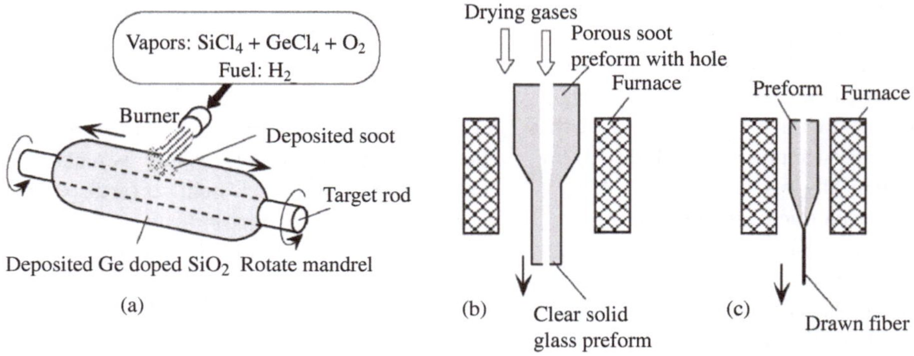

Schematic illustration of **OVD** and the preform preparation for fiber drawing. (a) Reaction of gases in the burner flame produces glass soot that deposits on to the outside surface of the mandrel. (b) The mandrel is removed and the hollow porous soot preform is consolidated; the soot particles are sintered, fused, together to form a clear glass rod. (c) The consolidated glass rod is used as a preform in fiber drawing.

where m_0 is the mass of the electron in vacuum. Using such an expression one may easily express the absorption coefficient $\alpha(\omega)$ for a bulk semiconductor (for $\hbar\omega \geq E_g$) as:

$$\alpha(\omega) = \frac{e^2(2m_r^*)^{3/2}}{4\pi\varepsilon_0 cn(\omega)\hbar^2} f_{cv}\sqrt{\hbar\omega - E_g}$$

where $1/m_r^*$ is the reduced effective mass (i.e. $1/m_r^* = 1/m_{dh}^* + 1/m_{de}^*$), c the speed of light in vacuum, and $n(\omega)$ the refractive index of the semiconductor as a function of the photon energy $\hbar\omega$. See *Lorentz oscillator, sum rule*.

Outside vapor deposition (OVD) is one of the vapor deposition techniques used to produce the rod preform used in fiber drawing. The OVD process has two stages. The first laydown stage involves using a fused silica glass rod (or a ceramic rod such as alumina) as a target rod. This acts as a mandrel and is rotated. The required glass material for the preform with the right composition is grown on the outside surface of this target rod by depositing glass soot particles. The deposition is achieved by burning various gases in an oxyhydrogen burner (torch) flame where glass soot is produced as reaction products.

Oxidation is the term used to refer to the production of silicon dioxide on silicon. Typically, silicon is exposed to either water vapor or oxygen at high temperatures to controllably produce a layer of oxidized silicon, or silicon dioxide, on the surface of a Si crystal.

P

Packet switching see *optical network*.

Parabolic mirror is a concave mirror that has the shape of parabolic surface and is able to bring all rays parallel to its axis to a single focal point without spherical aberration. Parabolic mirrors are widely used for detecting TV signals sent by a satellite, or by radio-astronomers for scanning some regions of the electromagnetic spectrum. It is also the best shape for a one-mirror telescope (Newton's telescope), since the parabola ensures perfect stigmatism between the on-axis object (say a distant star) and the main focal point.

Parallel ray is a ray that is parallel to the optical axis on the object side, but, after being refracted by the lens, it passes through the image focal point on the image side.

Parametric amplification refers to the amplification of optical waves whereby an intense coherent pump wave interacts with a nonlinear optical crystal to produce amplification at two other optical wavelengths.

Parametric oscillator is a device comprising a *parametric amplifier* located within a resonant optical cavity to generate a frequency-tunable coherent beam of light from an intense laser beam of fixed frequency. This device may be tuned by varying the phase-matching properties of the nonlinear material.

Paraxial chief ray is the ray, among a pencil of rays, passing through the very center of the entrance pupil.

Paraxial optics refers to the range of validity of the Gaussian approximation for geometric optics, that is for rays that are not far from the principal axis and only slightly inclined with respect to it (typically, with an angle α such as $\tan\alpha \approx \alpha$).

Paraxial ray equation is a limiting case of the ray equation in the paraxial approximation.

Parity selection rule states that, for a permitted atomic (or ionic) transition that involves electromagnetic radiation, wavefunctions of the initial and final states must have different *parity*, that is, the parity must change in a radiation absorbing or emitting electronic transition. The parity is a property of any wavefunction (or quantum mechanical state) that describes the function after a mirror reflection; that is, in one dimension, the x-coordinate $+x$ in $\psi(x)$ replaced by $-x$, the change in the symmetry of the wavefunction is examined to define its parity. Wavefunctions with *even* parity (states) are symmetric, identical after reflection (like a cosine function), while wavefunctions that have *odd* parity are antisymmetric (like a sine function). For an ion (or an atom) which is embedded in a host material, the parity selection rule may be partially removed due to the action of the *crystal field* or *ligands* giving rise to "forbidden lines" in the absorption–emission spectra.

Partial coherence is a notion experimentally related to the contrast of interference fringes. If two sources of equal luminous intensity are completely coherent, both spatially and spectrally (point sources, narrow line, constant phase relationship), then the interference fringes exhibit a high contrast. On the other hand, if they are completely incoherent, then the intensity resulting from their superposition is only the sum of their individual contributions, with no fringes at all. In-between, the sources are said to be *partially coherent*; the contrast of the fringes provides a quantitative measurement of their *degree of coherence*. See *Young's interference fringes*.

Passivation refers to the process of chemically protecting a semiconductor surface from degradation. For example, for bare gallium arsenide surfaces, surface charge builds up on exposure to air due to chemical oxidation. Chemical passivation with, for example, sulfur-containing solutions has been shown to be able to compensate this charge and thus avoid deleterious fields within a device structure. Another level of passivation is physical passivation, that is, encapsulating a semiconductor surface with, for example, a protective layer of silicon nitride or silicon dioxide.

Passive pixel sensor see *active pixel sensor, CMOS image sensor*.

Pauli exclusion principle excludes any two electrons from having the same set of quantum numbers, n, ℓ, m_ℓ, m_s in a given system of confined electrons, where n is the *principal quantum number*, ℓ is the orbital angular momentum quantum number, m_ℓ is the magnetic quantum number and m_s is the magnetic spin quantum number. There can be no two electrons occupying a given state $\psi(n, \ell, m_\ell, m_s)$, where ψ is the electron wavefunction including spin. Equivalently, up to two electrons with opposite spins can occupy a given orbital, $\psi(n, \ell, m_\ell)$.

Penetration depth of an electromagnetic wave into a medium is the distance over which the electric field drops by a factor of e^{-1}. The penetration depth δ is the reciprocal of the attenuation coefficient α in the medium; $\delta = 1/\alpha$. See *absorption coefficient*.

Penta prism, named after its pentagonal cross-section, is designed to deviate an incident beam by 90° without inverting or reversing the image. The 90° deflection does not require a precise alignment of the prism (orienting the prism) with respect to the incident beam, which is its virtue.

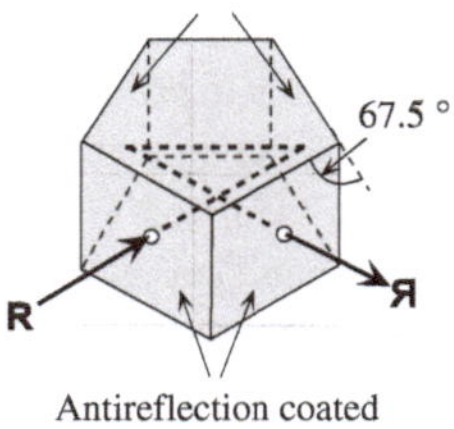

Penta prism.

Perturbation theory in quantum mechanics refers to approximation methods for finding allowed states of electronic systems subject to external perturbation. The external perturbation can include, for example, an applied field. Two distinct cases are normally considered – stationary perturbation and time-dependent perturbation methods. In the former case, the perturbation is considered to be time-independent, whereas in the latter case, the perturbation is assumed to be a function of time. As an example of the use of perturbation theory, stationary perturbation theory is discussed below. Consider the perturbed system to be described by the Hamiltonian, $H = H^0 + \lambda H'$, where H^0 represents the Hamiltonian of the unperturbed system, H′ is the perturbation such that $H' < H^0$, and λ is an arbitrary parameter that is a measure of the strength of the perturbation. The allowed energy states, or eigenstates, ϕ_n, of the unperturbed system can be found directly by solving the Schrödinger equation for the unperturbed system,

$$H^0\phi_n = E_n\phi_n$$

where E_n are the eigenenergies of the unperturbed system, while the perturbed eigenstates, Ψ_n, are given by

$$H\Psi_n = W_n\Psi_n$$

where W_n are the eigenenergies of the perturbed system, with the assumption that as $\lambda \to 0$, $W_n = E_n$. It can be shown that by expanding the perturbed wavefunction as a superposition of unperturbed state wavefunctions, W_n can be written as

$$W_n = W_{n,0} + \lambda W_{n,1} + \lambda^2 W_{n,2}$$

and

$$\Psi_n = \psi_{n,0} + \lambda\psi_{n,1} + \lambda^2\psi_{n,2}$$

where $W_{n,0} = E_n$ and $\psi_{n,0} = \phi_n$. The first-order correction to the energy (i.e. the λ term) is thus,

$$W_{n,1} = H'_{nn} = \langle\phi_n|H'|\phi_n\rangle.$$

Note that the notation $H'_{ji} = \langle\phi_j|H'|\phi_i\rangle$ is introduced for the matrix elements between two states ϕ_i and ϕ_j. Higher order corrections to the energy can be found in a similar fashion (e.g. using the λ^2 term for the second-order correction) as shown below:

$$W_{n,0} = E_n$$
$$W_{n,1} = H'_{nn}$$
$$W_{n,2} = \sum_{i\neq n}(H'_{ni}H'_{in})/(E_n - E_i).$$

Each of these terms represents a successive correction to the approximation for the energy eigenvalue. Thus, if λ is small, a first-order correction or a first-order stationary perturbation approximation $W_{n,1}$ may be sufficient as λ^2 is so small. If λ is a little larger, then the second-order stationary perturbation approximation $W_{n,2}$ may need to be added to achieve a better approximation for the eigenvalue.

Petermann mode radius of a propagating mode in a fiber defines the extent of the light intensity in a mode from the center of the fiber axis, based on two definitions, labelled as Petermann I and II mode radii. The Peterman I mode radius w_{I} is defined so that it represents the radial distance squared, r^2, weighted by the light power $P(r)$ in a narrow annular ring of radius r, i.e.

$$w_{\text{I}}^2 = \frac{\int_0^\infty r^2 P(r) 2\pi r\, dr}{\int_0^\infty P(r) 2\pi r\, dr} = \frac{\int_0^\infty r^3 E^2(r)\, dr}{\int_0^\infty r E^2(r)\, dr}.$$

For a Gaussian distribution, at $r = w_{\text{I}}$, the power density has fallen by a factor $1/e$. In Petermann II mode radius, the definition involves averaging $2r$,

$$w_{\text{II}}^2 = \frac{2\int_0^\infty r E^2(r) 2\pi r\, dr}{\int_0^\infty r(dE/dr)^2 dr}$$

so that at $r = r_{\text{II}}$, the power density fallen to $1/e^2$ and the field is down to $1/e$. Thus for a Gaussian distribution, $w_{\text{II}} = w_{\text{I}}\sqrt{2}$. (K. Petermann, *Electron. Letts.*, **12**, 102, 1976; **19**, 712, 1983.)

Petzval condition see *field curvature*.

Petzval lens is a lens system that is designed to reduce or eliminate the field curvature by using a combination of converging and diverging lenses, for example, two achromatic lenses with an aperture stop between them. It is used as an objective lens in certain cameras and in projectors where the image needs to be formed on a flat surface (screen, films, etc.). A negative lens (f negative, i.e. diverging), called a *field flattener lens*, can also be used near the focal plane of the projector's positive lens to compensate for the inward curvature without affecting other aberrations.

Petzval surface see *field curvature*.

Phase of a traveling wave is the quantity $(kx - \omega t)$ which determines the amplitude of the wave at position x and at time t given k $(= 2\pi/\lambda)$ and ω. In three dimensions it is the quantity $(\mathbf{k}\cdot\mathbf{r} - \omega t)$ where $\mathbf{k}$ is the wavevector and $\mathbf{r}$ is the position vector; the amplitude is measured at time t.

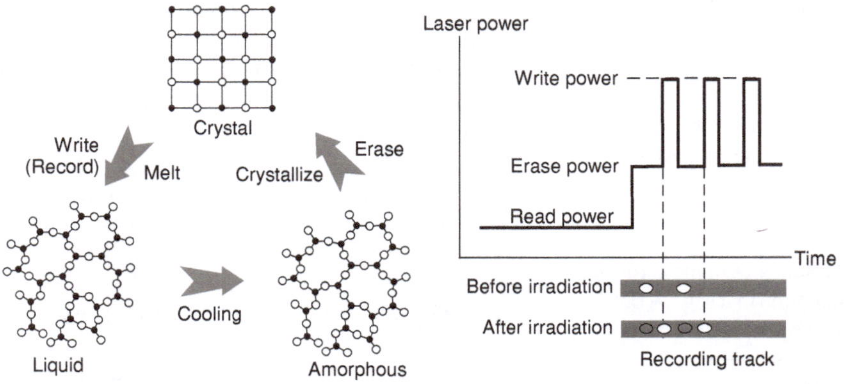

Schematic illustration of **phase change recording**. Left, the changes in the phase as a result of laser irradiation. Right, the laser power levels for erasing and writing information. Read power is much smaller than the erase power.

Phase change recording is an optical data recording process in which a "phase-change alloy glass" (such as a thin film of $Ge_2Sb_2Te_5$ deposited on a disk substrate) is used in such a way that a local change in the structure is induced (by heating from a laser diode) so that this local region undergoes a phase change from the glass or an *amorphous* phase to the *crystalline* phase, and vice versa. The two phases have different optical properties (i.e. different reflectances) and hence can represent different bits of information. This type of optical storage is used in RW (rewriteable) and RE (recordable-erasable) disks. The Ge-Sb-Te alloy for optical storage was discovered by Yamada and co-workers in Japan in 1987. The Ge-Sb-Te (so-called GST) alloys are used in DVD-RAM (digital versatile disk-random access memory) disks. A small focused laser spot is used to induce the phase change. If the laser pulse is sufficiently strong to take an amorphous region above the glass transition region but below the melting temperature T_m, then the region crystallizes. This *crystallization* corresponds to the "erase" operation. On the other hand, if a crystalline region receives an intense laser pulse so that the local temperature increases above T_m, then this region melts, and upon rapid cooling, when the laser pulse is off, it forms an amorphous phase. This *amorphization* corresponds to the "write" or "record" operation. Another useful alloy is the Ag-In-Sb-Te glass, known as AIST.

Phase conjugate mirrors differ from ordinary mirrors by reversing the wavefronts. As a consequence, an incident wave is back-reflected onto itself. This wave process, which can be thought of as a kind of time-reversal, is also found in acoustics. The first figure shows the difference between an ordinary mirror and a phase conjugate mirror (PCM) in terms of their effects on the wavefront.

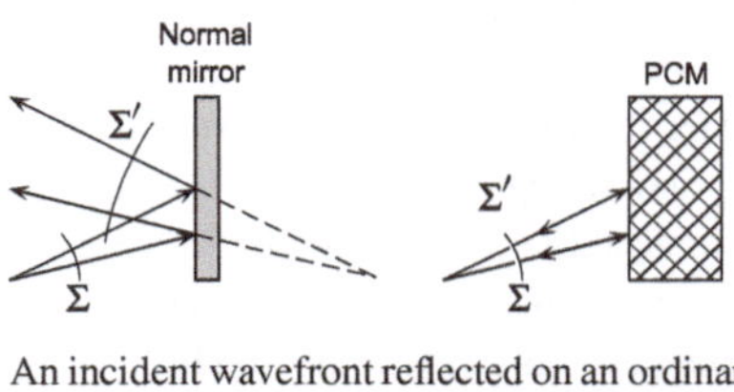

An incident wavefront reflected on an ordinary mirror as compared to the effect of a phase-reversal as provided by a **phase conjugate mirror** (PCM). Σ and Σ' are the incident and reflected wavefront surfaces.

The above PCM property can be used to correct the wavefront deformation provided by the propagation through a refractive object. It should be noted that although the physical mechanism involves real-time nonlinearity, the effect is formally similar to phase-reversal holography.

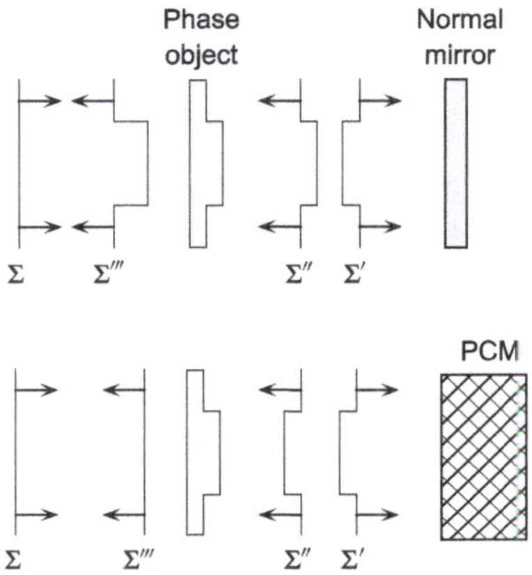

Phase conjugated mirrors The amplitude of the wavefront deformation caused by propagation through a phase plate (object) is doubled after reflection on an ordinary mirror. By contrast, the deformation is corrected after reflection from a phase conjugate mirror (PCM). Σ, Σ', Σ'' and Σ''' are the successive wavefronts as the wave passes through the phase plate, progresses towards the mirror and becomes reflected and passes through the phase plate again. Notice the distinct difference in Σ''' between the two cases.

Phase conjugation see also *Brillouin scattering*, *dynamic optical grating*.

Phase contrast method (Zernike) see *spatial filtering*.

Phase gratings see *diffraction grating*.

Phase matching see *second harmonic generation*.

Phase matching angle see *second harmonic generation*.

Phase mismatch between two propagating waves is simply the difference between their propagation constants β, that is $\Delta\beta$.

Phase modulator is a Pockels effect modulator in which an applied voltage, and hence the applied electric field, modulates the refractive index and hence the phase of a light wave propagating through the crystal. For example, in the transverse phase modulator, the applied electric field $E_a = V/d$, is applied in a transverse direction to the propagation of the light wave. Suppose that E_a is parallel to the y-direction, normal to the direction of light propagation along z. Suppose that the incident beam is linearly polarized with **E** say at 45° to the y-axis. We can represent the incident light in terms of polarizations ($\mathbf{E}_x$ and $\mathbf{E}_y$) along the x- and y-axes. These components $\mathbf{E}_x$ and $\mathbf{E}_y$ experience refractive indices n'_1 and n'_2, respectively. Thus, when $\mathbf{E}_x$ traverses the distance L, its phase changes by ϕ_1:

$$\phi_1 = \frac{2\pi n'_1}{\lambda} L = \frac{2\pi L}{\lambda}\left(n_o + \tfrac{1}{2} n_o^3 r_{22} \frac{V}{d}\right)$$

where n_o is the refractive index in the absence of the applied field and r_{22} is the appropriate Pockel's coefficient. When the component $\mathbf{E}_y$ traverses the distance L, its phase changes by ϕ_2, given by a similar expression except that r_{22} changes sign. Thus, the phase change $\Delta\phi$ between the two field components is

$$\Delta\phi = \phi_1 - \phi_2 = \frac{2\pi}{\lambda} n_o^3 r_{22} \frac{L}{d} V.$$

The applied voltage thus inserts an adjustable phase difference $\Delta\phi$ between the two field components. The polarization state of the output wave can therefore be controlled by the applied voltage and the Pockels cell is a *polarization modulator*. We can change the medium from a *quarter-wave* to a *half-wave* plate by simply adjusting V. The voltage $V = V_{\lambda/2}$, the *half-wave voltage*, corresponds to $\Delta\phi = \pi$ and generates a half-wave plate. The advantage of the transverse Pockels effect is that we can independently reduce d, and thereby increase the field, and increase the crystal length L, to build-up more phase change; $\Delta\phi$ is proportional to L/d. This is not the case in the longitudinal Pockels effect. If L and d were the same, typically $V_{\lambda/2}$

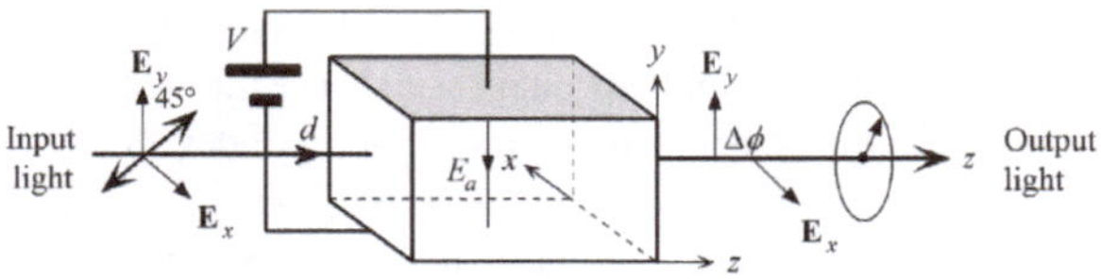

Transverse Pockels cell **phase modulator**. Linearly polarized input light entering an electrooptic crystal emerges as a circularly polarized light. E_a is the applied field.

would be a few kilovolts but tailoring *d/L* to be much smaller than unity would bring $V_{\lambda/2}$ down to desirable practical values. Phase modulators are extremely useful in phase shift key (PSK) modulation in communications. $LiNbO_3$ phase modulators can provide excellent PSK modulation.

Phase noise is the fluctuations in the frequency of oscillations of a wave that is caused by instabilities on the time domain. Suppose that we wish to generate a perfect sine wave at a frequency ν_0, but generate many independent truncated sine waves one after another with each train containing oscillations at ν_0, and each train having a certain duration Δt of its own. The generated wave will then have average oscillations at ν_0 but also oscillations at frequencies neighboring ν_0. Thus there is "noise" in the frequency of the generated wave, which is due to sudden changes in the phase from one wave train to another even though each is oscillating at ν_0. Thus phase noise is due to fluctuations in the phase of a wave (due to its generation or other processes) that appears as unwanted frequency components adjacent to the average, desired frequency ν_0. Jitter in the time domain can lead to phase noise in the frequency domain. Phase noise is usually quoted in terms of the power that appears at some offset frequency from the center or carrier frequency at ν_0 with respect to the power at the carrier frequency as dBc/Hz. Phase noise can be viewed as a "random source," with a certain probability density distribution, that modulates the phase of the oscillations; the result is the appearance of side bands around ν_0, and these side bands represent phase noise. In many cases, the phase noise has a *Lorentzian spectrum*.

Phase object (esp. in microscopy) is an object that alters the phase, but not the amplitude of the transmitted wave with respect to the incident wave; for instance, a nonabsorbing transmission grating, or a thin transparent object with nonuniform optical thickness. See also *amplitude object*.

Phase-squeezed state see *squeezed states*.

Phase velocity is the rate at which a given phase on a traveling wave advances. It represents the velocity of a given phase rather than the velocity at which information is carried by the wave. Two consecutive peaks of a wave are separated by a wavelength λ and it takes a time period $1/\upsilon$ for one peak to reach the next (or time separation of two consecutive peaks at one location); then the phase velocity is defined as $v = \lambda/(1/\upsilon) = \lambda\upsilon$.

Phillips–Van Vechten model is an approximate approach for calculating allowed energy states (i.e. E_{bonding} and $E_{\text{antibonding}}$) and wavefunctions for the interaction between two orbitals, with energies E_A and E_B, located on two atomic sites, A and B, having an interaction energy V_{AB}. It can be shown that the allowed energies $E_{+,-}$ are given by

$$E_{+,-} = \frac{E_A - E_B}{2} \pm \sqrt{V_{AB}^2 + \frac{(E_A - E_B)^2}{4}}$$

where $E_+ = E_{\text{antibonding}}$ and $E_- = E_{\text{bonding}}$, giving a binding energy, $\Delta E = E_+ - E_-$ given by

$$\Delta E = \sqrt{E_I^2 + E_C^2}$$

where $E_I = E - E_B$ and $E_C = 2V_{AB}$ are the ionic and covalent contributions to the binding energy, respectively, and the fractional ionicity of the bond $f_I = E_I/\Delta E$.

Phonon is a quantum of energy associated with the vibrations of the atoms in a crystal, analogous to the photon. It is often referred to as a quantum of lattice vibration. A phonon has an energy of $\hbar\Omega$ where Ω is the angular frequency of the lattice vibration, and $\hbar = 2\pi/h$, h = Planck's constant. If K is the wavevector of a lattice vibration wave, a phonon, then the momentum of the phonon is $\hbar K$.

Phonon echoes originate from acoustic nonlinear phenomena and may be observed in certain materials such as glasses at very low temperatures. In a typical backward-wave echo experiment, two microwave pulses, separated by a fixed delay time τ, are applied to a sample of interest. Owing to a transducer bonded to the sample, for example, a forward-propagating hypersonic wave is generated by the first pulse. This wave and the microwave electric field may then interact in a material as a result of intrinsic nonlinearities in the material, or from impurities, causing a reversal of the propagating wavevector. For example, in glasses, the amplitude of the resulting backward-propagating wave was found to be proportional to the concentration of OH^- ions. At some time 2τ, this backward-propagating wave reaches the transducer and gives rise to an echo. In the case of chalcogenide glasses such as doped As_2S_3, the nonlinear coupling between the acoustic wave and electric field is due to tunneling systems present in the material. Experimentally, the intensity of the echo increases with increasing acoustic wave intensity until the tunneling systems become saturated. Such phonon echo experiments have proven to be very important in glasses providing direct measurements of acoustic properties of these materials, including deformation potentials and the dynamics of tunneling systems.

Phosphorescence see *luminescence*, *photoluminescence*.

Phosphors see *luminescence*.

Photobleaching is the shift of short-wavelength absorption edge, α vs. $h\nu$ behavior, to higher photon energies or shorter wavelengths. For example photobleaching has been observed in as-evaporated films of Ge-based chalcogenides, GaLaS glass, oxide glasses containing colloid Ag particles, sol-gel-derived germanium oxide glass thin films. It has also been observed in glasses that have been photodarkened to saturation, when illuminated with subbandgap light ($h\nu < E_g$).

Photocarrier is a charge carrier in a semiconductor (or any device) that has been generated by the absorption of light.

Photochromatic glass is glass that undergoes reversible photodarkening; it becomes darker in the presence of stronger light, and reverses to its original transparent state when left in the dark. One type of chromatic glass has Ag^+Cl^- particles distributed in the volume. Ag^+Cl^- particles can be dissociated by the absorption of light to form Ag particles that scatter light and cause attenuation in the direction of light propagation.

Photoconductive detectors have simple metal-semiconductor-metal structures, that is two electrodes are attached to a semiconductor that has the desired absorption coefficient and quantum efficiency over the wavelengths of interest. Incident photons become absorbed in the semiconductor and photogenerate electron–hole pairs (EHPs). The result is an increase in the conductivity of the semiconductor and hence an increase in the external current which constitutes the photocurrent I_{ph}. The actual response of the detector depends on whether the contacts to the semiconductor are ohmic or blocking (for example Schottky junctions that do not inject carriers) and on the nature of carrier recombination kinetics.

The photoconductivity of the sample $\Delta\sigma$ under uniform illumination is given by

$$\Delta\sigma = \frac{e\eta I\lambda\tau(\mu_e + \mu_h)}{hcd}$$

where e is the electronic charge, η is the quantum efficiency, μ_e and μ_h are the electron and hole drift mobilities, τ is the mean recombination time

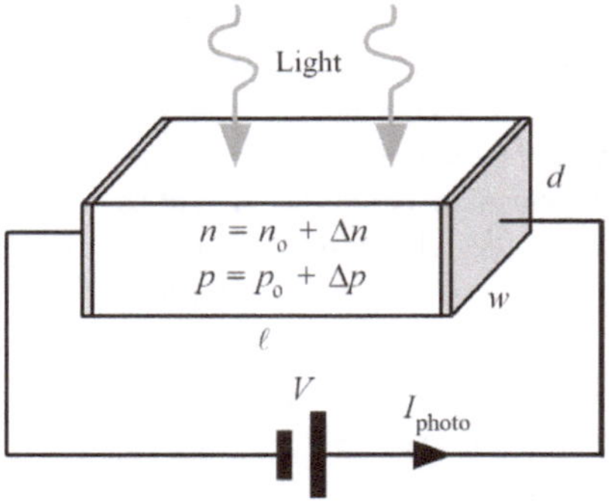

Photoconductive detector A semiconductor slab of length ℓ, width w and depth d is illuminated with light of wavelength λ and intensity is I. Equilibrium carrier concentrations are n_0 and p_0. Under steady state illumination these become $n = n_o + \Delta n$ and $p = p_o + \Delta p$. The change in the conductivity $\Delta\sigma$ is the photoconductivity of the sample.

(lifetime), h is Planck's constant, c is the speed of light and d is the thickness of the photoconductive slab. It is assumed that all the incident radiation is absorbed, otherwise this expression must be multiplied by $[1 - \exp(-\alpha d)]$, where α is the absorption coefficient at the wavelength of interest.

Photoconductive gain occurs in a photoconductor when the *external* photocurrent is due to more than one electron flowing per absorbed photon. The photoconductor must have ohmic contacts to allow carriers to enter the semiconductor. The photoconductive gain is then simply

$$G = \frac{\text{Rate of electron flow in external circuit}}{\text{Rate of electron generation by light absorption}}$$

which can be shown to be

$$G = \frac{\tau}{t_e}\left(1 + \frac{\mu_h}{\mu_e}\right)$$

where μ_e and μ_h, are the drift mobilities of the electrons and holes respectively, t_e is the electron *transit time*, $t_e = \ell/(\mu_e E)$, where ℓ is the photoconductor length, E is the electric field and τ is the mean recombination time. The photoconductive gain can be quite high if τ/t_e is kept large which means a long recombination time and a short transit time. The transit time can be made shorter by applying a greater field but this will also lead to an increase in the dark current and more noise. The speed of response of the device is limited by the recombination time of the injected carriers. A long τ means a slow device.

Photoconductivity is the increase in the conductivity of a sample resulting from illumination, $\sigma_{light} - \sigma_{dark}$.

Photodarkening is an apparent "darkening" of a material, usually a glass, due to changes in its optical absorption characteristics induced usually by

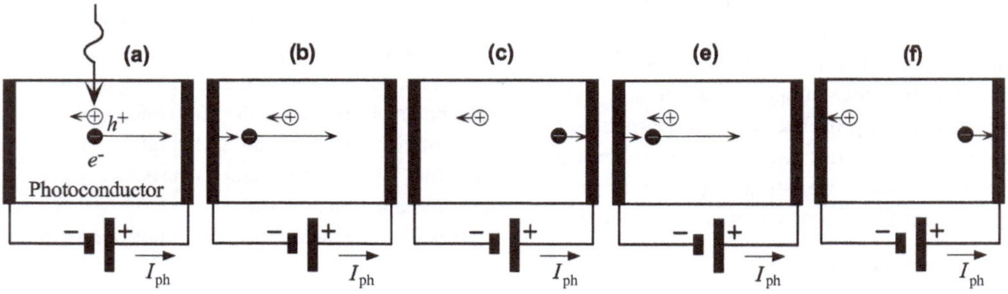

Photoconductive gain A photoconductor with ohmic contacts (contacts not limiting carrier entry) can exhibit gain. As the slow hole drifts through the photoconductors, many fast electrons enter and drift through the photoconductor because, at any instant, the photoconductor must be neutral. Electrons drift faster which means as one leaves, another must enter.

illumination, when it is illuminated. Photodarkening is a photoinduced effect in which a glass that is exposed to near-bandgap illumination (photon energy $\geq E_g$, bandgap) has its optical absorption coefficient vs. wavelength, α vs. $h\nu$, characteristic behavior near band edge absorption shifted to lower photon energies, i.e. longer wavelengths, hence the appearance of darkening. This photodarkening corresponds to a reduction in the optical bandgap. Typically maximum shift is reached for exposure with light having photon energy near the optical bandgap of the exposed glass. Chalcogenide glasses exhibit substantial photodarkening which has led to their potential use in various photonics applications, e.g. photoinduced fabrication of waveguides, diffraction gratings, etc. Photodarkening has been observed in both vitreous thin films and melt-quenched bulk chalcogenide glasses but not in crystalline chalcogenides. Photoinduced changes are favorable in chalcogenide glasses because of two important factors: the low coordination of chalcogens results in a "flexible" (not very rigid) glass structure, and the chalcogens have lone pair electrons that make up the valence band. There are both reversible and irreversible photodarkening effects. Irreversible photodarkening occurs in as-deposited (freshly evaporated) chalcogenide films. It can be induced by illumination, or by annealing at high temperatures. This irreversible component is caused by thermal or photoinduced permanent changes in the structure, e.g. conversion of wrong bonds to normal bonds, photopolymerization, etc. Reversible photodarkening is typically observed on previously annealed films and melt quenched glasses. The photostructural change results in a metastable state, which can be reversed (restored to the original state) by annealing at a temperature near to the glass transition temperature. During illumination itself, photodarkening can have both *transient* and *metastable* components. The transient part decays rapidly after the illumination is switched off and only normally observed metastable effect remains. Transient photodarkening could be a significant fraction of the total changes. (K. Shimakawa, A. Kolobov and S. R. Elliott: *Adv. Phys.*, **44**, 475, 1995.)

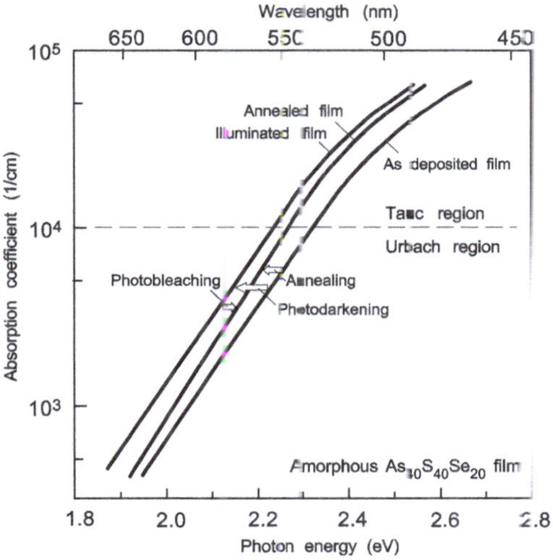

Illustration of **photodarkening** in an amorphous chalcogenide film (E. Márquez, J. M. González-Leal, R. Prieto-Alcón, R. Jiménez-Garay and M. Vleck, $S_{40}Se_{20}$, *J. Phys. D: Appl. Phys.*, **32**, issue 24, 1999, IOP Publishing.)

Photodetector is a device that converts a light signal to an electrical signal such as a voltage or current. In many photodetectors such as photoconductors and photodiodes this conversion is typically achieved by the creation of free electron–hole pairs by the absorption of photons, that is, the creation of electrons in the conduction band and holes in the valence band. In some devices, such as pyroelectric detectors, the energy conversion involves the generation of heat which increases the temperature of the device which changes its polarization and hence its relative permittivity.

Photodiode is a semiconductor diode (a *pn* junction) that generates a photocurrent upon illumination with a suitable range of wavelengths. *pn* junction based photodiodes are small and have high speed and good sensitivity for use in various optoelectronics applications, the most important of which is in optical communications.

Photoelastic birefrigence in optical fibers is an induced birefringence in the optical fiber due to internal stresses in an optical fiber that have resulted either in the production or during handling, or splicing. See *photoelastic effect.*

Photoelastic effect is the phenomenon in which the induced strain, S, in a crystal changes its refractive index n. The strain changes the density of the crystal and distorts the bonds (and hence the electron orbits) which leads to a change in the refractive index n. If we were to examine the change in $1/n^2$ instead of n, then we would find that this is proportional to the induced strain S and the proportionality constant is the photoelastic coefficient p, i.e.

$$\Delta\left(\frac{1}{n^2}\right) = pS.$$

The relationship is not as simple as stated above since we must consider the effect of a strain S along one direction in the crystal on the induced change in n for a particular light propagation direction and some specific polarization. The above equation in reality is a tensor relationship,

$$\Delta\left(\frac{1}{n^2}\right)_i = \sum_{k=1}^{6} p_{ik} S_j$$

where $j = 1, 2, 3$ corresponds to x, y and z, and $k = 1, \ldots, 6$ and p_{ik} are the tensor elements. The photoeleastic effect can also be attributed to stresses σ_j, and written as

$$\Delta\left(\frac{1}{n^2}\right)_i = \sum_{k=1}^{6} \pi_{ik}\sigma_j$$

where π_{ik} are the *piezooptical coefficients*. In the presence of stresses within a medium, light will experience different refractive indices depending on its polarization, or its displacement vector **D**. It has been found that one only needs to consider the principal stresses within a sample that has stresses. If σ_p and σ_q are the *principal stresses*, which are in orthogonal directions, the refractive indices n_p and n_q corresponding to light with **D** parallel to σ_p and σ_q are given by

$$n_p - n_q = C(\sigma_p - \sigma_q)$$

where C is a photoelastic constant called the *Brewster constant* and has the units of Brewsters (B), in which $1\ \mathrm{B} = 10^{-12}\ \mathrm{m^2\ N^{-1}}$. Since light with orthogonal polarizations propagates with different refractive indices, a *birefringence* of $n_p - n_q$ has been induced by $\sigma_p - \sigma_q$. Built-in stresses (or strains) can result in inherent birefringence in optical materials such as glasses and transparent plastics, etc.; they can usually be reduced by annealing at a high temperature. For glasses, $C \approx 2.6$ B. The photoelastic effect can be used with optical measurements (such as interference) to study stress distributions in various (usually) transparent objects.

Photoelectric effect is the emission of electrons from a metal upon illumination with a radiation of frequency υ above a critical frequency that depends on the material. The kinetic energy E_{K} of the emitted electron is independent of the light intensity but dependent on the energy $h\upsilon$ of the photons (where h is the Planck constant) as well as on the work function Φ of the metal. If $h\upsilon < \Phi$, no emission occurs. If $h\upsilon > \Phi$, the kinetic energy of the photoemitted electron is

$$E_{\mathrm{K}} = h\upsilon - \Phi.$$

This interpretation of the photoelectric effect by Einstein has led to the concept of the photon as a quantum of the electromagnetic field. As an essential contribution to quantum mechanics, a Nobel Prize subsequently acknowledged it in 1921.

Photogeneration is the excitation of an electron into the conduction band by the absorption of a photon. If the photon is absorbed by an electron in the valence band then its excitation to the conduction band will generate an electron–hole pair.

Photoinduced changes (effects) in glasses encompass a wide range of physical and/or chemical changes that are induced by illumination and involve structural changes. The photochanges that are induced in the material depend not only on the rate of light exposure (or irradiance, incident optical power per unit area), duration of exposure, the wavelength, polarization, but also on the properties of the irradiated glass. Typical photoinduced effects encompass photocrystallization, photopolymerization, photodecomposition, photodissolution, photovaporization, photoexpansion, photocontraction, photodarkening, photobleaching, photodichroism, and other possibilities such as photoamorphization (e.g. by a suitable laser pulse) of a devitrified glass. One particularly important example of photoinduced effects is the change in the refractive index n or the absorption coefficient α by the suitable illumination of a glass; the effect is particularly strong in chalcogenide glasses, if the absorption coefficient increases. Furthermore, the photoinduced changes may be *reversible* or *irreversible*, and in some cases the change may have both characteristics. Reversible changes are usually observed in well annealed glasses, and the photoinduced change can be reversed (original glass property restored) by simply annealing the glass near its glass transition temperature. Irreversible changes on the other hand are permanent changes in the glass properties associated with permanent structural changes, and usually are observed in either as-prepared glasses (not annealed) or in films. Photoinduced changes in the optical properties, n and α, have been extensively studied, especially in chalcogenide glasses, and have been fruitfully used to fabricate photoinduced diffraction gratings and waveguides for potential use in photonics. If the absorption coefficient increases with illumination, the photoinduced effect is *photodarkening*, and if it decreases with illumination, it is called *photobleaching*. *Photoinduced anisotropy* has also been observed in which the illumination of a glass with linearly polarized high-intensity light leads to the glass exhibiting optical anisotropy, that is, *photoinduced birefringence* and *photoinduced dichroism*.

Photoinduced vectoral change, or **photoinduced anisotropy (PA) in glasses** is an optical anisotropy that is induced by illumination with plane polarized light, called the pump light. In some glasses such as chalcogenide glasses, linearly polarized pump light photoinduces dichroism (absorption coefficient along and perpendicular to the optical field in the pump are not the same; $\alpha_{//} \neq \alpha_{\perp}$) and birefringence (refractive index $n_{//} \neq n_{\perp}$). Because of the directional properties of the induced anisotropies (they depend on the polarization state of the pump light) such photoinduced effects have been termed *vectoral changes*. Photoinduced *scalar changes* are those induced changes in the optical properties that depend on the pump light intensity and not on the pump beam polarization. Vectoral effects are much smaller in magnitude than scalar effects, and are reversible by annealing at a suitable temperature below or near T_g; vectoral effects tend to anneal at a lower temperature than scalar effects. (See *photodarkening*.) PA has been observed to be particularly pronounced in glasses that exhibit smaller scalar effects at the same temperature (e.g. Se, Sb_2S_3). Photoinduced dichroism is usually studied by irradiating a thin glass film with an intense linearly polarized light, and then measuring the transmittance anisotropy by using a low-intensity probing light beam. Some of the most important vector effects are light-induced mass transport, photocrystallization and optomechanical effects. These phenomena indicate that there is a change in the atomic structure that depends on the polarization state of the pump light, e.g. anisotropic local expansion around the Se atom in As_2Se_3 glass. Photoinduced anisotropy (birefringence) has also been observed in silica. (G. J. Adriaenssens, V. K. Tikhomirova, S. R. Elliott, *J. Non-Crystal. Solids*, **227**, 688, 1998.)

Photoinjection is the photogeneration of carriers in the semiconductor by illumination. Photogeneration may involve the excitation of an electron from the valence band to the conduction band in which case electrons and holes are generated in pairs.

Photolithography uses light in the lithographic process to define a pattern on the surface of a wafer.

Photoluminescence is the *luminescence* obtained from a material (or an atom or an ion in a medium) that has been excited by external illumination. Photoluminescence should not be confused with reflection and light scattering, which are also caused by external illumination. The criterion that distinguishes these phenomena is the characteristic decay time after the cessation of the incident light. While reflection and scattering would occur within $\sim 10^{-14}$ s (characteristic time of electromagnetic oscillations of incident light wave), the photoluminescence would persist at least more than $\sim 10^{-12}$ s (characteristic time of atomic vibrations of the luminescent material). The photoluminescence is called *fluorescence* if the characteristic decay time is about 10^{-8} s or less. The emission of light from a fluorescent tube is actually a *fluorescence* process. The tube contains a gas mixture of argon and mercury. The Ar and Hg gas atoms become excited by the electrical discharge process and emit light mainly in the ultraviolet region which is absorbed by the fluorescent coating on the tube. The excited electrons in the fluorescent coating material then recombine emitting light in the visible spectrum. There are also materials from which light emission may continue for milliseconds to hours after the cessation of excitation. These slow luminescence processes are normally referred to as *phosphorescence*. The photoluminescence may be characterized by its *quantum yield* or *photoluminescence efficiency*, which is the ratio of the number of emitted to absorbed photons. The quantum yield is defined by the competition of radiative recombination with the nonradiative recombination and *energy transfer or/and energy migration* from the luminescent center. While the influence of nonradiative processes is always adverse, the transfer of energy may play a positive role as in the case of energy transfer from a *sensitizer* to an *activator*. Very often, the luminescence is connected with the presence of specific luminescent centers embedded in the host, which are called *activators*. The examples of activators, which are widely used in various applications, are trivalent ions Cr^{3+} and Eu^{3+} in Al_2O_3:Cr^{3+} (ruby) and Y_2O_3:Eu^{3+}. The concentration of these activators normally cannot be more than 1 at.%. Another example of an activator with a concentration of 100% is the WO_4^{2-} ionic group in $CaWO_4$ compound, which can absorb x-rays and UV radiation and re-emit visible light. The addition of additional centers called *sensitizers* may greatly influence the photoluminescence. The task of a sensitizer is to absorb energy and to transmit it to an activator. Thus, in a lamp with phosphor $Ca_5(PO_4)_3$ doped with Sb^{3+} and Mn^{2+}, the ultraviolet energy is absorbed only by Sb^{3+} ions and is partially transmitted to Mn^{2+} ions. The resulting emission is nearly white because it consists of a mixture of Sb^{3+} blue emission and Mn^{2+} yellow emission. A *photoluminescence emission spectrum* characterizes the spectral distribution of emitted energy. The emission spectrum may consist of broad bands (lines) and/or narrow lines. The overall tendency is as follows: the emission line tends to be broader if the radiative transition is allowed, otherwise the line is narrow. A typical example is the behavior of *rare-earth* ions in many hosts, where divalent ions produce wide bands, while trivalent ions generate narrow lines. Thus, an Eu^{2+} ion in SrB_4O_7 produces broadband emission due to allowed radiative transitions $4f^6 5d \rightarrow 4f^7$, while the trivalent ion Eu^{3+} embedded in Y_2O_3 would produce a series of narrow lines due to *parity* forbidden transitions inside the 4f shell. Usually, photoluminescence may be excited within a definite range of excitation wavelengths. The energy difference between the maximum of the (lowest) excitation band and that of the emission is called the *Stokes shift*. The physical reason for the Stokes shift is the relaxation of the luminescence center and host within the time interval after excitation and before radiative recombination. The stronger the electron lattice

Table: Photometric units

Quantity	Usual Symbol	Luminous units	Radiant units	Photonic units
Flux	Φ	lumen [lm]	W	s^{-1}
Intensity	I	candela [cd]	W sr^{-1}	s^{-1} sr^{-1}
Luminance	L	nit [cd m^{-2}]	–	s^{-1} m^{-2} sr^{-1}
Radiance		–	W m^{-2} sr^{-1}	
Emittance	M	[lm m^{-2}]	–	s^{-1} m^{-2}
Excitance		–	W m^{-2}	
Illuminance, Irradiance	E	lux [lx]	W m^{-2}	s^{-1} m^{-2}
Energy		lm.s	J	number of photons

coupling, the greater the Stokes shift. See *luminescence, Stokes shift, Frank–Condon principle.*

Photometric units are the lumen (lm), candela (cd) and the nit, and are the counterparts of the radiometric units of W, W sr^{-1}, and W sr^{-2} m^{-2} respectively. See table.

Photometry is a subset of radiometry that deals with the measurements of radiation that is within the visible spectrum, and weighted by the spectral responsivity of the eye. See *radiometry, luminous flux.*

Photon is the quantum of electromagnetic radiation. A monochromatic field of wavevector $\mathbf{k}$ at frequency υ can be described as a stream of quanta of light, each carrying a discrete amount of electromagnetic energy $E = h\upsilon$ and momentum $\mathbf{p} = \hbar\mathbf{k}$, where h is the Planck constant and $\hbar = h/(2\pi)$. The absolute momentum can also be expressed as $p = |\mathbf{p}| = h/\lambda$. The photon has no rest mass (hence the expression *travel light*), its velocity in vacuum is a constant c, independent of its frequency. Its spin number is 1, which can be related to the states of polarization. The concept of the photon is especially useful at low levels of illumination, when the granularity property of light is more important. *Single-photon experiments* have shown photon localization at the time and/or place of detection. Einstein interpreted the photoelectric effect in terms of particle interaction. Exchange of momentum is used for laser cooling and trapping of atoms. The limit of sensibility for the human eye to detect a photon flux in the dark, in the visible range of the spectrum, is estimated to be a few hundred photons per second. An optical power of 0.1 mW at wavelength $\lambda = 1.5$ μm carries about 7.5×10^{14} photons per second. For most practical applications that do not involve very

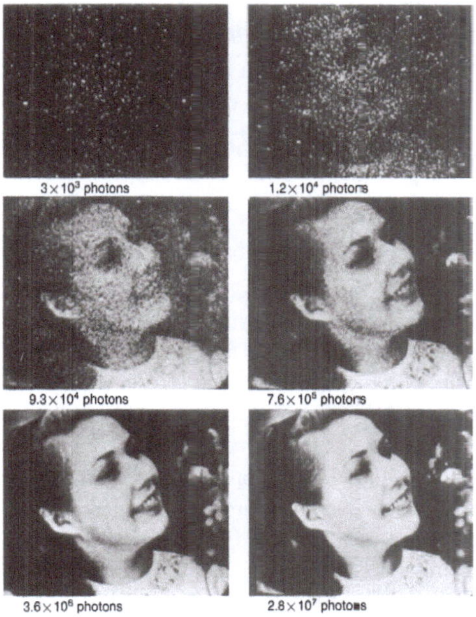

Photon These electronic images were made with the number of photons indicated. The discrete nature of photons means that a large number of photons are needed to constitute an image with satisfactorily discernable details. (A. Rose, *J. Opt. Soc. Amer.*, **43**, 715, 1953, Courtesy of OSA.)

low light intensity detection, and outside quantum optics, the semiclassical description of the electromagnetic radiation in terms of the optical field is quite sufficient. See *single photon interference*.

Photon diffusion describes the case where light propagating through a random medium undergoes multiple scattering events, and the transport of light intensity through the medium is described by a diffusion equation based on radiative transfer. After only a few scattering events in such a medium, photons become directionally randomized resulting in the photons performing a random walk, and the transport mean free path is the distance over which directional memory is lost. Typically, the assumption of diffusive photon transport is a good approximation when the thickness of the medium exceeds about eight times the step size of the photon random walk. The governing equation is the *time-dependent photon diffusion equation*:

$$\nabla^2 U(\mathbf{r}, t) - \frac{v\mu}{D} U(\mathbf{r}, t) - \frac{1}{D}\frac{\partial U(\mathbf{r}, t)}{\partial t} = -\frac{1}{D} S(\mathbf{r}, t)$$

where $U(\mathbf{r}, t)$ is the photon density at position $\mathbf{r}$ and time t, v the speed of light in the medium, D the diffusion coefficient, and μ the absorption coefficient, and $S(\mathbf{r}, t)$ the source term. To a good approximation, when $\mu < 0.1\mu_s, D = 3v/\mu_s$, where μ_s is the medium reduced scattering coefficient. This would be typical of the case for propagation of light in infinite, homogeneous, and isotropic media with optical properties resembling those of human tissues. (A. Ishimaru, *Wave Propagation and Scattering in Random Media*, Academic, New York, 1978, Vol. 1.)

Photon drag is an effect in which intense radiation passing through a semiconductor is able to transfer momentum to free charge carriers, and thereby cause a drift of these carriers in the direction of the photon flux. Photon flux drags the carriers and builds up an internal field, and hence a voltage difference along the sample. See *photon drag detector*.

Photon drag detector relies on the *photon drag effect* in which photoinduced carriers acquire a net momentum from the photon flux and in the direction of the photon field. This results in the appearance of an electromotive force or a current in the semiconductor which may be used to measure the photon flux. Such detectors, based on bulk semiconductors such as p-type Ge, are commercially available for use as sensors of CO_2 laser radiation. The photon drag effect is related to the momentum relaxation time in the material. In the case of bulk p-type semiconductors, absorbed photons cause transitions between heavy and light hole states, with only a very small fraction of the carriers resonantly interacting with the photons of a given energy. In the case when the energy bands of the ground and excited states are parallel to each other, such as is the case for the subbands of a two-dimensional electron gas, or the Landau states for a semiconductor in an applied magnetic field, the photon drag effect as described above occurs, together with new features which arise when the exciting photon energy is tuned over the resonant transition energy. The new features occur due to the difference in the momentum relaxation rates for ground and excited subband states. Photon drag detectors are inherently fast owing to subpicosecond momentum relaxation times being accessible in semiconductors. Moreover, quantum well-based photon drag detectors offer both fast response times and enhanced sensitivity due to the large absorption occurring in quantum well systems. One type of commercial photon drag detector (Hamamatsu B749) for use at a wavelength of 10.6 μm (CO_2 laser wavelength) has a peak photosensitivity of 1 mV/kW, that is 1 mV is induced from an incident radiation of power 1 kW, and the rise time is ~1 ns. However, the noise equivalent power (NEP) is usually higher than other types of detectors; for the latter detector it is quoted as 4×10^{-3} W Hz$^{-1/2}$. If a photon flux of intensity I

results in the generation of a field E, due to the photon drag effect, then the photon drag coefficient S is defined by $S = E/I$, i.e. field generated per unit light intensity. If P_o is the incident optical power which generates a photon drag voltage of V, then the voltage responsivity or photosensitivity R_V is defined by V/P_o. At long wavelengths, the classical theory predicts that

$$\begin{aligned} R_V &= \frac{V}{P_o} \\ &= S\frac{1-R}{A\alpha}\exp(-\alpha l)\frac{1-\exp(-\alpha L)}{1+R\exp(-\alpha L)} \end{aligned}$$

where e is the elementary charge, α is the absorption coefficient and R is the reflectance of the semiconductor at the wavelength of interest, l is the distance between the front surface (receiving radiation) and the front contact, and L and A are the length (between contacts, along the photon flux) and cross-sectional area of the sample. (K. Cameron, A. F. Gibson, J. Giles, C. B. Hatch, M. F. Kimmith and S. Shafik, *J. Physics C*, **8**, 3137, 1975.)

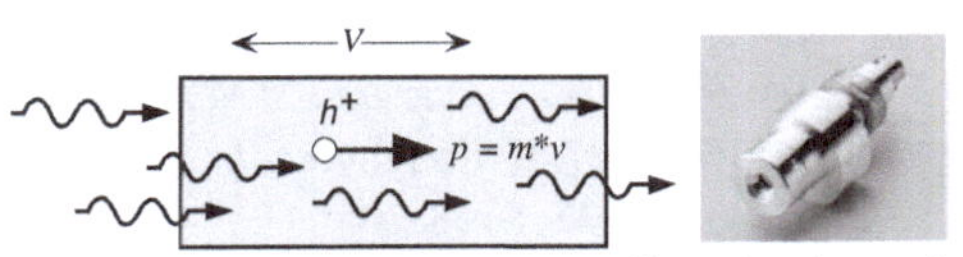

Semiconductor

Photon drag detector for use at 10.6 μm (CO_2 laser radiation) by Hamamatsu.

Photon drag detector Left: The photon drag effect in which a current is generated owing to the transfer of photon momentum ***p*** to the hole carrier system. In the open circuit configuration, a voltage V is built up in the sample along the direction of the incident light. Right: Photon drag detector for use at 10.6 μm (CO_2 laser radiation) by Hamamatsu.

Photon noise see *quantum noise*.

Photon or **optical confinement** is the restriction of optical power, photons, to a certain region of a device by virtue of refractive index profiling; for example, by constructing a waveguide structure.

Photon reabsorption broadening is simply the broadening of an intrinsic emission spectrum as the emitted photons from emission centers (such as Er^{3+} ions) are reabsorbed during their propagation towards the observer. Consider the emission of photons from excited Er^{3+}. Since the intrinsic spectrum has photons of different frequency, their removal from the spectrum by absorption depends on their frequency. As an example consider the observed photoluminescence (PL) spectrum when dopants such as Er^{3+} ions in a medium (such as glass) are excited from one side with an excitation beam of wavelength λ_{ex} as shown in the figure. The excitation photons are absorbed by Er^{3+} which then emit, by photoluminescence, photons of wavelength λ. If the individual emission spectrum from a dopant is $\Phi_i(\lambda)$ then the spectrum of the photon flux observed on the right-hand side, the "front" (F), is

$$\Phi_F(\lambda) = \Phi_i(\lambda)\int_{x=0}^{x=L}\exp[-\alpha(\lambda_{ex})x] \times \exp[-\alpha(\lambda)(L-x)]dx.$$

On the other hand, the spectrum of the photon flux exiting from the excitation side, the "back" (B), is

$$\Phi_B(\lambda) = \Phi_i(\lambda)\int_{x=0}^{x=L}\exp[-\alpha(\lambda_{ex})x] \times \exp[-\alpha(\lambda)x]dx$$

where $\alpha(\lambda_{ex})$ is the absorption coefficient of the medium with Er^{3+} ions at the excitation wavelength, and $\alpha(\lambda)$ is the absorption coefficient at the wavelength of interest recorded by the detector. The spectrum depends on the sample size. In fact, spectra of different sized samples with the same amount of Er^{3+} doping can be different because of the photon reabsorption effect; small particles show less reabsorption broadening and hence exhibit the $\Phi_i(\lambda)$ shape in their observed PL spectrum. One should also consider the possibility of re-emission from the Er^{3+} ions, which complicates the analytical treatment. Indeed, if one considers multiple

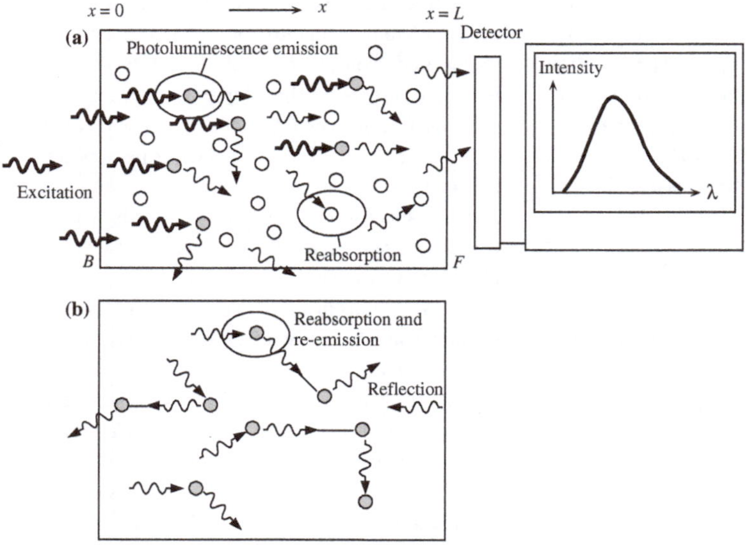

Photon reabsorption broadening (a) The reabsorption of photoluminescence photons on their way to the detector modifies the observed spectrum. (b) In general there could be multiple reabsorption and reemission processes that affect the observed spectrum.

reabsorption and re-emission processes as the luminescent photons propagate in the medium, the spectrum becomes modified by these processes; one possible method of calculating the spectrum is to use Monte Carlo methods to describe the multiple reabsorption and reemission processes. The reabsorption and reemission processes also affect the observed photoluminescence decay time, the effective PL lifetime, since the processes cause a delay in the escape of the photons and hence give the appearance of a longer lifetime. Reflections from the surface add to the photon path since they cause "photon trapping" inside the medium. The severity of photon reabsorption effect on the spectrum depends on the absorption coefficient. The spectral broadening due to photon reabsorption and reemission process is similar to *photon recycling*, and many authors refer to it as *photon trapping* since the PL photon does not immediately escape the medium and is trapped in it by the absorption and reemission processes. See *photon trapping*. (M. Mattarelli *et al.*, *Europhys. Lett.*, **71**, 394, 2005, S. O. Kasap *et al.*, *J Non-Cryst Solids*, **353**, 1364, 2007; K. Koughia *et al.*, *J. Mater. Sci.: Mater. Electron*, **18**, S153, 2007.)

Photon recycling is a term commonly used in semiconductor luminescent devices, in particular in heterojunction devices, and describes a process in which luminescent photons are reabsorbed in the material and then reemitted. The absorbed and the reemitted photons usually have different wavelengths, or energies. In effect, it is the reabsorption and reemission of a photoluminescent photon, which is accompanied by a redshift of the wavelength; a decrease in the photon energy.

Photon trapping refers to the trapping of photons in a medium due to various processes. For example, photoluminescence (PL) photons that have been emitted by dopants can be trapped in a medium by self-absorption, that is, dopants reabsorb the

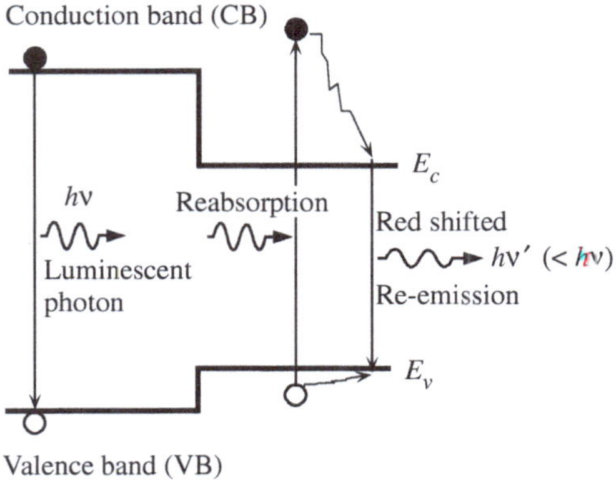

An example of **photon recycling**.

PL photons. Any process that delays or prevents the escape of photons generated in a system lead to photon trapping. (D. S. Sumida, T. Y. Fan, *Opt. Lett.*, **19**, 1343, 1994.) See *photon reabsorption broadening*.

Photonic bandgap see *photonic crystals*.

Photonic condensation see *threshold-crossing*.

Photonic crystals are artificial materials (or structures) where there is a periodic modulation of refractive index or dielectric constant along a given direction. For appropriate wavelength electromagnetic radiation, such structures can strongly influence the propagation and confinement of light in the structures. Analogous to semiconductor crystals, forbidden energy states arise over a range of energies resulting in a so-called *photonic bandgap* in such crystals. Similar to the semiconductor case, these are nonpropagating states for light, arising for wavelengths around the Bragg wavelength of the crystal. Photonic crystals can have periodic dielectric contrast in one, two or three dimensions and are thus termed 1D, 2D or 3D photonic crystals, respectively. (See first figure.) The 1D counterpart of a photonic crystal is the well-known multilayer distributed Bragg reflector (DBR), composed of quarter-wavelength stacks of alternating high-index and low-index materials. Radiation with wavelengths in a band (called the stop-band) close to the Bragg wavelength is subject to high reflection, but the stop-band varies with incidence angle and polarization and does not cover all directions.

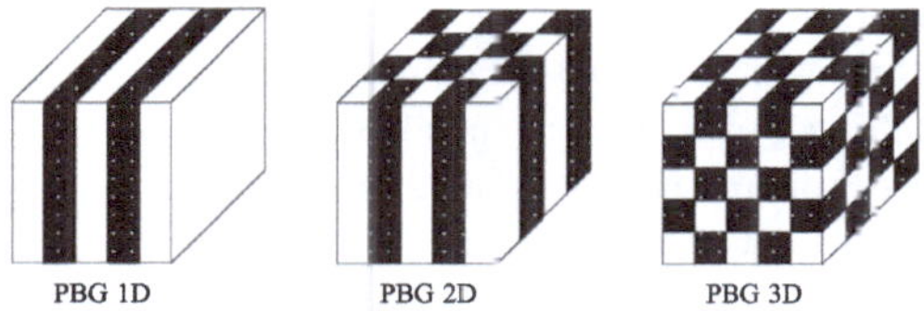

Photonic crystals Photonic bandgaps: one-dimensional, two-dimensional and three-dimensional.

So-called full photonic bandgap structures – ones in which the photonic bandgap exists for all spatial directions and for all polarizations – can exist. However, the details of the structure, including the strength of the refractive index contrast, structure symmetry and crystal perfection, are critical aspects. Another important practical aspect of photonic crystals, analogous to semiconductor crystals, is the importance of defects. Point defects, for example, can be formed by removing or substituting an "atom" or introducing "defects" (doping), enabling localized electromagnetic modes to be formed, with energy states within the stop-band. Owing to the potential to use such defects to tightly confine a mode of light in a very small cavity volume, such artificial structures appear promising for controlling spontaneous emission (by inhibition or enhancement), and thus possibly reducing the threshold current of lasers. Linear defects, where a row of "atoms" is missing, for example, can provide propagating states within the stop-band and hence represent near loss-free waveguides for guiding light, as shown in the figure below.

Moreover, by designing such linear defects as curved waveguides, light can be guided with almost

 Photonic crystals

90° bends, as compared with traditional waveguides, where large bend radii dictate the need for much larger spaces. An important application of this technology is in so-called *holey fibers* – a central core region of photonic crystal coaxial with the fiber acting as an engineered waveguide.

Photonic crystals have also been shown to exhibit a so-called *superprism effect*. Under appropriate conditions, the dispersion of light by a prism-shaped photonic crystal is considerably enhanced over that corresponding to a homogeneous prism having the same average refractive index as the photonic crystal. This originates from the strong curvature of the frequency-wavevector curve near the edge of the Brillouin zone. This may also be viewed as a high refractivity variation with wavelength. Researchers have shown that the superprism effect can result in as much as a threefold increase in the dispersion angle for selected wavelengths of light in appropriately designed photonic crystals. This effect can be very important for telecommunications applications. For example, in wavelength division multiplexing, different wavelength channels can be separated or combined in a much more compact space than would be the case for traditional demultiplexing or multiplexing systems.

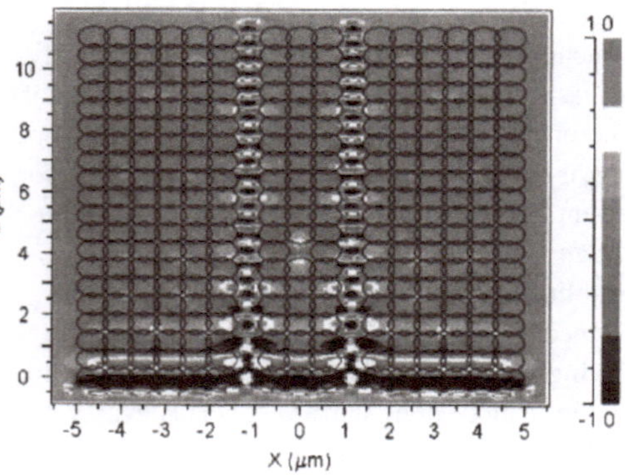

Photonic crystals TE simulation for two-channel triangular lattice waveguide with 0.7 μm wide channels surrounded on both sides by a triangular air-in-dielectric lattice. A plane wave source of width 4 μm was used. Notice how the two waveguides can guide light in the two channels without interaction or lateral loss, offering the prospect of dense lossless photonic waveguide circuits. (Courtesy of Shaniya Mathews and Harry Ruda, University of Toronto.)

One potentially attractive field for using photonic crystal structures is in fabricating planar lightwave circuits. This technology is analogous to integrated circuit technology where multiple electronic functions are integrated together on a chip to make an electronic circuit – in the case of the planar lightwave circuit, optical waveguides play the role of wires, guiding light between optical circuit elements. Photonic crystals are very attractive for implementing such schemes as they offer the possibility of highly compact circuits with additional functionality over their traditional optical counterparts. Such optical chips have applicability in advanced telecommunications systems. To implement such circuits, 2D-photonic crystal structures that can be manufactured using standard integrated circuit processing technology are actively being pursued. The two main types of structure are 2D arrays of solid elements (e.g. rods) and 2D arrays of holes in a thin film. The latter has the advantage that a suspended film or air-bridge acts as a waveguide with the periodic contrast superimposed on it. Considerations of out-of-plane losses are important in these structures and often control the overall optical performance of the structures.

Photonic bandgap (PBG) structures have proved to be of special interest in the microwave regime, where the dimensional requirements make them easier to fabricate; they have been used to improve radiation patterns of antennas, increase the output power and efficiency of power amplifiers, as well as with the design of reflectors, broadband absorbers, frequency selective surfaces or integrated filters. For instance, a microstrip line can be deposited on a two-dimensional PBG substrate, obtained by periodically etching the metal of the ground plane as illustrated in the figure.

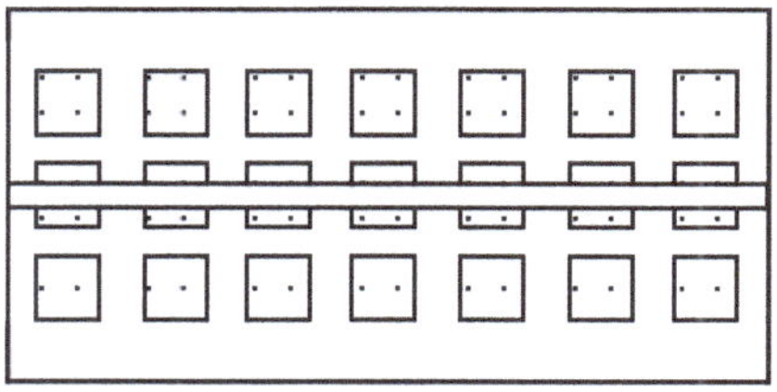

Photonic crystal Microstrip line on PBG substrate.

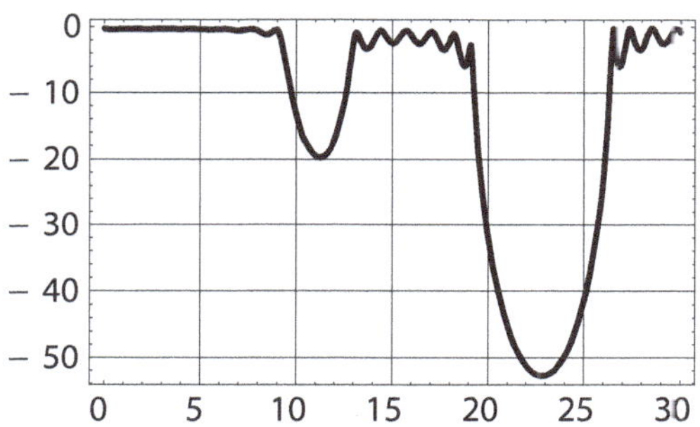

Photonic crystal Typical transmission spectrum of a microstrip line on PBG substrate in the microwave range. The vertical axis is the transmittance, $T(\nu)$ in dB, and the horizontal axis is the frequency ν in GHz.

The typical optical response of such a structure is presented in the third figure as transmittance vs. frequency: the first two stop-bands are well defined. The precise location and width of the stop-bands are found to depend strongly on the periodicity and duty cycles (or lattice constants) of the etched motifs. Applications to spectral filtering have been developed.

Photonic network see *optical network.*

Photonics means "the technology of generating and harnessing light and other forms of radiant energy whose quantum unit is the photon. The science includes light emission, transmission, deflection, amplification and detection by optical components and instruments, lasers and other light sources, fiber optics, electrooptical instrumentation, related hardware and electronics, and sophisticated systems. The range of applications of photonics extends from energy generation to detection to communications and information processing." (From *Photonics Dictionary* at http://photonics.com by Laurin Publishing, 2009)

Photopic is an adjective for the response of an average human eye when it is light adapted. See *luminous efficiency, eye.*

Photoreflectance spectroscopy measures the change in a sample's reflectivity $\Delta R(\lambda)$ for a weak probe beam of wavelength λ when the sample is pumped using modulated above-bandgap excitation, as compared with the unperturbed reflectivity, $R(\lambda)$. The photoreflectance spectrum $\Delta R/R$ versus probing photon energy or wavelength, is a measure of the derivatives of the unperturbed dielectric function, and directly provides information on critical points in the band structure, and has also been used to provide information on internal fields and stresses. Normally, phase-locked loop-type detection techniques are used and very small changes can be measured, providing resolution as low as 10^{-7} for $\Delta R/R$. Moreover, since the technique reflects higher order derivatives of the unperturbed spectrum, features are naturally sharp even at room temperature and using standard analysis of the lineshape function for standard critical points enables room temperature measurement of critical points with high precision, typically within a few meV, normally only feasible using liquid helium temperature photoluminescence measurements.

When the modulation amplitude is small, the photoreflectance changes can be related to changes in the complex dielectric function, $\varepsilon = \varepsilon_1 + i\varepsilon_2$, through the following relationship

$$\Delta R/R = \alpha\Delta\varepsilon_1 + \beta\Delta\varepsilon_2$$

where α and β are the *Seraphin coefficients*, related to the unperturbed dielectric function, ε_1 and ε_2 are the real and imaginary parts, and $\Delta\varepsilon_1$ and $\Delta\varepsilon_2$ are

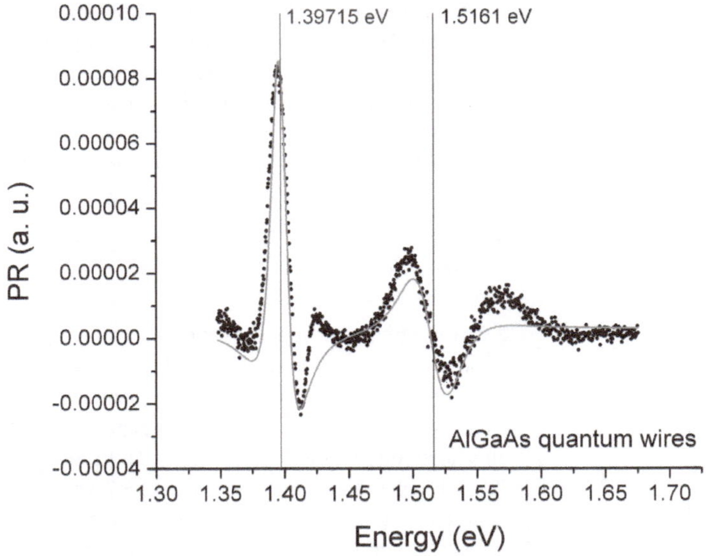

Photoreflectance spectroscopy Room temperature photoreflectance (PR) spectrum (PR vs. photon energy) of AlGaAs nanowires grown on a GaAs substrate, showing clear third derivative line shapes for the nanowires and substrate, with transition energies as indicated, separated by ~119 meV. The solid line shows a third derivative lineshape fit to the spectrum, enabling the transitions at 1.3972 eV and 1.5161 eV to be identified. (Courtesy of Christina de Souza and Harry Ruda.)

the changes in the complex dielectric function due to the perturbation. The *Seraphin coefficients* can be written as

$$\alpha = \frac{1}{R}\frac{\partial R}{\partial \varepsilon_1} \quad \text{and} \quad \beta = \frac{1}{R}\frac{\partial R}{\partial \varepsilon_2}.$$

Near the fundamental energy gap, $\beta \sim 0$, so that $\Delta R/R \sim \Delta\varepsilon_1$ is the only significant term. The quantities $\Delta\varepsilon_1$ and $\Delta\varepsilon_2$ are related by *Kramers–Kronig inversion*. The functional form of $\Delta\varepsilon_1$ and $\Delta\varepsilon_2$ can be calculated for a given perturbation provided that the dielectric function and critical points are known. For example, assuming weak modulation and a *Lorentzian lifetime* broadening Γ (i.e. due to interactions with phonons), it can be shown that

$$\Delta R/R = \mathrm{Re}[Ce^{i\phi}(E - E_g + i\Gamma)^{-m}]$$

where E is the photon energy, C is the amplitude, ϕ is a phase angle accounting for the mixture of real and imaginary components in ε, and the influence of nonuniform fields, and E_g is the band gap energy. The parameter m depends on the type of critical point. The figure shows the usefulness of photoreflectance for providing a direct measure of the critical point energy for AlGaAs nanowires prepared on a GaAs substrate, at room temperature. The clear bandgap signature from each is visible in the spectrum in two separate spectral ranges, and shows how a third derivative lineshape function provides an excellent fit to the experimental data. Moreover, if we consider in general the effects of perturbations to the unperturbed dielectric function induced by a weak modulation field, F_{ac}, $\Delta\varepsilon$ can be expressed as:

$$\Delta\varepsilon = \left[\frac{\partial\varepsilon}{\partial E_g}\frac{\partial E_g}{\partial F_{ac}} + \frac{\partial\varepsilon}{\partial\Gamma}\frac{\partial\Gamma}{\partial F_{ac}} + \frac{\partial\varepsilon}{\partial I}\frac{\partial I}{\partial F_{ac}}\right] F_{ac}$$

where $\partial E_g/\partial F_{ac}$ is the change in energy gap due to the *Stark effect,* $\partial\Gamma/\partial F_{ac}$ is the change in the broadening parameter due to a variation of the lifetime, and $\partial I/\partial F_{ac}$ is the change in the intensity due to the redistribution of charges. (F. H. Pollak, *Proc. SPIE,* **3944**, 408, 2000.)

Photorefraction, discovered by Ashkin *et al.* in 1966, is a spatially or locally varying change in refractive index of a material due to the absorption of light whose intensity varies spatially. That is, spatially varying light intensity induces a spatially varying refractive index, through a nonlinear optical effect. The absorption of light in the high-intensity (bright) region generates electrons (photoelectrons), which diffuse away from this bright region to the neighboring darker region where they recombine with empty donors. The migration of photogenerated electrons exposes charged dopants, and thereby creates an internal electric field. This space charge field, in turn, modifies the refractive index through the (Pockels) electrooptic effect. Thus, photorefraction occurs in photoconductive and electrooptic materials. The final result is a photoinduced imprinted space charge distribution. The photorefractive effect can be used to store and detect information. If $E(x)$ is the spatially varying electric field, and n and r are values of the refractive index and relevant electrooptic coefficient for a given material, the spatially varying refractive index change associated with the internal field modulation is,

$$\Delta n(x) = -(1/2)n^3 r E(x).$$

In the absence of an applied field, the change in internal field $E(x)$ can be associated with the incident light intensity $I(x)$,

$$E(x) = (k_B T/e)[d \ln I(x)/dx]$$

where k_B is the Boltzmann constant and e the elementary charge. The above equation is similar to an internal field generated by nonuniform doping if one replaces $I(x)$ with the donor concentration $N_d(x)$. This expression provides a relationship between the refractive index change and the illumination intensity

$$\Delta n(x) = -(1/2)n^3 r(k_B T/e)[d \ln I(x)/dx].$$

The processes leading to the photorefractive effect include free carrier generation by photoionization of impurity levels at a rate proportional to the optical power. Photogenerated carriers subsequently diffuse from where they were generated leaving behind ionized impurity atoms from which they were generated, as immobile ionized centers. The result is an inhomogeneous internal electric field distribution that can remain in place beyond the period of irradiation of the material. As discussed above, this field results in a modulation of the refractive index. The charge image may be revealed by monitoring the spatial pattern of the refractive index using an optical probe beam. One can restore the material to its original condition by uniform irradiation with light or by heating. This approach thus permits such materials to be used to record and store images. Some important photorefractive materials include barium titanate, bismuth silicon oxide, lithium niobate ($LiNbO_3$), gallium arsenide and strontium barium niobate. These photorefractive materials have been widely used in high-sensitivity dynamic holographic gratings (i.e. volume phase holograms), optical data storage, information processing, and fiber optic devices. Such dynamic photorefractive holograms do not need chemical development and are particularly convenient because the hologram that is generated by interference (of waves from the object and reference) within the photorefractive medium can diffract the reference beam at the same time to obtain the image or diffraction pattern immediately.

Recently there has been an interest in photorefractive quantum wells (QWs). Quantum confined excitons in such structures give rise to large electrooptic and nonlinear optical effects at room temperature. Photorefractive effects are introduced by applying an external electric field. The orientation of the electric field with respect to the

QWs determines the nature of the resonant photorefractive effect. If the electric field is parallel to the QWs, the *Franz–Keldysh effect* occurs, while the *quantum confined Stark effect* (QCSE) occurs for a perpendicular field. Typically, the QCSE produces the larger resonant photorefractive effect and higher diffractive efficiencies. Moreover, the shorter carrier transit distances for QCSE devices provide for faster response times. (A. Ashkin *et al.*, *Appl. Phys. Lett.*, **9**, 72, 1966; S. I. Stepanov, *Rep. Prog. Phys.*, 39, 1994; P. M. Johansen, *J. Opt. A*, **5**, S398, 2003; Gu *et al.*, *J. Opt. A*, **5**, S420, 2003.)

Phototransistor is a bipolar junction transistor (BJT) that operates as a photodetector with a photocurrent gain. Typically the base terminal is normally open and there is a voltage applied between the collector and emitter terminals just as in the normal operation of a common emitter BJT. An incident photon is absorbed in the space charge layer (SCL) between the base and collector to generate an electron–hole pair (EHP). The electric field E in the SCL separates the electron hole pair and drifts them in opposite direction. This is the primary photocurrent and effectively constitutes a base current even though the base terminal is open circuit (current is flowing into the base from the collector). Since the photon generated primary photocurrent I_{pho} is amplified as if it were a base current (I_{B}), the photocurrent flowing in the external circuit is

$$I_{\text{ph}} \approx \beta I_{\text{pho}}$$

where β is the current gain (or h_{FE}) of the transistor. The phototransistor construction is such that incident radiation is absorbed in the base–collector junction SCL. It is possible to construct a *heterojunction phototransistor* that has different bandgap materials for the emitter, base and collector. For example, if the emitter is InP ($E_g = 1.35$ eV) and the base is an InGaAsP alloy (for example, $E_g \approx 0.85$ eV) then photons with energies less than 1.35 eV but more than 0.85 eV will pass through the emitter and become absorbed in the base. This means the device can be illuminated through the emitter.

Photovoltaic devices or **solar cells** convert the incident solar radiation energy into electrical energy. Incident photons are absorbed to photogenerate charge carriers that pass through an external load to do electrical work. Photovoltaic devices may be metal–semiconductor Schottky junctions, *pn* junction or *pin* devices. For example, a crystalline Si *pn* junction solar cell may have a thin n-type semiconductor layer on a thick p-type substrate. The electrodes attached to the n-side must allow illumination to enter the device and at the same time result in a small series resistance. They are deposited on to the n-side to form an array of *finger electrodes* on the surface. The n-side is very narrow to allow most of the photons to be absorbed within the depletion region and within the neutral p-side. These photons photogenerate electron–hole pairs (EHPs) in these regions. EHPs photogenerated in the depletion region are immediately separated by the built-in field E_o which drifts them apart. The electron drifts and reaches the neutral n^+ side whereupon it makes this region relatively more negative by an amount of charge $-e$. Similarly the hole drifts and reaches the neutral p-side and thereby makes this side positive. Consequently an open circuit voltage develops between the terminals of the device with the p-side positive with respect to the n-side. If an external load is connected then the excess electron in the n-side can travel around the external circuit, do work, and reach the p-side to recombine with the excess hole there. There will be an external photocurrent. It is important to realize that without the internal field E_o it is not possible to drift apart the photogenerated EHPs and accumulate excess electrons on the n-side and excess holes on the p-side. A thin antireflection coating on the surface reduces reflections and allows more light to enter the device.

Photovoltaic effect is the generation of an emf (electromotive force) as a result of illumination. See *photovoltaic devices*.

 Photovoltaic or **solar cell *I-V* characteristics**

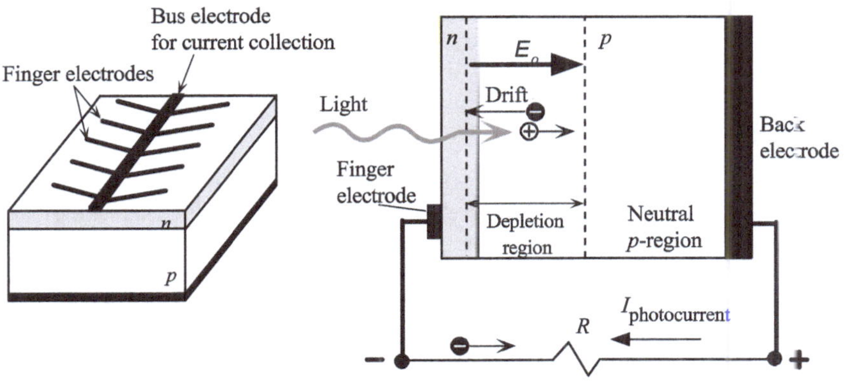

Photovoltaic device Left: finger electrodes on the surface of a solar cell reduce the series resistance. Right: the principle of operation of the solar cell.

Photovoltaic device Honda's two-seated Dream car is powered by photovoltaics. The Honda Dream was first to finish 3010 km in four days in the 1996 World Solar Challenge. (Courtesy of Photovoltaics Special Research Centre, University of New South Wales, Sydney, Australia.)

Photovoltaic or **solar cell *I-V* characteristics**

are the current vs. voltage characteristics of the device under different intensities of illumination within the spectral response curve of the device. Under illumination, the dark I–V characteristics become shifted so that there is a short circuit current I_{sc} and an open circuit voltage V_{oc}; that is, a photocurrent I_{sc} when the solar cell is shorted and a voltage V_{oc} when it is in open circuit. I_{sc} represents the conventional current that would normally flow in a forward biased diode. However, the actual photocurrent I_{ph} is in the opposite direction to I_{sc}, thus $I_{sc} = -I_{ph}$.

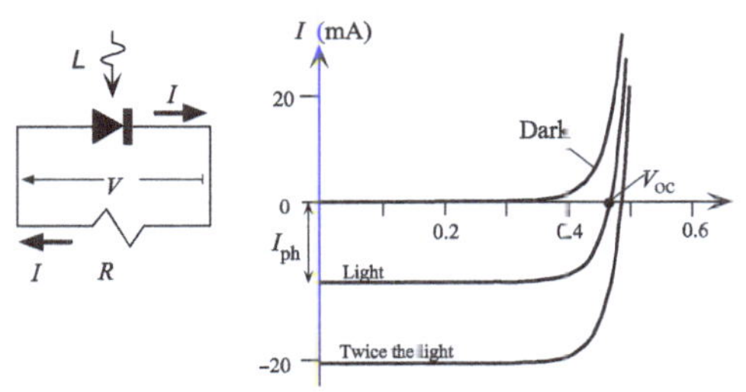

Photovoltaic *I–V* characteristics Typical I–V characteristics of a Si solar cell. The short circuit current is $-I_{ph}$ and the open circuit voltage is V_{oc}. The I–V curves for positive current require an external bias voltage. Photovoltaic operation is always in the negative current region.

When a solar cell drives a load R, R has the same voltage as the solar cell but the current through it is in the opposite direction to the convention that current flows from high to low potential.

The total current I through a pn junction solar cell can normally be written as

$$I = -I_{\text{ph}} + I_{\text{o}}\left[\exp\left(\frac{eV}{nk_{\text{B}}T}\right) - 1\right]$$

where V is the voltage across the solar cell, I_{ph} is the short circuit photocurrent, I_{o} is the pn junction reverse bias saturation current (in dark), k_{B} is the Boltzmann constant, T is temperature (K) and n is the ideality factor (1 to 2) for the pn junction. The open circuit output voltage, V_{oc}, of the solar cell is given by the point where the I–V curve cuts the V-axis ($I = 0$). It is apparent that although it depends on the light intensity, its value for Si solar cells typically lies in the range 0.4–0.6 V. When the solar cell is connected to a load R, the load has the same voltage as the solar cell and carries the same current. But the current I through R is now in the opposite direction to the convention that current flows from high to low potential. Thus,

$$I = -\frac{V}{R}.$$

The actual current I' and voltage V' in the circuit must satisfy both the I–V characteristics of the solar cell and that of the load. We can find I' and V' by solving these two equations through a graphical solution; called a load line construction. The I–V characteristics of the load is a straight line with a negative slope $-1/R$. This is called the *load line*. The load line cuts the solar cell characteristic at a point P called the *operating point*. At P, the load and the solar cell have the same current and voltage I' and V'. The current I' and voltage V' in the circuit are hence given by point P. The power delivered to the load is $P_{\text{out}} = I'\,V'$, which is the area of the rectangle bound by the I- and V-axes and the dashed lines in the figure. Maximum power is delivered to the load when this rectangular area is maximized (by changing R or the intensity of illumination), when $I' = I_m$ and $V' = V_m$. Since the maximum possible current is I_{ph} and the maximum possible voltage is V_{oc}, then $I_{\text{ph}}V_{\text{oc}}$, represents the desirable goal in power delivery for a given solar cell. It is therefore useful to compare the maximum power output, I_mV_m, with $I_{\text{sc}}V_{\text{oc}}$. The fill factor FF, which is a figure of merit for the solar cell, is defined as

$$\text{FF} = \frac{I_mV_m}{I_{\text{sc}}V_{\text{oc}}}.$$

FF is a measure of the closeness of the solar cell I-V curve to the rectangular shape (the ideal shape). It is clearly advantageous to have FF as close to unity as possible but the exponential pn junction I-V properties prevent this. Typically FF values are in the range 70–85% and depend on the device material and structure. The maximum power voltage V_m and current I_m can be calculated approximately from

$$\frac{V_m}{nV_T}\exp\left(\frac{V_m}{nV_T}\right) \approx \frac{I_{\text{ph}}}{I_0}$$

and $$I_m = -I_{\text{ph}}\left(1 - \frac{nV_T}{V_m}\right)$$

where $V_T = k_{\text{B}}T/e$ is the "thermal voltage."

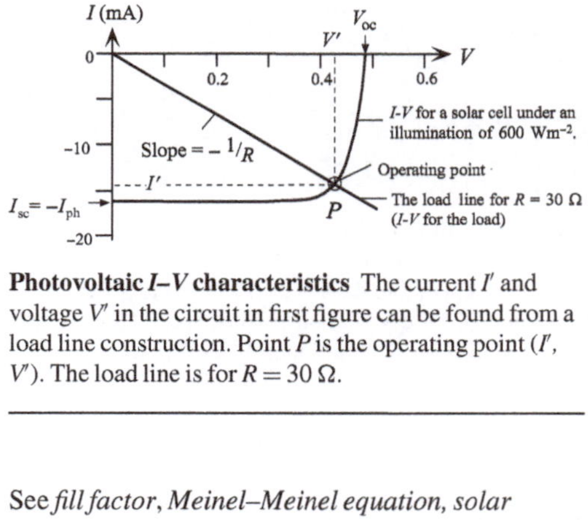

Photovoltaic I–V characteristics The current I' and voltage V' in the circuit in first figure can be found from a load line construction. Point P is the operating point (I', V'). The load line is for $R = 30\ \Omega$.

See *fill factor, Meinel–Meinel equation, solar radiation.*

Physical vapor deposition (PVD) refers to processes for depositing thin films of material onto a substrate. PVD includes evaporation and sputtering processes. PVD is commonly used to deposit metals on semiconductors.

Table: Pikhtin-Yaskov dispersion relation Constant in the Pikhthin–Yaskov equation from the *Handbook of Optical Constants of Solids*, Edited by E. D. Palik, Academic Press, New York, 1985.

	E_0(eV)	E_1(eV)	E_2(eV)	E_3(eV)	A	G_3(eV^2)	G_1(eV^2)	G_1(eV^2)
CdTe	1.500	3.50	5.50	0.0174	0.857321	0.001080	64.375000	17.664
GaAs	1.428	3.00	5.10	0.0174	0.585779	0.002180	39.194000	136.08
InAs	0.356	2.20	4.90	0.0174	1.173203	0.002010	28.748000	79.354
InP	1.345	3.20	5.10	0.0174	0.811819	0.003920	57.889000	65.937

Piezoelectric material has a noncentrosymmetric crystal structure which leads to the establishment of a *polarization vector P*, or charges on the crystal surfaces, upon the application of a mechanical stress, T. When strained, a piezoelectric crystal develops an internal field and therefore exhibits a voltage difference between two of its faces. Equivalently, a piezoelectric crystal when strained develops an internal field and therefore exhibits a voltage difference between two of its faces. The mechanical stress, T_i, in one direction (along i) can generate polarization, P_j, along a different direction (along j) and are related by $P_i = d_{ij}\, T_j$ where d_{ij} is the piezoelectric coefficient.

Piezooptical effect see *photoeleastic effect*.

Pigtail is a short optical fiber that is directly coupled to an optical component, for example, an LED. A pigtailed emitter, for example, comes with this short fiber already attached to it so that one can simply use this short fiber to couple the emitter to another component (e.g. a long-haul fiber by splicing, etc.).

Pigtail A 1550 nm MQW-DFB InGaAsP laser diode pigtail-coupled to a fiber. (Courtesy of Alcatel-Lucent.)

Pikhtin–Yaskov dispersion relation is a refractive index n vs. photon $h\nu$ energy relation for semiconductors of the form

$$n^2 = 1 + \frac{A}{\pi}\ln\left[\frac{E_1^2-(h\nu)^2}{E_0^2-(h\nu)^2}\right] + \frac{G_1}{E_1^2-(h\nu)^2} + \frac{G_2}{E_2^2-(h\nu)^2} + \frac{G_3}{E_3^2-(h\nu)^2}$$

in which $E_0, E_1, E_2, E_3, A, G_1, G_2$ and G_3 are all constants specific to the material as listed in the table. (A. N. Pikhtin and A. D. Yas'kov, *Sov. Phys. Semicond.*, **12**, 622, 1978.)

***pin* diode** is a semiconductor diode that has the structure, p^+/ intrinsic /n^+ with a relatively thick intrinsic region that is depleted of free carriers. The intrinsic layer has a much smaller doping than both p^+ and n^+ regions and it is much wider than these regions, typically 5–50 μm depending on the particular application. In the idealized *pin* photodiode, the i-Si region is simply taken to be truly intrinsic whereas in practice it is slightly doped. When the structure is first formed, holes diffuse from the p^+-side and electrons from the n^+-side into the i-Si layer where they recombine and disappear. This leaves behind a thin layer of exposed negatively charged acceptor ions in the p^+-side and a thin layer of exposed positively charged donor ions in the n^+-side. The two charges are separated by the i-Si layer of thickness W. The depletion layer widths of the thin sheets of acceptor and donor charges in the p^+ and n^+ sides are negligible compared with W. There is a uniform built-in field E_o in i-Si layer from the exposed positive donor ions to the exposed acceptor negative ions. In contrast, the built-in field in the depletion layer of a *pn* junction is not uniform.

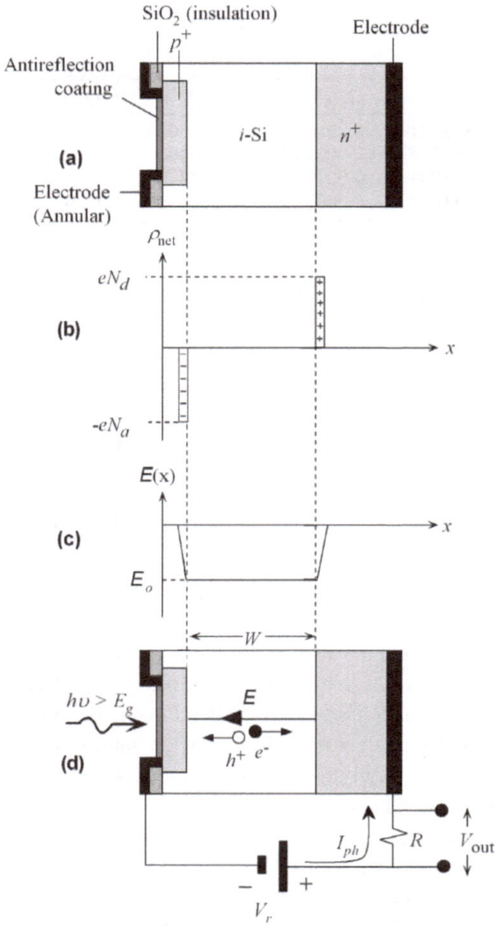

***pin* diode** (a) The schematic structure of an idealized *pin* photodiode. (b) The net space charge density across the photodiode. (c) The built-in field across the diode. (d) The *pin* photodiode in photodetection is reverse biased.

With no applied bias, equilibrium is maintained by the built-in field E_0 which prevents further diffusion of majority carriers into the *i*-Si layer. The separation of two very thin layers of negative and positive charges by a fixed distance, width *W*, of the *i*-Si, is the same as that in a parallel plate capacitor. The junction capacitance (the depletion layer capacitance C_{dep}) tends to be very small, in the picofarad range, and independent of the bias voltage. The *pin* is normally reverse biased with a voltage V_r which drops almost entirely across the width of *i*-Si layer. The field *E* in the *i*-Si layer is still uniform and increases to $E \approx V_r/W$. The *pin* structure is designed so that photon absorption occurs over the *i*-Si layer. The photogenerated electon–hole pairs (EHPs) in the *i*-Si layer are then separated by the field *E* and drifted towards the n^+ and p^+ sides respectively. The drift of the photogenerated carriers through the *i*-Si layer gives rise to an external current (photocurrent) which is detected as a voltage across a small sampling resistor *R*. The response time of the *pin* photodiode is determined by the transit times of the photogenerated carriers across the width *W* of the *i*-Si layer. Increasing *W* allows more photons to be absorbed which increases the quantum efficiency but it slows down the speed of response inasmuch as carrier transit times become longer. The intrinsic response time of the *pin* is normally determined by the drift time of the slowest carriers across the depletion region. This response time can be shortened by increasing the drift velocity of the carrier, that is, by increasing the field (i.e. V_r) but eventually the carrier drift velocity saturates and the drift time cannot be shortened further. For an *i*-Si layer of width 10 μm, with carriers drifting at saturation velocities, the drift time is about 0.1 ns which is longer than typical RC_{dep} time constants. The speed of *pin* photodiodes is invariably limited by the transit time of photogenerated carriers across the *i*-Si layer.

Pixel is a small element of a sensor whose signal makes up a small piece of the image that the sensor is able to capture. Stated differently, a pixel is a tiny individual detector which itself is part of a large array of detectors that make up an image sensor. A pixel can also refer to an element of an image; an image is made of numerous pixels each of which carries the local image properties.

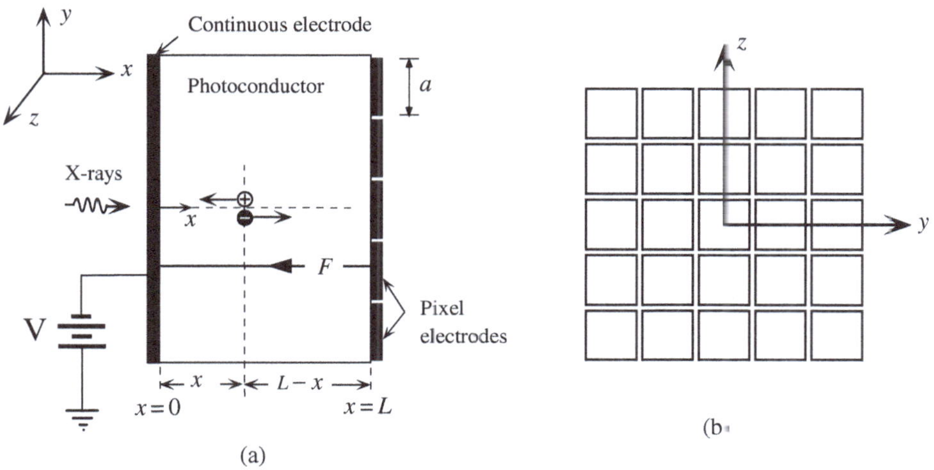

Pixellated detector A pixellated x-ray image detector: (a) a cross-section; (b) two-dimensional array of pixel electrodes. An electron and a hole are generated at x and drift under the influence of the electric field F.

Pixellated detector is the semiconductor detector configuration in which a photoconductor layer is sandwiched between two electrodes; the electrode at one side is a continuous metal plate and the electrode on the other side of the photoconductor is segmented into a 1D or 2D array of individual pixels. There is a gap of negligible dimension between the pixel electrodes. The lateral dimension of the photoconductor slab is much greater than the detector thickness L. The most common pixellated detector consists of a 2D array of individual square pixels of size $a \times a$ as shown in figure. Separate electrical contacts are generally made to each pixel electrode. The continuous electrode (top electrode) is biased with a voltage V to establish an electric field F in the photoconductor. Since the spacing between pixels is negligible compared to the pixel dimensions, the effects of the small gaps are negligible in determining the actual electric potential and field distribution throughout the detector volume. Therefore, the detector geometry is just like a parallel plate configuration from the electrostatic point of view, except that the currents through individual pixels are integrated separately. The pixellated detectors have been primarily designed for imaging applications, e.g. direct conversion flat-panel x-ray image detector.

Planar dielectric waveguide see *slab waveguide*.

Planck's black body radiation law see *black body radiation*.

Planck's constant h, is a universal constant that is defined as the smallest quantum of radiation energy (at one frequency) per unit radiation frequency; $h = 6.626 \times 10^{-34}$ J s. The quantity $h/2\pi$ is denoted as $\hbar$.

Plane electromagnetic wave is a traveling wave with time varying electric and magnetic fields, that is E_x and B_y respectively, which are propagating through space in such a way that they are always perpendicular to each other and the direction of propagation taken to be along z. At any instant, the plane wave has constant electric field and a constant magnetic field in any given infinite plane perpendicular to the direction of propagation, **k**. The field in successive planes, however, oscillates with time. The field is said to have a constant phase anywhere on a given plane perpendicular to **k**. The simplest traveling wave is a sinusoidal wave which, for propagation along z, has the general mathematical form,

$$E_x = E_0 \cos(\omega t - kz + \phi_0)$$

where k is the propagation constant, or wavenumber, given by $2\pi/\lambda$, where λ is the wavelength, ω is the angular frequency, E_0 is the amplitude of the wave and ϕ_0 is a phase constant which accounts for the fact that at $t = 0$ and $z = 0$, E_x may or may not necessarily be zero depending on the choice of origin. The argument $(\omega t - kz + \phi_0)$ is called the *phase* of the wave and denoted by ϕ. In any plane perpendicular to the direction of propagation (along z), the phase of the wave is constant which means that the field in this plane is also constant. A surface over which the phase of a wave is constant is referred to as a *wavefront*. A wavefront of a plane wave is obviously a plane perpendicular to the direction of propagation. The plane wave is one particular solution of the electromagnetic (EM) wave equation. A perfect plane wave can only be generated by an infinitely extended source, though the concept is

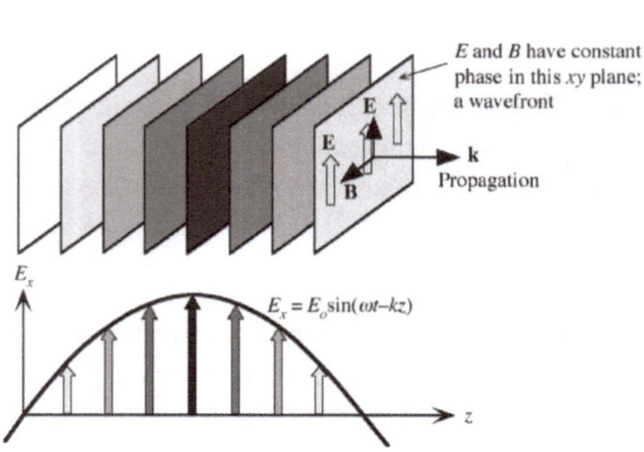

An **plane electromagnetic wave** is a traveling wave which has time varying electric and magnetic fields which are perpendicular to each other and the direction of propagation z.

very useful in the discussion and analysis of light waves propagating in a particular direction in a uniform and isotropic medium when the divergence is small or practically negligible. The magnetic and electric fields **E** and **B** in the wave are related by

$$\mathbf{k} \times \mathbf{E} = \omega \mathbf{B}.$$

The ratio of **E** to the magnetic field intensity **H** is called the *impedance* Z of the medium; $Z = 1/(\varepsilon_r^{1/2}\varepsilon_0 c)$. For free space, $Z = Z_o = 377\ \Omega$. See *electromagnetic wave, electromagnetic wave equation, spherical wave, Gaussian wave, impedance of a medium.*

Plane of incidence see *reflection.*

Plane wave see *plane electromagnetic wave.*

Planoconvex lens, or **planar convex lens**, is a converging lens that has one surface flat and the other surface curved to reduce spherical aberration in focusing a collimated beam incident on its curved surface; or in generating a nearly collimated beam from a point source placed at its focal point and illuminating the flat surface. While the planoconvex lens has reduced spherical aberration for either an object at infinity (incident parallel light rays on curved surface), or image at infinity, this is not so in optical applications where the object and image distances are comparable. Nonetheless, they are useful for reducing spherical aberration in applications which involve nearly collimated light or in lens combinations where the image and object are on the same side of the lens.

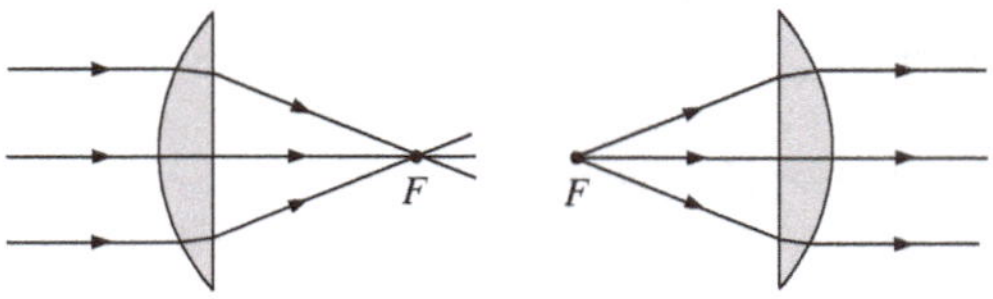

Planoconvex lens and its uses.

Plasma displays are based on generating a plasma by electrical breakdown of a low-pressure inert gas confined within a "pixel" or small volume. Light emission results from the ultraviolet light of the plasma exciting a thin film of phosphor which reemits visible light. A display comprises cells, each containing three subcells – containing red, green and blue phosphor films, respectively. To generate color shades, each color subcell is controlled independently using so-called pulse code modulation. In this scheme, the address signal pulse widths are modulated in 256 step units. Since the human eye has a much slower response than typical modulation times, the eye integrates these changes to provide 256 different intensity levels. Since such displays are emissive and use phosphors, they have excellent viewing angle and color performance. In addition, these displays typically offer contrast ratios (on:off) of 400:1. A main problem with the plasma display technology is reducing the pixel size below about 300 μm. For this reason, this technology is unlikely to play a significant role in the desktop computer market and will remain best suited to TV and large screen (up to 70 inch) presentation formats.

Plasma-enhanced chemical vapor deposition (PECVD) is a variation of the chemical vapor deposition (CVD) process where a gas plasma is used to lower the temperature required to obtain a chemical reaction and achieve film deposition.

Plasmons are quanta of plasma oscillations occurring in metals and semiconductors. A simple description of plasmons in a metal can be developed using a free electron gas model – the positive ion cores are replaced with a uniform positive charge density equal to the mean electron charge density, $\rho_0 = nq$, where n is the average electron concentration. If the actual charge density if $\rho(r,t)$, and we neglect electron–electron collisions since plasmon oscillations occur on a much shorter time scale than such collision processes, one can write an equation of motion:

$$\partial^2(\rho - \rho_0)/\partial t^2 + \omega_p^2(\rho - \rho_0) = 0$$

with $\omega_p = nq^2/m^*\varepsilon_0$ where m^* is the effective electron mass. Note that this is the equation of the harmonic oscillator with angular frequency ω_p. Substituting typical values for metals, gives values of $\omega_p \approx 10^{16}\ s^{-1}$ and characteristic plasmon energies of about 10 eV. Note that because this energy is so high, plasmons are not normally excited in metals. Also, note that it was justified to neglect electron–electron collisions as the typical time τ between with such collisions is about 10^{-12} s, and hence $\omega_p\tau \gg 1$. Thus, the collective oscillations of the free carrier gas occur without damping from carrier–carrier collisions. Such plasmons may also be excited by light – it is found that for $\omega < \omega_p$, the propagation wavevector for the light k is imaginary and the corresponding wave is evanescent (i.e. light is totally externally reflected as in the case of Reststrahlen). Thus, all frequencies below ω_p are totally reflected while above ω_p, k is real. Thus for frequencies above ω_p, the material should be transparent if the electrical resistance of the material is neglected – if this is taken into account, then such an electromagnetic wave would be damped. Indeed, for sufficiently thin films of metals, they are in fact transparent in the ultraviolet spectral region. Interestingly, recently there has been strong interest in nanoscale metallic dielectric composites such as gold-coated silica nanospheres which offer tunable optical response (i.e. from the blue to the near-infrared spectral regions) based on plasmon resonance.

Platinum point black body (at the freezing temperature of platinum, or 2042 K) was adopted in 1948 as a standard of the luminous intensity of light by the Comité International des Poids et Mesures (CIPM), and used to define the *candela*; finally abandoned in 1979.

Pockels cell intensity modulator is a Pockels phase modulator with a polarizer P and an analyzer A before and after the phase modulator such that they are cross-polarized, i.e. P and A have their transmission axes at 90° to each other. The

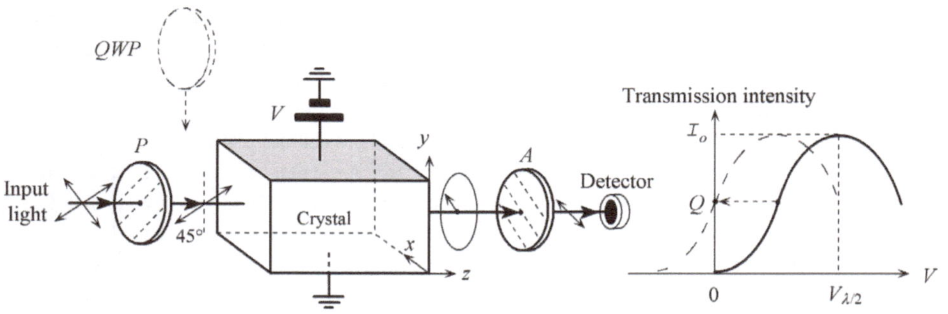

Pockels cell intensity modulator Left: A transverse Pockels cell intensity modulator. The polarizer *P* and analyzer *A* have their transmission axis at right angles and *P* polarizes at an angle 45° to *y*-axis. Right: Transmission intensity vs. applied voltage characteristics. If a quarter-wave plate (*QWP*) is inserted after *P*, the characteristic is shifted to the dashed curve.

transmission axis of *P* is at 45° to the *y*-axis (hence *A* also has its transmission axis at 45° to *y*) so that the light entering the crystal has equal $\mathbf{E}_x$ and $\mathbf{E}_y$ components. In the absence of an applied voltage, the two components travel with the same refractive index and polarization; the output from the crystal is the same as its input. There is no light detected at the detector as *A* and *P* are at right angles ($\theta = 90°$ in Malus's law). An applied voltage inserts a phase difference $\Delta\phi$ between the two electric field components. The light leaving the crystal now has an elliptical polarization and hence a field component along the transmission axis of *A*. A portion of this light will therefore pass through *A* to the detector. The transmitted intensity now depends on the applied voltage *V*. The field components at the analyzer will be out of phase by an amount $\Delta\phi$. Only the component of the field along the transmission axis of *A* can pass through. The intensity *I* of the detected beam (output beam) is then

$$I = I_o \sin^2\left(\frac{\pi}{2} \cdot \frac{V}{V_{\lambda/2}}\right)$$

where I_o is the light intensity under full transmission, and an applied voltage of $V_{\lambda/2}$ is needed to allow full transmission. In digital electronics, we would switch a light pulse on and off so that the nonlinear dependence of transmission intensity on *V* would not be a problem. However, if we wish to obtain a linear modulation between the intensity *I* and *V* we need to bias this structure about the apparent "linear region" of the curve at half-height. This is done by inserting a *quarter-wave plate* after the polarizer *P* which provides *circularly polarized light* as input. That means $\Delta\phi$ is already shifted by $\pi/4$ before any applied voltage. The applied voltage then, depending on the sign, increases or decreases $\Delta\phi$. The new transmission characteristic is shown as a dashed curve. We have, effectively, optically biased the modulator at point *Q* on this new characteristic.

Pockels effect is a change in the refractive index of a crystal due to the application of an external electric field, other than the field of the light wave. Suppose that *x*, *y* and *z* are the principal axes of a crystal with refractive indices n_1, n_2 and n_3 along these directions. For optically isotropic crystals, these would be the same whereas for a uniaxial crystal $n_1 = n_2 \neq n_3$. Suppose that we suitably apply a voltage across a crystal and thereby apply an external dc field E_a along the *z*-axis. In the Pockels effect, the field will modify the optical indicatrix. The exact effect depends on the crystal structure. For example, a crystal like GaAs, optically isotropic with a spherical indicatrix, becomes birefringent,

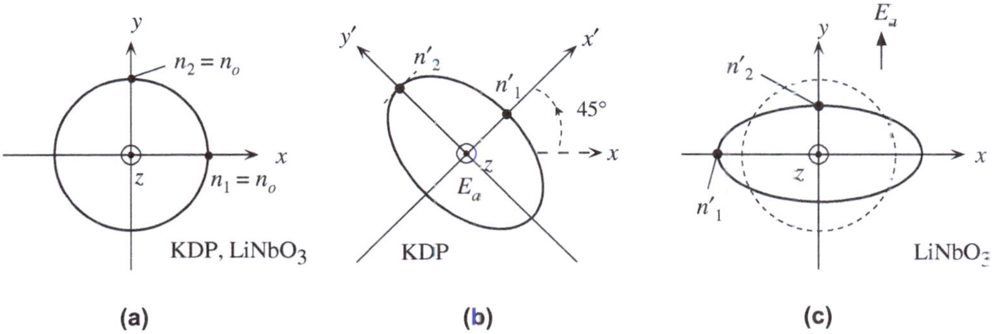

Pockels effect (a) Cross-section of the optical indicatrix with no applied field, $n_1 = n_2 = n_o$. (b) The applied external field modifies the optical indicatrix. In a KDP crystal, it rotates the principal axes by 45° to x' and y' and n_1 and n_2 change respectively to n'_1 and n'_2. (c) Applied field along y in $LiNbO_3$ modifies the indicatrix and changes n_1 and n_2 change to n'_1 and n'_2.

Friedrich Carl Alwin Pockels (1865–1913), son of Captain Theodore Pockels and Alwine Becker, was born in Vincenza, Italy. He obtained his doctorate from Göttingen University in 1888. From 1900 until 1913, he was a professor of theoretical physics in the Faculty of Sciences and Mathematics at the University of Heidelberg where he carried out extensive studies on electrooptic properties of crystals – the Pockels effect is the basis of many practical electrooptic modulators (Courtesy of the Department of Physics and Astronomy, University of Heidelberg, Germany.)

and a crystal like KDP (KH_2PO_4 – potassium dihydrogen phosphate) that is uniaxial becomes biaxial. In the case of KDP, the field E_a along z rotates the principal axes by 45° about z, and changes the principal indices. The new principal indices are now n'_1 and n'_2 which means that the cross-section is now an ellipse. Propagation along the z-axis under an applied field now occurs with different refractive indices n'_1 and n'_2. The applied field induces new principal axes x', y' and z' for this crystal (though in this special case $z' = z$). In the case of $LiNbO_3$ (lithium niobate), an optoelectronically important uniaxial crystal, the field E_a along the y-direction does not significantly rotate the principal axes but rather changes the *principal refractive indices* n_1 and n_2 (both equal to n_o) to n'_1 and n'_2. As an example consider a wave propagating along the z-direction (optic axis) in a $LiNbO_3$ crystal. This wave will experience the same refractive index ($n_1 = n_2 = n_o$) whatever the polarization. However, in the presence of an applied field E_a parallel to the principal y-axis, the light propagates as two orthogonally polarized waves (with fields parallel to x and y) experiencing different refractive indices n'_1 and n'_2. The applied field thus *induces a birefringence* for light traveling along the z-axis. The field induced rotation of the principal axes in this case, though present, is small and can be neglected. Before the field E_a is applied, the refractive indices n_1 and n_2 are both equal to n_o.

Table: Pockels effect Low- and high-frequency values of elements of the electrooptic tensor for various crystals. Combined from C. C. Davis, *Lasers and Electro-Optics*, Cambridge University Press, 1996, p. 478, and A. Yariv, *Optical Electronics in Modern Communications, 5th Edition*, Oxford University Press, 1997, p. 334. λ is in μm.

Material	Symmetry	n	Electrooptic coefficients (10^{-12} m/V) at specific wavelength (μm); low frequency	Electrooptic coefficients (10^{-12} m/V) at specific wavelength (μm); high frequency
KDP	$\bar{4}2$ m	$n_o = 1.5074$ $n_e = 1.4669$	$r_{63} = -10.5$; $r_{41} = 8.8(\lambda = 0.546)$	$r_{63} = 8.15$
ADP	$\bar{4}2$ m	$n_o = 1.5220$ $n_e = 1.4773$	$r_{63} = -8.5; r_{41} = 8.6$	$r_{63} = 5.5$
CDA (CsH_2AsO_4)	$\bar{4}2$ m	$n_o = 1.572$ $n_e = 1.550$ $(\lambda = 0.55)$	$r_{63} = 8.6$	
$LiNbO_3$	3m	$n_o = 2.286$ $n_e = 2.200$	$r_{33} = 32.2; r_{13} = 10$; $r_{22} = 7; r_{51} = 32(\lambda = 0.633)$	$r_{33} = 30.8; r_{13} = 8.6$; $r_{22} = 3.4; r_{51} = 28(\lambda = 0.633)$
$LiTaO_3$	3m	$n_o = 2.176$ $n_e = 2.180$		$r_{33} = 35.8; r_{13} = 7$; $r_{22} \approx 1; r_{51} = 20$ $(\lambda = 0.633)$
$BaTiO_3$	4mm	$n_o = 2.437$ $n_e = 2.365$ $(\lambda = 0.546)$	$r_{51} = 1640(\lambda = 0.546)$	$r_{51} = 820(\lambda = 0.546)$; $r_{33} = 28; r_{13} = 8(\lambda = 0.633)$
$PbTiO_3$	4mm			$r_{33} = 5.9; r_{13} = 13.8(\lambda = 0.633)$
$Ba_2NaNb_5O_{15}$	mm2		$r_{33} = 48; r_{13} = 15$; $r_{23} = 13; r_{42} = 92$; $r_{51} = 90(\lambda = 0.633)$	$r_{33} = 29; r_{13} = 7$; $r_{23} = 8; r_{42} = 75$; $r_{51} = 88(\lambda = 0.633)$
GaAs	$\bar{4}3$m	$n = 3.3(\lambda = 10.6)$	$r_{41} = 1.6(\lambda = 10.6)$	$r_{41} = 1.5(\lambda = 10.6)$
GaP	$\bar{4}3$m	$n = 3.32$	$r_{41} = -1.0$ $(\lambda = 0.55\text{–}1.3)$	$r_{41} = -0.97(\lambda = 0.633)$
CdTe	$\bar{4}3$m	2.60	$r_{41} = 6.8(\lambda = 10.6)$	
ZnSe	$\bar{4}3$m	$n = 2.60$	$r_{41} = 2.0(\lambda = 0.546)$	$r_{41} = 2.0(\lambda = 0.633)$
ZnS (α-ZnS, wurtzite)	6mm	$n_o = 2.347$ $n_e = 2.360$		$r_{33} = 1.8; r_{13} = 0.9(\lambda = 0.633)$
ZnTe	$\bar{4}3$m	$n = 2.99$	$r_{41} = 4.51(\lambda = 0.589)$; $r_{41} = 3.9(\lambda = 10.6)$	$r_{41} = 4.3(\lambda = 0.633)$
SiO_2(quartz)			$r_{11} = -0.47(\lambda = 0.5)$; $r_{41} = 0.2$	$r_{11} = 0.29(\lambda = 0.633)$
CuCl	$\bar{4}3$m		$r_{41} = 3.6(\lambda = 0.633)$; $r_{41} = 3.2(\lambda = 10.6)$	$r_{41} = 2.35(\lambda = 0.633)$; $r_{41} = 2.2(\lambda = 3.39)$

The Pockels effect then gives the new refractive indices n'_1 and n'_2 in the presence of E_a as

$$n'_1 \approx n_1 + \frac{1}{2}n_1^3 r_{22} E_a \quad \text{and}$$
$$n'_2 \approx n_2 - \frac{1}{2}n_2^3 r_{22} E_a$$

where r_{22} is a constant, called a *Pockels coefficient*, that depends on the crystal structure and the material. The reason for the seemingly unusual subscript notation is that there is more than one constant and these are elements of a tensor that represents the optical response of the crystal to an applied field along a particular direction with respect to the principal axes (the exact theory is more mathematical than intuitive). We therefore have to use the correct Pockels coefficients for the refractive index changes for a given crystal and a given field direction. If the field were along z, the Pockels coefficient above would be r_{13}. The Pockels effect involves examining the effect of the field on the indicatrix, that is, the index ellipsoid, and requires the electrooptic tensor. For example, the change in the principal refractive index n_1 along the principal axis x (where **D** and **E** are parallel) of the indicatrix is written as

$$\Delta n_1 = -\frac{1}{2}n_1^3 r_{11} E_1 - \frac{1}{2}n_1^3 r_{12} E_2 - \frac{1}{2}n_1^3 r_{13} E_3$$

where E_j is the field along j, j = 1,2,3 corresponding to x, y and z, and r_{ij} are the elements of the electrooptic tensor $\mathbf{r}$, a 6×3 matrix. It is clear that the control of the refractive index by an external applied field (and hence a voltage) is a distinct advantage which enables the phase change through a Pockels crystal to be controlled or modulated; such a phase modulator is called a *Pockels cell*. In the *longitudinal Pockels cell phase modulator* the applied field is in the direction of light propagation whereas in the *transverse phase modulator*, the applied field is transverse to the direction of light propagation. See *electrooptic effects*.

Point source is an ideal approximation for a "very small" luminous source, at the scale of a given optical system, that ideally emits spherical wavefronts. If d_x and d_y represent the lateral dimensions of a source at a distance L from the observer, the source would be considered a point source if $d_x d_y \ll L^2$. For instance, a distant star as seen through a telescope is very close to an ideal point source. The total power radiated by an oscillating dipole that has a maximum magnitude p_{max} is given by

$$P = \frac{p_{max}^2 \omega^4}{12\pi\varepsilon_0 c^3}$$

where ω is the angular frequency and other symbols have their usual meaning. The above equation is useful for approximately calculating the total power emitted by a point source and indicates that the total power is proportional to ω^4. It can be used to explain that short wavelength light is scattered more than long wavelength light by a small dielectric particle (smaller than the wavelength of light). When a passing electromagnetic wave excites a dipole oscillation in a small dielectric particle (or a molecule), the latter will emit like a quasipoint source with a power proportional to ω^4 or λ^{-4}. The result is a partial energy transfer from the propagating wave to the radiation emitted by the dielectric particle.

Point spread function see *impulse response function*.

Poisson distribution in photonics is a particular probability distribution of quanta, such as photons or electrons, that allows the number of quanta to be found in a certain time interval. Suppose that the average rate of quanta generation, or detection, is r. The probability $P(n, t)$ of finding n quanta in a time interval t is given by *Poisson statistics* as

$$P(n,t) = \frac{(rt)^n e^{-rt}}{n!}.$$

The average number of quanta generated or detected $\bar{n}$ is then the rate r times the time interval t; $\bar{n} = rt$ The variance σ^2, or the mean standard deviation σ, of a Poisson process (that which obeys the Poisson statistics) is given by

$$\sigma^2 = \overline{(n^2 - \bar{n}^2)} = \bar{n}.$$

If several processes contribute to a process, and if each of the constituent processes follows the

Poisson distribution, then the process itself follows the Poisson distribution,

$$\Sigma\text{Poisson process} = \text{Poisson process}$$

and the overall rate r is the sum of all individual rates r_i, i.e. $r = \Sigma r_i$.

Polariton is a quantum of a coupled phonon–photon field in a material. Coupling occurs predominantly for small wavevectors, q (as compared with the zone boundary), where $\omega_{\text{photon}} = cq_{\text{photon}} = \omega_{\text{phonon}} \approx 10^{13}\ \text{s}^{-1}$, and thus giving $q \approx 300\ \text{cm}^{-1}$. Photons and optical phonons interact strongly near the crossing of their respective $\omega(q)$ dispersion curves, which become modified into *polariton dispersion curves*. As seen in the figure, this occurs for uncoupled transverse and longitudinal optical phonons in the vicinity of q_{TO} and q_{LO}, respectively. This results in the polariton dispersion curve having two branches – upper and lower branches. Light dispersion curves are straight lines going through the origin, and in the case of the upper polariton branch become the limit curves for large q-values described by $\omega_{\text{photon}}(q) = (c/\sqrt{\varepsilon_\infty})q$ – that is, the upper polariton dispersion curve becomes photon-like for high q-values, and phonon-like for low q-values (approaching the constant value ω_{LO} for $q = 0$). A line defined by the slope of the lower polariton curve and passing through the origin defines another photon-like curve, but of lower slope value, corresponding to $\omega_{\text{photon}}(q) = (c/\sqrt{\varepsilon_s})q$. ($\varepsilon_\infty$ and ε_s are the high and low frequency relative permittivities.) Thus, the lower polariton dispersion curve is photon-like for small q-values, becoming phonon like in the limit of high q (saturating at the constant value of ω_{TO} for $q = \infty$). An estimate of the q-value for which the lower polariton changes from being photon-like to phonon-like, q_{TO}, with increasing q-value, is given by $\omega_{\text{photon}}(q_{\text{TO}}) = (c/\sqrt{\varepsilon_s})q_{\text{TO}} = \omega_{\text{TO}}$ or $q_{\text{TO}} = (\sqrt{\varepsilon_s}/c)\omega_{\text{TO}}$. Similarly, an estimate of the q-value for which the upper polariton changes from being phonon-like to photon-like, q_{LO}, with increasing q-value, is given by $\omega_{\text{photon}}(q_{\text{LO}}) = (c/\sqrt{\varepsilon_\infty})q_{\text{LO}} = \omega_{\text{LO}}$ or $q_{\text{TO}} = (\sqrt{\varepsilon_\infty}/c)\omega_{\text{LO}}$. Also note that for $\omega_{\text{TO}} < \omega < \omega_{\text{LO}}$ it can be shown that $q^2 < 0$, or that q is imaginary. This means that no wave can propagate – indeed, this describes an evanescent wave that decays exponentially into the crystal corresponding to total external reflection of such incident radiation. This is often referred to as Reststrahlen reflection.

The above discussion explains how optical phonons and photons with approximately the same energy and q-value can become strongly coupled. A similar coupling can occur between an exciton and a photon, and is commonly termed an exciton-polariton. An important case of such an excitation occurs in low-dimensional structures in which the exciton binding energy is enhanced within such structures due to the confinement of the respective charges as well as the peculiarities of screening in low-dimensional structures. Finally, cavity related effects which condition the optical field can also influence the polariton behavior.

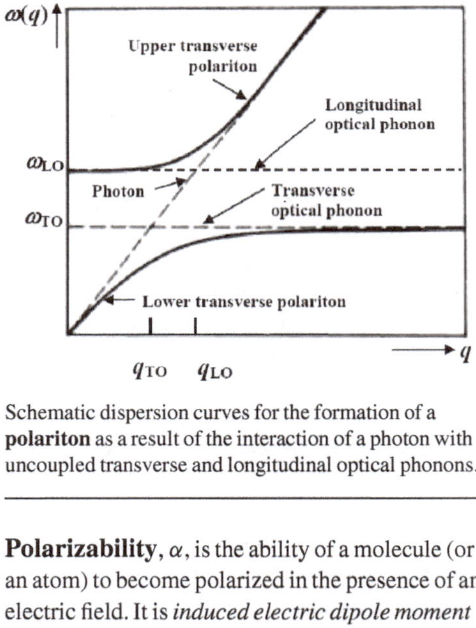

Schematic dispersion curves for the formation of a **polariton** as a result of the interaction of a photon with uncoupled transverse and longitudinal optical phonons.

Polarizability, α, is the ability of a molecule (or an atom) to become polarized in the presence of an electric field. It is *induced electric dipole moment* in the molecule per unit electric field acting on the molecule. Suppose that $\mathbf{E}_{\text{local}}$ is the actual, local

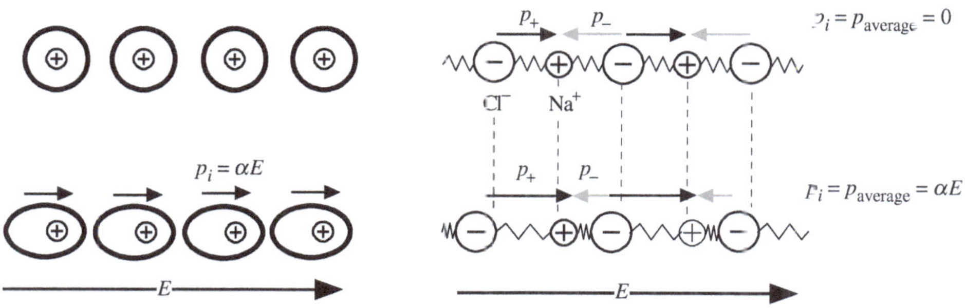

Electronic polarization Ionic polarization

Electronic and ionic polarization and different origins of **polarizability**.

field acting on a molecule in a material. $\mathbf{E}_{local}$ is usually not the same as the macroscopic field in the material but somewhat higher. In response to this local field, the molecule becomes polarized and develops an induced *dipole moment* $\mathbf{p}_i$

$$\mathbf{p}_i = \alpha \mathbf{E}_{local},$$

where α is the *polarizability* of the molecule; there is an induced displacement of the centers of the negative and positive charges in the molecule. The *polarization* **P** induced in a medium is simply $\mathbf{P} = N\mathbf{p}_i$ where N is the number of molecules or atoms per unit volume. In *electronic polarizability*, the charge displacement is due to electrons being displaced from their equilibrium distribution in the molecule. In a crystal, valence electrons can also become displaced in so-called bond polarization. In *ionic polarizability,* the ions in the crystal are displaced with respect to their equilibrium positions. There are other polarization mechanisms but electronic and ionic are usually the most important contributions in insulators in many photonics applications. (In a conducting medium, such as a conductor or a highly doped semiconductor, the displacement of the free electrons with respect to the positive ionic cores can also give rise to the polarization of the medium but this is excluded in insulators and high-resistivity semiconductors). Once the polarizability per atom, per molecule or per ion in a material is known, the relative permittivity ε_r and the refractive index n can be found by using the Clausius–Mossotti and the Lorentz–Lorenz equations; the latter relate the microscopic description of polarization embedded in α to the macroscopically measured quantities ε_r and n. For example, one can use the *Lorentz oscillator model* to describe α, and then use this model to predict n. The definition of polarizability is the same in both CGS and SI but the conversion is $\alpha(\mathrm{SI}) = 4\pi\varepsilon_0(10^{-6})\alpha(\mathrm{CGS})$. In general α is a 3×3 matrix but becomes a scalar quantity for gases, liquids, noncrystalline solids (e.g. glasses) and cubic crystals. Further, the dependence of p_i on E_{local} may not be linear at high fields, in which case higher terms in the field such as $\beta E_{local}^2 + \gamma E_{local}^3$ are also needed to describe p_i. The constants β and γ in these terms are called hyperpolarizabilities and represent nonlinear effects. See *hyperpolarizability, Lorentz oscillator model, Clausius–Mossotti equation, Lorentz–Lorenz equation, electric dipole moment.*

Polarization, or **state of polarization**, in optics normally refers to the behavior of the electric field vector **E** as light propagates a distance equivalent to a wavelength. A propagating electromagnetic

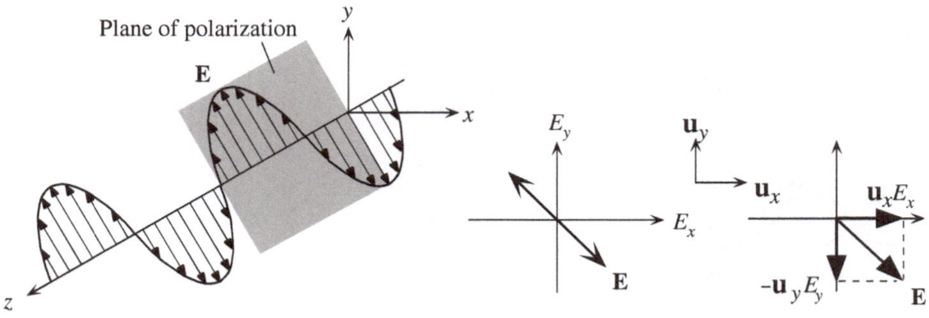

Polarization (a) A linearly polarized wave has its electric field oscillations defined along a line perpendicular to the direction of propagation, z. The field vector **E** and z define the plane of polarization. (b) The **E**-field oscillations are contained in the plane of polarization. (c) Linearly polarized light at any instant can be represented by the superposition of two fields $\mathbf{E}_x$ and $\mathbf{E}_y$ with the right magnitude and phase.

(EM) wave has its electric and magnetic field at right angles to the direction of propagation. If we place the z-axis along the direction of propagation, then the electric field can be in any direction in the plane perpendicular to the z-axis. The term *polarization* of an EM wave describes the behavior of the electric field vector in the EM wave as it propagates through a medium over a distance equivalent to a whole wavelength. If the oscillations of the electric field at all times are contained within a well-defined line then the EM wave is said to be *linearly polarized* as in the figure. The field vibrations and the direction of propagation (z) define a plane of polarization (plane of vibration) so that linear polarization implies a wave that is *plane-polarized*. By contrast, if a beam of light has waves with the **E**-field each in a random direction but perpendicular to z, then this light beam is *unpolarized*. A light beam can be linearly polarized by passing the beam through a *polarizer*, such as a polaroid sheet, a device that only passes electric field oscillations lying on a well-defined plane at right angles to the direction of propagation.

Suppose that we arbitrarily place the x and y axes and describe the electric field in terms of its components E_x and E_y along x and y (E_x and E_y are perpendicular to z). To find the electric field in the wave at any space and time location, we add E_x and E_y *vectorially*. Both E_x and E_y can individually be described by a wave equation which must have the same angular frequency ω and wavenumber k. However, we must include a phase difference ϕ between the two so that $E_x = E_{xo}\cos(\omega t - kz)$ and $E_y = E_{yo}\cos(\omega t - kz + \phi)$ where ϕ is the phase difference between E_y and E_x; ϕ can arise if one of the components is delayed (retarded). In the figure, the linearly polarized wave has the **E** oscillations at $-45°$. We can generate this field by choosing $E_{xo} = E_{yo}$ and $\phi = \pm 180°$ ($\pm\pi$), i.e. E_x and E_y have the same magnitude but they are out of phase by 180°. If $\mathbf{u}_x$ and $\mathbf{u}_y$ are the unit vectors along x and y, using $\phi = \pi$, the field in the linearly polarized wave is

$$\begin{aligned}\mathbf{E} &= \mathbf{u}_x E_x + \mathbf{u}_y E_y \\ &= \mathbf{u}_x E_{xo}\cos(\omega t - kz) \\ &\quad - \mathbf{u}_y E_{yo}\cos(\omega t - kz) \\ &= \mathbf{E}_o\cos(\omega t - kz)\end{aligned}$$

where $\mathbf{E}_o = \mathbf{u}_x E_{xo} - \mathbf{u}_y E_{yo}$, that is, $\mathbf{E}_o$ is at $-45°$ to the x-axis (since $E_{xo} = E_{yo}$) and propagates along the z-direction. There are many choices for the behavior of the electric field besides the simple linear polarization in this example. In *circular polarization* there is a phase difference of $\pm\pi/2$, between E_x and E_y and $|E_x| = |E_y|$. A more general state of polarization can be described in terms of the *polarization ellipse* in which **E** traces out an ellipse as the wave propagates. See *elliptical polarization.*

Polarization of a dielectric is the separation of positive and negative charges in a system so that there is a *net* electric dipole moment per unit volume. Put differently, it is a phenomenon in which, when a dielectric material is placed in an electric field **E**, it develops polarization, that is, the surface of the dielectric acquires positive charge on one side and negative charge on the other side; these are bound charges, but separated so that there is a net dipole for the whole dielectric. In insulators, the most important polarization mechanisms are electronic and ionic polarization. A free-carrier contribution to polarization can occur in highly doped semiconductors. Polarization is measured by the *polarization vector* **P**. See *polarization vector, polarizability.*

Polarization angle see *Brewster's angle.*

Polarization by scattering is a phenomenon in which the polarization of the scattered light wave depends on the direction of scattering and the polarization of the incoming light wave. An oscillating electric dipole emits electromagnetic radiation which has a nonspherical electric field (E) pattern around it as shown in the first figure for an oscillating *electric dipole moment* $p(t)$ parallel to the y-axis. There is no field radiation along the dipole axis y. The irradiance I of the radiation along a direction at an angle θ to the perpendicular to the dipole axis is proportional to $\cos^2\theta$. The electric field in an incoming electromagnetic wave will induce dipole oscillations in a molecule of the medium. The scattering of an incident unpolarized electromagnetic wave by this molecule leads to waves with different polarizations along the x and y axes.

A passing light beam through the molecule can be decomposed into the two fields, E_x and E_y as in the second figure. E_x polarizes the molecule along x and E_y along y. The field at any distance along y is due to E_x and is parallel to y, and the field at any distance along x is due to E_y and is parallel to y. Thus, the EM waves scattered along the x- and y-directions are polarized. The fractional extent of polarization, $\eta_{\text{polarization}}$, for a spherical molecule (or a tiny spherical particle) is given by

$$\eta_{\text{polarization}} = \frac{1 - \cos^2\theta}{1 + \cos^2\theta}$$

where θ is the angle between the incident and scattered waves. The transmitted beam for which $\theta = 0$, has $\eta_{\text{polarization}} = 0$, no polarization, whereas the 90° scattered light has $\eta_{\text{polarization}} = 1$, full polarization.

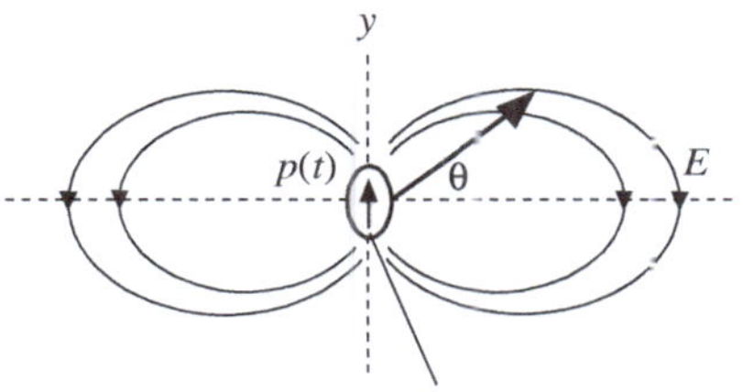

Polarization by scattering A snapshot of the field pattern around an oscillating dipole moment in the y-direction. Maximum electromagnetic radiation is perpendicular to the dipole axis and there is no radiation along the dipole axis. This is a polar plot.

Polarization ellipse see *elliptically polarized light.*

Polarization-maintaining fiber (PMF) is an optical fiber in which the polarization state of propagating light is maintained along the length of the fiber. A main cause for the loss of the polarization state is perturbing birefringence which may occur due to internal perturbations from defects including non-ideal core and/or cladding geometry, or externally applied bends, twists and other distortions of the fiber. This effect results in difficulties in fiber

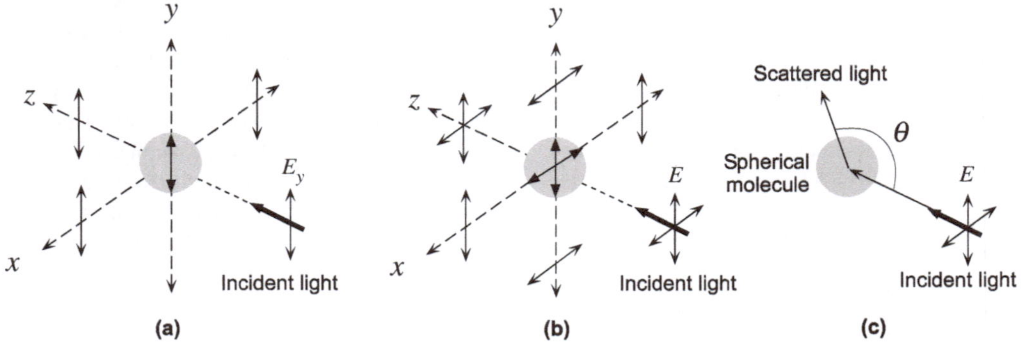

Polarization by scattering (a) There is no field anywhere along the dipole axis if the dipole oscillates along the y-axis. (b) Polarization by scattering. (c) Definition of the scattering angle θ.

pigtailing for integrated optical devices, in sensors, as well as polarization fading in interferometers, for example. A number of fiber structures have been developed to preserve the polarization state. The principle of these approaches is to limit the propagation of one polarization state. The latter is achieved by making the PMF as a birefringent fiber where the difference in refractive indices between orthogonal polarization states has been purposely made very large. There are two basic types of polarization-maintaining fibers (PMFs). First is the geometrical or form birefringence, in which the fiber core has been made to be elliptical or has core-vicinity structures as in the first figure. A purely geometrical birefringence type fiber is very difficult to manufacture because there are always some non-symmetrical stress distributions when there is axial nonsymmetry. Maximum birefringence for a geometric type fiber corresponds to the ratio is $(a/b - 1)$ being approximately 5, where a and b denote the ellipse's axes. The second type is stress induced birefringence, in which the birefringence is generated by nonsymmetrical stresses acting on the circular core. This is achieved by the selected regions applying a large stress (or strain) to the core. For example, diametrically opposite regions are silica doped with B_2O_3 glasses which have large thermal expansion coefficients and therefore apply a large tensile force on the axis with the doped cladding and a compressive force on the orthogonal axis after the fiber has been drawn and cooled. This causes internal mechanical stresses and therefore birefringence inasmuch as the refractive index of the glass is changed by the *photoelastic* effect. There are several types of these fibers with four of the most common ones used in optical communications depicted in the second figure. Two successful structures that maintain polarization are the so-called PANDA and Bow-Tie fibers, which can provide a high birefringence, without compromising on loss or cross-talk. In this case, strain generating rods of material are introduced aligned along the diameter of the fiber to generate a stress pattern across the core, as shown in the figure. This stress pattern causes a natural separation between the slow and fast axes for the propagating light. Inasmuch as polarization-maintaining fibers are axially birefringent, the refractive indices are different for two directions along the fiber. By design, these refractive indices of the core are orthogonal to each other and therefore have a distinct fast and slow principal optical axis. A polarization-maintaining fiber is

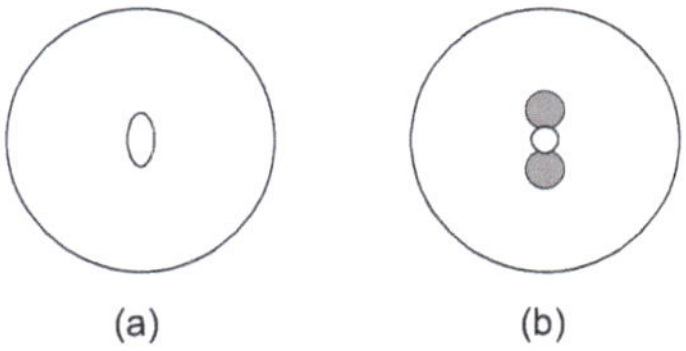

Geometrical **polarization-maintaining fibers**: (a) elliptical core and (b) core-vicinity structures that have a low refractive index.

typically used to guide linearly polarized light from point to point. Only when the electric field of the light is entirely aligned with the slow or fast optical axis is a polarization-maintaining fiber able to maintain polarization. A polarization-maintaining fibers is characterized by its beat length, which corresponds to the separation between two points along the fiber for which the state of polarization repeats itself. Typical single-mode fibers have a beat length that ranges from a few centimeters to a few meters, whereas polarization-maintaining fibers can have a beat length as short as a few millimeters. The *polarization beat length* quantifies the linear birefringence in fibers and is defined through

$$L_{\mathrm{B}} = \frac{2\pi}{|\beta_x - \beta_y|}$$

where β_x and β_y are the orthogonal propagation constants. Polarization cross-talk is a figure of merit for the confinement of light to a single axis of a PMF and is given by

$$\text{Polarization cross-talk} = 10 \log\left(\frac{P_{\min}}{P_{\max}}\right).$$

The power level $P_{\min}$ indicates the amount of light that has coupled into the undesired axis of the polarization-maintaining fiber and $P_{\max}$ indicates the amount of light that has been coupled into the desired axis. (Dennis Derickson, *Fiber Optic Test and Measurement*, Prentice-Hall, 1997; Chin-Lin Chen, *Elements of Optoelectronics and Fiber Optics*, McGraw-Hill, 1996).

Polarization mode dispersion arises when the fiber is not perfectly symmetric and homogeneous, that is, the refractive index is *not isotropic*. When the refractive index is different for different directions of the electric field, the propagation constant of a given mode depends on its polarization. Due to various variations in the fabrication process, such as small changes in the glass composition, geometry and induced local strains (either during fiber drawing or cabling), the refractive indices n_1 (core) and n_2 (cladding) may not be isotropic. Suppose that n_1 has the values n_{1x} and n_{1y} when the electric field is parallel to the x- and y-axes, respectively. Then the propagation constant for fields along x and y would be different, $\beta_x(01)$ and $\beta_y(01)$ (01 represents the fundamental mode.), leading to different *group delays* and hence to dispersion even if the source wavelength is monochromatic. The situation in reality is more

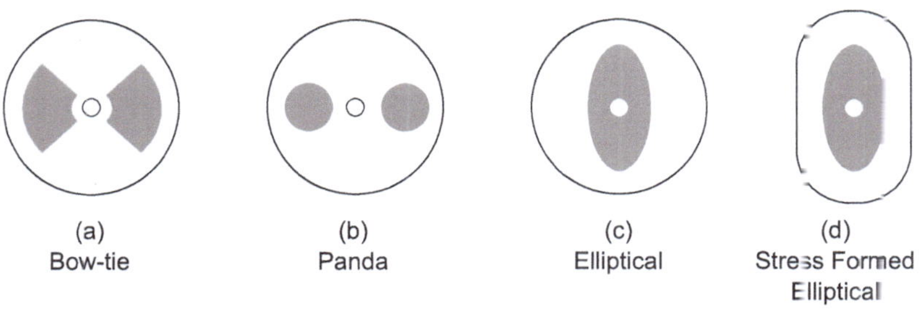

Stressed induced **polarization-maintaining fibers**.

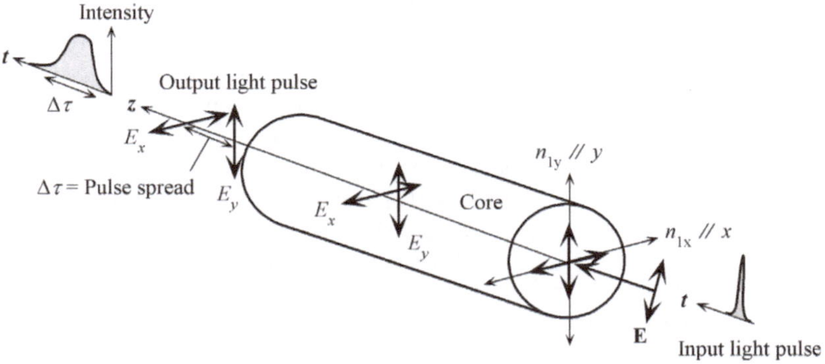

Polarization mode dispersion The core refractive index has different values along two orthogonal directions corresponding to two orthogonal electric field oscillation directions (polarizations). Take the *x*- and *y*-axes along these directions. Input light will travel along the fiber with E_x and E_y polarizations having different group velocities and hence arrive at the output at different times which leads to the dispersion of the light pulse. (The explanation is over simplified.)

complicated because n_{1x} and n_{1y} would vary along the fiber length and there will be an interchange of energy between these modes as well. The final dispersion nonetheless depends on the extent of anisotropy, $n_{1x}-n_{1y}$, which is kept minimum by various manufacturing procedures (such as rotating the fiber during drawing, etc.). Typically polarization mode dispersion is less than a fraction of a picosecond per kilometer of fiber. However, this type of dispersion does not scale linearly with fiber length L; in fact, the dispersion scales roughly as $L^{1/2}$ and the polarization mode dispersion in current fibers is typically 0.05–0.5 ps km$^{-1/2}$.

Polarization prisms are polarizing optical components that normally use two prisms, made from birefringent crystals, to split an incident unpolarized light beam into two differently and linearly polarized beams. The beams come out in different directions and hence the incident beam has been effectively split into two differently polarized beams. One of the beams is usually fully polarized and has the electric field oscillations along a direction that depends on the type of prism that has been used and the alignments of the optic axes. The other beam may or may not be fully polarized, e.g. in the Wollaston prism, the two beams are orthogonally polarized. See *Glan–Foucault prism, Nicol prism, Wollaston prism.*

Polarization vector, P, in a dielectric measures the extent of polarization in a unit volume of dielectric matter. It is the vector sum of dielectric dipoles per unit volume. If **p** is the average dipole moment per atom or per molecule and N is the number of atoms or molecules per unit volume that can be polarized (or have permanent dipole moments), then by definition $\mathbf{P}=N\mathbf{p}$. In polarized dielectric matter (e.g. in an electric field), the bound surface charge density σ_p due to polarization is equal to the normal component of **P** at that point, $\sigma_p=\mathrm{P}_{\text{normal}}$. The polarization **P** can also be induced by the application of an electric field; the field induces an average dipole moment of **p** in each molecule and the induced polarization $\mathbf{P}=N\mathbf{p}_i$. The electric susceptibility χ of the medium relates the induced polarization **P** to the electric field **E** in the medium. For isotropic linear media (gases, liquids, noncrystalline solids and cubic crystals),

$$\mathbf{P}=\varepsilon_0\chi\mathbf{E}.$$

In an anisotropic crystal, χ is a tensor or a 3×3 matrix. See *electric susceptibility, polarizability, relative permittivity*.

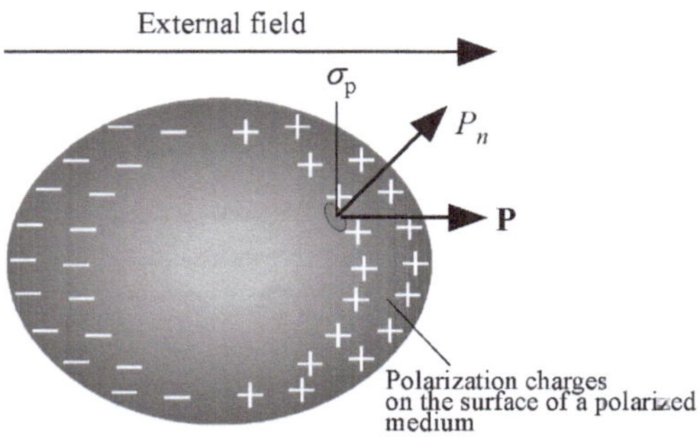

Polarization vector A dielectric medium in an electric field become polarized, and develops polarization charges on its surfaces.

Polarized EM wave (polarized light) is an EM (electromagnetic) wave which has the oscillations of the electric field in a well-defined direction (but still at right angles to the direction of propagation). In unpolarized light the field oscillations are in random directions but once polarized the field oscillations are along a well-defined direction. If this direction does not change as the wave propagates it is said to be *plane polarized*.

Polarizing prism is a composite prism made of birefringent prisms to separate an incident beam into two orthogonally polarized beams, or to pass or deflect an incident beam of particular polarization. See *Glan–Foucault, Wollaston, Nicol, Fresnel prism*.

Polaroid sheet see *linear polarizer*.

Polysilicon (Poly) is a type of silicon which is not a single-crystal, but consists of randomly oriented grains. Such material is typically produced by chemical vapor deposition (CVD) from a silicon gas source such as silane, on to silicon dioxide or silicon nitride.

Popcorn noise see *excess noise*.

Population inversion is the phenomenon of having more atoms (or ions) in a given material system occupy an excited energy level E_2, than a lower energy level E_1, so that the normal equilibrium distribution is reversed, i.e. $N(E_2) > N(E_1)$. Population inversion occurs as a result of the excitation of a medium, which is called *pumping*. If left on its own the medium will eventually, however long, return to its equilibrium population distribution with more atoms at E_1 than at E_2. For gas atoms, this means $N(E_2)/N(E_1) = \exp[-(E_2-E_1)/k_B T]$, where k_B is the Boltzmann constant and T is the temperature (kelvins). In a medium with population inversion, the temperature T becomes "negative," which actually indicates population inversion. When population inversion is achieved, the absorption of photons of energy $h\nu_0 = E_2 - E_1$ that excites atoms from E_1 to E_2 is highly diminished. If the pumped medium is now subject to a flux of photons of energy $h\nu_0 = E_2 - E_1$, there will be negligible absorption but, instead, the atoms at E_2 will undergo *stimulated emissions*, and fall down to E_1 by the emission of photons in phase with incoming stimulating photons. The pumped medium will then exhibit an optical gain. Population inversion is an essential requirement for the operation of optical amplifiers and lasers both of which must be suitably pumped. See *LASER, stimulated emission, threshold gain*.

Porous silicon is a form of silicon in which pores are intentionally introduced. The field grew from early work on electrochemically etched silicon in which luminescence was observed, but translating these findings into device applications has proven difficult, and current interest in porous silicon relates to the ability to tune the refractive index of the material by adjusting pore size and spacings, and to forming ordered porous structures as photonic crystals. The first figure shows an example of the use of macroporous silicon (MPSi) membranes for use as UV filters. This is an area where current multilayer interference stacks perform poorly as

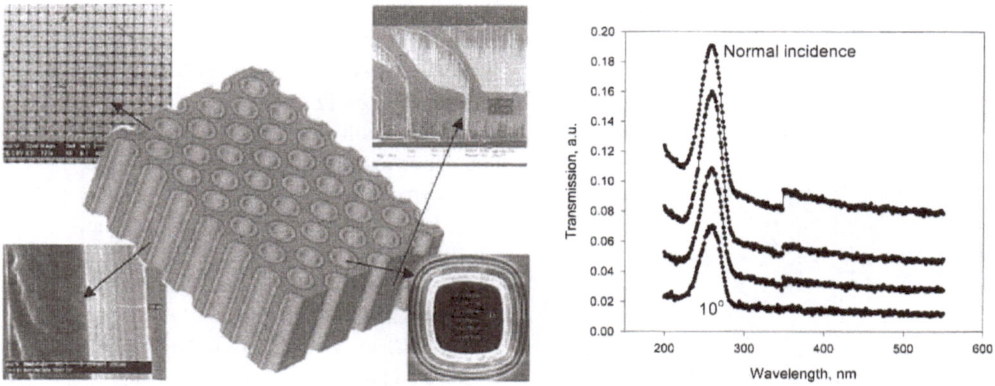

Porous silicon Left: an illustration of the structure of a macroporous silicon (MPSi) UV filter. Right: experimental transmission spectra through the MPSi membrane with multilayer pore wall coating taken at different incidence angles. UV pass is apparent. (After V. Kochergin and H. Foell, *Phys. Stat. Sol.* (c), **4**(6), 1933, 2007.)

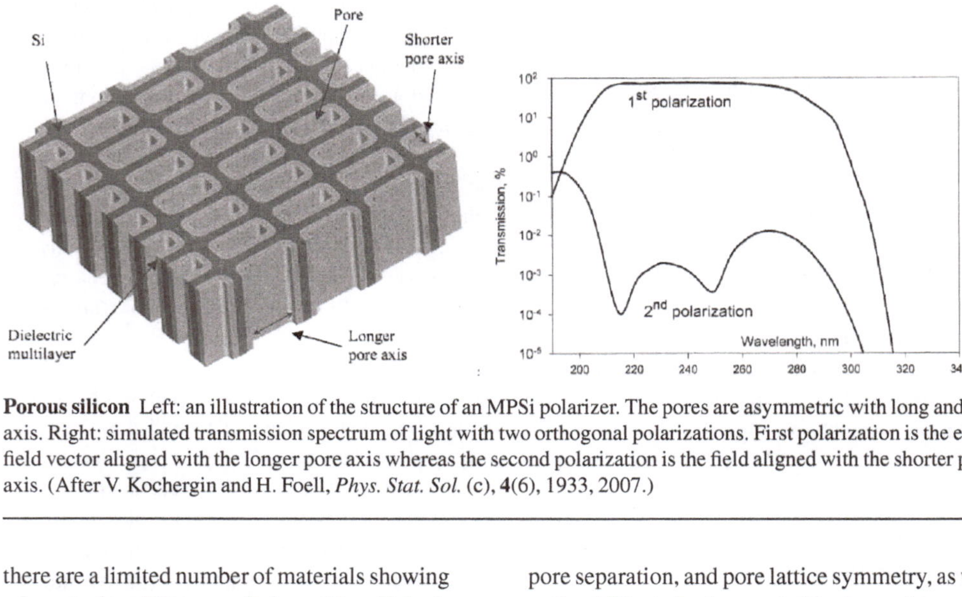

Porous silicon Left: an illustration of the structure of an MPSi polarizer. The pores are asymmetric with long and short axis. Right: simulated transmission spectrum of light with two orthogonal polarizations. First polarization is the electric field vector aligned with the longer pore axis whereas the second polarization is the field aligned with the shorter pore axis. (After V. Kochergin and H. Foell, *Phys. Stat. Sol.* (c), **4**(6), 1933, 2007.)

there are a limited number of materials showing adequate deep UV transmission with sufficient refractive index contrast and environmental stability to provide a high transmission passband with sufficient sideband rejection. Ordered porous silicon membrane structures permit engineered optical response by control over the pore diameter, inter-pore separation, and pore lattice symmetry, as well as the ability to further control leaky mode propagation in such structures by coating pore walls with a multilayer dielectric stack. The first figure shows that such an approach can enable a very wide and deep rejection band to be achieved. Another interesting application of macroporous silicon is

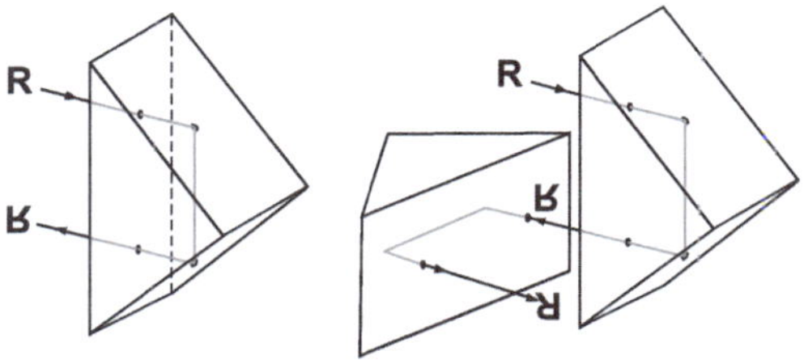

The **Porro prism** flips the image by 180°. A double Porro prism can flip the image both vertically and horizontally.

for optical polarizers in the deep UV where there are again few choices currently. The second figure shows how using similar ordered pore arrays with multilayer coated pores, in which the pores are highly asymmetric in their cross-sectional shape, one can provide for a high polarization selectivity.

Porro prism is a special type of prism that is able to reflect the image and at the same time rotate it through 180°, that is flip the image vertically (or horizontally); it is named after Ignazio Porro, an Italian engineer (1801–1875). The light beam enters the prism at normal incidence through the large face of the prism and undergoes two total internal reflections and exits out from the large face but displaced with the respect to the incoming beam. It is used in various optical instruments, including binoculars to invert the image. Two Porro prisms can be used to fully flip the image vertically and horizontally. See *Abbe–Porro prism*.

Posener width see *Voigt lineshape*.

Positive uniaxial see *principal refractive indices*.

Potassium dihydrogen phosphate, KH_2PO_4, see *KDP*.

Potassium titanyl phosphate see *KTP*.

Power spectral density see *noise*.

Poynting vector, S, represents the instantaneous energy flow per unit time per unit area in a direction determined by $\mathbf{E} \times \mathbf{B}$ (direction of propagation), which in free space is given by $\mathbf{S} = (c^2\varepsilon_0)\mathbf{E} \times \mathbf{B}$. Its magnitude, power flow per unit area, is called the instantaneous *irradiance* or intensity of the wave. In a linear and isotropic medium of relative permittivity ε_r and refractive index n ($\varepsilon_r = n^2$), the Poynting vector becomes $\mathbf{S} = v^2\varepsilon_0\varepsilon_r\mathbf{E} \times \mathbf{B}$ where v is the phase velocity in the medium. A more convenient and simpler notation uses **H**, the magnetizing field or the magnetic field intensity, instead of **B**, and expresses **S** simply as

$$\mathbf{S} = \mathbf{E} \times \mathbf{H}.$$

See *irradiance*.

Preform is a glass rod which has an appropriate *core* and *cladding* from which an optical fiber is drawn by a fiber drawing process at an elevated temperature.

Pressure broadening see *impact broadening*.

Primary color is a fundamental or essential color that cannot be simply obtained by mixing other primary colors. For example, red R, green G and blue B are three primary colors that can be added in various proportions to match a wide range of colors that are contained within the triangle RGB in the chromaticity diagram. See *color, color matching functions*.

Primary or **unmultiplied photocurrent** is normally the photocurrent in an avalanche photodiode just below the onset of avalanche multiplication, i.e. when the reverse bias is sufficient to reach through (the depletion region reaches through the device) but not sufficiently high to cause avalanche breakdown.

Principal axes of a crystal are three appropriate orthogonal directions, say x, y and z, corresponding to the three principal refractive indices n_1, n_2 and n_3.

Principal or chief ray is the ray that goes from the object to image through the center of the lens or the center of the optical aperture.

Principal points see *cardinal point*.

Principal quantum number, n, is a quantum number with integer values $1, 2, 3, \ldots$, which characterizes the total energy of the electron. The energy increases with n. It determines, with the other quantum numbers ℓ and m_ℓ, the orbital of the electron in an atom, $\psi_{n,\ell,m_\ell,m_s}(\rho,\theta,\phi)$. $n = 1, 2, 3, 4, \ldots$ are labeled as $K, L, M, N, \ldots$ shells within each of which there may be subshells based on $\ell = 0, 1, 2, \ldots, (n-1)$ and corresponding to *s, p, d,*... states.

Principal refractive indices of a crystal, according to all experiments and theories, are a set of three refractive indices n_1, n_2 and n_3 (along three mutually orthogonal directions in the crystal, say x, y and z called the principal axes) which are necessary to characterize light propagation even in the most anisotropic crystal. These indices correspond to the polarization state of the wave along these axes. Crystals that have three distinct principal indices also have *two* optic axes and are called *biaxial crystals*. On the other hand, *uniaxial crystals* have two of their principal indices the same ($n_1 = n_2$) and only have *one* optic axis. Uniaxial crystals, such as quartz, that have $n_3 > n_1$ and are called *positive*, and those such as calcite that have $n_3 < n_1$ are called *negative uniaxial* crystals.

Prism is an optical component that is generally made of a transparent material and shaped in such a way to deflect an incident beam of light, either through refractions, or by one or more total internal reflections (TIRs), or both. The deflection depends not only on the incidence angle of the light beam on the prism, the prism material and geometry, but also on the wavelength and the polarization state of the incident light, because the refractive index n of the prism material normally depends on the wavelength, and further, for certain materials (e.g. quartz, calcite, etc.), it depends on the polarization of light (with respect to the optic axis of the prism material). Prisms are usually classified into *refracting* type or *reflecting* type, depending on whether refractions or TIRs are the main deflection mechanism, though some prisms, usually composite prisms, need both refractions and TIR to fulfill their purpose as in the case of the Glan–Foucault prism which passes through linearly polarized light. For a refractive prism, the minimum angle of deflection δ depends on the angles of incidence and refraction, θ_i, and θ_r respectively

$$\delta_{\min} = 2\theta_i - 2\theta_r.$$

The minimum deflection condition can also be expressed in terms of the prism angle α, and the refractive index n,

$$n \sin(\alpha/2) = \sin[(\alpha/2) + (\delta_{\min}/2)]$$

which can be solved to find $\delta_{\min}$; or as is more common, n from the measurement of $\delta_{\min}$. Under minimum deflection, the light passes through the prism

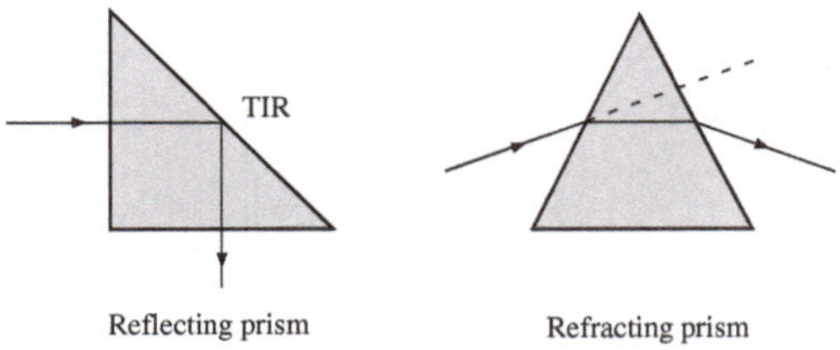

Basic types of **prism**.

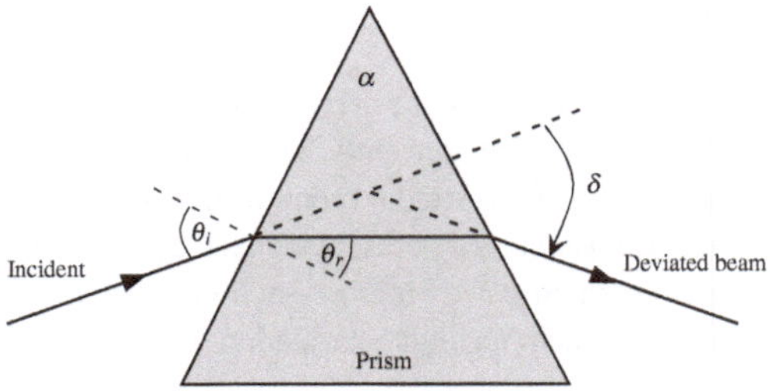

Deviation, or deflection, of a light beam by a **prism**. The angle of deviation is δ.

symmetrically. If $\theta_i < 0.1$ rad, and $\alpha < 0.1$ rad, then the deviation angle δ takes the simple form,

$$\delta = (n - 1)\alpha$$

and shows that all rays with small angles of incidence suffer the same deflection.

The power of a prism, Δ, is defined as

$$\Delta = 100 \tan \delta$$

and is measured in *diopters*, D, where 1 D is a deflection of 1 cm at a distance of 100 m. The resolving power of a prism of base length b is given by

$$\frac{\lambda}{\Delta\lambda} = b \left| \frac{dn}{d\lambda} \right| .$$

See *polarizing prisms*, *Glan–Foucault prism*.

Profile dispersion is the broadening of a propagating optical pulse in a fiber as a result of the group velocity $v_g(01)$ of the fundamental mode (01) also depending on the refractive index difference Δ, i.e. $\Delta = \Delta(\lambda)$. If Δ changes with wavelength, then different wavelengths from the source would have different group velocities and experience different group delays leading to pulse broadening. It is part of chromatic dispersion because it depends on the width of the input spectrum $\Delta\lambda$. The profile dispersion coefficient D_p is the time spread of a propagating optical pulse in an optical guide per unit length and per unit spectral wavelength width due to the wavelength dependence of the refractive index difference. Typically, D_p is less than 1 ps $nm^{-1}km^{-1}$.

Profile index or **index grading coefficient,** γ, is a parameter that describes the variation of the refractive n with radial distance r from the fiber axis assuming that for $r < a$, the refractive index decreases with r as $n = n_1[1 - 2\Delta(r/a)^\gamma]^{1/2}$, where a is the fiber core region radius, Δ is the normalized refractive index difference and n_1 is the maximum refractive index at the fiber axis. At $r = a$, $n = n_2$ and is constant.

Propagation constant, β, is a property of a wave that, for a given frequency and at a given time, represents the change in the phase per unit distance. Since this phase change is 2π over a wavelength λ, $\beta = 2\pi/\lambda$. The propagation constant β and wavevector **k** represent essentially the same property except that **k** has a direction associated with it. See *complex propagation constant*.

Pumping see *LASER, population inversion*.

Pumping in lasers involves exciting atoms from their ground states to states at higher energies.

Purkinje shift refers to the shift in the eye's spectral peak response towards the blue (shorter wavelengths) when the illumination is dimmed. It is named after a Czech physiologist who reported that red colors fade faster than blue colors when the ambient light fades. The shift in the eye's responsivity as the lights are dimmed is due to the difference in the spectral luminous efficiency of the light- and dark-adapted eye, which have their peaks at 555 nm and at about 507 nm respectively (photopic responsivity vs. scotopic responsivity).

Pyroelectric coefficient see *pyroelectricity, pyroelectric detector*.

Pyroelectric detector is an *infrared detector* that uses the pyroelectric effect (*pyroelectricity*) to detect and measure radiation by virtue of the heating effect of an absorbed radiation. Any infrared radiation that can raise the temperature of a pyroelectric crystal, even by a thousandth of a degree, can in principle be detected. For example, many intruder alarms use pyroelectric detectors because as the human or animal intruder passes by the view of the detector the infrared radiation from the warm body raises the temperature of the pyroelectric detector which generates a voltage that actuates an alarm. In a typical pyroelectric radiation detector, the detecting element, labeled as A in the figure, is actually a thin crystal or ceramic (or even a polymer) of a pyroelectric material that has electrodes on opposite faces. Pyroelectric materials are also piezoelectric and therefore also sensitive to stresses. Thus pressure fluctuations, e.g. vibrations from the detector

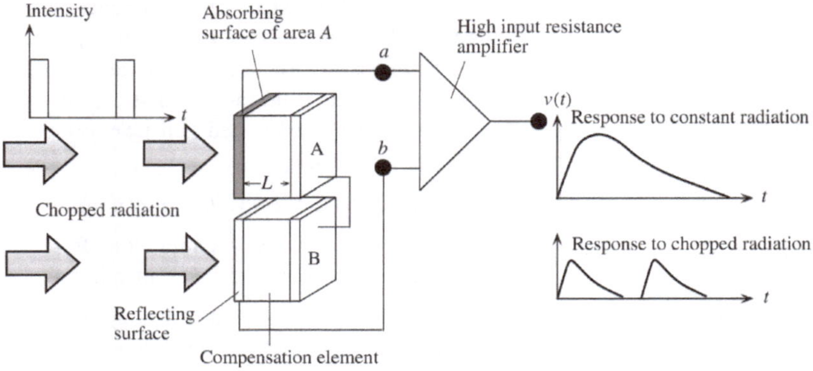

A simple **pyroelectric detector** concept.

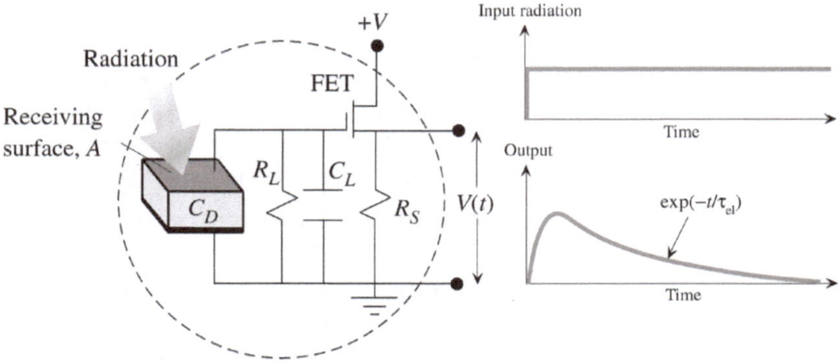

A single-element **pyroelectric detector** with an FET voltage follower circuit.

mount or sound waves, interfere with the response of the detector to radiation alone. These can be compensated for by having a second dummy detector (*compensation* element), B, which has a reflecting coating and is subjected to the same vibrations (air and mount). Thus there are two elements in the detector, one with an absorbing surface, detecting element (A), and the other with a reflecting surface, compensating element (B). Stress fluctuations give rise to the same piezoelectric voltage in both, which then cancel each other between a and b at the input of the amplifier. When radiation is incident then only the detecting element absorbs the radiation, becomes warmer and hence generates a pyroelectric voltage. This voltage appears directly across a and b. As the incident radiation warms the detecting element (A) and increases its temperature, the pyroelectric voltage increases with time. Eventually the temperature reaches a steady state value determined by heat losses from the element. We therefore expect the pyroelectric voltage to reach a constant value as well. However, a constant pyroelectric voltage cannot be sustained because the surface charges slowly become neutralized or leak away. The constant radiation is therefore normally chopped to subject the detector to periodic bursts of radiation as shown in the figure. The pyroelectric voltage is then a changing function of time which is readily measured and relates to the power in the incident radiation. If, upon the absorption of radiation, the temperature of the pyroelectric crystal changes by ΔT, then the induced polarization is $\Delta P = p\Delta T$, where p is the *pyroelectric coefficient*. ΔP induces a voltage ΔV across the electrodes of the detector. Suppose that the radiation of intensity I is received during a time interval Δt, and delivers an amount of energy ΔH to the pyroelectric detector. ΔH, in the absence of any heat losses, increases the temperature by ΔT. The *pyroelectric voltage* ΔV across the electrodes of the detector of thickness L is

$$\Delta V = \frac{L}{\varepsilon_r \varepsilon_o} p\Delta T \approx \left(\frac{p}{\rho c \varepsilon_r \varepsilon_o}\right) \eta I \Delta t$$

where ε_r is the relative permittivity, c is the specific heat capacity, ρ is the density, all of the detector material, and η is the emissivity of the radiation receiving surface (only ηI will be absorbed). The above equation applies during the initial heating-up of the crystal by the incident radiation. If $C_D = \varepsilon_o \varepsilon_r A/L$ (A is the surface area of the receiving crystal) is the detector capacitance, the quantity $C_D \Delta V/\Delta t$ is the *pyroelectric* detector current I_p, or

$$I_p = A\frac{d\Delta P}{dt} = Ap\frac{d\Delta T}{dt}.$$

The *voltage responsivity* R_V of the detector is the maximum voltage generated (output) per unit incident radiation energy (input),

$$R_V = \frac{\Delta V}{I\Delta t} \approx \frac{p\eta}{\rho c \varepsilon_r \varepsilon_o}.$$

The quantity p/ε_r is a useful *pyroelectric figure of merit* for comparing various pyroelectric crystals.

Many pyroelectric detectors are sold with a single-crystal element coupled to an FET voltage follower, the output of which is usually coupled to an amplifier. The gate bias (load) resistance R_L, the effective input capacitance of the FET, C_L, together with the pyroelectric detector element capacitance, C_D, form an *electrical time constant* $\tau_{el} = R_L(C_L + C_D)$. The output signal $V(t)$ is the voltage across the source resistance R_S. Even if the electrical circuit could respond instantaneously, there is still an effective thermal time constant τ_{th} associated with the heating up (or cooling) of the detector which involves its effective heat capacity and the effective thermal losses. The signal $V(t)$ in response to a step radiation input would be

$$V(t) = V_0\left[\exp\left(-\frac{t}{\tau_{th}}\right) - \exp\left(-\frac{t}{\tau_{el}}\right)\right]$$

where V_0 represents a constant determined not only by the responsivity of the crystal but also by the thermal factors that control the rate of change of detector temperature, and electrical factors that control the rate of change of voltage in the input circuit. The output peaks and then decays, because R_L leaks the induced charge on the pyroelectric sample to ground at a rate determined by the electrical time constant τ_{el}. Due to these time constants, and other loading effects such as the circuit to which the FET output is connected, the pyroelectric detector has *low* and *high cut-off frequencies*, f_L and f_H. The detector is usually operated at a radiation chopping frequency f_T that is between f_L and f_H, e.g. $f_T = (f_L f_H)^{1/2}$. In general, higher f_H can only be obtained at the expense of voltage responsivity.

Pyroelectric detectors are widely used in laser power measurements, noncontact temperature measurements, IR spectroscopy, solar instrumentation, security and surveillance sensors, industrial process control, x-ray calorimetry, gas analyzer instrumentation, etc. A particular $LiTaO_3$ pyroelectric detector with an FET, marketed as P1–73 (by Coherent), has the following reported characteristics: Infrared response up to 1000 μm, diameter = 3 mm; NEP ≈ 1×10^{-9} W $Hz^{-1/2}$; $D^* \approx 3\times10^{8}$ cm $Hz^{1/2}$ W^{-1}; $R_V \approx 90$ V W^{-1}; current responsivity = 0.5 μA W^{-1}; $C_D = 54$ pF; $R_L = 10^{11}$ Ω; $R_s = 5$ kΩ; $f_L =$ 1 Hz; $f_H \approx 3$ kHz for external load of 1 MΩ. (R. W. Whatmore, *Rep. Prog. Phys.*, **49**, 1335, 1986; P. Muralt, *Rep. Prog. Phys.*, **64**, 1339, 2001.) See *pyroelectricity*.

Table: Pyroelectric detector Properties of pyroelectric detector materials (approximate)

	ε_r'	Pyroelectric coefficient $\times 10^{-6}$ $C\,m^{-2}\,K^{-1}$.	Density $g\,cm^{-3}$	Heat capacity $J\,K^{-1}\,g^{-1}$ ($J\,K^{-1}\,cm^{-3}$)	Thermal conductivity $W\,m^{-1}\,K^{-1}$	Curie temperature °C
PZT	290	380	7.7	0.3 (2.4)	1–1.2	230
PVDF	12	27	1.8	1.3 (2.4)	0.13	80
$LiTaO_3$	44 (ε_{r33})	180–230	7.45	0.43 (3.2)	4.4	610
Lithium tantalate [trigonal]	52 (ε_{r11})					
TGS	22–32	280–300		1 (2.5)	0.7	49
$BaTiO_3$		20		$110\,J\,mol^{-1}K^{-1}$	4–5	130

PZT is lead zirconate titanate ($PbZrTiO_3$); TGS is triglycine sulfate [$(NH_2CH_2\text{-}COOH)_3H_2SO_4$]

Pyroelectricity is a phenomenon observed in certain dielectric materials (called *pyroelectric materials* such as barium titanate, $BaTiO_3$) in which a temperature change δT induces a proportional change δP in the polarization, i.e. $\delta P = p\,\delta T$ where p is the *pyroelectric coefficient* of the crystal, and measures the strength of the effect. The change δP gives rise to a change δV in the voltage which can be measured. Certain crystals, called *ferroelectric*, are permanently polarized even in the absence of an applied field. The crystal already possesses a finite polarization vector **P** due to the separation of positive and negative charges in the crystal. $BaTiO_3$ below 130 °C is a tetragonal crystal in which the Ti^{4+} atom is not located at the center of mass of the negative charges. The crystal is therefore polarized by the separation of the centers of mass of negative and positive charges, and hence possesses a finite polarization vector **P**. The critical temperature, above which ferroelectric property is lost, in this case 130 °C, is called the *Curie temperature* (T_C). Below the Curie temperature, the whole crystal becomes spontaneously polarized. A change in the temperature displaces the Ti^{4+} ion and hence causes a change δP in P, which means the crystal is pyroelectric. An important and technologically useful characteristic of a ferroelectric crystal is its ability to be *poled*. Above 130 °C there is no permanent polarization in the crystal. If we apply a temporary field E and let the crystal cool to below 130 °C we can induce the spontaneous polarization **P** to develop along the field direction. This process is called *poling*. The axis along which **P** is developed is also called the *ferroelectric axis*. Very small temperature changes, even thousandths of degrees, in ferroelectric material can develop voltages that can be readily measured. For example, for a PZT type pyroelectric ceramic, taking $\delta T = 10^{-3}$ K, and $p \approx 380 \times 10^{-6}$, we find $\delta P = 3.8 \times 10^{-7}$ C m^{-2}, and from $\delta P = \varepsilon_0(\varepsilon_r - 1)\delta E$ with $\varepsilon_r = 290$ we find $\delta E = 148.5$ V m^{-1}. If the distance between the faces of the ceramic where the charges are developed is 0.1 mm then $\delta V = 0.014$ V or 14 mV which can be readily measured.

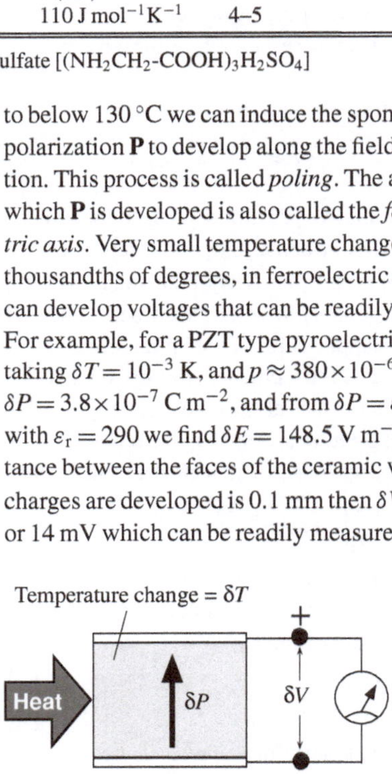

Pyroelectricity is the generation of a voltage across a crystal (due to a change in the *polarization*) whose temperature has been changed by the absorption (or loss) of heat.

PZT is a general acronym for the lead zirconate titanate ($PbZrO_3$-$PbTiO_3$ or $PbTi_{0.48}Zr_{0.52}O_3$) family of crystals.

Q

Q-factor or quality factor of an impedance is the ratio of its reactance to its resistance. The Q-factor of a capacitor is X_c/R_p where $X_c = 1/\omega C$ and R_p is the equivalent parallel resistance that represents the dielectric and conduction losses. The Q-factor of a resonant circuit measures the circuit's peak response at the resonant frequency and also its bandwidth. The greater is Q, the higher is the peak response and the narrower is the bandwidth. For a series RLC resonant circuit, $Q = \omega_0 L/R = 1/(\omega_0 CR)$ where ω_0 is the resonant angular frequency, $\omega_0 = 1/\sqrt{LC}$. The width of the resonant response curve between the half-power points is $\Delta\omega = \omega_0 Q$.

Q-switched laser is a laser whose optical cavity is switched from a low Q to a high Q to generate an intense laser pulse. While the optical cavity has low Q, lasing is suppressed and the active medium can be pumped to high energies to achieve a large population inversion. When the Q of the optical cavity is switched to a high value, then an intense lasing emission is generated. While the Q is low, the pumped lasing medium is effectively a very high-gain photon amplifier. There is too much loss in the optical resonator (i.e. no optical feedback) to achieve a lasing oscillation. As soon as the Q is switched to a high value, the low loss in the optical resonator allows lasing oscillations to occur which depletes the population inversion and decreases the gain until the population inversion falls below the threshold value and lasing oscillations cease. The Q-switching can be conveniently, for example, implemented by using an electrooptic switch.

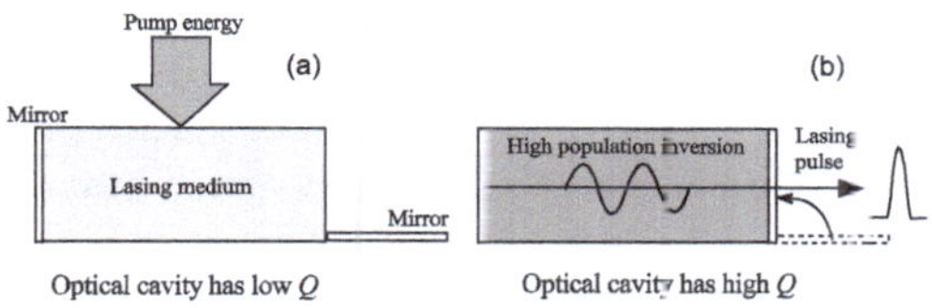

Q-switched laser (a) The optical cavity has a low Q so that pumping takes the medium to a very high degree of population inversion by the excitation of atoms; lasing is prevented by not having a right-hand mirror. (b) The right mirror is flung to make an optical resonator, Q is switched to a high value which immediately encourages lasing emissions. There is an intense pulse of lasing emission which brings down the excess population inversion.

Quadratic Stark effect see *Stark effect*.

Quantization noise represents the error involved in the analog-to-digital conversion process due to round-up or truncation. When an analog signal A is digitized to an N-bit digital number D (N-number of zeros and ones) then D does not exactly represent A because there are insufficient numbers of bits, and the difference between D and A is the error, and can be viewed as "noise" in the conversion process. For example, suppose we convert a 0–15 V analog signal, A = 0–15, to a four-bit digital signal. Then there are only 16 values for D, from 0000 corresponding to 0 to 1111 corresponding to 15, but almost an infinite number of values for A. $A = 14$ and 15 would correspond to 1110 and 1111, respectively, but $A = 14.5$ would have to be represented either by 1110 or 1111, and thus there is a quantization error of 0.5. The quantization noise is the noise associated with the quantization error.

Quantum (plural quanta) designates the smallest quantity of some physical property (energy, momentum, etc.) that a system can possess, according to the quantum theory of physics, for which this property can manifest itself only in discrete amounts. It can also designate the particle itself.

Quantum bits or qubits are the fundamental units of information in quantum computing systems. Consider a two-level quantum system having two observable orthogonal states, $|0\rangle$ and $|1\rangle$. The state of the system is represented by a coherent representation or qubit $|q\rangle$ with arbitrary coefficients c_0 and c_1,

$$|q\rangle = c_0|0\rangle + c_1|1\rangle$$

where c_0 and c_1 satisfy the normalization condition $|c_0|^2 + |c_1|^2 = 1$. A *Bloch sphere* provides a useful method for representing a qubit. If one considers two qubits, $|q_1\rangle$ and $|q_2\rangle$, they will exist in a *coherent superposition* of four orthonormal states,

$$\begin{aligned}|q_1\rangle|q_2\rangle &= (a_1|0_1\rangle + b_1|1_1\rangle)(a_2|0_2\rangle + b_2|1_2\rangle)\\ &= |q_1q_2\rangle\end{aligned}$$

with $$\begin{aligned}|q_1q_2\rangle &= a_1a_2|0_10_2\rangle + a_1b_2|0_11_2\rangle\\ &\quad + b_1a_2|1_10_2\rangle + b_1b_2|1_11_2\rangle\end{aligned}$$

where the subscripts 1 and 2 distinguish the two qubits. The normalization conditions become $|a_1|^2 + |b_1|^2 = |a_2|^2 + |b_2|^2 = |a_1a_2|^2 + |a_1b_2|^2 + |b_1a_2|^2 + |b_1b_2|^2 = 1$. This two-qubit system carries information on four measurable input and output numbers. Thus, a computation being performed on two qubits is actually computing the four different values in parallel. For a system of three qubits, there are 8 or 2^3 superposed values for the superposition state upon which calculations can be performed, in parallel. So when the number of qubits increases linearly, the information-bearing capacity of the system increases exponentially. Thus, with only 500 qubits it is possible to superpose the 2^{500} or about 10^{150} measurable outcomes – a number far greater than the total estimated number of atoms in the universe. Such an increase in information capacity is what allows a quantum computer to solve certain problems that are considered intractable for classical computers. A number of practical approaches have been developed to implement a physical realization of qubits – these include ion traps, nuclear magnetic resonance, quantum dots, cavity quantum electrodynamics, nonlinear optics, linear optics and superconducting quantum interference devices.

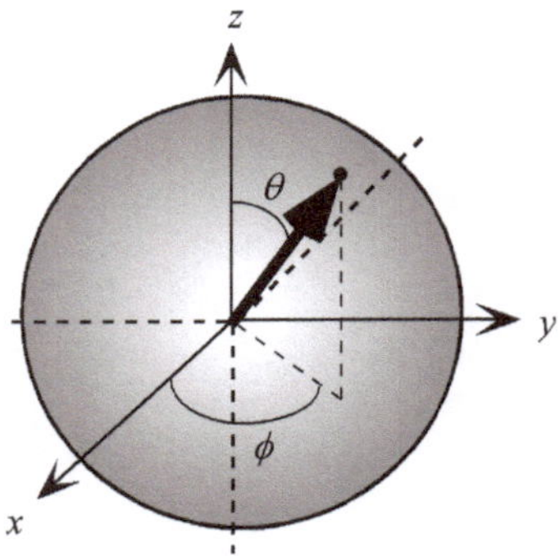

Quantum bits The Bloch sphere represents the state of a qubit. When $z = +1, \theta = 0$ and the qubit is in the state $|0\rangle$. If $z = -1, \theta = \pi/2$ and the qubit is in the state $|0\rangle$. The state $|q\rangle = (1/2^{1/2})[|0\rangle + e^{j\phi}|1\rangle]$ is a superposed state existing in the xy plane at an angle ϕ from the x-axis.

The *Bloch sphere* provides a useful approach for visualizing the state of a qubit. The complex coefficients c_0 and c_1 in the expression $|q\rangle = c_0|0\rangle + c_1|1\rangle$ for a qubit $|q\rangle$ may be rewritten in terms of real polar, azimuthal and global angles θ, ϕ, and γ as

$$|q\rangle = e^{j\gamma}\cos(\theta/2)|0\rangle + e^{j(\gamma+\phi)}\sin(\theta/2|1\rangle.$$

Because the global angle γ does not contribute to any observable effect, this factor can be dropped from the equation giving,

$$|q\rangle = \cos(\theta/2)|0\rangle + e^{j\phi}\sin(\theta/2|1\rangle.$$

The θ and ϕ define the direction of a unit vector from the origin of a three-dimensional space and ending at the surface of a unit sphere known as the *Bloch sphere*. The unit vector is termed the Bloch vector with x, y and z Cartesian coordinates of $[\cos(\phi)\sin(\theta), \sin(\phi)\sin(\theta), \cos(\theta)]$.

Quantum computing is based on harnessing quantum mechanical phenomena for information processing. Whereas current computers are based on the manipulation of a classical bit of information that can assume one of two possible states 0 or 1, the fundamental unit of information for quantum computing is the *qubit* or the *quantum bit*. The qubit can exist in a coherent superposition of the two possible classical states – that is in the state 1, the state 0. Such a superposition may be viewed as a vector composed of a weighted sum of two basis vectors. One example of a practical implementation of a qubit is the use of an *ion trap* where the nuclear spin states of the ions represent the states of the qubit. Only when a measurement on a given two-state system is performed will the system emerge in one of the two classically allowed basis states. More interestingly, a series of such qubits can be entangled – that is, the individual qubits exist in a joint state with strong correlations (as compared to any classical system) between the qubits. This results in the whole system being in a joint state and enables information to be encoded in a completely new way – an N-qubit system can now hold 2^N superposed values upon which calculations can be performed, in parallel. Such a capacity for handling information in parallel allows a quantum computer to solve certain problems that are considered intractable for classical computers. One example of such a problem is the use of quantum computers to break cryptographic codes – such codes underlie internet, banking and even military data security. However, the effects which make quantum computation so powerful, also makes a quantum computing system extremely delicate. Any observation or measurement of such a system destroys the superposition of the qubits and the interference between the states can no longer be used for quantum computation once a measurement is made. This means that the qubits must be well isolated during a computation, and the outcome of a computation must involve observing the results of the interference effects *without* making a direct measurement of the qubits. In addition, one has to design systems against the natural processes of decoherence due to coupling of the system to its environment. The concept of realizing a computer whose functionality was based on quantum mechanics dates back to work of Bennet, Benioff, Deutsch and Feynman in the 1970s, driven by discussions concerning the fundamental limits of conventional computers based on silicon integrated circuits, with shrinking dimensions. The latter discussions naturally touched upon the concept of atomic-scale logical elements, at which scale the physical laws that govern the behavior and properties of the circuit become inherently quantum mechanical and not classical. This naturally raised the question of whether a new kind of computer could be devised based on the principles of quantum mechanics. In 1982 Feynman answered the question by developing a model showing how a quantum system can indeed act as a computer. Shor then showed the importance of such a so-called quantum computer, as providing a means to crack an important problem in number theory – namely, factorization – and how on such a machine, huge numbers could in principle be factored extremely rapidly, much faster than is possible on conventional computers. Indeed, current computers are limited in the size of numbers that can be practically factored in any reasonable time period – while factoring such numbers would still be quite feasible with a quantum computer, and thus having significant practical implications. In terms of practical realization of a quantum computer, there has been progress on their demonstration for few-qubit systems capable of performing some simple arithmetic functions and sorting. However, these studies highlighted some serious challenges that will need to be overcome in order to build practical larger scale systems

Among these, are error correction, decoherence, and hardware architecture. Error correction refers to correcting errors that arise from decoherence – that is, the fact that in real quantum systems, interaction with their environment is inevitable, and this interaction acts to cause the quantum system to decay from a given quantum state into an incoherent state, leading to a loss of information and leading to computational errors. In order to reap the benefits of quantum computating a most serious obstacle is to develop appropriate hardware – something that is still in its infancy. *Nuclear magnetic resonance* has become the most popular component in quantum hardware architecture, although designs involving *ion traps*, cavity quantum electrodynamics, NMR and quantum dots are also being pursued. Of these approaches the latter are among the most interesting as they potentially offer scalable (multiqubit) architectures. See *quantum bits, nuclear magnetic resonance*.

Quantum dot is a crystal that is so small in all dimensions that the conduction electrons are confined in the *x*, *y*- and *z*-directions. The electron energy in the quantum dot is quantized in a similar fashion to the electron energy in a finite three-dimensional potential energy (PE) well that is well described in quantum mechanics. Quantum effects become apparent when the crystal size is several nanometers, typically less than 10 nm. The 0D (zero-dimensional) electronic density of states (DOS, number of allowed electronic states per unit energy) is made of Dirac singularities. Electron quantization effects associated with microcrystals of II-VI semiconductors (such as CdS, CdSe, ZnS, etc.) in glasses are well documented. InAs quantum dots of the order of 10 nm in dimension can form in a self-organized fashion when InAs is being grown on GaAs substrates. The strain mismatch between the materials and diffusion cause self-assembled arrays of InAs islands to form, representing an array of quantum dots. The preparation of technologically useful quantum dots is still under intense research. QD lasers have a number of distinct theoretical advantages. Compared with quantum well lasers, QD lasers have the lowest threshold current densities and have the narrowest emission spectrum. Further, the threshold current for a QD laser is expected to show much less temperature dependence compared with quantum well lasers. An InAs QD laser emitting near 1.3 μm with a threshold current as low as 24 A cm^{-2} has already been demonstrated; present extensive research and development will undoubtedly bring QD lasers into viable commercialization in the future. Despite the technological difficulties, various QD structures have been developed and currently there is much interest in developing QD lasers due to their promising features in terms of increased optical gain, carrier dynamics and lower laser threshold. The most promising techniques involve the self-organized growth of dense arrays of uniform islands of narrow bandgap material inside a defect-free semiconductor matrix. (S. Reimann, *Rev. Mod. Phys.*, **74**, 1283, 2002; A. D. Yofee, *Adv. Physc.*, **50**, 1, 2001.)

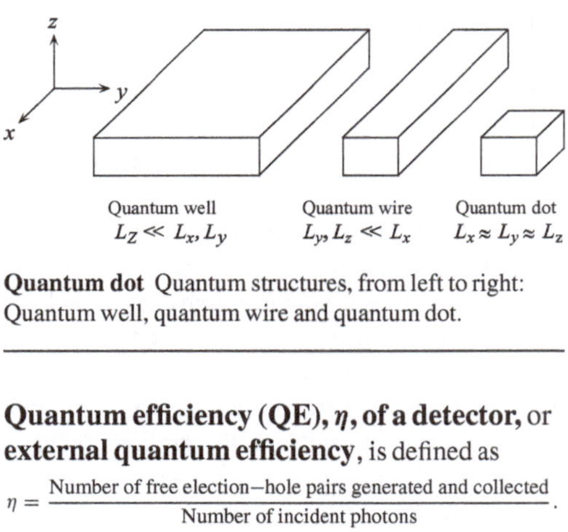

Quantum dot Quantum structures, from left to right: Quantum well, quantum wire and quantum dot.

Quantum efficiency (QE), η, of a detector, or **external quantum efficiency**, is defined as

$$\eta = \frac{\text{Number of free election–hole pairs generated and collected}}{\text{Number of incident photons}}.$$

Not all the incident photons are absorbed to create *free* electron–hole pairs (EHPs) that can be collected and give rise to a photocurrent. The efficiency of the conversion process of received photons to free EHPs is measured by the above quantum efficiency definition. The measured photocurrent I_{ph} in the external circuit is due to the flow of electrons per second to the terminals of the photodiode. The number of electrons collected per second is I_{ph}/e. If P_o is the incident optical power then the number of photons arriving per second is $P_o/h\upsilon$. Then the QE, η, can also be defined by

$$\eta = \frac{I_{ph}/e}{P_o/h\upsilon}$$

where h is the Planck constant, e is the electronic charge and υ is the frequency of light. Not all of the absorbed photons may photogenerate free EHPs that can be collected. Some EHPs may disappear by recombination without contributing to the photocurrent or become immediately trapped. Further if the semiconductor length is comparable with the *penetration depth* ($1/\alpha$ where α is the absorption coefficient) then not all the photons will be absorbed. The device QE, or the external QE, is therefore always less than unity. It depends on the *absorption coefficient* α of the semiconductor at the wavelength of interest and on the structure of the device. QE can be increased by reducing the reflections at the semiconductor surface, increasing absorption within the depletion layer and preventing the recombination or trapping of carriers before they are collected. The above QE is for the whole device. More specifically, it is known as the *external quantum efficiency*. *Internal quantum efficiency*, on the other hand, is the number of free EHPs photogenerated per *absorbed photon* and is typically quite high for many devices (see also *internal quantum efficiency*). The external quantum efficiency incorporates internal quantum efficiency because it applies to the whole device See *absorption coefficient, optical properties of semiconductors.*

Quantum efficiency, in medical physics as used in x-ray detectors, and as distinct from the definition used in semiconductor physics, is the fraction of the incident photons that are attenuated by the detecting medium. If μ is the linear attenuation coefficient, and L is the thickness of the medium, then the quantum efficiency η is

$$\eta = 1 - \exp(-\mu L).$$

Quantum noise or **photon noise in photodetectors** is the fluctuations in the photocurrent due to the quantum or discrete (i.e photon) nature of light. Incident light on a photodetector is a flux of photons, discrete entities, which means that the photogeneration process is *not* a smooth continuous process without fluctuations; photons arrive like discrete particles. Quantum noise is the same as shot noise because both electrons and photons are discrete entities that are involved in the flow: discrete electrons in an electric current and discrete photons in a beam of light. The arrival of photons at a detector follows the Poisson distribution. If σ is the standard deviation in the collected number of photons and if N is the average number of photons collected, then for a Poisson distribution $\sigma^2 = N$ and hence the signal-to-noise ration is given by

$$\text{SNR} = \frac{\text{Signal power}}{\text{Noise power}} = \frac{N^2}{\sigma^2} = N$$

which shows that the quantum noise is significant when the number of photons is small.

Quantum noise is carried by the electric field of a propagating light wave due to the quantum nature of light. This arises directly from Heisenberg's uncertainty relation. Just as one cannot simultaneously measure the position and momentum of a particle with arbitrary precision, the more precisely the phase of a light wave is measured, the less determined becomes its amplitude and vice versa. Thus, the product of phase and amplitude uncertainty has a fixed lower limit. States of the light field with the minimum amount of quantum noise are called minimum uncertainty states. An example of such a state is the coherent state – the state closely corresponds to that of a classical sinusoidally propagating light wave, the classical harmonic oscillator, with the amount of quantum noise distributed evenly in time (being the same for any phase angle of the wave). The coherent state is the state which is emitted by an ideal laser. To understand this, consider a simplistic model for a laser as a box containing the light field in the ground state. If the electric field within the box is perturbed by an external force (e.g. by optical or electrical pumping), quantum mechanical perturbation theory predicts that after a given time interval, the most likely state of the light field inside the box will be a coherent state. For such a state, the quantum noise depends on both the phase and amplitude of the wave. Thus, with increasing amplitude the overall amount of quantum noise remains constant. However the relative signal-to-noise ratio (i.e. when considering the light wave itself as the signal) improves. This explains why quantum effects of the light field are essentially unobservable for classical light waves, or waves with large amplitudes – for such waves, the influence of the very small amount of quantum noise can be neglected in most cases. Note that when the amplitude goes to zero there is still a finite amount of quantum noise. This is the vacuum noise of the light's electrical field, corresponding to the *zero point energy* of the harmonic oscillator. By sampling the quantum noise at a series of time intervals, one can obtain information on the wave packet at different times and thus observe the time evolution of the state's wave packet. A coherent state's wave packet oscillates evenly, similar to the movement of a classical particle in a harmonic potential. However, unlike a classical particle, there will always be a certain width to the wave packet, or an uncertainty in its exact position. For a coherent state, this width remains constant throughout the whole oscillation period. Most instructively, the state of the light field may be described by its *Wigner function*. See *Poisson distribution*.

Quantum optics is concerned with understanding and harnessing the quantum mechanical properties of light. Some important concepts include the uncertainty relation, wave packets, *quantum noise*, *Wigner functions*, and *photon number distributions*. Quantum optics is applied to quantum state engineering for light, atoms and ions, including entangled, squeezed and other interesting states – thus quantum optics also touches on quantum information processing, including quantum computing, quantum teleportation, quantum cryptography and quantum memory. See *quantum bits, quantum computing*.

Quantum well device is a *heterostructure device* that has an ultrathin, typically less than 50 nm, narrow bandgap semiconductor, such as GaAs, sandwiched between two wider bandgap semiconductors, such as AlGaAs. We assume that the two semiconductors are lattice matched in the sense that they have the same *lattice parameter a*. This means that

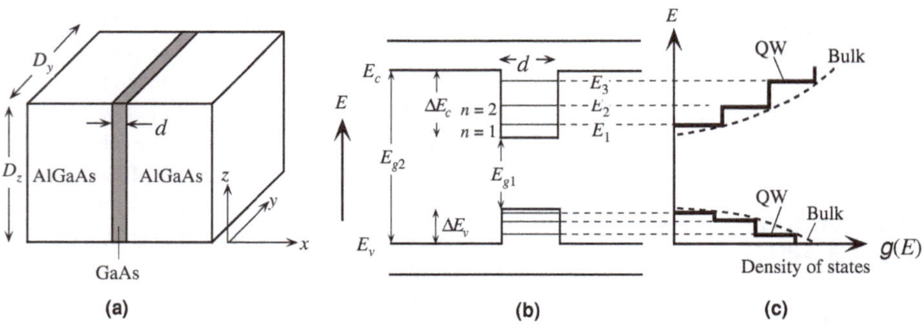

A **quantum well** (QW) device. (a) Schematic illustration of a quantum well structure in which a thin layer of GaAs is sandwiched between two wider bandgap semiconductors (AlGaAs). (b) The conduction electrons in the GaAs layer are confined (by ΔE_c) in the *x*-direction to a small length *d* so that their energy is quantized. (c) The density of states of a two-dimensional QW. The density of states is constant at each quantized energy level.

interface defects due to mismatch of crystal dimensions between the two semiconductor crystals are minimal. Since the bandgap, E_g, changes at the interface, there are discontinuities in E_c and E_v at the interfaces. These discontinuities, ΔE_c and ΔE_v, depend on the semiconductor materials and their doping. In the case of a GaAs/AlGaAs heterostructure shown in the figure, ΔE_c is greater than ΔE_v. Very approximately, the change from the wider E_{g2} to narrower E_{g1} is proportioned 60% to ΔE_c and 40% to ΔE_v. Because of the potential energy barrier, ΔE_c, conduction electrons in the thin GaAs layer are confined in the x-direction. This confinement length d is so small that we can treat the electron as in a one-dimensional potential energy (PE) well in the x-direction but as if it were free in the yz plane.

We can appreciate the confinement effect by considering the energy of the conduction electron that is bound by the size of the GaAs layer which is d along x and D_y and D_z along y and z. The energy of the conduction electron will be the same as that in a three-dimensional PE well of size d, D_y and D_z and given by

$$E = E_c + \frac{h^2 n^2}{8m_e^* d^2} + \frac{h^2 n_y^2}{8m_e^* D_y^2} + \frac{h^2 n_z^2}{8m_e^* D_z^2}$$

where n, n_y and n_z are quantum numbers having the values 1,2,3,.... The reason for the E_c in this equation is that the potential energy (PE) barriers are defined with respect to E_c. These PE barriers are ΔE_c along x and the *electron affinity* (energy required to take the electron from E_c to vacuum) along y and z. But D_y and D_z are orders of magnitude greater than d so that the minimum energy, denoted as E_1, is determined by the term with n and d, the energy associated with motion along x. The minimum energy E_1 corresponds to $n = 1$ and is above E_c of GaAs. The separation between the energy levels identified by n_y and n_z and associated with motion in the yz plane is so small that the electron is free to move in the yz plane as if it were in the bulk semiconductor. We therefore have a two-dimensional electron gas which is confined in the x-direction. The holes in the valence band are confined by the potential energy barrier ΔE_v (in an energy band diagram, hole energy is in the opposite direction to the electron energy) and behave similarly. The density of electronic states for the two-dimensional electron system is not the same as that for the bulk semiconductor. For a given electron concentration n, the *density of states* $g(E)$, the number of quantum states per unit energy per unit volume, is constant and does not depend on the energy. The density of states for the confined electron and that in the bulk semiconductor are shown schematically in the first figure. $g(E)$ is constant at E_1 until E_2 where it increases as a step and remains constant until E_3 where again it increases as a step by the same amount and at every value of E_n. The density of states in the valence band behaves similarly.

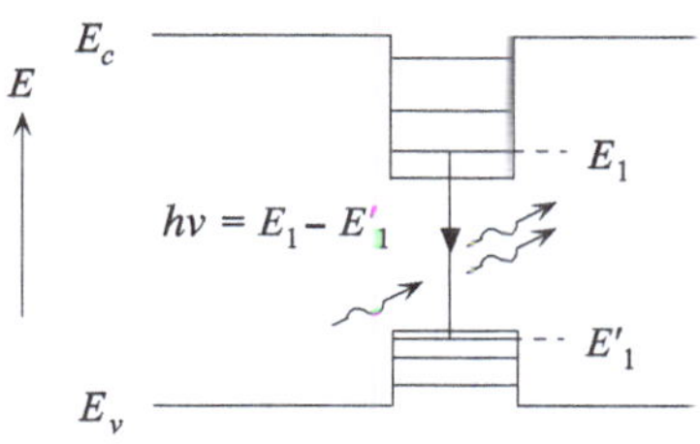

In single **quantum well** (SQW) lasers electrons are injected by the forward current into the thin GaAs layer which serves as the active layer. Population inversion between E_1 and E'_1 is reached even with a small forward current which results in stimulated emissions.

Since at E_1 there is a finite and substantial density of states, the electrons in the conduction band do not have to spread far in energy to find states. In the bulk semiconductor, on the other hand, the density of states at E_c is zero and increases slowly with energy (as $E^{1/2}$) which means that the electrons are spread more deeply into the conduction band in search of states. A large concentration of electrons can easily occur at E_1 in a QW whereas this is not the case in the bulk semiconductor. Similarly, the majority of holes in the valence band will be around E'_1 since there are sufficient states at this energy. Under a forward bias electrons are injected into the conduction band of the GaAs layer which serves as the active layer. The injected electrons readily populate the large number of states at E_1 which means that the electron concentration at E_1 increases rapidly with the current and hence population inversion occurs quickly without the need for a large current to bring in a great number of electrons. Stimulated transitions of electrons from E_1 to E'_1 leads to a lasing emission. The *threshold current* for population inversion and hence lasing emission is markedly reduced with respect to that for bulk semiconductor devices. For example in a *single quantum well* (SQW) laser this is typically in the range 0.5–1 mA whereas in a double heterostructure laser the threshold current is in the range 10–50 mA.

Quantum wire (QW) is made of a thin layer of lower bandgap semiconductor (e.g. GaAs) surrounded or confined in two dimensions by another semiconductor of wider bandgap (e.g. AlGaAs); if the *y*- and *z*-axes represent the spatial confinement dimensions, and if the narrower bandgap crystal has lengths L_y and L_z along *y*- and *z*-directions, then typically, $L_y \approx L_z \approx 10$ nm, comparable or less than the De Broglie wavelength. The electron energy states become quantized with electron motion described by standing waves in the two directions (*y* and *z*), and are described by quasifree Bloch wave states in the third direction (*x*), hence the term *quantum wire*. A quantum wire can be regarded as a two-dimensional potential well with boundaries defined by the dimensions of the quantum wire. Similar to the case of quantum dots, the density of states distribution is more concentrated than that for bulk material, in the vicinity of the quantized energies. The electron energies in a quantum wire are defined by two quantum numbers. The simultaneous confinement of electronic wavefunctions along two orthogonal directions makes the (1D, one-dimensional) electronic density of states (DOS, number of electronic states per unit energy) vary as hyperbolas, which leads to characteristic (or specific) forms for the spectral properties of absorption, emission and gain.

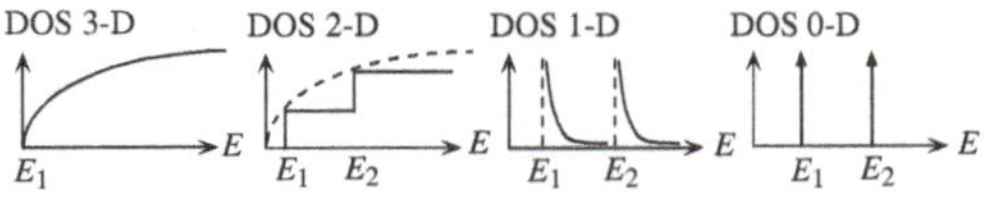

Quantum wire Electronic density of states vs. energy in the conduction band for semiconductor structures (schematic): bulk (3D) and reduced dimensionalities, quantum well (2D), quantum wire (1D), quantum dot (0D).

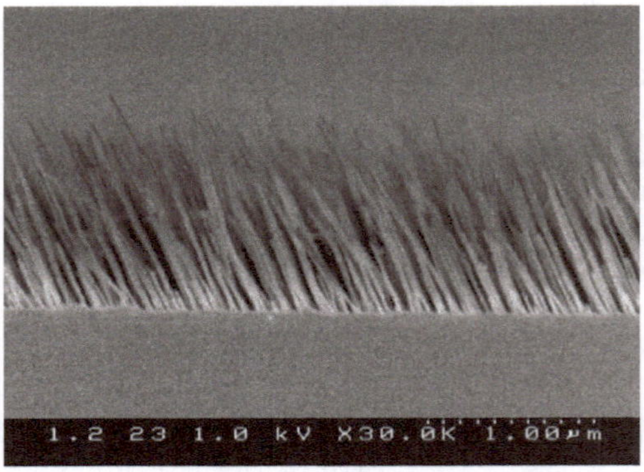

GaAs **quantum wires** grown using molecular beam epitaxy, Ruda and Wu. (Courtesy of H. Ruda and Z. Wu, University of Toronto.)

One common method for fabricating semiconducting quantum wires is to grow over the cleaved edge of a *superlattice structure*, revealing an array of long parallel quantum wires. It should be noted that as the wires become narrower, confinement energies rise and at appropriately low temperatures, phase coherent ballistic transport can occur, revealing a class of devices based on electron interference in waveguide structures.

Quantum yield see *photoluminescence*.

Quarter-wave plate is a retarder plate (made from a birefringent refracting crystal) that results in a relative phase change of a quarter of a wave ($\pi/4$) between the *ordinary* and *extraordinary waves* propagating inside the crystal from entry to exit. A *linearly polarized* light incident with its **E**-field at an angle α to the optic axis leaves the crystal either as *elliptically* polarized light or as *circularly* polarized light ($\alpha = 45°$).

Quartz (SiO_2) is a crystalline optical material that has good transmission from about 0.15 to 2.5 μm, which includes the UV down to about 150 nm. Single crystals are grown using a hydrothermal process. It is a uniaxial positive ($n_e > n_o$) birefringent crystal (trigonal 32) with the optic axis along the crystal's *c*-axis. $\rho = 2.650$ g cm^{-3}; $T_{\text{melt}} \approx 1700$ °C; $\nu_d = 969.87$; $\alpha_L = 7.1 \times 10^{-6}$ // to *c*-axis K^{-1}, 13.2×10^{-6} ⊥ to *c*-axis K^{-1}; $\kappa = 10.4$ // to *c*-axis, 6.2 ⊥ to *c*-axis, W cm^{-1} K^{-1}; $c_s = 0.71$ J g^{-1} K^{-1}; $E = 97$ // to *c*-axis, 76.5 ⊥ to *c*-axis GPa; $K = 36.4$ GPa; $G = 31.1$ GPa; $H_{\text{Knoop}} = 740$ kg mm^{-2}. Its refractive indices over the 200–3000 nm wavelength range are given in terms of Cauchy's equation with λ in μm as (to be rounded to 5 decimals):

$$n_0 = 1.535509 + \frac{3.32873 \times 10^{-3}}{\lambda^2} + \frac{5.02005 \times 10^{-5}}{\lambda^4} - 3.86816 \times 10^{-3}\lambda^2$$

$$n_e = 1.544490 + \frac{3.39158 \times 10^{-3}}{\lambda^2} + \frac{5.43370 \times 10^{-5}}{\lambda^4} - 4.14222 \times 10^{-3}\lambda^2.$$

See *optical materials*.

Quaternary (III–V) alloys have four different elements from the III and V groups. For example, $In_{1-x}Ga_xAs_{1-y}P_y$ is a quaternary alloy having Ga and In from group III and As and P from group V. There are as many group III elements as there are group V elements and the formula reflects this requirement. Of the III–V compound semiconductors, commonly used systems include $Ga_xIn_{1-x}As_yP_{1-y}$, $In_{1-x-y}Al_xGa_yAs$ and $In_{1-x-y}Al_xGa_yP$ – these alloys all have applications in the optoelectronics industry, with the two former alloys being used to fabricate lasers, photodiodes, and high-power and high-frequency electronics for the telecommunications industry. The latter alloys have applicability for developing high-power, high-brightness light-emitting diodes with red, orange, green and yellow emission wavelengths. The table shows some of the pertinent properties of the former two alloys. The figure shows the variation of energy bandgap versus lattice constant. The clear relationship between substrate choice and alloy system for given bandgap is evident. Thus, III-V materials for UV light emitters and detectors, tend to be designed

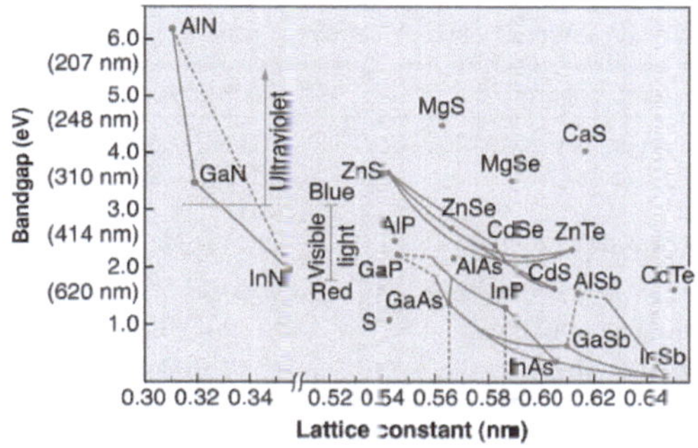

Quaternary alloys Bandgap energy E_g and lattice constant for various semiconductor materials. The solid lines between binary compounds indicate regions of direct bandgap, while the dashed lines represent the regions of indirect bandgap. Loci enclosed by a series of such lines define multicomponent alloy systems showing complete solid solubility of the end-point alloys. Examples of these include GaP-GaAs-InAs-InP, defining the quaternary alloy system $Ga_xIn_{1-x}As_yP_{1-y}$. Note how such alloys offer the possibility of a tunable direct bandgap lattice matched to either GaAs or InP. (From *Laser Focus World*, **41**(2), February, 2005. © 2005 Penn Well Corp., used by permission)

Table: Quaternary (III–V) alloys Properties of quaternary alloys. m_o is the electron mass in vacuum.

System/Property	$Ga_xIn_{1-x}As_yP_{1-y}$	$In_{1-x-y}Al_xGa_yAs$
Direct energy gap at 300 K (eV)	$1.35 + 0.668x - 1.068y - 0.758x^2 - 0.078y^2 - 0.069xy - 0.322x^2y + 0.03xy^2$	$0.36 + 2.093x + 0.629y + 0.577x^2 + 0.436y^2 + 1.013xy - 2.0xy(1 - x - y)$
Crystal structure	Zinc blende	Zinc blende
Properties of alloys lattice matched to InP:		
x as a function of y:	$0.1894y/[0.4184 - 0.013y]$	$0.4761–1.0173y$
Electron effective mass. m^*/m_o:	$0.08–0.039y$	$0.0427 + 0.068x$

around nitride materials with lattice constant around 0.30–0.34 nm, while telecom applications favor materials with lattice constant around 0.55–0.58 nm. Recently, however, such nitride based materials have started to have an impact in optoelectronic device development at telecommunications wavelengths – in particular GaInNP and GaInNAs quaternary alloys show remarkable bandgap bowing, offering a narrow bandgap region suitable for developing telecommunications wavelength lasers, as well as high-power high-frequency electronic devices.

R

Radiance or **radiant sterance** [W m^{-2} sr^{-1}] is emitted radiant flux density per unit solid angle, or stated differently, radiant intensity emitted per unit area of source. Quantitatively, if the elementary radiant flux $d^2\Phi_e$ is emitted by an elementary surface area dS (e.g. a source or a reflecting plane) into an elementary solid angle $d\Omega$, in the direction θ (direction of interest or observation looking at the surface) with respect to the normal to the surface, then the *radiance* L_e is defined by

$$L_e = \frac{d^2\Phi_e}{dS\cos\theta d\Omega} = \frac{dI_e}{dS\cos\theta}$$

where the quantity $L_e = L_e(\theta)$ is the *radiance*, I_e is the radiant intensity, defined by $I_e = d\Phi_e/d\Omega$ (W sr^{-1}), and $dS\cos\theta$ is the projected area that is normal to the direction of interest or observation looking at the surface. Radiance in a linear optical system, in which there are negligible transmission losses, is conserved. If L_o and L_i are the radiance of the object and image respectively in an optical system, then $L_i = L_o$. See *luminance*.

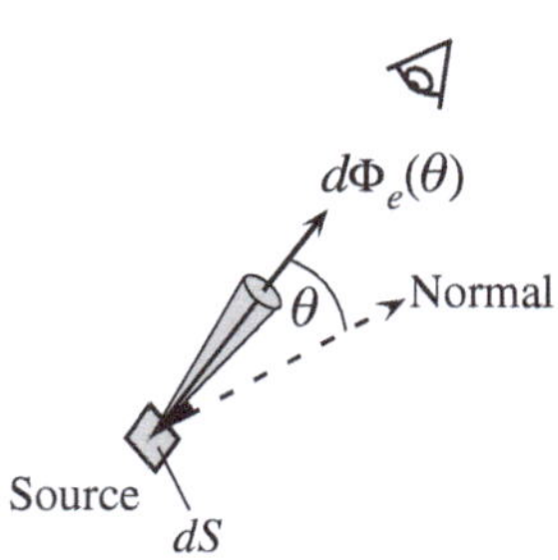

Definition of **radiance**.

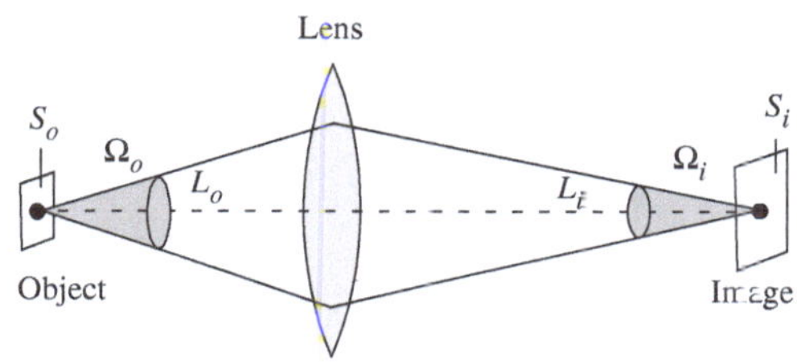

Conservation of **radiance** in a lossless linear optical system.

Radiance conservation see *radiance*.

Radiant is a common adjective used to imply the involvement of radiation, that is, electromagnetic waves, in the noun that it qualifies; e.g. *radiant energy* is the energy transmitted by radiation.

Radiant exitance or emittance, M_e, see *exitance*.

Radiant flux density see *irradiance*.

Radiant incidence see *irradiance*.

Radiant intensity is the radiant flux (radiant energy flowing per unit time) per unit solid angle; it is a radiometric quantity with the units of W sr^{-1}. It is a particularly useful quantity for point sources. If $d\Phi_e$ is the radiant flux within a solid angle $d\Omega$ in a particular direction θ, ϕ, as defined on spherical coordinates placed at the source, then

$$I_e = \frac{d\Phi_e}{d\Omega} = r^2 \frac{d\Phi_e}{dS}.$$

Radiant intensity usually has an angular dependence (on the spherical coordinates θ and ϕ). If it is symmetric, with no angular dependence, then the source is *isotropic*. The counterpart of radiant intensity in photometry (over visual wavelengths) is *luminous intensity*, measured in lm sr^{-1}. See *radiometry, radiant flux, luminous intensity*.

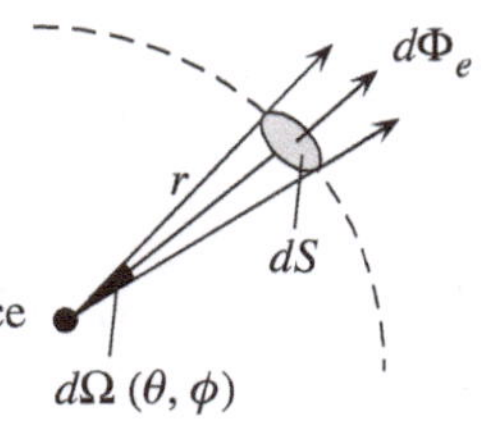

Radiant intensity definition using the spherical coordinates.

Radiant power is radiant flux. See *flux*.

Radiant sensitivity see *responsivity*.

Radiation energy conservation or flux conservation law as used in radiation science relates the reflectivity, ρ, the transmittivity, τ, and absorptivity, α, of a thin substance (a "surface") as $\alpha(\lambda) + \rho(\lambda) + \tau(\lambda) = 1$ at every wavelength (assuming no nonlinear phenomena occurring within the material). Suppose that $\Phi_{A\lambda}$ is the spectral radiant flux absorbed at a surface, i.e. energy absorbed per unit time per unit wavelength, $\Phi_{R\lambda}$ is the spectral flux reflected at a surface, and $\Phi_{T\lambda}$ is the spectral flux transmitted at (through) a surface. Spectral absorptivity α_λ characterizes the fraction of radiation absorbed with respect to the spectral flux $\Phi_{\text{blackbody},\lambda}$ of a black body, $\alpha_\lambda = \Phi_{A\lambda}/\Phi_{\text{blackbody},\lambda}$, *spectral reflectivity* ρ_λ characterizes the fraction of radiation reflected with respect to $\Phi_{\text{blackbody},\lambda}$, of a black body, $\rho_\lambda = \Phi_{R\lambda}/\Phi_{\text{blackbody},\lambda}$, and spectral transmittivity τ_λ characterizes the fraction of radiation transmitted with respect to $\Phi_{\text{blackbody},\lambda}$ of a black body, $\tau_\lambda = \Phi_{T\lambda}/\Phi_{\text{blackbody},\lambda}$. Energy conservation requires that $\alpha_\lambda + \rho_\lambda + \tau_\lambda = 1$. For an opaque substance there is no transmission, and $\alpha_\lambda + \rho_\lambda = 1$. In some applications $\alpha_\lambda, \rho_\lambda, \tau_\lambda$ are defined in terms of the incident spectral radiant flux, but the energy conservation relationship $\alpha_\lambda + \rho_\lambda + \tau_\lambda = 1$ remains valid. See *emissivity, Kirchhoff law*.

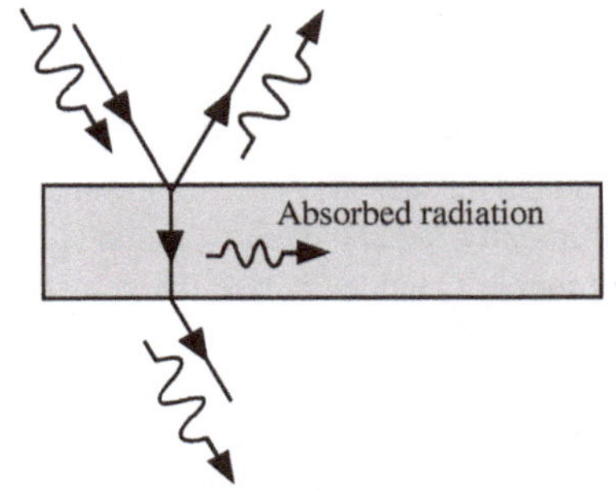

Radiative energy conservation.

Radiation thermometer see *infrared thermometer*.

Radiationless recombination, a nonradiative transition, is that in which an electron and a hole recombine through a recombination center such as a crystal defect or an impurity and emit phonons (lattice vibrations). There is no electromagnetic radiation emitted in such a recombination process.

Radiative recombination center see *luminescence*.

Radiative recombination lifetime see *direct recombination coefficient, recombination lifetime*.

Radiometry is the science of radiation measurement, for example, the measurement of emitted, absorbed, reflected, transmitted radiation energy; radiation is understood to mean electromagnetic energy normally in the optical frequency range (ultraviolet, visible and infrared). Photometry, on

Table: Radiometry Summary of radiometric quantities, A = area through which flux crosses; S = surface area emitting or receiving radiation.

Radiometric units			Spectral radiometric units		
Quantity	Usual symbol	Units	Quantity	Symbol	Units
Radiant energy	Q_e	Joule, J			
Radiant flux or radiant power	$\Phi_e = \frac{dQ_e}{dt}$	Watt, W	Spectral radiant flux; radiant power (flux) per unit wavelength	$\Phi_\lambda = \frac{d\Phi_e}{d\lambda}$	W nm^{-1}
Radiant flux density: Flux per unit area	$\frac{d\Phi_e}{dA}$	W m^{-2}			
Radiant exitance or emittance: exiting flux per unit surface area	$M_e = \frac{d\Phi_e}{dS}$	W m^{-2}	Spectral exitance or emittance	$M_\lambda = \frac{dM_e}{d\lambda}$	W m^{-2} nm^{-1}
Radiant intensity: flux per unit solid angle	$I_e = \frac{d\Phi_e}{d\Omega}$	W sr^{-1}	Spectral radiant intensity	$I_\lambda = \frac{dI_e}{d\lambda}$	W m^{-2} sr^{-1}
Radiance: radiant power per unit solid angle and unit projected area	$L_e = \frac{d\Phi_e}{d\Omega dS\cos\theta}$ $L_e = \frac{dI_e}{dS\cos\theta}$	W sr^{-1} m^{-2}	Spectral radiance	$L_\lambda = \frac{dI_e}{d\lambda}$	W sr^{-1} m^{-2} nm^{-1}

the other hand, is a subset of radiometry in which radiation is measured with respect to the *spectral responsivity* of the eye, that is over the visible spectrum and by taking into account the spectral visual sensitivity of the eye under normal light adapted conditions, i.e. *photopic* conditions. Radiometric quantities, such as *radiant flux*, Φ_e, radiant energy flow per unit time, usually have a subscript e, and invariably involve energy or power. Radiometric *spectral* quantities, such as *spectral radiant flux*, Φ_λ, refer to the radiometric quantity per unit wavelength, i.e. $\Phi_\lambda = d\Phi_e/d\lambda$. See *flux, photometry*.

Radiosity represents all the radiation power (in Watts) that leaves a surface; it includes the radiation that is reflected and also the radiation that is emitted as a result of the surface being at a finite temperature (thermal radiation). If G is the irradiation of a surface with a reflectivity ρ, then a portion ρG will be reflected. The surface will also have an emissive radiation power E leaving due to its finite temperature, as determined by Stefan's law. The radiosity J is given by $J = E + \rho G$. *Spectral radiosity* J_λ refers the total power emitted per unit wavelength range: $J_\lambda = E_\lambda + \rho_\lambda G_\lambda$, where E_λ and G_λ are the spectral emissive power and spectral irradiation respectively and ρ_λ is the monochromatic reflectivity.

Raman fiber amplifier is a fiber amplifier that is based on the stimulated Raman scattering effect. In Raman scattering, a light wave of frequency υ_P is scattered from the molecular vibrations in a medium as a result of which its frequency is downshifted to υ_S. This is an inelastic interaction and depends on the nonlinear dielectric behavior of the medium. The change in the photon energy $h\upsilon_P - h\upsilon_S$ excites the molecular vibrations in the interaction region by a phonon energy, that is $h\upsilon_R$, where υ_R is the Raman shift frequency, a resonance frequency associated with molecular oscillations of the medium. The photon $h\upsilon_S$ is called a Stokes photon, and the downshift in the frequency is called the Stokes shift. For silica, υ_R is not actually a single frequency but a range of frequencies, a "band," with a peak at about 13–14 THz. Suppose that we send a pump laser beam at frequency υ_P into a fiber to excite the molecular vibrations, and suppose that the pump beam is very intense. (In practice this would be something like 1 W fed into a single-mode fiber.) If the incoming relatively weak signal has a frequency υ_S then it can *stimulate* the Stokes shift and hence extract energy from the pump beam to add to itself. The result is an optical amplification of the signal at υ_S. Since the discovery of the EDFA, interest in Raman fiber amplifiers has somewhat waned due to long fiber lengths and high pump powers that are required to achieve acceptable optical gains. One of their advantages is that they have a wide optical bandwidth, and unlike ERDFA they are not restricted to the 1550 nm range.

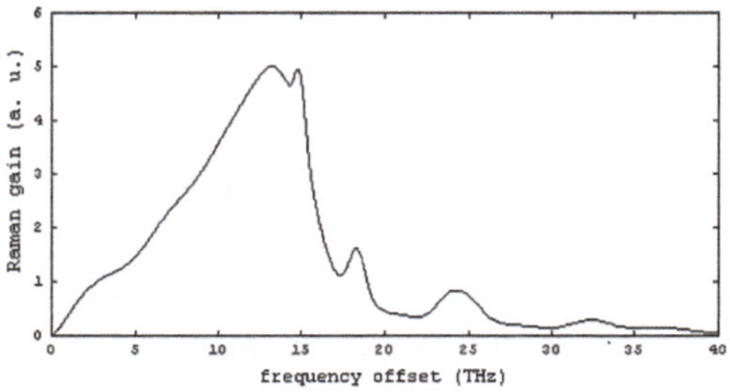

Raman fiber amplifier Raman gain spectrum of silica. The frequency offset of the signal is measured with respect to the pump. (D. Hollenbeck and C. D. Cantrell, *J. Opt. Soc. Am.*, **B19**, 2886, 2002.)

Raman–Nath effect is based on the diffraction of a beam of light by a dynamic acoustic grating. The propagation of an acoustical wave of frequency f in an optical medium causes a periodic change in the refractive index of the medium. The origin of the refractive index change is the formation of a dynamic strain caused by the propagation of the wave traveling at the speed of sound υ_S in the medium, resulting in a polarization of the medium in some direction x. The dynamic acoustic grating thus has a wavelength $\Lambda = \upsilon_s/f$, and an associated wavevector of magntiude $|\boldsymbol{q}| \equiv 2\pi/\Lambda$. Consider the optical beam as a plane wave traveling in the xz plane in this medium with frequency ν, free space wavelength $\lambda_0 = c/\nu$ and wavelength $\lambda = \lambda_0/n$ in the unperturbed medium, where c is the speed of light in vacuum and n is the unperturbed refractive index of the medium. The optical beam has an associated wavevector $|\mathbf{k}| \equiv 2\pi/\lambda$ making an angle θ with the z-axis. Assume that the acoustic wave is constrained in a thin beam and diffracts the incident optical beam, resulting in diffracted waves with a wavevector $|\mathbf{k}_r| \equiv 2\pi/\lambda$ appearing as shown in the figure – conservation of momentum requires that $\mathbf{k}_r = \mathbf{k} \pm \mathbf{q}$.

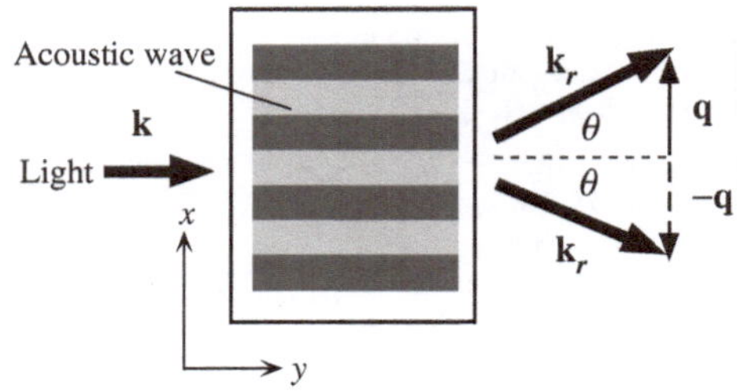

In the **Raman–Nath effect**, a thin (in the z-dimension) acoustic beam acts as a diffraction grating for an optical beam. The wavevectors for the incident and diffracted optical beams are **k** and $\mathbf{k}_r$, respectively, with the grating wavevector as **q**. θ is the angle between $\mathbf{k}_r$ and **k**.

Thus, the Raman–Nath diffraction condition becomes $\sin(\theta/2) = \lambda/2\Lambda$. For typical acoustic waves in solids, the velocity of sound in these materials is about 3000 m/s at a frequency of 100 MHz giving an acoustical wavelength $\Lambda \sim 30$ μm. For a visible-wavelength optical beam (e.g. a He-Ne laser) the wavelength is ~ 0.63 μm and thus the diffraction angle is a few tenths of a degree. Thus, the Raman–Nath condition may be simplified to $\theta \sim \lambda/\Lambda$. That is, the incident optical beam is deflected into directions making angles $\pm\theta$ as shown above, for the acoustical standing wave.

Raman–Nath diffraction see *acoustooptic (AO) modulator.*

Raman scattering is the scattering of light due to its interactions with the vibrations of the molecules of the material system, or the vibrations of the

Chandrasekhara Venkata Raman (1888–1970, India), won the 1930 Nobel Prize in physics for the discovery of Raman scattering while at Calcutta University. Subramanyan Chandrasekhar is Raman's nephew and also a Nobel Laureate (1983). He joined the Indian Institute of Science at Bangalore in 1933 and then in 1948 became the Director of the Raman Institute of Research at Bangalore, established and endowed by himself. The *Indian Journal of Physics* was founded by Raman in 1926 (AIP Emilio Segre Visual Archives, Born Collection.)

crystal lattice, in such a way that the scattered light has been shifted in wavelength; the photon energy has been changed by the scattering process by an amount that corresponds to a change in the vibrational energy of the molecule or the lattice vibrations. Suppose that $h\vartheta$ is the phonon energy associated with molecular or lattice vibrations of frequency ϑ, $h\upsilon$ is the incident photon energy that has a frequency υ, then the scattered photon energy is

$$h\upsilon' = h\upsilon - h\vartheta.$$

Normally Raman scattering is weak. The scattered photon has less energy because it generates a phonon, or excites a molecular vibration, therefore it has a lower frequency but a longer wavelength. The shift $\upsilon' - \upsilon$ (or its corresponding shift in wavelength) is called the *Stokes shift*. If the incident photon interacts with a molecule that is already excited so that the photon absorbs energy from the molecular vibration, the scattered photon has a higher energy, higher frequency and therefore a shorter wavelength. This is called the *anti-Stokes shift*. The latter process is, in the crystal, the absorption of a phonon by a photon. Raman scattering results from the interaction of light with *optical phonons* in a semiconductor, as distinct from *Brillouin scattering* involving *acoustic phonons*. Similar to the latter process, under strong coherent light excitation, stimulated Raman scattering can occur, with the material acting as a *parametric amplifier*.

Ramo's theorem, also known as the **Shockley–Ramo theorem**, relates the external photocurrent $i(t)$ due to a charge q being drifted with a velocity $v_{\mathrm{d}}(t)$ by a field between two biased electrodes separated by L, that is

$$i(t) = \frac{q v_{\mathrm{d}}(t)}{L}; \qquad t < t_{\mathrm{transit}}.$$

Consider a semiconductor material with a negligible dark conductivity that is electroded and biased to generate a field E in the sample that is uniform and is given by V/L. This situation is almost identical to the intrinsic region of a *reverse biased pin* photodiode. Suppose that a single photon is absorbed at a position $x = l$ from the left electrode and instantly creates an electron–hole pair (EHP). The electron and the hole drift in opposite directions with respective drift velocities $v_e = \mu_e E$ and $v_h = \mu_h E$, where μ_e and μ_h are the electron and hole drift mobilities respectively. The *transit time* of a carrier is the time it takes for a carrier to drift from its generation point to the collecting electrode. The electron and hole transit times t_e and t_h are

$$t_e = \frac{L - l}{v_e} \quad \text{and} \quad t_h = \frac{l}{v_h}.$$

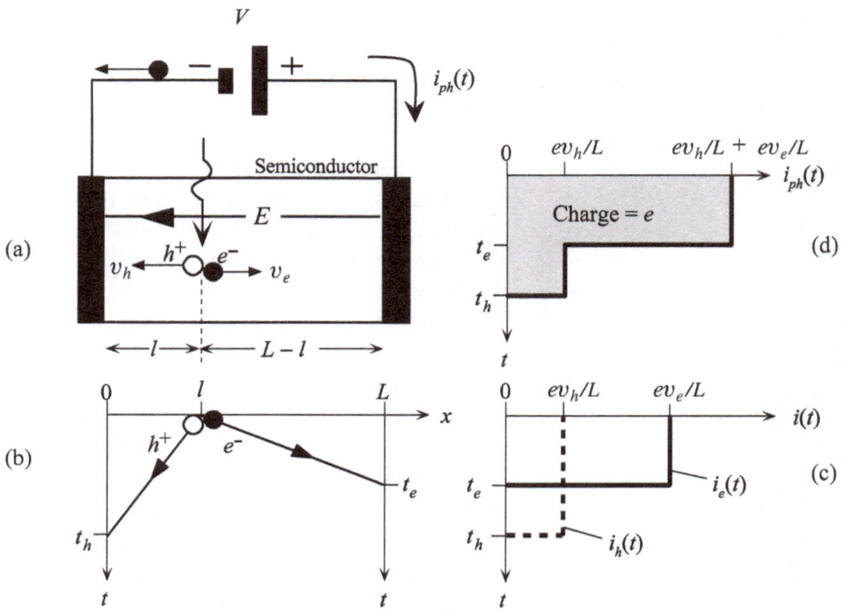

Ramo's theorem (a) An EHP is photogenerated at $x = l$. The electron and the hole drift in opposite directions with drift velocities v_h and v_e. (b) The electron arrives at time $t_e = (L - l)/v_e$ and the hole arrives at time $t_h = l/v_h$. (c) As the electron and hole drift, each generates an external photocurrent shown as $i_e(t)$ and $i_h(t)$. (d) The total photocurrent $i_{ph}(t)$ is the sum of hole and electron.

Consider first only the drifting electron. Suppose that the external photocurrent due to the motion of this electron is $i_e(t)$. The electron is acted on by the force eE of the electric field. When it moves a distance dx, work must be done by the external circuit. In time dt, the electron drifts a distance dx and does an amount of work $eEdx$ which is provided by the battery in time dt as $Vi_e(t)dt$. Thus,

$$\text{Work done} = eEdx = Vi_e(t)dt.$$

Using $E = V/L$ and $v_e = dx/dt$ we find the electron photocurrent

$$i_e(t) = \frac{ev_e}{L}; \quad t < t_e.$$

It is apparent that this current continues to flow as long as the electron is drifting (has a velocity v_e) in the sample. It lasts for a duration t_e at the end of which the electron reaches the battery. Thus, although the electron has been photogenerated instantaneously, the external photocurrent is *not* instantaneous and has a *time spread*. We can apply similar arguments to the drifting hole as well, which will generate a hole photocurrent $i_h(t)$ in the external circuit given by

$$i_h(t) = \frac{ev_h}{L}; \quad t < t_h.$$

The total external current, the photocurrent, $i_{ph}(t)$, will be the sum of $i_e(t)$ and $i_h(t)$. If we integrate the external current $i_{ph}(t)$ to evaluate the collected charge $Q_{\text{collected}}$ we would find,

$$Q_{\text{collected}} = \int_0^{t_e} i_e(t) + \int_0^{t_h} i_h(t) = e.$$

Thus, the collected charge is not $2e$ but just one electron.

The above statement of Ramo's theorem is applicable for a single-element detector (two biased electodes), which is a special case of the generalized Ramo's theorem. The generalized Ramo's theorem provides a convenient way to calculate the induced current flowing through an electrode of multielectrode detectors (e.g. pixellated image detectors) due to the motion of charge carriers in the detector. The induced current and charge on an electrode *j* by a moving point charge *e* at *x* are given by

$$I_j = e\mathbf{v} \cdot \mathbf{E}_{wj} \quad \Phi_j = -eV_{wj}(x)$$

where $\mathbf{v}$ is the instantaneous velocity of charge *e*, $\mathbf{E}_{wj}(x)$ and $V_{wj}(x)$ are the weighting field and potential of electrode *j*. $E_{wj}(x)$ and $V_{wj}(x)$ are the electric field and potential that would exist at charge *e*'s instantaneous position *x* if the electrode of interest (*j*) is raised to unit potential and all other electrodes kept at zero potential and all charges removed. For a large-area single-element detector the weighting field $E_w(x) = 1/L$ and thus the current is $i = ev/L$ as shown above. W. Shockley, *J. Appl. Phys.*, **9**, 635, 1938; S. Ramo, *Proc. I.R.E.*, **27**, 584, 1939.

Random telegraphic noise see *excess noise*.

Rare earth (RE) is the common name for the elements from lanthanum (La) to lutetium (Lu). They have atomic numbers from 57 to 71 and form a separate group in the Periodic Table. A specific feature of these elements is the incompletely filled 4f shell.

Table: Rare earth Occupation of outer electronic shells for RE elements.

57	**La**	$4s^2$	$4p^2$	$4d^{10}$	–	$5s^2$	$5p^6$	$\mathbf{5d^1}$	$6s^2$
58	**Ce**	$4s^2$	$4p^2$	$4d^{10}$	$\mathbf{4f^1}$	$5s^2$	$5p^6$	$\mathbf{5d^1}$	$6s^2$
59	**Pr**	$4s^2$	$4p^2$	$4d^{10}$	$\mathbf{4f^3}$	$5s^2$	$5p^6$	–	$6s^2$
60	**Nd**	$4s^2$	$4p^2$	$4d^{10}$	$\mathbf{4f^4}$	$5s^2$	$5p^6$	–	$6s^2$
...									
68	**Er**	$4s^2$	$4p^2$	$4d^{10}$	$\mathbf{4f^{12}}$	$5s^2$	$5p^6$	–	$6s^2$
...									
70	**Yb**	$4s^2$	$4p^2$	$4d^{10}$	$\mathbf{4f^{14}}$	$5s^2$	$5p^6$	–	$6s^2$
71	**Lu**	$4s^2$	$4p^2$	$4d^{10}$	$\mathbf{4f^{14}}$	$5s^2$	$5p^6$	$\mathbf{5d^1}$	$6s^2$

A RE may be embedded in different host materials in the form of divalent or trivalent ions. As divalent ions, REs exhibit broad absorption–emission lines related to the allowed 4f → 5d transitions. In the trivalent form (e.g. Er^{3+}), the RE loses two 6s electrons and one 4f or 5d electron. As a result of Coulombic interaction of the 4f electron(s) with the positively charged core, the 4f level becomes split into a complicated set of manifolds whose positions, in the first approximation, are relatively independent of the host matrix because the 4f level is well screened by the 5s and 5p shells from outer influences. In the second approximation, the exact construction and precise energy positions of manifolds depend on the host material via the *crystal field* and via the covalent interaction with *ligands* surrounding the RE ion. (A ligand is an atom, molecule, radical or ion, with one or more unshared pairs of electrons that can attach to a central metallic ion (or atom) to form a coordination complex. Examples of ligands are ions, F^-, Cl^-, Br^-, I^-, S^{2-}, CN^-, NCS^-, OH^-, NH_2^-, or molecules, NH_3, H_2O, NO, CO, that donate a pair of electrons to a metal atom or ion. Some ligands that share electrons with metals form very stable complexes.) The optical transitions between 4f manifold levels are forbidden by the *parity transition rule*, which is somewhat relaxed by the admixture of 5d state to 4f and by the disturbance of RE ion symmetry due to the influence of the host, which increases with the growth of *covalency*. A higher covalency implies a stronger sharing of electrons between the RE ion and the ligands an effect that is known as the *nephelauxetic effect*. The resulting absorption–emission lines are characteristic for individual RE ions and are quite narrow because they are related to forbidden inner-shell 4f transitions. (W. T. Carnall, G. L. Goodman, K. Rajnak, and R. S. Rana, *J. Chem. Physics*, **90**, 3443, 1989.) See *Judd-Ofelt analysis*, *rare-earth doped fiber amplifiers*.

Rare earth doped fiber see *erbium (rare earth) doped fiber amplifier (EDFA)*.

Rare earth luminescence see *photoluminescence*.

Rare gas halide lasers see *excimer*.

Ray equation presents another formulation of the *Eikonal equation* in an inhomogeneous medium with a refractive index $n(\mathbf{r})$ where $\mathbf{r}$ is the position vector. If ds denotes a differential length along a ray trajectory Γ, s being the curvilinear coordinate, the calculus of variations leads eventually to the following set of partial differential equations in Cartesian coordinates $x(s), y(s), z(s)$:

$$\frac{d}{ds}\left(n\frac{dx}{ds}\right) = \frac{dn}{dx},$$
$$\frac{d}{ds}\left(n\frac{dy}{ds}\right) = \frac{dn}{dy},$$
$$\frac{d}{ds}\left(n\frac{dz}{ds}\right) = \frac{dn}{dz}.$$

Alternatively, with $\mathbf{r}(s)$ being the position vector, the vectorial ray equation can be compactly written as

$$\frac{d}{ds}\left(n\frac{dr}{ds}\right) = \nabla n.$$

The trajectory Γ can be represented either by the 3D vectorial set $\{x(s), y(s), z(s)\}$, or by the Cartesian relationship $\{x(z), y(z)\}$. Consider a paraxial ray in a graded-index medium, that is, a ray whose trajectory is almost parallel to one of the axes, taken as the z-axis. In that case, $ds \approx dz$ and the paraxial ray equation reduces to

$$\frac{d}{dz}\left(n\frac{dx}{dz}\right) = \frac{dn}{dx}, \quad \frac{d}{dz}\left(n\frac{dy}{dz}\right) = \frac{dn}{dy}.$$

As an example, in a graded-index optical fiber, exhibiting cylindrical symmetry with a parabolic index $n = n(\rho) = {n_0}^2[1 - \alpha^2\rho^2]$, with $\rho^2 = x^2 + y^2$ and $\alpha^2\rho^2 \ll 1$, the results are harmonic functions of z with spatial periodicity $2\pi/\alpha$. The general solution is therefore *helicoidal*, and the paraxial ray remains confined inside the optical fiber close to its main axis.

Rayleigh criterion for the resolving power of an imaging system with an aperture of diameter D specifies the minimum angular separation $\Delta\theta_{\text{min}}$ of the two point light sources (S_1 and S_2) of wavelength λ such that their images can still be just distinguished, where

$$\Delta\theta_{\text{min}} = \frac{1.22\lambda}{D}$$

and the corresponding separation s of the point sources can be found from $s \approx L\Delta\theta_{\text{min}}$. The equation applies when $D \gg \lambda$. Each source generates an Airy diffraction pattern and the two Airy patterns A_1 and A_2, corresponding to S_1 and S_2, would be separate and distinct if the two sources are well separated. As the points get closer, their angular

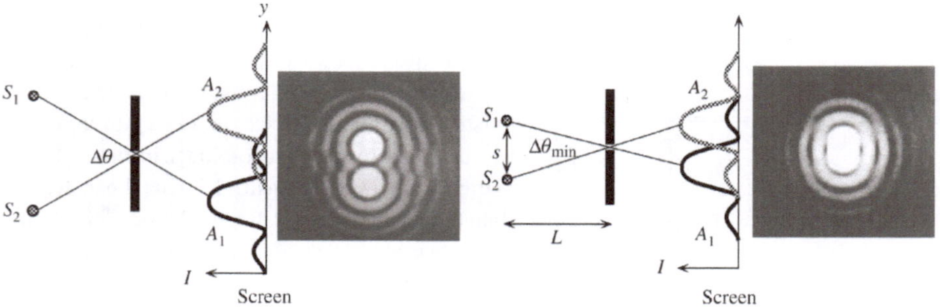

Rayleigh criterion Resolution of imaging systems is limited by diffraction effects. As point sources S_1 and S_2 get closer, eventually the Airy patterns overlap so much that the resolution is lost. The Rayleigh criterion allows the minimum angular separation of the two point sources to be determined. (Schematic illustration inasmuch as the side lobes are actually much smaller than the central peak.)

separation becomes narrower and the diffraction patterns begin to overlap. According to the Rayleigh criterion, the two spots are just resolvable when the principal maximum of one diffraction pattern coincides with the minimum of the other, which leads to the above equation. *Dawes*, studying various stars, was able to show that the resolving power of a telescope is actually better than the Rayleigh criterion, and can be written empirically as $\Delta\vartheta_{min} = 116/D$ for green light, where D is in mm and the angle in arc sec. According to the *Sparrow criterion*, the two points can still be distinguished even if two Airy patterns are closer than the Rayleigh criterion, until their superposition or their sum is such that the top of the combined pattern is nearly flat.

Rayleigh scattering refers to the scattering of light from scatterers having a size of about a tenth of the wavelength of the light, or smaller. Although Rayleigh scattering off the molecules of the air can be used to explain the blue colour of the sky, this effect was first explained by Tyndall and is referred to as *Tyndall scattering*. Rayleigh calculated the scattered intensity from such dipole scatterers to be

$$I = I_0(8\pi^4 N\alpha^2/\lambda^4 R^2)(1 + \cos^2\theta)$$

where I_0 is the incident light intensity of wavelength λ, θ is the angle the observer subtends with respect to the incident light, R is the distance of the observer from the scatterer and N is the number of scatterers with polarizability α. It should be noted that Rayleigh scattering can be considered to be elastic scattering since the energy of a scattered photon is unchanged. For a spherical dielectric particle of radius a and refractive index n_{sph} in a surrounding medium of refractive index n_{med} (1 for air and 1.33 for water) the differential cross-section per unit solid angle is,

$$\frac{d\sigma}{d\Omega} = \frac{1}{2}\left(\frac{2\pi n_{med}}{\lambda_o}\right)^4 \left(\frac{n^2-1}{n^2+1}\right) a^6(1+\cos^2\theta)$$

where λ_o is the wavelength in vacuum, $n = n_{sph}/n_{med}$, and the total scattering cross-section is

$$\sigma = \frac{8\pi}{3}\left(\frac{2\pi n_{med}}{\lambda_o}\right)^4 \left(\frac{n^2-1}{n^2+1}\right) a^6.$$

In its modern formulation, the Rayleigh scattering cross-section for a single molecule is written as

$$\sigma = \frac{24\pi^3\nu^4}{N^2}\left(\frac{n_r^2-1}{n_r^2+2}\right) F$$

where ν is the frequency, N the molecular density, n_r the refractive index, the term in brackets is the Clausius–Mossotti or Lorentz–Lorenz factor (which is proportional to N) and F is the so-called *King correction factor* defined as:

$$F = \frac{6+3\rho}{6-7\rho}$$

where ρ is the depolarization ratio of unpolarized light and accounts for the anisotropy of nonspherical molecules. The modern equation for σ accounts for the effects of nonresonant scattering that were not known to Rayleigh – for example, the fine structure on the Rayleigh line related to vibrational and rotational Raman scattering and the effects of collisional redistribution that produce Rayleigh wing scattering.

Rayleigh scattering is particularly important in optical fiber communications because it sets the lowest limit to the achievable attenuation. In a glass fiber in which refractive index fluctuations give rise to Rayleigh scattering, the attenuation is

$$\alpha_{Rayleigh} = \frac{8\pi}{3\lambda^4}(n^2-1)^2\beta_T k_B T_f$$

where n is the refractive index of the fiber core at the wavelength of interest, β_T is the isothermal compressibility (at T_f) of the glass, k_B is the Boltzmann constant and T_f is a quantity called the *fictive temperature* (roughly the *glass transition temperature*) where the liquid structure during the cooling of the fiber is frozen to become the glass structure. Fiber is drawn at high temperatures and, as the fiber cools, eventually the temperature drops sufficiently for the

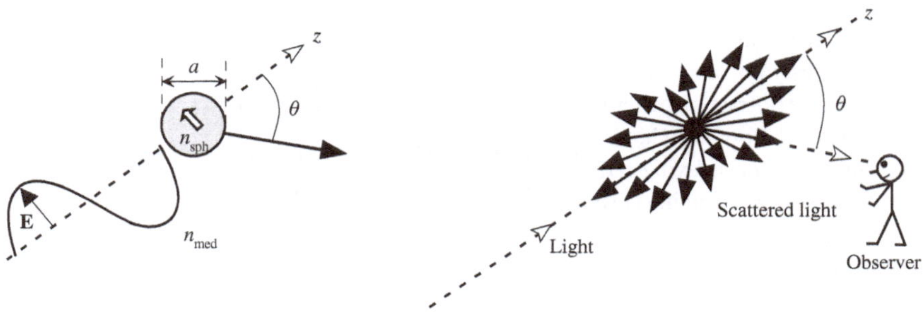

Rayleigh scattering Right: Schematic of geometry for sunlight being scattering from molecules following the theory of Rayleigh. The strong wavelength dependence (i.e. scattering intensity $\propto\lambda^{-4}$) of the Rayleigh scattering at short wavelengths results in the blue appearance of the sky. Note that the intensity of scattered light orthogonal to the incident light is half that of the forward scattered light intensity. Left: Rayleigh scattering from a spherical dielectric particle.

Lord Rayleigh (John William Strutt) was an English physicist (1877–1919) and a Nobel Laureate (1904) who made a number of contributions to wave physics of sound and optics. (Courtesy of AIP Emilio Segrè Visual Archives, *Physics Today* Collection.)

atomic motions to be so sluggish that the structure becomes essentially "frozen-in" and remains like this even at room temperature; α_{Rayleigh} is typically about 0.14 dB km^{-1} at 1.55 μm.

Rayleigh scattering loss in silica-based optical fibers can be estimated approximately from

$$\alpha_{\text{Rayleigh}} \approx \frac{B}{\lambda^4};\; B = 0.9\ \text{dB km}^{-1}\,\mu\text{m}^4.$$

(A. J. Cox, A. J. DeWeerd, J. Linden, *Am. J. Phys.*, **70**, 620, 2002.)

Reabsorption broadening see *photon reabsorption broadening*.

Reach-through avalanche photodiode (APD) is an avalanche photodiode device with a structure $n^+ - p - \pi - p^+$, where π represents very slight p-type doping. The n^+ side is thin and it is the side that is illuminated through a window. There are three p-type layers of different doping levels next to the n^+ layer to suitably modify the field distribution across the diode. The first is a thin p-type layer and the second is a thick lightly p-type doped (almost intrinsic) π-layer and the third is a heavily doped p^+ layer. The diode is reverse biased to increase the fields in the depletion regions. Under zero bias the depletion layer in the p-region (between n^+p) does not normally extend across this layer, but when a sufficient reverse bias is applied, the depletion region in the p-layer widens to *reach-through* to the π-layer (and hence the name *reach-through*). The field extends from the exposed positively charged donors in the thin depletion layer in the n^+ side, all the way to the exposed negatively charged acceptors in the thin depletion layer in the p^+-side. The electric field is given by the integration of the net space charge density ρ_{net} across the diode subject to an applied voltage V_r across the device. The variation in the field across the diode is such that the field lines start at positive ions and end at negative ions which exist through the p, π and p^+ layers. This means that E is maximum at the n^+p junction, then decreases slowly through the p-layer. Through the π-layer it decreases only slightly as the net space charge density here is small. The field vanishes at the end of the narrow depletion layer in the p^+ side. The absorption of photons and hence photogeneration takes place mainly in the long π-layer. The nearly uniform field here separates the electron–hole pairs (EHPs) and drifts them at velocities near saturation towards the n^+ and p^+ sides, respectively. When the drifting electrons reach the p-layer, they experience even greater fields and therefore acquire sufficient kinetic energy (greater than the bandgap E_g) to *impact-ionize* some of the Si covalent bonds and release EHPs. These generated EHPs themselves can also be accelerated by the high fields in this region to sufficiently large kinetic energies to further cause *impact ionization* and release more EHPs which leads to an *avalanche of impact ionization processes*. Thus from a single electron entering the p-layer one can generate a large number of EHPs all of which contribute to the observed photocurrent. The photodiode possesses an *internal gain mechanism* in that a single photon absorption leads to a large number of EHPs generated. The photocurrent in the APD in the presence of avalanche multiplication therefore corresponds to an *effective* quantum efficiency in excess of unity. The reason for keeping the photogeneration within the π-region and reasonably separate from the avalanche p-region is that avalanche multiplication is a statistical process and hence leads to carrier generation fluctuations which leads to *excess noise* in the avalanche multiplied photocurrent. This is minimized if impact ionization is restricted to the carrier with the highest impact ionization efficiency which in Si is the electron.

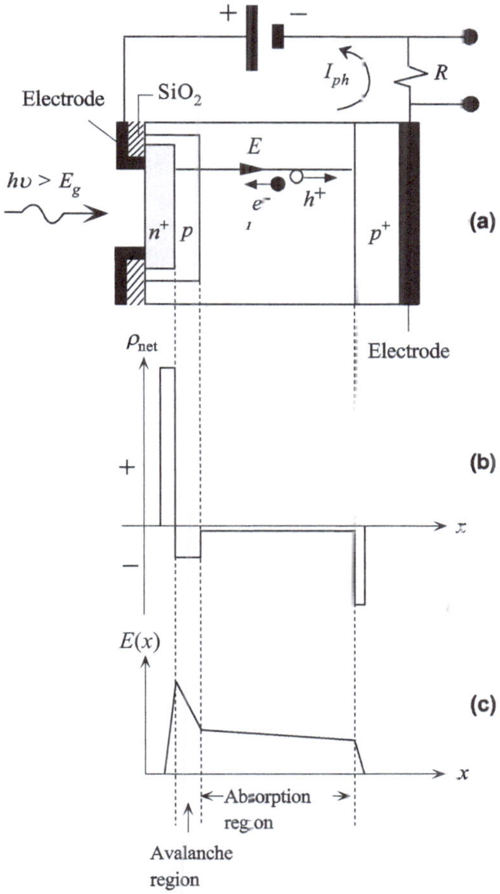

Reach-through APD (a) A schematic illustration of the structure of an avalanche photodiode (APD) biased for avalanche gain. (b) The net space charge density across the photodiode. (c) The field across the diode and the identification of absorption and multiplication regions.

Real image see *image*.

Real refractive index see *complex propagation constant*.

Reciprocity is a term that is typically used to describe what happens to a physical phenomenon, or to observable effects, when two quantities (that are part of describing the effect) are interchanged. For example, an oscillating current I at a point A will generate an electric field E at a point B. What happens to the E-I relationship when we interchange the two quantities and place the oscillating current at B and measure the field at A? In the *Lorentz reciprocity theorem*, the relationship between I and E remains unchanged when we interchange the two locations. Consider harmonic oscillations at a single frequency of ω. Consider two locations $\mathbf{r}_1$ and $\mathbf{r}_2$. A localized current density $\mathbf{J}(\mathbf{r}_1)$ at $\mathbf{r}_1$ generates a field $\mathbf{E}(\mathbf{r}_2)$ at $\mathbf{r}_2$. Then the Lorentz theorem can be written in terms of volume integrals as

$$\int \mathbf{J}(\mathbf{r}_1) \cdot \mathbf{E}(\mathbf{r}_2) dV = \int \mathbf{J}(\mathbf{r}_2) \cdot \mathbf{E}(\mathbf{r}_1) dV.$$

Rayleigh originally derived a similar reciprocity theorem for sound waves, which was later applied to radio waves by Carson, and came to be known as the *Rayleigh–Carson reciprocity* theorem; Lorentz and Rayleigh–Carson reciprocity theorems are similar.

Maxwell equations in a lossless linear medium are time-invariant, which means that replacing time t with $-t$ does not change the wave equation but simply reverses the direction of electromagnetic wave propagation; the amplitude does not change. In such a medium, Maxwell's equation are said to have *time reversal symmetry.* (R. J. Potton, *Rep. Prog. Phys.*, **67**, 717, 2004; H. A. Lorentz, *Amsterdammer Akademie der Wetenschappen*, **4**, 176, 1896.)

Recombination of an electron–hole pair involves an electron in the conduction band (CB) falling in energy down into an empty state (hole) in the valence band (VB) to occupy it. The result is the annihilation of the electron–hole pair. Recombination is direct when the electron falls directly down into an empty state in the VB as in GaAs. The recombination results in the emission of a photon, an electromagnetic radiation and the recombination process is called *radiative recombination*. Recombination is *indirect* if the electron is first captured locally by a defect or an impurity, called a *recombination center*, and from there it falls down into an empty state (hole) in the VB as in Si and Ge. In indirect recombination, the energy of the electron

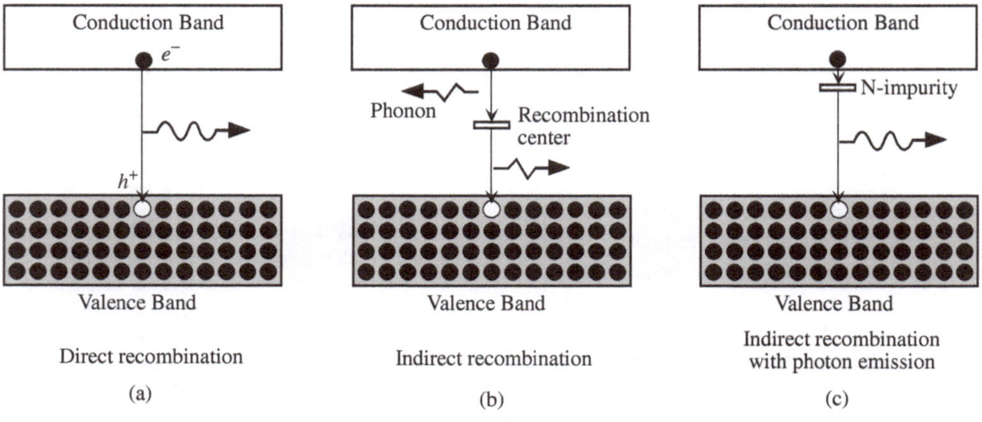

Recombination (a) Direct recombination. (b) Indirect radiationless recombination through a recombination center. The electron's energy is lost through the emission of phonons (lattice vibrations). (b) Indirect recombination in GaP through an isoelectronic N impurity with photon emission.

is usually lost by the emission of phonons but, in certain semiconductors, the recombination center can allow the emission of a photon. For example, in N-doped GaP, the electron is first capture in the N impurity and the transition involves the emission of phonons. The electron at the N-center then recombines with a hole radiatively, and emits a photon. N-doped GaP is an LED material. See *direct recombination, direct recombination capture coefficient, recombination centers.*

Recombination centers are defects or impurities in a crystal that facilitate the recombination of electrons and holes. Usually, the recombination process is nonradiative with the excess energy being dissipated through interaction with the lattice (i.e. phonons). Radiative recombination can also occur, and involves excess energy being emitted in the form of light. An example of radiative recombination is for GaP co-doped with Zn and O to form a so-called isoelectronic center. Recombination involving these centers results in efficient red luminescence under injection, and is used for making inexpensive red light-emitting diodes. In general, all recombination processes that do not result in the emission of photons are termed *nonradiative*. Several processes are included in this category – these include Auger recombination, surface recombination, and phonon emission. In the Auger effect, the energy released by a recombining carrier is transferred to another carrier which subsequently dissipates energy by phonon emission. A large number of possible Auger processes can occur including those involving two electrons and a single hole or two holes and a single electron, for example. The probability of such processes will depend on carrier concentration, temperature and the semiconductor bandgap. In the case of surface recombination, carriers within a diffusion length of a semiconductor surface can recombine nonradiatively through this band of noncrystal states. See *recombination.*

Recombination current flows under forward bias to replenish the carriers recombining in the space charge (depletion) layer (SCL). Typically the recombination current is described by

$$I = I_{ro}[\exp(eV/k_BT) - 1]$$

where I_{ro} is a constant, e is the electronic charge, k_B is the Boltzmann constant and T is the temperature (kelvins). The constant I_{ro} contains the details of the carrier recombination process in the space charge layer, for example, the mean recombination times in the SCL in the p and n-sides, e.g.

$$I_{ro} = \frac{en_i}{2}\left(\frac{W_p}{\bar{\tau}_e} + \frac{W_n}{\bar{\tau}_h}\right)$$

where n_i is the intrinsic concentration, W_p and W_n are the widths of the depletion region on the p- and n-side (under the bias V), and $\bar{\tau}_e$ and $\bar{\tau}_h$ are the mean recombination times in W_p and W_n, respectively

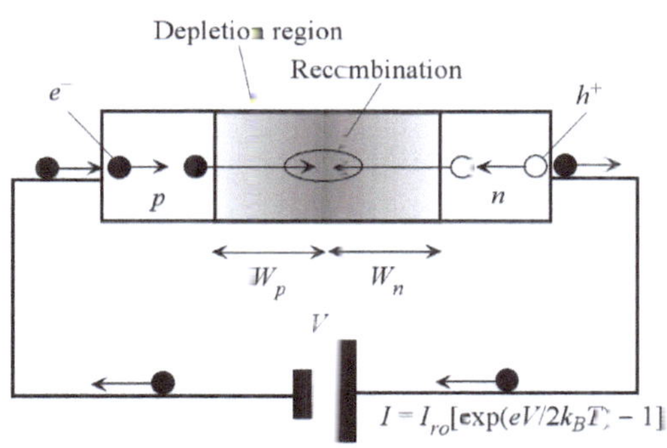

Recombination current in a *pn* junction is due to the recombination of injected electrons and holes in the depletion region.

Reddy equation for the refractive index of silicates see *optical electronegativity.*

Reduced rate (Statz–De Mars) equations for steady state solutions and stability conditions of a single-mode laser can be described by considering an ideal (open) two-level model as a starting point. N_1 and N_2 [m^{-3}] denote the populations of the lower, 1, and upper, 2, levels, of energy E_1 and E_2, respectively; Λ_i [$m^{-3}s^{-1}$] ($i = 1$ or 2) is the pumping rate into, and γ_i [s^{-1}] the relaxation rate from

Reduced rate (Statz–De Mars) equations

level i; $W\,[\mathrm{s}^{-1}]$ is the rate of stimulated processes, which is proportional to the optical intensity. The gain coefficient $[\mathrm{m}^{-1}]$ is $\alpha = \sigma\Delta N$, in which σ is the gain cross-section $[\mathrm{m}^2]$ and $\Delta N = N_2 - N_1$ the inversion population. For simplicity, suppose $h\nu = E_2 - E_1$, $\gamma_1 = \gamma_2 = \gamma$ and neglect the rate A of spontaneous emissions compared with W and γ. No other transition mechanisms are considered.

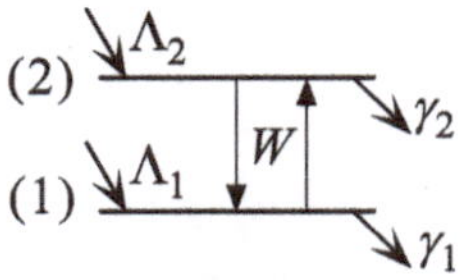

Reduced rate equations An open two-level system with pumping Λ_i, relaxation γ_i and stimulated processes W, neglecting spontaneous emission.

The rate equations are

$$dN_1/dt = +\Lambda_1 - (W+\gamma)N_1 + WN_2,$$
$$dN_2/dt = +\Lambda_2 + WN_1 - (W+\gamma)N_2.$$

Let $\tilde{N}_i = \Lambda_i/\gamma$ denote the steady state value of the population of level i for a vanishingly small laser field ($W = 0$), and $\Delta\tilde{N} = \tilde{N}_2 - \tilde{N}_1$. In the laser cavity, where c is the light velocity and α_L is the cavity losses, the optical intensity and population inversion are coupled as

$$dW/dt = cW(\sigma\Delta N - \alpha_L),$$
$$d\Delta N/dt = -\gamma(\Delta N - \Delta\tilde{N}) - 2W\Delta N.$$

The inversion population at threshold is $\Delta N_\mathrm{th} = \alpha_\mathrm{L}/\sigma$. We introduce the dimensionless parameters $Y_\mathrm{P} = \Delta\tilde{N}/\Delta N_\mathrm{th}$ (reduced pumping), $D = \Delta N/\Delta N_\mathrm{th}$ (reduced population inversion), $\tau = \gamma t$ (reduced time), $I = 2W/\gamma$ (reduced laser intensity) and $k = c\alpha_\mathrm{L}/\gamma$. The reduced rate equations, as derived by Statz and deMars in 1960, read:

$$dI/d\tau = kI(D-1),$$
$$dD/d\tau = Y_\mathrm{P} - D(1+I).$$

The steady state behavior is obtained by putting $(d/d\tau) = 0$. At threshold, $Y_\mathrm{P} = 1$. Below threshold, $Y_\mathrm{P} < 1, D = Y_\mathrm{P}, I = 0$. Above threshold, $Y_\mathrm{P} > 1, D = 1, I = Y_\mathrm{P} - 1$.

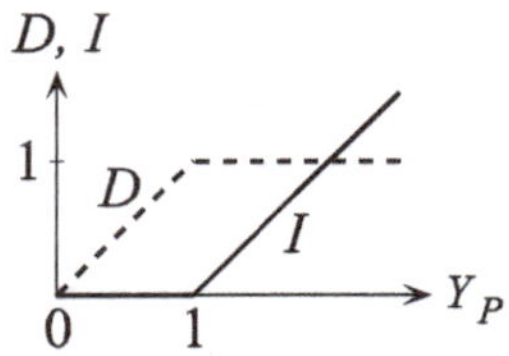

Reduced rate equation Dependence of reduced population inversion, D, and reduced laser intensity, I, on reduced pumping Y_P. The steady state evolution of reduced laser intensity I (straight line) and population inversion D (dashed line) as a function of reduced pumping Y_P. Above threshold, the steady state gain is "clamped" (through D) to its threshold value.

Linear stability analysis consists in following the response of the laser to a small perturbation: the system is stable if it returns to its initial state. To do so, we linearize the reduced rate equations in the vicinity of a steady state operating point. Consider $Y_\mathrm{P} > 1$, and the solution $\{I_0 = Y_\mathrm{P} - 1, D_0 = 1\}$. The perturbed intensity and population inversion are $I(\tau) = I_0 + \delta I(\tau)$ and $D(\tau) = D_0 + \delta D(\tau)$, with $\delta I \ll I_0$ and $\delta D \ll D_0$. By neglecting the second-order term $\delta I \delta D$, we get

$$\delta I' = k\delta D(Y_\mathrm{P} - 1),$$
$$\delta D' = -\delta I - Y_\mathrm{P}\delta D,$$

where the prime (′) stands for the derivative with respect to the (reduced) time τ. An intensity perturbation δI is a solution to a second-order differential equation with constant coefficients, quite similar to that of a damped oscillator,

$$\delta I'' + Y_\mathrm{P}\delta I' + k(Y_\mathrm{P} - 1)\,\delta I = 0,$$

admitting exponential solutions as $\delta I = \delta I_0 \exp(\mu\tau)$. The laser behavior depends on two possible eigenvalues for μ:

$$\mu^{(\pm)} = -Y_\mathrm{P} \pm \left[Y_\mathrm{P}^2 - 4k(Y_\mathrm{P} - 1)\right]^{1/2}.$$

The key parameter is k, whereas Y_P remains adjustable by pumping. If $k \ll 1$, both roots are real and negative, the oscillation is *overdamped* and the system is called a *class A laser*. If not, the system is a *class B laser*. Around threshold, for sufficiently high k such as $k(Y_P - 1) \gg Y_P^2/4$, we get $\mu^{(\pm)} \approx -(Y_P/2) \pm j\Omega'$, with $\mathrm{Re}[\mu] < 0$ and $\Omega' = [k(Y_P - 1)]^{1/2}$. The laser oscillates at eigenfrequency Ω' while returning to its initial state in an *underdamped* way.

Reduced slope fiber see *dispersion*.

Reflectance is the fraction of power in the reflected electromagnetic wave with respect to the incident power. For a material whose complex refractive index is $n - jK$, where n is the refractive index and K the extinction coefficient, for normal incidence, the reflectance R is given by

$$R = \left|\frac{1 - n + jK}{1 + n - jK}\right|^2 = \frac{(1 - n)^2 + K^2}{(1 + n)^2 + K^2}$$

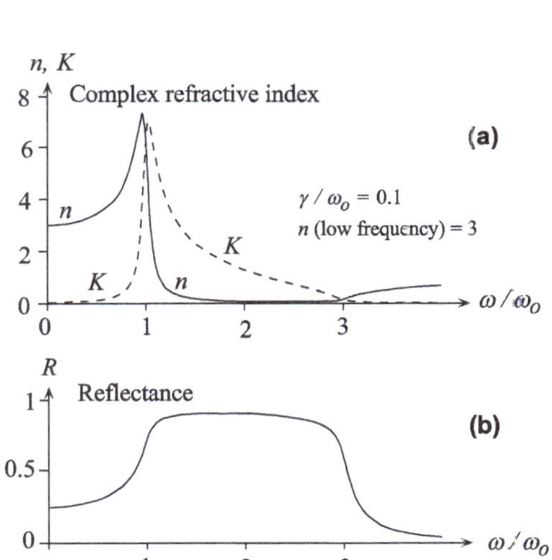

Reflectance (a) Refractive index and extinction coefficient vs. normalized frequency for a single-frequency dipole oscillator model for $\gamma/\omega_o = 0.1$ and low frequency $n = 3$. (b) Reflectance vs. normalized frequency. γ represents the losses and ω_o is the resonant frequency.

which reduces to the usual form when the extinction coefficient $K = 0$. Notice that whenever K is large, for example over a range of wavelengths, the absorption is strong, and the reflectance is almost unity. The light is then reflected, and any light in the medium is highly attenuated. The figure shows the dependence of R on the frequency for a single resonant frequency (ω_o) dipole oscillator model; that is when the electromagnetic wave induces forced dipole oscillations in which the single resonant frequency is ω_o and the damping factor is γ. It is apparent that R reaches its maximum value at a frequency slightly above $\omega = \omega_o$, and then remains high until ω reaches nearly $3\omega_o$; the reflectance is substantial while absorption is strong. See *complex propagation constant, complex refractive index, Lorentz oscillator*.

Reflecting or **reflective prism** see *prism*.

Reflection is an electromagnetic phenomenon in which an electromagnetic wave (the incident wave) that is traveling in a particular medium (with refractive index n_1) becomes incident at an interface with a second medium (n_2) so that some or all of the wave is "returned" back into the first medium (n_1); the returned wave is the "reflected wave." The n_1/n_2 interface constitutes the surface on which reflection takes place. The two laws of reflection are: (1) If we use rays to represent waves, then the incident ray, the reflected ray and the normal to the surface are all in one plane, called the *plane of incidence*. (2) If we measure the angle of incidence θ_i as the angle between the wavevector of the incident beam and the normal to the surface, and the angle of reflection θ_r as the angle between the reflected wave's wavevector and the normal again, then $\theta_r = \theta_i$. Reflection from surfaces is classified as either *specular* or *diffusive*. If reflection occurs from a surface that is smooth, that is, the surface roughness features (i.e. irregularities) are much less than the wavelength λ of light, then the reflection is *specular*. On the other hand, if the surface irregularities are greater than λ but less than the extent of the incident light beam, the reflection is *diffuse*. The rays become incident at different incidence angles on various differently shaped irregular surfaces and hence the reflected rays do not appear to follow the law of reflection with respect to the *whole* surface; the rays appear to be diffusing in various seemingly random directions (reminiscent to being scattered).

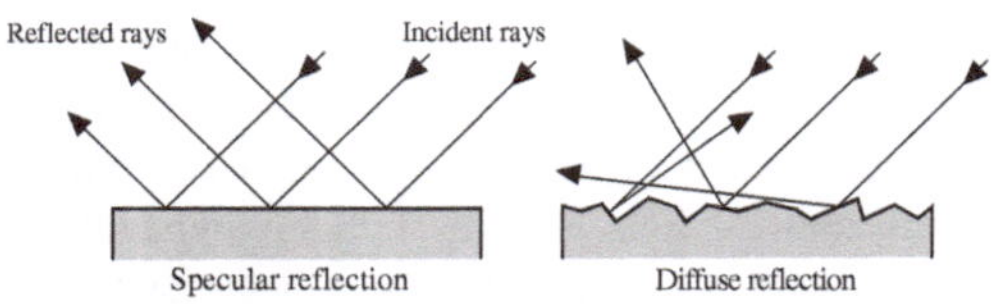

Reflection Specular and diffuse reflections from optically smooth and rough surfaces.

Reflection coefficient is the ratio of the amplitude of the reflected EM (electromagnetic) wave to that of the incident wave, both in the same polarization state. It can be positive, negative or a complex number which then represents a phase change. Suppose that $E_{io\perp}$ and $E_{ro\perp}$ are the electric field amplitudes of the incident and reflected waves, respectively, where the fields are perpendicular ($\perp$) to the plane of incidence; then the reflection coefficient $r_\perp$ for this particular polarization state is defined by $r_\perp = E_{ro\perp}/E_{io\perp}$. Similarly, one can define a reflection coefficient $r_{//}$ for the field components $E_{io//}$ and $E_{ro//}$ parallel (//) to the plane of incidence by $r_{//} = E_{ro//}/E_{io//}$. The reflection coefficients can be found by using the boundary conditions, and lead to the well-known *Fresnel equations*. See *Fresnel's equation*.

Refracting or **refractive prism** see *prism*.

Refraction is a change in the direction of a propagating wave when it enters a medium with a different refractive index. A wave that is incident at a boundary between two media with different refractive indices experiences refraction and changes direction in passing from one to the other medium. The angles of incidence and refraction obey *Snell's law*. If light traveling in a medium with index n_1 is incident on a medium of index n_2, and if the angles of incidence and refraction (transmission) are θ_i and θ_t, then according to Snell's law:

$$\frac{\sin\theta_i}{\sin\theta_t} = \frac{n_2}{n_1}.$$

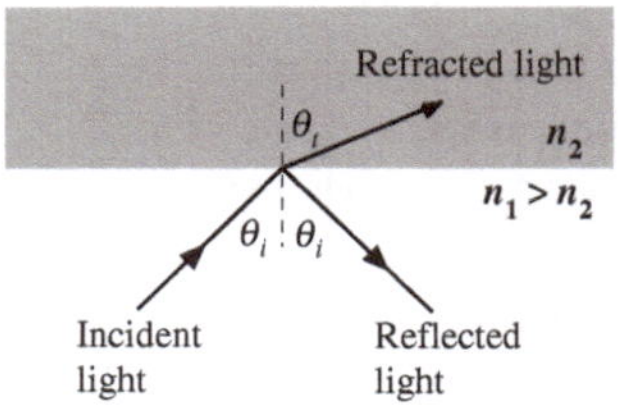

Refraction In this example, a light wave traveling in a more dense medium strikes a less dense medium. The angle of incidence is less than the critical angle for total internal reflection, and the incident wave experiences refraction. There is also a reflected wave.

Refractive index, n, of an optical or dielectric medium is the ratio of the velocity of light c in vacuum to its velocity v in the medium; $n = c/v$. If ε_r is the relative permittivity at the frequency at which n is of interest, then $n = (\varepsilon_r)^{1/2}$. The refractive index depends on the wavelength of light by virtue of the polarization (ε_r) depending on the frequency of the electric field; for example, in glasses n decreases with increasing λ. The dependence of n on the wavelength is called *dispersion*. An intuitive guide to explaining dispersion in insulators is based on a single-oscillator model in which the electric field in the light induces forced dipole oscillations in the material (displaces the electron shells to oscillate about the positive nucleus) with a single resonant frequency ω_0. Neglecting damping (losses), at frequencies $\omega < \omega_0$, if ε_{DC} is $\varepsilon_r(0)$, the DC permittivity, then a simple Lorentz oscillator model predicts that

$$n^2 \approx 1 + (\varepsilon_{DC} - 1)\frac{\omega_0^2}{\omega_0^2 - \omega^2}; \qquad \omega < \omega_0.$$

Since, $\lambda = 2\pi c/\omega$, defining $\lambda_o = 2\pi c/\omega_o$ as the resonance wavelength, then

$$n^2 \approx 1 + (\varepsilon_{DC} - 1)\frac{\lambda^2}{\lambda^2 - \lambda_o^2}; \qquad \lambda > \lambda_o$$

The normal dispersion region is the frequency range below ω_o where n falls as the frequency decreases, that is, n decreases as the wavelength λ increases. The *anomalous dispersion* region is the frequency range above ω_o where n decreases as ω increases. While intuitively useful, the above dispersion relation is far too simple. More rigorously, we have to consider the dipole oscillator quantum mechanically which means a photon excites the oscillator to a higher energy level. The result is that we would have a series of $\lambda^2/(\lambda^2 - \lambda_i^2)$ terms with various weighting factors f_i that add to unity, where λ_i represent different resonance frequencies. The weighting factors f_i involve quantum mechanical matrix elements. In practice the dispersion relationships can be quite complicated. For example, the *Sellmeier dispersion* equation is an empirical expression for the wavelength λ dependence of n that is in the form of a series of $\lambda^2/(\lambda^2 - \lambda_i^2)$, where λ_i is a constant, i.e.

$$n^2 = 1 + \frac{A_1\lambda^2}{\lambda^2 - \lambda_1^2} + \frac{A_2\lambda^2}{\lambda^2 - \lambda_2^2} + \frac{A_3\lambda^2}{\lambda^2 - \lambda_3^2} + \cdots$$

where A_1, A_2, A_3 and λ_1, λ_2 and λ_3 are constants, called *Sellmeier coefficients*, that are determined by fitting this expression to the experimental data. Higher terms involving A_4 and higher A coefficients can generally be neglected in representing n vs. λ behavior over typical wavelengths of interest. Such dispersion relationships are essential in designing photonic devices such as waveguides.

The refractive index of a semiconductor material typically decreases with increasing energy bandgap E_g. There are various empirical and semiempirical rules and expressions that relate n to E_g. In *Moss's rule*, n and E_g are related by $n^4E_g = K =$ constant (roughly ~100 eV). In the Hervé–Vandamme relationship,

$$n^2 = 1 + \left(\frac{A}{E_g + B}\right)^2$$

where A and B are constants ($A \approx 13.6$ eV and $B \approx 3.4$ eV and $dB/dT \approx 2.5 \times 10^{-5}$ eV K^{-1}). The refractive index typically increases with increasing temperature. The temperature coefficient of refractive index TCRI of semiconductors can be found from the *Hervé–Vandamme relationship*

$$\text{TCRI} = \frac{1}{n}\cdot\frac{dn}{dT} = \frac{(n^2-1)^{3/2}}{13.6n^2}\left[\frac{dE_g}{dT} + \frac{dB}{dT}\right].$$

TCRI is typically in the range 10^{-6}–10^{-4}. See *Sellmeier dispersion equation, Lorentz oscillator, Lorentz-Lorenz refractive index.*

Refractometer is an optical instrument that allows the measurement of the refractive index n of a substance. Various optical principles have been used to determine the refractive index but two most common methods involve interference techniques and critical angle measurements between the material of interest and a medium whose index is known. If a material can be shaped into a prism, then a spectrometer can be used to measure the minimum angle of deflection δ_{min} when light is passed through a prism, and then, using the prism angle, one can calculate the refractive index of the prism material from δ_{min}. The *Abbe refractometer* is a widely used instrument for measuring the refractive index n of various samples, usually liquids. The liquid is placed on a prism that has a high and known refractive index n_G. The light that is reflected from the liquid–glass interface is examined to find the border between the bright and dark regions, which identifies the critical angle. The light that experiences total internal reflection from the liquid–glass interface makes up the bright region, and the dark region corresponds to very little reflection inasmuch as at those angles the light at the interface refracts

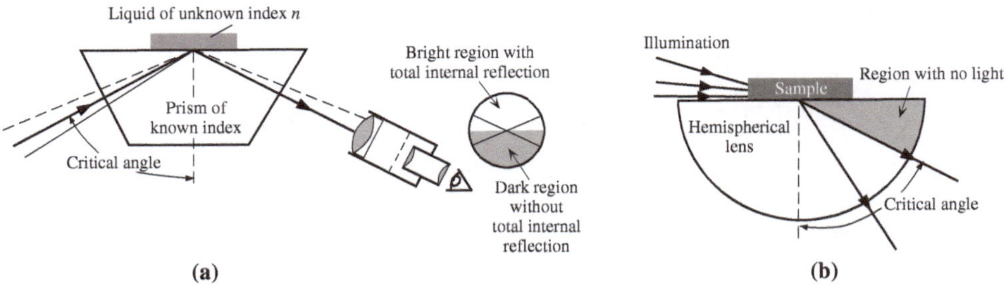

Refractometer (a) The principle of the Abbe refractometer for measuring the refractive index of a sample such as a liquid. The prism has a large known refractive index. The telescope or the sample is rotated until the border of the total internal reflection region lies on cross-hairs, and the precalibrated instrument then uses the critical angle to calculate and display the refractive index of the sample. (b) An Abbe refractometer using a hemispherical glass lens. The border of the dark region identifies the critical angle.

into the liquid. The critical angle θ_c is then given by $\sin\theta_c = n/n_G$. The instrument can be calibrated against known refractive index materials at various wavelengths, and n can be calculated by a microprocessor and displayed as in digital refractometers. The prism can be a hemispherical lens and the illumination can be done from the side as well, as shown in the figure. In interference techniques, an interferometer is used in which a monochromatic beam is split into two paths and then the waves are made to interfere on a screen and form an interference pattern such as Young's fringes. The unknown index material of length L is then placed in one of the paths, which modifies the phase change in this path by nL, and thereby shifts the interference pattern by a certain amount d. The shift d in the pattern is used to find n given L. The *Rayleigh refractometer* is essentially based on this principle. See *spectrometer, Swanepoel technique.* (S. Singh, *Phys. Scr.*, **65**, 167, 2002.)

Relative aperture see *f-number*.

Relative permittivity, ε_r, is a measure of the polarizability of a dielectric medium. In an applied field **E**, a medium becomes polarized and develops a polarization **P**. The polarization of the medium depends on **E** or dielectric constant of the dielectric. In electrostatics the standard definition of ε_r is the fractional increase in the stored charge per unit voltage on the capacitor plates due to the presence of the dielectric between the plates (the whole space between the plates is assumed to be filled). It is also defined as the fractional increase in the capacitance of a capacitor when the insulation between the plates is changed from a vacuum to a dielectric material, keeping the geometry the same. While these standard definitions are based on using the medium to store charge (electrostatic definitions), we can also view ε_r to represent the fractional increase in the stored energy per unit volume for a given field when free space is replaced by a dielectric medium; $\varepsilon_r = E_{volume}(\text{medium})/E_{volume}(\text{vacuum})$ at the same field value where E_{volume} is the energy stored per unit volume at a location where the magnitude of the field is E. We know from electrostatics that $E_{volume}(\text{vacuum}) = (^1/_2)\varepsilon_o E^2$ and $E_{volume}(\text{medium}) = (^1/_2)\varepsilon_o\varepsilon_r E^2$. The energy storage is due to the polarization of the medium, which may be electronic, ionic or dipolar polarization in origin. For an isotropic medium, a given field **E** induces polarization **P** in the dielectric, which would be in the same direction. ε_r relates the "effect" P to the "cause" E by $P = \varepsilon_o(\varepsilon_r - 1)E$. In a linear medium, ε_r is independent of the field. If χ_e is the *electric susceptibility*, then $\varepsilon_r = 1 + \chi_e$. See *electric susceptibility*.

Relaxation oscillations in a laser occur as it is turned on, or after a perturbation of the pumping power has caused it to depart from a steady-state operation. It is an essentially transient behavior mostly observed when the recovery time of the population inversion is substantially longer than the laser cavity decay time. The output of a continuously pumped solid state laser typically consists of irregular spikes, constantly excited by mechanical and thermal fluctuations. Relaxation oscillations are a fundamental characteristic of Fabry–Perot lasers. The oscillations can vary in frequency, amplitude, and duration. Relaxation oscillations have not been a serious problem for low-speed optical links, but are becoming a major issue for gigabit links such as the gigabit Ethernet. The high-speed modulation response of such lasers is directly related to ultrafast carrier dynamics in the laser medium The transient response of any laser to a change in pumping conditions (e.g. the drive current in a semiconductor laser) is determined by the interplay between the dynamics of population inversion and that of the laser field. In particular, both quantities respond to such a change by undergoing *coupled* damped oscillations towards steady state values, with the damping mechanism provided by stimulated emission and gain saturation. These relaxation oscillations produce a resonance enhancement of the laser modulation response at their characteristic frequency f_c. The modulation response becomes increasingly weak as the drive frequency is increased above this resonance, so that f_c provides a measure of the laser intrinsic modulation bandwidth. An interesting solution for ultrafast optical links is to use *quantum cascade lasers*. These lasers are mid-infrared semiconductor light sources based on inter-subband electron transitions in coupled quantum-well systems – a unique feature of these devices is their ultrafast carrier relaxation lifetime. Electron equilibration between energy subbands is dominated by longitudinal optical phonon relaxation which occurs on a picosecond time-scale. By comparison, in conventional semiconductor lasers based on inter-band transitions the intrinsic carrier lifetime is on the order of a few nanoseconds, and even slower relaxation time constants are found in most other laser systems. In the case of quantum cascade lasers, since relaxation of the carrier density to equilibrium is dominated by phonon scattering, which is an ultrafast mechanism, transient oscillations of the population inversion and hence of the photon density are overdamped and no resonance will therefore appear in the frequency response – hence there is the characteristic absence of relaxation oscillations in these lasers. The modulation dynamics of quantum cascade lasers are, to a large extent, simply limited by their picosecond carrier lifetime. Thus, intrinsic bandwidths in excess of 100 gigahertz are feasible.

Repeater, optical repeater, is a device or an electronic circuit that restores the received optical signal back to its original shape with amplification so that it can be sent again along a transmission line.

Resolution see *resolving power*.

Resolvance see *diffraction grating*.

Resolving power of an optical system or an imaging system (such as an image detector or sensor), qualitatively, is its ability to generate an image in which as fine features in the object as possible can be resolved; or two neighboring points in the object with different contrasts can be resolved in the image. It can be defined quantitatively in a number of ways depending on the particular imaging task and the system used for that task. The resolution of a TV picture has quite a different meaning from the inherent resolution in the image generated by a simple lens. There is also a distinct difference in the resolving power of a lens and that of a photographic emulsion. The resolving power of an optical system such as a telescope is the minimum angular separation $\Delta\theta_{\min}$ of two points of equal luminance that can be distinguished. $\Delta\theta_{\min}$ is proportional to the wavelength λ of light but inversely proportional to the aperture diameter D, e.g. the objective lens or the primary mirror diameter. Larger aperture telescopes have a better resolution, and resolution increases with decreasing wavelength. Another measure of an imaging system's resolving capability is its *modulation transfer function*, MTF, which measures the ability of the system to faithfully reproduce the different spatial frequency components that make up the object's true image (the true image would contain all the necessary spatial frequencies that represent an accurate image of the object). MTF is taken as 100% at the zero spatial frequency and then it decreases with the spatial frequency f. The *spatial resolution limit* of an imaging system is defined as that critical spatial frequency f_r (line pairs per mm or cycles per mm) beyond which the MTF falls below some threshold value, $MTF_{threshold}$. In many practical applications, $MTF_{threshold}$ is taken as 20%, whereas in others the threshold is set to the noise equivalent modulation, NEM, which is defined as the required modulation depth that would generate a signal-to-noise ratio of unity. (MTFs below NEM are not useful.) The uppermost spatial frequency where MTF = 0 (the ultimate resolution limit) is usually avoided in resolution specifications since MTFs are difficult to measure and the high spatial frequency region has more measurement errors, as well as possible Fourier transform errors. See *diffraction grating, Rayleigh criterion, modulation transfer function.*

Resonance, in general, is a "state" of a system in which the smallest driving oscillations can most easily build up into large-amplitude oscillations of the system. For example, a parallel inductor-capacitor, (LC), circuit is in resonance when the driving oscillator frequency is $f = 1/[2\pi(LC)^{1/2}]$ at which the circuit oscillates naturally and consumes the minimum energy from the driving oscillator.

Resonant frequency, in general, is the frequency of an allowed wave or oscillation within a given structure; a natural frequency of the structure. A *resonant frequency of an optical cavity* is a frequency of an allowed mode of electromagnetic oscillations within the optical cavity. The resonant frequency depends on the cavity properties: the cavity size, reflectances and the refractive index.

Response time is the duration of time it takes for a device to produce the required output in response to a step input.

Responsivity is the photocurrent generated by a photodiode per unit incident optical power. It depends on the quantum efficiency (QE) and the wavelength (λ) of the incident radiation. The responsivity R of a photodiode characterizes its performance in terms of the photocurrent generated (I_{ph}) per incident optical power (P_o) at a given wavelength by

$$R = \frac{\text{Photocurrent (A)}}{\text{Incident Optical Power (W)}} = \frac{I_{ph}}{P_o}.$$

From the definition of *quantum efficiency* (QE)

$$R = \eta \frac{e}{h\upsilon} = \eta \frac{e\lambda}{hc}$$

where η is the quantum efficiency which depends on the light wavelength λ, h is Planck's constant, υ is the light frequency and e is the electronic charge. The responsivity therefore clearly depends on the wavelength. R is also called the *spectral responsivity* or *radiant sensitivity*. The R vs. λ characteristics represent the spectral response of the photodiode and are generally provided by the manufacturer. Ideally with a quantum efficiency of 100% ($\eta = 1$), R should increase with λ up to λ_g. In practice, QE limits the responsivity to lie below the ideal photodiode line with upper and lower wavelength limits as shown for a typical Si photodiode in the figure. The QE of a well-designed Si photodiode in the wavelength range 700–900 nm can be close to 90–95%. Operation under maximum QE corresponds to the condition

$$\frac{dR}{d\lambda} = \frac{R}{\lambda}$$

which corresponds to the point where a straight line through the origin just touches (it is a tangent to) the R vs. λ responsivity curve.

Reststrahlen absorption

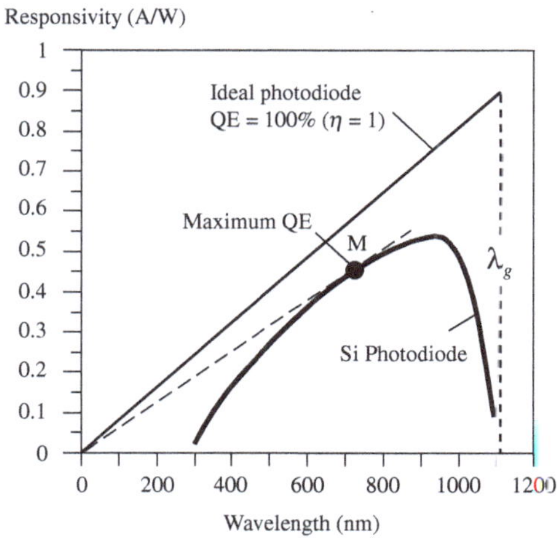

Responsivity (R) vs. wavelength (λ) for an ideal photodiode with QE $= 100\%$ ($\eta = 1$) and for a typical inexpensive commercial Si photodiode. Point M represents operation at the maximum QE.

Reststrahlen absorption is the absorption of radiation by a material in the infrared wavelength region due to the interaction of the electromagnetic waves with optical lattice vibrations. In the infrared wavelength region, ionic crystals reflect and absorb light strongly due to the resonance interaction of the electromagnetic (EM) wave field with the optical phonons. A simple single-dipole oscillator model assumes that ions in the crystal execute forced oscillations driven by the electric field in the EM wave so that

$$\varepsilon_r = \varepsilon_r' - j\varepsilon_r''$$

$$= \varepsilon_{r\infty} + \frac{\varepsilon_{r\infty} - \varepsilon_{ro}}{\left(\dfrac{\omega}{\omega_T}\right)^2 - 1 + j\dfrac{\gamma}{\omega_T}\left(\dfrac{\omega}{\omega_T}\right)}$$

where ε_{ro} and $\varepsilon_{r\infty}$ are the relative permittivity at $\omega = 0$ (very low frequencies) and $\omega = \infty$ (very high frequencies) respectively, γ is the loss coefficient per unit reduced mass (see below) representing the rate of energy transfer from the EM wave to optical phonons and ω_T is a transverse optical phonon frequency, that is related to the nature of bonding between the ions in the crystal, i.e. $\omega_T^2 = \omega_0^2(\varepsilon_{r\infty} + 2)/(\varepsilon_{ro} + 2)$ in which $\omega_0^2 = \beta/M_r$, β is the force constant in "restoring force $= -\beta \times$ displacement" and M_r is the reduced mass of the negative and positive ions in the crystal. The loss, ε_r'' and the absorption are maxima when $\omega = \omega_T$, and the wave is attenuated by the transfer of energy to the transverse optical phonons, thus the EM wave couples to the transverse optical phonons. At $\omega = \omega_L$, the wave couples to the longitudinal optical phonons. The refractive index n from ionic polarization vanishes, and the reflectance is a minimum. The figure shows the optical properties of AlSb in terms of n, K and R vs. wavelength. The extinction coefficient K and reflectance R peaks occur over about the same wavelength region, corresponding to the coupling of the EM wave to the transverse optical phonons. At a wavelength close to $\lambda_T = 2\pi/\omega_T$, n and K peak, and there is a strong absorption of light which corresponds to the EM wave resonating (coupling) with the TO lattice vibrations, then R rises sharply. ω_T and ω_L are related through

$$\omega_T^2 = \omega_L^2(\varepsilon_{r\infty}/\varepsilon_{ro})$$

which is called the *Lyddane–Sachs–Teller relation*. It relates the high- and low-frequency refractive indices $n_\infty = \sqrt{\varepsilon_{r\infty}}$ and $n_o = \sqrt{\varepsilon_{ro}}$ to ω_T and ω_L. The complex refractive index $N = n + jK$ is then

$$N^2 = (n + jK)^2 = \varepsilon_r' - j\varepsilon_r''$$

$$= \varepsilon_{r\infty}\left[1 + \frac{\omega_L^2 - \omega_T^2}{\omega_T^2 - \omega^2 + j\gamma\omega}\right].$$

Taking CdTe as an example, and substituting the values for $\varepsilon_{r\infty}$, ω_T, ω_L and γ from the table into the above expression, at $\lambda = 70$ μm or $\omega = 2.6909 \times 10^{12}$ rad s^{-1}, $n = 3.20$ and $K = 0.00235$. The above expression is usually sufficient to predict N in the infrared for many compound semiconductors and ionic crystals, though for low bandgap semiconductors there may also be a significant contribution arising from photogeneration.

Table: Restrahlen absorption Reststrahlen absorption and refractive index in the IR for selected crystals. ω_T, ω_L and γ are in rad s^{-1}. Data from various sources including E. D. Palik, *Handbook of Optical Constants of Solids* (Academic Press, San Diego, 1985); R. S. Knox and K. J. Teegarden, *Physics of Color Centers* (Academic Press, San Diego, 1968); C. Kittel, *Introduction to Solid State Physic, Sixth Edition* (Wiley, New York, 1986); S. Adachi, *Properties of Group IV, III–V and II–V Semiconductors* (Wiley, Chichester, 2005).

	CdTe	GaAs	InAs	InSb	InP	SiC	α-GaN ($\perp c$)	α-GaN ($//c$)	GaP	GaSb	InAs
ε_{ro}	10.20	12.999	14.181	17.721	12.401	10.004	9.435	10.432	10.7	16.1	14.9
$\varepsilon_{r\infty}$	7.10	11.000	11.700	15.680	9.610	6.700	5.350	5.350	8.5	14.4	12.3
$\omega_T \times 10^{13}$	2.658	5.065	4.109	3.376	5.725	14.948	10.550	9.983	6.9	4.3	4.1
$\omega_L \times 10^{13}$	3.186	5.506	4.524	3.589	6.503	18.265	14.010	13.940	7.6	4.6	4.5
$\gamma \times 10^{13}$	0.124	0.045	0.075	0.054	0.066	0.090	0.113	0.131	–	–	–
	Si	Ge	LiH	LiF	LiCl	LiBr	NaF	NaCl	NaBr	KF	KCl
ε_{ro}	11.70	15.80	12.90	8.90	12.00	13.20	5.10	5.90	6.40	5.50	4.85
$\varepsilon_{r\infty}$	11.70	15.80	3.60	1.90	2.70	3.20	1.70	2.25	2.60	1.50	2.10
$\omega_T \times 10^{13}$	9.900	5.700	11.000	5.800	3.600	3.000	4.500	3.100	2.500	3.600	2.700
$\omega_L \times 10^{13}$	9.900	5.700	21.000	12.000	7.500	6.100	7.800	5.000	3.900	6.100	4.000
$\gamma \times 10^{13}$	–	–	–	–	–	–	–	–	–	–	–
	KI	RbF	RbI	CsCl	CsI	TlCl	TlBr	AgCl	AgBr	MgO	CsF
ε_{ro}	5.10	6.50	5.50	7.20	5.65	31.90	29.80	12.30	13.10	9.80	–
$\varepsilon_{r\infty}$	2.70	1.90	2.60	2.60	3.00	5.10	5.40	4.00	4.60	2.95	2.16
$\omega_T \times 10^{13}$	1.900	2.900	1.400	1.900	1.200	1.200	0.810	1.900	1.500	7.500	–
$\omega_L \times 10^{13}$	2.600	5.400	1.900	3.100	1.600	3.000	1.900	3.400	2.500	14.000	–
$\gamma \times 10^{13}$	–	–	–	–	–	–	–	–	–	–	–

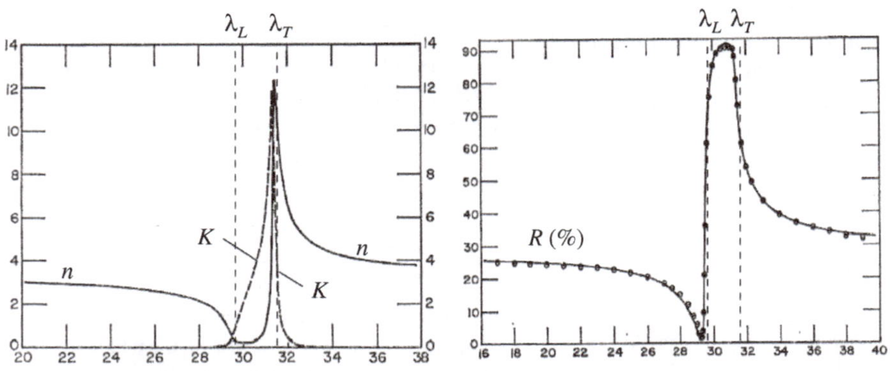

Reststrahlen absorption Infrared refractive index *n*, extinction coefficient *K* (left), and reflectance *R* (right) of AlSb. Note: the wavelength axes are not identical, and wavelengths λ_T and λ_L corresponding to ω_T and ω_L respectively, are shown as dashed vertical lines (data from W. J. Turner and W. E. Reese, *Phys. Rev.*, **127**, 126, 1962).

Retardation see *retarding plates.*

Retarding plates are optical devices that change the state of polarization of an incident light beam. For example, consider a positive uniaxial crystal such as a quartz ($n_e > n_o$) plate that has the optic axis (taken along z) parallel to the plate faces. Suppose that a *linearly polarized* wave is incident at normal incidence on a plate face. If the optical field **E** is parallel to the optic axis ($E_{//}$), then this wave will travel through the crystal as an extraordinary wave (e-wave) with a velocity c/n_e slower than the ordinary wave (o-wave) since $n_e > n_o$. Thus, the optic axis is the "slow axis" for waves polarized parallel to it. If **E** is at right angles to the optic axis ($E_\perp$) then this wave will travel with a velocity c/n_o, which will be the fastest velocity in the crystal. Thus, the axis perpendicular to the optic axis (say x) will be the "fast axis" for polarization along this direction. When a light ray enters a crystal at normal incidence to the optic axis and plate surface, then the o- and e-waves travel along the same direction. We can resolve a linear polarization at an angle α to z into $E_\perp$ and $E_{//}$. When the light comes out at the

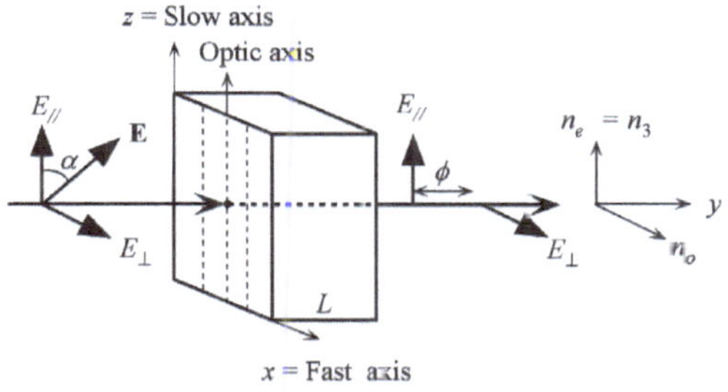

A **retarding plate**. The optic axis is parallel to the plate face. The o- and e-waves, corresponding to ordinary and extraordinary waves, travel in the same direction but at different speeds.

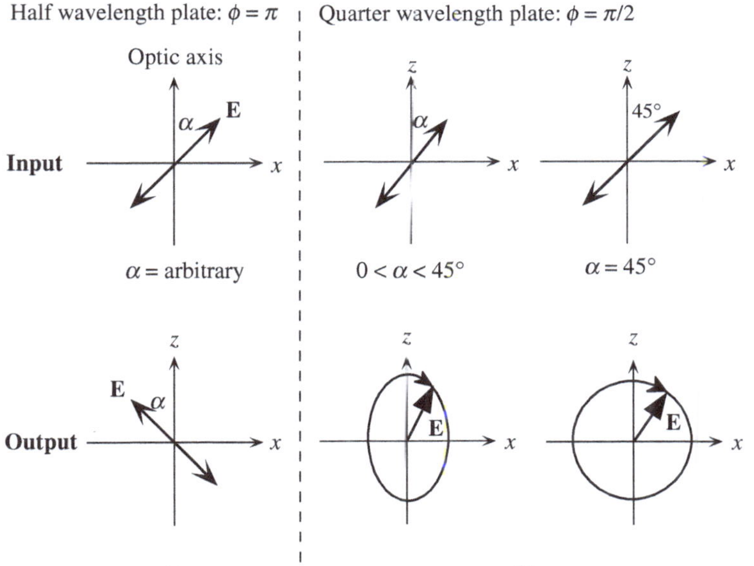

Retarding plate Input and output polarizations of light through (a) a half-wavelength (half-wave) plate and (b) through a quarter-wavelength (quarter-wave) plate.

opposite face these two components would have been phase-shifted by ϕ (one component retarded relative to the other). Depending on the initial angle α of **E** and the length of the crystal, which determines the total phase shift ϕ through the plate, the emerging beam can have its initial linear polarization rotated, or changed into an elliptically or *circularly polarized* light. If L is the thickness of the plate then the o-wave experiences a phase change $k_{o\text{-wave}}L$ through the plate where $k_{o\text{-wave}}$ is the wavevector of the o-wave: $k_{o\text{-wave}} = (2\pi/\lambda)n_o$, where λ is the free space wavelength. Similarly, the e-wave experiences a phase change $(2\pi/\lambda)n_eL$ through the plate. Thus, the phase difference ϕ between the orthogonal components $E_\perp$ and $E_{//}$ of the emerging beam is

$$\phi = \frac{2\pi}{\lambda}(n_e - n_o)L.$$

The phase difference ϕ expressed in terms of full wavelengths is called the *retardation* of the plate. For example, a phase difference ϕ of 180° is a half-wavelength retardation. The polarization of the through-beam depends on the crystal-type, $(n_e - n_o)$, and the plate thickness L. We know that depending on the phase difference ϕ between the orthogonal components of the field, the EM wave can be *linearly*, *circularly* or *elliptically* polarized. A *half-wave plate* retarder has a thickness L such that the phase difference ϕ is π or 180°, corresponding to a half of wavelength ($\lambda/2$) of retardation. The result is that $E_\perp$ is delayed by 180°. If we add the emerging $E_\perp$ and $E_{//}$ with this shift ϕ, **E** would be at an angle $-\alpha$ to the optic axis and still linearly polarized. **E** has been rotated counterclockwise through 2α. A *quarter-wave plate* retarder has a thickness L such that the phase difference ϕ is $\pi/2$ or 90°, corresponding to a quarter of wavelength. If we add the emerging $E_\perp$ and $E_{//}$ with this shift ϕ, the emerging light will be elliptically polarized if $0 < \alpha < 45°$ and circularly polarized if $\alpha = 45°$.

Return-to-zero (RZ) is a digital coding system in which information in binary form is represented by pulses that return to zero, for a duration equal to the pulse width, before the next bit of information. Suppose that T is the information bit period, that is information is sent every T seconds. In RZ coding, the pulse width for information is $T/2$ and is zero for the next half-period $T/2$. In *nonreturn to zero* (NRZ) the information is represented by pulses that do not have to return to zero. The pulse width is the same as the information period T.

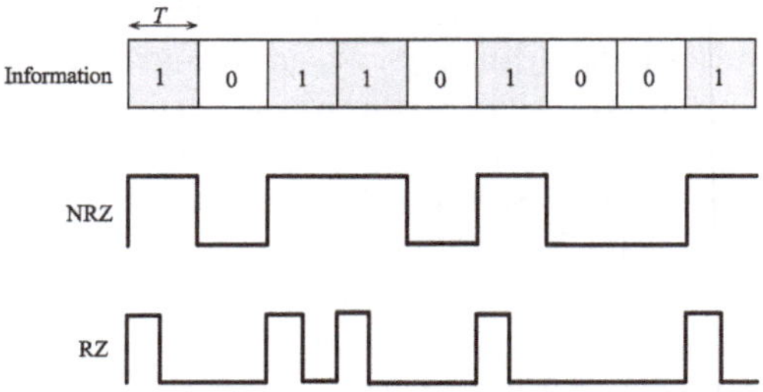

Return to zero digital data.

Reverse bias is the application of an external voltage to a *pn* junction such that the positive terminal is connected to the n-side and negative to the p-side. The applied voltage increases the built-in potential and hence the internal field in the space charge layer. Many *pn* junction based photodiodes are used in reverse bias. See *photodiode*, *pn junction*.

Reverse diode current is the diode current in the dark when the diode is reverse biased. Under reverse bias, the hole concentration in the n-side just outside the space charge layer (SCL) is nearly zero by the law of the junction, whereas the hole concentration in the bulk (or near the negative terminal) is the equilibrium concentration p_{no}, which is small. There is therefore a small concentration gradient and hence a small hole diffusion current towards the SCL.

Similarly, there is a small electron diffusion current from the bulk p-side to the SCL. Within the SCL, these carriers are drifted by the field. This reverse current is basically the Shockley equation under reverse bias and is called the reverse saturation current, I_{so}. This saturation current depends

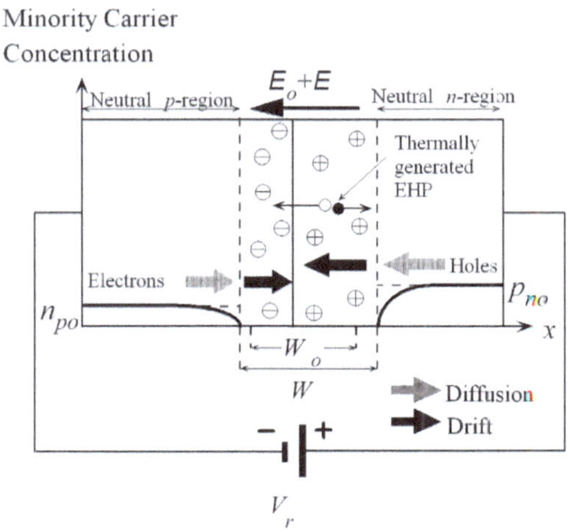

Reverse biased *pn* junction. Minority carrier profiles and the origin of the **reverse diode current**.

on the square of the intrinsic concentration, that is n_i^2, but not on the voltage. The thermal generation of electron–hole pairs (EHPs) in the space charge layer (depletion region) can also contribute to the observed reverse current since the internal field in this layer will separate the electron and hole and drift them towards the neutral regions. This drift will result in an external current $I_{generation}$ due to EHP generation in the SCL. $I_{generation}$ increases slightly with the reverse bias V_r because the SCL width increases with V_r. $I_{generation}$ is proportional to n_i. The total *reverse current* $I_{reverse}$ is the sum of the diffusion and generation components. The relative importance of these terms depends not only on the semiconductor properties but also on the temperature since $n_i \sim \exp(-E_g/2k_BT)$, where E_g is the semiconductor bandgap, k_B is the Boltzmann constant and T is the temperature. For example, the reverse dark current $I_{reverse}$ in a Ge *pn* junction (a photodiode) when plotted as $\ln(I_{reverse})$ vs. $1/T$ as in the figure shows two regions. Above 238 K, $I_{reverse}$ is controlled by n_i^2 because the slope of $\ln(I_{reverse})$ vs. $1/T$ yields an E_g of approximately 0.63 eV, close to the expected E_g of about 0.66 eV in Ge. In this range, the reverse current is due to minority carrier diffusion in neutral regions. Below 238 K, $I_{reverse}$ is controlled by n_i because the slope of $\ln(I_{reverse})$ vs. $1/T$ is equivalent to an $E_g/2$ of approximately 0.33 eV. In this range, the reverse current is due to EHP generation in the SCL via defects and impurities (recombination centers).

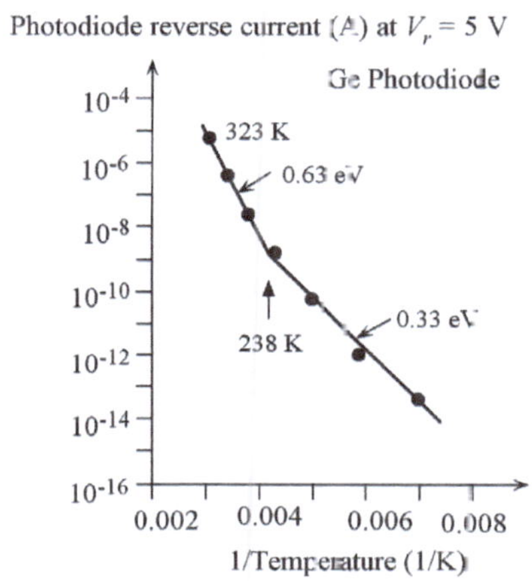

Dark current, i.e. **reverse diode current**, in a Ge *pn* junction as a function of temperature in a $\ln(I_{reverse})$ vs. $1/T$ plot. Above 238 K, $I_{reverse}$ is controlled by n_i^2 and below 238 K it is controlled by n_i. The vertical axis is a logarithmic scale with actual current values. (From D. Scansen and S.O. Kasap, *Cnd. J. Physics*. **70**, 1070, 1992.)

Reverse saturation current is the reverse current that would flow in a reversed biased ideal *pn* junction obeying the Shockley equation. See *reverse diode current*.

RGB see *color*.

Rib or a ridge waveguide is a three-dimensional dielectric structure formed from a layer of thin film on a substrate by partially removing material from the uppermost region adjacent to a patterned strip, resulting in a light-guiding region beneath the strip defined by the dimensions of the strip (namely, its width and height) and the refractive indices of the

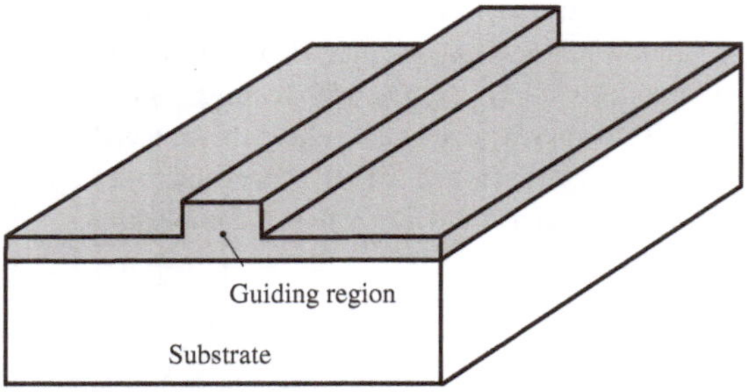

Rib or ridge waveguide

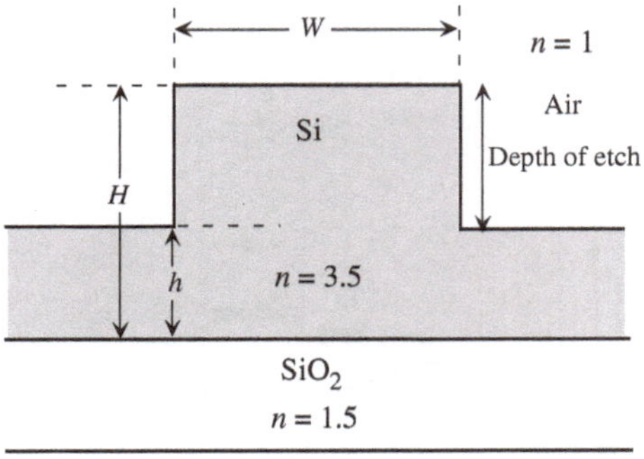

A **rib waveguide** with a width w and height h.

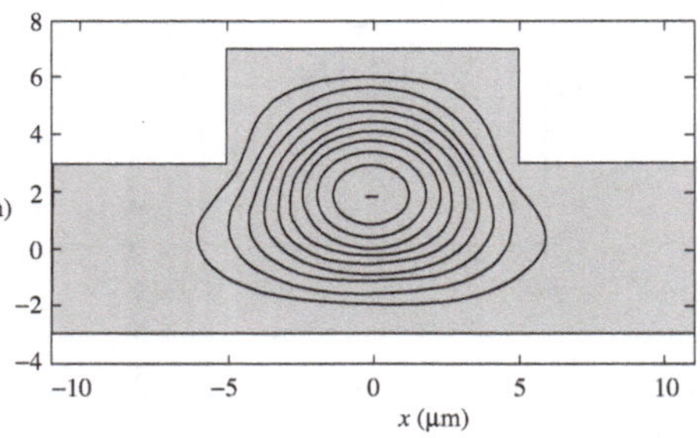

Calculated intensity distribution in cross-section of a Si **ridge waveguide** on an oxide base, where the thickness of the silicon outside the rib region is 6 µm (i.e. for $x < -5$ µm, $x > 5$ µm), and 10 µm in the rib region (i.e. for $-5 < x\,(\mu m) < 5$), and with a rib width of 10 µm. (Timo T. Aalto, Mikko Harjanne, Markku Kapulainen, Paivi Heimala, and Matti J. Leppihalme, *Development of silicon-on-insulator waveguide technology*, in *Integrated Optics: Devices, Materials, and Technologies VIII Proceedings of the SPIE 5355*, 2004.)

respective materials used, as shown in the first figure. Such structures are commonly fabricated using thin film processing, typically photolithography, to pattern and etch the resulting rib waveguide. The waveguide is inherently asymmetric, owing to its geometry. While the rib waveguide appears to be similar to the asymmetric slab waveguide, it is nonetheless a different type of guide inasmuch as optical confinement occurs in two independent dimensions. The second figure shows the intensity distribution across the cross-section of a rib waveguide for the fundamental mode, showing how such guides provide effective confinement for light guiding. (See *strip waveguides*.)

The most common materials that have been used to fabricate such waveguides for planar lightwave circuits include silicon, GaAs, InP, silicon oxide, and polymide. There is considerable interest in *silicon-on-insulator* (SOI) structures for optical waveguides owing to their compatibility with CMOS electronics. SOI waveguides present an interesting optical structure in which there is a large refractive index contrast between the silicon core ($n \approx 3.46$) and the silicon oxide cladding ($n \approx 1.46$). If a rectangular section waveguide is desired, transverse dimensions should be of submicron dimensions to act as a single-mode guide – however, this will result in very poor coupling efficiency to a single mode fibre. Thus, SOI rib waveguides with large (several micron) cross-sections have been developed. In these devices, when the higher order modes propagate down the waveguide, they leak out into the surrounding slab regions either side of the rib, and consequently, near-single-mode propagation in the rib region is realized in practice.

For a rib waveguide with a large cross-section ($h > \lambda$), a key criterion is the *single-mode condition* which can be given for SOI waveguides, due to Soref, as

$$t \leq \alpha + \frac{r}{\sqrt{1-r^2}}; \alpha = 0.3; r \geq 0.5$$

where $t = W/H$ (ratio of rib full width to total height from the Si/SiO_2 interface) and $r = h/H$ are defined in the first figure and the inset of the third figure. This approximation is excellent for $r > 0.45$ as shown by more exact beam propagation method calculations. (R. A. Soref, J. Schmidtchen, K. Petermann, *IEEE Journal of Quantum Electronics*, **27**, 1971, 1991; O. Powell, *J. Lightwave Technol.*, **20**, 1851, 2002.)

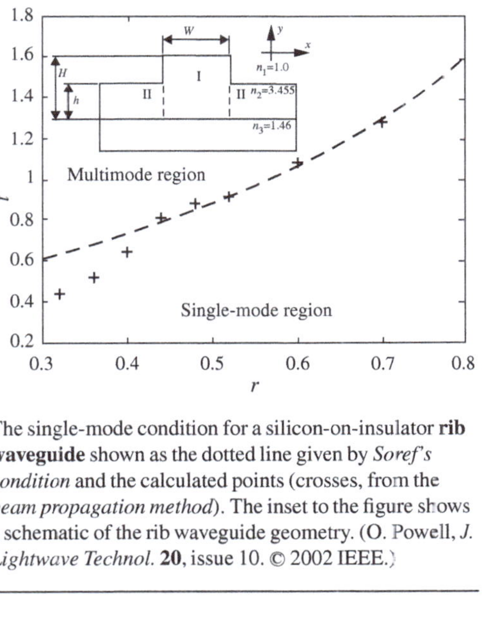

The single-mode condition for a silicon-on-insulator **rib waveguide** shown as the dotted line given by *Soref's condition* and the calculated points (crosses, from the *beam propagation method*). The inset to the figure shows a schematic of the rib waveguide geometry. (O. Powell, *J. Lightwave Technol.* **20**, issue 10. © 2002 IEEE.)

Ridley's lucky drift see *impact ionization*.

Ring resonators are waveguides formed into closed loops or rings coupled to another waveguide by a waveguide coupler, the latter being a region where the curved ring guide is brought proximate to another waveguide over a coupling length *l*. Such ring resonators behave like wavelength filters having a finesse that is directly related to the waveguide loss encountered in one round-trip through the loop. Ring resonators are thus promising candidates for the creation of wavelength selective filters with small channel spacings. Rings can be highly selective and have high out-of-band rejection, suitable for large-scale integration, and have been demonstrated in many technologies and on different materials. These include glass, glass polymer, Si-silicon dioxide, silicon oxynitride, GaAs and GaInAsP/InP. Free spectral ranges may range from a few GHz to some THz, depending on the technology used. Such resonators can serve as *channel-drop filters* in WDM systems – input light comprised of several wavelengths will normally bypass the resonator and exit in the same waveguide. Only wavelengths that satisfy the cavity resonance condition will be "dropped" through the resonator to the other output waveguide. The free-space wavelengths λ_0 are given by the resonance condition $L_c = m\lambda_0/n_{effective}$ where L_c is the cavity length, *m* is an integer and $n_{effective}$ is the effective refractive index of the cavity. The free spectral range FSR of the filter is determined by the cavity round-trip length, FSR $= \lambda_0{}^2/n_{effective}L_c$. In order to pick out a particular wavelength, the FSR must exceed the WDM signal bandwidth which is typically about 10 nm. For an FSR of 10 nm, the resonator round-trip needs to be of the order of $50\lambda_0$, assuming $\lambda_0 \approx 1550$ nm. Such a small radius ring can only be realized by using strongly confined waveguides. The main constraint on the ring design is the choice of waveguide index contrast. The higher the index contrast, the lower the minimum bending radius supported by the waveguide and the higher the maximum achievable free spectral range. However, in using high index contrast waveguides it is often critical to control scattering losses, fiber-to-waveguide coupling losses and tolerance sensitivity.

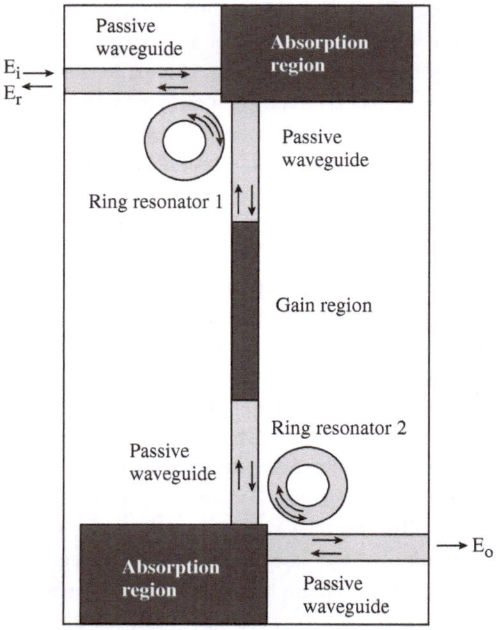

An example of a **ring resonator** structure involving two ring resonators coupled to a laser structure. E_i, E_r and E_o are the incident, reflected and output optical signals. (B. Liu, *IEEE Photonics Technol. Lett.*, **14**, 600, 2002, (© IEEE 2002).)

Rise time is the time it takes for the photocurrent to rise from 10% to 90% of the steady state value when the photodetector is suddenly illuminated as a step function.

Ritland's approach to the refractive index of a glass is to represent the local field E_{local} acting at an ion as being given by bP, where b is a variable parameter, and to represent the refractive index through the Lorentz-Lorenz equation so that in CGS units

$$\frac{n^2-1}{4\pi+b(n^2-1)}=\sum_i N_i\alpha_i$$

where N_i and α_i is the atomic concentration and polarizability of type i ions in the glass. The best b is found by fitting the equation to experimental data. (H. N. Ritland, *J Am Ceram Soc.*, **38**, 86, 1955.)

Rochon prism see *Wollaston prism*.

Roentgen is an ionizing radiation exposure unit that is widely used in x-ray applications; one roentgen (1 R) has been traditionally defined as the amount of radiation that will impart sufficient energy to 1 cm^3 of air to ionize an amount of charge that is 0.334×10^{-9} C at STP (standard temperature and pressure). 1 R ionizes 1 cm^3 of air to generate 0.33 nC of charge. A more recent definition is such that, at STP, 1 R releases 2.58×10^{-4} C of charge per kg of air. The amount of energy that 1 R deposits into a volume of 1 cm^3 of air is 1.14×10^{-8} J/cm^3 or, stated differently, the energy deposited by 1 R into a 1 kg of air is 8.77×10^{-3} J/kg. The definitions based on charge and energy can be related by noting that the average ionization energy of air is 34 eV per ion pair at STP.

Table: Roentgen Definition of 1 roentgen at STP.

1 roentgen on	1 cm^3 of air	1 kg of air
Charge released	0.334×10^{-9} C / cm^3	2.58×10^{-4} C / kg
Energy deposited	1.14×10^{-8} J / cm^3	8.77×10^{-3} J / kg

The *energy fluence* Ψ of incident ionizing radiation is the incident beam energy per unit area, and it increases linearly with the exposure X (roentgens)

$$\frac{\Psi}{X}=\frac{2.58\times10^{-4}W_{\text{air}}}{e(\mu_{\text{en,air}}/\rho_{\text{air}})}$$

$$=\frac{0.00873}{(\mu_{\text{en,air}}/\rho_{\text{air}})}\ \text{J cm}^{-2}\,\text{R}^{-1}$$

where the density of air is ρ_{air} and its *mass energy coefficient* is $\mu_{\text{en,air}}/\rho_{\text{air}}$ at the photon energy of interest, and $\mu_{\text{en,air}}/\rho_{\text{air}}$ is in g cm^{-2}. If E_{photon} is the photon energy, the *photon fluence* Φ, the number of photons incident per unit area, is

$$\Phi=\Psi/E_{\text{photon}}.$$

(H. E. Johns and J. R. Cunningham, *The Physics of Radiology*, Fourth Edition, Springfield, IL: Charles C Thomas Publishers, 1983.)

Rowland ghosts see *diffraction grating*.

Rydberg constant, R_∞, is a constant that is used in the study of atomic spectra and appears in the *Rydberg* or *Rydberg–Ritz* equation which relates the wavelength of emission for the electronic transition from an energy level with a quantum number n_2 to that with n_1 in the hydrogen atom,

$$\lambda = R_\infty\left(\frac{\overline{m}_e}{m_e}\right)\left(\frac{1}{n_1^2}-\frac{1}{n_2^2}\right)$$

where $\overline{m}_e$ is the reduced mass of the electron in the atom ($m_eM/(M+m_e)$; M = proton mass) and R_∞ is the Rydberg constant given by

$$R_\infty = \frac{m_e e^4}{8h^3\varepsilon_0^2 c} = 1.097373\ldots\times 10^7\text{m}^{-1}.$$

One rydberg represents 13.60569 . . . eV, and sometimes the rydberg is used as an energy unit in spectral physics.

S

Sagittal plane is a plane in an optical system that is at right angles to the tangential or meridional plane; it has the off-axis object point, and the principal ray of the meridional plane. See *astigmatism* for an illustration.

Sagnac effect is used in an interferometer in which the phase difference between two light signals, one traveling clockwise and the other counterclockwise in an optical loop, depends on the angular rotation of the loop. The effect is illustrated in the figure in which light from a monochromatic source (such as a laser) is split into two by a beam splitter. One beam is made to travel in a clockwise and the other in a counterclockwise direction in an optical loop, in a ring, by using corner mirrors. The two beams are combined again at the beam splitter and form fringes (an interference pattern) on a fringe detector. When the whole interferometer, that is the whole experiment with the detector and light source as well, is rotated, it is found that the fringes shift with respect to that for the stationary interferometer. This is the Sagnac effect. The shift ΔZ in the fringes is given by

$$\Delta Z = 4\boldsymbol{\Omega} \cdot \mathbf{A}\lambda_0 c$$

where $\mathbf{A}$ is the area enclosed by the light path (the optical loop), λ_0 is the vacuum wavelength and c is the free space velocity of light. The area defined by the optical circuit or the ring resonator is a vector $\mathbf{A}$ with a direction normal to the area. Thus, the shift depends on the cosine of the angle between the two vectors $\boldsymbol{\Omega}$ and $\mathbf{A}$. The effect does not depend on the shape of the optical loop or the center of rotation, which can be outside the loop. The shift in the fringes is a result of the *phase shift* introduced between the clockwise and counterclockwise beams when the whole interferometer rotates. Thus, the phase shift depends on $\boldsymbol{\Omega}\cdot\mathbf{A}$. In fiber optics, the *Sagnac interferometer* involves a fiber loop into which light is fed into both ends through a double Y-junction splitter. Light waves are thus coupled into both ends of the fiber loop and hence there are two waves in the loop traveling in clockwise and counterclockwise directions. The two waves are combined at the splitter again and form the output in the output fiber. The rotation-induced phase difference $\Delta\phi$ between the clockwise and counterclockwise waves is given by

$$\Delta\phi = \frac{8\pi\,\Omega A N}{\lambda_0 c}$$

where N is the number of turns in the fiber loop, and A is the loop area. The above examples represent passive Sagnac interferometers. It is also possible to construct *a ring laser Sagnac interferometer*; an *active interferometer*, in which the light is generated and self-sustained inside a ring cavity. If the ring laser is rotated around an axis normal to the plane of the ring, then photons traveling along the direction of rotation would experience a shorter optical path length than those traveling in the direction of rotation, leading to two different standing waves in

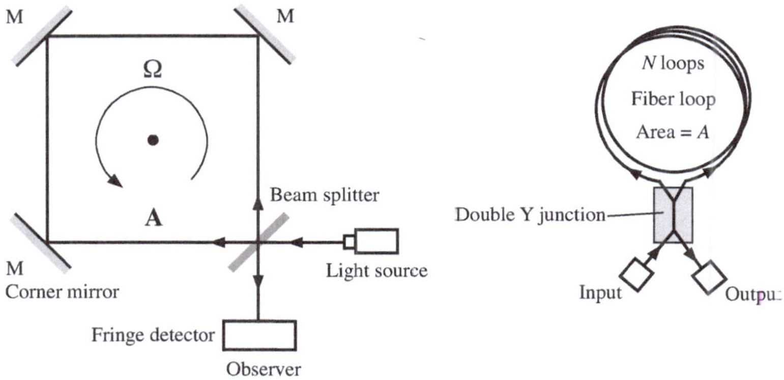

Sagnac interferometer being rotated   Fiber optic Sagnac interferometer

Schematic illustration of the **Sagnac effect**, and the schematic structure of a fiber optic Sagnac interferometer.

the ring, i.e. two different modes. These two cavity modes would be separated in frequency by an amount Δf that depends on the rotational velocity Ω. By tapping out the radiation from the ring cavity and measuring the beat frequency Δf, we can determine Ω. (E. J. Post, *Rev. Mod. Phys.*, **39**, 475, 1967; G. Sagnac, *Comptes Rendus de l'Academie des Sciences (Paris)*, **157**, 708 and 1410, 1913.)

Saturable absorption occurs in a medium at a sufficiently high level of optical injection. The mechanism is exactly the same as *gain saturation*, without the population inversion. Consider the simplest case of a collection of open two-level systems. The higher the incident intensity, the greater the population in the upper state, 2, and the smaller in the lower state, 1. The transition probability from 1 to 2 therefore diminishes accordingly. At very high levels of incident intensity, the two populations are equal and the medium becomes transparent (*self-induced transparency*).

Saturation see *color*.

Saturation intensity of an optical amplifier is the intensity (strictly irradiance) at which the optical gain of the amplifying medium is halved with respect to the small signal input gain. Correspondingly, the *saturation output power* is the optical output power level at which the optical gain has been reduced to half its small signal gain value.

Saturation output power see *saturation intensity of an optical amplifier*.

Saturation parameter see *gain saturation in an active optical medium*.

Scalar changes see *photoinduced vectoral change*.

Scanning probe based nanofabrication relies on the strong interactions that occur between an atomically sharp tip and a substrate. Several tip–substrate interactions can be employed to fabricate nanostructures, including attractive and repulsive forces, confined electron currents, and electric fields. A *scanning tunneling microscope* (STM) can operate in either the tunneling mode or the field-emission mode. In the former case, when the tip is brought sufficiently close to the surface with an appropriately high positive bias, electrons can tunnel from surface bound atoms to the tip, permitting selective removal of surface atoms. Alternatively, in the field-emission mode with an appropriate negative bias, tip-adsorbed atoms can be transferred to the surface. An STM can also provide a finely focused (i.e. nanometer size) beam of low-energy electrons, permitting processes such as bond breaking, surface migration, and chemical reactions.

Scanning probe microscopy see *nanoelectronics.*

Scanning tunneling microscopy see *nanoelectronics.*

Scattering is a process by which some of the power in a propagating EM (electromagnetic) wave is redirected as secondary EM waves in various directions away from the original direction of propagation. There are a number of scattering processes. In *Rayleigh scattering*, the scattering particle size is much smaller than the wavelength of light; fluctuations in the refractive index, inhomogeneities, etc. lead to the scattering of light that decreases with the wavelength as λ^4. For an EM wave propagating along an optical fiber, scattering is a cause of attenuation along the fiber (direction of propagation). Rayleigh scattering is responsible for the blue sky and the red sunset because blue wavelengths are scattered more in the atmosphere. At sunset, the light travels the longest distance, and suffers substantial scattering out of the blue light, and hence the red sunset. In *Mie scattering,* the scattering particle can be of any size but typically comparable or larger than the wavelength of light. Mie scattering is responsible for the "blue moon" or the "green sun" that is often observed after volcanic explosions and forest fires; the scattering is due to particles larger than the light wavelength [. . . virtually every aspect of physical optics is an example of light scattering." Mark P. Silverman, *Waves and Grains*, Princeton University Press, 1998.]

See *Rayleigh scattering, Mie scattering, Tyndall scattering.*

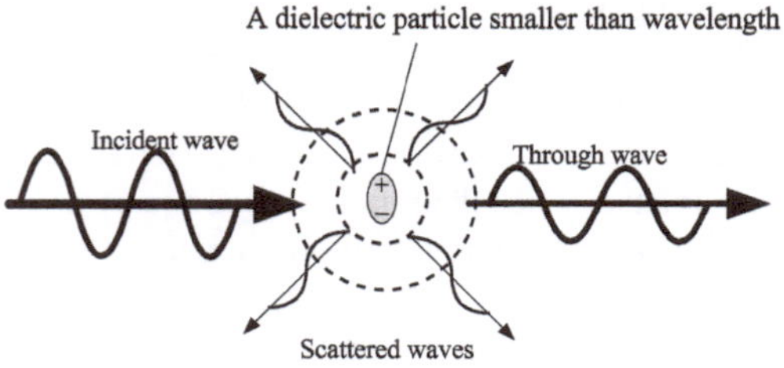

Rayleigh **scattering** involves the polarization of a small dielectric particle or a region that is much smaller than the light wavelength. The field forces dipole oscillations in the particle (by polarizing it) which leads to the emission of EM waves in "many" directions so that a portion of the light energy is directed away from the incident beam.

Scattering cross-section is the effective cross-sectional area of an impurity, dopant or an imperfection in a solid that is involved in scattering light. Suppose that there N_s number of scatterers per unit volume. The attenuation in the light intensity I in the direction of propagation z can be written as

$$I(z) = I(0)\exp(-\sigma N_s z)$$

where σ is the scattering cross-section; thus if α is the attenuation, $\alpha = \sigma \mathrm{N_s}$. The above equation can be used to measure σ if $\mathrm{N_s}$ is known. See *cross-section.*

Schadt–Helfrich effect see *liquid crystal display.*

Schawlow–Townes linewidth limit represents the theoretical limit to narrowest linewidth that one can expect to have in the output spectrum of a laser in terms of the cavity losses, Γ, and the total radiation output power P_o,

$$\Delta\nu = \frac{h\nu}{4\pi}\frac{\Gamma^2}{P_o}$$

in which $\Delta\nu$ is the spectral width, P_o is the output power through both mirrors of the cavity, and Γ is the cavity loss rate (per unit time). In an ideal Fabry–Perot cavity, losses are due to the end mirrors R_1 and R_2 only; $\Gamma = -(c/2L)\ln(R_1R_2)$ represents the so-called "cold cavity loss rate." The limit grossly underestimates the true linewidth in practical lasers. The theory assumes that the laser gain curve is homogeneously broadened, the output is a single-mode, and the angular frequency gain bandwidth is greater than the cavity loss rate, i.e. it applies to good cavities with low losses. The Shawlow–Townes limit has been discussed by various researchers and has been modified to take into account various other factors contributing to the linewidth. (A. L. Schawlow and C. H. Townes, *Phys. Rev.*, **112**, 1940, 1958; S. J. M. Kuppens *et al.*, *Phys. Rev. Letts.*, **72**, 3815, 1994.)

Schrödinger equation is a fundamental equation in nature whose solution describes the wave-like behavior of a particle. The equation cannot be derived from a more fundamental law. Its validity is based on its ability to predict any known natural phenomena. The solution requires, as input, the potential energy (PE) function, $V(x,y,z,t)$, of the electron and the boundary and initial conditions. The PE function, $V(x,y,z,t)$, describes the interaction of the particle with its environment. The time-independent Schrödinger equation describes the wave behavior of a particle under steady state conditions, i.e. when the PE is time independent, $V(x,y,z)$. If E is the total energy, $\nabla^2 = (d^2/dx^2 + d^2/dy^2 + d^2/dz^2)$, m_e is the electron mass and ψ is the wavefunction of the electron then

$$\nabla^2\psi + (2m_e/\hbar^2)(E - V)\psi = 0.$$

The solution of the time-independent Schrödinger equation gives the wavefunction $\psi(x,y,z)$ of the electron and its energy E. The wavefunction, $\psi(x,y,z)$, has the interpretation that $|\psi(x,y,z)|^2$ is the probability of finding the electron per unit volume at point (x, y, z).

Schubweg is the average distance a charge carrier drifts under the action of an applied field before it is permanently captured by a trap (a deep trap) or a recombination center, and hence becomes unavailable for conduction over the time-scale of measurement. If μ is the drift mobility of the carrier, τ is its *lifetime* (e.g. trapping time if the lifetime is controlled by deep trapping), and E is the applied field, then the carrier schubweg is defined as $\mu\tau E$. Schubweg per unit field is often termed the carrier *range*. Schubweg is an important quantity in characterizing photoconductors. See *Hecht collection efficiency*.

Scintillator is a material that gives off light when a charged particle or high-energy quantum (like an x-ray photon) passes through it. When fast moving, charged particles, such as cosmic rays or high-energy quanta, enter the scintillator they excite the scintillator atoms by giving them some or all of their energy, as a consequence of which the cosmic ray then slows down, or becomes absorbed as in the case of an x-ray photon. The excited atoms then lose this additional energy by emitting photons of light. In the case of an x-ray photon, it becomes absorbed and generates an energetic photoelectron (the photoelectric effect) in the scintillator. The photoelectron then excites electron hole pairs as it loses its energy in the solid. The light is usually detected by a *photomultiplier*, or a similar highly sensitive detector. The choice of materials for scintillators depends on the application. Thus, the alkali halides NaI and CsI, activated by a small amount of either thallium or sodium impurity, are used in x-ray astronomy because these materials can be made into large area crystals, have good x-ray sensitivity and are efficient light producers. For a material thickness of about 5 mm, for both NaI and CsI, the detection efficiency of particles with energy 20–100 keV is close to unity. Other materials, such as plastics (or organic solids) and bismuth germanate (BGO), can also be used as x-ray scintillators, though BGO is difficult to manufacture in large areas, while plastics have a lower sensitivity. The advantage of plastics is that they are obviously relatively cheap and may be easily shaped so that much of the light produced inside it is reflected off its surfaces and back into the material, and is eventually funneled into a photomultiplier. Usually the decay constant for the optical emission lies in the microsecond range for most inorganic scintillators, and in the nanosecond range for plastics.

Scotopic is an adjective for the response of an average human eye when it is dark adapted. See *luminous efficiency, human eye*.

Second-harmonic generation (SHG) occurs when an intense light beam of angular frequency ω passing through an appropriate crystal (e.g. quartz) generates a light beam of double the frequency, 2ω. If E is the electric field in the light wave, then the induced polarization P becomes a function of E and can be written as

$$P = \varepsilon_0 \chi_1 E + \varepsilon_0 \chi_2 E^2 + \varepsilon_0 \chi_3 E^3$$

where χ_1, χ_2 and χ_3 are the linear, second-order and third-order susceptibilities. SHG is based on a finite χ_2 coefficient in which the effect of χ_3 is negligible. Consider a beam of monochromatic light, with a well-defined angular frequency ω, passing through a medium. The optical field E at any point in the medium will polarize the medium at the same point in synchronization with the optical field oscillations. An oscillating dipole moment is well known as an electromagnetic emission source (just like an antenna). These secondary electromagnetic emissions from the dipoles in the medium interfere and constitute the actual wave traveling through the medium (Huygen's construction from secondary waves). Suppose that the optical field is oscillating sinusoidally between $\pm E_o$ as shown in the figure. In the linear regime (E_o is "small"), P oscillations will also be sinusoidal with a frequency ω. If the field strength is sufficiently large, the induced polarization will not be linear and will not oscillate in a simple sinusoidal fashion. The polarization now oscillates between P_+ and P_- and is not symmetrical. The oscillations of the dipole moment P now emit waves not only at the frequency ω but also at 2ω. In addition there is a dc component (light is "rectified," i.e. gives rise to a small permanent polarization). The *fundamental*, ω, and the *second harmonic*, 2ω, components, along with the dc, are shown in the figure. If we write the optical field as $E = E_o \sin(\omega t)$ and substitute into the expression for P we would find the induced P as

$$P = \varepsilon_0 \chi_1 E_o \sin(\omega t) - {}^1/_2 \varepsilon_0 \chi_2 E_o \cos(2\omega t) + {}^1/_2 \varepsilon_0 \chi_2 E_o.$$

The first term is the fundamental, the second is the second harmonic and the third is the dc term. The second-harmonic (2ω) oscillation of local dipole moments generates secondary second-harmonic (2ω-) waves in the crystal. It may be thought that these secondary waves will interfere constructively and result in a second harmonic beam just as the fundamental (ω-) secondary waves interfere and give rise to the propagating light beam. However, the crystal will normally possess different refractive indices $n(\omega)$ and $n(2\omega)$ for frequencies ω and 2ω which means that the ω- and 2ω- waves propagate with different phase velocities v_1 and v_2, respectively. As the ω-wave propagates in the crystal it generates secondary 2ω waves along its path, S_1, S_2, S_3, ... When wave S_2 is generated, S_1 must arrive there in phase which means S_2 must travel with the same velocity as the fundamental wave; and so on. It is apparent that only if these 2ω waves are in phase, that is they propagate with the same velocity as the ω-wave, can they interfere constructively and constitute a second-harmonic beam. Otherwise, the S_1, S_2, S_3 ... will eventually fall out of phase and destroy each other and there will be either no or very little second-harmonic beam. The condition that the second-harmonic waves must travel with the same phase velocity as the fundamental wave to constitute a second-harmonic beam is called *phase matching* and requires $n(\omega) = n(2\omega)$. For most crystals this is not possible as n is dispersive; it depends on the wavelength. SHG efficiency depends on the extent of phase matching, $n(\omega) = n(2\omega)$. One method is to use a birefringent crystal as these have two refractive indices: an ordinary index n_o and extraordinary index n_e. Suppose that along a certain crystal direction at an angle θ to the optic axis, $n_e(2\omega)$ at the second harmonic is the same as $n_o(\omega)$ at the fundamental frequency:

 Second-harmonic generation (SHG)

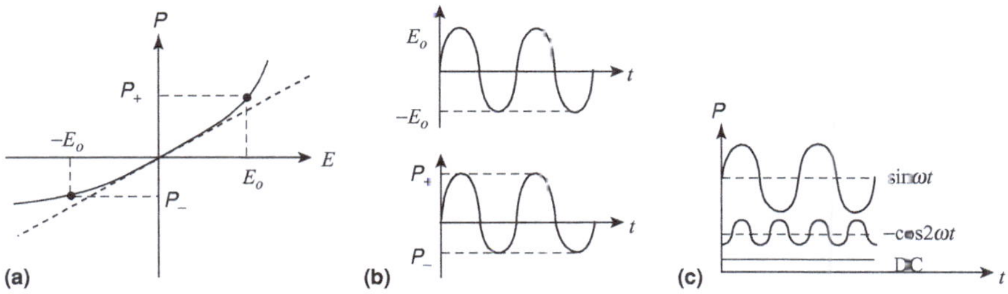

Second-harmonic generation (a) Induced polarization vs. optical field for a nonlinear medium. (b) Sinusoidal optical field oscillations between $\pm E_o$ result in polarization oscillations between P_+ and P_-. (c) The polarization oscillation can be represented by sinusoidal oscillations at angular frequencies ω (fundamental), 2ω (second harmonic) and a small dc component.

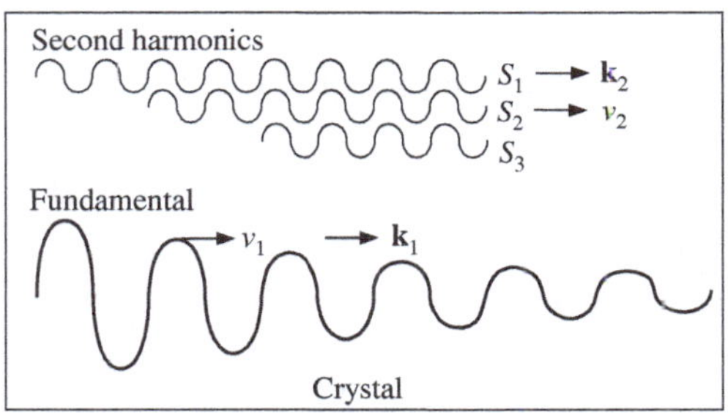

Second-harmonic generation As the fundamental wave propagates, it periodically generates second harmonic waves (S_1, S_2, S_3, ...) and if these are in phase then the amplitude of the second harmonic light builds up.

$n_e(2\omega) = n_o(\omega)$. This is called *index matching* and the angle θ is the *phase matching angle*. Thus, the fundamental would propagate as an ordinary wave and the second harmonic as an *extraordinary* wave and both would be in phase. This would maximize the *conversion efficiency*, though this would still be limited by the magnitude of the second-order term with respect to the first. To separate the second-harmonic beam from the fundamental beam, something like a diffraction grating, a prism or an optical filter will have to be used at the output. The phase matching angle θ depends on the wavelength (or ω) and is sensitive to temperature. Second-harmonic generation was first demonstrated by P. A. Franken, A. E. Hill, C. W. Peters, and G. Weinreich at the University of Michigan, Ann Arbor, in 1961 (*Phys. Rev. Lett.* **7**, 118, 1961).

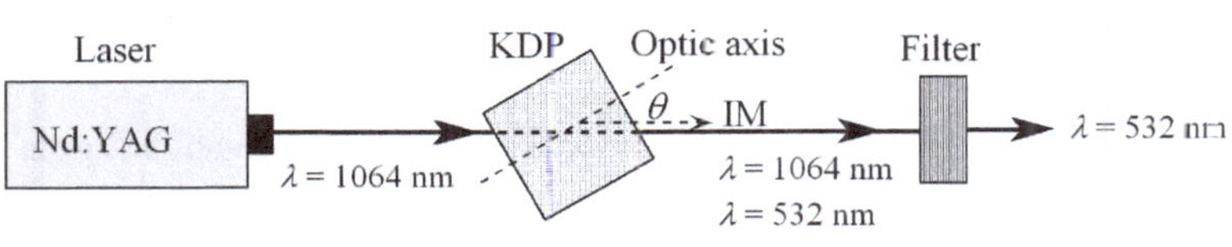

Second harmonic generation A simplified schematic illustration of optical frequency doubling using a KDP (potassium dihydrogen phosphate) crystal. IM is the index matched direction at an angle θ (about 35°) to the optic axis along which $n_e(2\omega) = n_o(\omega)$. The focusing of the laser beam onto the KDP crystal and the collimation of the light emerging from the crystal are not shown.

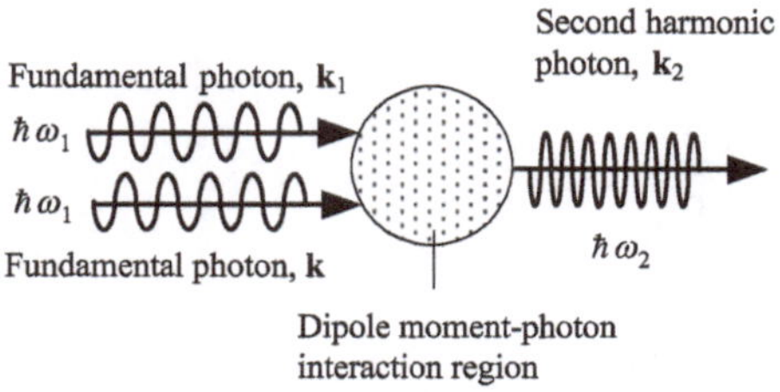

Second harmonic generation Photonic interpretation of second harmonic generation (SHG).

SHG in terms of the photonic interpretation involves two fundamental mode photons that interact with the dipole moments to produce a single second-harmonic photon. The photon momentum is $\hbar\mathbf{k}$ and the energy is $\hbar\omega$. Suppose that subscripts 1 and 2 refer to fundamental and second-harmonic photons.

In general terms, we can write the following two equations. Conservation of momentum requires that

$$\hbar\mathbf{k}_1 + \hbar\mathbf{k}_1 = \hbar\mathbf{k}_2.$$

The conservation of energy requires that

$$\hbar\omega_1 + \hbar\omega_1 = \hbar\omega_2.$$

We tacitly assumed that the interaction does not result in phonon (lattice vibration) generation or absorption. We can satisfy the second equation by taking $\omega_2 = 2\omega_1$, so that the frequency of the second harmonic is indeed twice the fundamental. To satisfy the first equation we need $\mathbf{k}_2 = 2\mathbf{k}_1$. The phase velocity v_2 of the second-harmonic waves is

$$v_2 = \frac{\omega_2}{k_2} = \frac{2\omega_1}{2k_1} = \frac{\omega_1}{k_1} = v_1.$$

Thus, the fundamental and the second-harmonic photons are required to have the same phase velocity which is tantamount to a phase matching criterion in terms of pure waves. If $\mathbf{k}_2$ is not exactly $2\mathbf{k}_1$, i.e. $\Delta k = k_2 - 2k_1$ is not zero, i.e. there is a mismatch, then SHG is only effective over a limited length l_c which can be shown to be given by $l_c = \pi/\Delta k$. This length l_c is essentially the coherence length of the second harmonic (depending on the index difference, this may be quite short, e.g. $l_c \approx$ 1–100 μm). If the crystal size is longer than this, the second harmonics will interfere randomly with each other and the SHG efficiency will be very poor, if not zero. Phase matching is therefore an essential requirement for SHG. The conversion efficiency depends on the intensity of the exciting laser beam, the material's χ_2 coefficient and the extent of phase matching and can be substantial (as high as 70–80%) if well-engineered by, for example, placing the converting crystal into the cavity of the laser itself.

Second-order electrooptic effect coefficients see *electrooptic effects*.

Secondary spectrum see *chromatic dispersion*.

Seidel aberrations see *aberrations*.

Seitz–Mark model of ionic crystals allows the bulk and surface bandgaps, and surface states of ionic crystals such as alkali halides (e.g. NaCl) to be calculated from the average Madelung energy for the crystal structure (different in the bulk and at the surface), the ionization potential I_m of the cation forming atom (Na), the electron affinity A_x of the anion forming atom (Cl), and the surface structure or geometry. The bulk bandgap E_{gb} and the surface bandgap E_{gs} are given by

$$E_{gb} = 2V_b - I_m + A_x \quad \text{and}$$
$$E_{gs} = 2V_s - I_m + A_x$$

where V_b and V_s are the average Madelung energy in the bulk and surface of the crystal, respectively. The key concept is that in ionic crystals such as NaCl the valence band is formed from orbitals from Cl^- and the conduction band from orbitals from Na^+. The valence and conduction bands in the model are treated as "narrow bands."

(P. Mark, *J. Phys. Chem. Solids*, **29**, 689, 1968; J. D. Levine and P. Mark, *Phys. Rev.*, **144**, 751, 1966.)

Selection rules for an atom determine what values of ℓ and m_ℓ are allowed for transitions involving the emission and absorption of electromagnetic radiation, i.e. photons. For a hydrogen-like atom, the emission or absorption of a photon must have $\Delta\ell = \pm 1$ and $\Delta m_\ell = 0, \pm 1$. The spin of the electron, m_s, remains unchanged, $\Delta m_s = 0$. These rules can also be extended to certain simple many-electron atoms and molecules. In all photon involving transitions, the parity of the electron wavefunction must change. The transition of the electron within an atom from one state to another, due to collisions with other atoms or electrons, does not necessarily obey the selection rules. There are also selection rules for photon absorption and emission for electronic transitions in a quantum well. The electron energy in a quantum well is quantized by the quantum number n for states in the conduction band well, and by n' for states in the valence band well. Photon absorption and emission by an electron in an infinite quantum well must obey $\Delta n = n' - n = 0$. In finite quantum wells, the $\Delta n \neq 0$ transition is quite weak, and forbidden for Δn being odd.

Self-assembly is a process in which atoms and molecules arrange themselves into ordered (i.e. *organized*) structures driven by thermodynamic considerations. The term *self-assembly* is used for describing various processes. It often refers to a nucleation phase for adsorbed species on a surface forming the basis for a subsequent growth process defined by the underlying pattern. The final pattern depends on interactions between the species themselves, and between the species and the surface. Self-assembly can also occur on *vicinal surfaces* (i.e. those formed by cleavage of a single-crystal structure at a small angle to a high-index crystal plane). Such surfaces comprise an array of terraces and steps that can provide favorable nucleation sites suitable for forming *quantum wires*. In heavily lattice-mismatched epitaxial growth, after a critical amount of material is deposited (typically on the scale of a monolayer), subsequent deposition results in three-dimensional island formation (typically with dimensions on the order of tens of nanometers, constituting *quantum dots*). The morphology of these dots depends on the thermodynamics and kinetics of their formation, as well as on their growth by the coarsening process. This, in turn, determines their electronic properties. Typical examples of such strained systems are InAs/GaAs and GeSi/Si. Self-assembled nanocrystals can also be prepared by controlled precipitation from solution.

Self-coherence function see *coherence*.

Self-focused light see *optical Kerr effect*.

Self-phase modulation is the modulation of the phase of a wave as a result of time-dependent changes in the refractive index of the medium due to the instantaneous intensity of the wave; see *optical Kerr effect*.

Self-phase modulation dispersion is the modulation of the phase of a wave by its own magnitude. At sufficiently high light intensities, the refractive index of glass n' can be written as $n' = n + n_2 I$ where n_2 is a nonlinearity constant, and I is the light intensity. The intensity of light modulates its own phase.

Sellmeier dispersion equation relates the refractive index n of a substance to the wavelength λ of light through a semiempirical dispersion relation of the form

$$n^2 - 1 = \frac{A_1\lambda^2}{\lambda^2 - \lambda_1^2} + \frac{A_2\lambda^2}{\lambda^2 - \lambda_2^2} + \frac{A_3\lambda^2}{\lambda^2 - \lambda_3^2}$$

where A_1, A_2, A_3 and λ_1, λ_2 and λ_3 are constants (called Sellmeier coefficients) that are determined by fitting this expression to the experimental data. The actual Sellmeier formula has more terms in the right-hand summation of the same type, e.g. $A_i\lambda^2/(\lambda^2 - \lambda_i^2)$ where $i = 4, 5, \ldots$ but these can generally be neglected in representing n vs. λ behavior

Table: Sellmeier dispersion equation Sellmeier coefficients. The λ_1, λ_2, λ_3 are in μm.

Material	A_1	A_2	A_3	λ_1	λ_2	λ_3
SiO_2 (fused silica)	0.696749	0.408218	0.890815	0.0690660	0.115662	9.900559
86.5%SiO_2–13.5%GeO_2	0.711040	0.451885	0.704048	0.0642700	0.129408	9.425478
GeO_2	0.80686642	0.71815848	0.85416831	0.068972606	0.15396605	11.841931
Barium fluoride	0.63356	0.506762	3.8261	0.057789	0.109681	46.38642
Sapphire	1.023798	1.058264	5.280792	0.0614482	0.110700	17.92656
Diamond	0.63306	4.3356		0.175	0.106	
Quartz, n_o	1.35400	0.010	0.9994	0.092612	10.700	9.8500
Quartz, n_e	1.38100	0.0100	0.9992	0.093505	11.310	9.5280
KDP, n_o	1.2540	0.0100	0.0992	0.09646	6.9777	5.9848
KDP, n_e	1.13000	0.0001	0.9999	0.09351	7.6710	12.170

over typical wavelengths of interest and ensuring that three terms included in the Sellmeier equation correspond to the most important or relevant terms in the summation. Sellmeier coefficients for pure silica (SiO_2) and 86.5 mol.%SiO_2–13.5 mol.% GeO_2 are given in the table. There are two methods to find the refractive index of silica-germania glass $(SiO_2)_{1-x}(GeO_2)_x$: (a) First is a simple, but approximate, linear interpolation of the refractive index between known compositions, e.g. $n(x) - n(0.135) = (x - 0.135)[n(0.135) - n(0)]/0.135$, where $n(x)$ is for $(SiO_2)_{1-x}(GeO_2)_x$; $n(0.135)$ is for 86.5 mol.%SiO_2–13.5 mol.% GeO_2; $n(0)$ is for SiO_2. (b) Second is an interpolation for coefficients A_i and λ_i (J. W. Fleming, *Appl. Optics*, **23**, 4486, 1984) between SiO_2 and GeO_2, i.e

$$n^2 - 1 = \frac{\{A_1(S) + x[A_1(G) - A_1(S)]\}\lambda^2}{\lambda^2 - \{\lambda_1(S) + x[\lambda_1(G) - \lambda_1(S)]\}^2} + \cdots$$

where S and G in parentheses refer to silica and germania. The theoretical basis of the Sellmeier equation lies in representing the solid as a sum of N lossless (frictionless) Lorentz oscillators that each have the $\lambda^2/(\lambda^2 - \lambda_i^2)$ form with different λ_i and each have different strengths, weighting functions (A_i, $i = 1, \ldots, N$). The Sellmeier equation with only the first term is equivalent to the Wemple–DiDomenico dispersion relation. See *Wemple–DiDomenico dispersion, optical glasses, refractive index*.

(K. L. Wolf and K. F. Herzfeld, *Handbuch der Physik*, H. Geiger and K. Scheel (Edition), Vol. 20 (Springer, 1928), ch. 10; M. Herzberger, *Opt. Acta*, **6**, 197, 1959.)

Semiconductor is a nonmetallic element (e.g. Si or Ge) that contains both electrons and holes as charge carriers in contrast to the enormous number of conduction electrons in metals. A hole is essentially a "half-broken" covalent bond which has a missing electron and therefore behaves effectively as if positively charged. Under the action of an applied field the hole can move by accepting an electron from a neighboring bond thereby passing on the "hole." Electron and hole concentrations in a semiconductor are generally many orders of magnitude less than those in metals thus leading to much smaller conductivities.

Semiconductor laser is a solid state *diode* that has a special single or double heterojunction structure made from differently doped semiconductors in such a way that under sufficiently large forward bias, a large number of carriers are injected into a confined region in the structure. These injected excess carriers then recombine by stimulated emission, which leads to optical gain. The structure also has an optical cavity to build up electromagnetic oscillations and hence achieve a lasing output – coherent radiation. The forward current needs to exceed a certain threshold current to provide the

Semiconductor optical (laser) amplifier

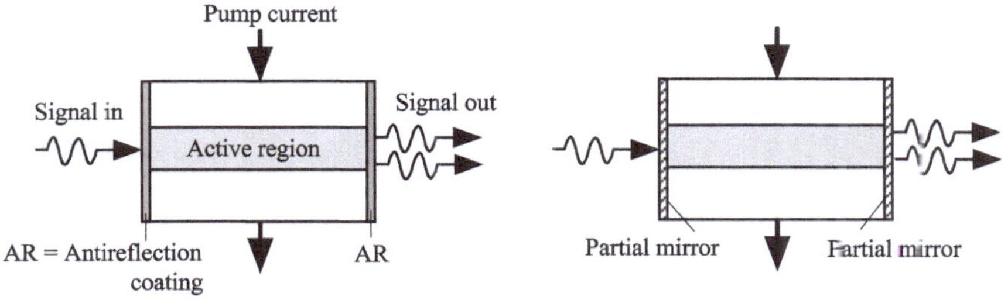

(a) Traveling wave amplifier (b) Fabry-Perot amplifier

Semiconductor optical amplifier (a) Traveling wave amplifier. (b) Fabry–Perot amplifier. Simplified schematic illustrations of two types of laser amplifiers.

necessary conditions for generating coherent radiation as output. In quantum-well semiconductor lasers, stimulated recombination in a quantum well inside the semiconductor structure is used to obtain optical gain. See *laser diode, heterostructure laser diodes, multiple quantum well (MQW) lasers.*

Semiconductor optical (laser) amplifier is a semiconductor laser structure that can be used as an optical amplifier that amplifies light waves passing through its active region. Light to be amplified is made to enter the active region of the laser semiconductor diode and the active region is pumped by a sufficiently large diode current just as in a normal laser diode. The wavelength of radiation to be amplified must be within the optical gain bandwidth of the laser. Such a device would not be a laser oscillator, emitting lasing emission without an input, but an optical amplifier with input and output ports for light entry and exit. In the *traveling wave semiconductor laser* amplifier the ends of the optical cavity have antireflection (AR) coatings so that the optical cavity does not act as an efficient *optical resonator*, a condition for laser oscillations. Light, for example, from an optical fiber, is coupled into the active region of the laser structure. As the radiation propagates through the active layer, optically guided by this layer, it becomes amplified by the induced stimulated emissions, and leaves the optical cavity with a higher intensity. The *Fabry–Perot laser amplifier* is similar to the conventional laser oscillator, but is operated below the threshold current for lasing oscillations; the active region has an optical gain but is not sufficient to sustain a self-lasing output. Light passing through such an active region will be amplified by stimulated emissions but, because of the presence of an optical resonator, there will be internal multiple reflections. These multiple reflections lead to the gain being highest at the resonant frequencies of the cavity within the optical gain bandwidth. Optical frequencies around the cavity resonant frequencies will experience higher gain than those away from resonant frequencies. Although the Fabry–Perot laser amplifier can have a higher gain than the traveling wave amplifier, it is less stable.

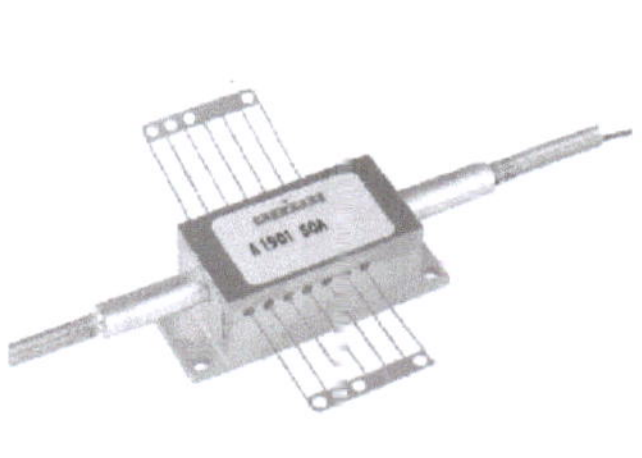

A 1550 nm **semiconductor optical amplifier** using an InGaAsP chip. (Courtesy of Alcatel-Lucent.)

Semiconductor laser rate equations describe the rate of change with time of the number of carriers (*dn/dt*) and photons (*dN/dt*) in a laser cavity. In general, the change in the carrier population can be written as

$$dn/dt = G - R$$

where *G* is the net generation rate of carriers which may be written as $G = \eta_i J/qd$ where η_i is the injection efficiency, *I* the injection current and *d* the cavity thickness; *R* is the net rate of carrier recombination given by $R = R_{sp} + R_{st} + R_{nr}$, where R_{sp} is the spontaneous emission rate, R_{st} is the stimulated emission rate, and R_{nr} is the nonradiative recombination rate. $R_{sp} = Bn^2$ where *B* is a constant, and $R_{nr} = An + Cn^3$ with *A* and *C* as constants. An effective spontaneous recombination lifetime, τ, can be defined as $1/(A+Bn+Cn^2)$, enabling the rate of change in the carrier population to be written as

$$dn/dt = (\eta_i J/qd) - (n/\tau) - R_{st}.$$

R_{st} depends on the number of photons *N*, the gain coefficient, $g(n) = a(n - n_{tr})$, where n_{tr} is the transparency concentration when the gain is zero and *a* is the differential gain, and the group velocity $v_g = c/n_g$ with n_g as the (group) refractive index; that is $R_{st} = av_g(n - n_{tr})N$. The rate equation for photons in the cavity can be written based on the arguments presented above, adjusting for the fact that the generated photons can occupy a volume larger than that occupied by the electrons which are confined to the active medium. The ratio of the volume of the cavity to the volume occupied by the photons in the cavity is called the *confinement factor* and is given the symbol Γ. Also note that photon loss occurs within the cavity due to optical absorption and scattering out of the mode, and also at the output coupling mirror where a fraction of the resonant mode is usefully coupled to some output medium. These losses can be characterized by the photon lifetime τ_p. The photon rate equation is then

$$dN/dt = \Gamma(\beta_{sp} R_{sp} + R_{st}) - (N/\tau_p)$$

where β_{sp} is the spontaneous emission factor. For uniform coupling to all modes, β_{sp} is the reciprocal of the number of optical modes within the bandwidth of the spontaneous emission. At steady state, both *dn/dt* and *dN/dt* equal zero. Also, the term $\Gamma \beta_{sp} R_{sp}$ representing spontaneous emission into the lasing mode in the equation for the photon generation rate can generally be neglected. This gives

$$N = (\Gamma \eta_i \tau_p/qd)[J - J_{th}]$$

where we define the threshold current density for lasing J_{th} as $J_{th} = (qd/\eta_i\tau)[n_{tr} + (1/\Gamma a v_g \tau_p)]$. This expression may be transformed into one for the optical output power *P* as a function of the injection current *I* by recognizing that $P \propto (h\nu)N(Ad)$, where $h\nu$ is the energy of the emitted photon, *N* the photon density and *Ad* the volume of the active region (defined by an area *A* and thickness *d*). This gives

$$P \propto (h\nu\Gamma\eta_i\tau_p/q)[I - I_{th}].$$

That is, the output power of the laser at steady state rises linearly with the injection current above threshold. One can express the so-called differential external quantum efficiency of the laser η_e as $(q/h\nu)(dP/dI)$, or $\eta_e \propto \Gamma\eta_i\tau_p$. See *Laser diode equation.*

Sensitizer see *photoluminescence.*

Separate absorption and multiplication (SAM) avalanche photodiode (APD) is typically a heterostructure based detector, e.g. InGaAs-InP, with different bandgap materials to separate absorption and multiplication (SAM). InP has a wider bandgap than InGaAs and the *p* and *n* type doping of InP is indicated by capital letters, *P* and *N*. The main depletion layer is between P^+-InP and *N*-InP layers and it is within the *N*-InP. This is where the field is greatest and therefore it is in this *N*-InP layer where avalanche multiplication takes place. With sufficient reverse bias the depletion layer in the *n*-InGaAs reaches through to the *N*-InP layer. The field in the depletion layer in *n*-InGaAs is not as great as that in *N*-InP. Although the long-wavelength photons are incident on the InP side, they are not absorbed by InP since the photon energy is less than the bandgap energy of InP ($E_g =$ 1.35 eV). Photons pass through the InP layer and become absorbed in the n-InGaAs layers. The field in the n-InGaAs layer drifts the holes to the multiplication region where *impact ionization* multiplies the carriers. The real device is more complicated than this simple description. Photogenerated holes drifting from n-InGaAs to N-InP become trapped at the interface because there is a sharp increase in the bandgap and a sharp change ΔE_v in E_v (valence band edge) between the two semiconductors and holes cannot easily surmount the potential energy barrier ΔE_v. This problem is overcome by using thin layers of n-type InGaAsP with intermediate bandgaps to provide a graded transition from InGaAs to InP. Effectively ΔE_v has been broken up into two steps. The hole has sufficient energy to overcome the first step and enter the InGaAsP layer. It drifts and accelerates in the InGaAsP layer to gain

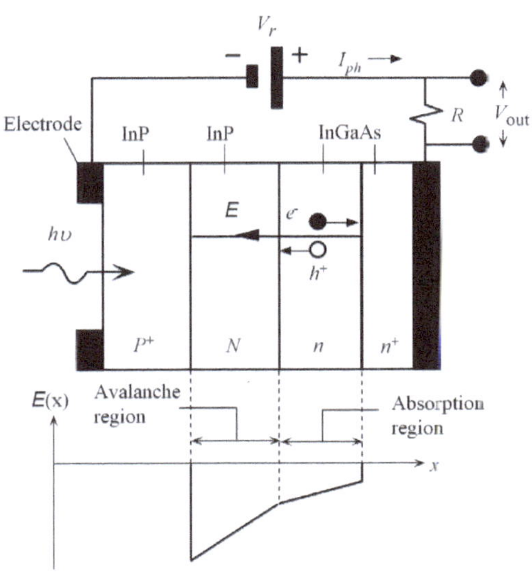

Simplified schematic diagram of a **separate absorption and multiplication** (SAM) APD using a heterostructure based on InGaAs-InP. *P* and *N* refer to *p*- and *n*-type wider-bandgap semiconductors.

sufficient energy to surmount the second step. These devices are called *separate absorption, grading and multiplication* (SAGM) APDs. Both the InP layers are grown epitaxially on an InP substrate. The substrate itself is not used directly to make the *P*-*N* junction to prevent crystal defects (for example, dislocations) in the substrate appearing in the multiplication region and hence deteriorating the device performance.

Separate absorption, grading and multiplication (SAGM) APDs see *separate absorption and multiplication (SAM) avalanche photodiode.*

Seraphin coefficients see *photoreflectance spectroscopy.*

SHB see *spectral hole burning.*

Shear strain see *strain.*

Sheik-Bahae equations are useful equations that relate the nonlinear refractive index n_2 (or the optical Kerr effect coefficient), and the nonlinear absorption coefficient β, to the bandgap E_g of the material and its normal refractive index n_0,

$$n_2 = A\frac{\overline{G}_2(\hbar\omega/E_g)}{n_0^2 E_g^4}$$

where $A \approx 3\times10^{-35}$ (MKS units), and $\overline{G}_2$ in $\overline{G}_2(\hbar\omega)$ in a normalized spectral function that describes the dispersion lineshape of n_2 (see the references for its exact form, and the figure) and

$$\beta[\mathrm{m}^2/W] = B\frac{\overline{F}_2(\hbar\omega/E_g)}{n_0^2 E_g^3}$$

where $B \approx 5.67\times10^{-66}$ (SI units), and $\overline{F}_2$ in $\overline{F}_2(\hbar\omega)$ is the spectral function given by $F(x) =$ $[2(2x-1)^{3/2}]/x^5$; see the figure.

(M. Sheik-Bahae *et al.*, *Phys. Rev. Letts.*, **65**, 96, 1990; M. Sheik-Bahae and P. Hasselbeck, Chapter 17 in the *OSA Handbook of Optics*, Vol. IV, Opt. Soc. Am. 2000.)

Shimakawa's rule for the bandgap of amorphous (a-) semiconductors is an empirical relationship that allows the bandgap of certain alloys in this class of

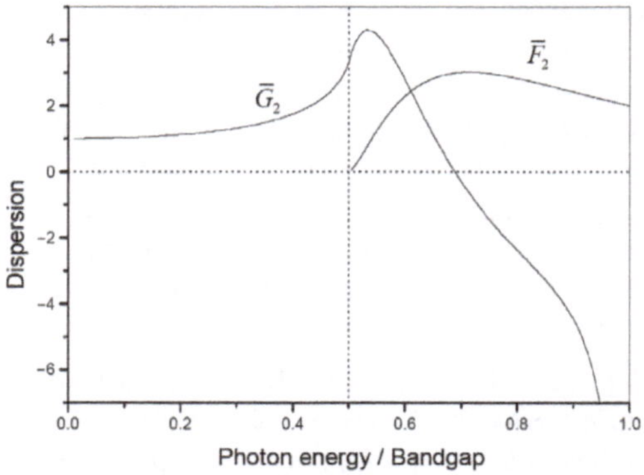

Sheik-Bahae equations The spectral dispersion functions $\overline{F}_2$ and $\overline{G}_2$ for β and n_2. (Data from M. Sheik-Bahae and P. Hasselbeck, Chapter 17 in the *OSA Handbook of Optics*, Vol. IV, Opt. Soc. Am. 2000.)

materials to be calculated from the bandgaps and volume fractions of the constituents. In the random covalent bond network model of the rule, the bandgap E_g of the alloy AB is simply given by

$$E_g = YE_{gA} + (1 - Y)E_{gB}$$

where Y is the volume fraction of A, and $E_{g\mathrm{A}}$ and $E_{g\mathrm{B}}$ are the bandgaps of A and B. There is a linear change in E_g with Y, and the rule therefore differs from the case for crystalline semiconductors in which the bandgap is often represented by a parabolic dependence on the composition. Amorphous Si_xGe_{1-x}, Sb_xSe_{1-x}, As_xTe_{1-x} fall into this relatively simple category. In the chemically ordered bond network model of the rule, the structure is assumed to have molecular constituents. For example, in a-$As_{1-x}Se_x$, it is assumed that As_2Se_3 units exist in the structure so that, on the Se rich side, we should take the alloy to be a mixture of (As_2Se_3) and Se, yielding $E_g = YE_g(As_2Se_3) + (1-Y)E_g(Se)$. Although the rule does not apply to all classes of amorphous semiconductors, it can, nonetheless, be used to find E_g given the composition for those alloys that follow the rule. The rule can be viewed as an effective medium approximation to the optical gap. (K. Shimakawa, *J. Non-Cryst. Solids*, **43**, 229, 1981.)

Shockley diode equation relates the current through a *pn* junction diode to the voltage across the diode through $I = I_0[\exp(eV/k_\mathrm{B}T) - 1]$, where I_0 is the reverse saturation current. It is based on the injection and diffusion of injected minority carriers in neutral regions of the *pn* junction by the application of a forward bias. See *pn junction*.

Shockley–Ramo theorem see *Ramo's theorem*.

Shockley–Reed–Hall (SRH) generation and recombination processes are electron and hole capture and emission processes that involve a localized state in the bandgap (a trap or a recombination center), as illustrated in the figure (a)–(d). Electron and hole emission processes can both be classified as carrier generation processes, providing an increase in the carrier concentration as a result of carriers bound at the center being released from the center. If n is the electron concentration in the conduction band, N_t is the concentration and f_t is the occupation probability of the centers at E_t (i.e. N_tf_t is the concentration of traps occupied by electrons), then for electron capture (a), the rate is

$$R_n = c_nnN_t(1 - f_t)$$

where c_n is a proportionality constant. The rate is proportional to the concentration of empty traps $N_t(1-f_t)$ and also to the number of electrons able to fill them, n. For the case of recombination of holes (c), the rate is given by

$$R_p = c_ppN_tf_t$$

following the same logic used to derive the expression for electrons. For electron emission (b), the generation rate is

$$G_n = e_nN_tf_t$$

where e_n is the emission rate for electrons from the trap. The generation rate for holes (d), is

$$G_p = e_pN_t(1 - f_t),$$

where all terms are defined analogously to that for G_n, except that a factor of $(1-f_t)$ appears as this corresponds to the occupation probability for holes and e_p is the hole emission rate. Based on the principle of detailed balance, there should be no net generation of electrons or holes, respectively, at thermal equilibrium. That is, $R_n - G_n = 0$ and $R_p - G_p = 0$ when all terms assume their thermal equilibrium values; that is $n = n_0, p = p_0$, and $f_t = f_{t0}$ where the subscript "0" denotes thermal equilibrium value. Defining $n_1 = n_0(1-f_{t0})/f_{t0}$ and $p_1 = p_0 f_{t0}/(1-f_{t0})$, and the principle of detailed balance, gives $e_n = c_n n_1$ and $e_p = c_p p_1$. Note that the law of mass action applies to n_1 and p_1, such that $n_1 p_1 = n_0 p_0 = n_i^2$. The net recombination rate can thus be written as

$$R = R_n - G_n = R_p - G_p$$
$$= (np - n_i^2)/(\tau_p(n + n_1) + \tau_n(p + p_1))$$

where the lifetime for holes is defined by $\tau_p = 1/(c_p N_t)$ and for electrons, $\tau_n = 1/(c_n N_t)$. If Δn excess carriers are injected, the net recombination rate R, under low level injection conditions, is

$$R = \Delta n/\tau_0$$

in which

$$\tau_0 = \frac{\tau_p(n + n_1) + \tau_n(p + p_1)}{n_0 + p_0}.$$

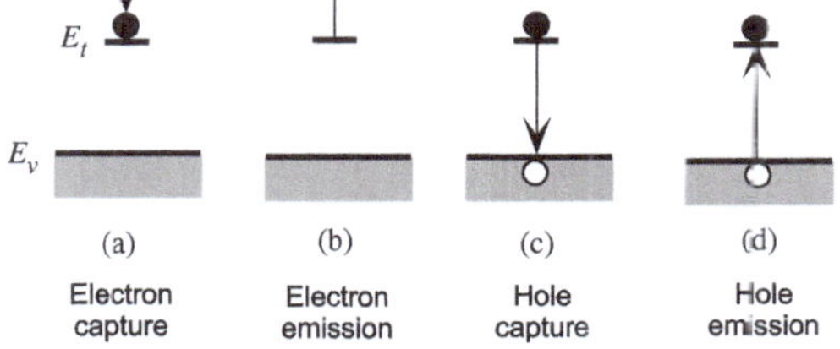

SRH generation and recombination Various electron and hole capture processes. The arrows refer to electron transitions.

Short diode is a *pn* junction in which the neutral regions are shorter than the minority carrier diffusion lengths.

Shot noise is the fluctuations in the current due to the discrete nature of charge. Electrons in a current flow as discrete charges, as particles with each one carrying a charge of $-e$. If the current were a continuous flow of "gaseous charge" without any fluctuations then the shot noise would be zero. Each time an electron arrives at an electrode, there is a discrete charge of $-e$ in the collected charge. It is analogous to randomly dropping ball bearings on a drum. The balls are discrete (each the same mass) but they arrive at random times, like shots, and the sound they make is called shot noise. If the current were a continuous flow of "gaseous charge" without any fluctuations then the shot noise would be zero. The root mean square fluctuations in the current about its mean value I is often simply called the *noise current*, i_n, and is given by

$$i_n = \sqrt{\overline{[i(t) - I]^2}} = \sqrt{2eIB}$$

where I is the average current (or the dc value of the current), and B is the bandwidth of the circuit or the system in which the shot noise is measured. Shot noise follows *Poisson statistics*. For a large number of electrons, the Poisson distribution approaches the normal distribution.

Signal-to-noise ratio (SNR, *S/N*) in a photodetector is defined as the ratio of the power in the photogenerated current to the noise power (power in the random fluctuations) in the detector. When the photodetector is connected to a receiving circuit, the noise power must include that generated in the circuit as a whole:

$$\text{SNR} = \frac{\text{Signal Power}}{\text{Noise Power}}.$$

Silica glass or fused silica glass is the noncrystalline, or amorphous, form of SiO_2. It is similar to melt silica but, due to a very high viscosity at room temperature, behaves like a solid (glass). The crystalline form is called *quartz*. UV grade fused silica has very high purity, and transmits UV wavelength ranges down to 180 nm. Fused silica can be manufactured to better optical standards than fused quartz. Fused silica has a low thermal expansion coefficient and good resistance to laser damage, and is used in transmission and reflection optics. It is preferred to BK7 (borosilicate crown optical glass) for use over a wider wavelength range. $\rho = 2.202$ g cm^{-3}; $\nu_d = 67.7$; $\alpha_L = 0.52\times10^{-6}$ K^{-1}; $\kappa = 1.38$ W cm^{-1} K^{-1}; $c_s = 0.75$ J g^{-1} K^{-1}; $E = 73$ GPa; $H_{Knoop} = 600$ kg mm^{-2}. For refractive index n vs. λ behavior, see *optical materials, Sellmeier dispersion equation.*

Silicon carbide (SiC) is a IV-IV compound semiconductor that has applications in optoelectronics and high-temperature or radiation resistant high-speed electronics. SiC has more than a hundred crystalline modifications. For example, 3C-SiC or β-SiC has the zinc blende, 2H-SiC or α-SiC has the wurtzite (hexagonal), 15R-SiC has rhombohedral unit cells. C, H and R stand for cubic, hexagonal and rhombohedral.

3C-SiC: $a = 0.43596$ nm (297 K); sublimes; 3C to 6H transformation at $T > 215$ °C; $d = 3.116$ g cm^{-3} (293 K); E_g(I) $= 2.2$ eV (300K): 2.417 eV (2 K); $\mu_e \approx 900$ cm^2 V^{-1}s^{-1}; $\mu_h \approx 20$ cm^2 V^{-1}s^{-1}; electron drift velocity $= 2.5\times10^7$ cm s^{-1} at $E > 2\times10^5$ V cm^{-1}; $\alpha_L = 4\times10^{-6}$ K^{-1} (300 K); 7.8×10^{-6} K^{-1} (373 K); 8.3×10^{-6} K^{-1} (473 K); $\kappa = 500$ W m^{-1} K^{-1}; $C_P = 28.5$ J K^{-1} mol^{-1}; $c_s = 0.34$ J K^{-1} g^{-1}; $T_{Debye} = 1270$ K; $H_{Vickers} = 112$ kg mm^{-2} (1 kg indenter) $H_{Knoops} = 120$ kg mm^{-2}; $E = 448$ GPa; $K = 249$ GPa; $\upsilon = 0.28$; $\varepsilon_r(DC) = 9.52$; $\varepsilon_r(\infty) = 6.38$; Reststrahlen peak at $\lambda = 45.7$ µm; $dn/dT \approx 6\times10^{-5}$ K^{-1} at 10.6 µm; n in the 1–11 µm range described by a Cauchy expression (λ in µm); breakdown field $= 4\times10^6$ V cm^{-1}; for $\lambda = 467 - 691$ nm,

$$n = 2.55387 + (3.417 \times 10^4)/\lambda^2.$$

Silicon nitride, hydrogenated and amorphous, a-Si$_{1-x}$N$_x$:H, film is an amorphous semiconductor film that is usually fabricated by chemical vapor deposition (CVD) or plasma enhanced chemical vapor deposition, etc. Alloying a-Si:H with nitrogen increases the optical bandgap, and decreases the refractive index. Its high bandgap with high N content, as shown in the figure, means that it has a high electrical resistivity as well. It has a number of useful applications such as a passivation layer in microelectronics, an "oxidation mask," dopant diffusion barrier, interlevel dielectric in semiconductor fabrication, gate-dielectric in thin film FETs, antireflection coatings on photovoltaic devices and detectors since the refractive index can be tailored with the N content. It has also been used in multilayer structures to fabricate, for example, avalanche detectors as well as optical components such as distributed Bragg reflectors.

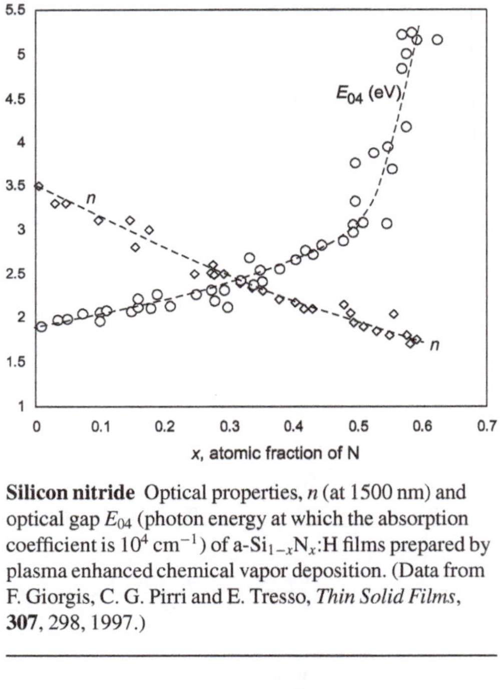

Silicon nitride Optical properties, n (at 1500 nm) and optical gap E_{04} (photon energy at which the absorption coefficient is 10^4 cm^{-1}) of a-Si$_{1-x}$N$_x$:H films prepared by plasma enhanced chemical vapor deposition. (Data from F. Giorgis, C. G. Pirri and E. Tresso, *Thin Solid Films*, **307**, 298, 1997.)

Silicon photonic modulator is a modulator based on the change in the refractive index n and/or the absorption coefficient of silicon. If the Si sample whose n is modulated is in one of the arms (paths)

Silicon photonic modulator

of a Mach–Zehnder interferometer, then the output light intensity can be readily modulated. The change in the refractive index of Si due to changes in the free carrier concentration can be explained by the Drude (free carrier) model. Suppose that ΔN_e and ΔN_h represent the injected free electron and hole concentrations in an undoped Si crystal (for example, injection can be done by doping the sample or by the application of a bias to a *pn* junction, etc.), which decrease the refractive index of the undoped crystal by Δn at wavelengths longer than the bandgap absorption. In simple terms, the change in the refractive index Δn and the absortion coefficient $\Delta\alpha$ can be written as

$$\Delta n = -(e^2\lambda^2/8\pi^2c^2\varepsilon_0^2 n)[(\Delta N_e)/m_{ce}^* + (\Delta N_h)/m_{ch}^*]$$

and

$$\Delta\alpha = -(e^3\lambda^2/4\pi^2c^3\varepsilon_0^2 n)[(\Delta N_e)/\mu_e m_{ce}^{*\,2} + (\Delta N_h)/\mu_h m_{ch}^{*\,2}]$$

where $m_{ce}{}^*$ and $m_{ch}{}^*$ are the conductivity effective masses of electrons and holes, and μ_e and μ_h are the electron and hole drift mobilities. The actual experimental data do not follow simple linear relationships as is apparent in the figures for Δn vs. N_e and N_h. The Soref–Bennett empirical relations relate Δn to changes ΔN_e and ΔN_h in the electron and hole concentrations at various wavelengths; for example, the best fits in the figure give

$$\begin{aligned}\Delta n &= \Delta n_e + \Delta n_h \\ &= -[6.80 \times 10^{-23}(\Delta N_e)^{1.053} \\ &\quad + 1.53 \times 10^{-18}(\Delta N_h)^{0.83}] \text{ at } 1.3\ \mu\text{m}\end{aligned}$$

and

$$\begin{aligned}\Delta n &= \Delta n_e + \Delta n_h \\ &= -[2.66 \times 10^{-22}(\Delta N_e)^{1.029} \\ &\quad + 4.22 \times 10^{-18}(\Delta N_h)^{0.816}] \text{ at } 1.5\ \mu\text{m}.\end{aligned}$$

(R. A. Soref and B. R. Bennett, *J. Quantum Electron.* 23, issue 1. © 1987 IEEE.)

Silicon photonic modulator

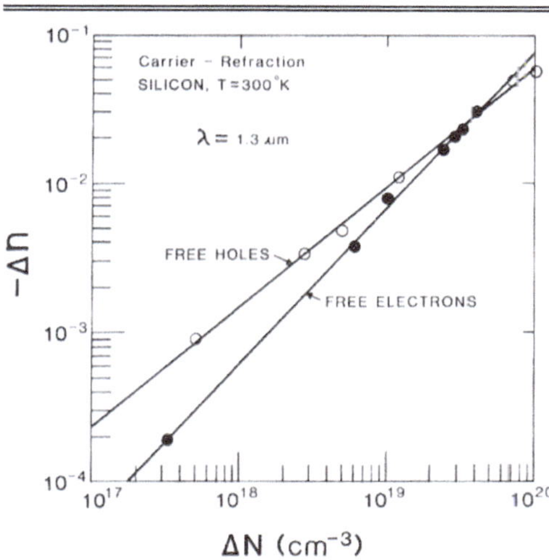

Δn at 1.3 μm vs. injected free carrier concentration
Both from R. A. Soref and B. R. Bennett, *IEEE J. Quantum Elecron.*, **QE23**, 123, 1987 (Figures 10 and 11)

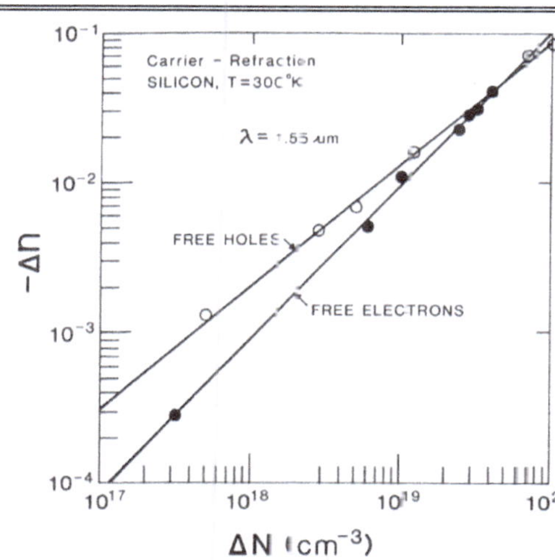

Δn at 1.5 μm vs. injected free carrier concentration

(a)

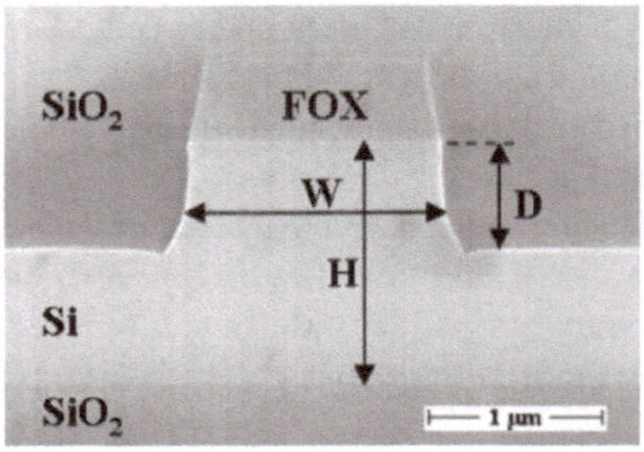

(b)

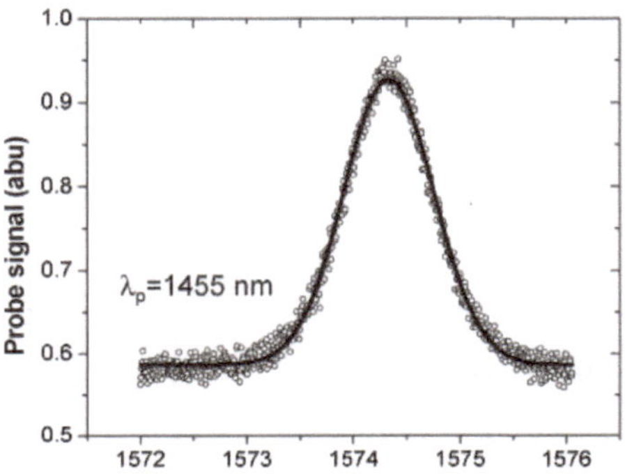

(c)

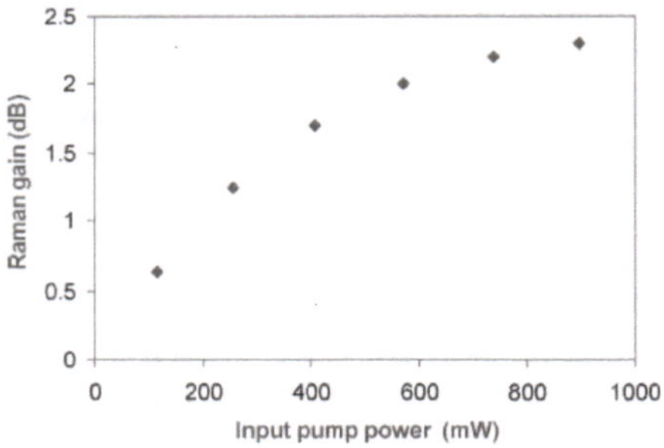

Silicon photonic Raman amplifier (a) Scanning electron micrograph of waveguide structure used for Raman gain measurement shown in (b) and (c). (b) Spontaneous Raman emission spectrum for a pump power of 700 mW at 1455 nm, and (c) measured peak gain as a function of input power for a pump laser wavelength of 1455 nm. (Ansheng Liu; Haisheng Rong; R. Jones, O. Cohen, D. Hak, and M. Paniccia, *Journal of Lightwave Technology* 24 issue 3, 1440 © 2006 IEEE.)

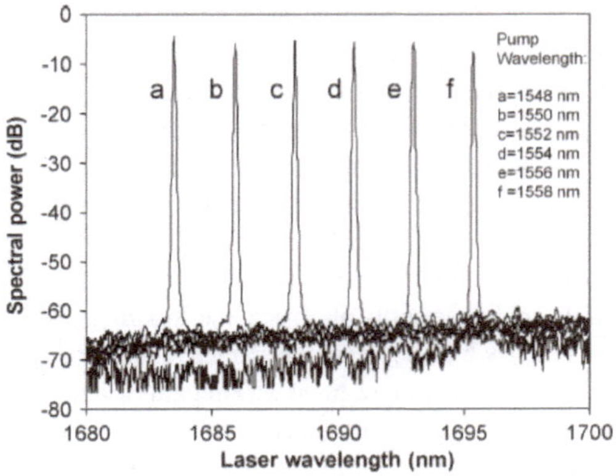

Silicon photonic Raman amplifier Output spectra of CW all-silicon Raman laser at different pump wavelengths. The laser was fabricated from a low-loss single-mode rib waveguide structure containing a reverse biased *p-i-n* diode structure, and having a high reflectivity (90% at both pump and Raman/Stokes wavelength) multilayer facet coating on one facet. (Ansheng Liu; Haisheng Rong; R. Jones, O. Cohen, D. Hak, and M. Paniccia, *Journal of Lightwave Technology* 24 issue 3, 1440 © 2006 IEEE.)

Silicon photonic Raman amplifier is an optical amplifier based on Raman gain in a silicon waveguide device. In such devices, an input signal is amplified in the Raman-active silicon medium while co- or counter-propagating with a pump beam whose wavelength is typically a few tens of nanometers shorter in wavelength. Typically, when such amplifiers are used for telecommunications systems, they are realized using erbium-doped fiber as the Raman-active medium for amplification. Such amplifiers can offer relatively high output power and low noise as compared with silicon counterparts. However, the silicon-based technology is compatible with planar lightwave circuit development and was recently shown to provide the possibility of a silicon based laser.

Silicon quantum dot is a nanoscale (i.e. a few nanometers) Si single crystal which can be excited

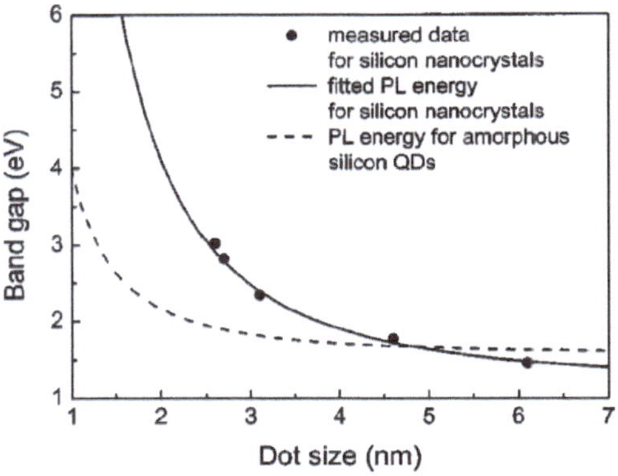

Silicon quantum dot Photoluminescence peak energy ("Bandgap (eV)") for silicon nanocrystals as a function of their size. The solid line is calculated using effective mass theory for three-dimensionally confined silicon nanocrystals. The dashed line was obtained from effective mass theory for an amorphous silicon quantum dot structure. (Reprinted with permission from: Tae-Youb Kim, Nae-Man Park, Kyung-Hyun Kim, Gun Yong Sung, Young-Woo Ok, Tae-Yeon Seong, and Cheol-Jong Choi, *Appl. Phys. Lett.*, **85**, 5355, 2004. © 2004, American Institute of Physics.)

to emit light. Making use of size-dependent quantum confinement effects, the peak wavelength of luminescence can span from the visible to near infrared, and with the incorporation of rare earth impurities (such as erbium) can be extended out to longer wavelengths. Quantum efficiencies of around 10% are feasible, but the emission spectrum is typically quite broad, and application to devices is also limited by relatively low modulation speed capability. Spectral narrowing can be achieved by embedding the quantum dot layer in a photonic cavity to achieve linewidths on the order of 1 nm. Aside from light-emitting diodes, a major thrust in this field is for silicon-based optoelectronics relying on developing lasers. An encouraging development in this area has been to observe amplified spontaneous emission in various structures incorporating silicon quantum dots under optical excitation. Future work will focus on demonstrating lasing under current injection.

(P. M. Fauchet, *Materials Today*, 25 January 2005).

Singer–Kuzyk–Sohn (SKS) model relates the electrooptic effect coefficient of a dye-doped polymer to the hyperpolarizability (nonlinear susceptibility) and the number density of dyes in the material. The material has to be poled, that is an electric field has to be applied so that the dyes are aligned in order to impart an electrooptic coefficient to the material. If β^* is the second-order (electrooptic) hyperpolarizability, N is the number density, m^* is the ground-state dipole moment, E_p is the poling-electric field, and T is the poling temperature, the electrooptic coefficient r in the SKS model, using the authors' original notation, is given by

$$r = \frac{1}{n_0^4} N\beta^* \left(\frac{2m^* E_p}{5kT}\right)$$

where n_0 is the bulk refractive index, and the term in parentheses is dimensionless, so each parameter can be evaluated in SI units, while keeping β and N in Gaussian units. (K. D. Singer, M. G. Kuzyk, and J. E. Sohn, *J. Opt. Soc. Amer.*, **4**, 968, 1987.)

Single-mode step index fiber is a step index fiber that allows only the fundamental mode to propagate at the wavelength of interest. To allow the fundamental mode to propagate, the fiber normally has a thin core (e.g. ≈ 5–10 μm in diameter) and a very small refractive index difference between the core and cladding (0.2–0.3%). A *single-mode fiber* is normally a step index fiber (unless otherwise defined). Suppose that the core radius is a, the core and cladding refractive indices are n_1 and n_2, respectively, and λ is the free-space wavelength. The V-number of the SMF, defined as $V = (2\pi a/\lambda)[n_1{}^2 - n_2{}^2]^{1/2}$, must be less than 2.405. Below a cut-off wavelength λ_c, the single-mode fiber becomes multimode. If a is the core radius, and $\mathrm{NA} = (n_1{}^2 - n_2{}^2)^{1/2}$ is the numerical aperture, then

$$0.766\lambda_c = (2a)\mathrm{NA}.$$

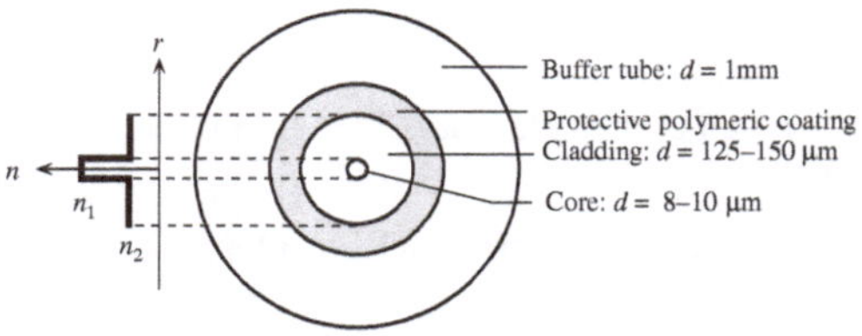

The cross-section of a typical **single-mode step index fiber** with a tight buffer tube (d = diameter).

Single-mode waveguide is a waveguide that can only carry the *fundamental mode* (lowest mode) of radiation within the wavelength range of interest, that is wavelengths longer than the critical cut-off wavelength. See *cut-off wavelength, single mode step index fiber*.

Single-photon detector (SPD) is a detector capable of detecting a single photon (used throughout the gamma-ray, x-ray, ultraviolet, visible and infrared spectral regimes). A single-photon detector is the most sensitive radiation detector. In most SPDs, the amount of current generated by a single-photon is very small and well below the noise of the detector, and thus not measurable at all. A single-photon detector requires a certain amplification mechanism, as noiseless as possible, in which a primary generated electron or electron–hole pair can generate a number of secondary electrons or electron–hole pairs. The resulting instantaneous current should be quite large and becomes easily detectable by additional electronics. The single-photon counting detectors detect and count individual photons. Si avalanche photodiodes (APDs) and photomultiplier tubes (PMTs) are the best known and most widely used SPDs for the detection of visible and infrared light. The operation of photomultiplier and avalanche devices is based on the electron cascade and multiplication effect, which significantly amplifies the response and allows for an easy measurement of response pulses. SPDs can also be used as x-ray imaging detectors such as single-photon counting hybrid pixel detectors (a silicon- and a CdTe detector are bump bonded onto the multipicture element counter chip), and superconducting tunnel junction (STJ) detectors. STJ single-photon detectors have a large variety of configurations and have been operated at energies ranging from the x-ray down to the near-infrared (NIR).

Single-photon interference illustrates the wave–particle "duality" by illuminating a classical system with a source of photons emitted one at a time. For instance, a beam splitter splits the incoming flux toward two single-shot detectors D_1 and D_2; each detection event corresponds to one photon only. The series of detection events on both outputs is highly correlated: although the classical continuous wave is partially transmitted and partially reflected, a single photon is either totally transmitted or totally reflected, with an equal probability. This is typical of a particle-like behavior. On the other hand, when a Mach–Zehnder interferometer is illuminated with an input stream of single photons, a truly particle-like behavior would lead us to expect an equal repartition of events on both output detectors; as a matter of fact, all photons are sequentially detected on the same detector D_1 corresponding to constructive interference, none of them on the complementary path. Only by repeating the experiment over a large number of shots can we obtain the classical wave-like behavior.

Young's interference fringes are another example of spatial photon localization: each detection event involves a single pixel of the detector plane; by integrating over a large number of events, we recover classical interference fringes.

These experiments show that neither the concept of particle or wave is sufficient to describe completely a photon (nor any quantum object): the notion of well-defined trajectory loses its meaning, only the probability wavefunction is relevant. The same observations have been successfully obtained with single electrons.

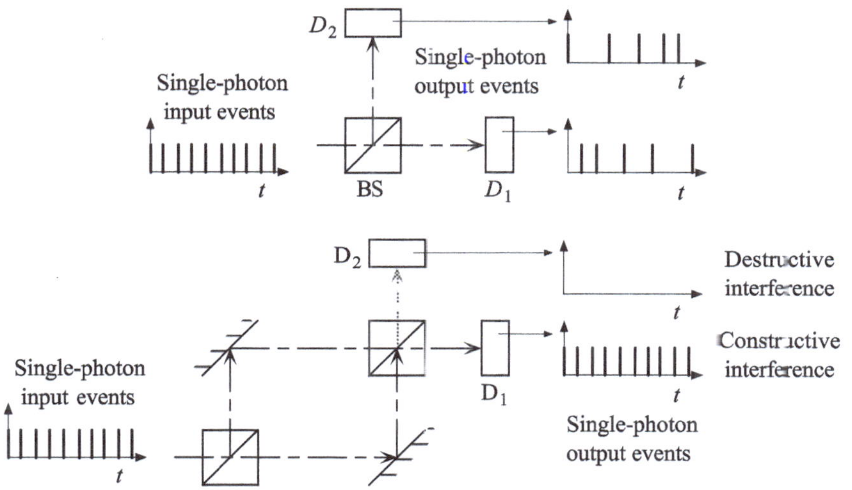

Single-photon interference Experimental set-up and chronograms of input and output events. BS is a beam splitter, D_1 and D_2 are detectors.

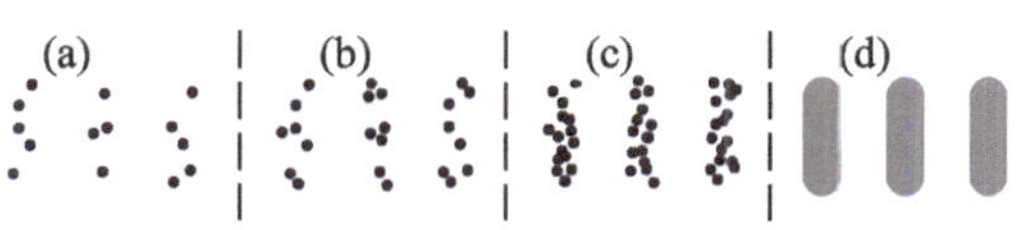

Single-photon interference Progressive formation of Young's interference fringes in a single-photon experiment. Each impact is determined in a probabilistic way only: (a) to (c); temporal integration gives rise to classical fringes (d).

Single quantum dot spectroscopy refers to probing the photoluminescence (PL) response from a single quantum dot at low temperature. Such spectroscopy is warranted since the PL from a given quantum dot is sensitive to its size, with discrete transitions for a single dot being broadened into an emission band when light is collected from a number of quantum dots having a distribution in their respective sizes. The size-broadened spectrum can be a limiting factor in many applications including, for example, quantum dot lasers where a sharp luminescence linewidth is required. Probing of the PL from a single quantum dot requires the use of an ultrasensitive detection system, appropriate immersion lenses, and samples with relatively widely separated quantum dots in relation to the resolution of the optical technique used. Spectroscopic studies of single quantum dots have revealed unexpected new physics including blinking and spectral diffusion which are not observable in samples comprising an ensemble of such quantum dots. Furthermore, very high quantum efficiencies have been demonstrated for individual quantum dots even at room temperature.

Singlet and triplet states are quantum mechanical states that have *total spin quantum number S* of 0 and 1, respectively. An individual electron has an intrinsic angular momentum of its own called electron spin **S** that has a quantized magnitude of $\hbar[s(s+1)]^{1/2}$ where s is the spin quantum number. In an atom comprising several electrons, the individual spins $\mathbf{S}_i$ of the electrons are coupled yielding a resultant overall spin **S** for the atom whose

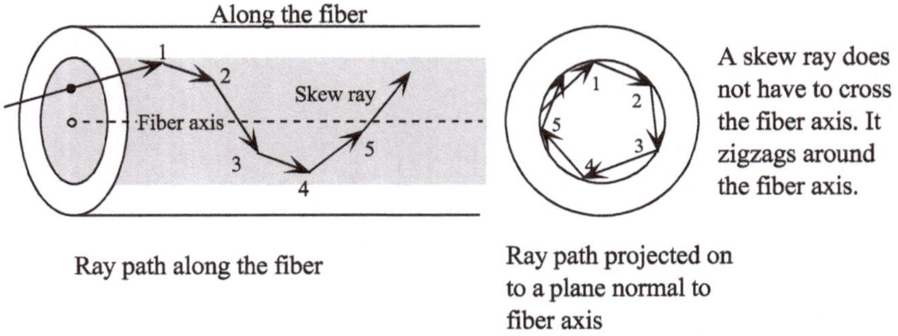

Skew ray.

magnitude is quantized in terms of the atomic total spin quantum number S,

$$S = \hbar\sqrt{S(S+1)}.$$

Skew ray is a light ray that enters the fiber off the fiber axis and zigzags down the fiber without crossing the axis and when viewed looking down the fiber (its projection in a plane normal to the fiber axis) it traces out a polygon around the fiber axis. A skew ray therefore has a *helical path* around the fiber axis. In a step index fiber both meridional and skew rays give rise to guided modes (propagating waves) along the fiber each with a propagation constant β along z, as in the case of the planar waveguide. Skew rays give rise to modes that have both E_z and B_z (or H_z) components along the guide axis z and they are therefore not TE or TM waves.

Slab waveguide is a planar dielectric thin film (sometimes also called the *core*) of thickness t and index n_f sandwiched between a substrate and a "cover" (or superstrate, or cladding) with lower indices n_s and n_c, respectively. In the simplest case, the guide is a thin film deposited on some substrate (such as Si or SiO_2) so that n_c is simply 1, the refractive index of the surrounding medium. The guiding mechanism is provided by total internal reflections along the vertical dimension (see also *integrated optics*). The slab is called *symmetric* if $n_s = n_c$, *asymmetric* otherwise.

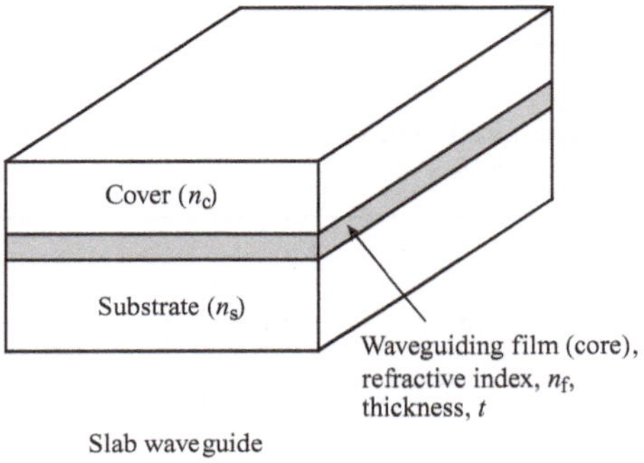

Defining characteristics of the **slab waveguide**.

The basic mechanism involved in guiding the light along the film is the confinement of the optical wave by internal total reflection. The guided rays correspond to incidence angles at the core-substrate and core-cover boundaries greater than the critical angle θ_c for total internal reflection. The principle of guiding can be explained in terms of waves. Only waves that can interfere constructively can be propagated along the guide. This implies that after two reflections, the total phase change (including phase change due to internal reflection at the boundaries) must be 2π. For a given geometry, only a discrete set of *transverse modes* can be supported by the waveguide. The field distribution of the transversally confined modes is sinusoidal-like inside the film, and evanescent in the surrounding dielectrics (decaying exponentially with distance into the surrounding dielectric).

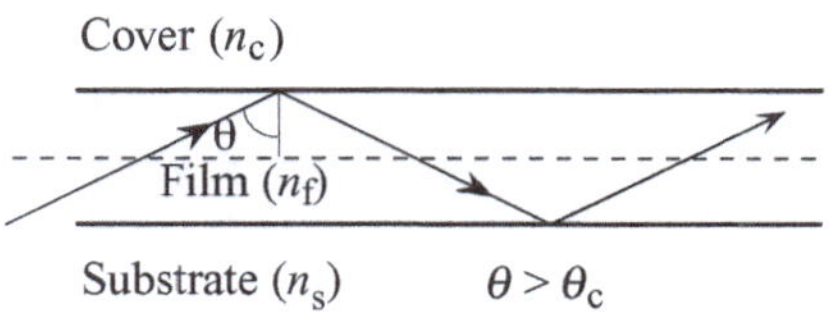

The light ray model for the propagation of light along the **slab waveguide**.

The field distribution inside an asymmetric slab waveguide is not symmetric. The minimum refractive index contrast, $\Delta n = n_f - n_s$, to support guiding propagating modes, can be found by solving Maxwell's equations under the appropriate boundary conditions, providing the so-called cut-off condition. At a vacuum wavelength of λ, for an asymmetric waveguide of thickness t on a planar substrate, the Δn that should be chosen to permit waveguiding of a particular TE mode, m, is given by

$$\Delta n > \left[\frac{(2m+1)\lambda}{4t}\right]^2 \cdot \frac{1}{n_f + n_s}.$$

A single-mode propagation requires $[3\lambda/4t]^2(n_f + n_s)^{-1} > \Delta n > [\lambda/4t]^2(n_f + n_s)^{-1}$.

Slit scan is a particular x-ray imaging system where images of objects are obtained by scanning with slits in front of the object and detector to collimate the beam to the detector size, and moving the collimated scanning set across the object during exposure to acquire the whole image with the aim of reducing x-ray scattering effects in the image. The slit scan system has a linear detector array covering only one pixel row. The collimator shaping the x-ray beam to match the detector is a vital part of

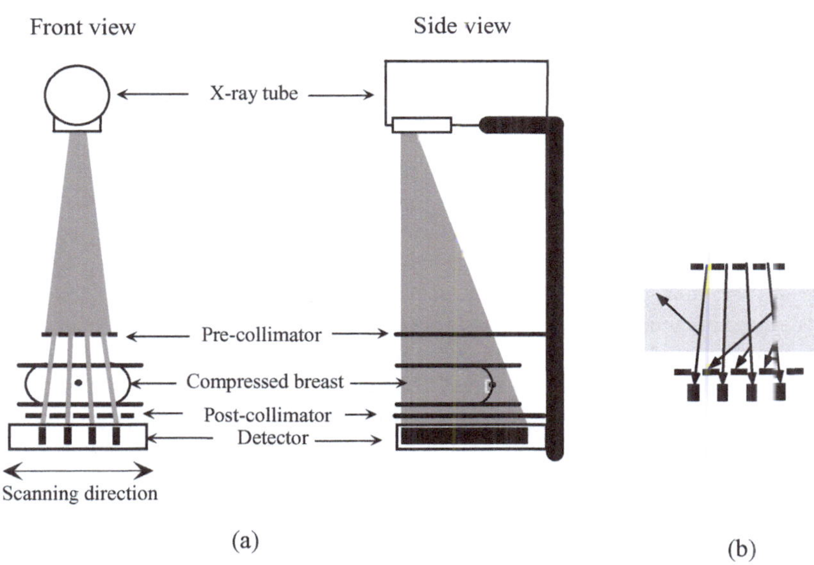

Slit scan (a) Schematic of a scanning multi-slit system. The x-ray source, collimators (slits) and detector lines are mounted on a common mechanical structure that performs the scanning motion relative to the object. (b) Scatter reduction mechanism in a scanning multi-slit system. Scatter from adjacent beams is controlled by the post-collimator slits so that very little scatter is imaged.

this scanning system. This system has the benefit of reducing the amount of scattered radiation and also the reduced detector area decreases the detector price. But precision mechanics is needed to correctly align the x-ray beam, detector and slits. There are some drawbacks with the scanned systems such as a relatively low fraction of the photons emitted from the x-ray tube is used and a longer exposure time. However, it is not the total exposure time that is relevant. In fact, most detector lines will image a specific point in the patient body during a time which is significantly smaller compared to the total passage time. The pixel width in the scanning direction is never more than 100 μm. A scanned slit system based on one detector line could never acquire an image fast enough, but several lines are stacked together to form a multi-slit system. This x-ray imaging system has been successfully applied to mammography. The principle of the slit scan system is illustrated in the figure. There is an array of slits (pre-collimator) between the focal spot and patient, and an array of slits (post-collimator) between the patient and image receptor. Usually 30–50 slits are used for mammographic applications. The pre-collimator slits and the post-collimator slits are aligned during an exposure. The volume of tissue irradiated by a single beam at any instant is small because of the long narrow beam geometry, so little scatter is produced and little of this scatter reaches the image plane in the area defined by the beam. Scatter from adjacent beams is controlled by the post-collimator slots so that very little scatter is imaged. The x-ray source, collimators (slits) and detector lines are mounted on a common mechanical structure that performs the scanning motion relative to the object. Depending on the spacing between the slits, scanning over a relatively small distance can acquire the full area of the breast. Thus this multi-slit scanning can greatly reduce the exposure time and at the same time reduce scattering.

Slope efficiency of a laser diode determines the optical power, P_o, in the lasing emission in terms of the diode current above the threshold current I_{th}. If I is the diode current, the slope efficiency η_{slope} (in watts per ampere) is defined by

$$\eta_{slope} = \frac{P_o}{I - I_{th}}.$$

Slot scan is an x-ray imaging system where the radiation beam is confined to a "slot," and images of objects are obtained by moving the slot across the object during exposure to acquire the whole image. The detector covers the entire image field (~ 20 cm) in one direction and only around ~1 cm wide in the scanning direction. This system has the benefit of reducing the amount of scattered radiation, and also the reduced detection area decreases the detector price. There are some drawbacks with the scanned systems such as a relatively low fraction of the photons emitted from the x-ray tube is used and a longer exposure time is required. The slot scan imaging system has been successfully applied to mammographic imaging. The x-ray beam is shaped to match the rectangular detector by a collimator. The detector structure can be (i) a phosphor/CsI layer that is fiber (optically) coupled to a number of CCDs, or (ii) a photodiode array bonded to a CCD by indium bump bonds. The images are acquired using a time delay integration (TDI) technique. The x-ray beam is activated continuously during the image scan and the collected charges in the pixel rows are shifted down CCD columns at an equal speed but in the opposite direction to the motion of the x-ray beam and detector assembly across the breast. Thus, the collected charge packets remain stationary with respect to a given x-ray projection through the breast. In this way the charge corresponding to a pixel value in the image is integrated during the passage of the detector. This TDI method increases the sensitivity of the slot-scan detector system (and thus reduces x-ray intensity) by the ratio of the slot width to the pixel width, a substantial factor of 80. When the charge packet reaches the final element of the CCD, the charge signals in the columns are read out on a transfer register and

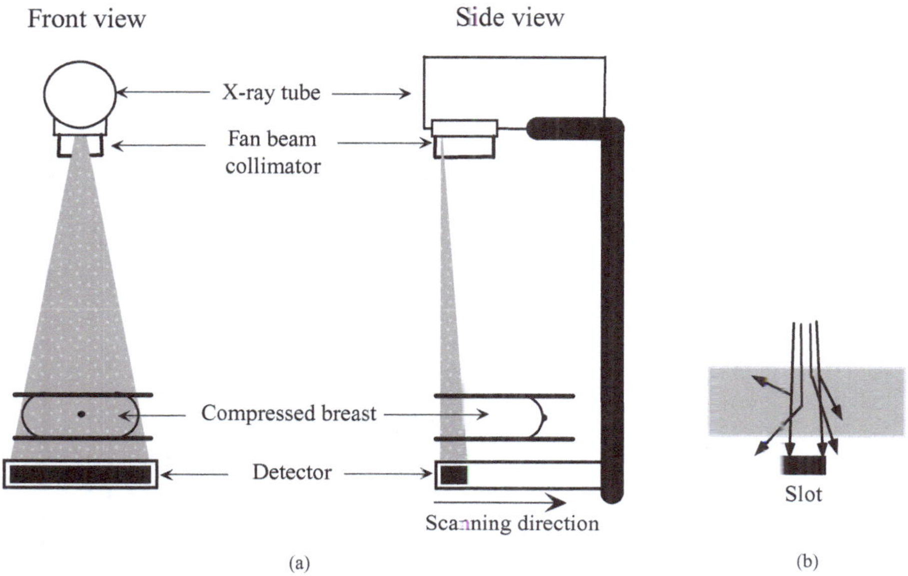

Slot scan (a) Schematic of a slot scanned digital mammography system. The x-ray source, collimator (slot) and detector are mounted on a common mechanical structure that performs the scanning motion relative to the object. (b) Illustration of the scatter reduction mechanism in a scanning multi-slit system. Compared to an areal detector, the probability of a scattered photon hitting the slot detector is very low.

digitized. The slot-scanned x-ray imaging system is shown in (a) in the figure. The scatter reduction mechanism is illustrated in (b).

Small pixel effect is the phenomenon in semiconductor pixellated detectors in which the collected charge depends mainly on the motion of the charge carriers that move towards the pixel electrodes. The motion of charge carriers close to the pixel electrodes (near-field region) can produce almost full signal (collected charge) strength in the external circuit and the motion of charge carriers outside this region induces very little signal, because the induced charge on a particular pixel electrode by a moving charge is very small when the charge is far away (the distance to the pixel is much greater than the pixel dimension) due to charge sharing among many pixel electrodes, and increases rapidly when the charge is in the vicinity of this pixel. As a result, the change in induced charges in pixel electrodes due to the carrier motion through the near-field region is much higher than their motion in the outside of the near-field region (this change in induced charge produces electric current in the external circuit and the integration of this current is the collected charge). The extent of the near-field region decreases with decreasing pixel dimensions. Therefore, this effect is more pronounced in detectors with pixel dimensions that are small compared with the detector thickness. Since the charge collection efficiency of pixellated detectors depends mainly on the transport properties (mobility and lifetime) of the charges that move towards the pixel electrodes, the effects of poor transport properties of other types of carriers can be eliminated by having smaller pixels. Some of the

Willebrord Snellius (Willebrord Snel van Royen, 1580–1626) was a Dutch astronomer and a mathematician who was a professor at the University of Leiden. He discovered his law of refraction in 1621which was published by Réne Descartes in France 1637; it is not known whether Descartes knew of Snell's law or formulated it independently. (AIP Emilio Segre Visual Archives, Brittle Books Collection.)

René Descartes (1596–1650) was a French philosopher who was also involved with mathematics and the sciences. He has been called the "Father of Modern Philosophy." Descartes was responsible for the development of Cartesian coordinates and analytical geometry. He also made significant contributions to optics, including reflection and refraction. (AIP Emilio Segre Visual Archives.)

potential photoconductors (e.g. *a*-Se, CdZnTe and HgI_2) for x-ray image detectors have asymmetric transport properties (mobility-lifetime product) for electrons and holes. Therefore, the charge collection efficiency of pixellated detectors can be improved by ensuring that the carrier with the higher mobility-lifetime product is drifted towards the pixellated electrodes.

Smectic phase see *liquid crystals*.

Snell's law describes the motion of electromagnetic waves through a medium with different refractive indices (e.g. traveling from air to glass). Suppose that light traveling in a medium with refractive index n_1 is incident on a medium of refractive index n_2, and if the angles of incidence and refraction (transmission) are θ_i and θ_t with respect to the normal to the surface between the media at the point of incidence, then according to Snell's law, $n_1 \sin\theta_i = n_2 \sin\theta_t$. See *refraction*.

Softening point of glass is the temperature at which the viscosity of a glass is about 10^3–10^4 Pa s. For example, for fused silica glass this is about 1600–1730 °C. The viscosity of glass is strongly temperature dependent and decreases steeply as the temperature increases. At the softening point, the viscosity is sufficiently low to allow the glass to flow under its own weight. Very roughly, the softening point is the glass transformation point where the rigid glass (solid) "transforms" to a supercooled liquid that can flow under a small shear stress (just like honey).

Sol gel is a process used primarily for the chemical synthesis of ceramic and glass materials. The process generally involves the transformation of a colloidal liquid known as a "sol" into a solid or "gel" phase. The sol is usually prepared from a precursor comprising inorganic metal salts or organic compounds such as metal alkoxides. Typically, the precursor then undergoes a series of hydrolysis or polymerization reactions, finally resulting in the formation of a colloidal suspension or the sol. At this point in the process, there are three main routes usually followed to fabricate the final materials – these are coating, gelling and spinning. Coating processes include both spin coating and dip coating. These enable thin films of the desired material to be prepared on a substrate, which are then heat treated to form dense films. In the case of gelling, a typical approach is to cast the sol into a mold to form a wet gel which with further drying and heat treating is converted into a dense ceramic or glass bulk material. If the liquid (i.e. solvent) is extracted from the wet gel under supercritical conditions (i.e. with respect to temperature and pressure), a material known as an aerogel can be obtained. Drying is normally performed inside an autoclave enabling one to exceed the critical point (P_C, T_C) of the solvent. Different schedules can be used to reach the critical point of the solvent, with careful attention to appropriate choice of solvent. Strong inorganic solids are commonly dried using alcohol (or acetone) as solvent, while organic solids susceptible to decomposition at temperatures exceeding 100 °C are commonly dried using CO_2 as the solvent. The process of spinning is performed to form fibers of ceramic or glass material. In this case, careful attention to control the viscosity of the sol is paid, enabling fibers to be drawn and used after an appropriate heat treatment. Sol gel films have found extensive application in the field of optical systems for guiding light as well as for nonlinear optics, as well as for sensor applications, while sol gel processing has been used to form artificial opals and inverse opals in both thin film and bulk form, possessing promising photonic crystal properties. There is extensive work on fabricating optical fiber for telecommunications, for example, using sol gel processing. These are just a few of the applications of sol gel processing.

Solar cell see *photovoltaic devices*.

Solar constant is the integrated intensity above Earth's atmosphere, that is, the total power flow through a unit area perpendicular to the direction of the Sun. This quantity is called the *solar constant* or *air-mass zero* (AM0) radiation and it is approximately constant at a value of 1.353 kW m^{-2}. See *solar radiation*.

Solar radiation generally refers to the electromagnetic radiation energy flux that we receive from the Sun. The solar radiation is characterized by two important quantities: *air mass* and *spectrum*. *Solar radiation air mass m* (AM*m*) is defined as the ratio of the actual Sun's radiation path h to the shortest path h_0, that is $m = h/h_0$. Since $h = h_0 \sec\theta$, AM*m* is AM $\sec\theta$. This spectrum refers to incident energy on a unit area *normal* to Sun's rays (which have to travel the atmospheric length h.) The actual intensity spectrum on Earth's surface depends on the absorption and scattering effects of the atmosphere and hence on the atmospheric composition and the radiation path length through the atmosphere. These atmospheric effects depend on the wavelength. Clouds increase the absorption and scattering of sunlight and hence substantially reduce the incident intensity. On a clear sunny day, the light intensity arriving on Earth's surface is roughly 70% of the intensity above the atmosphere. Absorption and scattering effects increase with the Sun beam's path through the atmosphere. The shortest path through the atmosphere is when the Sun is directly above that location and the received spectrum is called air mass one (AM1). All other angles of incidence increase the optical path through the atmosphere, and hence the atmospheric losses. In addition, atmospheric molecules and dust particles scatter the sunlight. Scattering not only reduces the

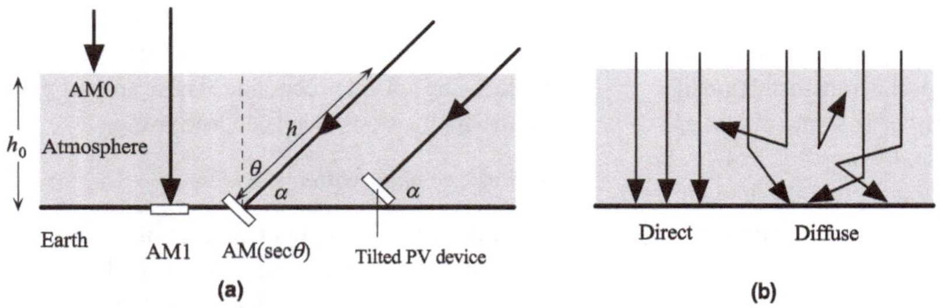

Solar radiation (a) Illustration of the effect of the angle of incidence θ on the ray path length and the definitions of AM0, AM1 and AM(sec θ). The angle α between the sunbeam and the horizon is the solar latitude. (b) Scattering reduces the intensity and gives rise to a diffused radiation.

intensity in the direction towards the Earth but also gives rise to the Sun's rays arriving at random angles. Consequently, the terrestrial light has a *diffuse* component in addition to the *direct* component. The diffuse component increases with cloudiness and the Sun's position, and has a spectrum shifted toward the blue light. The scattering of light increases with decreasing wavelength so that shorter wavelengths in the original sunbeam experience more scattering than longer wavelengths. On a clear day, the diffuse component can be roughly 20% of the total radiation, and significantly higher on cloudy days.

Solar spectrum is the spectral intensity (intensity per unit wavelength, also know as *spectral flux*) I_λ vs. wavelength λ characteristic of the solar radiation that Earth receives. $I_\lambda \delta\lambda$ is the intensity in a small interval $\delta\lambda$. Integration of I_λ over the whole spectrum gives the integrated or *total intensity*. The integrated intensity above Earth's atmosphere gives the total power flow through a unit area perpendicular to the direction of the Sun. This quantity is called the *solar constant* or air-mass zero (AM0) radiation and it is approximately constant at a value of 1.353 kW m^{-2}. The solar spectrum that is received at sea level is radiation that has passed through the atmosphere and hence suffered absorption and scattering. While the overall gross features of the sea-level spectrum resemble those of black body radiation, the solar spectrum has many sharp absorption losses (troughs) where the radiation has been absorbed by atmospheric molecular species, primarily by water vapor molecules, H_2O, and then by CO_2 and O_2. There is also a reduction in the solar flux at short wavelengths, typically less than 320 nm, due to absorption by ozone (O_3) and scattering by molecules, aerosols and small particles in the atmosphere; scattering increases with decreasing wavelength. See also *solar radiation*

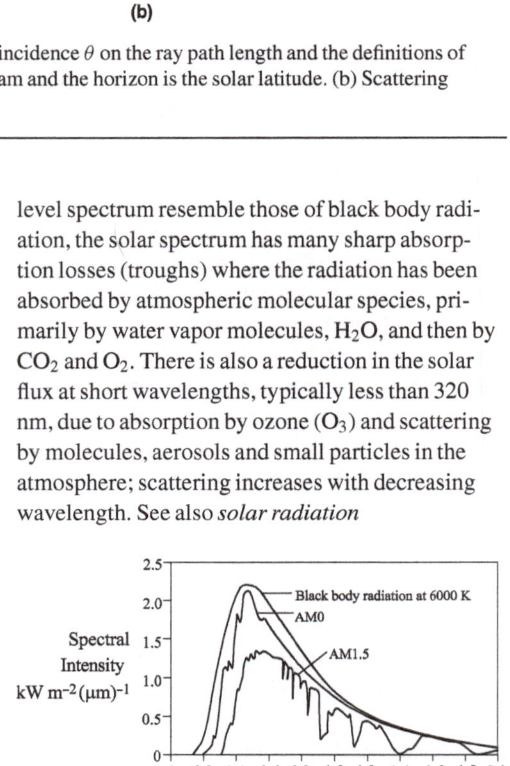

Solar spectrum The spectrum of the solar energy represented as spectral intensity (I_λ) vs. wavelength above the Earth's atmosphere (AM0 radiation) and at the Earth's surface (AM1.5 radiation). Black body radiation at 6000 K is shown for comparison.

(K. R. Lang, *The Cambridge Encyclopedia of the Sun*, Cambridge University Press, 2001.)

Soleil–Babinet compensator is an optical device that allows one to control the retardation (i.e. the phase change) of a wave passing through it. In a wave plate retarder such as the half-wave plate, the relative phase change ϕ between the *ordinary* and *extraordinary waves* depends on the plate thickness and cannot be changed. In compensators, ϕ is adjustable. The Soleil–Babinet compensator has two quartz wedges touching over their large faces to form a "block" of adjustable height d. Sliding one wedge over the other wedge alters the "thickness" d of this block. The two-wedge block is placed on a parallel plate quartz slab with a fixed thickness D. The slab has its optic axis parallel to its surface face. The optical axes in the wedges are parallel but perpendicular to the optic axis of the slab. Suppose that linearly polarized light is incident on this compensator at normal incidence. We can represent this light by field oscillations parallel and perpendicular to the optic axis of the two-wedge block; these fields are E_1 and E_2, respectively. The polarization E_1 travels through the wedges (d) experiencing a refractive index n_e (E_1 is along the optic axis) and travels through the plate (D) experiencing an index n_o (E_1 perpendicular to the optic axis). Its phase change is

$$\phi_1 = \frac{2\pi}{\lambda}(n_e d + n_o D).$$

But the E_2 polarization wave first experiences n_o through the wedges (d) and then n_e through the plate (D) so that its phase change is

$$\phi_2 = \frac{2\pi}{\lambda}(n_o d + n_e D).$$

The phase difference ϕ $(= \phi_2 - \phi_1)$ between the two polarizations is

$$\phi = \frac{2\pi}{\lambda}(n_e - n_o)(D - d).$$

It is apparent that as we can change d continuously by sliding the wedges (by using a micrometer screw), we can continuously alter the phase difference ϕ from 0 to 2π. We can therefore produce a quarter-wave or half-wave plate by simply adjusting this compensator. It should be emphasized that this control occurs over the surface region that corresponds to both the wedges in contact and in practice this is a narrow region.

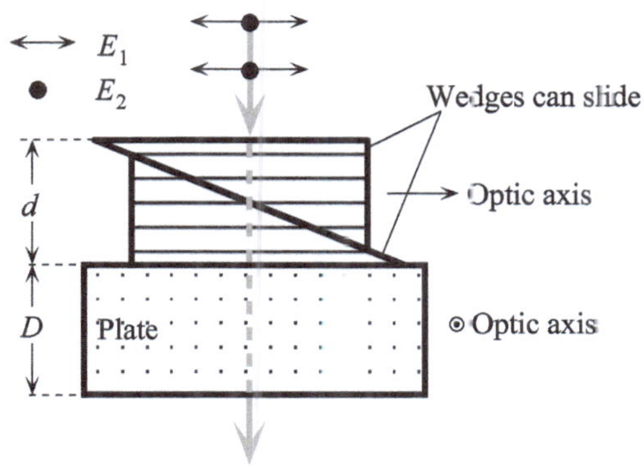

Soleil-Babinet compensator.

Solid angle, qualitatively, is the angle Ω in three dimensions that an area S on a sphere subtends at the center of the sphere; its magnitude is given by $\Omega = S/r^2$. Its units are steradians. More rigorously, the solid angle is defined in terms of an *element of solid angle* $d\Omega$ by $d\Omega = dS/r^2$, where dS is the elemental area on a sphere of radius r. Using spherical coordinates, the area $dS = r^2 \sin\theta d\phi d\theta$ so that

$$d\Omega = \sin\theta d\phi d\theta.$$

For a whole sphere, $S = 4\pi r^2$ so that $\Omega = 4\pi$, and for a hemisphere Ω is 2π. The solid angle that corresponds to a cone of apex angle 2θ is of particular importance, especially when LED emission distribution is considered,

$$\Omega = \int_0^{2\pi}\int_0^{\theta} \sin\theta d\phi d\theta = 2\pi(1 - \cos\theta) \approx \pi\theta^2$$

for small angles (θ in radians).

Soliton is an isolated wave (a solitary wave), in the form of a "pulse of vibration" with a certain shape and intensity profile, in a nonlinear medium that is

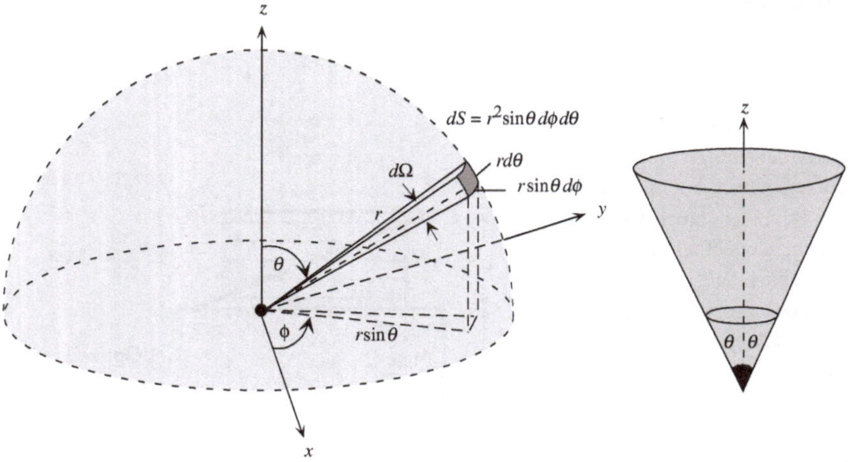

Definition of a **solid angle** and the solid angle of a cone.

able to propagate without suffering dispersion. A soliton can propagate without experiencing a net dispersion because the dispersive effect, due to the refractive index depending on the wavelength, is balanced by the nonlinearity of the medium in which the refractive index depends on the intensity. The overall result is as if the wave were propagating in a nondispersive linear medium. Such solitons have interesting properties. For example, two solitons do not interfere as normal waves do. They pass right through each other without changing their shapes as if they were independent entities; only a phase shift is induced as a result of the interaction. If ψ is the disturbance or perturbation that is propagating, such as displacement, one possible soliton shape (wave envelope) that is propagating along the z-direction is a sech function,

$$\psi = A\,\text{sech}^2[(vt - z)/l]$$

where v is the local signal velocity and A and l are constants; l is a constant that represents the spatial extent of the soliton. The above is only one possible soliton from a family of solitons that can exist in a given nonlinear medium. Solitons satisfy a special

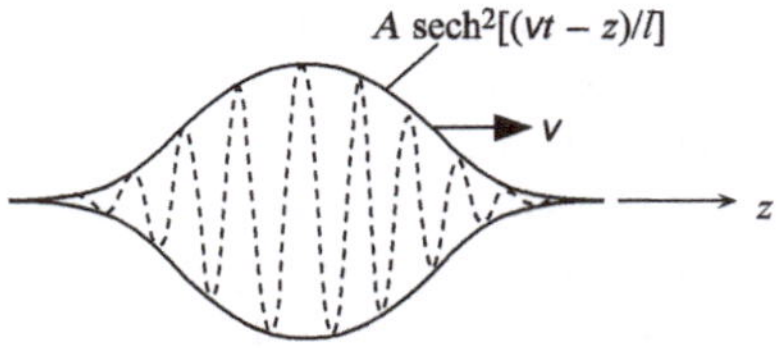

A visualization of a **soliton** wave packet.

equation called the *nonlinear Schrödinger equation*. A soliton pulse can only be excited and propagated if the input pulse has the right shape and the amplitude × width product; otherwise an ordinary optical pulse will be excited. A *spatial soliton* is an intense beam that can propagate without diverging because the diffraction effects (that lead to divergence) have been balanced by the nonlinear effects of the medium. One can intuitively appreciate this as follows. All propagating beams of finite cross-sectional area are known to diverge. If the intensity is sufficiently large, the intensity profile of the beam will modify the refractive index of the medium due to the nonlinear response of the medium (polarization is not linearly related to the field). If the beam has the right intensity profile such that the altered index variation results in an induced waveguide, and the beam is a mode of this waveguide, then the beam becomes *self-guided*; it no longer diverges.

SONET see *synchronous optical networking*.

Sonoluminescence is the emission of light from a material in a particular state, as a result of excitation with acoustic radiation such as ultrasonic waves, for example. Typically, this phenomenon is associated with acoustic excitation of small bubbles (containing gas) in a liquid. That is, under intense compression associated with collapse of such bubbles, bubbles radiate light corresponding to an effective black body temperature of the order of tens of thousands of kelvin, with spectral broadening reflecting the pressure broadening of the atomic transitions for the hot compressed gas. Typically, light emission occurs in the ultraviolet spectral region with emission occurring on time-scales of tens of picoseconds following the excitation.

Soref's single-mode rib waveguide equation is a condition that applies to large cross-section rib waveguides fabricated using the Si on oxide technology (Si/SiO_2 waveguide) to determine whether the rib waveguide is a single or a multimode guide. The condition is given in *rib waveguides*. (R. A. Soref et al, *IEEE J. Quantum Electron.*, **27**, 1971, 1991.)

Sound velocity in a crystal refers to the group velocity of lattice waves within the crystal, that is, the propagation of coupled atomic vibrations as a wave where the frequency of the oscillations is typically less than 10^{12} Hz. These waves are called acoustic *phonons*. The frequency ω vs. wavevector K at such low frequencies is a straight line so that the waves have a group velocity v_g independent of the frequency. The acoustic or the sound velocity depends on the crystal direction and the direction of the oscillations of the atomic displacements. If the atomic displacements are along the direction of propagation then the wave is a *transverse acoustic* (TA) wave, whereas if the displacements are transverse, then it is a *transverse acoustic wave* (TA). The sound velocity can be calculated from the elastic constants (or elastic moduli). For a cubic crystal, along the [111] direction, $v_{LA} = (C_{11}/\rho)^{1/2}$ and $v_{TA} = (C_{44}/\rho)^{1/2}$ where ρ is the density, and C_{11} and C_{44} are two of the elastic coefficients (usually listed).

Sparrow criterion of resolving power see *Rayleigh criterion*.

Spatial coherence describes the extent of coherence between waves radiated from different locations on a light source. If the waves emitted from two locations P and Q on the source are in phase then P and Q are spatially coherent. A spatially coherent source emits waves that are in phase over its entire emission surface. These waves however may have partial *temporal coherence*. A light beam emerging from a spatially coherent light source will hence exhibit spatial coherence across the beam cross-section. A mostly incoherent beam will contain waves that have very little correlation with each other.

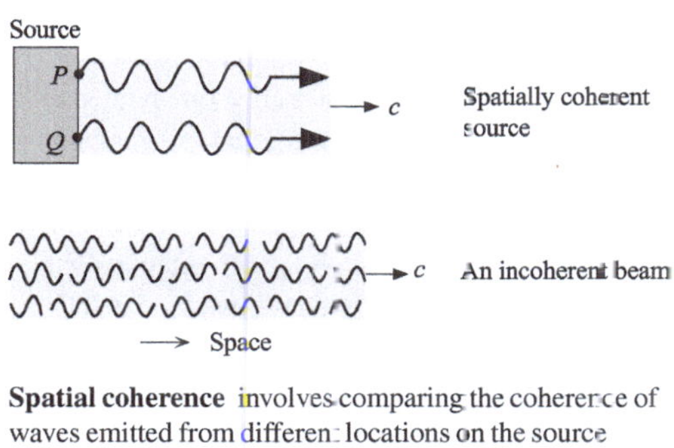

Spatial coherence involves comparing the coherence of waves emitted from different locations on the source

More generally, two points P_1 and P_2 separated in space at respective positions $\boldsymbol{r}_1$ and $\boldsymbol{r}_2$ are said to be mutually coherent if the vibrations of the electromagnetic field at P_1 and P_2 are correlated (constant

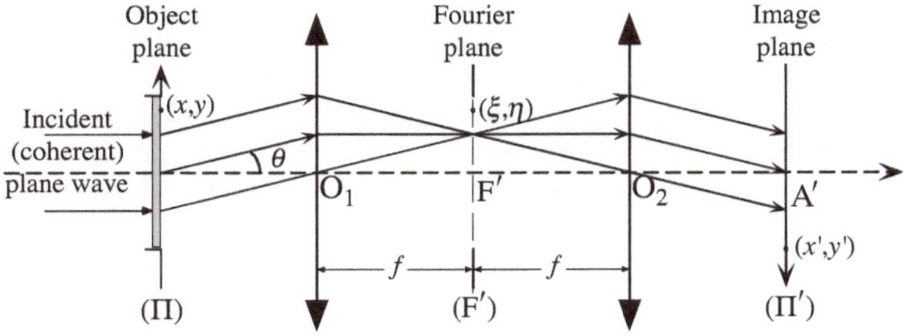

Illustration of **spatial filtering**.

phase difference). For instance, two thin pinholes in Young's experiment, illuminated by a single point source located at equal distance from both of them, are reached by the primary wave exactly in phase, and thus form completely coherent secondary point sources.

Spatial cut-off frequency see *spatial frequency*.

Spatial filtering in coherent illumination enables one to filter out some spatial frequencies by positioning a suitable mask in the Fourier plane, F′, of the transparent, grating-like object in the Π (object) plane. Two objective lenses are placed along a common axis in telescope fashion; the image focal plane of the first coincides with the object focal plane of the second. The final image is observed in the Π′ plane, in which the coordinates (x',y') are related to (x,y) in the Π plane through the geometric magnification coefficient. See also *Abbe's theory of microscope imaging*.

For instance, a low-pass filter will select only the low spatial frequencies, which is illustrated in the figure.

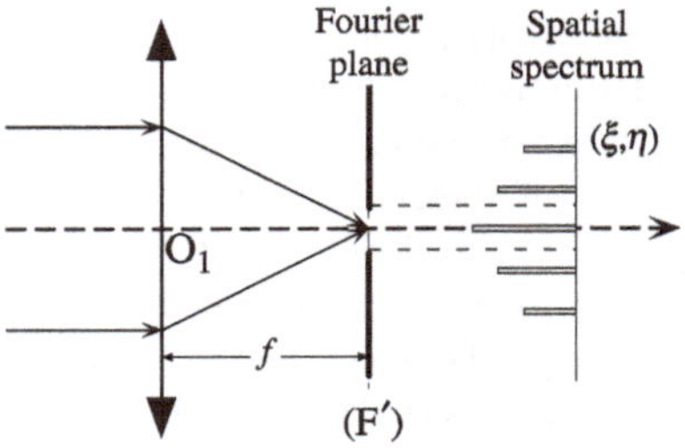

Illustration of **spatial filtering**.

On the other hand, in the so-called *strioscopy* technique, a central mask is used to filter out the zero-order diffraction spot only, enabling one to see the details of a *phase object* on dark background. For this reason, it is also known as the *central dark ground method* of observation. Zernike's *phase contrast method* is even more sensitive for observing a *phase object* of complex transmittance $t(x,y) = \exp[j\Phi(x,y)]$. As long as the phase variation $\Phi(x,y)$ across the object plane remains small, $t(x,y) \approx 1 + j\,\Phi(x,y)$. Without any filtering, the image in the Π′ plane would be identical to the object, characterized by an amplitude $a(x',y') = 1 + j\,\Phi(x',y')$, or an intensity proportional to $|a|^2 = 1 + \Phi^2(x',y')$, that is essentially 1 to first order in Φ. The details would be lost on the continuous (zero-order) background. Suppose now that the central order is phase-shifted by a quarter of a wavelength through a phase plate in the F′ plane, whereas the higher-order components remain unaffected. The amplitude becomes $a'(x',y') = j\,[1 + \Phi(x',y')]$ in the Π′ plane, giving an intensity $|a'|^2 \approx 1 + 2\,\Phi(x',y')$. The intensity variations in the restituted image are now directly proportional to the phase variations in the original object. For this method, first described in 1935, Frits

Zernike was awarded the Nobel Prize in physics in 1953.

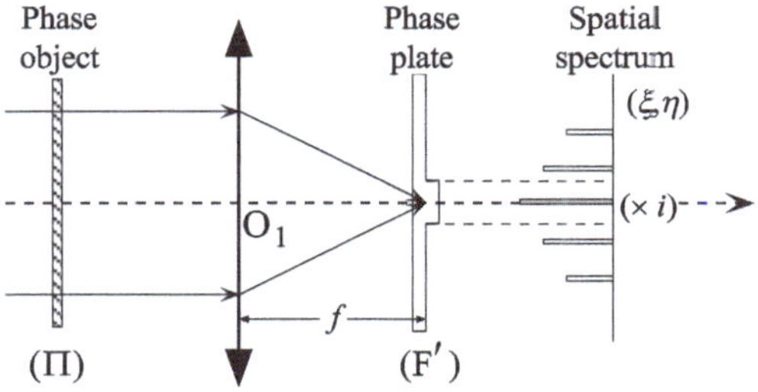

Spatial filtering Zernike's phase contrast method.

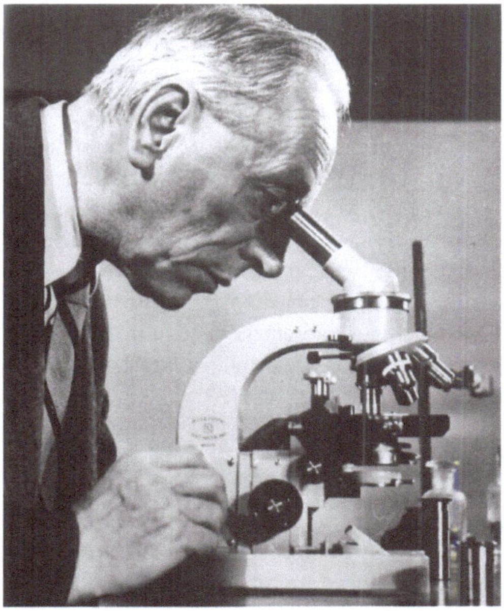

Spatial filtering Frits Zernike (1888–1966) was a Dutch physicist, who received the Nobel Prize in physics in 1953 for his phase contrast microscope. (AIP Emilio Segrè Visual Archives.)

Spatial frequency represents the frequency of a sinusoidal oscillation of a spatially dependent quantity, such as the electric field, that varies periodically with distance (instead of time as in a sinusoidal time signal). Suppose the field $E(x)$ varies sinusoidally with distance along some line, taken arbitrarily to be along the x-axis. Then $E(x) = E_0\sin(2\pi x/\lambda)$ where x is distance, and λ is the *spatial period* ("repetition distance") on x. The reciprocal of the spatial period, $1/\lambda$, is the *spatial frequency*, denoted by ν (or f), of this harmonic spatial variation just as the sinusoidal oscillation $\sin(2\pi t/T)$, where T is the period of the oscillations, has a frequency $\nu = 1/T$. Thus, spatial frequency $\nu = 1/\lambda$. It is customary to use the subscript to indicate the coordinate axis on which the spatial variation takes place, in this case the x-axis, so that the spatial frequency on x is ν_x. In analogy with the k-vector of a traveling wave, $2\pi/\lambda$ is the *angular spatial frequency* and usually denoted by k_x (sometimes by K_x or even q_x). It is measured in cycles per millimeter. Most quantities do not have a perfect sinusoidal variation with distance, which means that they cannot be represented with a single spatial frequency ν_x, and need a range of spatial frequencies to be representable. For example, a square wave pattern along the x-axis with a spatial period λ needs spatial frequencies $\nu_x = 1/\lambda, 3/\lambda, 5/\lambda, \ldots$ to describe it. A two-dimensional image that has a brightness or luminance distribution $l(x,y)$ on the xy-plane can be described by (or can be constructed from) summing up properly weighted contributions of harmonic (sinusoidal-type) intensity patterns that have different spatial frequencies ν_x and ν_y along x and y,

$$l(x, y) = \int\int_{-\infty}^{+\infty} L(\nu_x, \nu_y) \times \exp[-j(2\pi\nu_x x + 2\pi\nu_y y)]d\nu_x d\nu_y$$

where $L(\nu_x,\nu_y)$ is the Fourier transform of the image's brightness pattern $l(x,y)$ and describes the weighting (as a complex amplitude) of the harmonic variation $\exp[-j(2\pi\nu_x x + 2\pi\nu_y y)]$ at the spatial frequencies ν_x and ν_y that has to be used in the construction of $l(x,y)$. The *spatial cut-off frequency* $\nu_{\text{cut-off}}$ is defined as the highest spatial frequency

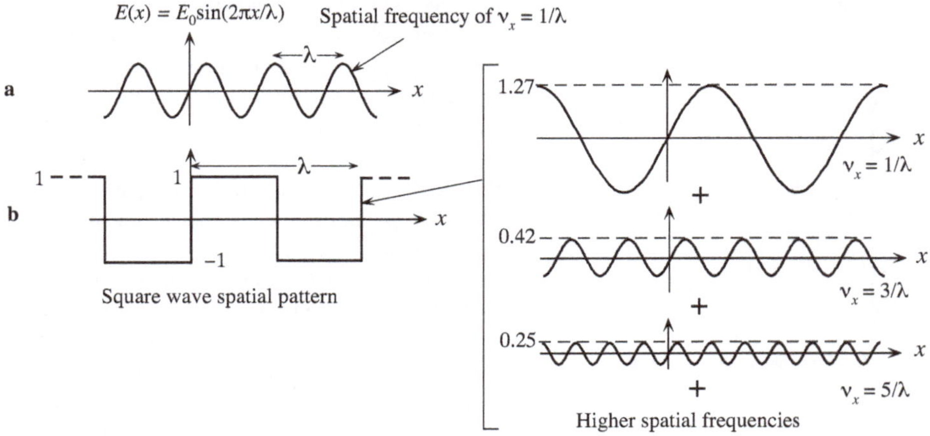

Spatial frequency (a) Schematic illustration of the meaning of spatial frequency. (b) A square wave pattern with a spatial period needs an infinite number of spatial frequencies, $\nu_x = 1/\lambda, 3/\lambda, 5/\lambda \ldots$ to describe it.

an optical system can image or distinguish. Small or sharp objects contain high spatial frequencies, which may fall beyond $f_{\text{cut-off}}$. In those cases, the optical instrument cannot resolve a sharp edge or a small object. For a lens, due to diffraction effects, the upper (best case) spatial cut-off frequency is

$$\nu_{\text{cut-off}} = \frac{1}{\lambda F}$$

where λ is the wavelength of light used in imaging, and F is the *f*-number of the lens, $F = f/D$, f is the focal length and D is the aperture or pupil diameter of the lens. See *Fourier analysis of diffraction*.

Spatial resolution see *modulation transfer function, resolving power*.

Spatial soliton see *soliton*.

Specific rotatory power see *optical activity*.

Speckle noise see *modal noise*.

Spectra is a general term used to describe the emitted or absorbed radiation at various wavelengths. Spectra of gases have discrete wavelengths because they correspond to electronic transitions between various quantized energy levels. In an absorption process, an atom with an energy E_1 is excited to a higher energy level E_2 by absorbing a photon of energy $h\upsilon = E_2 - E_1$. The absorption spectrum is observed when white light, light of all wavelengths, is passed through the medium. Those wavelengths that are missing represent the photons that have been absorbed from white light in exciting the gas atoms. The emission process is the reverse of the absorption process, except that there must be atoms at the energy level E_2 that can become de-excited to E_1 by emitting photons of energy $h\upsilon = E_2 - E_1$. Emission spectra can be obtained by first exciting the atoms in the gas, for example, by an electric discharge. The excited atom with an energy E_2 can become de-excited to a lower energy level E_1 by emitting a photon that has an energy $h\upsilon = E_2 - E_1$, provided that the transition obeys the quantum mechanical *selection rules*. Typically the excitation of gas atoms involves accelerating electrons (or gas ions) in the discharge colliding with the gas atoms.

Spectral color see *color*.

 Spectral hole burning

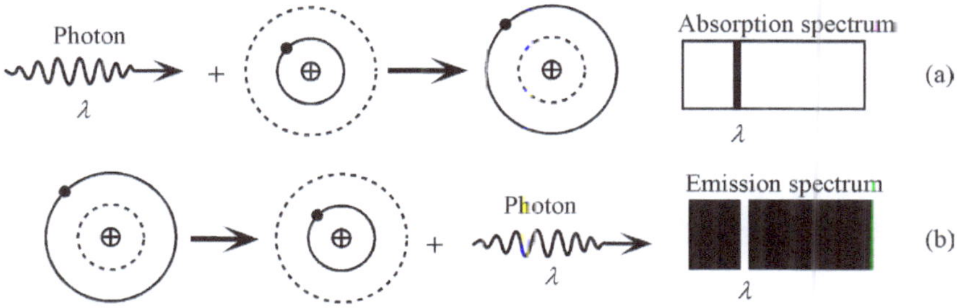

Spectra The physical origin of (a) absorption spectra and (b) emission spectra.

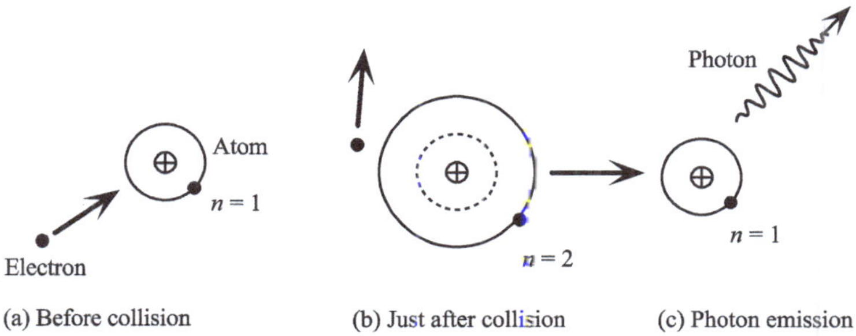

Spectra An atom in a gas discharge tube can become excited by a projectile electron colliding with it.

Table: Spectra Common nomenclature assigned to various wavelength ranges.

Terminology	Typical approximate range (nm)
Extreme ultraviolet (UV)	10–200
Far UV	200–300
Near UV	300–380
Visible	380–700
Near IR (infrared)	770–1500
Middle IR	1500–6000
Far IR	6000–40,000
Far-far IR	40,000–1,000,000

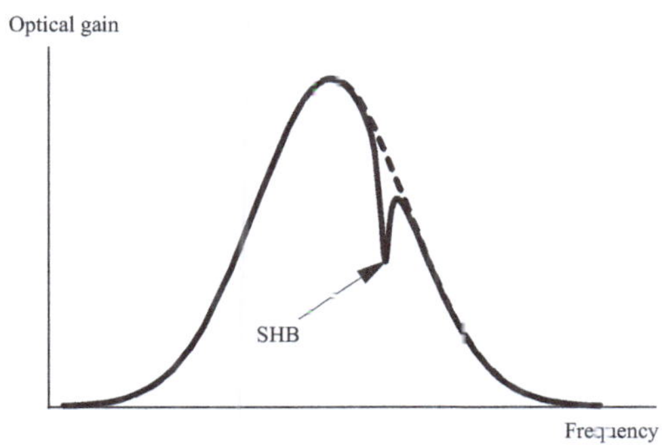

Optical gain vs. frequency of an inhomogeneously broadened medium in which there is **spectral hole burning** (SHB); a hole has been burnt around the frequency of the optical field. Both the depth and the width of the hole increase with its saturating intensity.

Spectral hole burning (SHB) is caused by stimulated emission creating a depletion, "burning a hole," in the distribution of excited atoms (or, in semiconductors, in the carrier distribution), thus causing a deviation from the unsaturated steady state distribution (e.g. a Gaussian lineshape for Doppler-broadened gas lasers, Fermi–Dirac statistics for semiconductors). The time constant associated with this process is the time needed to restore the steady state distribution. In semiconductors, the restoration mechanism is carrier–carrier scattering, with a typical time of several tens to hundreds of femtoseconds.

Spectral intensity, I_λ, is the intensity per unit wavelength so that $I_\lambda \delta\lambda$ is the intensity in a small interval $\delta\lambda$. Integration of I_λ over the whole spectrum gives the integrated or *total intensity*, *I*. See *solar spectrum.*

Spectral responsivity see *responsivity*.

Spectral width, $\delta\upsilon_m$, of the Fabry–Perot etalon is the *full width at half maximum* (FWHM) of an individual mode intensity and depends on the *finesse factor* of the cavity. The greater the finesse, the narrower the width of an individual mode. See *optical resonator.*

Spectral width is the frequency (or wavelength) width of the spectrum of a signal that contains the majority of the power.

Spectrometer is an optical instrument that uses a suitably designed prism or a diffraction grating to generate, view or take measurements on a pure spectrum from a light source (e.g. to measure the wavelength of light), or to measure the refractive index of a material in the form of a prism (e.g. by measuring the minimum deviation angle) or to study the performance of diffraction gratings (e.g. by measuring its various diffraction angles, etc). There are three basic components as shown in the first figure: (a) A *collimator* that has an *adjustable slit* and an achromatic lens to produce a collimated light beam, (b) A *spectrometer table* which is a graduated turntable whose rotations around the vertical axis through its center can be measured. It serves as a *goniometer* to measure angles. The turntable carries a prism or a diffraction grating which can therefore be rotated with respect o the collimated incident beam. The turntable can usually be leveled by three adjustable screws. (c) A *telescope* which can be rotated around the turntable's vertical axis, and the relative rotation can be measured. The eyepiece is normally focused onto the crosswires in the telescope to facilitate measurements. These three simple basic components enable the spectrometer to be a highly versatile optical instrument; it is widely used in light wavelength, refractive index and diffraction grating measurements. The second figure illustrates the measurement of the refractive index of a prism by first measuring the prism angle α and then the minimum deflection $\delta_{\min}$ from which we can calculate the refractive index. See *monochromator*, *refractometer*.

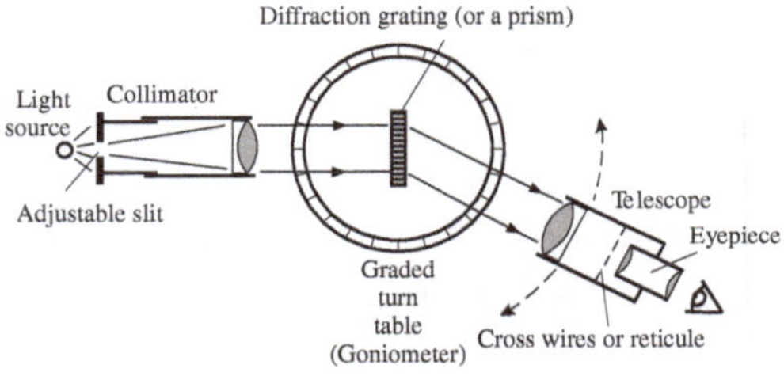

A **spectrometer**.

Spectroscopy or **spectrometry** is the study and measurement of absorbed, transmitted or emitted radiation with reference to its spectrum. Spectroscopy includes methods for generating, detecting and analyzing spectra. It is very useful in identifying atomic and molecular energy levels responsible for the observed emission or absorption spectra, and hence in chemical analysis (determination of composition, etc.).

Spectrum of an optical signal is generally understood to be the "relative optical power" vs. frequency distribution that constitutes the optical light signal. It represents all the necessary frequencies each with the correct optical power level that is needed to constitute the signal. The relative optical

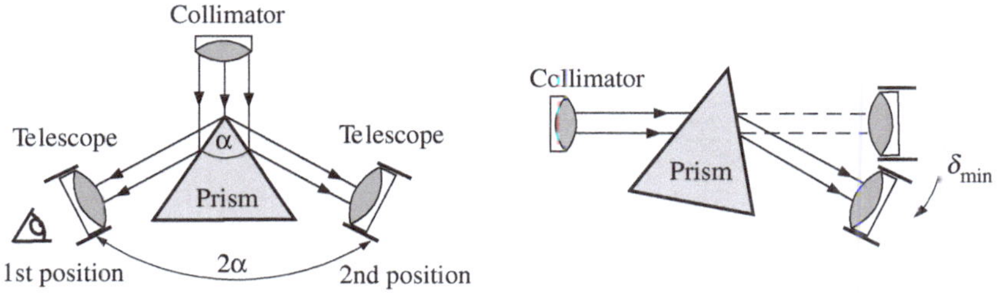

Measurement of the prism angle α
a

Measurement of the minimum deflection δ_{min}
b

Spectrometer Schematic illustration of the measurement of the prism angle α as in (a) and then the minimum deflection δ_{min} as in (b) to extract the prism refractive index.

power vs. frequency spectrum is represented in such a way that its integration over all the frequencies represents the total optical power of the signal. Thus, "relative optical power" implies optical power per unit frequency. The spectrum can also be represented in terms of wavelength. The necessary phase information is typically not included in spectrum discussions. A better term is "spectral optical power".

Spectrum of a laser diode output is the optical power density vs. wavelength behavior of the lasing emission from the diode. This spectrum may be multimode or single mode depending on the optical resonator structure and the pumping current level. (See figure.)

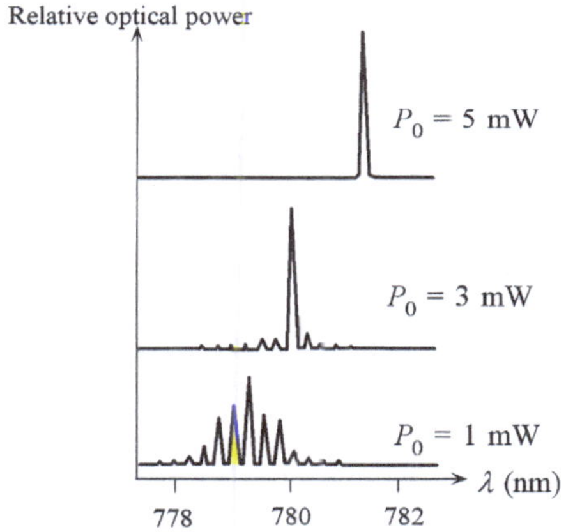

Spectrum of a laser diode output Output spectra of lasing emission from an index guided laser diode. At sufficiently high diode currents corresponding to high optical power, the operation becomes single mode. (Note: the relative power scale applies to each spectrum individually and not between spectra.)

Spherical aberration (SA) is a third-order aberration in which the focal point of a lens, or mirror, depends on the height of the incident ray from the lens axis, which is due to the failure of the paraxial approximation. In an ideal lens, parallel light rays, whether traveling parallel or obliquely to the lens axis, will be brought to a focus on a flat surface called the *paraxial focal plane* σ_{paraxial}. SA causes blurring and reduces contrast, and hence reduces the modulation transfer function (MTF). In the figure, the marginal rays are refracted more and have a shorter focal length (this is called *positive* SA). The *longitudinal SA* (LSA) is the distance between the paraxial focal point and the focal point of marginal rays (F_m) on the lens axis. *Transverse (lateral) SA* (TSA) is the distance from the paraxial lens axis focal point F and the point where the marginal ray cuts the paraxial focal plane (M). The refracted rays on the image side of the lens intersect each other and define an *envelope* called the *caustic*; it is a surface. All the refracted light is inside the caustic. The

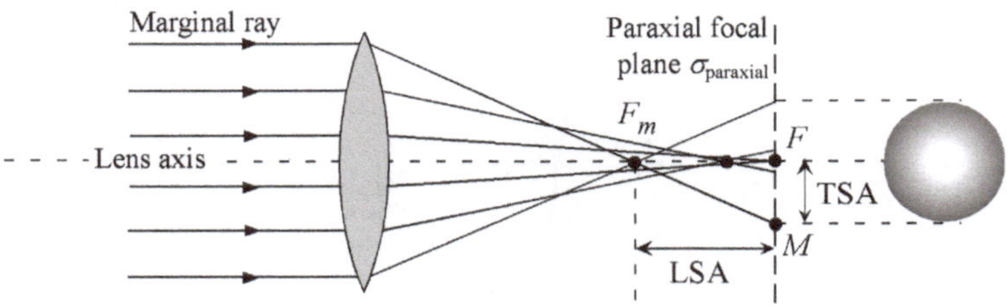

Spherical aberration (SA) in a lens. LSA, longitudinal SA, TSA, transverse SA.

intersection of marginal rays with the caustic surface identifies a *circle of least confusion*, which is the best location to place a screen to observe the image. SA can obviously be reduced by reducing the lens aperture (i.e. limiting marginal rays) but that undesirably reduces the total light used in forming the image and reduces the image illuminance. SA depends on the object distance and the lens shape. It is possible to shape the lens, i.e. vary the radius of curvature on the lens surface, in such a way as to minimize the SA for a given aperture and known object distance. For a planoconvex lens with the object nearly at infinity and flat side facing the object, the spot size is given by $0.067f/(f/\#)^3$.

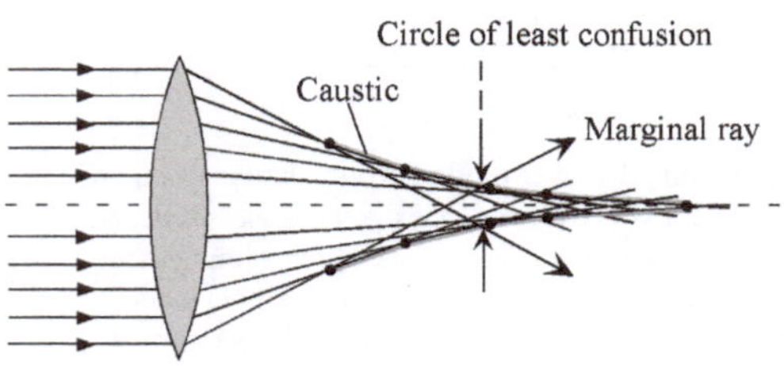

Spherical aberration The caustic and the circle of least confusion.

Spherical lens, or **ball lens**, is a simple glass sphere that has good focusing ability for coupling light from one optical fiber to another, between a fiber and an emitter, and a fiber and a detector. Some ball lenses are so small (diameters much less than

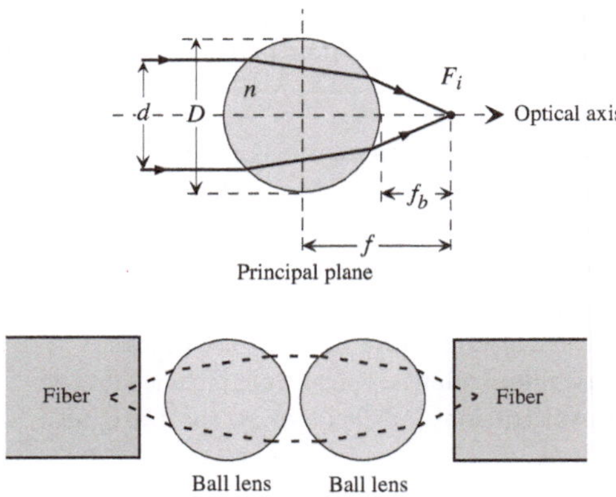

Spherical lens definitions and use in fiber coupling (schematic only).

1 mm) that they are called *microball lenses*. A ball lens is characterized by its diameter D and refractive index n. The effective focal length f of a ball lens is measured from the center of the ball whereas the so-called *back focal length* f_b is measured from the surface of the ball.

$$f = \frac{nD}{4(n-1)} \quad \text{and} \quad f_b = \frac{nD}{4(n-1)} - \frac{D}{2}.$$

If the diameter of the beam impinging on the ball lens is d, then the numerical aperture NA is given by

$$\text{NA} = \frac{2d(n-1)}{nD}.$$

For efficient coupling of light into a fiber, the NA of the ball lens should be smaller than the NA of the fiber. Ball lenses suffer from spherical aberration. (A. Nicia, *Applied Optics*, **20**, 3136, 1981; Edmund Optics website, 2004.)

Spherical mirror is a mirror of spherical form, with center C and summit S. In terms of the *paraxial approximation*, its equation of conjugation between object A and image A' states is

$$\frac{1}{\overline{SA'}} + \frac{1}{\overline{SA}} = \frac{1}{\overline{SF}} = \frac{1}{\overline{SF'}} = \frac{2}{\overline{SC}}.$$

Both focal points F and F' are located at mid-distance from S and C. See also *imaging equations*.

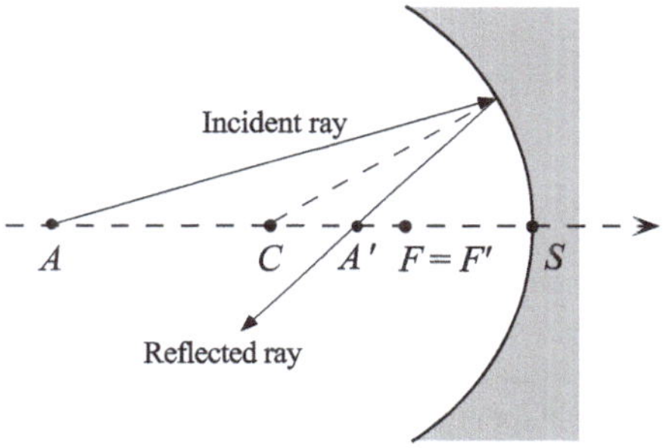

Spherical mirror with center C, summit S, and focal points F and F'; within the paraxial approximation, any incident ray from the object point A is reflected toward image point A'.

Spherical wave is described by a traveling field that emerges from a point electromagnetic source (EM) source and whose amplitude decays with distance r from the source. At any point r from the source, the field at time t can be described by

$$E = \frac{A}{r}\cos(\omega t - kr)$$

where A is a constant, ω is the angular frequency and k is the wavevector. In spherical coordinates, the wave propagation can be described by a relatively simple equation of the form

$$\nabla^2 E = \frac{1}{r}\frac{d^2(rE)}{dr^2}.$$

The sinusoidal (or harmonic) spherical wave equation above is one possible solution, but other solutions (other types of spherical waves) can also exist. The irradiance (EM power flowing per unit area) of a spherical wave decreases as $1/r^2$. (This is constant in an ideal plane wave.)

Spin of an electron, S, is its intrinsic angular momentum (analogous to the spin of Earth around its own axis) which is space quantized to have two possibilities. The magnitude of the electron's spin is constant, $\hbar\sqrt{3}/2$, but its component along a magnetic field in the z-direction is $m_s\hbar$, where m_s is the spin magnetic quantum number, $+^1/_2$ or $-^1/_2$. The quantity $\hbar$ is Planck's constant divided by 2π.

Splice is a permanent joint between two fibers, usually achieved by aligning the fibers and then fusing the two glasses to each other at a high temperature in an electrical arc between two electrodes. Whenever two fibers are needed to be connected permanently, for example, when laying down additional fiber or repairing a broken fiber, they are always spliced. Splicing is achieved by an instrument called the *fusion splicer*. Splice losses tend to be small, usually not greater than 0.1 dB.

Splice loss see *fiber-to-fiber attenuation*.

SPM, self-phase modulation, see *optical Kerr effect*.

Spontaneous emission in an atom is the phenomenon by which a photon is emitted when an electron in a high-energy state, $\psi(n, \ell, m_\ell, m_s)$, with an energy E_2, spontaneously makes a transition to a lower energy state, $\psi(n', \ell', m_\ell', m_s')$ with energy E_1 that is not occupied; the transition does not need another photon with the same energy to stimulate it. The emitted photon energy $h\nu_o$ is the change in the energy; $\Delta E = (E_2 - E_1)$. Since the emitted photon has an angular momentum, the orbital quantum number, ℓ, of the electron must change, i.e. $\Delta\ell = \ell' - \ell = \pm 1$.

In more general terms, spontaneous emission occurs whenever an electronic system is in

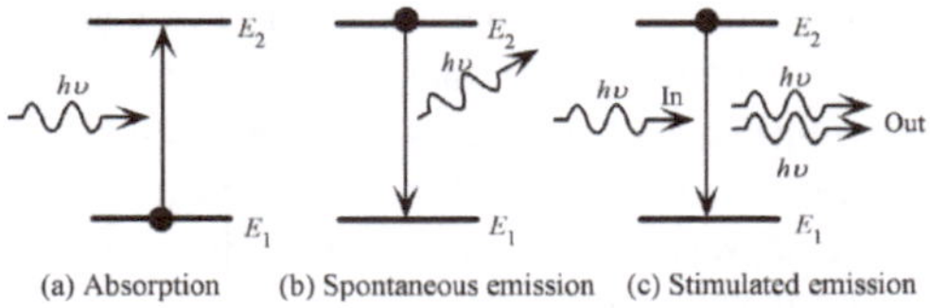

Spontaneous emission Absorption, spontaneous (random photon) emission and stimulated emission.

an excited state and lowers its energy by an electron making a radiative transition to an available lower energy state. Normally the lower and upper energy levels E_1 and E_2 are not well-defined discrete levels but either distributions centered at E_1 and E_2 or correspond to two sets of energy manifolds. Consequently ν_o is some dominant frequency in the spontaneous emission spectrum, i.e. the central frequency or the peak intensity frequency. Consider a material that has been doped with a concentration N_d of dopants; these may be luminescent centers in a host material, e.g. Ce^{3+} in $Y_3Al_5O_{12}$ or Er^{3+} in silica glass. Suppose that, of these, N_1 is at energy E_1 and N_2 is at energy E_2 at some instant. The rate of spontaneous transitions per unit volume is proportional to the concentration of excited dopants N_2 and independent of the number of photons already created in this mode in the system (or in an optical cavity). Let dN_{ph}/dt be the number of photons emitted per unit time, and τ_{sp} be the lifetime of the excited state before it makes a spontaneous transition, or the mean time for a spontaneous emission. The mean probability of spontaneous transition per unit time W_{sp} is simply $1/\tau_{sp}$. There are $N_2 V$ dopants in the sample, and dN_{ph}/dt is given by

$$\frac{dN_{ph}}{dt} = \frac{N_2 V}{\tau_{sp}} = \sigma_{sp}(c/n)N_2$$

where c/n is the speed of light in the medium, n is the refractive index, and σ_{sp} is known as the *spontaneous emission cross-section*, and given by $1/\tau_{sp} = (c/n)\sigma_{sp}/V \cdot \sigma = \sigma(\nu)$ depends on the frequency, and is a narrow function centered on the transition resonance frequency ν_o. Suppose that at some instant there are N_2 excited oscillators and there is no ongoing excitation mechanism present. Then $\delta N_2 = (W_{sp}\delta t)N_2$ will undergo a transition in a very small time interval δt, which means that $dN_2/dt = -W_{sp}N_2$, giving the number of oscillators at any time $N_2(t)$, as

$$N_2(t) = N_2(0)\exp(-W_{sp}t)$$

decaying exponentially with a time constant $1/W_{sp}$ or τ_{sp}. The cross-section $\sigma(\nu)$ describes the character of the interaction between the oscillator and the radiation – the integral of this function is known as the *oscillator strength* S:

$$S = \int_0^\infty \sigma(\nu)d\nu.$$

$\sigma(\nu)$ and S also define a so-called lineshape function $g(\nu) = \sigma(\nu)/S$. The function $g(\nu)d\nu$ is the probability that a spontaneously emitted photon will appear at a frequency in the interval between ν and $\nu + d\nu$. In general, the full width half maximum for the spectrum of the spontaneously emitted light, $\Delta\nu$, is proportional to $1/g(\nu)$.

Spot size see *Gaussian beam.*

Squeezed states of a light field are generated when the quantum noise is concentrated at a specific phase angle, or amplitude. This may be achieved by reducing the quantum noise in one quadrature of the light field (e.g., the phase) at the expense of enhancing it in the complementary observable (i.e. the amplitude, or vice versa). This may be accomplished using nonlinear optics – for example, parametric amplification and de-amplification. In such a case, the wave packets oscillate in time, as for the case of the coherent state, but also get wider and narrower – the corresponding phase space distribution has an elliptical shape. Squeezed states have been studied extensively over the past decades owing to their providing a means for reducing the amount of noise in specially designed precision optical measurements. The amount of squeezing, the amplitude of the coherent excitation, as well as the relative angle between the squeezed quadrature and the coherent excitation of the light can vary. Thus, there is a whole family of such squeezed states of light. By knowing the evolution of the wave packet or the Wigner function, one can deduce the distribution of energy quanta or photons that constitute the light wave. However, having knowledge of the distribution of photons does not necessarily provide information on the state of the light field, since this contains no information on the state's phase. The photon number distribution is given by the diagonal elements of another complete description of the state of the light field – the density matrix in the Fock (i.e. number state) basis. The diagonal elements of the density matrix contain the information on the intensity distribution of the state, while the phase information is contained within the off-diagonal elements. Analyzing the diagonal elements of these matrices shows that the photon number distribution of the coherent state is *Poissonian* – thus the mean photon number equals the spreading of the photon number distribution, or the variance. For the other states, the more enhanced the state's amplitude noise, the more spread out is its photon number distribution, or the more enhanced its intensity noise. Thus, for the *amplitude-squeezed* state, the photon number variance is *less* than the mean photon number. Such light is called *sub-Poissonian* light. This light played an important role in the history of quantum optics, since it was the first type of nonclassical light that was detected. However, note that not all amplitude-squeezed light is sub-Poissonian (e.g. states with very small amplitude) – also, sub-Poissonian light may very well not be amplitude-squeezed (e.g. if it is classically not coherent, phase diffused). *Phase-squeezed* light, however, has a much broader *super-Poissonian* photon number distribution than the coherent state. The phase distribution provides another noncomplete representation of the quantum state, complementary to the photon number distribution. Thus, the phase distribution of a coherent state will be narrower the farther away the state is from the origin, or the larger its amplitude. Therefore, when comparing phase distributions of different kinds of states, the states should have the same amplitude. There are a variety of methods that may be used to produce squeezed light. The most successful approach is using parametric down-conversion. In this case, a coherent state of very small amplitude (i.e. the signal input) is overlapped within a nonlinear crystal with a light wave of twice its frequency (i.e. the pump wave). Owing to the polarizability of the nonlinear crystal, the pump wave induces a periodic change in the medium for propagation of the signal wave. The effect on the signal wave depends on its relative phase with respect to the pump wave. Experimentally this phase is controlled by a mirror mounted on a piezoelectric transducer in the beam path of the signal wave. If the signal wave is in phase with the pump wave, it is amplified, while if it is out of phase, it is attenuated. This behavior may be visualized in phase space – the light field's state is depicted by its uncertainty area in phase space, which is the contour (i.e. a horizontal cut) of its Wigner function. The action of the pump wave in the nonlinear medium (mathematically denoted as the *squeezing operator*) is a distortion of the phase space along hyperbolic lines. The phase angle of

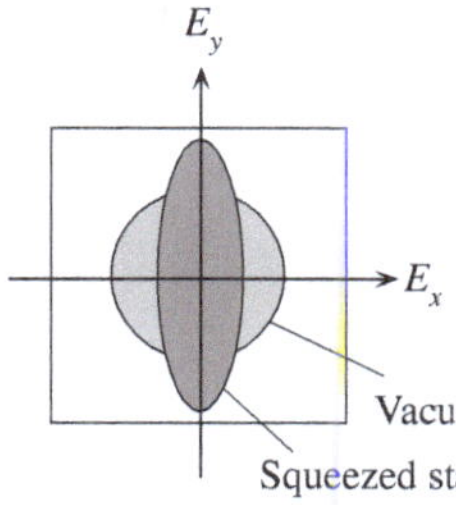

A representation of the probability distribution of a vacuum and a **squeezed state**. Both distributions are minimum uncertainty states and thus they both have the same area, but the squeezed state has less noise in one phase quadrature as compared with the other.

the input states is given by the relative phase angle between the pump and signal waves. All kinds of squeezed states can be generated by altering the phase or the amplitude of the input state.

Table: Standard wavelengths

Fraunhofer designation	Wavelength (nm)	Color	Element
i	365.01	UV	Hg
h	404.6561	Violet	Hg
g	435.8343	Blue	Hg
F'	479.9914	Blue	Cd
F	486.1327	Blue	H
e	546.0740	Green	Hg
d	587.5618	Yellow	He
D_2	588.9953	Yellow	Na
D	589.2938	Yellow	Na
D_1	589.5923	Yellow	Na
C'	643.8469	Red	Cd
C	656.2725	Red	H
r	706.5188	Red	He
A'	768.19	IR	K
s	852.11	IR	Cs
t	1013.98	IR	Hg

Squeezing operator see *squeezed states*.

Standard wavelengths are those light wavelengths that have been chosen for refractive index measurements so that the wavelength is accurate to within almost four decimal places and reproducible. These standard wavelengths were originally chosen to be prominent lines in the solar spectra (Fraunhofer lines), though in the laboratory they are normally obtained as light emitted from various gaseous arc discharges that have narrow emission spectra.

Star coupler see *optical networks*.

Stark effect results from the interaction of an external electric field with the atoms or molecules of a medium, as distinct from the **Zeeman effect** which involves the interaction with an external magnetic field. Generally, the effect is observed by examining the spectral properties of the medium such as changes in the absorption and emission of electromagnetic (EM) radiation. An atom with an electric dipole moment **p** placed in an electric field **E** interacts with the field and possesses an electrostatic potential energy (PE) given by

$$E_{\text{PE}} = -\mathbf{p} \cdot \mathbf{E}.$$

The energy of the atom must be calculated by using the Schrödinger equation with this additional electrostatic PE term. The result is that the applied field perturbation of the system can remove part of the degeneracy in the system, causing the previously degenerate energy level to be split into two states equally separated from the initial unperturbed state as illustrated in the figure. There are two types of Stark effect. In the *linear Stark effect*, the atom or the molecule possesses a permanent dipole moment by virtue of having a nonsymmetric electronic charge distribution. The magnitude of the field-induced splitting of an energy level, the Stark effect, is *linearly* proportional to the applied field. In the *quadratic Stark effect*, the atom or the molecule has no permanent dipole moment, but acquires an induced dipole moment when it is placed in an external field, that is, the field induces a dipole moment p_{induced} that is proportional to the field E. The PE term is then proportional to E^2 and the Stark effect is hence proportional to the square of the field.

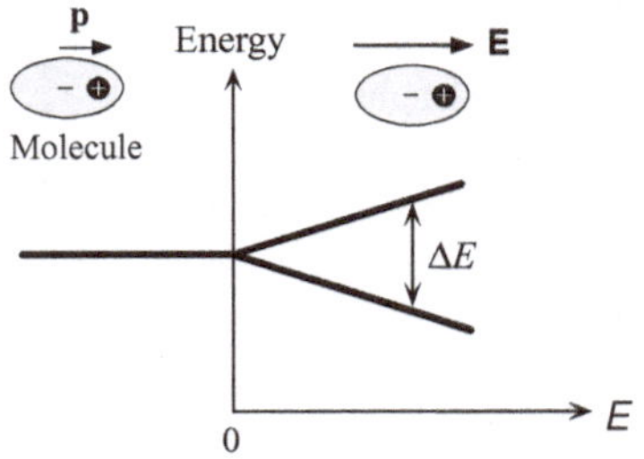

An example of the **Stark effect**. In the presence of an applied electric field, the energy of a molecule splits into two levels, one below and the other above the original level.

The *quantum confined Stark effect* is the shift in the electron energy levels inside a quantum well.

Photon absorption in a quantum well generates a bound electron–hole pair since the electron and hole are confined to the quantum well. The electron and hole pair effectively form an exciton (called the *quantum confined exciton*). Consider a quantum well of width d. The applied field pushes the electron and hole apart and thereby modifies the energy; it lowers the energy. The change in the $n = 1$ energy level for a field applied along z, the confinement direction, is

$$\Delta E = -24\left(\frac{2}{3\pi}\right)^6 \frac{e^6 m_r^* d^4}{\hbar^2} E_z^2$$

where $m_r{}^*$ is the reduced effective mass, where $1/m_r^* = 1/m_e^* + 1/m_h^*$ with $m_e{}^*$ and $m_h{}^*$ as the electron and hole effective masses, respectively.

State is a possible wavefunction for the electron that defines its spatial (orbital) and spin properties, e.g. $\psi(n,\ell,m_\ell,m_s)$ is a state of the electron. From the Schrödinger equation, each state corresponds to a certain electron energy E. We thus speak of a state with energy E as state of energy E, or even an *energy state*. Generally there may be more than one state ψ, with the same energy E.

Station see *optical network*.

Statz–De Mars equations see *reduced rate equations*.

Stefan's law, or **Stefan–Boltzman law**, states that the total net radiation power (i.e. rate of flow of electromagnetic energy) emitted or absorbed per unit surface area of a black body is proportional to the difference between the fourth powers of the temperature of the surface and that of the ambient, with the constant of proportionality σ_S being a universal constant ($= 5.670 \times 10^{-8}$ W m^{-2} K^{-4}),

$$P_{\text{blackbody}} = \sigma_S S\left(T^4 - T_o^4\right)$$

where $P_{\text{blackbody}}$ is the net radiation power emitted or absorbed by the black body, T and T_o are the temperatures of the black body and the ambient respectively, S is the surface area, and σ_S is the Stefan constant. A black body, in simple terms, is defined as a surface that absorbs all radiation that falls on it. For a surface that is not a black body, then the net radiated power can be written as

$$P_{\text{blackbody}} = \varepsilon\sigma_S S\left(T^4 - T_o^4\right)$$

where ε is the *total emissivity* of the surface; for real surfaces $\varepsilon < 1$ and for a black body $\varepsilon = 1$. See *black body, emissivity*.

Steinhart–Hart equation see *bolometer*.

Step index fiber has a central core of constant refractive index and a surrounding layer of cladding of smaller refractive index. There is a step decrease in the refractive index from the core to the cladding.

Stepped taper or **staircase concatenation method** see *beam propagation method*.

Sterance, as in *radiant sterance*, is the quantity of interest per unit solid angle. Radiant sterance is simply *radiance*.

Stimulated Brillouin scattering see *Brillouin scattering*.

Stimulated or induced emission from an atom is the phenomenon by which an incoming photon with frequency ν_o and energy $h\nu_o = E_2 - E_1$ interacts with an electron in a high-energy state, $\psi(n,\ell,m_l,m_s)$, at E_2, and induces it to make a transition to a lower-energy state, $\psi(n',\ell',m_l',m_s')$, at E_1, that is not occupied. The photon that is emitted by stimulation has the same energy and phase as the incoming photon and, further, it is in the same direction. Consequently, stimulated emission results in two coherent photons, with the same energy, traveling in the same direction. The stimulated emission process obeys the selection rule (just as spontaneous emission); $\Delta\ell = \ell' - \ell = \pm 1$. In more general terms, stimulated emission in a system refers to the fact that the rate of transition of an electron from an excited state does not need to wait for spontaneous emission to occur as under appropriate conditions such an electron can be stimulated to its ground state in a much shorter time. The stimulus is provided by the presence of another photon of appropriate frequency, direction of propagation and polarization. This photon stimulates the emission of an identical photon as a result of the electronic transition. Consider a material that contains dopants (e.g. Er^{3+}) of which N_1 per unit volume are at E_1 and N_2 per unit volume are at E_2 where E_1 and E_2 are the lower and upper energy levels. The energy levels E_1 and E_2 are likely to represent two manifolds of energy or, as in some cases, energy distributions centered at E_1 and E_2. (E_1 is not necessarily the ground energy but it is lower than E_2.) A photon of energy $h\nu_o = E_2 - E_1$ can either be absorbed by a dopant at E_1 or it can stimulate a dopant at E_2 to emit a stimulated photon. Let Φ be the stimulated photon flux flow in a certain direction in a medium in which such transitions take place. Then $d\Phi/dt$ is the rate at which this photon flux grows with time. This growth will be due to E_2-to-E_1 transitions, proportional to Φ and N_2, exceeding E_1-to-E_2 absorption processes, proportional to ΦN_1. The *stimulated emission cross-section,* σ, can be defined by

$$\frac{d\Phi}{\Phi dt} = \sigma(c/n)(N_2 - N_1)$$

where c/n is the speed of light in the medium, and n is the refractive index. If g_2/g_1 is the ratio of the degeneracies of the upper and lower levels then we must replace $N_2 - N_1$ by $[N_2 - (g_2/g_1)N_1]$. Alternatively, if the probability of stimulated emission per unit time is $W_{\text{stimulated}}$, then

$$W_{\text{stimulated}} = \frac{\sigma c}{nV}$$

where V is the volume of the sample so that $W_{\text{stimulated}}(N_2 - N_1)V$ gives the total rate of stimulated emissions. It must be emphasized that $\sigma = \sigma(\nu)$, that is, it depends on the frequency ν through the broadening mechanism that gives rise to the observed emission spectrum. $\sigma(\nu)$ can also be defined through the *stimulated emission gain* $g = (1/\Phi)(d\Phi/dx)$,

$$g = \frac{d\Phi}{\Phi dx} = \sigma[N_2 - (g_2/g_1)N_1]$$

in which $g = g(\nu)$ because $\Phi = \Phi(\nu)$ does not occur at one discrete wavelength but has a frequency distribution; and thus $\sigma = \sigma(\nu)$. The maxium stimulated cross-sections σ_H and σ_D in the presence of homogeneous and Doppler broadening, respectively, are

$$\sigma_H = \frac{1}{4\pi^2}\left(\frac{\lambda_o^2 A_{21}}{n^2 \Delta\nu}\right) \quad \text{and}$$

$$\sigma_D = \frac{1}{4\pi}\left(\frac{\ln 2}{\pi}\right)^{1/2}\left(\frac{\lambda_o^2 A_{21}}{n^2 \Delta\nu}\right)$$

where A_{21} is the spontaneous radiative transition probability per unit time (rate) from the upper to lower level, λ_o is the central wavelength in the emission spectrum, and $\Delta\nu$ is the spectral width (full width half maximum) in frequency for the particular broadening mechanism. It should be noted that in an optical cavity in which there is both stimulated and spontaneous emission, at steady state, the sum of these rates equals the rate of absorption. In order to influence the relative rates of stimulated to spontaneous emission, in practice one designs an optical cavity that permits a sufficiently high photon density to be established at frequencies near the transition frequency $\nu = \nu_o$ through multiple reflections and constructive interference of such photons. Moreover, by maximizing the rate of stimulated emission to absorption in such a structure, one can maintain an excess population

Table: Stimulated emission Stimulated emission cross-sections in various laser systems. (From W. T. Silfvast, *Laser Fundamentals, Second Edition,* Cambridge University Press, 2004, p. 246.)

Laser	λ (nm)	σ (m^2)
He-Ne	632.8	3.0×10^{-17}
Argon	488.0	2.5×10^{-16}
He-Cd	441.6	9.0×10^{-18}
Copper (CVL)	510.5	8.6×10^{-18}
CO_2	10,600.0	3.0×10^{-22}
Excimer	248.0	2.6×10^{-20}
Dye (Rh6G)	577.0	2.5×10^{-20}
Semiconductor	800.0	1.0×10^{-22}
Nd:YAG	1064.1	6.5×10^{-23}
Nd:Glass	1062.3	3.0×10^{-24}
$Ti:Al_2O_2$	800.0	3.4×10^{-23}
Cr:LiSrAlF	850.0	4.8×10^{-24}

of electrons in the excited state compared with the ground state and thereby achieve a medium that has an optical gain.

See *cross-section*.

Stimulated Raman scattering is the *Raman scattering* of electromagnetic waves in such a way that the scattered waves are in phase and reinforce each other to generate a strong scattered radiation. This can be achieved by "pumping" the system with an intense radiation at the right frequency, for example, at the frequency that excites the required molecular vibrations.

Stokes scattering is the inelastic scattering of light in such a way that the frequency of light is shifted down with respect to the original incident light. (The wavelength is increased). See *Stokes shift*.

Stokes shift refers to the process in which light (electromagnetic radiation) incident on a material at a given frequency emerges from the material with a *lower* frequency – this so-called *down-shifted* light is said to have undergone *Stokes-shifted scattering*. In the case when the emerging light appears from the material with a higher frequency than the incident frequency, it is said to have undergone *anti-Stokes shifted scattering*. In the case of semiconductors, light can interact with lattice vibrational modes or *phonons* to produce Stokes shifted (anti-Stokes shifted) light as a result of phonon emission (absorption).

Stokes' theorem was derived by Sir George Stokes in 1850 and permits any surface integral to be reduced to a line integral following the rim of the surface – indeed, Stokes' theorem allows one to transform from one surface integral to any other, so long as they both share a common rim, c:

$$\oint_c \mathbf{F} \cdot d\mathbf{r} = \int \int_S (\nabla \times \mathbf{F}) \cdot \mathbf{n} dS.$$

Note that the path c is defined in any anticlockwise sense in this equation, and $\mathbf{n}$ is a unit vector pointing out from the surface S, as defined by the right-hand rule. Stokes' theorem has been very significant in the history of science – it not only provides an interpretation of the curl of a vector field, but together with the divergence theorem, gives a direct connection between macroscopic integral laws and differential equations; the latter have played a pivotal role in the discovery of the laws of electrodynamics and the development of radio and wireless communication.

Sir George Gabriel Stokes (1819–1903) was a mathematician and physicist who made a number of contributions in both fields. He held a Lucasian professorship of mathematics at Cambridge and for some time was the President of the Royal Society. (AIP Emilio Segre Visual Archives, E. Scott Barr Collection.)

Straddled alignment see *heterojunction*.

Strain is a measure of the deformation a material exhibits under an applied stress. It is expressed in normalized units. Under an applied tensile stress, strain (ε) is the change in the length per unit original length, $\Delta L/L_0$. When a shear stress is applied, the resulting deformation involves a shear angle. *Shear* *strain* is defined as the tangent of the shear angle that is developed by the application of a shearing force.

Strain point of glass is the temperature at which the viscosity of a glass is about 10^{13} Pa s. For example, for fused silica glass this is about 1050 °C. The viscosity of glass is strongly temperature dependent and decreases sharply as the temperature increases. Below the strain point, the glass can be cooled rapidly without introducing unwanted internal stresses.

Stress is force per unit area, F/A. When the applied force is perpendicular to the area it leads either to a tensile or compressive stress, $\sigma = F/A$. If the applied force is tangential to the area then it leads to a shear stress, $\tau = F/A$.

Stress birefringence of an optical glass represents the extent of birefringence induced in a material due to an applied mechanical stress. The *stress optical constant* is a measure of this stress induced birefringence. Suppose that light travels a distance $\Delta\ell$ in a sample that is under stress σ, and the induced optical path difference between the propagation of the two orthogonal fields is Δs. Then

$$\text{Stress optical constant} = \frac{\Delta s}{\sigma \Delta\ell}.$$

Strioscopy or **central dark ground method of observation**, see *spatial filtering*.

Strip or strip loaded waveguide is a three-dimensional waveguide in which confinement is not necessarily achieved by surrounding the waveguide material with another material of lower refractive index, as shown for various possibilities in the figure, where n_s, n_f and n_c represent the refractive indices of the "substrate," "film" (the guide) and "cover" or surrounding material. The simplest structure is a simple strip on the surface of a substrate. The embedded strip waveguide has the strip embedded from the surface of the substrate. The buried strip waveguide is essentially a channel buried inside the substrate. In the case of strip loaded waveguides, the presence of a loading strip on top of a planar waveguide layer causes the effective refractive index of the layer under the strip to be larger than that in adjacent regions, improving lateral confinement. When a strip loaded waveguide is made with the index of the loading strip equal to that of the waveguide, the waveguide is referred to as a *rib or ridge waveguide*. Another interesting case is that where two parallel metallic loading strips are placed on a planar waveguide layer surface – such a structure provides excellent lateral confinement in the aperture between the two strips owing to the reduced penetration of guided waves beneath the strips. In principle, strip loaded waveguides should have lower optical loss than an embedded strip, because the scattering due to side wall roughness is reduced.

Stripe geometry laser diode is a double heterostructure laser device in which one of the electrodes has a stripe geometry, narrow width, to increase the current density in the active region, and which results in a lower threshold current for lasing emissions.

Strong injection is usually a condition in a semiconductor device in which the injected minority carrier concentration is much greater than the equilibrium majority carrier concentration in the same region; injection may be due to photogeneration or voltage application to forward bias a *pn* junction. See *minority carrier lifetime*.

STS-1, STS- *N* see *synchronous optical networking*.

Sub-Poissonian see *squeezed states*.

Substrate is a single mechanical support which carries active and passive devices. For example in integrated circuit technology, typically, many integrated circuits are fabricated on a single silicon crystal wafer which serves as the substrate.

Sum rules are useful summation rules in the science of optical properties of materials that sum the optical constants, or their functions, over all

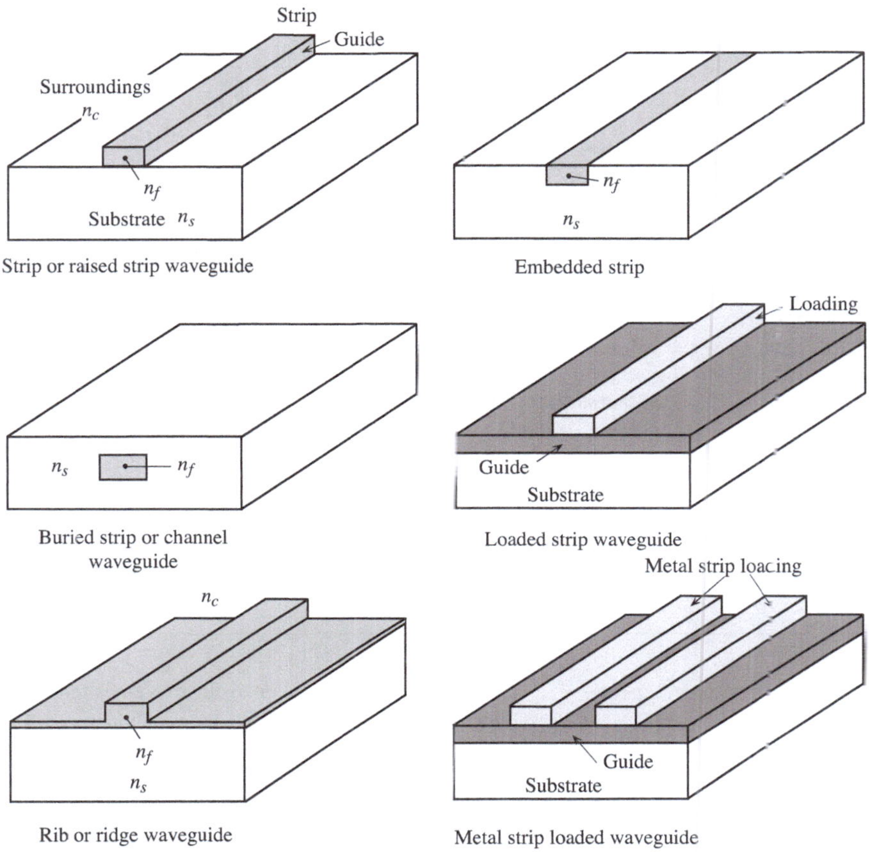

Strip waveguides Various types of strip waveguides.

frequencies (or wavelengths) sum to certain finite values. For example, if $n(\omega)$ is the refractive index, then the integration of $[n(\omega) - 1]$ over all frequencies must be zero. Two important so-called *f-sum rules* are the following. If $K(\omega)$ is the extinction coefficient, then the integration of $\omega K(\omega)$ over all frequencies gives $(\pi/4)\omega_p{}^2$, where $\omega_p = \hbar(4\pi NZe^2/m_e)^{1/2}$ is the free electron plasma frequency in which N is the atomic concentration, Z is the total number of electrons per atom, and e and m_e are the charge and mass of the electron. Further, the integration of $\omega\mathrm{Im}[-1/\varepsilon_r(\omega)]$, where Im represents the imaginary part, and $\varepsilon_r(\omega)$ is the relative permittivity of the medium, over all frequencies must be $(\pi/2)\omega_p{}^2$. The sum rules provide a consistency check and enable various constants to be interrelated.

Super CCD sensor is a CCD image sensor that has its pixel area in octagonal geometry arranged in a honeycomb structure to increase the horizontal and vertical resolution, to which our eyes are more sensitive than the diagonal resolution. See *image sensor, charge coupled device*.

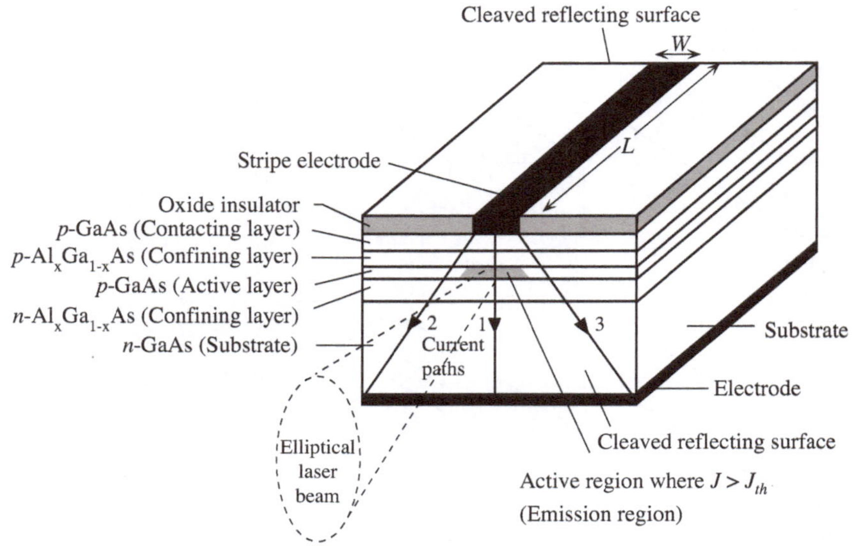

Stripe geometry laser diode Schematic illustration of the structure of a double hererojunction stripe contact laser diode.

Super-large area (SLA) fiber see *dispersion compensation.*

Super-Poissonian state see *squeezed states.*

Superconducting quantum interference devices (SQUIDs) may be used to perform *quantum computing.* In this case, small gaps are made in a superconducting loop of metal creating a so-called island. A *qubit* is represented by the electric charge on the island. The rate at which electron tunneling occurs across the gaps and onto the island is controlled by external electric and magnetic fields.

Supercontinuum process see *holey fibers.*

Superlattice avalanche photodiode (APD) is a photodiode that minimizes excess noise in the photocurrent by limiting avalanche multiplication to a single type of carrier by the use of a superlattice structure which is formed by alternating thin layers of different bandgap semiconductors, as in multiple quantum well (MQW) device structures. The multilayered structure consisting of many alternating layers of different bandgap semiconductors is called a *superlattice.* Typically, the bandgap in each layer changes from a minimum E_{g1} to a maximum E_{g2} which is more than twice E_{g1}. There is a step change ΔE_c in the conduction band edge between two neighboring graded layers that is greater than E_{g1}. In very simple terms, the photogenerated electron initially drifts in the graded layer conduction band. When the electron drifts into the neighboring layer, it now has a kinetic energy ΔE_c above E_c in this layer. It therefore enters the neighboring layer as a highly energetic electron and loses the excess energy ΔE_c by impact ionization. The process repeats itself from layer to layer leading to an avalanche multiplication of the photogenerated electron. Since the impact ionization is primarily achieved as a result of transition over ΔE_c, the device does not need the high fields typical of avalanche multiplication in a bulk semiconductor;

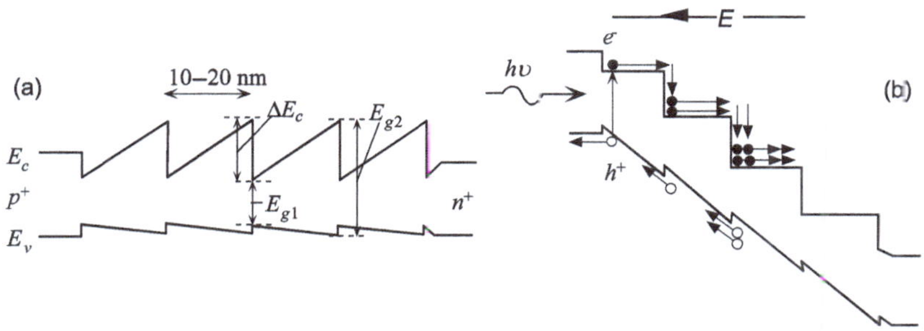

Superlattice APD Energy band diagram of a staircase superlattice APD: (a) no bias; (b) with an applied bias.

it can operate at lower fields. Further, the impact ionized holes experience only a small ΔE_{v} which prevents hole multiplication (ΔE_{v} involves the hole gaining energy). Thus, effectively, only electrons are multiplied and the device is a solid state photomultiplier. Such staircase superlattice APDs are difficult to fabricate and involve varying the composition of a quaternary semiconducting alloy (such as AlGaAsSb) to obtain the necessary bandgap grading. Superlattice structures that are simply alternating layers of low and high bandgap semiconductor layers, that is, layers that do not have a graded bandgap, are easier to fabricate and constitute multiple quantum well (MQW) detectors. Typically *molecular beam epitaxy* (MBE) is used to fabricate such multilayer structures.

Superlattices are artificial structures comprising a sequence of semiconductor materials differing in composition or doping, so as to create a series of potential wells and barriers. For the case where the structure is defined by a repeat unit recurring a sufficiently large number of times, an artificial one-dimensional lattice is created. On the other hand, if the repeat unit only recurs a small number of times, the associated structure is termed a *short-period superlattice*. Superlattices may be fashioned from both type I and type II heterostructures, resulting in either spatially coincident or spatially separated lattices of electron and hole populations, respectively. Superlattice structures rely on the ability to prepare semiconductor layers with monolayer precision using available epitaxial growth technologies. The latter include molecular beam epitaxy and metalorganic chemical vapor deposition. The theory of the band structure and properties of real superlattices is not simple as the wells and barriers in real structures possess complex profiles. Moreover, the effective masses in the different wells and barriers can be different. In addition, when the thickness of the respective layers approaches a few monolayers, the commonly used effective mass approximation is inapplicable. For superlattices comprising a large number of identical structural units, approximating an infinite one dimensional lattice, the so-called *Kronig–Penney model* can be used with appropriate effective masses. From this model, an energy state of an isolated quantum well splits into a series of N states when N wells of the superlattice are included. The origin of this is that the N-fold degeneracy is lifted by the occurrence of tunneling through the associated barriers between the respective wells of the superlattice. With superlattices possessing periodicity (e.g. for a lattice of identical quantum wells) the periodicity leads to the condition that the physical properties of the system should be invariant to translating an electron through one or more integer periods of the superlattice, giving a condition that is a direct analog of Bloch's theorem – namely

$$\Phi(z + nd) = e^{iqnd}\Phi(z)$$

where $\Phi(z)$ is the wavefunction of the electron at location z in the superlattice, and the superlattice has a period, d, with n being some integer number of repeat units (where $n \leq N$). The one-dimensional superlattice wavevector is taken as q. The wavefunctions corresponding to different states in the same superlattice energy band are characterized by different q values, and thus q acts as a new quantum number for the system. Moreover, applying a cyclic boundary condition for a superlattice of length $L = Nd$, enables one to express q in terms of integer values $m = 1, 2, 3, \ldots, N$:

$$q = (2\pi/Nd)m.$$

In addition, using the reduced zone scheme, q values can be shown to lie in the range $\{-\pi/d, +\pi/d\}$, which is the first Brillouin zone of the superlattice. We note that as the electron energy increases the tunneling probability increases and so as the band index i for a given state $E_i(q)$ increases, the band width of successively higher index bands will increase. Important features of the band structure of these systems are the ability to artificially engineer bandgaps and dispersion curves, $E_i(q)$ vs. q, in these artificial structures. Such structures thus afford an opportunity to study and realize new phenomena such as, for example, *Bloch oscillations*. See *quantum well, Bloch oscillations*.

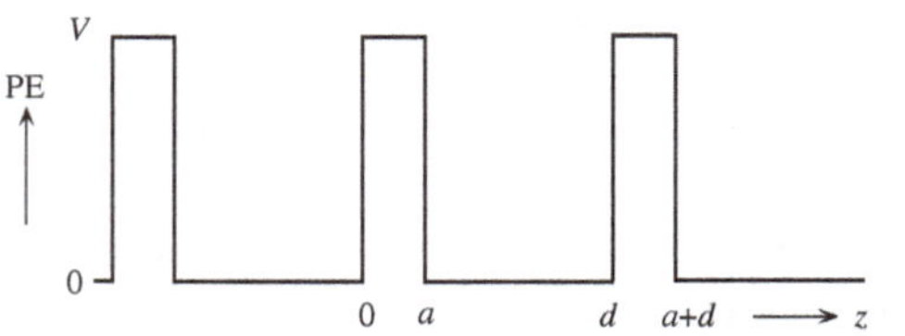

A schematic of a one-dimensional **superlattice** in which the potential energy (PE) varies periodically from 0 to V over a spatial period of d in the z-direction. In the case shown, there are two distinct regions: those regions where the potential is zero, in the so-called well-regions, adjacent to regions where the potential is finite, the so-called barrier regions. A unit cell of the structure can be considered to be composed of one barrier region and one well region. In an infinite superlattice, there would be an infinite repetition of the unit cell, while in a finite superlattice, this structure repeats a finite number of times.

Superlens is a planoconcave photonic crystal lens that is apparently able to focus light without being limited by diffraction effects; the lens is able to overcome the diffraction limit. The superlens is still a theoretical concept though it is expected that a practical device will eventually be built. (N. Fang, H. Lee, C. Sun, X. Zhang, *Science*, **308**, 534, 2005; H. Lee, Y. Xiong, N. Fang, W. Srituravanich, S. Durant, M. Ambati, C. Sun and X. Zhang, *New J. Phys.*, **7**, 255, 2005.) See *left-handed materials, negative refractive index*.

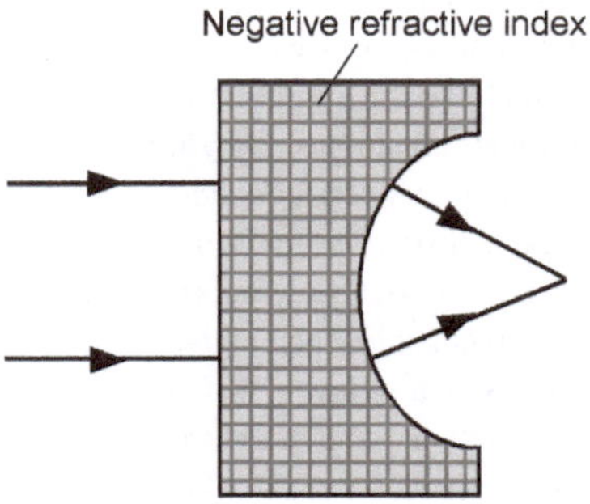

A **superlens** from a planoconcave periodic structure based on using a negative effective refractive index material (conceptual).

Supraluminal velocity refers to the propagation of a light pulse by optical tunneling (*frustrated total internal reflection*), which is sometimes interpreted in terms of quicker-than-light velocity, because the transit time across the barrier is found to be shorter than the time it would take to cross the same distance by free propagation in a vacuum.

This effect is mainly due to the phase redistribution of the spectral components of the pulse across the barrier. Consider a quasi monochromatic wave packet $\Psi(z, t)$ with spectral extension $\delta\omega$ around a central angular frequency ω_0, such as $\delta\omega \ll \omega_0$. A

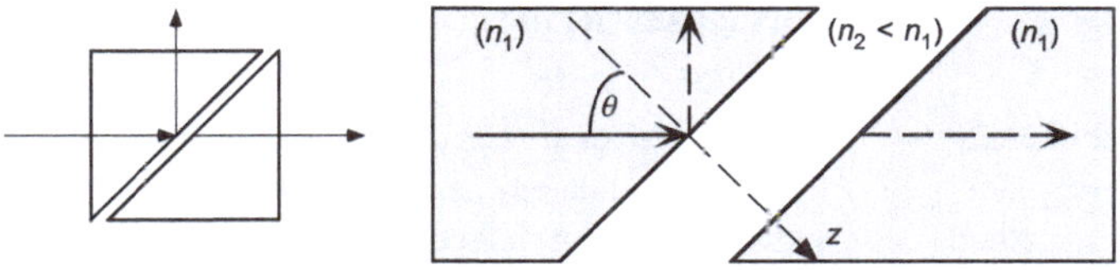

Supraluminal velocity Frustrated total reflection (optical tunneling): the angle θ is greater than the critical angle of total reflection $\theta_c = \arcsin(n_2/n_1)$.

plane wave decomposition enables one to write,

$$\Psi(z,t) = [2\pi]^{-1} \int B(\omega) \exp[j(\omega t - kz)] d\omega,$$

where $B(\omega)$ is the spectral amplitude and each wavevector k is related to ω through the dispersion relation $\omega(k)$. Due to the small spectral extension, we have $\omega(k) - \omega_0 \approx (k - k_0)\, v_g$, with v_g the group velocity. $\Psi(z, t)$ appears as the product of a high-frequency carrier (ω_0) propagating at phase velocity ω_0/k_0 by a low-frequency envelope $F(z, t)$ propagating at group velocity:

$$\Psi(z,t) = F(z,t) \exp[j(\omega_0 t - k_0 z)],$$
$$F(z,t) = [2\pi]^{-1/2} \int B(\omega) \exp[j(\omega - \omega_0) \times (t - z/v_g)] d\omega.$$

Consider now a linear filter between abscissas $z = 0$ and $z = L$, characterized by its spectral transfer function $t(\omega)$. The incident Ψ_1 and transmitted Ψ_2 wave packets read:

$$\Psi_1(0,t) = \exp[j\omega_0 t][2\pi]^{-1/2} \int B(\omega) \times \exp[j(\omega - \omega_0)t] d\omega,$$
$$\Psi_2(L,t) = \exp[j\omega_0 t][2\pi]^{-1/2} \int t(\omega) B(\omega) \times \exp[j(\omega - \omega_0)t] d\omega,$$

For a free propagation over distance L in vacuum, $t(\omega) = \exp(-j\omega L/c)$, whereas in optical tunneling, the transfer function is well approximated by:

$$t(\omega) = K \exp[-j(\omega - \omega_0) d/v_g],$$

where K is a complex constant ($|K| < 1$) and d the penetration depth of the evanescent wave inside the barrier. If we are interested only in the intensity, the transit time of the envelope is L/c in the free-space case, d/v_g across the tunnel. As soon as $L > d$, the pulse seems to have jumped across the tunnel quicker than it would have traveled in vacuum: hence the apparent "*supraluminal*" velocity.

It is not the light velocity that is greater than in vacuum; it is rather the effective length d of the tunneling barrier that appears shorter than its actual length L. Such an effect can be universally associated with any kind of evanescent wave, and can be observed for instance inside the forbidden band of a photonic crystal. Besides, it should be noted that no fundamental rule of relativity is broken; it has been shown for instance that "quicker-than-light" transmission of information cannot be achieved in such a configuration.

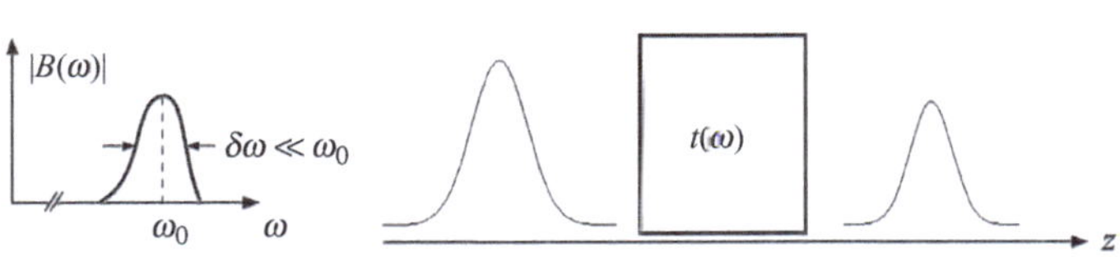

Supraluminal velocity Quasi-monochromatic spectrum of a wave packet; incident and transmitted wave packets across a linear filter.

Surface acoustic wave see *acoustooptic modulator.*

Surface emitting LED (SLED) has its emitted radiation emerging from an area in the plane of the recombination layer. The simplest method of coupling the radiation from a surface emitting LED into an optical fiber is to etch a well in the planar LED structure and lower the fiber into the well as close as possible to the active region where emission occurs. This type of structure is called a *Burrus* type device (after its originator). An epoxy resin is used to bond the fiber and provide refractive index matching between the glass fiber and the LED material to capture as much of the light rays as possible. Note that in the double heterostructure LED used in this way, the photons emitted from the active region (e.g. *p*-GaAs) do not get absorbed by the neighboring layer (AlGaAs) which has a wider bandgap. Another method is to use a truncated *spherical lens* (a microlens) with a high refractive index ($n = 1.9$–2) to focus the light into the fiber. The lens is bonded to the LED with a refractive index matching cement and, in addition, the fiber can be bonded to the lens with a similar cement.

Surface micromachining see *MEMS.*

Surface scattering in waveguides see *interface scattering*.

Surface states are electronic states originating from surface-related phenomena. These include surface relaxation and reconstruction processes, as well as states formed as a result of chemical processes on surfaces. The latter includes surface oxidation, and the interaction of surfaces with adsorbed species. The surface state related energy bands would not exist in the absence of these phenomena, resulting in only a bulk band structure. It should be noted that a so-called ideal surface, in which atoms assume their bulk relationships, can still result in a surface band structure owing to the severing of bonds at the surface resulting in a redistribution of charge, distinct from that of the bulk. The occupation of surface bands is responsible for a surface charge, and this, in turn, results in a near-surface field. The insensitivity of Schottky barrier heights to the metal work function, for certain combinations of metals deposited on semiconductor surfaces, has been attributed to a pinning of the Fermi energy within such surface states.

Swanepoel's technique allows the determination of the optical constants (refractive index n and absorption coefficient α) and the thickness d of a

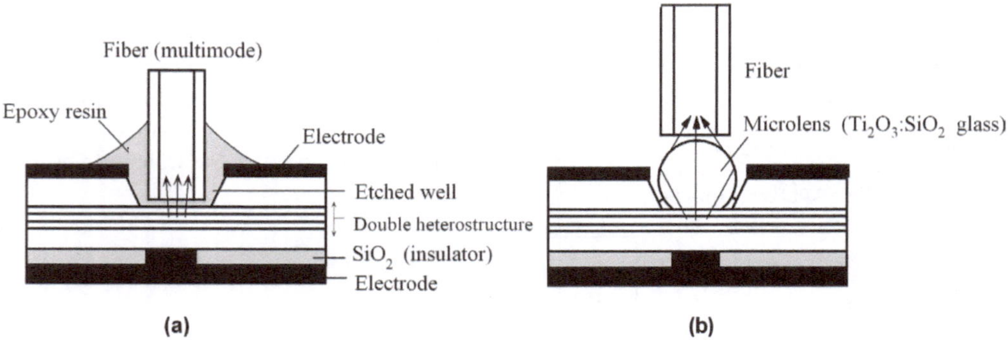

Surface emitting LED (a) Light is coupled from a surface emitting LED into a multimode fiber using an index matching epoxy. The fiber is bonded to the LED structure. (b) A microlens focuses diverging light from a surface emitting LED into a multimode optical fiber.

thin film of a transparent material deposited on a thick transparent glass or other (e.g. sapphire) substrate from the measurement of the transmission spectrum, i.e. transmittance $T(\lambda)$ vs. λ; see the first figure. The amount of light that gets transmitted through a thin film material depends on the amount of reflection and absorption that takes place along the light path. If the material is a thin film with a moderate absorption coefficient α there will then be multiple interferences at the transmitted side of the sample, and the transmission spectrum $T(\lambda)$ vs. λ exhibits oscillations in the wavelength due to interference effects within the thin film. As the wavelength becomes shorter, the transmittance decreases due to absorption of light in the bulk of the film. The minima and the maxima form two envelopes which are determined by $\alpha = \alpha(\lambda)$ and $n = n(\lambda)$; see the second figure. According to Swanepoel,

$$T = T(\lambda) = \frac{Ax}{B - Cx\cos\varphi + Dx^2}$$

where $A = 16n^2 s$, s = refractive index of the substrate, $B = (n+1)^3(n+s^2)$, $C = 2(n^2-1)(n^2-s^2)$, $D = (n-1)^3(n-s^2)$, $\varphi = 4\pi nd/\lambda$, $x = \exp(-\alpha d)$ is the so-called "absorbance," and n, s, and α are all functions of the wavelength, λ. There are two envelopes that have to be constructed before any of the expressions derived from the Swanepoel equation can be used to extract the optical properties. This can be done by locating all the extreme points of the interference fringes in the transmission spectrum and then making sure that the respective envelopes, $T_M(\lambda)$ for the maxima and $T_m(\lambda)$ for the minima, pass through the corresponding extremes, the maxima and minima, of $T(\lambda)$ tangentially. From the Swanepoel equation at $\cos\varphi = \pm 1$, the envelopes for maxima and minima are

$$T_M = \frac{Ax}{B - Cx + Dx^2}$$

$$T_m = \frac{Ax}{B + Cx + Dx^2}.$$

The second figure shows two envelopes constructed for a transmission spectrum of an a-Se thin film.

The transmission spectrum is normally divided into three special regions according to their transmittance values: the *transparent region* is where $T(\lambda) \geq 99.99\%$ of the substrate's transmittance value of $T_s(\lambda)$; the *strong absorption region* is where $T(\lambda)$ is typically smaller than 20%; and the *absorption region* is between the two latter regions.

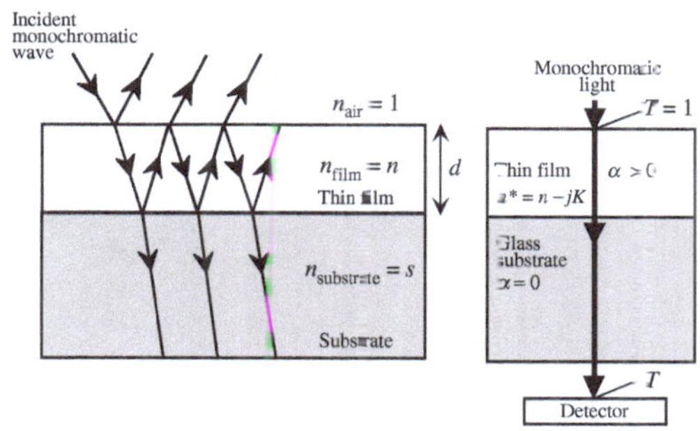

Swanepoel's technique Schematic sketch of the typical behavior of light passing through a thin film on a substrate. On the left, oblique incidence is shown to demonstrate multiple reflections. In most measurements, the incident beam is nearly normal to the film as shown on the right. The detector measures the transmittance as the wavelength is scanned.

The refractive index of the thin film can be calculated from the two envelopes, $T_M(\lambda)$ and $T_m(\lambda)$, and the refractive index of the substrate s through

$$n = \left[N + (N^2 - s^2)^{1/2}\right]^{1/2};$$

$$N = 2s\left[\frac{T_M - T_m}{T_M T_m}\right] + \frac{s^2+1}{2}$$

where N is defined by the second equation above. The equation is not valid in the strong absorption region where there are no maxima and minima, which means that the calculated refractive index has to be fitted to a well-established dispersion model for extrapolation to shorter wavelengths before it can be used to obtain other optical constants. Usually either the Sellmeier or the Cauchy dispersion equation is used to fit n vs. λ experimental data. With the refractive index of the thin film corresponding to two adjacent maxima (or minima)

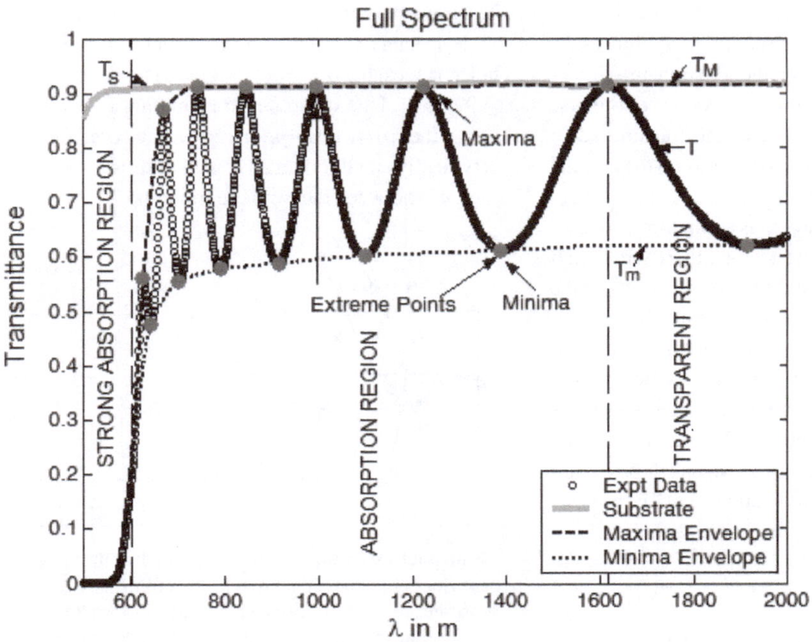

Swanepoel's technique The construction of envelopes in the transmission spectrum of a thin a-Se film, 0.969 μm thick amorphous Se thin film that has been vacuum coated on a glass substrate held at a substrate temperature of 50 °C during the deposition. (Courtesy of Wee C. Tan, 2005).

at points 1 and 2 given as n_1 at λ_1 and n_2 at λ_2, the thickness can be easily calculated from the basic interference equation of waves

$$d_{12} = \frac{\lambda_1 \lambda_2}{2(\lambda_1 n_2 - \lambda_2 n_1)}$$

where d_{12} refers to the thickness obtained from the maxima (minima) at points 1 and 2. As other adjacent pairs of maxima or minima points are used, more thickness values can be deduced, and hence an average value calculated. It is assumed that the film ideally has a uniform thickness. The absorption coefficient α can be obtained once the absorbance x is extracted from the envelope of the maxima inasmuch as

$$\alpha = -\frac{\ln(x)}{d_{\text{ave}}}$$

where

$$x = \frac{E_M - \sqrt{E_M^2 - (n^2-1)^3(n^2-s^4)}}{(n-1)^3(n-s^2)}$$

$$E_M = \frac{8n^2 s}{T_M} + (n^2-1)(n^2-s^2)$$

and d_{ave} is the average thickness, T_M is the envelope of the maxima. The first step in improving the accuracy is to determine a new set of interference order number, m', for the interference fringes from the basic interference equation of waves, that is,

$$m' = \frac{2n_e d_{\text{ave}}}{\lambda_e}$$

where n_e and λ_e are values taken at any extreme points, and m' is an integer if the extremes taken are maxima or a half-integer if the extremes taken are minima. The second step is to get a new corresponding set of thicknesses, d', from this new set of order numbers m', by

$$d' = \frac{m'\lambda_e}{2n_e}.$$

From this new set of thicknesses, d', a new average thickness, d_{new}, must be calculated before it can be applied to improve the refractive index. With this new average thickness, a more accurate refractive index can be obtained from the same equation,

$$n_e = \frac{m'\lambda_e}{2d_{new}}.$$

This new refractive index can then be fitted to the previous dispersion model again so that an improved absorption coefficient α can be calculated from x. All these parameters can then be used to regenerate a transmission spectrum $T_{cal}(\lambda)$ so that the root mean square error (RMSE) can be determined from the experimental spectrum T_{exp}. The RMSE can be minimized by fine tuning the optical constants until the regenerated transmission spectrum converges with the experimental spectrum. While originally derived for a thin film of uniform thickness, the Swanepoel technique has been extended to cover nonuniform thickness films that are wedge-like. The equations are more complicated.

(R. Swanepoel, *J. Phys. E. Sci. Instrum.*, **16**, 1214, 1983; *J. Phys. E. Sci. Instrum.*, **17**, 896, 1984; E. Marquez, J. B. Ramirez-Malo, P. Villares, R. Jimenez-Garay and R. Swanepoel, *Thin Solid Films*, **254**, 83, 1995.]

See *thin film optics*.

Swank factor is the quantitative description of noise associated with the conversion gain (x-ray to charge carriers or light) in x-ray image sensors. The conversion gain is described by a random process with a mean gain $\bar{g}$ and a variance σ_g^2. Then, the swank factor, A_s, is given by

$$A_s(E) = \frac{\bar{g}^2(E)}{\bar{g}^2(E) + \sigma_g^2(E)}$$

where E is the energy of a single x-ray photon. In the ideal case, conversion gains for all x-rays are equal and the swank factor is unity. But in practice, the swank factor is less than but close to unity.

Synchronous digital hierarchy (SDH) see *synchronous optical networking*.

Synchronous optical networking (SONET) is a communication standard that is used for the transmission and multiplexing of optical signals on optical networks. See *optical network*.

Synchronous transport module or signal see *synchronous optical networking*.

Szigetti equation see *Born–Szigetti equation*.

T

Talbot effect, discovered by Talbot in 1836, refers to the formation of an image of a diffraction grating that is formed when the grating is illuminated by a coherent, monochromatic collimated beam. When a "plane" wave passes through a diffraction grating, an image of the grating is formed at a distance z_T, the *Talbot distance*, which is given by

$$z_T = \frac{2d^2}{\lambda}$$

where d is the periodicity of the diffraction grating, and λ is the wavelength. The image of the grating is the Talbot image. The Talbot effect allows the imaging of a diffraction grating without the use of a lens.

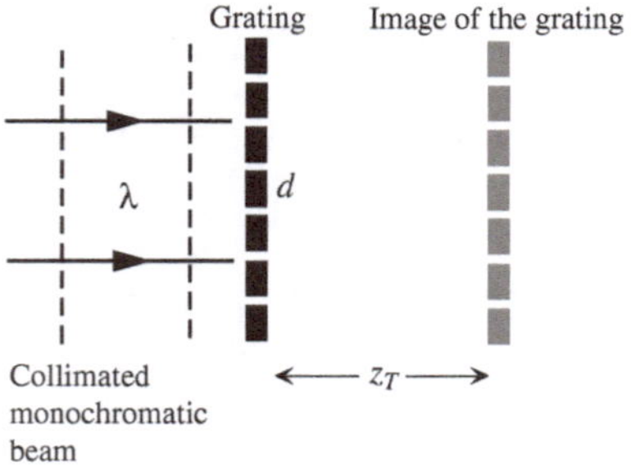

A schematic illustration of the **Talbot effect**.

Tangential plane see *meridional plane*.

Tauc's law is an empirical equation that is used to describe the optical absorption edge in amorphous semiconductors and glasses where the absorption coefficient $\alpha > 10^4$ cm^{-1},

$$(h\nu)\alpha = B(h\nu - E_g)^2$$

where h is Planck's constant, ν is the photon frequency, B is a constant, and E_g is the "optical bandgap", sometimes called the Tauc optical gap. Typically experimental α vs. ν data are plotted as $(h\nu\alpha)^{1/2}$ vs. $h\nu$ and a best straight line fit is drawn through the points; the intercept is used to find E_g. (J. C. Tauc, *Optical Properties of Solids*, North-Holland, Amsterdam, 1972.)

Temporal coherence measures the extent to which two points P and Q on a given waveform separated in time at a given location in space can be correlated, that is one can be reliably predicted from the other. For a pure sine wave, at any given location, any two points P and Q separated by any time interval are always correlated because we can predict the phase of one (Q) from the phase of the other (P) for any temporal separation. Consider a quasi-monochromatic plane wave propagating along the z-axis, with a rectilinear polarization parallel to $\mathbf{u}_x$, the unit vector along the x-direction

$$\mathbf{E}(\mathbf{r}, t) = A\exp[j(\omega t - kz + \varphi)]\mathbf{u}_x.$$

If the phase φ were constant in time, then the vibration would take the form of a pure sine wave. In reality, an actual optical wave is never purely monochromatic, but exhibits a finite bandwidth $\Delta\upsilon$. As a result, it can be described by a randomly time-varying phase $\varphi(t)$, where the phase variation is statistically distributed piecewise-like throughout the whole interval $[0, 2\pi]$. The field is equivalent to a succession of sine waves of finite duration with random phases. The average duration of each successive constant value taken by $\varphi(t)$ is the *coherence time* T_C, which is inversely proportional to the bandwidth $\Delta\upsilon$. The corresponding *coherence length* is $L_C = cT_C$ in vacuum, or $L_C = (c/n)T_C$ in a refractive medium of index n.

Consider the superposition of two quasi-monochromatic plane waves sharing the same state of polarization (for instance rectilinear parallel to $\mathbf{u}_x$):

$$\mathbf{E}_1(\mathbf{r}, t) = A_1 \exp[j(\omega t - kz + \varphi_1(t))]\mathbf{u}_x,$$

$$\mathbf{E}_2(\mathbf{r}, t) = A_2 \exp[j(\omega t - kz + \varphi_2(t))]\mathbf{u}_x.$$

The total field is $\mathbf{E}(\mathbf{r},t) = \mathbf{E}_1(\mathbf{r},t) + \mathbf{E}_2(\mathbf{r},t)$. The resulting *intensity* is

$$I = (1/2)\varepsilon_0 cn\langle|\mathbf{E}(\mathbf{r}, t)|^2\rangle,$$

where $\langle f(t)\rangle$ denotes the time average of any time-varying function $f(t)$ over the integrating time T_D of the detector. Using $\Delta\varphi(t) = \varphi_2(t) - \varphi_1(t)$,

$$I = (1/2)\varepsilon_0 cn\{A_1^2 + A_2^2 + 2A_1A_2 \times\langle\cos[\Delta\varphi(t)]\rangle\},$$

$$I = I_1 + I_2 + 2(I_1I_2)^{1/2}\langle\cos[\Delta\varphi(t)]\rangle.$$

If the time variations of $\varphi_1(t)$ and $\varphi_2(t)$ are correlated in such a way that, at least at the scale of T_D, $\Delta\varphi(t) = \Delta\varphi_0 =$ constant, then the intensity of the sum can be different from the sum of the intensities. The two fields are then said to be *mutually coherent*. See *Young's interference fringes*.

Ternary III-V alloys have three different elements from the III and V groups. For example GaAs and InAs are III-V compounds individually but they can be alloyed to produce a ternary alloy, $In_{1-x}Ga_xAs$. There are as many group III elements as there are group V elements. See *ternary semiconducting alloy*.

Ternary semiconducting alloy has three different elements in its composition, for example, $Al_xGa_{1-x}As$ has three elements Al, Ga, As. This $Al_xGa_{1-x}As$ alloy is composed of III (Al and Ga) and V (As) groups only and the molar amounts from each group are the same; hence it is a *III-V ternary alloy*. The bandgap E_g of the ternary alloys $Al_xGa_{1-x}As$ follows the empirical expression,

$$E_g(\text{eV}) = 1.424 + 1.266x + 0.266x^2.$$

This system is one of the most important III-V ternary systems owing to the ability to form near-lattice matched heterostructures over almost the entire compositional range. It has found extensive use in laser development, and in fabrication of high electron mobility transistors. Other important ternaries include InGaAs which is also used in high-power and high-frequency electronics because of its superior electron velocity with respect to the more common semiconductors such as silicon and GaAs. The ability to tune the alloy bandgap to cover the important 1300–1550 nm spectral window makes it an excellent system for telecommunications detector development. These alloys are usually grown on InP substrates – the lattice matched composition, $In_{0.53}Ga_{0.47}As$, provides devices with a *cut-off wavelength* of 1.68 μm. By further increasing the In/Ga ratio, it is possible to extend this wavelength to about 2.6 μm – however, in this case special precautions need to be taken to avoid dislocation and defect formation due to elastic mismatch.

Another important ternary III-V alloy system is $In_xGa_{1-x}N$, a direct bandgap semiconductor which is normally grown on optically transparent substrates (e.g. sapphire and silicon carbide). These alloys have very high *heat capacity* and low sensitivity to *ionizing radiation*, making them excellent choices for solar cell materials for deployment in space – for example, on satellites. Interestingly, in these alloys, segregation of indium occurs resulting in defects which act as quantum wells – these then form the basis for efficient light emitters in the green, blue and ultraviolet. Such structures have been exploited to form efficient LEDs and diode lasers. Depending on the Ga/In ratio, x,

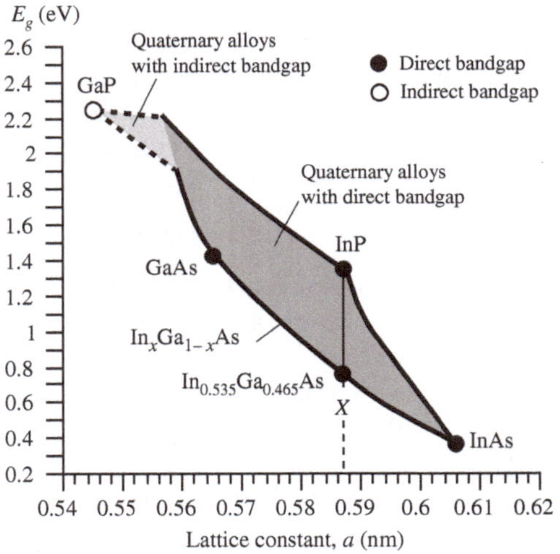

Ternary semiconductor alloy Bandgap energy E_g and lattice constant a for various III–V alloys of GaP, GaAs, InP and InAs. A line represents a ternary alloy formed with compounds from the end points of the line. Solid lines are for direct bandgap alloys whereas dashed lines are for indirect bandgap alloys. Regions between lines represent quarternary alloys. The line from X to InP represents quarternary alloys $In_{1-x}Ga_xAs_{1-y}P_y$ made from $In_{0.535}Ga_{0.465}As$ and InP which are lattice matched to InP.

emission from the near ultraviolet ($x \sim 0.02$) to blue ($x \sim 0.3$) and even red is possible. As solar cells, with high immunity to ionizing radiation, two junction cells with bandgaps of 1.1 eV and 1.7 eV can attain a theoretical efficiency of ~50%.

One of the earliest examples of using ternary semiconducting alloys was that of *GaAsP*. Using liquid phase epitaxy, and later, chemical vapour deposition, it proved to be an excellent material for manufacturing red, orange and yellow LEDs. It is often grown on a gallium phosphide substrate to form a GaP/GaAsP heterostructure. Doping with nitrogen is used to fabricate bright green LEDs.

There are many other important ternary semiconducting alloys in the groups other than III-V and these include II-VI alloys – the latter includes HgCdTe which is widely used as a material for developing infrared photodiode arrays in the first and second atmospheric windows.

Table: Ternary Compounds Compositional dependence of room temperature direct energy bandgap for standard III–V ternary alloys. Data combined from J. W. Mayer and S. S. Lau, *Electronic Materials Science*, Macmillan Publishing Company, New York, 1990; H. C. Casey Jr, M. B. Panish, *Heterostructure Lasers, Part A, Fundamental Principles, Part B, Materials and Operating Characteristics*, Academic Press, New York, 1978.

Compound	Energy Bandgap, E_g (eV)
$Al_xIn_{1-x}P$	$1.351 + 2.230x$
$Al_xGa_{1-x}As$	$1.424 + 1.266x + 0.266x^2$
$Al_xIn_{1-x}As$	$0.360 + 2.120x + 0.698x^2$
$Al_xGa_{1-x}Sb$	$0.726 + 1.129x + 0.368x^2$
$Al_xIn_{1-x}Sb$	$0.172 + 1.621x + 0.430x^2$
$Ga_xIn_{1-x}P$	$1.351 + 0.643x + 0.786x^2$
$Ga_xIn_{1-x}As$	$0.360 + 1.064x$
$Ga_xIn_{1-x}Sb$	$0.172 + 0.139x + 0.415x^2$
$GaAs_{1-x}P_x$	$1.424 + 1.150x + 0.176x^2$
$GaSb_{1-x}As_x$	$0.726 - 0.502x + 1.200x^2$
$InAs_{1-x}P_x$	$0.360 + 0.891x + 0.101x^2$
$InSb_{1-x}As_x$	$0.180 - 0.410x + 0.580x^2$

TFT see *thin film transistor*.

Thermal broadening see *Doppler broadening*.

Thermal equilibrium carrier concentrations are those electron and hole concentrations that are solely determined by the statistics of the carriers and the density of states in the band under thermal equilibrium and in the absence of excitation. Thermal equilibrium concentrations obey the mass action law, $np = n_i^2$ where n and p are thermal equilibrium electron and hole concentrations and n_i is the intrinsic concentration.

Thermal generation current is the current that flows in a reverse biased *pn* junction as a result of the thermal generation of electron–hole pairs in the depletion layer which become separated and swept across by the built-in field. It is usually the most significant contribution to the reverse diode current, in particular, in Si diodes.

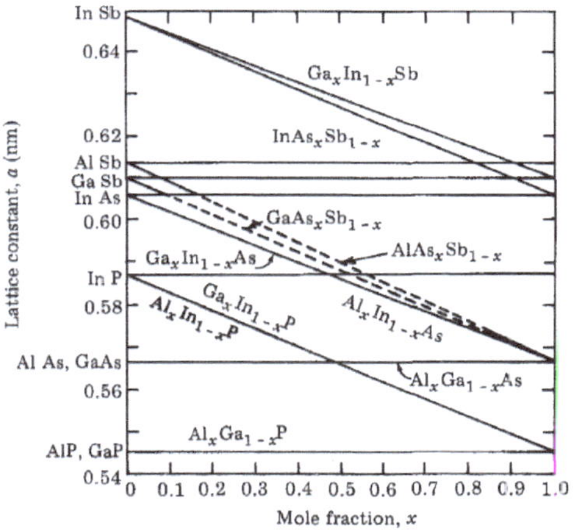

Ternary semiconductor alloy Dependence of lattice parameter on alloy composition for a selection of III-V ternary alloys. (This figure was published in H. C. Casey Jr and M. B. Panish, *Heterostructure Lasers, Part A, Fundamental Principles, part B, Materials and Operating Characteristics*, © Elsevier (for Academic Press, NY) 1978.)

Thermal shock is the fracture of a brittle component, such as glass, due to induced stresses that result from a sudden change in the temperature. Rapid cooling, for example, will result in tensile stresses on the surface of a glass sample because the inner portion would still be hot; glass has poor thermal conductivity. A crack on the surface can then lead to brittle fracture. If the induced stress is sufficiently large, then the sudden temperature is sufficiently large to propagate the crack. *Thermal shock resistance* (TSR) is a relative figure of merit that gauges the resistance of a sample to withstand sudden temperature changes:

$$\mathrm{TSR} = \frac{\sigma_{\mathrm{FS}}\kappa}{E\alpha_{\mathrm{L}}}$$

where σ_{FS} is the fracture strength of the sample, κ is the thermal conductivity, E is Young's modulus, and α_{L} is the linear thermal expansion coefficient. To obtain good TSR we need large σ_{FT} and κ but small E and α_{L}. For example, the ratio of TSR of MgF_2 to that of CaF_2 is about 3; MgF_2 is well known to have better TSR than CaF_2. See *optical materials*.

Thermalization is a process in which an excited electron in the conduction band (CB) loses excess energy as it is collides with lattice vibrations, and falls close to E_{c}, that is, until its average energy is $(3/2)k_{\mathrm{B}}T$ above E_{c}. The same concept also applies to holes in the valence band.

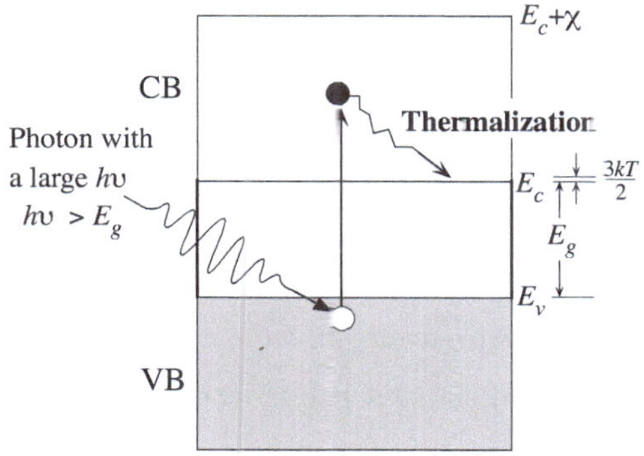

Thermalization Optical absorption generates electron–hole pairs. Energetic electrons must lose their excess energy to lattice vibrations until their average energy is $(3/2)k_{\mathrm{B}}T$ in the CB. (CB = Conduction Band, VB = Valence Band, E_g = Bandgap energy.) We assumed the effective mass of the electron is much smaller than that of the hole.

Thermionic emission is the emission of electrons from the surface of a heated metal. When a metal is heated, the proportion of electrons having energy greater than the work function (Φ) plus Fermi energy (E_{F}) increases due to the tailing of the Fermi–Dirac function to higher energies. Consequently, some of the electrons can be emitted over this potential barrier into vacuum and become free. The emission probability depends exponentially on the temperature. If the emitted electrons are collected and the metal replenished with electrons, as in the vacuum tube, then the thermionic current density is described by the *Richardson–Dushman equation*.

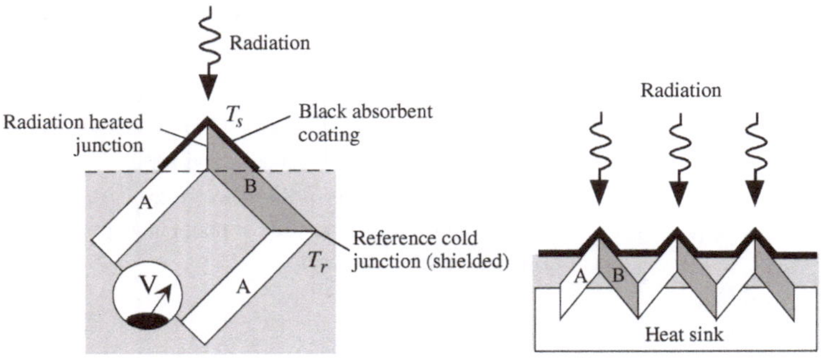

Thermocouple principle Thermopile

Schematic illustration of the principle of operation of the **thermopile detector**.

Thermographic camera see *infrared camera.*

Thermoluminescence see *luminescence.*

Thermomagnetic recording see *magnetooptic recording.*

Thermopile detector is a thermal detector, used primarily for infrared detection, that utilizes the *thermocouple effect* to generate a voltage difference when one junction of the thermocouple is heated by incident radiation with respect to the other junction. A thermocouple is made from two different metals A and B (e.g. antimony and bismuth) connected as A-B-A to form two AB junctions. The thermocouple has a "cold" reference junction (T_r) and a sensing junction (T_s) that is coated with a black radiation-absorbing material. The reference junction has to be heat-sunk and isolated from the sensing junction so that its temperature T_r is unaffected by the radiation. When radiation is absorbed by the sensing junction, its temperature T_s rises with respect to T_r, and the temperature difference $\Delta T = T_s - T_r$ generates a voltage at the output terminals of the thermocouple. Two different metals, A and B, with different Seebeck coefficients are needed for the thermoelectric couple to generate a voltage from ΔT. The single thermocouple voltage is usually very small, less than a microvolt, and a number of thermocouples are used in series to generate a higher output voltage. For example in one particular type of bismuth–antimony thermopile detector 72 junctions are connected in series to generate 0.1 mV from an incident flux of 1 μW. Thermopile detectors tend to be slow with response times typically 10–100 ms range; usually speed comes at the expense of responsivity. Advances in microelectronic thin film fabrication, and silicon micromachining, have led to the development of efficient and compact thermopile detectors. Differently doped semiconductors or semiconductor–metal (silicon–metal) junctions have been used for A and B with good responsivity. Their distinct advantage is their wide flat spectral response from almost 0.2 μm to 1000 μm, and the absence of flicker (1/*f*) noise. Typical responsivities are 0.02–0.1 mV/μW, specific detectivity (D^*) is $(1-3)\times 10^8$ cm Hz$^{1/2}$ W^{-1}.

Thick lens is a lens whose thickness cannot be neglected in the formation of an image. In the paraxial approximation (light rays making small angles with the optical axis of the lens), the optical properties of a thick lens may be described by using an *object (first) principal plane H* and an *image (second) principal plane H′*. Rays from the object to the lens are drawn to the *H* plane and rays leaving the lens to the image start from the *H′* plane. For

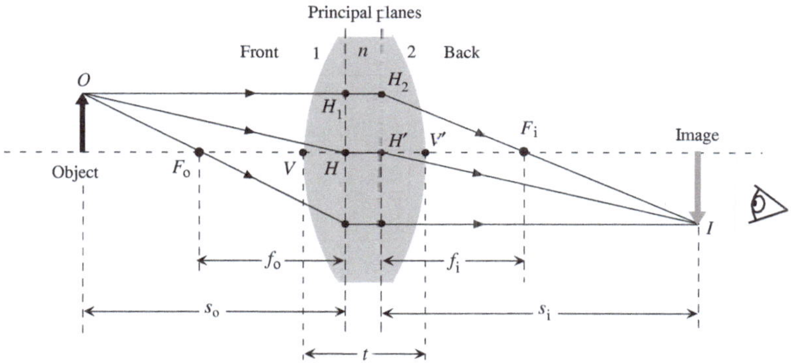

Thick lens H and H' are the principal planes.

example, the ray OH_1 from the object O is drawn to H_1 on H. The point H_2 is opposite to H_1 on H'. The ray from the lens to the image, starts at H_1, crosses F_i to make it to I on the image. The object distance s_o and image distance s_i are normally measured to the vertices of the lens V and V'. Thick lens effective focal length f is

$$\frac{1}{f} = (n-1)\left(\frac{1}{R_1} - \frac{1}{R_2}\right) + \left[\frac{(n-1)^2}{n}\left(\frac{t}{R_2 R_2}\right)\right]$$

where R_1 and R_2 are the radii of curvature of surfaces 1 (object side) and 2 (image side), respectively, n is the lens refractive index, and $t = V' - V$. With the focal length f calculated from above, the normal thin lens equations can be used for object and image distances from V and V'; $1/s_o + 1/s_i = 1/f$. Other important thick lens parameters are

$$\frac{1}{h} = \frac{1}{VH} = \frac{n(R_1 - R_2) - (n-1)t}{R_1 t}$$

$$\frac{1}{h'} = \frac{1}{H'V'} = \frac{n(R_1 - R_2) - (n-1)t}{R_2 t}$$

and

$$HH' = \frac{(R_1 - R_2 - t)(n-1)t}{n(R_1 - R_2) - (n-1)t}.$$

Gullstrand's equation gives the effective power of a thick lens at the H' plane as

$$P = P_1 + P_2 - \frac{t}{n} P_1 P_2$$

where P_1 and P_2 are the surface powers of the surfaces 1 and 2 (see figure). Front and back adjectives refer to the object and image sides of the lens, *back focal length* f_i being the distance from V' to F_i, and F_i being the *back focal point*.

See *cardinal points, lens, Newton's lens equation, paraxial optics.*

Thin film optics usually involves multiple reflections of light entering a thin film dielectric (typically on a substrate) so that the reflection and transmission coefficients are determined by multiple wave interference phenomena.

Consider a thin film coating on an object. Suppose that the incident wave has an amplitude of A_0. Then there are various transmitted and reflected waves as shown in the figure. We then have the following amplitudes based on the definitions of the reflection and transmission coefficients:

$A_1 = A_0 r_{12}$ $\quad B_4 = A_0 t_{12} r_{23} r_{21} r_{23}$
$A_2 = A_0 t_{12} r_{23} t_{21}$ $\quad B_5 = A_0 t_{12} r_{23} r_{21} r_{23} r_{21}$
$A_3 = A_0 t_{12} r_{23} r_{21} r_{23} t_{21}$ $\quad B_6 = A_0 t_{12} r_{23} r_{21} r_{23} r_{21} r_{23}$
$B_1 = A_0 t_{12}$ $\quad C_1 = A_0 t_{12} t_{23}$
$B_2 = A_0 t_{12} r_{23}$ $\quad C_2 = A_0 t_{12} r_{23} r_{21} t_{23}$
$B_3 = A_0 t_{12} r_{23} r_{21}$ $\quad C_3 = A_0 t_{12} r_{23} r_{21} r_{23} r_{21} t_{23}$

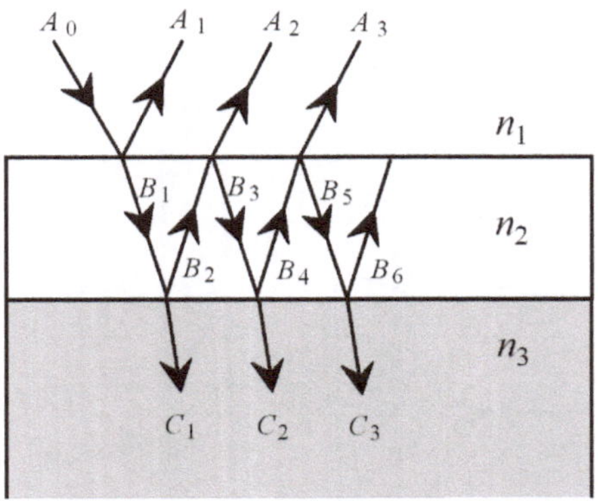

Thin film optics Thin film coating on a substrate and multiple reflections.

and so on, where r_{12} is the reflection coefficient of a wave in medium 1 incident on medium 2, and t_{12} is the transmission coefficient from medium 1 into 2. For simplicity, we will assume normal incidence. The phase change in traversing the coating thickness d is $\phi = (2\pi/\lambda)n_2 d$ where λ is the free-space wavelength. The wave has to be multiplied by $\exp(-j\phi)$ to account for this phase difference through d.

The reflection and transmission coefficients are given by

$$r_1 = r_{12} = \frac{n_1 - n_2}{n_1 + n_2} = -r_{21},$$
$$r_2 = r_{23} = \frac{n_2 - n_3}{n_2 + n_3}$$

and

$$t_1 = t_{12} = \frac{2n_1}{n_1 + n_2}, \qquad t_2 = t_{21} = \frac{2n_2}{n_1 + n_2},$$
$$t_3 = t_{23} = \frac{2n_3}{n_2 + n_3}$$

where

$$1 - t_1 t_2 = r_1^2.$$

The overall reflection coefficient is then

$$r = \frac{A_{\text{reflected}}}{A_0} = r_1 - \frac{t_1 t_2}{r_1}\sum_{k=1}^{\infty}\left(-r_1 r_2 e^{-j2\phi}\right)^k$$

which can be summed to

$$r = \frac{r_1 + r_2 e^{-j2\phi}}{1 + r_1 r_2 e^{-j2\phi}}.$$

The overall transmission coefficient t is

$$t = \frac{C_{\text{transmitted}}}{A_0} = -\frac{t_1 t_{23} e^{j\phi}}{r_1 r_2}\sum_{k=1}^{\infty}\left(-r_1 r_2 e^{-j2\phi}\right)^k$$
$$= \left(\frac{t_1 t_3 e^{j\phi}}{r_1 r_2}\right)\frac{r_1 r_2 e^{-j2\phi}}{1 + r_1 r_2 e^{-j2\phi}}$$

which can be simplified to

$$t = \frac{t_1 t_3 e^{-j\phi}}{1 + r_1 r_2 e^{-j2\phi}}.$$

See *Swanepoel's technique* for the determination of the optical constants of thin transparent films.

Thin film transistor is a *field effect transistor* (FET) that has been fabricated by the deposition of various thin film layers (such as metals, semiconductors and insulators), using special thin film deposition techniques (usually vacuum deposition process) onto a suitable substrate, usually a glass substrate, so that the transistor is almost flat on the substrate and can be interconnected by using thin film techniques to other TFTs or to external electronics. The TFTs are widely used in *flat panel displays* because they can be readily fabricated on the surface of a large glass substrate and millions of them can be placed to form an array, called an *active matrix array* (AMA), and suitably interconnected. Each pixel of an AMA has one or more TFTs as active elements. Usually the electrodes and interconnects of the TFT can be made transparent, while the TFT itself (unless made from a wide bandgap semiconductor such as ZnO) is not transparent. The channel of the TFT FET is a semiconductor that can be deposited as a thin film and over a wide area (for displays), and obviously allows the field effect to operate (the field is able to generate a "conducting" channel). Many TFTs use hydrogenated amorphous silicon (a-Si:H) or nanocrystalline or microcrystalline a-Si:H because of the highly developed large area thin film technology for this material.

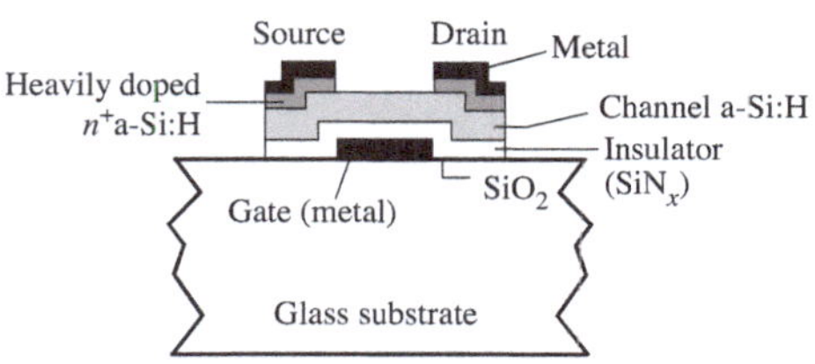

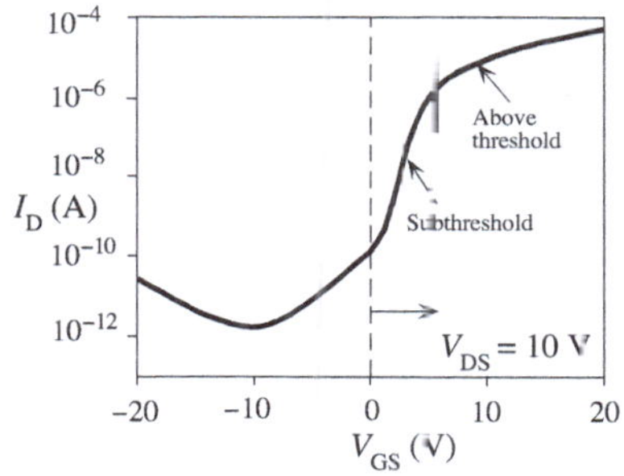

Left: schematic sketch of a typical **thin film transistor** cross-section from a-Si:H technology. Right: typical drain current I_D vs. gate voltage V_{GS} characteristics with $V_{DS} = 10V$ constant.

Other semiconductors have also been used including polycrystalline CdSe and organic materials. Organic TFTs are cheaper and easier to fabricate, e.g. using spin coating. In the a-Si:H TFT, the channel is undoped a-Si:H, and heavily doped n^+ a-Si:H layers are used to make the source (S) and drain (D) contacts, onto which metals are deposited for the actual S and D electrodes, as shown in the simplified diagram. (A. Nathan *et al., IEE Proc. Cirt. Dev. Syst.*, **150**, 329, 2003; M. Powell, *IEEE Trans Elec. Dev.*, **36**, 2753, 1989.) See *active matrix array*.

Thin lens see *lens*.

Third-order harmonic generation is a nonlinear optical process in which three photons of energy $\hbar\omega$ are absorbed by a material resulting in the emission of a single photon of energy $3\hbar\omega$. In the case of centrosymmetric media, second-order nonlinearity is absent since the polarization must reverse itself when the electric field is reversed. In this case, the dominant nonlinearity is of third order and the medium is called a *Kerr medium*. Kerr media respond to applied optical fields by generating third harmonics and sums and differences of triplets of frequencies. It should be noted that typical values of the third order susceptibility $\chi^{(3)}$ are orders of magnitude smaller than $\chi^{(2)}$ coefficients for typical nonlinear conversion crystals – this means that the incident laser intensity needs to be proportionately higher and can often lead to material damage. In addition, third-harmonic generation in crystals can be hard to phase-match in order to build up a reasonable net third-harmonic signal. For this reason, cascading of two second-harmonic effects in separate crystals is the preferred method for obtaining higher multiples of the pump laser frequency. Unlike second-harmonic generation, however, third-harmonic generation can be obtained in practically all materials, as long as they are transparent at the wavelengths involved. However, under strong focusing conditions, the harmonic radiation produced ahead of the focus is reabsorbed coherently after the focus as a result of the *Gouy phase shift* (i.e. the phase shift for such radiation passing through a focus – this has been observed using THz spectroscopy) and hence third-harmonic generation is often "background free" and highly sensitive to interfaces between materials, differing in dispersion or $\chi^{(3)}$. This has an important application as a microscopic spectroscopic tool.

Third-order nonlinear processes are those processes that involve the third-order *nonlinear optical properties*, that is the third-order susceptibility $\chi^{(3)}$, that include *third-order harmonic generation*, sum frequency generation, frequency mixing, parametric amplification, coherent Stokes and anti-Stokes Raman scattering processes, the bound-electronic optical Kerr effect, the Raman induced Kerr effect, the molecular orientation Kerr effect, *two photon absorption*, the ac Stark effect, stimulated Raman scattering, and stimulated *Rayleigh–Wing scattering*.

Third-order susceptibility, $\chi^{(3)}$, see *electric susceptibility third order harmonic generation, third order nonlinear processes*.

Thomas–Reiche–Kuhn sum rule states that the summation of oscillator strengths f_{ij} for an atom in an initial state i is equal to the number of electrons in the atom; $\sum_j f_{ij} = 1$. Oscillator strengths are involved in calculating various optical properties. See *sum rules*.

Thomson scattering is the scattering of electromagnetic radiation, usually low energy photons, by nonrelativistic free electrons. In classical Thomson scattering, the oscillating electric field in the incident radiation causes the free electron to oscillate at the same frequency, and thereby emit radiation at the same frequency so that the scattered radiation and the incident radiation essentially have the same frequency. Thus, the scattering is *elastic* or *coherent*. Further, the scattered radiation has preferential polarization depending on the observation position. Thomson scattering can be viewed as Compton scattering in which the scattered photo has a frequency very close to the incident photon, the elastic limit of Compton scattering. The relative intensity (radiation power flowing per unit solid angle) of the scattered radiation in a direction at an angle θ to the incident radiation depends on $r^2(1 + \cos^2\theta)$ where r is the classical radius of the electron, given as $e^2/(m_e c^2)$. See *Compton scattering*.

Three CCD color image sensor see *imaging sensor*.

Three-Five and III-V binary, ternary and quaternary compounds are semiconductors from group III and V elements such as GaP, GaAs, GaSb, InP, InAs, InSb, $Al_xGa_{1-x}As$, $In_{1-x}Ga_xAs_yP_{1-y}$ and are important optoelectronic materials. The bandgap covers a wide range, e.g. $E_g = 0.18$ eV for InSb, but 2.2 eV for GaP. Their optical properties can be found in S. Adachi, *J. Appl. Phys.* **66**, 6030, 1989.

Threshold crossing in a single-mode laser is characterized by a qualitative change of the field behavior that can be understood with slightly modified *reduced rate equations*, which couple the intensity I to the material gain D. Let us add a source term (spontaneous emission) of the form βD:

$$dI/d\tau = kI(D-1) + \beta D,$$
$$dD/d\tau = Y_P - D(1+I),$$

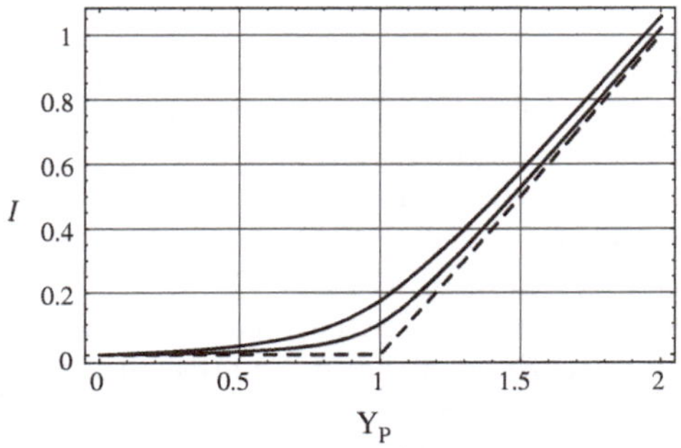

Threshold crossing I vs. Y_P; from top to bottom, $(\beta/k) = 3\times10^{-2}, 10^{-2}, 10^{-5}$. The smaller the spontaneous contribution, the closer the curve to the classical one.

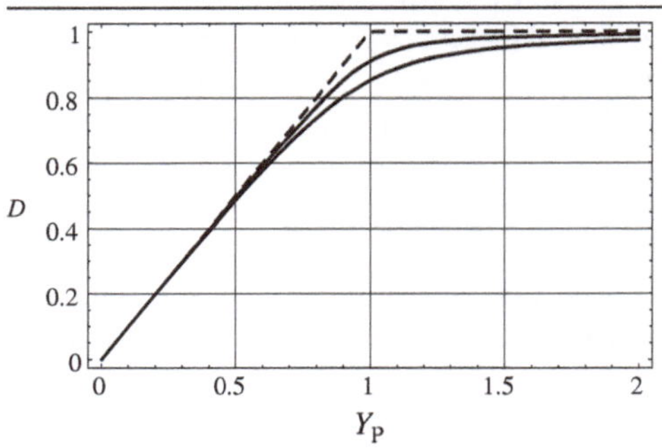

Threshold crossing D vs. Y_P; from bottom to top, $(\beta/k) = 3\times10^{-2}, 10^{-2}, 10^{-5}$. The smaller the spontaneous contribution, the closer the curve to the classical one.

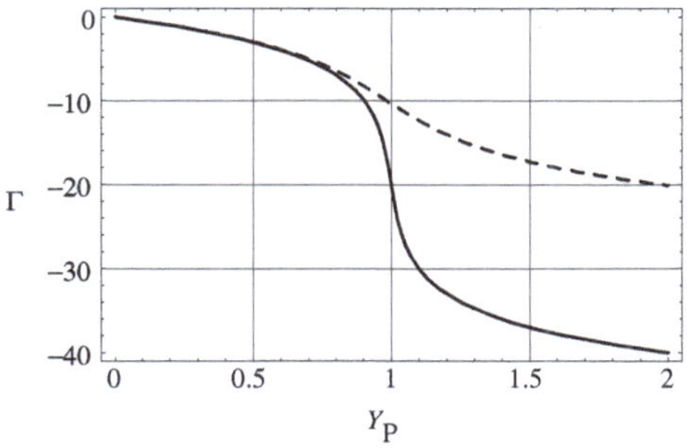

Threshold crossing Γ vs. Y_P; radiation linewidth of the emitted field as compared to the cold-cavity resonance [dB] as a function of pumping, for $(\beta/k) = 10^{-2}$ (upper curve) and 10^{-4} (lower curve).

where Y_P is the pumping rate, β a dimensionless constant ($\beta \ll 1$) and βD the fraction of spontaneous emission projected into the laser mode. In the steady state regime, $d/d\tau = 0$, so that:

$$I = (\beta/k)D/(1 - D),$$
$$D = Y_P/(1 + I).$$

The intensity I appears as the solution to a simple second-order equation:

$$I^2 + I(1 - Y_P) - (\beta Y_P/k) = 0.$$

The transition from the below-threshold behavior to the above-threshold behavior is actually continuous. Gain clamping is also progressive.

The linewidth Γ is also subject to a dramatic narrowing across threshold, as long as the field becomes highly coherent; this is sometimes referred to as *photonic condensation*.

Threshold current, I_{th}, of a laser diode is the minimum current that is needed for achieving a lasing emission from the device. At this current value, the optical gain in the active region is the threshold gain. Lasing radiation is only obtained when the optical gain in the medium can overcome the photon losses from the cavity, which requires the diode current I to exceed a *threshold value* I_{th}. Below I_{th}, the light from the device is due to spontaneous emission and not stimulated emission. The light output is then composed of incoherent photons that are emitted randomly and the device behaves like an LED. Threshold current is strongly temperature dependent.

Threshold gain is the critical optical gain, g_{th}, that an active medium in an optical resonator must have to achieve steady state (self-sustained) lasing radiation as output from this resonator; this

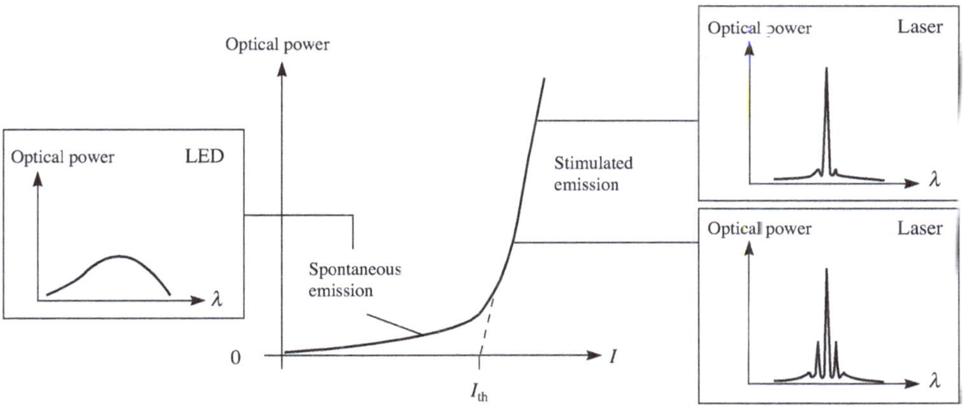

Threshold current Typical output optical power vs. diode current (I) characteristics and the corresponding output spectrum of a laser diode.

threshold gain just overcomes the losses in the active medium and also the losses from the optical cavity and the reflectors. When the optical gain of the medium is equal to the threshold gain, g_{th}, the device becomes a self-sustained optical oscillator in which some of the oscillations in the cavity are allowed to escape as radiation output by making one of the mirrors partially transmitting. The term laser is understood to refer to a device that is an optical oscillator with an output radiation. Energy is continuously pumped into the active medium to maintain its gain and hence maintain the optical oscillations. Consider an optical cavity with mirrors at the ends, such as the Fabry–Perot cavity. The cavity contains a laser medium so that lasing emissions build up to a steady state, that is we have continuous operation, or steady state and we have stationary electromagnetic (EM) oscillations in the cavity. Consider an EM wave with an initial optical power P_i starting at some point in the cavity and traveling towards the cavity face 1 as shown in the figure. It will travel the length of the cavity, become reflected at face 1, travel back the length of the cavity to face 2, become reflected at 2 and arrive at the starting point with a final optical power P_f. Under steady state conditions, oscillations do not build up and do not die out which means that P_f must be the same as P_i. Thus there should be no optical power loss in the round trip which means that the *net round-trip optical gain* G_{op} must be unity,

$$G_{op} = P_f/P_i = 1.$$

Let γ be the loss coefficient of the medium, which represents all losses in the cavity and its walls and scattering, but excludes light transmission losses though the end mirrors and absorption across the energy levels involved in stimulated emissions. If L is the length of the optical cavity, and R_1 and R_2 are the reflectances of the end-reflectors of the cavity, then the optical gain of the medium for lasing oscillations (i.e. for continuous wave lasing emission from the device) is

$$g_{th} = \gamma + \frac{1}{2L}\ln\left(\frac{1}{R_1 R_2}\right).$$

As the medium is pumped the population inversion increases and eventually reaches the *threshold population inversion*, which is that population inversion $(N_2 - N_1)_{th}$ between the energy levels E_2 and $E_1 (E_2 > E_1)$ that corresponds to the threshold optical gain (g_{th}),

$$(N_2 - N_1)_{th} = g_{th}\frac{c\Delta\nu}{B_{21} n h \nu_o} = g_{th}\frac{8\pi n^2 \nu_o^2 \tau_{sp}\Delta\nu}{c^2}$$

where c is the speed of light, n is the refractive index of the medium, B_{21} is the Einstein B_{21} coefficient, $\Delta\nu$ is the frequency linewidth of the optical gain curve (lineshape function) of the lasing medium which has its peak at the frequency ν_o, and $\tau_{sp} = 1/A_{21}$ is the mean time for spontaneous emissions from E_2 to E_1. The examination of the steady state continuous wave coherent radiation output power P_o and the population difference $(N_2 - N_1)$ in a laser as a function of the pump rate would reveal the simplified behavior shown in the second figure. Until the pump rate can bring $(N_2 - N_1)$ to the threshold value $(N_2 - N_1)_{th}$, there would be no coherent radiation output. When the pumping rate exceeds the threshold value, then $(N_2 - N_1)$ remains clamped at $(N_2 - N_1)_{th}$ because this controls the optical gain g which must remain at g_{th}. Additional pumping increases the *rate of stimulated transitions* and hence increases the optical output power P_o. Also note that we have not considered how pumping actually modifies N_1 and N_2 except that $(N_2 - N_1)$ is proportional to the pumping rate. Note that, as in all oscillators, the initiation and build-up of oscillations to the steady state initially involves a net round-trip optical gain that is greater than unity ($G_{op} > 1$), which becomes unity ($G_{op} = 1$) when steady state is reached. See *laser*.

Table: Threshold population inversion Threshold and other properties of selected laser systems. Data from W. T. Silfvast, *Laser Fundamentals*, second Edition, Cambridge University Press, Cambridge, 2004.

	He-Ne laser (Gas: Ne lases)	Ar ion laser (Gas)	Nd:YAG laser (Solid state)	Ruby laser Solid state
$\lambda_o = c/\upsilon_o$ (nm)	632.8	488.0	1064	694.3
$\Delta\upsilon$ (GHz)	1.5	2.7	120	330
g_{th} (m^{-1})	0.15	0.5	5	20
$(N_2 - N_1)_{th}$ (cm^{-3})	5×10^9	2×10^9	1.6×10^{17}	8×10^{18}
τ_{sp}	30 ns	10 ns	230 μs	3 ms
n	1.0	1.0	1.82	1.76
Single pass gain	1.015–1.16	1.05–1.65	2–20	7.5

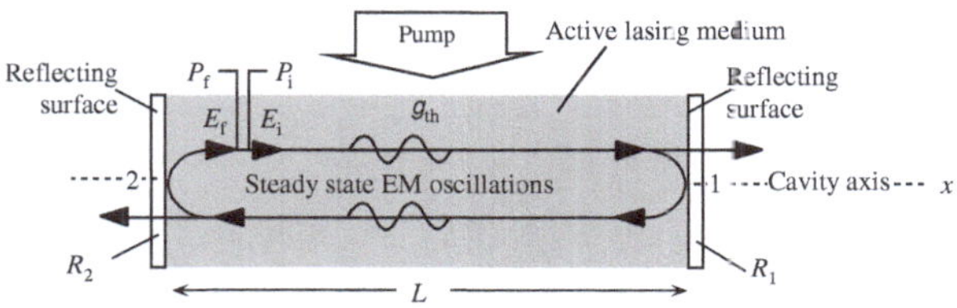

Optical cavity resonator

The **threshold gain** in a pumped active medium in a Fabry–Perot cavity that leads to self-sustained electromagnetic oscillations inside the cavity.

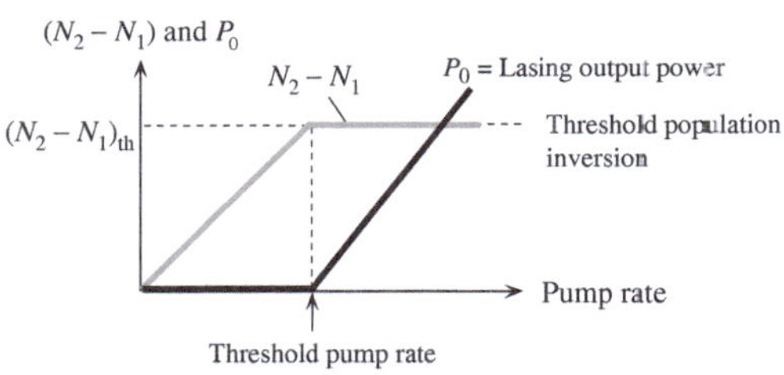

Threshold gain Simplified description of a laser oscillator. $(N_2 - N_1)$ and coherent output power (P_o) vs. pump rate under continuous wave steady state operation.

Threshold population inversion see *threshold gain*.

Tien's equation allows the scattering loss to be determined in a thin film optical guide in which the core-cladding interface roughness causes the scattering of light. The attenuation coefficient depends on the variance of the interface roughness and the fraction of the light intensity at the interface, among other factors. Consider a slab, e.g. a thin film, optical waveguide in which the upper and lower interfaces have some roughness. Suppose W is the guide thickness, n_1 is the core refractive index, k is the free space wavenumber $(2\pi/\lambda)$, λ is the free space wavelength, σ_{12} and σ_{10} are the RMS surface roughnesses of the top and bottom interfaces as illustrated in the figure in Tien's original formulation and θ_1 is the angle of incidence for the mode in question at the interface. Then, the attenuation coefficient due to interface scattering is

$$\alpha = \left(\frac{4n_1\pi}{\lambda}\right)^2 \left(\frac{\cos^3\theta_1}{2\sin\theta_1}\right) \times \left[\frac{1}{W + (2/p)}\right] (\sigma_{12}^2 - \sigma_{10}^2)$$

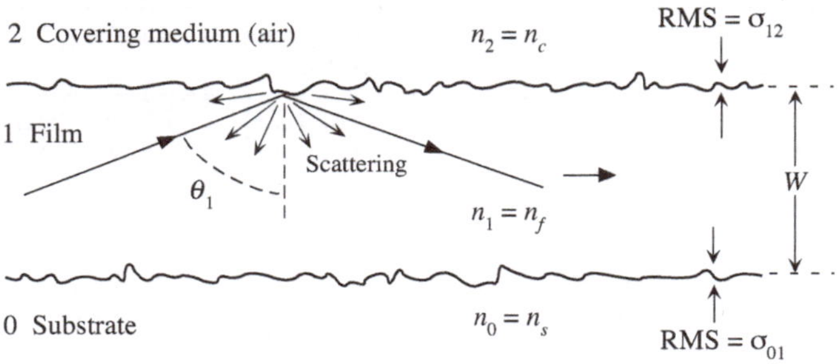

Tien's equation A schematic illustration of interface scattering induced attenuation of a propagating mode in a planar waveguide.

in which $\beta = n_1 k \sin\theta_1$ is the propagation constant, and $p = (\beta^2 - n_1^2 k^2)^{1/2}$. It is assumed that σ_{12} and σ_{10} are less than λ. There have been other formulations of interface scattering but most are quite complicated whereas Tien's equation is relatively easy to use. (P. K. Tien, *Appl. Opt.*, **10**, 2395, 1971).

Time Division Multiplexing (TDM) is a technique used in optical communications to transmit different signals on the same channel, by interweaving in a synchronous manner the time-slots allotted to each of them.

Total internal reflection (TIR) is the total reflection of a wave traveling in a medium of refractive index n_1 when it is incident at a boundary with another medium of lower refractive index n_2. The angle of incidence must be greater than the critical angle θ_c which depends on the refractive indices since $\sin\theta_c = n_2/n_1$.

Transfer matrix formalism is a useful technique for describing various optical transfer phenomena, and stems from the linearity of the electromagnetism equations. Usually 2 × 2 transfer matrices were originally developed for isotropic layered media, made of several homogeneous layers separated by parallel interfaces, all perpendicular to a common axis conventionally oriented in the z-direction from the left to the right, and devoid of any internal source.

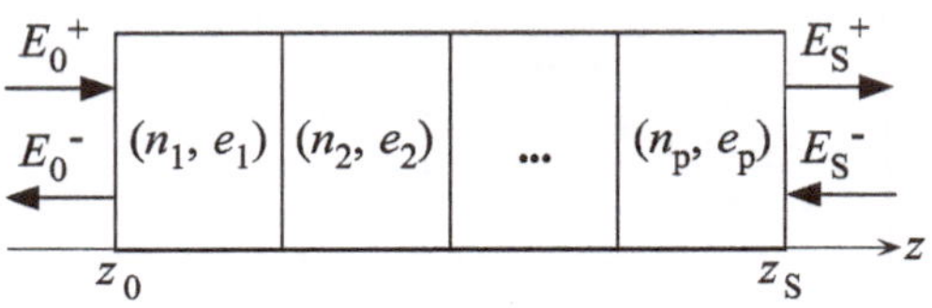

Transfer matrix Multilayer structure, made of uniform layers of width $(e_1, e_2, \ldots)$ and index $(n_1, n_2, \ldots)$, separated by plane parallel interfaces, between two semi-infinite media of respective index n_0 and n_s. $(E_0{}^+, E_0{}^-)$ represent the propagative and contrapropagative components of the electric field at abscissa z_0, $(E_s{}^+, E_s{}^-)$ their counterpart at abscissa z_s.

A. (2×2) transfer matrix

We are interested only in longitudinal effects along the propagation axis. For a given frequency ν (angular frequency $\omega = 2\pi\nu$), a given angle of incidence θ and a given eigenstate of polarization (either s or p), the electric component (E) of the electromagnetic field is completely determined by its two propagative (E^+) and contrapropagative (E^-) parts. The time dependence is $\exp(+j\omega t)$. If we work in the spectral domain, then all fields are expressed in $\mathrm{V\,m^{-1}\,Hz^{-1/2}}$. Throughout its propagation along the structure, the frequency, the state of polarization as well as the transverse component $k_{//}$ of the wavevector are conserved. Only the longitudinal component

Transfer matrix formalism

$k_z = [\varepsilon_r(\omega/c)^2 - k_{//}{}^2]^{1/2}$ is affected, where ε_r is the relative dielectric permittivity. Because of the sign convention, $\mathrm{Im}[\varepsilon_r] > 0$ in the presence of optical gain. For instance, under normal incidence, $k_z = n(\omega/c) + j(g/2)$, where n is the refractive index and g (in m^{-1}) the intensity gain coefficient.

The same formalism remains valid for any one-dimensional multisection structure. In an active integrated waveguide or optical fiber, k_z is to be replaced by the modal wavevector $\beta = n(\omega/c) + j(g/2)$, with n the effective index and g the net gain. Each section of length L is completely defined by n and g. In the spectral domain, chromatic dispersion is as easily taken into account as spectral gain behavior.

The state of the field at any abscissa z is related to its state at any other abscissa $z' > z$ by a linear relationship. As a result, the fields between the two end-facets located at abscissas z_0 and z_s are related by a transfer matrix [M] such as:

$$\begin{pmatrix} E_0^+ \\ E_0^- \end{pmatrix} = \begin{pmatrix} M_{11} & M_{12} \\ M_{21} & M_{22} \end{pmatrix} \begin{pmatrix} E_S^+ \\ E_S^- \end{pmatrix}.$$

Any one-dimensional complex structure can thus be decomposed into simpler subsections. On the other hand, strictly periodic structures such as distributed-Bragg reflectors (DBR) or distributed-feedback (DFB) lasers can be described, in the frame of coupled-wave equations (CWE), by one transfer matrix only.

B. Usual (2×2) transfer matrices

We give the explicit form of the transfer matrix associated to: (i) a pure propagation inside a linear, isotropic, homogeneous medium; (ii) an interface between two such media.

(i) Propagation over a distance L:

$$M_{11} = \exp(+jk_zL), \quad M_{12} = 0,$$
$$M_{21} = 0, \quad M_{22} = \exp(-jk_zL),$$

where k_z is the component of the wavevector in the z-direction.

(ii) Interface between two linear, homogeneous, isotropic media of relative (complex) permittivity ε_1 and ε_2; for $i = 1$ or 2, we note $k_{iz} = [\varepsilon_i(\omega/c)^2 - k_{//}{}^2]^{1/2}$, $p_i = (k_{iz}/\varepsilon_i)$. For s polarization (TE):

$$M_{11} = (k_{1z} + k_{2z})/(2k_{1z}),$$
$$M_{12} = (k_{1z} - k_{2z})/(2k_{1z}),$$
$$M_{21} = (k_{1z} - k_{2z})/(2k_{1z}),$$
$$M_{22} = (k_{1z} + k_{2z})/(2k_{1z}).$$

For p polarization (TM):

$$M_{11} = [\varepsilon_2/\varepsilon_1]^{1/2}(p_1 + p_2)/(2p_1),$$
$$M_{12} = [\varepsilon_2/\varepsilon_1]^{1/2}(p_2 - p_1)/(2p_1),$$
$$M_{21} = [\varepsilon_2/\varepsilon_1]^{1/2}(p_2 + p_1)/(2p_1),$$
$$M_{22} = [\varepsilon_2/\varepsilon_1]^{1/2}(p_1 - p_2)/(2p_1).$$

From these equations, it is straightforward to derive the Fresnel transmission and reflection coefficients on the interface, for both states of polarization.

C. Spectral transfer functions

The transfer matrix holds whatever the boundary conditions. The transmission and reflection coefficients in complex amplitude are, from left to right ($E_s{}^- = 0$):

$$t = \left.\frac{E_S^+}{E_0^+}\right|_{E_S^-=0} = \frac{1}{M_{11}},$$

$$r = \left.\frac{E_0^-}{E_0^+}\right|_{E_S^-=0} = \frac{M_{21}}{M_{11}}$$

and, from right to left ($E_0{}^+ = 0$):

$$t' = \left.\frac{E_0^-}{E_S^-}\right|_{E_0^+=0} = \frac{\det(M)}{M_{11}},$$

$$r' = \left.\frac{E_S^+}{E_S^-}\right|_{E_0^+=0} = \frac{-M_{12}}{M_{11}}.$$

For simplicity, we limit ourselves to the case where the two semi-infinite media surrounding the structure are identical ($n_0 = n_s$), in which case, $\det(M) = 1$. From a physical point of view, this restriction doesn't change anything, but presents the advantage of keeping formal expressions simpler to handle. The transmittance and reflectance for the intensity are, accordingly,

$$T = T' = |t|^2 = |1/M_{11}|^2,$$
$$R = |r|^2 = |M_{21}/M_{11}|^2,$$
$$R' = |r'|^2 = |M_{12}/M_{11}|^2.$$

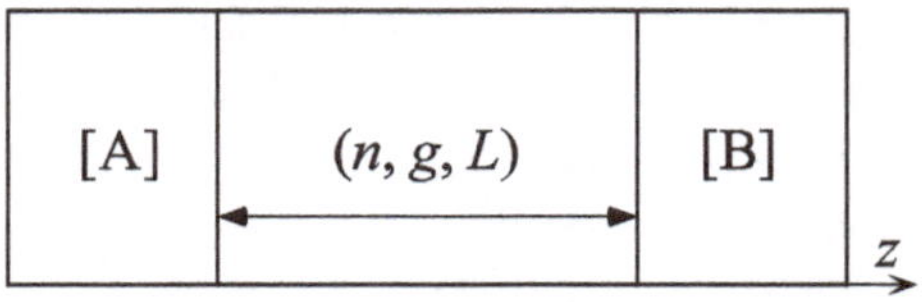

Resonant cavity made of one homogeneous section of length L, index n, gain g between two passive reflectors [A] and [B].

It should be noted that, in the linear regime, the overall transfer function is the same in both directions ($T = T'$). It is frequency-dependent: $T = T(\nu)$.

Consider the typical example of a resonator made of an active zone of (effective) index n and (net) gain g, sandwiched between two passive reflectors represented by their matrix [A] and [B]. Its transfer matrix reads:

$$[\mathrm{M}] = \begin{bmatrix} A_{11} & A_{12} \\ A_{21} & A_{22} \end{bmatrix} \begin{pmatrix} \exp(+j\beta L) & 0 \\ 0 & \exp(-j\beta L) \end{pmatrix} \times \begin{bmatrix} B_{11} & B_{12} \\ B_{21} & B_{22} \end{bmatrix},$$

so that

$$\begin{aligned} M_{11} &= A_{11}B_{11}\exp(+j\beta L) + A_{12}B_{21}\exp(-j\beta L), \\ M_{12} &= A_{11}B_{12}\exp(+j\beta L) + A_{12}B_{22}\exp(-j\beta L), \\ M_{21} &= A_{21}B_{11}\exp(+j\beta L) + A_{22}B_{21}\exp(-j\beta L), \\ M_{22} &= A_{21}B_{12}\exp(+j\beta L) + A_{22}B_{22}\exp(-j\beta L). \end{aligned}$$

As in any classical Fabry–Perot-like resonator, the transfer function takes the form of the so-called Airy function:

$$T(v) = T_0/[1 + m\sin^2(\Phi/2)],$$

where T_0 is the maximum transmission,

$$T_0 = \frac{T_A T_B \exp(gL)}{[1 - \rho_A\rho_B\exp(gL)]^2},$$

m is the *selectivity parameter*

$$m = \frac{4\rho_A\rho_B\exp(gL)}{[1 - \rho_A\rho_B\exp(gL)]^2},$$

and Φ is the round-trip phase change inside the cavity

$$\Phi = 2nL\omega/c - \varphi_A - \varphi_B,$$

and $\rho_{\mathrm{A}}\exp(j\varphi_{\mathrm{A}}) = -A_{12}/A_{11}$ and $\rho_{\mathrm{B}}\exp(j\varphi_{\mathrm{B}}) = B_{21}/B_{11}$ are the reflection coefficients, in complex amplitude, of both reflectors seen from inside the active zone, with $(\rho_{\mathrm{A}}, \rho_B)$ real. T_A, T_B are the transmission coefficients in power of the reflectors. Note that the amplitudes $(\rho_{\mathrm{A}}, \rho_{\mathrm{B}})$ as well as the phases $(\varphi_{\mathrm{A}}, \varphi_{\mathrm{B}})$ are allowed to vary with frequency ν. See *extented transfer matrix formalism.*

Transmission axis of a polarizer identifies the electric field direction that is allowed to pass through the polarizer.

Transmission coefficient is the ratio of the amplitude of the transmitted wave to that of the incident wave when the incident wave traveling in a medium meets a boundary with a medium of different refractive index. See also *transmission matrix formalism.*

Transmission grating see *diffraction grating.*

Transmission matrix see *Jones matrices.*

Transmittance is the fraction of transmitted intensity when a wave traveling in a medium is incident at a boundary with a different medium (different refractive index).

Transparency current, I_{trans}, is the diode current that provides just sufficient injection to lead to stimulated emissions just balancing absorption so that there is then no net photon absorption and the medium is transparent. Above I_{trans} there is a net optical gain in the medium though the optical output is not yet a continuous wave coherent radiation.

Transverse electric field (TE) mode has the electric field oscillations at right angles to the direction of propagation of the mode along the waveguide axis (see figure). A *transverse magnetic field* (TM) mode has the magnetic field oscillations at right angles to the direction of propagation of the mode along the waveguide axis. *Transverse electric and magnetic field modes* (TEM) have both the electric and magnetic field at right angles to the direction of propagation.

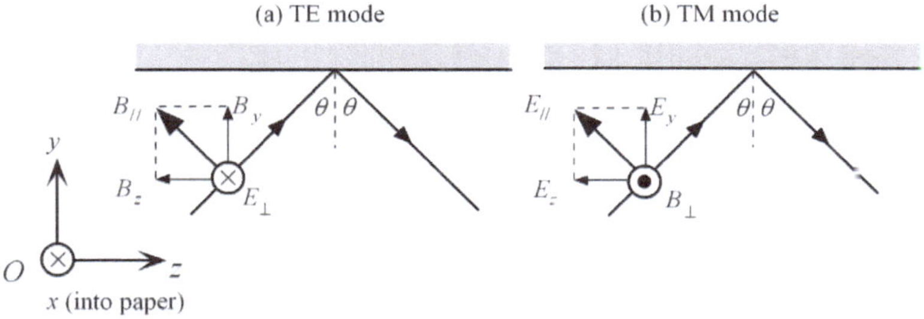

Transverse electric field Possible modes can be classified in terms of (a) transelectric field (TE) and (b) transmagnetic field (TM). Plane of incidence is the paper.

Transverse holographic diffraction gratings see *fiber Bragg grating*.

Trap in a semiconductor is a localized electronic state located within the energy bandgap of the semiconductor which may act to bind ("trap") either an electron or hole, termed an electron or hole trap, respectively. For the case of an electron trap, the energy $E_c - E_t$ (E_c is the conduction band edge energy and E_t is the energy of the trap in the bandgap) required to ionize an electron from the trap is relatively large compared with $k_B T$ such that the electron is said to be tightly bound in the trap. Typically, the potential energy associated with traps, having a deep binding energy, will be of relatively short range, and thus the trap will act to strongly localize the bound carrier. Associated with a trap are characteristic properties including its capture cross-section σ_n, or the *effective area* that it presents to a mobile carrier over which it acts to effectively capture the carrier. An argument based on detailed balance may be used to relate the electron emission rate e_n from an electron trap as a function of the thermal velocity of carriers v_{th}, and conduction band effective density of states N_c:

$$e_n = \sigma_n v_{th} N_c \exp[-(E_c - E_t)/k_B T].$$

If m^* is the electron's effective mass, the thermal velocity is then $v_{th} = (3k_B T/m^*)^{1/2}$, and since $N_c = 2(2\pi m^* k_B T)/\hbar^2)^{3/2}$, the emission rate can be written,

$$e_n/T^2 = \sigma_n \gamma \exp[-(E_c - E_t)/k_B T];$$
$$\gamma \equiv 2m^* k_B^2 (2\pi/\hbar^2)^{\frac{3}{2}} \sqrt{3}$$

Trapping of atoms requires a cavity in which the atoms are confined; the cavity is called an *atomic trap*. Some of the most important traps include magnetic traps (which use the force of an inhomogeneous magnetic field on the atomic magnetic dipole moment) and optical dipole traps (which use the force exerted on atoms by focused laser beams). Confinement of atoms between two atomic mirrors has also been suggested and is analogous to a Fabry–Perot type cavity for light. When a gas of atoms is sufficiently cooled and sufficiently confined in a trap, all the atoms occupy a single quantum state known as a *Bose–Einstein condensate* (BEC). This state was first observed in 1995 using both magnetic fields and laser beams to slow and trap the atoms – Eric A. Cornell of the National Institute of Standards and Technology, Carl E. Wieman of the University of Colorado at Boulder and Wolfgang Ketterle, professor of physics at the Massachusetts Institute of Technology, were awarded the 2001 Nobel Prize in physics for the discovery of the BEC. Not all atoms can be confined in a magnetic trap – for example atoms without an unpaired outer electron are not influenced by magnetic fields. Optical traps can be used to form condensates of

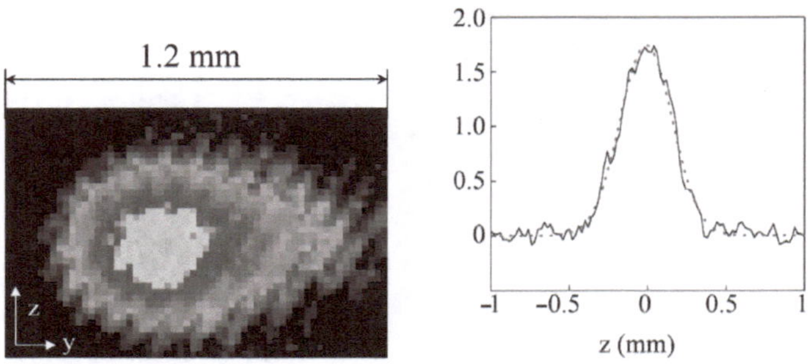

(a) (b)

Trapping of atoms A Bose–Einstein condensate in rubidium 87 achieved by Professor William van Wijngaarden, York University, Canada, using a so-called quadruple plus Ioffe coil (QUIC) trap. The image (a) was acquired for a final evaporation frequency of 2.400 MHz, and 18 msec after the magnetic fields were turned off. A probe laser passes through the cloud and is imaged with a CCD camera. Note from the line profile of optical density shown in (b) that the profile is nearly a perfect Gaussian fit indicating that a thermal cloud of cold atoms is present. The cloud corresponds to 3.4×10^6 atoms with corresponding temperature of 0.9 μK. (Courtesy of William van Wijngaarden, York University, Canada.)

nonmagnetic atoms, of molecules, and of several spin states of the same atom. Typically, the first step in most methods of generating a BEC is to reduce the temperature of the gas sufficiently and raise its density sufficiently to load it into the optical trap. This initial cooling is done in a magnetooptic trap (MOT) featuring a weak magnetic field and three orthogonal pairs of laser beams. Using this approach, an atom moving in any direction always has a velocity component pointed to at least one of the six lasers. These cooling lasers are tuned slightly below the atomic resonance, to ensure that the atoms move toward them and see the light as Doppler-shifted into resonance. Atoms absorb these oppositely directed photons and get a momentum kick that reduces their velocity. Typically the second step is evaporative cooling, performed in an optical dipole trap. The key to success in this approach may be attributed, in part, to the high atomic densities, typically exceeding 10^{14} atoms cm^{-3}, achieved in practice as atoms are loaded from the MOT into the optical dipole trap, before final evaporative cooling. See *atom trapping*.

Traveling wave semiconductor laser see *optical semiconductor amplifier*.

Triboluminescence see *luminescence*.

Tristimulus values are three fundamental "colors" (or variables), such as, red, green, and blue or, *X*, *Y* and *Z*, that define a three-dimensional color space (RGB or XYZ color space) that can be used represent the color of an object or the perception of almost any color. See *color*, *color matching functions*.

Trunk see *optical network*.

Tunabe–Sugano diagram see *crystal field*.

Tunable lasers emit a wavelength that can be adjusted at will, over a definite spectral range. The first requirement for tunability is an active medium with a broad enough spectral gain (such as organic dye, semiconductor or titanium:sapphire). The second condition lies with the resonance frequency of the resonator: it can be shifted by modifying the cavity length, or by replacing one reflector by a mobile diffraction grating, or by any suitable dispersive

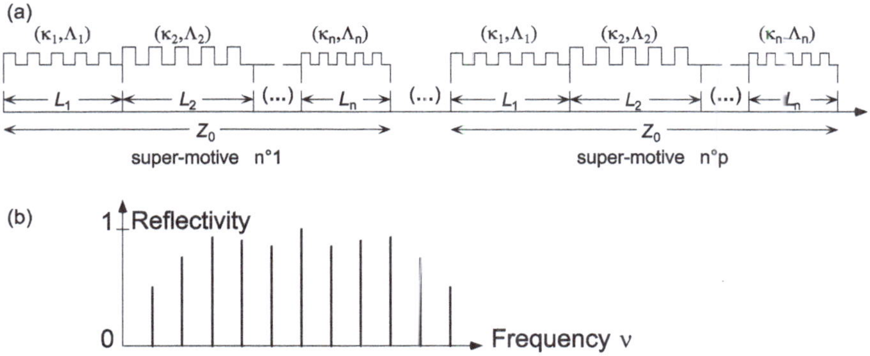

Tunable laser A super-structure grating (SSG) is a complex reflector where a "super-motive" made of different distributed feedback sections (with various pitches and/or coupling constants) is periodically repeated (a). It exhibits a "comb-like" reflectivity spectrum (b).

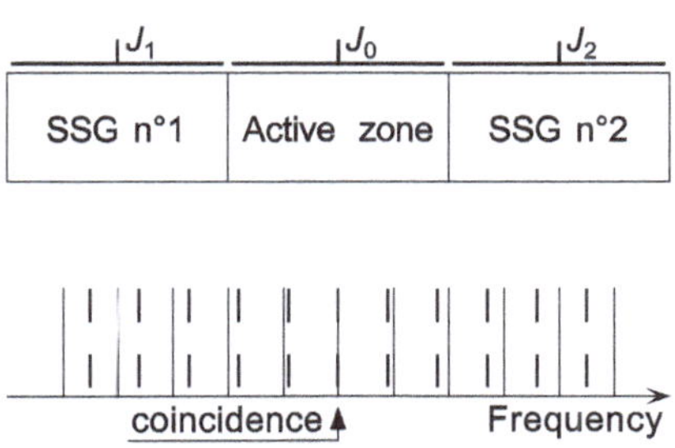

Tunable laser An active zone sandwiched between two homothetic SSG reflectors. The reflectivity spectra coincide at at most one peak (vernier effect), fixing the emission wavelength. Injection currents (J_1, J_2) can shift the spectra, thus the coincidence.

element that helps control the round-trip phase change (see *optical resonator*).

Tunability can be continuous or discrete, or both. In semiconductor lasers provided with super-structured gratings (SSG) (see the figure), broad tunability (over a few tens of nanometers) can be achieved thanks to the vernier effect.

The reflectivity spectrum of a SSG structure exhibits a comb-like series of equidistant narrow peaks, distributed over a few tens of nanometers. The whole SSG is controlled by an electrode that can shift the spectrum by injection of carriers. Consider now an active zone between two SSG reflectors, with three independent electrodes (current J_0 controls the optical gain, while J_1 and J_2 control the two comb-like reflectors). If the two spectra are not identical but homothetic, they coincide at most at one peak.

The variation of only one current will shift that coincidence (thus the laser wavelength) from that peak to the next (discrete tuning by vernier effect), while a correlated control of both J_1 and J_2 will shift both coincident peaks simultaneously (continuous tuning).

Tunability refers to the potential (or possibility), for a resonant system, to shift the spectral position of its resonance frequency, in the same manner that one can tune a radio receiver or a violin string. For instance, a passive Fabry–Perot filter (or an active laser cavity) exhibits well-defined peaks of transmission (or emission) at longitudinal mode frequencies depending upon its optical length. By translating a mirror with a piezoelectric displacement control, or by injecting carriers in a semiconducting zone (thus modifying its refractive index), the resonance condition is shifted. This is the basic

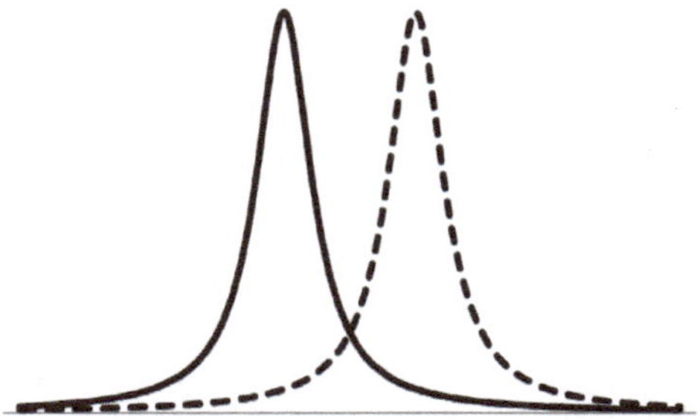

Tunability.

principle of a Fabry–Perot interferometer used as a device for scanning wavelengths.

Two-photon absorption (TPA) consists of the excitation of a system such as an atom, molecule or solid, from a lower quantum state to an excited state of the same parity as the lower state in a single step. In this case, the two states cannot be connected by an electric-dipole based transition, and thus parity conservation implies that the two light quanta must be absorbed simultaneously. The two photons need not necessarily have the same frequency, with neither photon being able to complete the transition alone – thus TPA is observed in a spectral range where the material is normally transparent. TPA can be considered to occur as a result of a first photon undergoing a transition to some intermediate virtual state with very short lifetime – only if the second photon arrives within the lifetime of this excited state can the transition to the final state be completed. Otherwise, the system relaxes to its ground state and no absorption occurs. Thus, to have an appreciable rate of TPA, high photon intensities are required to ensure a reasonable probability that both photons will be present during the lifetime of the virtual state – since the lifetime is so short, photon fluxes need to be very high. Thus, in TPA the absorbed power is proportional to the square of the incident radiation field intensity and is only observed at high excitation intensities. If I is the intensity of light passing through a medium, then the rate of decay of the light intensity can be written as

$$dI/dz = -\alpha I - \beta I^2$$

where α (cm^{-1}) is the usual one-photon or linear absorption coefficient, and β is the two-photon absorption coefficient with units of cm W^{-1}. In the same way as α is related to the imaginary part of the refractive index, β can be related to the imaginary part of the nonlinear refractive index n_2 by

$$\beta = 2k\mathrm{Im}[n_2]$$

or

$$\beta = \frac{3k}{2\varepsilon_0 cn^2}\mathrm{Im}[\chi^{(3)}]$$

where $k = 2\pi/\lambda$ is the wavevector of the propagating light in the medium, c is the speed of light, n is the refractive index, and $\chi^{(3)}$ is the third-order susceptibility. A propagating light wave subject to two-photon absorption (loss) due to the optical Kerr effect (i.e. from n_2), results in a change in the refractive index $\Delta n = n_2 I(z)$ and suffers a phase change $\Delta\Phi$ due to propagating over a distance from $z = 0$ to $z = \Delta z$ given by

$$\Delta\Phi = (2\pi k n_2/\beta)\ln[1 + \beta I_0 \Delta z]$$

where I_0 is the intensity at $z = 0$. Consider light

Table: Two photon absorption Table of TPA coefficients. PPV: Poly (p-phenylenevinylene); DANS (di-alkyl-amino-nitro-stilbene), sidechain polymer.

Material	Si crystal	Fused silica	Fuse silica + 3.5 mol%Ge	GaSb crystal	As_2S_3 glass	PPV	DANS
β, cm/GW (at λ)	0.45 (1.54 μm)	0.02 (264 nm)	0.42 (264 nm)	380 (3 μm)	0.03 (1.24 μm) 20 (0.62 μm)	3–50 (800 nm)	5.5 (920 nm)

incident on a sample of thickness L. The transmittance T in the presence of TPA is given by

$$\frac{1}{T} = \frac{\exp(\alpha L)}{(1-R)^2} + \beta I_0 \frac{\exp(\alpha L) - 1}{\alpha(1-R)}$$

where R is the reflectance, which depends on the refractive index of the sample. Moreover, α and β depend on the photon energy; for example, for glasses $\beta \propto \exp(\hbar\omega/E_\beta)$, where E_β is a constant. Typical values for β range from 10^{-2} to 100 cm/GW. TPA is an important tool in laser spectroscopy, as it enables transitions between states unallowed through an electric-dipole interaction. Although such transitions can be studied using resonant one-photon processes through an intermediate level, in that case the measured linewidth of the process is increased by the linewidths of the two successive one-photon absorption processes. TPA also permits coherent excitation of molecules with energies in the far-UV to be made using visible radiation, for which high-intensity coherent sources are readily available. The importance of TPA may be gauged by using a figure of merit (FOM) for TPA, defined by,

$$\mathrm{FOM} = 2\beta\lambda/n_2.$$

(K. Tanaka, *Appl. Phys. Letts.*, **80**, 177, 2002.) See *optical Kerr effect*.

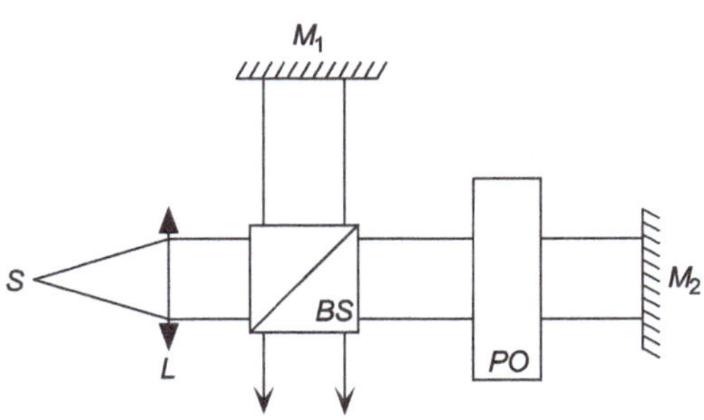

Twyman–Green interferometer Modified Michelson configuration with a monochromatic point source, S, at the focal point of a convergent lens, L, used to analyze the phase distortion introduced by a transparent phase object (PO) inserted into one arm after the beamsplitter (BS).

John Tyndall (1820–1893), who was able to explain the blue color of the sky in terms of light scattering, known as the Tyndall effect. (Courtesy of AIP Emilio Segre Visual Archives, Zeleny Collection.)

Twyman–Green interferometer is an adaptation of the Michelson scheme where a transparent object is inserted into one of the arms. By analyzing the interference fringes, one can characterize the phase distribution of the object.

Tyndall scattering is named after John Tyndall (1820–1893) and refers to the first explanation for the blue color of the sky, which was later refined by Rayleigh. The blue color originates from the scattering of sunlight from molecules in the atmosphere. The scattering is more effective at short wavelengths (i.e. the blue rather than red end of the *visible spectrum*). A quantitative relationship between the intensity of the light scattering, I, and the light wavelength, λ, was derived by Rayleigh and is commonly known as *Rayleigh's law of scattering*. The law gives $I \propto N(n_r - 1)^2/\lambda^4$, where N is the concentration of scattering centers per unit volume and n_r is the refractive index. This relation is applicable to scattering centers with size of about $\lambda/10$. The scattering intensity is strongly dependent on the wavelength of the light and is proportional to λ^{-4}. Because for a given size of scattering center, it presents a more severe obstacle to shorter wavelength light as such centers correspond to a larger fraction of the wavelength. Since red light (720 nm) has a wavelength 1.8 times that of blue light (400 nm), the law predicts $(1.8)^4$ or about 10 times greater scattering of the blue light, assuming scattering centers with size 40 nm or less. This effect is often used as a measure of the existence of a colloid. The effect is measurable in colloids with concentrations as low as 0.1 ppm (parts per million). It should be noted that both Tyndall and Rayleigh thought that the blue color of the sky was due to either small particles of dust or droplets of water vapor in the atmosphere. Even today, this is still a common misconception. It was soon realized that if this were true, there would be much stronger variations in the color of the sky with changes in humidity than are actually observed; thus it was realized that molecules of oxygen and nitrogen in the air can correctly account for the observed light scattering. This issue was finally settled by Einstein in 1911, who calculated a detailed formula for light scattering from molecules in close agreement with experiment. Einstein was even able to use these calculations to further verify Avogadro's number.

U

Ulbricht sphere see *integrating sphere*.

Ultraviolet (UV) radiation is an electromagnetic wave with a wavelength typically in the range 400 nm–4 nm, shorter than violet light but longer than x-rays. When the wavelength is in the 400–300 nm range it is called near-UV and in the 300–200 nm range it called far-UV. Ultraviolet radiation with a wavelength shorter than 200 nm is known as extreme or vacuum UV.

Uncertainty principle states that the uncertainty, Δx, in the position of a particle and the uncertainty, Δp_x, in its momentum in the x-direction obey $(\Delta x)(\Delta p_x) > \hbar$. This is a consequence of the wave nature of matter and has nothing to do with the precision of measurement. If ΔE is the uncertainty in the energy of a particle during a time, Δt, then according to the uncertainty principle, $(\Delta E)(\Delta t) > \hbar$. To measure the energy of a particle without any uncertainty means that we will need an infinitely long time, $\Delta t \to \infty$.

Uniaxial crystals see *anisotropy*.

Up-conversion laser involves a mechanism whereby the active medium emits radiation at a wavelength λ_L shorter than the pump wavelength λ_P. This is because the upper level is not excited directly, but through intermediary levels. Consider for instance the simplified energetic diagram of erbium, Er^{3+}, emitting around $\lambda_L \approx 523$ nm, with a double pumping scheme at $\lambda_{P1} \approx 840$ nm and $\lambda_{P2} \approx 1$ μm, as shown in the figure.

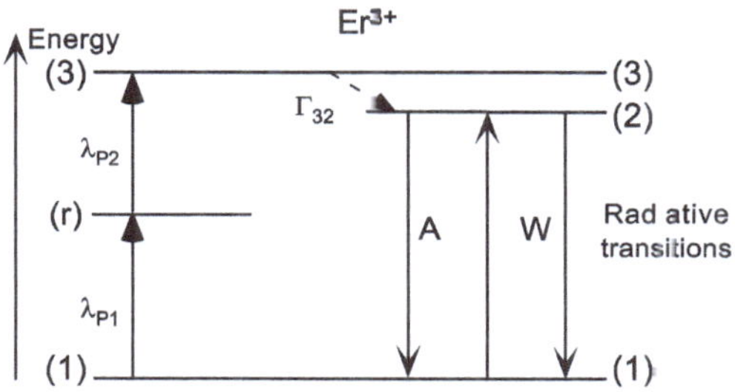

Up-conversion laser A first pump beam at $\lambda_{P1} \approx 840$ nm (1.48 eV) excites atoms from the lower level $^4I_{15/2}$ (1) into the intermediary level $^4I_{9/2}$ (r); a second pump beam at $\lambda_{P2} \approx 1$ μm (1.24 eV) excites metastable levels (3), which rapidly relax to the upper level $^2H_{11/2}$ (2). The laser emission takes place between levels (2) and (1) at $\lambda_L \approx 523$ nm (2.37 eV).

Other mechanisms are possible. Consider now the Tm^{3+} system as schematically depicted in the second figure, emitting around $\lambda_L \approx 455$ nm, with a main pumping at $\lambda_{P1} \approx 645$ nm and an optional additional pumping at $\lambda_{P2} \approx 1.064$ μm.

Urbach (or Urbach–Martienssen) rule describes the exponential dependence of the absorption coefficient α on the photon energy $\hbar\omega$ as observed for a diverse spectrum of materials in terms of a simple expression of the form

$$\alpha = A\exp[(\hbar\omega - E_1)/\Delta E]$$

where $\hbar\omega$ is the photon energy, A is a constant, E_1 is the photon energy when $\alpha = A$ (usually,

Urbach (or Urbach–Martienssen) rule

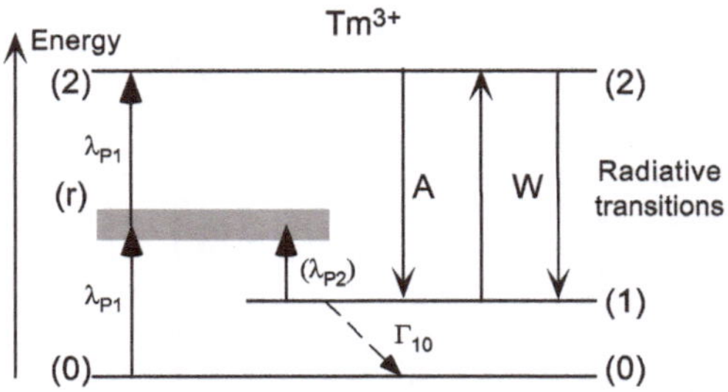

Up-conversion laser A pump beam at $\lambda_{P1} \approx 645$ nm (1.92 eV) excites atoms from the ground state (0) into an intermediary continuum of levels (r), and simultaneously excites atoms from that very continuum (r) into the upper level (2). The laser transition takes place between levels (2) and (1), around $\lambda_L \approx 455$ nm (2.73 eV). In order to improve the overall efficiency, a second (optional) pump beam at $\lambda_{P2} \approx 1.064$ mm (1.17 eV) can help recycle atoms from lower level (1) into the continuum (r), thus enhancing the pumping and maintaining a higher population inversion.

$A = 10^3\ \text{cm}^{-1}$) and $\Delta E = 1/\beta$ is the *Urbach absorption tail width*, or simply the *Urbach width*, and β is the Urbach parameter. Usually the Urbach rule is obeyed over a restricted range of photon energies that typically correspond to $1 < \alpha < 10^4\ \text{cm}^{-1}$, i.e. $\hbar\omega$ below fundamental absorption, e.g. for a-Si:H typically for α below $5\times10^3\ \text{cm}^{-1}$. In addition, at sufficiently low photon energies other absorption processes such as absorption due to defects begin to dominate over the Urbach absorption process. For many materials, ΔE increases with temperature; it is usually expressed as $\Delta E = k_B T/\sigma$ where σ is the *Urbach tail steepness* parameter. There are various theories that account for the ΔE-T dependence. For crystalline materials, $\sigma = \sigma_0(2k_B T/h\upsilon_p)\tanh(h\upsilon_p/2k_B T)$ in which σ_0 is a constant, and $h\upsilon_p$ is a phonon energy. On the other hand, for noncrystalline materials it has been suggested that

$$\Delta E = A + B[\exp(\theta_E/T) - 1]$$

where A and B are constants related to structural and thermal disorder and θ_E is the Einstein temperature.

While the origin of the Urbach rule is still debated, its wide applicability to many classes of

Table: Urbach rule Approximate Urbach tail energy widths ΔE at room temperature, and approximate optical gap; pc: polycrystalline.

Material	**a-Si:H**	As_2Se_3 **glass**	As_2Se_3 **crystal**	GeS_2 **glass**	$Ga_x(Ge_{30}Se_{70})_{100-x}$ **glass** ($x < 0.1$)	**Alkali halides**	**ZnO pc-film**
ΔE (meV)	50	50	28	130	70–100	$k_B T/0.8$	84 (as deposited) 44 (annealed)
E_g (eV)	1.8–1.9	1.9	1.9	3.2	2.1–2.3		3.2
Material	As_2S_3 glass	**a-GaAs**	**GaN film**	$GeSe_2$ **glass**	**Tetrahedral a-C**	**5%**WO_3 – – **95%**Pb_3O_4 **glass**	In_2O_3 **pc-film**
ΔE (meV)	67	120	400	82	600	90	400
E_g (eV)	2.39	1.1	3.5	2.24	3	2.3	3.6
Material	**PVC**	TeO_2-TiO_2 **film**	$CuGa_3Se_5$	Cu_7GeS_5I	**10%**Pb_3O_4– **90%**P_2O_5 **glass**	**40%**Pb_3O_4– **60%**P_2O_5 **glass**	**10%**Na_2O-**45%PbO-45%**P_2O_5 **glass**
ΔE (meV)	600	270	40–45	35	440	400	138
E_g (eV)	3.29	3.29	2.2	2.37	4.81	3.21	4.37

materials has proved to be useful in the comparative characterization of optical absorption. For many amorphous semiconductors, ΔE has been related to the width of the valence (or conduction) band tail states, and ΔE is typically 20–100 meV at room temperature for various classes of materials. ΔE also increases with disorder and depends on the preparation conditions, post-treatment such as annealing, etc. For a given class of materials, ΔE increases as the optical gap increases.

(F. Urbach, *Phys. Rev.*, **92**, 1324, 1953; H. W. Martienssen, *J. Phys. Chem.*, **2**, 257, 1957.)

UV grade fused silica see *silica*.

V

***V*-coating** see *antireflection coating*.

***V*-number** or **normalized frequency** is a dimensionless quantity that is a characteristic of a dielectric waveguide which determines the nature of propagation of EM (electromagnetic) waves along the guide. For a *step index fiber*, the transmission medium in optical communications, it is defined by

$$V = \frac{2\pi a}{\lambda}\left(n_1^2 - n_2^2\right)^{1/2}$$

where a is the core radius, λ is the free-space wavelength of the radiation to be guided and n_1 and n_2 are the refractive indices of the core and cladding, respectively. If the V-number is less than 2.405, then only the fundamental mode, the lowest mode, can propagate in a step index fiber, and the fiber is called a *single-mode fiber*. The propagation constant β depends on the V-number, which itself depends on wavelength λ. For an input light pulse that has a finite spectrum, the different wavelengths will propagate with different propagation constants, and hence lead to the spread of the light pulse, its dispersion. This type of dispersion is called *waveguide dispersion*. See *normalized propagation constant, waveguide dispersion*.

Table: *V*-number The V-number expressions for step index optical fibers.

V-number	
$V = (2\pi a/\lambda)\left(n_1^2 - n_2^2\right)^{1/2}$	λ = free-space wavelength; a = fiber core radius; n_1 and n_2 are the refractive indices of the core and cladding respectively
$V = ka\left(n_1^2 - n_2^2\right)^{1/2}$	$k = 2\pi/\lambda$, free-space wavenumber or wavevector
$V = (2\pi a/\lambda)(\text{NA}) = ka(\text{NA})$	NA = numerical aperture = $\left(n_1^2 - n_2^2\right)^{1/2}$
$V = (2\pi a n_1/\lambda)(2\Delta)^{1/2} = kan_1(2\Delta)^{1/2}$	Δ = normalized refractive index difference $\ll 1$ $\Delta = (n_1 - n_2)/n_1 \approx \left(n_1^2 - n_2^2\right)/2n_1^2$

V-alue of an optical medium see *Abbe number*.

Vacancy is a point defect in a crystal where a normally occupied lattice site has a missing atom.

Vacuum level is the energy level where the potential energy of the electron and the kinetic energy of the electron are both zero. It defines the energy level where the electron is just free from the solid. Typically at a distance of ~100 nm, the electron would be considered free from a solid.

Valence band (VB) normally refers to a band of energies for the electrons in bonds in a semiconductor. The valence band is made of all those states (wavefunctions) that constitute the bonding between the atoms in the crystal. At absolute zero of temperature the valence band is full of all the bonding electrons of the atoms. When an electron is excited to the conduction band, then this leaves behind an empty state which is called a hole. It carries positive charge and behaves as it were a "free" positively charged entity with an effective mass of $m_h{}^*$. It moves around the VB by having a neighboring electron tunnel into the unoccupied state.

Van Roosbroeck–Shockley relation is an expression derived by the consideration of detailed balance in a semiconductor in terms of the rates of absorption and spontaneous emission of photons. If the absorption rate for transitions between valence band and conduction band states, per photon of frequency ν, is $P_{\mathrm{vc}}(\nu)$, and the photon emission rate for the transition from conduction to valence band is $R_{\mathrm{cv}}(\nu)$, using the definition of the photon density ρ in the interval $d\nu$ about ν as

$$\rho(\nu)\frac{8\pi n_{\mathrm{r}}^3\nu^2}{c^3[e^{h\nu/k_{\mathrm{B}}T}-1]},$$

The van Roosbroeck–Shockley relation can be written as:

$$P_{\mathrm{cv}}(\nu)\rho(\nu) = R_{\mathrm{cv}}(\nu) \qquad (1)$$

where n_{r} refers to the refractive index for light in the material, h is Planck's constant, k_{B} Boltzmann's constant, T the absolute temperature and c the speed of light in vacuum. Using the definition for $\rho(\nu)$ and $P_{\mathrm{cv}} = \alpha(\nu)c/n_{\mathrm{r}}$, where $\alpha(\nu)$ is the absorption coefficient for light of frequency ν in the material, in Equation (1), gives the more usual form of the van Roosbroeck–Shockley relation:

$$R_{\mathrm{cv}}(\nu) = \frac{\alpha(\nu)8\pi n_{\mathrm{r}}^3\nu^2}{c^2}\cdot\frac{1}{e^{h\nu/k_{\mathrm{B}}T-1}}. \qquad (2)$$

The form of the equation relates the spontaneous emission probability to the absorption coefficient at a given frequency. Thus, one can calculate the photoluminescence spectrum from the measured absorption coefficient using this approach as shown in the figure. (W. van Roosbroeck and W. Shockley, *Phys. Rev.*, **94**, 1558, 1954.)

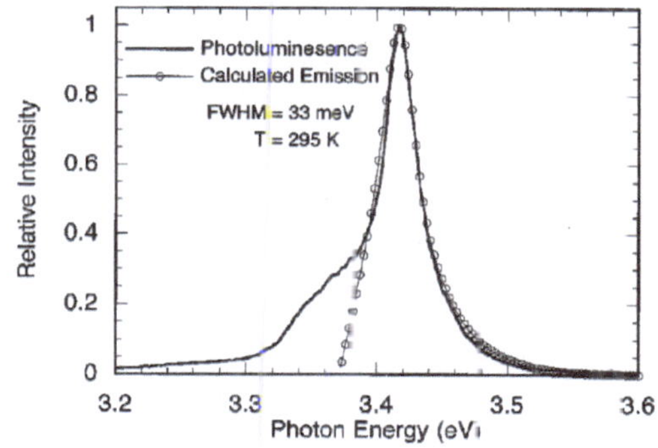

Comparison of measured photoluminescence spectrum (solid line) with spectrum calculated from the absorption coefficient using the **van Roosbroeck–Shockley relation** (open circles) for an epitaxial GaN layer. (From J. F. Muth, J. H. Lee, I. K. Shmagin, R. M. Kolbas, H. C. Kasey, B. P. Keller, U. K. Mishra and S. P. DenBaars, *Appl. Phys. Lett.*, **71**, 2572, 1997, (Figure 3).)

Varshni equation describes the change in the energy bandgap E_g of a semiconductor with temperature T in terms of

$$E_g = E_{g0} - \frac{AT^2}{B+T}$$

where E_{g0} is the bandgap at $T = 0$ K, and A and B are material-specific constants. For example, for GaAs, $E_{g0} = 1.519$ eV, $A = 5.405\times10^{-4}$ eV K^{-1}, $B = 204$ K, so that at $T = 300$ K, $E_g = 1.42$ eV. We can find the temperature coefficient of bandgap E_g' by differentiating the Varshni equation,

$$E_g' = \frac{dE_g}{dT} = -\frac{AT(T+2B)}{(B+T)^2}$$

which for GaAs, using the values above, gives $E_g{}' = -0.45$ meV K^{-1}. dE_g/dT can be used to calculate the shift in the peak emission wavelength of an LED with temperature or the cut-off wavelength of a detector.

Vectoral changes see *photoinduced vectoral change*.

Verdet constant see *Faraday effect*.

Table: Varshni equation

Semiconductor	E_{g0}, eV	A, eV K^{-1}	B, K	E_g (300K), eV	dE_g/dT (300K), meV K^{-1}
InAs (D)	1.519	5.405×10^{-4}	204	1.42	−0.45
InAs (D)	0.426	3.158×10^{-4}	93	0.354	−0.30
InP (D)	1.42	4.906×10^{-4}	327	1.35 eV	−0.36
AlAs (D)	2.239	6.0×10^{-4}	408	2.163	−0.40
GaP (I)	2.338	5.771×10^{-4}	372	2.261	−0.40
Si (I)	1.17	4.73×10^{-4}	636	1.12	−0.25
Ge (I)	0.744	4.56×10^{-4}	210	0.660	−0.25

Note: E_{g0}, A and B are combined from various sources. (Y. P. Varshni, *Physica*, **34**, 149, 1967.)

Vergence, as in *convergence* or *divergence*, represents the instantaneous *radius of curvature* of the wavefront of light that is either moving away from its source O (e.g. a point on the object), or moving towards an image point, I, or a focal point. If L is the distance from a point P on the wavefront to the source O, or the image point I, and the light is propagating in a medium of refractive index n, then, by definition, $V = n/L$, measured in diopters (D or m^{-1}). V is negative if P is moving away from O, it is positive if P is moving towards I, and L is increasing or decreasing, respectively. There is an alternative convention. Take x to be the direction of propagation. If PO is the vector from P to O, and if PO is in the negative x-direction, V is negative, as in divergent. If the vector PI is in the positive x-direction, V is positive, as in convergent. The change in vergence from V_1 to V_2 in traveling a distance l is,

$$V_2 = \frac{V_1}{1 - V_1(l/n)}.$$

Optical surfaces and lenses modify the light's wavefront and hence its vergence. When light is incident on a curved surface and suffers refraction going from one medium into another, there is a change in its vergence, given by,

$$V_2 = V_1 + P_s$$

where V_1 is the *entrance vergence*, V_2 is the *exit vergence*, P_s is the *surface power* (in changing light's vergence). For a spherical surface and for paraxial rays (near the optic axis OI) P_s is given by

$$P_s = (n_2 - n_1)/R$$

where R is the radius of curvature of the spherical surface whose center is at C. The surface power of the human eye is about +60 D.

Vertical cavity surface emitting lasers (VCSELs) have the optical cavity axis along the direction of current flow rather than perpendicular to the current flow as in conventional laser diodes.

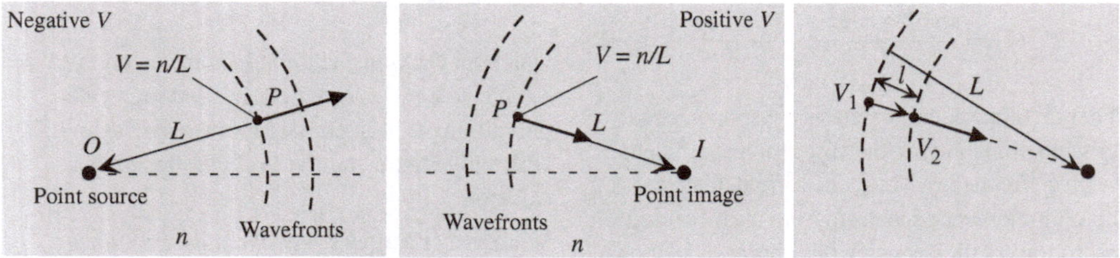

Definition of **vergence** for $V < 0$, and $V > 0$, and change in vergence.

 Vertical cavity surface emitting lasers (VCSELs)

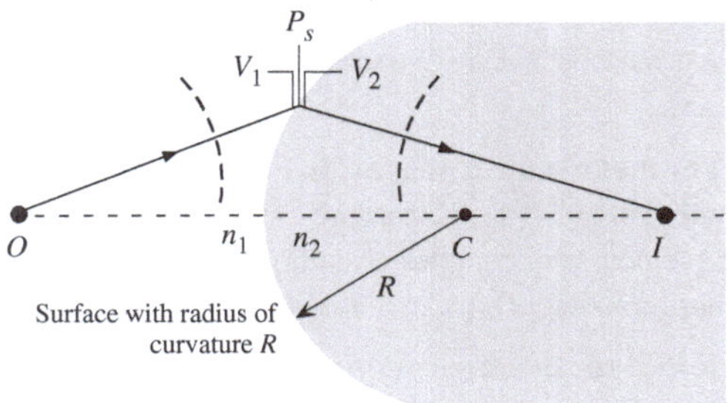

Change in **vergence** on refraction at a spherical surface of radius R and center at C. O is the object and I is the image.

The active region length is very short compared with the lateral dimensions so that the radiation emerges from the "surface" of the cavity rather than from its edge. The reflectors at the ends of the cavity are dielectric mirrors made from alternating high and low refractive index quarter-wave thick multilayers. Such dielectric mirrors provide a high degree of wavelength selective reflectance at the required free-surface wavelength λ if the thicknesses of alternating layers d_1 and d_2 with refractive indices n_1 and n_2 are such that

$$n_1 d_1 + n_2 d_2 = \tfrac{1}{2}\lambda$$

which then leads to the *constructive interference* of all partially reflected waves at the interfaces. Since the wave is reflected because of a periodic variation in the refractive index as in a grating, the dielectric mirror is essentially a *distributed Bragg reflector* (DBR). High reflectance end mirrors are needed because the short cavity length L reduces the optical gain of the active layer inasmuch as the optical gain is proportional to $\exp(gL)$ where g is the optical gain coefficient. There may be 20–30 or so layers in dielectric mirrors to obtain the required reflectance (~99%). The whole optical cavity looks "vertical" if we keep the current flow the same as in a conventional laser diode cavity. The active layer is generally very thin (< 0.1 μm) and is likely to

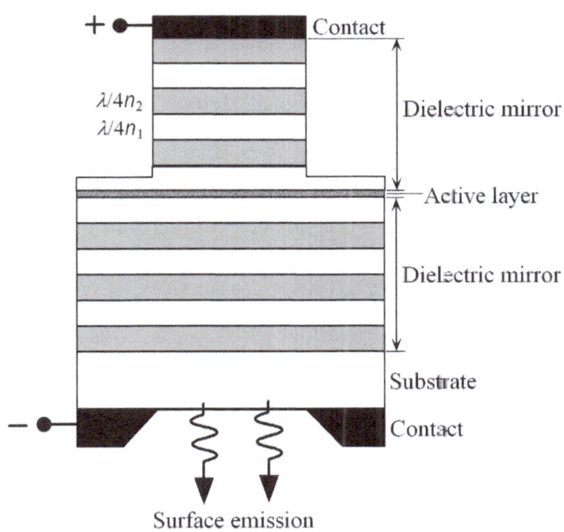

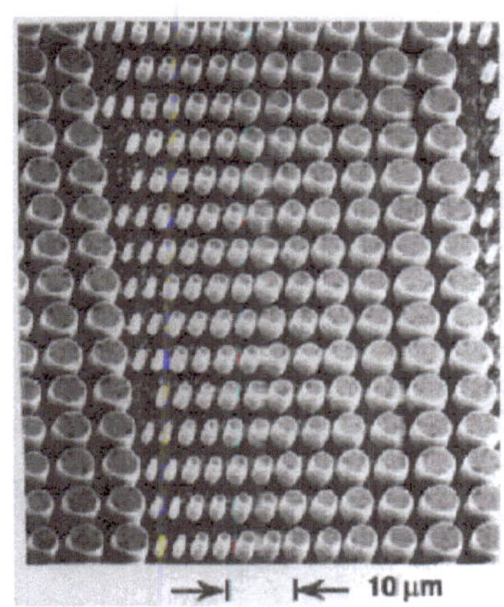

A simplified schematic illustration of a **vertical cavity surface emitting laser** (VCSEL). SEM (scanning electron microscope) of the first low-threshold VCSELs developed at Bell Laboratories in 1989. The largest device area is 5 μm in diameter (Courtesy of Axel Scherer, CalTech.)

be a *multiple quantum well* (MQW) for improved threshold current. The required semiconductor layers are grown by epitaxial growth on a suitable substrate which is transparent in the emission wavelength. For example, a 980 nm emitting VCSEL device has InGaAs as the active layer to provide the 980 nm emission, and a GaAs crystal is used as substrate which is transparent at 980 nm. The dielectric mirrors are then alternating layers of AlGaAs with different compositions and hence different bandgaps and refractive indices. The top dielectric mirror is etched after all the layers have been epitaxially grown on the GaAs substrate. In practice, the current flowing through the dielectric mirrors give rise to an undesirable voltage drop and methods are used to feed the current into the active region more directly, for example, by depositing "peripheral" contacts close to the active region. There are presently various sophisticated VCSEL structures. The vertical cavity is generally circular in its cross-section so that the emitted beam has a circular cross-section, which is an advantage. The height of the vertical cavity may be as small as several microns. Therefore the longitudinal mode separation is sufficiently large to allow only one longitudinal mode to operate. However, there may be one or more lateral (transverse) modes depending on the lateral size of the cavity. In practice there is only one single lateral mode (and hence one mode) in the output spectrum for cavity diameters less than ~8 μm. Various VCSELs in the market have several lateral modes but the spectral width is still only ~0.5 nm, substantially less than a conventional longitudinal multimode laser diode. With cavity dimensions in the microns range, such a laser is referred to as a *microlaser*. One of the most significant advantages of microlasers is that they can be arrayed to construct a *matrix emitter*, that is a broad area surface emitting laser source. Such laser arrays have important potential applications in optical interconnect and optical computing technologies. Further, such laser arrays can provide a higher optical power than that available from a single conventional laser diode. Powers reaching a few watts have been demonstrated using such matrix lasers.

Very large scale integration (VLSI) is the integration of 10^6–10^7 devices into a single Si chip to implement very complex circuits (for example, microprocessor chips, memories, etc.).

Vicinal surfaces see *self-assembly*.

Virtual image see *image*.

Visibility function see *luminosity*.

Visual acuity qualitatively is the sharpness of vision, and is normally tested on the Snellen chart. Put differently, it is the eye's ability to resolve fine details or the highest spatial frequency it can resolve. See *eye*.

Voigt effect see *Cotton–Mouton effect*.

Voigt lineshape profile is a lineshape function that is used in spectroscopy to describe the emission spectrum that has been broadened by two mechanisms, one due to homogeneous broadening (Lorentzian), and the other due to inhomogeneous broadening (Doppler broadening). For example, in a gas discharge, "foreign" atoms mixed into the gas (foreign to light emitting atoms) can act as perturbers and give rise to collision broadening, in addition to Doppler broadening. The Voigt lineshape function $V(\nu)$ is a probability density function, $\int V(\nu)d\nu = 1$, that is a *convolution* of the Lorentzian $L(\nu)$ and Gaussian $G(\nu)$ lineshapes so that

$$V(\nu) = \int_{-\infty}^{\infty} G(\nu')L(\nu - \nu')d\nu.$$

The Lorentzian and the Gaussian will each have their own widths and center frequencies. The actual expression for $V(\nu)$ is quite complicated. Various approximations have been developed for $V(\nu)$, one popular being the Whiting equation. A number of equations have been suggested for the width w_V of

Table: Voigt lineshape profile Approximations and semiempirical equations for the Voigt lineshape width in terms of Gaussian and Lorentzian linewidths.

R, d, a and W	$R = w_V/(w_L + w_D)$; $d = (w_L - w_D)/(w_L + w_D)$;
	$a = (\ln 2)^{1/2}(w_L/w_G)$; $W = (\ln 2)^{1/2}(w_V/w_G)$
Danos–Geshwind	$w_V^2 = w_G^2 + w_L^2$
Posener	$R^2 = \dfrac{(1-d)^2 + \ln 2(1+d)^2}{2\ln 2}$
Van de Hulst–Reesinck–Whiting	$w_V = \dfrac{w_L}{2} + \sqrt{\left(\dfrac{w_L^2}{4}\right) + w_G^2}$
Olivero–Longbothum	$W = 0.5346a + [0.2166a^2 + \ln(2)]^{1/2}$

the Voigt spectrum $V(\nu)$ in terms of the widths w_G and w_L of the Gaussian $G(\nu)$ and Lorentzian $L(\nu)$ constituents as summarized in the table. (E. E. Whiting, *J. Quant. Spect. Rad. Transf.*, **8**, 1379, 1968. J. J. Olivero and R. L. Longbothum, **17**, 233, 1977; R. J. Leiweke, **103**, 597, 2007.)

W

Waist radius see *Gaussian beam.*

Wang's rule is an empirical relation between the first- and third-order electric susceptibilities, $\chi^{(1)}$ and $\chi^{(3)}$, respectively, in isotropic media such as cubic crystals, glasses, and liquids, given in esu units as

$$\chi^{(3)} = \frac{1.2[\chi^{(1)}]^2}{N_{\text{eff}}\hbar\omega_o}$$

where N_{eff} is an oscillator strength, and ω_o is an average absorption frequency, and $\chi^{(3)}$ is at low frequencies (below ω_o). (C. Wang, *Phys. Rev. B*, **2**, 2045, 1970.)

Wannier–Mott exciton refers to excitons with radius far greater than the lattice constant of the material. This type of exciton was named after Sir Nevill Mott and Gregory Wannier. In the case of semiconductors, owing to the relatively high dielectric constant, screening tends to weaken the Coulomb interaction between the electron and hole in the exciton resulting in large exciton radii, or the formation of Wannier–Mott excitons. As a result, the effect of the lattice potential can be incorporated into the effective masses of both the electron and hole. See also *exciton.*

Wave is a periodically occurring disturbance, such as the displacements of atoms from equilibrium positions in a solid carrying sound waves, or a periodic variation in a measurable quantity, such as the electric field $E(x,t)$ in a medium or space. In a traveling wave, energy is transferred from one location to another by the oscillations. For example, $E_y(x,t) = E_{yo}\sin(kx - \omega t)$, where $k = 2\pi/\lambda$ and $\omega = 2\pi\upsilon$, is a traveling wave in the x direction. The electric field in the y-direction varies periodically along the x with a period λ called the wavelength and with time with a period $1/\upsilon$ where υ is the frequency. The wave propagates along the x-direction with a velocity of propagation, c, in vacuum. In a longitudinal wave, the displacement is in the direction of propagation as in the propagation of a sound wave, whereas in a transverse wave, the displacement is at right angles to the direction of propagation as in an electromagnetic wave. Electromagnetic waves are transverse waves in which the electric and magnetic fields $E_y(x,t)$ and $H_z(x,t)$ are at right angles to each other and also the direction of propagation, x. A traveling wave in the electric field must be accompanied by a similar traveling wave in the magnetic field $H_z(x,t) = H_{zo}\sin(kx - \omega t)$. Typical wave-like properties are *interference* and *diffraction*. See *electromagnetic wave.*

Wave equation is a second-order partial differential equation in space and time coordinates which is usually of the form

$$\frac{d^2u}{dx^2} + \frac{d^2u}{dy^2} + \frac{d^2u}{dz^2} = \frac{1}{v}\frac{d^2u}{dt^2}$$

whose solution describes the space and time dependence of a physical quantity $u = u(x, y, z, t)$ such as displacement, pressure, potential, field, etc.

from equilibrium or zero, given the boundary conditions applicable to the phenomena under consideration. The parameter v in the wave equation is the propagation velocity of the wave. The above wave equation is for a wave in a homogeneous, lossless and linear medium where the cause and the effect are linearly and simply related. In the case of electromagnetic waves in vacuum, the wave equation describes the variation of the electric (or magnetic) field E(or H) with space and time, e.g. for propagation along x, $d^2E/dx^2 = (1/c^2)d^2E/dt^2$ where $c = 1/(\mu_o\varepsilon_o)^{1/2}$ is the speed of light. The general solution is composed of two oppositely traveling waves; in one dimension along x, $u(x, t) = f_1(x - vt) + f_2(x + vt)$, where $f_1(x - vt)$ is a wave traveling along $+x$ and $f_2(x + vt)$ along $-x$. The functional forms and amplitudes of f_1 and f_2 depend on the boundary and initial conditions; if $f_2 = 0$, there is only one wave f_1 propagating along $+x$.

Wavefront is a surface where all the points associated with a wave have the same phase. A wavefront on a plane wave is an infinite plane perpendicular to the direction of propagation. For example, the wavefront of a plane electromagnetic field propagating along z is an infinite xy plane perpendicular to z, i.e. to the direction of propagation; the electric (or the magnetic) field in this xy plane has the same value everywhere at any given instant. See *plane wave*.

Wavefront reconstruction see *holography*.

Wavefunction, $\Psi(x,y,z,t)$, is a probability based function that is used to describe the wave-like properties of a particle. It is obtained by solving the Schrödinger equation which in turn requires knowledge of the potential energy of the particle and the boundary and initial conditions. $|\Psi(x,y,z,t)|^2$ is the probability per unit volume of finding the electron at (x,y,z) at time t. Put differently, $|\Psi(x,y,z,t)|^2dxdydz$ is the probability of finding the electron in the small volume $dxdydz$ at (x,y,z) at time t. Under steady state conditions, the wavefunction can be separated into space dependent, $\psi(x,y,z)$, and time-dependent components as $\Psi(x,y,z,t) = \psi(x,y,z)\exp(-jEt/\hbar)$ where E is the energy of the particle and $\hbar = h/(2\pi)$. The spatial part, $\psi(x,y,z)$, satisfies the time-independent Schrödinger equation.

Waveguide condition specifies the condition which an EM (electromagnetic) wave must satisfy to be propagated and hence be guided through a waveguide given the V-number of the waveguide. Each EM wave satisfying the waveguide condition is a *mode* of the waveguide as it can be propagated. The waveguide condition is a mathematical relationship for determining the allowed modes, particular EM radiation patterns that can propagate in a waveguide from the waveguide properties (thickness, refractive indices, etc.). See *optical fiber*.

Waveguide dispersion, as separate from material dispersion, is dispersion that arises because of the wavelength dependence of the V-number which determines the propagation constant inside the guide. As the radiation fed into the core has a finite range of wavelengths, $\Delta\lambda \neq 0$, the V-number is not constant which leads to fundamental modes with different wavelengths which propagate at different velocities. Waveguide dispersion results from the guiding properties of the dielectric structure and it has nothing to do with the frequency (or wavelength) dependence of the refractive index. Since increasing the wavelength decreases the V-number, a characteristic property of the guide, the dispersion can also be stated as being due to the wavelength dependence of the V-number.

Waveguide dispersion coefficient, D_w, is the time spread of a propagating optical pulse in an optical guide per unit length and per unit spectral wavelength width due to the wavelength dependence of the V-number, a waveguide property. If $\delta\lambda$ is the spread of the excitation wavelengths coupled into the fiber, $\delta\tau$ is the spread in propagation times of these different wavelengths due to waveguide dispersion, and L is the fiber length, then,

$$D_w = \frac{1}{L}\frac{\delta\tau}{\delta\lambda}.$$

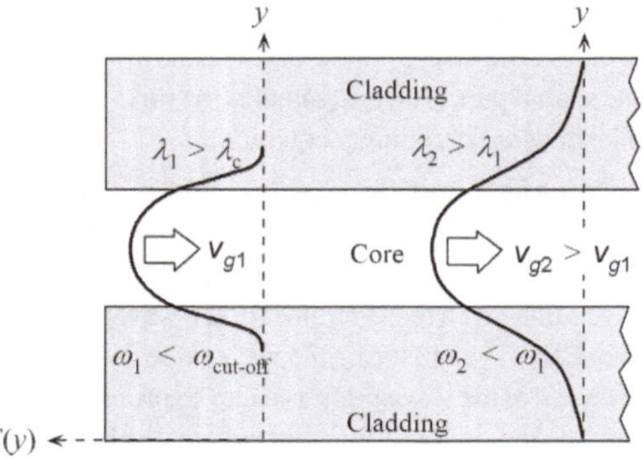

Waveguide dispersion The electric field of the TE_0 mode extends more into the cladding as the wavelength increases. As more of the field is carried by the cladding, the group velocity increases.

For a single-mode step index fiber, and for V-numbers in the range $1.5 < V < 2.4$,

$$D_w \approx -\frac{1.984 N_{g2}\lambda}{(2\pi a)^2 2cn_2^2}$$

where D_w is the waveguide dispersion coefficient in s m^{-2}, N_{g2} and n_2 are the group and refractive indices of the cladding, a is the core radius (m), λ is the free-space wavelength (m) and c (m s^{-1}) is the light velocity. Taking $N_{g2}/n_2^2 \approx 1/n_2$, the above expression can be simplified to

$$D_w \approx -\frac{0.025\lambda}{a^2 cn_2}$$

where D_w is in s m^{-2}, and to convert to ps nm^{-1} km^{-1}, multiply D_w by 10^6. If λ and a are both in μm, substituting for c as well,

$$D_w(\text{ps nm}^{-1}\text{km}^{-1}) \approx -\frac{83.76\lambda(\mu\text{m})}{[a(\mu\text{m})]^2 n_2}.$$

Wavelength add/drop multiplexer (WADM) or **optical add/drop multiplexer (OADM)** is a multiplexer that is able to selectively take out, that is remove, a signal at a specified wavelength such as λ_1 from wavelength division multiplexed signals propagating along a fiber, and then add new or different information at the same wavelength λ_1' into the fiber, in the same direction.

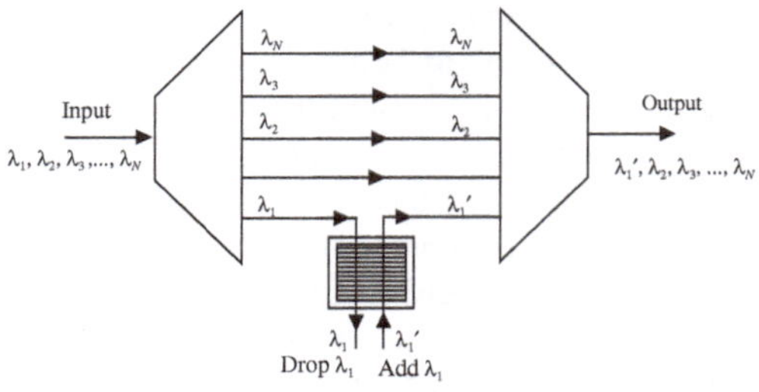

Wavelength add/drop multiplexer, **WADM**, or **OADM** schematic diagram.

Wavelength division multiplexing (WDM) is a method of transmitting information by breaking it into sets of information and then transmitting the sets at different wavelengths through the transmission medium, usually an optical fiber; it is the same as *frequency division multiplexing*. Thus, the communication system operates at more than one wavelength, and a specific carrier wavelength is assigned to each signal. WDM requires multiple sources with closely spaced emission wavelengths, provided they are all compatible with the bandwidth of the fiber.

At short distances, high bit-rate communication at 1 TBps has been reported. Around 1.55 μm for instance, a frequency spacing $\Delta\nu$ of 250 GHz is equivalent to a wavelength spacing $\Delta\lambda$ of 2 nm: thus 10 channels cover a band of 20 nm. This imposes severe requirements on the spectral purity of each laser source. For a better utilization of the available fiber bandwidth, several fixed-wavelength single-frequency lasers, or multiple-wavelength DFB (distributed feedback) laser arrays, are a possible solution. In the frame of ITU (International Telecommunications Union) wavelength standards around $\nu = 193.1$ THz (or $\lambda = 1552.52$nm), the frequency spacing is fixed to $\Delta\nu = 100$ GHz (or $\Delta\lambda \approx 0.8$ nm); the frequency accuracy is specified as 10% of the spacing, imposing a wavelength accuracy close to 0.08 nm. The thermal stability of usual laser sources is typically about 0.1 nm/K, which means

 Wavelength division multiplexing (WDM)

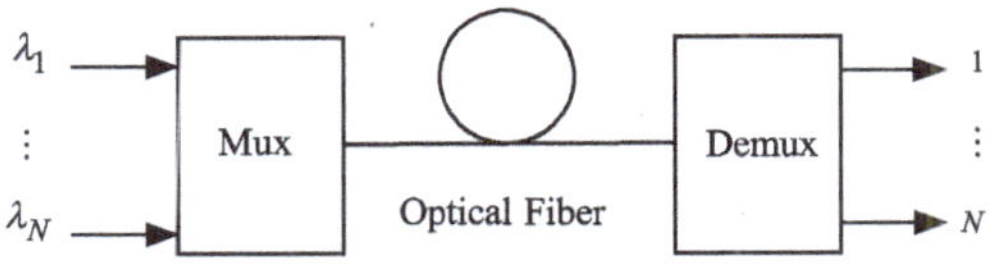

WDM Several optical signals at various wavelengths $\lambda_1, \lambda_2, \ldots, \lambda_N$ are mixed through a multiplexer into an optical fiber. At the other end of the transmission link, demultiplexing involves optical filters, tuned to each wavelength, that separate the different wavelengths toward different detectors.

temperature stabilization better than 1 K is required to maintain the laser output within specifications. Unequal frequency spacing may be preferable in order to avoid parasitic *four-wave mixing*. Inasmuch as several carriers, each modulated by its own signal, are being transmitted simultaneously, the degree of *cross-talk* should be limited to the smallest possible. An isolation of 30 dB between different channels is considered acceptable. Both unidirectional and bidirectional transmission schemes have been implemented. The increasing complexity of such systems imposes many technological requirements, such as the minimization of insertion losses.

WDM systems require a number of important basic functions: (a) Generation of different wavelengths of light each with a narrow spectrum to avoid any overlap in wavelengths; (b) modulation of light without wavelength distortion; i.e. without chirping (variation in the frequency of light due to modulation); (c) efficient coupling of different wavelengths into a single transmission medium; (d) optical amplification of all the wavelengths by an amount that compensates for attenuation in the transmission medium, which depends on the wavelength; (e) dropping and adding channels when necessary during transmission; (f) demultiplexing the wavelengths into individual channels at the receiving end; (g) detecting the signal in each channel. To achieve an acceptable bandwidth, we need to dispersion manage the fiber (use dispersion compensation fibers), and to reduce cross-talk and unwanted signals, we have to use optical filters to block or pass the required wavelengths. We need various optical components to connect the devices together and implement the whole system. Channel density is an important factor in designing WDM systems. It is normally measured in terms of frequency spacing between two adjacent optical channels. If the frequency spacing $\Delta\nu$ is less than 200 GHz then WDM is called *dense wavelength division multiplexing* and denoted as DWDM. At present, DWDM channel separation is at 100 GHz which is equivalent to a wavelength separation of 0.8 nm. DWDM has stringent requirements on lasers and modulators used for generating the optical signals. One cannot tolerate even slight shifts in the optical signal frequency when channels are spaced closely. As the channel spacing becomes narrower as in DWDM, one also encounters various other problems not previously present. For example, any nonlinearity in a component carrying the channels can produce intermodulation between channels, an undesirable effect. Thus, the total optical

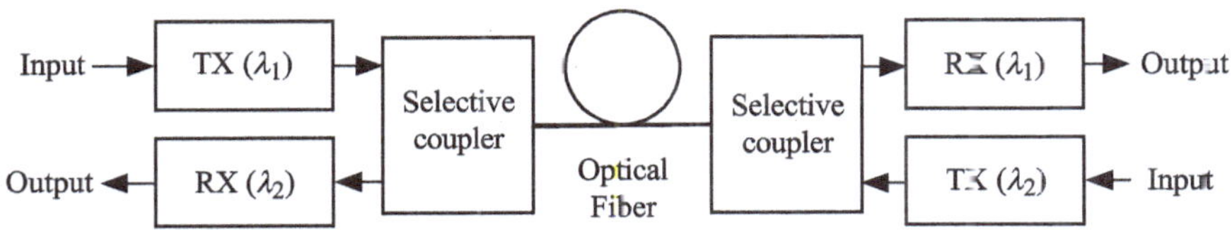

WDM Schematic representation of a two-channel bidirectional transmission, with a conveniently tuned transmitter and receiver systems at each end. TX (λ) represents a transmitter operating at λ, and RX(λ), a receiver operating at λ.

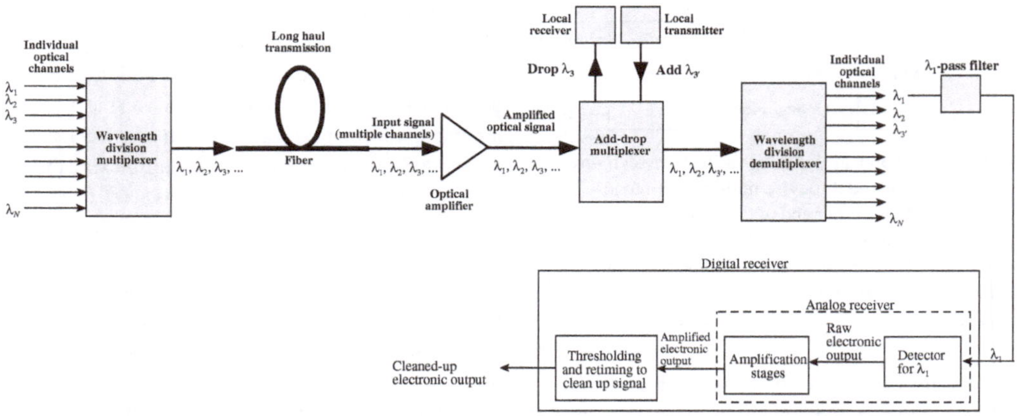

A **WDM** communications system. *N* optical information channels are generated by modulating *N* light waves, each with a different wavelength from λ_1 to λ_N. The *N* optical channels are then multiplexed into a single fiber for transmission.

power must be kept below the onset of nonlinearity in the optical amplifiers within the WDM system. Further the nonlinearity of optical fiber at high optical powers leads to a phenomenon called *four-photon mixing*. Consider three optical signals at ν_1, $\nu_2 = \nu_1 + 100$ GHz, $\nu_3 = \nu_1 + 200$ GHz. The nonlinearity in the optical fiber mixes signals to generate a new signal at $\nu_4 = \nu_1 + \nu_2 - \nu_3 = \nu_1 - 100$ GHz, on top of the channel below υ_1. Four-photon mixing thus adds noise into the communication system. There are limits to how much optical power can be efficiently transmitted in a fiber. The transmitted optical power through a fiber does not increase linearly with the input power when the latter is sufficiently high to cause *stimulated Brillouin scattering* (SBS). Interaction of the photons with acoustics phonons in the fiber generates a periodic variation in the refractive index which in turn reflects some of the power back. The effect increases as the input light power increases, and the spectral width of the input light becomes narrower. The onset of SBS depends not only on the fiber type and core diameter, but also on the linewidth $\Delta\lambda$ of the laser output spectrum. SBS is enhanced as the laser linewidth $\Delta\lambda$ is narrowed or the light pulse is lengthened.

Optical amplifiers are used at various points along the transmission medium to amplify the attenuated signal. If a particular information channel, for example the third channel at λ_3, needs to be dropped and another information channel (at the same wavelength) needs to be added, then this can be done by using a *wavelength add/drop multiplexer*. The new channel is labeled as λ'_3. At the destination, the optical signal has to be separated into its constituent *N* channels by a wavelength demultiplexer.

Wave number or **propagation constant** is defined as $2\pi/\lambda$. It is the phase shift in the wave over a distance of unit length.

Wavenumber of a monochromatic wave is defined as the reciprocal of its free-space wavelength, that is $1/\lambda$ [in m^{-1}] where λ is the wavelength of a propagating wave, i.e. the repeat distance along the direction of propagation at one instant. $\lambda = c/\nu$, and ν is the frequency. Spectroscopists usually express the wavenumber in cm^{-1}.

Wave packet is a group of waves with slightly different frequencies traveling together and forming a "group." This wavepacket travels with a *group velocity* v_g that depends on the slope of ω vs. k (ω is the angular frequency and k is the propagation constant) characteristics of the wavepacket, i.e. $v_g = d\omega/dk$.

Wavevector is a vector denoted as **k** that describes the *direction* of propagation of a wave and has the *magnitude* of the wavenumber, $k = 2\pi/\lambda$.

WDM see *wavelength division mutiplexing*.

Weak injection is usually a condition in a semiconductor device in which the injected minority carrier concentration is much less than the equilibrium majority carrier concentration in the same region; injection may be due to photogeneration or voltage application to a forward biased a *pn* junction. See *minority carrier lifetime*.

Weber contrast see *contrast*.

Wedge prism is a prism that is cut as a wedge (graded width) and, usually, with a round circumference. It can deviate a beam passing through it. If θ_w is the wedge prism apex angle, and θ_d is the minimum deflection angle of the beam through the prism, then

$$\theta_w = \arctan\left(\frac{\sin\theta_d}{n - \cos\theta_d}\right)$$

where n is the refractive index of the prism, assumed to be in air. Two wedge prisms of equal power and very close to each other with flat surfaces parallel can be used to steer a beam by rotating the prisms independently around the common central axis (normal to the common flat surfaces) with the input beam incident along the central axis. The deflected bean can be steered in any direction within the cone that has half an apex angle of $2\theta_d$ where θ_d is the deflection for one prism. They can also be used as an *anamorphic pair*. See *anamorphic prism pair*.

Weiner relationship for the refractive index of a mixture of say two components is

$$\frac{n^2 - n_1^2}{n^2 + 2n_1^2} = \phi_2 \frac{n_2^2 - n_1^2}{n_2^2 + 2n_1^2}$$

where n_1 and n_1 are the refractive indices of the components 1 and 2 in the mixture, and $1 - \phi_2$ and ϕ_2 the volume fractions of components 1 and 2 respectively. (O. Weiner, *Berichte* (Leipzig) **62**, 256, 1910.)

Wemple–DiDomenico dispersion relation is a useful semiempirical single oscillator based dispersion relationship for the refractive index of a variety of materials for photon energies usually below or near the interband absorption edge:

$$n^2 = 1 + \frac{E_0 E_d}{E_0^2 - (h\upsilon)^2}$$

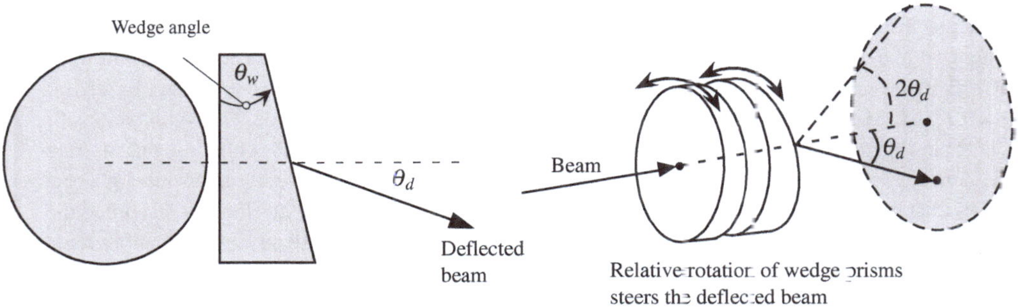

Wedge prism deflection and beam steering (Melles Griot website).

Table: Wemple–DiDomenico dispersion relation Examples of the Wimple–DiDomenico dispersion relatinship. S. H. Wemple and M. DiDomenico, *Phys. Rev. B* **3**, 1338, 1971.

Material	N_c	Z_a	N_e	E_o (eV)	E_d (eV)	β (eV)	β	Comment
NaCl	6	1	8	10.3	13.6	0.28	β_i	Halides, LiF, NaF, etc.
CsCl	8	1	8	10.6	17.1	0.27	β_i	CsBr, CsI, etc.
TlCl	8	1	10	5.8	20.6	0.26	β_i	TlBr
CaF_2	8	1	8	15.7	15.9	0.25	β_i	BaF_2, etc.
CaO	6	2	8	9.9	22.6	0.24	β_i	Oxides, MgO, TeO_2, etc.
Al_2O_3	6	2	8	13.4	27.5	0.29	β_i	
$LiNbO_3$	6	2	8	6.65	25.9	0.27	β_i	
TiO_2	6	2	8	5.24	25.7	0.27	β_i	
ZnO	4	2	8	6.4	17.1	0.27	β_i	
ZnSe	4	2	8	5.54	27	0.42	β_c	II-VI, Zinc blende, ZnS, ZnTe,CdTe
GaAs	4	3	8	3.55	33.5	0.35	β_c	III-V, Zinc blende, GaP, etc.
Si (Crystal)	4	4	8	4.0	44.4	0.35	β_c	Diamond, covalent bonding; C (diamond), Ge, β-SiC, etc.
SiO_2 (Crystal)	4	2	8	13.33	18.10	0.28	β_i	Average crystalline form
SiO_2 (Amorphous)	4	2	8	13.38	14.71	0.23	β_i	Fused silica
CdSe	4	2	8	4.0	20.6	0.32	$\beta_i - \beta_c$	Wurtzite

where υ is the frequency, h is the Planck constant, E_o is the single oscillator energy, E_d is the dispersion energy which is a measure of the average *strength* of interband optical transitions; $E_d = \beta N_c Z_a N_e$ (eV), where N_c is the effective coordination number of the cation nearest-neighbour to the anion (e.g. $N_c = 6$ in NaCl, $N_c = 4$ in Ge), Z_a is the formal chemical valency of the anion ($Z_a = 1$ in NaCl; 2 in Te; 3 in GaP), N_e is the effective number of valence electrons per anion excluding the cores ($N_e = 8$ in NaCl, Ge; 10 in TlCl; 12 in Te; $9^1/_3$ in As_2Se_3), and β is a constant that depends on whether the interatomic bond is ionic (β_i) or covalent (β_c): $\beta_i = 0.26 \pm 0.04$ eV (e.g. for halides NaCl, ThBr, etc. and most oxides, Al_2O_3, etc.), $\beta_c = 0.37 \pm 0.05$ eV (e.g. tetrahedrally bonded $A^N B^{8-N}$ zinc blende and diamond type structures, GaP, ZnS, etc.), respectively. (Wurtzite crystals have a β that is intermediate between β_i and β_c.) Further, empirically, $E_o = CE_g(D)$, where $E_g(D)$ is the *lowest* direct bandgap and C is a constant, typically $C \approx 1.5$. E_o has been associated with the main peak in the $\varepsilon_r''(h\upsilon)$ vs. $h\upsilon$ spectrum. The Wemple–DiDomenico dispersion relation is equivalent to a two–term Sellmeier equation $n^2 = 1 + A_1\lambda^2/(\lambda^2 - \lambda_1{}^2)$ in which $\lambda_1 = ch/E_o$, and $A_1 = E_d/E_o$. (S. H. Wemple and M. Di Domenico, *Phys. Rev. B*, **3**, 1338, 1971; S. H. Wemple, *Phys. Rev. B*, **7**, 3767, 1973.) See *Sellmeier dispersion relation.*

Whispering Gallery Mode (WGM) refers to a mode confined by total internal reflection in spherical or disk-like microresonators.

White LED is an LED, a *light-emitting diode,* emitting a mixture of blue and yellow light, or a mixture of red, green and blue light, which are registered visually by the eye as appearing white. The production of white LEDs became possible due to development of bright blue-emitting LEDs based on gallium-indium-nitride. The combination of red, green, and blue diode chips into one discrete package allows the generation of white light or any of 256 colors by utilizing circuitry that drives the three diodes independently. In applications requiring a full spectrum of colors from a single point source, this type of RGB diode format is the preferred technique. A cost-effective and more popular solution is a diode which employs a semiconductor

Table: White LED Complementary color wavelengths. In accordance with CIE 1931 the mixture of these colors in the given proportion produces white light.

Complementary Wavelengths		Power ratio
λ_1 (nm)	λ_2 (nm)	$P(\lambda_2)/P(\lambda_1)$
390	560.9	0.00955
410	561.3	0.356
430	562.2	1.42
450	564.0	1.79
470	570.4	1.09
480	584.6	0.562
484	602.1	0.440

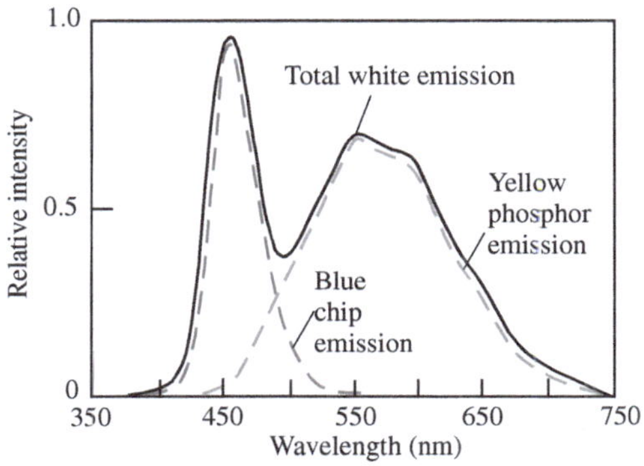

The spectral distribution of light emitted by a **white LED**. Blue luminescence is emitted by GaInN chip and "yellow" phosphorescence is produced by the phosphor. The combined spectrum looks "white."

chip emitting at a short wavelength (blue, violet or ultraviolet) and a wavelength converter, such as a *phosphor*, which absorbs light from the diode and undergoes secondary emission at a longer wavelength. Such diodes emit light of two or more wavelengths, that when combined, appear as white. The quality and spectral characteristics of the combined emission vary with different designs.

The *phosphor* in the LED exhibits *photoluminescence* which means that when it is photoexcited by a suitable radiation, such as blue light, it emits radiation of longer wavelength, such as yellow light. Typically utilized phosphors are composed of an inorganic host substance containing an optically active dopant. Yttrium aluminum garnet ($Y_3Al_5O_{12}$ or YAG) is a common host material, and for diode applications, it is usually doped with one of the rare-earth elements or a rare-earth compound. Cerium (Ce) is a common dopant element in YAG phosphors designed for white light-emitting diodes in so-called blue chip and yellow phosphor white LEDs. A brand new solution is white LEDs using quantum dots as light-emitting phosphors. The approach is based on encapsulating semiconductor quantum dots – nanoparticles approximately 10^{-9} m in size – and engineering their surfaces so they efficiently emit visible light when excited by near-ultraviolet (UV) LEDs. The quantum dots strongly absorb light in the near-UV range and re-emit visible light that has its color determined by both their size and surface chemistry. To date, quantum dot devices have largely been composed of cadmium sulphide. Another possible solution is presented by LEDs based on ZnSe where the chip itself emits two colors simultaneously, one peak at 484 nm (blue) and one peak at 605 nm (yellow) – mixing these two wavelengths inside an epoxy housing "generates" white light. The efficacy of present white LEDs is about 20–40 lm/W with a forecast as high as 100 lm/W, which make white LEDs a strong competitor to existing light sources.

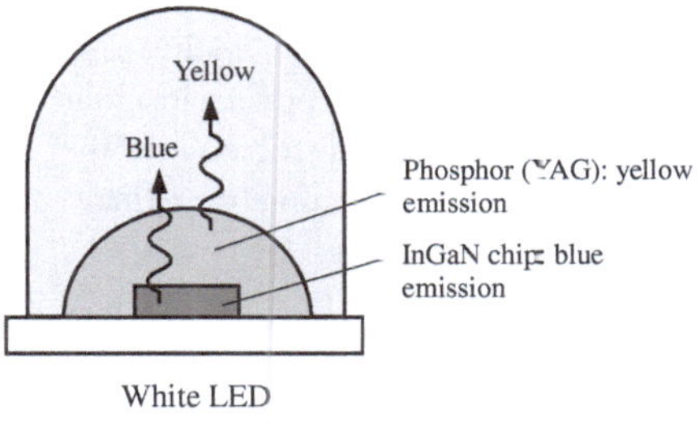

Schematic structure of a blue chip yellow phosphor **white LED**.

See *light emitting diode*, *luminous efficiency*.

Whiting equation see *Voigt lineshape.*

Wide area network (LAN) see *optical network.*

Wien's law is an approximation to the temperature dependence of the spectral luminance $L(\upsilon)$ [$\mathrm{W\,m^{-2}\,sr^{-1}\,Hz^{-1}}$] of a *black body*, valid at low frequencies. According to *Wien's displacement law,* the spectral energy density of radiation emitted by a body in thermal equilibrium exhibits a maximum at a specific wavelength λ_{max} which is inversely proportional to its absolute temperature T,

$$T\lambda_{\max} = \text{constant} = 2898\ \mu\text{m}K.$$

The emitted spectrum can be therefore defined by its color (around λ_{max}) or its temperature. For instance, the Sun has its maximum emission wavelength at around 480 nm which corresponds to a temperature $T \approx 6000$ K; a body at room temperature radiates around $\lambda_{max} \approx 10\ \mu$m (the infrared, also called "thermal" region of the spectrum).

Wiener–Khintchine theorem see *noise.*

Wigner function is a phase space distribution, similar to the Maxwell–Boltzmann distribution, of position and momentum of an ensemble of particles known from classical statistical mechanics. Wigner functions simultaneously describe the value of two conjugate variables, such as time and frequency, or space and momentum. For example, in the latter case the Wigner function can describe the joint transverse position–transverse momentum $(x\text{-}p)$ distribution for a backscattered light field. Measurements of $(x\text{-}p)$ distributions then provide imaging of both the spatial and angular distributions of the scattered light. The Wigner function $W(x,p)$ may be defined as follows:

$$W(x,p) = \frac{1}{2\pi}\int \exp(jp\varepsilon)\Big\langle E^*\left(x+\frac{\varepsilon}{2}\right) \times E\left(x-\frac{\varepsilon}{2}\right)\Big\rangle d\varepsilon$$

where x is the transverse position and p is the transverse momentum of the light field. The angled brackets indicate an ensemble averaging, with ε as an integration variable. $W(x,p)$ is a Fourier transform related to the mutual coherence function $\langle E^*(x_1)E(x_2)\rangle$. It is clear that Wigner functions contain phase information about the light field (as compared to intensity measurements), as is evident from its relationship to the mutual coherence function. Such functions also offer an attractive framework for studying the propagation of optical coherence through random media.

Windows in optoelectronic devices allow desired radiation to pass through, and at the same time isolate the device from its environment by excluding dust, dirt, vapors, humidity, etc. Window selection is critical for both light emitters and detectors if the device is to respond over its design wavelength range. In addition, the window is expected to protect the device from environmental effects. Window selection involves examining several factors: the transparency over the desired wavelength range; changes in the optical properties with temperature, humidity, and time (aging); mechanical properties such as elastic modulus and strength; thermal expansion, or thermal mismatch between the window and its holder; thermal shock (depends on the thermal conductivity and how it can withstand sudden temperature changes); wear and corrosion; cost. See *optical materials.*

Wire bonding refers to the attachment of aluminum or gold leads (via thermal compression or ultrasonic welding) to a die before it is packaged.

Wire-grid polarizer consists of closely spaced parallel thin conducting wires. The light beam passing through the wire-grid becomes linearly polarized at right angles to the wires. The E_x component along the wires drives the conduction electrons along the length of the wires and thereby generates a current. The collisions of these driven electrons with lattice vibrations leads to Joule loss, i.e. I^2R. Thus energy is absorbed from the E_x field and the wires "heat up" a little (negligible amount) just as energy would be absorbed from a battery driving a current through a wire. As the wires are

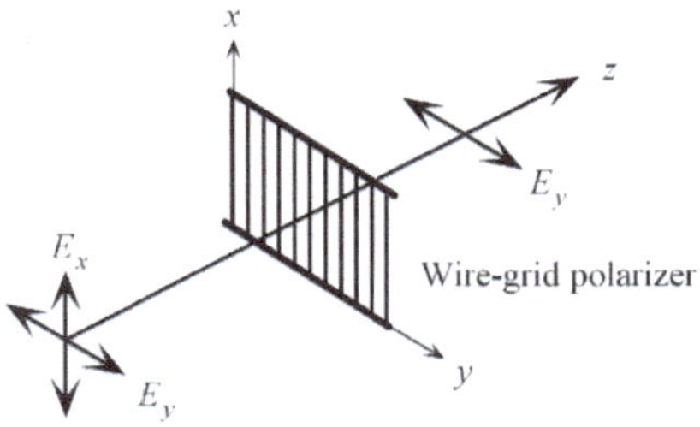

Wire-grid polarizer The wire grid acts as a polarizer.

very thin, E_y field cannot drive the electrons too far.

WKB (Wentzel–Kramers–Brillouin) approximation gives an approximate solution to the wave equation for the modes in a waveguide exhibiting a slowly varying index profile (at the scale of the wavelength). Let the index variation be $n(x)$ and the wave equation

$$d^2\psi/dx^2 = [\beta^2 - n^2(x)k^2]\psi,$$

in which ψ is the varying physical quantity, and $k = 2\pi/\lambda_0$ the wavevector in vacuum. We are looking for the effective index n_{eff} corresponding to a propagation constant $\beta = n_{\text{eff}}k$. Let x' and x'' denote the turning-point values of x where $n(x) = n_{\text{eff}}$, as shown in the figure.

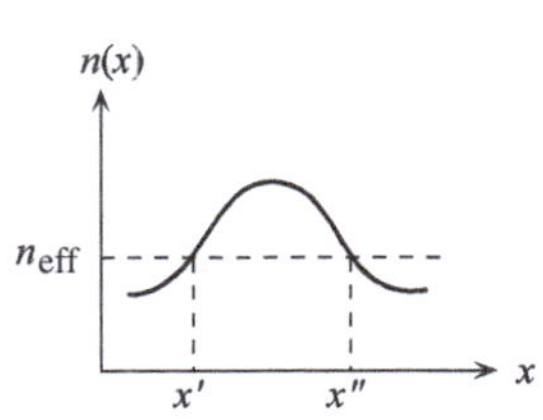

Illustration of the **WKB method**.

The values of n_{eff} predicted by the WKB method are such that:

$$\int_{[x',x'']} \left[n^2(x) - n_{\text{eff}}^2\right]^{1/2} dx = (p + 1/2)(\lambda_0/2),$$

or

$$\int_{[x',x'']} [n^2(x)k^2 - \beta^2]^{1/2} dx = \pi(p + 1/2),$$

where p is an integer. The modal field oscillates in the region $[x',x'']$ where $n(x) > n_{\text{eff}}$, and exponentially decays outside. The same method can be applied to the Schrödinger equation of a quantum wave in a medium with slowly-varying potential $V(x)$ with $k^2(x) = (2m_e/\hbar^2)[E - V(x)]$, where E is the electron energy, and m_e is the mass of the electron.

Wollaston prism is a *polarizing prism* in which the incident beam onto the prism is split into two diverging beams whose polarizations are orthogonal. The Wollaston prism is a polarization splitter in which the split beam has orthogonal polarizations. Two calcite (or quartz) right angle prisms A and B are placed with their diagonal faces touching to form a rectangular block (as in the figure). Looking at the cross-section of this block, the optic axis in A is in the plane of the paper and that in B coming out of the paper; the two prisms have their optic axes mutually orthogonal. Further, as shown, the optic axes are parallel to the prism sides. Consider a light wave of arbitrary polarization at normal incidence to prism A. The light beam entering prism A travels in this prism as two orthogonally polarized waves that have fields E_1 and E_2. E_1 (normal to the plane of paper) is orthogonal to the optic axis and corresponds to ordinary (o-) waves in A. E_2 (in the plane of the paper) is along the optic axis of A and corresponds to the extraordinary (e-) waves. E_1 has a refractive index n_o and E_2 has n_e. However, in prism B, E_1 is the e-wave. In going through the diagonal interface, E_1 experiences a *decrease* from n_o to n_e. On the other hand, the e-wave in A becomes the o-wave in B and experiences an *increase* from n_e to n_o. Notice that E_2 is now orthogonal to the optic axis in B. These refractive index changes are opposite which means that the two waves are refracted in opposite directions at the interface. The E_1-wave moves away from the normal to the diagonal face whereas the E_2-wave moves closer to this normal.

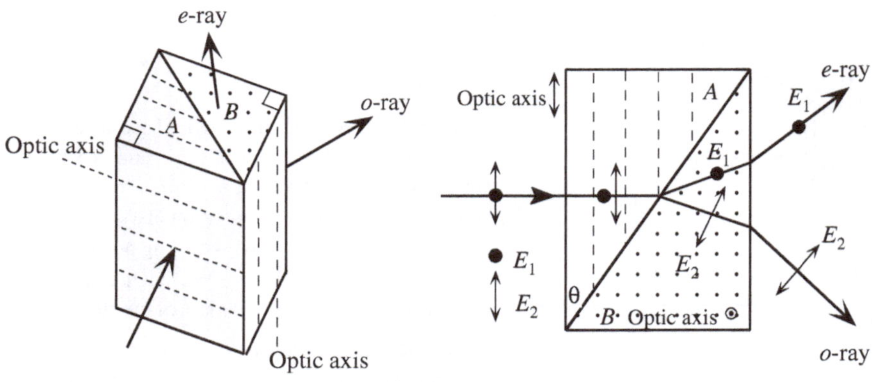

The **Wollaston prism** is a beam polarization splitter.

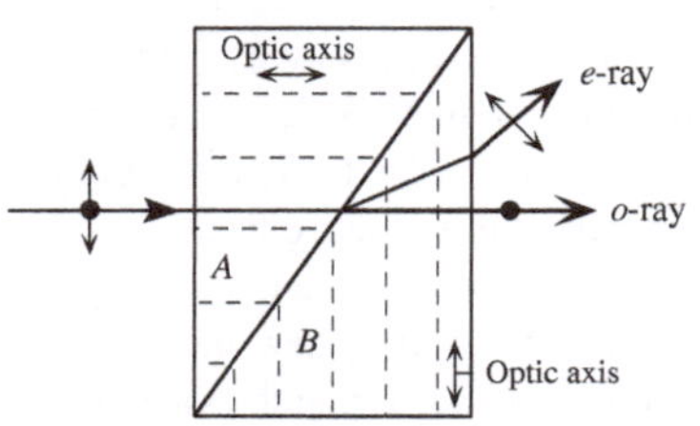

Wollaston prism The Rochon prism is a beam polarization splitter. Notice the difference in optic axes alignments between the Wollaston and Rochon prisms.

The two orthogonal polarizations are therefore angularly separated out by these oppositely sensed refractions. The divergence angle depends on the prism wedge angle θ. Various Wollaston prisms with typical beam splitting angles of 15–45° are commercially available. If we rotate the prism about the incident beam by 180°, we would see the two orthogonal polarizations E_1 and E_2 switch places, and if we use a quartz Wollaston prism, we would find E_1 and E_2 in the figure are again switched. The *Rochon prism* is a polarizing prism that is very similar to the Wollaston prism except the optic axes alignments are different. There are two prisms, and one of the prisms, *A*, has the optic axis along the incident beam. The optic axis in the other, *B*, is perpendicular to that in *A* and parallel to the exit face. The *o*-ray emerges nearly undeflected whereas the *e*-ray is deflected. The Wollaston prism was invented by William Hyde Wollaston (1766–1828), a well known British scientist.

Work function, Φ, is the minimum energy needed to free an electron from a metal at absolute zero of temperature. It is the energy separation of the Fermi level from the vacuum level. Thus, in general, it is the energy required to remove an electron from the solid to the vacuum level. The energy involved in freeing an electron from a metal surface depends on the crystal surface because different surfaces have different surface concentrations and arrangements of positive metal ions. Two examples are shown in

Table: Work functions of nickel and tungsten

Crystal plane	(100)	(110)	(111)	Typical
FCC relative planar concentration, atoms/nm^2 (1 is maximum)	0.87	0.61	1	
Nickel Φ (eV)	5.22	5.04	5.35	5.09
BCC relative planar concentration, atoms/nm^2 (1 is maximum)	0.71	1	0.41	
Tungsten Φ (eV)	4.63	5.25	4.47	4.57

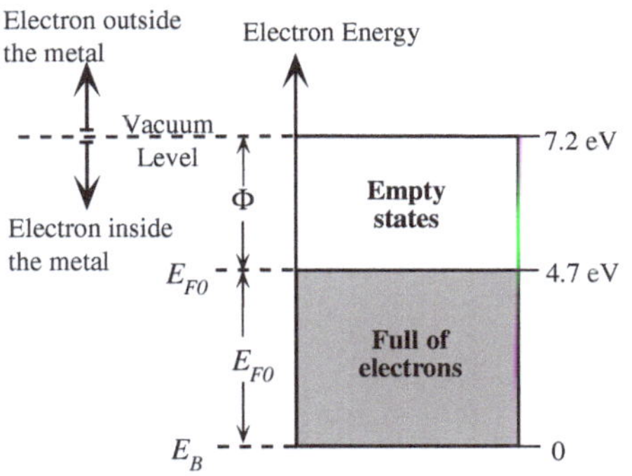

Work function Typical electron energy band diagram for a metal (Li). All the valence electrons are in an energy band which they only partially fill. The top of the band is the vacuum level where the electron is free from the solid (potential energy = 0). E_{F0} is the Fermi energy at 0 K. E_B is the bottom of the conduction band. Φ is the work function.

the table for an FCC and a BCC crystal. The vacuum level defines the energy where the electron is free from that particular solid (typically ~ 10 nm away from the surface) and where the electron has zero kinetic energy. The semiconductor work function $\Phi_{\text{semiconductor}}$ still represents the energy difference between the vacuum level and the semiconductor's Fermi energy E_F, even though there are normally no electrons at E_F. It may be thought that the minimum energy required to remove an electron from the semiconductor is simply the *electron affinity* χ but this is not so. Thermal equilibrium requires that only a certain fraction of all the electrons in the semiconductor should be in the conduction band at a given temperature. When an electron is removed from the conduction band, then thermal equilibrium is upset and some additional energy has to be supplied to maintain the thermal equilibrium. The effective energy $\Phi_{\text{semiconductor}}$ required to remove an electron, as for metals, is the energy from E_F to the vacuum level.

Working point of glass is the temperature at which the viscosity of a glass is about 10^7 Pa s. The viscosity of glass is strongly temperature dependent and decreases sharply as the temperature increases. At the working point, the viscosity is sufficiently low to allow the glass to be easily worked in various glass fabrication processes such as drawing, pressing, molding etc.

X

X-rays are electromagnetic waves of wavelengths typically in the range 1 pm–10 nm which are shorter than those for UV radiation. X-ray photons typically have high energies in the range 100 eV – 1 MeV. X-ray radiation is known as *ionizing radiation* since it is capable of ionizing the media into which it enters or passes through. X-rays can easily pass through various types of media in which its attenuation depends on the structure and composition of the medium – hence its use in imaging interiors of objects. Provided the wavelength fits into the Bragg diffraction condition, it is also diffracted by crystal planes, which has led to x-ray crystallography in which x-ray diffraction patters from crystals are used to extract the crystal structure of samples. X-rays were first discovered by Wilhelm Röntgen in 1895, soon after which x-rays were used in x-ray imaging – the first imaging experiments were crude and did not have necessary protection but nonetheless showed the usefulness of this radiation in medical imaging. X-rays can be *soft* with long wavelengths (photon energies 100–1000 eV), overlapping the short wavelength UV range, or they can be *hard* with short wavelengths (>10 keV) overlapping the γ-radiation range. X-rays are produced by accelerating and impinging electrons (in a vacuum tube) onto a metal anode target (such as tungsten), following which the rapid deceleration (or deflection) of the electrons in the presence of the positive nuclei of the target causes the emission of a continuous electromagnetic radiation, *Bremsstrahlung radiation*, that is in the x-ray range. In addition, the impinging electron can knock out an electron from the core of the metal ion, and when a higher energy level electron falls down to fill this vacancy, it emits a characteric x-ray that has a well-defined wavelength corresponding to the change in the energy. X-rays are usually characterized by their photon energy spectrum (which includes the average photon energy and the maximum photon energy or the maximum accelerating voltage kVp), their fluence (number of photons arriving per unit area), and their ionizing capability, which is measured in roentgens; one roentgen (1 R) is defined as the amount of radiation that will impart sufficient energy to 1 kg of air to ionize an amount of charge that is 2.58×10^{-4} C at STP (standard temperature and pressure).

X-ray image detector or **sensor, flat panel**, is a pixelated flat panel image sensor that captures the x-rays passing through an object to be imaged, and then converts the incident x-ray photons into electrical charges stored at the pixels of the detector. Such a flat panel detector, as in the first figure, has millions of pixels and the readout technology in most devices uses an *active matrix array* (AMA). The incident radiation is converted to charges and stored at the pixels for a frame period before being read out. These electrical signals from the different pixels are used to construct the x-ray image. There are two main types of x-ray image detectors: (i) *direct conversion detectors*; (ii) *indirect*

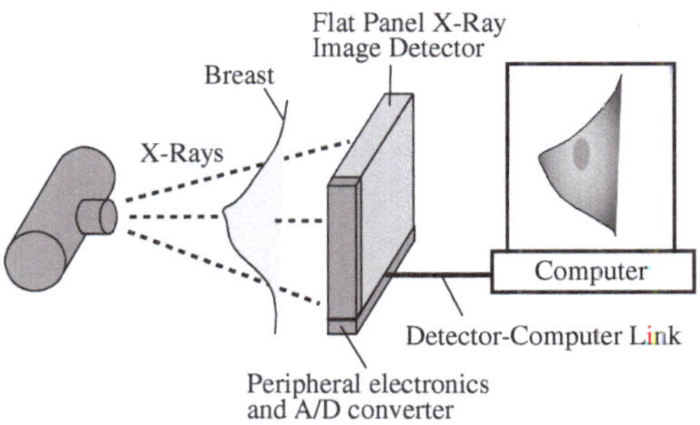

X-ray image detector Flat panel x-ray image detectors enable digital x-ray imaging. The basic read-out technology is an active matrix array. The detection can be direct or indirect.

conversion detectors. Most detectors today use the indirect approach. In direct conversion, an x-ray photoconductor, such as a-Se (amorphous selenium), converts the absorbed x-ray photons to electron–hole pairs, which separate and drift parallel to an applied field. In the example shown in the second figure, x-ray generated electrons are collected on a storage capacitor C at the pixel X, Y. When a signal is applied to the gate G of the thin film transistor (TFT) at this pixel (by applying an address signal to the gate line X), the TFT conducts (it is turned on) and transfers the charge on C to the data line Y; the signal at pixel X, Y is read. In indirect conversion, as shown in the second figure, a phosphor or a scintillator screen is used to convert the x-rays to light. These light photons are then detected by an array of photodiodes, one photodiode at each pixel. The photodiode in each pixel generates an electric charge whose magnitude is proportional to the total (integrated) light intensity emitted from the phosphor in the region close to the pixel. This charge is stored in the pixel capacitor until the corresponding row in the active matrix is read out. The light that is generated at the absorption point of the x-ray photon, spreads out from this point and therefore causes a marked spread in the information, and hence a loss of resolution. The lateral spreading of light can be contained substantially by using a columnar CsI scintillator in which the columns, just like optical fibers, channel or guide the light towards the detector pixels. The spread is reduced but not eliminated because some of the light will pass over to the neighboring columns. In a photoconductor, the x-ray generated electrons and holes follow the field lines as they drift. Their lateral spread is very small, and thus direct conversion detectors have an intrinsically higher resolution than indirect conversion detectors as illustrated in the third figure. (S. Kasap and J. A. Rowlands, *Proc. IEEE*, **90**, 591, 2002.) See *active matrix array*.

X-ray laser (**XASER**) is a laser emitting in the x-range of the electromagnetic spectrum. The active medium is ionized plasma, obtained by focusing amplified ultrashort light pulses on a metal target of high atomic number (typically palladium, selenium or titanium).

XASERs are more difficult to obtain than near-visible sources, for at least two reasons: First, the ratio of spontaneous upon stimulated transition rates varies as λ^{-3}; the shorter the wavelength, the more difficult it is to overcome spontaneous emission. Secondly, the high-reflection cavity mirrors remain difficult to build, due to the fact that all matter is almost transparent in the far-reaching range of the x-spectrum. Periodic structures with alternating layers of different absorption (loss-coupled *distributed Bragg reflectors*), as already used in astronomy, are the most effective, provided they resist the high-energy of the beam pulse. After each shot, the system must be allowed to cool down: the powerful Nova, at Lawrence Livermore National Laboratory (California), can be fired only about six times a day. Up to now, most high-power X-ray beams are not delivered by coherent "xasers," but by not-so-coherent synchrotron sources (particle accelerators).

X-ray sensor see *x-ray image detector*.

XASER see *x-ray laser*.

Xenon arc lamp see *arc lamp*.

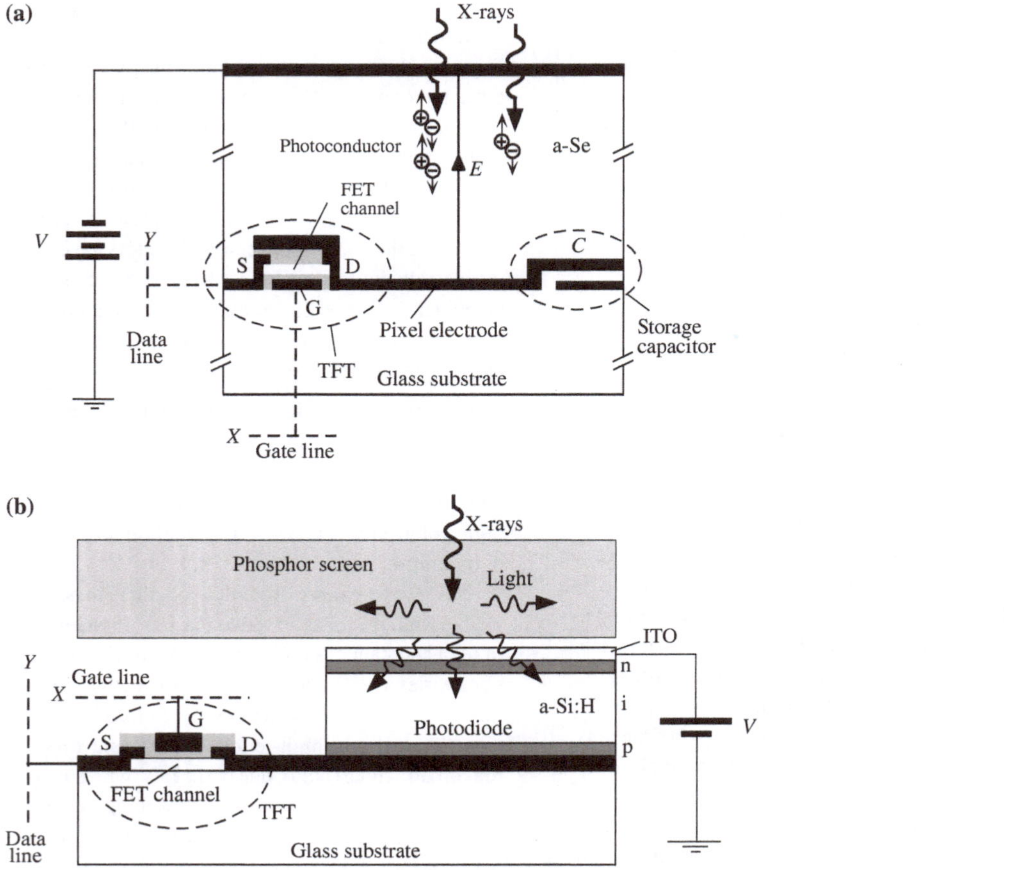

X-ray image detector (a) A cross-section of a pixel in a direct conversion x-ray image detector. (b) A cross-section of a pixel in an indirect detector. The phosphor screen absorbs x-rays and creates visible light photons. These visible photons create electron–hole pairs in the a-Si:H pin photodiode layer, and the charge carriers are subsequently collected. TFT = thin film transistor, G = gate, S = source, D = drain. The pixel at an arbitrary location *X*,*Y*.

Xenon flash lamp is a lamp that is widely used in many applications for generating a short light pulse by sparking an electric discharge in a xenon gas in a bulb. The electric discharge can be designed to generate a very bright flash that can be quite intense and/or of short or long duration, depending on the application. One of the distinct advantages of the xenon flash is its relatively broad light spectrum that stretches from the UV (120 nm) to infrared (1100 nm) through the visible spectrum as shown in the first figure. The exact spectrum depends on the discharge current density. The xenon flash bulb is not evacuated but has the xenon gas usually under pressure (e.g. up to 100 torr) and, depending on the lamp construction, the discharge current can be very large, e.g. up to 10 kA cm^{-2}. The xenon lamp has a

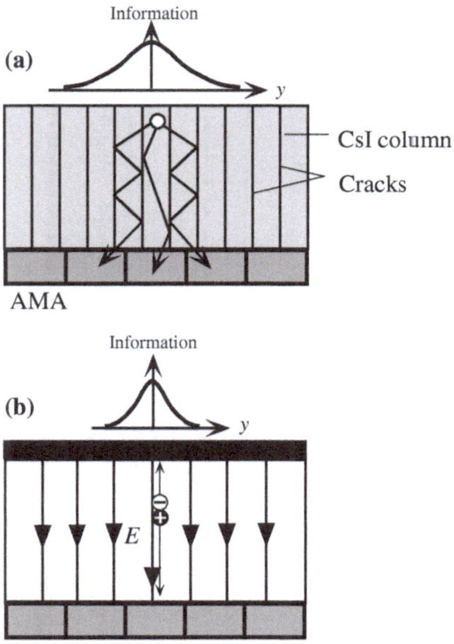

X-ray image detector (a) The spread of information in an indirect detector in which a columnar CsI phosphor (scintillator) is used to covert x-rays to light, which are then guided towards the active matrix array (AMA) that has photodiodes. (b) The charges generated by the x-rays drift along the field lines and the lateral spread is very small.

relatively high luminous efficacy (40 lm/W). The spectrum is made up of a black body type radiation that arises from the hot plasma generated in the discharge and numerous partially smeared spectral features that result from the transitions of the excited xenon atoms and ions as they become de-excited. There tends to be more "black body" radiation at higher current densities. Xenon flash lamps are used in many applications that require intense light pulses, such as strobe lights for high-speed photography and image processing, various applications in medical analysis, e.g. blood analyzers, chemical and environmental analysis, solar simulators, flash light pump for ruby and Nd:YAG lasers, etc.

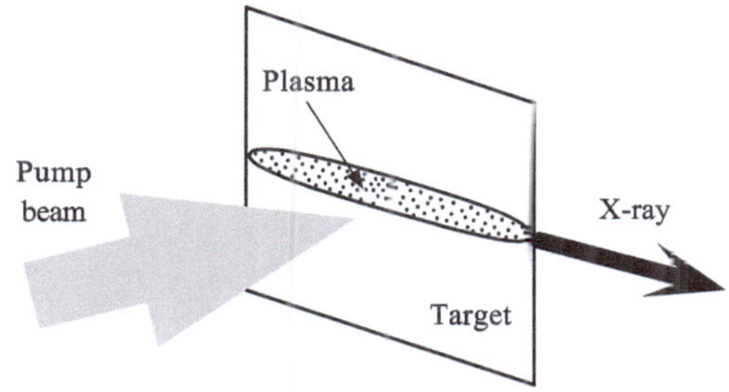

Schematic depiction of a **XASER**; the high-energy light pulse of the pump beam is focused on a line-like spot on the metal target, where it induces an ionized plasma. The latter forms an active medium for x-ray generation and stimulated amplification along its main axis.

A xenon flash lamp circuit is basically made out of two main sections: Charge storage circuit (with C_H) and triggering circuit (with C_T and a pulse transformer). The capacitors C_H and C_T are charged to a voltage of V_H and V_T volts, respectively; V_H is a high voltage (e.g. 1000 V) and V_T is a low voltage (e.g. 100 V). When a trigger pulse is applied to the gate of the silicon controlled rectifier (SCR), C_T suddenly becomes connected to a pulse transformer and discharges quickly through this pulse transformer, generating a large voltage spike in its secondary coil. Thus, a high voltage spike (5–7 kV) is applied to the trigger probe of the xenon flash. This high voltage spike ionizes the xenon gas in the spark gap of the flash bulb and reduces the impedance of the spark gap. The ionization of the spark gap initiates an arc discharge of the high voltage capacitor C_H. The duration and energy of the light pulse depends on the energy stored on C_H, $(1/2)(C_H V_H^2)$, as well as the impedance of the leads and the arc. After the discharge of C_H through the xenon arc, the bulb returns to its equilibrium and C_H is charged slowly through the resistor R_H for the next pulse. Shielding is important to reduce RF interference from the flash discharge. Typically light pulses as short as ~100 ns can be easily obtained by the right design.

XPM, see *cross-phase modulation*.

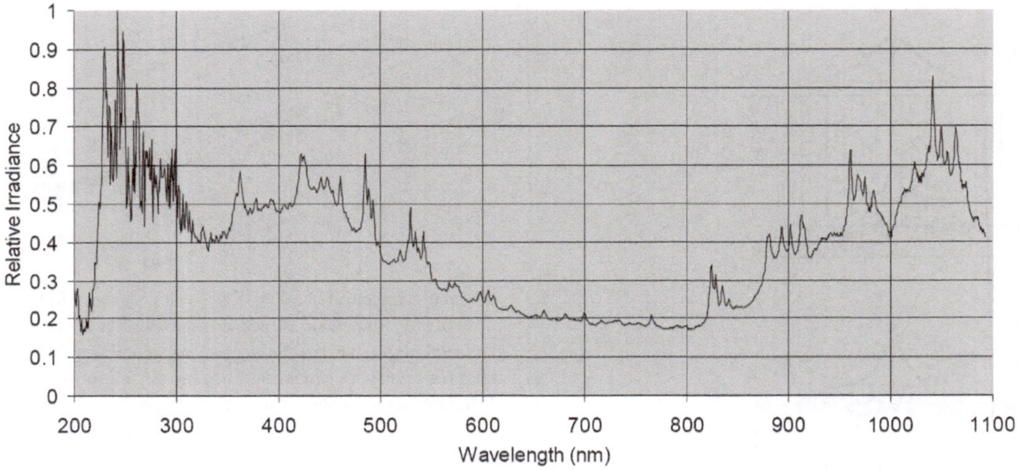

Xenon flash lamp Typical spectrum of light output from a xenon flash lamp (FX4400 from Perkin Elmer). Both short (UV) and IR wavelengths are present. Over the visible spectrum, the intensity distribution is relatively continuous and resembles black body radiation.

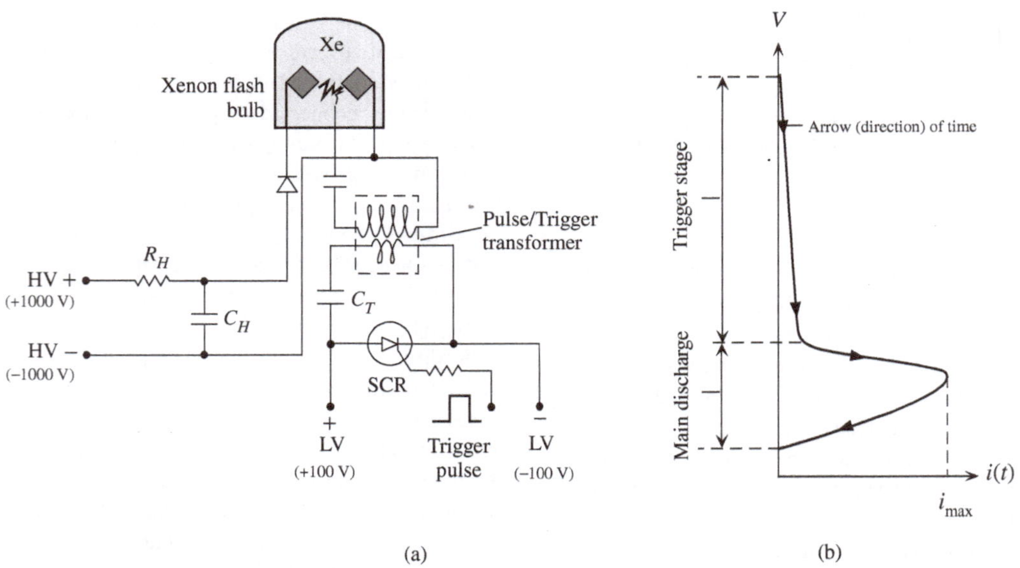

Xenon flash lamp (a) A typical circuit for triggering a xenon flash bulb. Many other circuits (simpler than this) are possible. (b) A simplified schematic sketch of the voltage–current (impedance) characteristics of the xenon arc.

Y

Young's interference fringes as encountered in simple optics experiments are alternating dark and bright bands of light obtained when two light waves of the same frequency, emanating from two coherent sources, interfere on a photographic (or photoluminescent) screen.

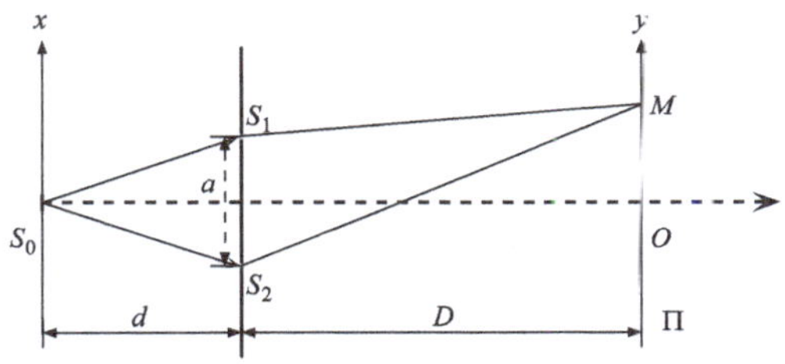

Young's interference fringes Definitions in Young's double slit experiment.

In a general description, we have to identify the following cases:

A. Far-field interference from two monochromatic point sources

Consider first a purely monochromatic point source of frequency $\nu_0 = c/\lambda_0$, located on the main symmetry axis of the system at point S_0 (see figure) at equal distance from pinholes S_1 and S_2. For simplicity, assume that the distance $a = S_1S_2$ is much smaller than the distance d separating S_0 from the plane of the pinholes, and the distance D separating the latter plane from the plane of observation Π. If the two (scalar) vibrations at S_1 and S_2 are written respectively as,

$$V_1(S_1) = A_{01} \exp[j(\omega t + \varphi_0)]$$

and

$$V_2(S_2) = A_{02} \exp[j(\omega t + \varphi_0)],$$

they become, respectively, at any point M of the observation plane,

$$V_1(M) = A_{01} \exp[j(\omega t + \varphi_0)] \times \exp[-(2\pi/\lambda_0)\, nS_1M]$$

and

$$V_2(M) = A_{02} \exp[j(\omega t + \varphi_0)] \times \exp[-(2\pi/\lambda_0)\, nS_2M],$$

where n is the refractive index of the medium (usually air where $n \approx 1$). We consider that the only effect due to propagation from S_1 (respectively S_2) to M is a pure phase change. If we restrict ourselves to the dimension Oy orthogonal to the symmetry axis Oz, the optical distances $\delta_1 = nS_1M$ and $\delta_2 = nS_2M$ can be obtained by a first-order expansion as

$$\begin{aligned}\delta_1 &= nD\{1 + [(y - a/2)^2/D^2]\}^{1/2} \\ &\approx nD\{1 + [(y - a/2)^2/(2D^2)]\}, \\ \delta_2 &= nD\{1 + [(y + a/2)^2/D^2]\}^{1/2} \\ &\approx nD\{1 + [(y + a/2)^2/(2D^2)]\}.\end{aligned}$$

The total vibration at point M is $V(M) = V_1(M) + V_2(M)$, and the corresponding observable intensity $I(M)$ is proportional to $\langle|V(M)|^2\rangle$, i.e.

$$I(M) = I_1 + I_2 + 2\sqrt{I_1 I_2}\cos\left(\frac{2\pi\delta}{\lambda_0}\right)$$
$$= I_1 + I_2 + 2\sqrt{I_1 I_2}\cos(2\pi p)$$

in which the quantity $\delta = \delta_2 - \delta_1 \approx (nay/D)$ is called the *optical path difference* and $p = (\delta/\lambda_0)$ is the *interference order*. The above is also the universal formula for describing the interference of two waves. The intensity can be rewritten as

$$I(M) = (I_1 + I_2)[1 + m\cos(2\pi p)],$$

where

$$m = 2(I_1 I_2)^{1/2}/(I_1 + I_2)$$

is the *modulation index*. If the two vibrations share the same amplitude, then $I_1 = I_2$, $m = 1$ and

$$I(M) = (2I_1)[1 + \cos(2\pi p)].$$

The maxima of the resulting intensity $I_{max} = (I_1 + I_2)[1 + m]$ are obtained for integer values of p such as $\cos(2\pi p) = +1$, whereas minima $I_{min} = (I_1 + I_2)[1 - m]$ correspond to half-integer values of p, such as $\cos(2\pi p) = -1$. In the two-pinhole experiment, the intensity along the vertical coordinate Oy is therefore sinusoidally modulated:

$$I(y) = I_1 + I_2 + 2\sqrt{I_1 I_2}\cos\left(\frac{2\pi\, nay}{\lambda_0\, D}\right)$$
$$= (I_1 + I_2)\left[1 + m\cos\left(\frac{2\pi\, nay}{\lambda_0\, D}\right)\right].$$

The spatial period of the modulation, called *interfringe separation*, is $\Delta y = (\lambda_0 D/na)$.

B. Interference from spectrally extended point sources

If the point sources are not rigorously monochromatic, each wavelength contributes to its own interference pattern. From the description above, the sine-like modulated patterns depend on the wavelength. If they coincide at $\delta = 0$, they become out of phase as soon as $\delta > 0$. For simplicity, consider first the case of a two-wavelength (λ_{01},λ_{02}) point source. The abscissas y_{p1} corresponding to maxima at wavelength λ_{01} are $y_{p1} = p_1\Delta y_1$, with p_1 being an integer and $\Delta y_1 = (\lambda_{01}D/na)$; similarly, the abscissa y_{p2} corresponding to maxima at wavelength λ_{02} are $y_{p2} = p_2\Delta y_2$, with p_2 being an integer and $\Delta y_2 = (\lambda_{02}D/na)$. If there exists a couple of integers (p_1,p_2) such as $y_{p1} = y_{p2}$, then the maxima coincide at this precise abscissa value, and the patterns are said to be in *coincidence*. On the other hand, if, at a given abscissa, one pattern reaches its maximum while the other is at its minimum, the patterns are said to be in *anticoincidence.*

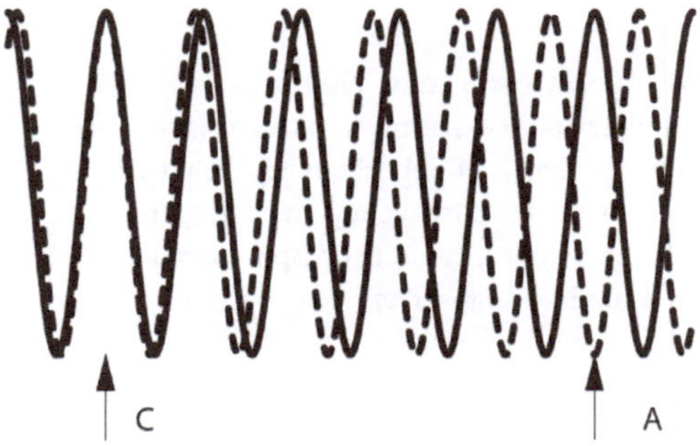

Young's interference fringes Superposition of two sine-like modulated patterns at slightly different periods. When the two patterns coincide, they are said to be in "coincidence" (C). When the maximum of one pattern coincides with a minimum of the other, they are in "anticoincidence" (A).

Consider now a continuously varying spectrum, centered on ν_0 and assumed uniform between $\nu_1 = \nu_0 - \gamma/2$ and $\nu_2 = \nu_0 + \gamma/2$, where γ is the width of the spectrum; the intensity pattern corresponding to frequency ν is best rewritten as,

$$I(\nu, y) = [I_1(\nu) + I_2(\nu)] \times \left[1 + m\cos\left(\frac{2\pi\nu\, nay}{cD}\right)\right],$$

where I_1 and I_2 are replaced by spectral densities $I_1(\nu) = I_{10}/\gamma$ and $I_2(\nu) = I_{20}/\gamma$. The total intensity pattern as a function of abscissa y is obtained by spectral integration,

$$I(y) = \int I(\nu, y)d\nu$$

or

$$I(y) = (I_{10} + I_{20}) \times \left[1 + m\,\mathrm{sinc}\left(\frac{\pi\gamma\, nay}{cD}\right)\cos\left(\frac{2\pi\nu_0\, nay}{cD}\right)\right],$$

where $\mathrm{sinc}(x) = \sin(x)/x$ is the "sinc" function.

The index of modulation appears reduced by the sinc function, limiting the visibility of the fringes to the smallest values of y. The first zero of the sinc function is reached for $y = (cD/\gamma na)$. Note that, between the first and the second zero, a *contrast inversion* is obtained: the sinc becoming negative, maxima become minima and *vice versa*. This explains why interference fringes are so hard to observe with ordinary white light. As an example, take $\nu_0 = 0.5\times10^{15}$ Hz ($\lambda_0 = 600$ nm) and $\gamma = 0.1 \times 10^{15}$ Hz or $\gamma = 0.4 \times 10^{15}$ Hz. Even with $m = 1$, only the central peak is distinct, as seen in the figure.

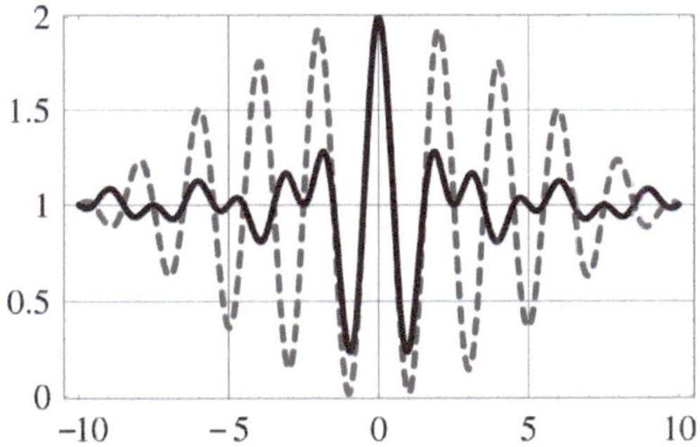

Young's interference fringes resulting from the superposition of two spectrally extended optical fields with the same intensity ($m = 1$), at central frequency $\nu_0 = 0.5\times10^{15}$ Hz, as a function of (nay/cD) [$\times10^{-15}$]. For dotted lines, $\gamma = 0.1\times10^{15}$ Hz; for straight lines, $\gamma = 0.4\times10^{15}$ Hz.

C. Interference from spatially extended monochromatic sources

Another limitation to fringe visibility comes from the spatial extension of the actual source. For simplicity, we will restrict ourselves to a one-dimensional monochromatic source with a wavelength λ_0, with an extension of total length s along the axis S_0x, parallel to Oy, and symmetrical with respect to S_0. We call d the distance between the plane source and the plane of the pinholes, assumed much greater than both s and a. For any secondary source S located at abscissa x, the optical path difference at point M between vibrations coming from S through S_2 and S_1 becomes

$$\delta = \delta_2 - \delta_1 \approx (nay/D) + (nax/d).$$

As a result, the interference pattern is shifted along the y-direction with respect to the primary pattern generated by S_0. The interfrange is still given by $\Delta y = (\lambda_0 D/na)$, but the center of the pattern, that is the abscissa corresponding to $\delta = 0$, is $y = -(xD/d)$. Consider first the case of a two-point source, S_0 at $x = 0$ and S_{01} at $x = x_1$. Different points from the same source are generally uncorrelated (or not coherent spatially), so that they add their own contribution in intensity. As soon as the shift reaches half an interfringe separation, that is for $(x_1 D/d) = \Delta y/2 = (\lambda_0 D/2na)$, the two patterns are in anti-coincidence at each point: it becomes impossible to distinguish any modulation at all. The second point source S_{01} should then not be distinct from S_0 by more than $(\lambda_0 d/2na)$, which gives the correct order of magnitude of the *spatial coherence* required for an extended source in order to actually obtain observable interference fringes in such an experiment.

Consider now a continuously extended source, centered on S_0 and assumed uniform between $x_1 = -s/2$ and $x_2 = +s/2$; the intensity pattern corresponding to abscissa x is

$$I(x, y) = (I_1(x) + I_2(x)) \times \left[1 + m\cos\left(\frac{2\pi nax}{\lambda_0 d} + \frac{2\pi nay}{\lambda_0 D}\right)\right],$$

where I_1 and I_2 are replaced by *spatial* densities $I_1(x) = I_{10}/s$ and $I_2(x) = I_{20}/s$.

The total intensity pattern as a function of abscissa y is obtained by spatial integration,

$$I(y) = \int I(x, y)\,dx,$$

or

$$I(y) = (I_{10} + I_{20}) \times \left[1 + m \,\mathrm{sinc}\left(\frac{\pi nas}{\lambda_0 d}\right) \cos\left(\frac{2\pi nay}{\lambda_0 D}\right)\right],$$

Once again, the index of modulation appears reduced by the sinc function, limiting the visibility of the fringes to the smallest values of s, for any y. For vanishingly small values of s, the source may be considered as a point source, and the visibility is not appreciably degraded. The first zero of the sinc function is reached for $s = (\lambda_0 d/na)$. The fringes are then completely washed out by mutual superposition. For greater values of s, a *contrast inversion* is once more obtained, but the visibility tends rapidly toward zero.

D. Interference from spatially extended, spectrally broadened sources.

The same principle can be applied to the more realistic case of a real source with finite dimension and broad spectrum. The contribution from an infinitesimal source of spatial and spectral extensions dx and $d\nu$, respectively, around abscissa x and frequency ν, is

$$I(\nu, x, y) = (I_1(\nu, x) + I_2(\nu, x)) \times \left[1 + m \cos\left(\frac{2\pi \nu nax}{cd} + \frac{2\pi \nu nay}{cD}\right)\right],$$

where $I_1(\nu, x) = (I_{10}/\gamma s)$ and $I_2(\nu, x) = (I_{20}/\gamma s)$. A double integration with respect to x over the interval $[-s/2, +s/2]$ and with respect to ν over the interval $[\nu_0 - \gamma/2, \nu_0 + \gamma/2]$ is necessary. Unfortunately, the result cannot be expressed in a simple explicit form (except with the help of so-called special functions), but the qualitative discussion of the two previous paragraphs still holds. It should be related to the conditions of *coherence* (both spatial and spectral) required for observing fringes.

Z

Z-scan is an optical measurement technique for measuring nonlinear absorption and refraction (both the sign and magnitude) of a material. The technique relies on sending an axially symmetric Gaussian laser beam through a converging lens onto a sample placed near the waist of the laser beam and finally through an aperture placed in front of a detector in the far field (see the figure). Translating the sample along the optical axis (z-direction) to one side of the beam waist results in the detected power increasing to a peak value. Translating the sample to the other side of the beam waist results in the detected power decreasing to a minimum value. The high intensity of the electromagnetic field in the sample induces a nonlinear lens of variable focal length. This is responsible for the refractive convergence or divergence of the laser beam, depending on the sample position relative to the beam focus and the sign of the optical nonlinearity. Measurements can be made with and without an aperture on the detector. When using the aperture, the difference in the measured power from peak to valley allows a determination to be made of both the sign and magnitude of the nonlinear index of refraction or Kerr coefficient, while measurements without the aperture allow a determination of the sign and magnitude of the nonlinear (two-photon) absorption. The transmittance $T(z)$ measured through an aperture of an input Gaussian beam is given by:

$$T(z) = \frac{\int_{-\infty}^{\infty} P_{\mathrm{T}}(\Delta\Phi_0(t))dt}{S\int_{-\infty}^{\infty} P_{\mathrm{i}}(t)dt}$$

where $\Delta\Phi_0 = k\Delta n_0 L_{\mathrm{eff}} = \gamma k L_{\mathrm{eff}} I_0(t)$ is the on-axis phase shift at the focus ($z = 0$), given in terms of the wavevector, $k = 2\pi/\lambda$, the on-axis nonlinear refractive index change Δn_0 at the focus, the on-axis irradiance at the focus, $I_0(t)$, and $L_{\mathrm{eff}} = (1 - \mathrm{e}^{-\alpha L})/\alpha$, where L is the sample thickness and α is the linear absorption coefficient, where $P_{\mathrm{i}}(t) = \pi w_0{}^2 I_0(t)$ is the instantaneous input power within the sample with w_0 as $\sqrt{2}$ times the on-axis beam radius at

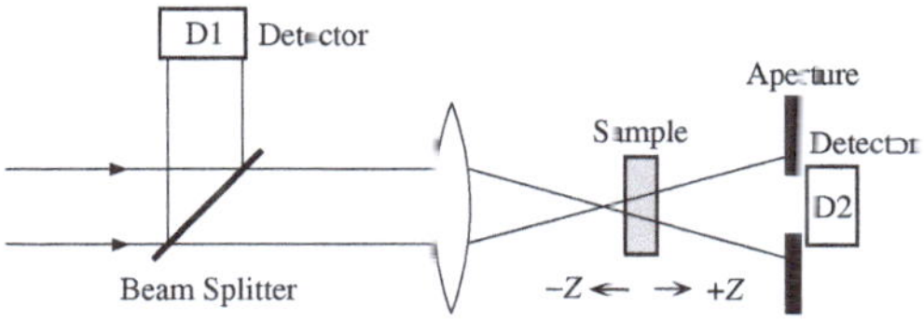

The **Z-scan technique** in which a sample is translated a distance Z along the optical axis from the focal point of a Gaussian beam, and the ratio of the light intensity incident on the sample (as measured by the detector D1) to that transmitted through the sample and a radial aperture of fixed diameter is recorded.

the focus, r_a is the beam radius at the aperture, $S = 1 - \exp[-2(r_a/w_0)^2]$, and $P_T(\Delta\Phi)$ is the transmitted power through the aperture. It can be shown that the peak-to-valley transmittance change is $\Delta T_{p\text{-}v} \sim 0.406(1 - S)^{0.25}|\Delta\Phi_0|$ and hence, knowing the aperture function, S, directly provides a value for $\Delta\Phi_0$ and thus the nonlinear refractive index change. (M. Sheik-Bahae, A. A. Said, T.-H. Wei, D. J. Hagan and E. W. van Stryland, *IEEE J Quam Electron*, **26**, 760, 1990.)

Zener breakdown is the enormous increase in the reverse current in a *pn* junction when the applied voltage is sufficient to cause the tunneling of electrons from the valence band in the *p*-side to the conduction band in the *n*-side. Zener breakdown occurs in *pn* junctions that are heavily doped on both sides so that the depletion layer width is narrow.

Zernike's phase contrast method see *spatial filtering*.

Zero dispersion shifted fiber see *dispersion shifted fiber*.

Zinc selenide (ZnSe) is a II-VI semiconductor and a very useful optical material used in various applications such as optical windows (especially high-power laser windows), lenses, prisms, etc. It transmits over 0.50–19 μm. ZnSe blue-green laser diodes have been demonstrated. ZnSe has the zinc blende crystal structure which is cubic. $a = 0.557$ nm; $T_m = 1798$; $d = 5.27$ g cm^{-3}; $E_g(\mathrm{D}) = 2.64$ eV (300 K): 2.80 eV (10 K); $\mu_e \approx 400$ cm^2 V^{-1}s^{-1}; $\mu_h \sim 100$ cm^2 V^{-1}s^{-1}; $\alpha_L = 7.1 \times 10^{-6}$ K^{-1} (273 K); 7.8×10^{-6} K^{-1} (373 K); 8.3×10^{-6} K^{-1} (473 K); $\kappa = 18$ W m^{-1} K^{-1}; $C_P = 52$ J K^{-1} mol^{-1}; $c_s = 0.34$ J K^{-1} g^{-1}; $T_{\mathrm{Debye}} = 339$ K; $H_{\mathrm{Vickers}} = 112$ kg mm^{-2} (1 kg indenter) $H_{\mathrm{Knoops}} = 120$ kg mm^{-2}; $E = 67.3$ GPa; $K = 40$ GPa; $\upsilon = 0.28$; $\varepsilon_r(\mathrm{DC}) = 8.6$; $\varepsilon_r(\infty) = 5.7$; Reststrahlen peak at $\lambda = 45.7$ μm; $dn/dT \approx 6 \times 10^{-5}$ K^{-1} at 10.6 μm; n in the 1–11 μm range described by a Cauchy-type expression (λ in μm),

$$n = 2.4365 + \frac{0.0485}{\lambda^2} + \frac{0.0061}{\lambda^4} - 0.0003\lambda^2.$$

"Physicists use the wave theory on Mondays, Wednesdays and Fridays, and the particle theory on Tuesdays, Thursday and Saturdays". Sir William Henry Bragg (1862–1942) (From *A Dictionary of Scientific Quotations*, Alan Mackay, IOP Publishing, Bristol, 1991.)

Appendix

Physical Constants

c	Speed of light in vacuum	299792458×10^{8} m s^{-1}
h	Planck's constant	6.6261×10^{-34} J s
$\hbar$	$\hbar = h/2$	1.0546×10^{-34} J s
e	Electronic charge	1.60218×10^{-19} C
ε_o	Absolute permittivity	8.8542×10^{-12} F m^{-1}
k_B	Boltzmann constant ($k_B = R/N_A$)	1.3807×10^{-23} J K^{-1}
$k_B T/e$	Thermal voltage at 300 K	0.02585 V
m_e	Mass of the electron in free space	9.10939×10^{-31} kg
μ_o	Absolute permeability	$4\pi \times 10^{-7}$ H m^{-1}
N_A	Avogadro's number	6.0221×10^{23} mol^{-1}
R	Gas constant ($N_A k$)	8.31457 J mol^{-1} K^{-1}

Useful Information

Wavelength ranges and colors as usually specified for LEDs

Color	λ (nm)
Blue	$\lambda < 500$
Emerald green	530–564
Green	565–579
Yellow	580–587
Amber	588–594
Orange	595–606
Red-Orange	607–615
Red	616–632
Deep red	632–700
Infrared	$\lambda > 700$

1 eV = 1.60218×10^{-19} J
1 Å = 10^{-10} m = 0.1 nm

Electrooptic properties

From C. C. Davis, *Lasers and Electro-Optics: Fundamentals and Engineering*, Cambridge University Press, Cambridge, 1996. Tables 19.1 and 19.2.

Table: Electrooptic properties *The electro-optic $\bar{\bar{r}}$ for the 32 crystal symmetry classes, zero element, • nozero element, •·········• equal nonzero elements, • · · · · · · o equal nonzero elements of opposite sign. Centrosymmetric classes ···· all elements of $\bar{\bar{r}}$ are zero.*

Triclinic

•	•	•
•	•	•
•	•	•
•	•	•
•	•	•
•	•	•

Example: Calcium thiosulphate ($CaS_2O_3.6H_2O$) strontium tartrate ($SrH_2(C_4H_4O_6)_2.4H_2O$)

Monoclinic

·	•	·
•	•	·
·	•	·
•	·	•
·	•	·
•	·	•

2 (symmetry axis parallel to y)

·	·	•
·	·	•
·	·	•
•	•	·
•	•	·
·	·	•

2 (symmetry axis parallel to z)

Examples: lithium sulphate ($LiSO_4.H_2O$), tartaric acid triglycine sulphate

•	·	•
•	·	•
•	·	•
·	•	·
•	·	•
·	•	·

m (perpendicular to y)

•	•	·
•	•	·
•	•	·
·	·	•
·	·	•
•	•	·

m (perpendicular to z)

Example: potassium nitrite (KNO_2)

Orthorhombic

·	·	·
·	·	·
·	·	·
•	·	·
·	•	·
·	·	•

222

Examples: α–iodic acid (α–HIO_3), magnesium sulphate ($MgSO_4.7H_2O$), Rochelle salt ($KNaC_4H_4O_6.4H_2O$)

·	·	•
·	·	•
·	·	•
·	•	·
•	·	·
·	·	·

mm2

Examples: Barium sodium niobate ($Ba_2NaNb_5O_{15}$), polyvinylidine fluoride (PVF), $(CH_2CF_2)_n$

Table: Electrooptic properties (*cont.*)

Tetragonal

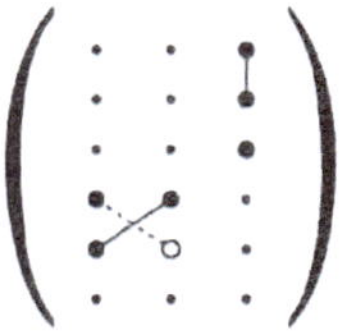

4

Example: iodo succinimide $(CH_2CO)_2NI$

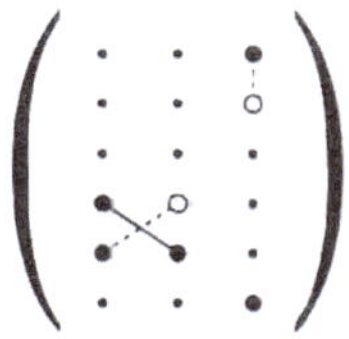

$\bar{4}$

Examples: pentaerythritol $(C(CH_2OH)_4)$, cahnite $(Ca_4B_2As_2O_{12}.4H_2O)$

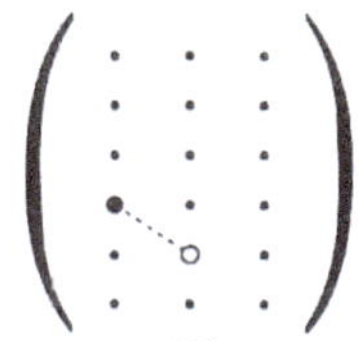

422

Example: nickel sulphate $NiSO_4.5H_2O$

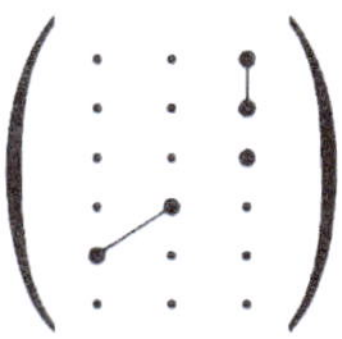

4 mm

Example: $BaTiO_3$

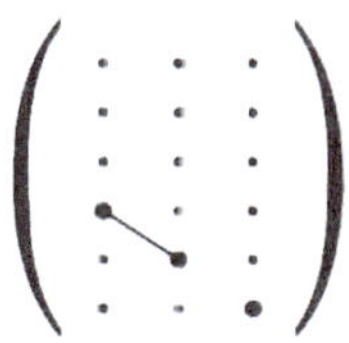

$\bar{4}$2 m (2-fold axis parallel to x)

Examples: KDP, ADP, $CdGeAs_2$, $AgGaSe_2$, $AgGaS_2$

Cubic

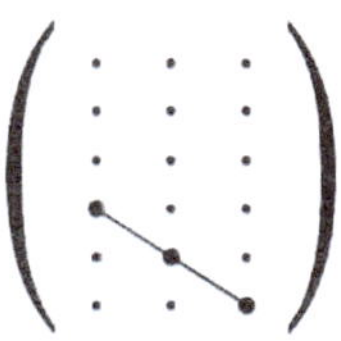

$\bar{4}$3 mm, 23

Examples: ($\bar{4}$3 m): GaAs, InAs, GaP, ZnSe, CdTe, InSb. (23): Sodium chlorate $(NaClO_3)$, sodium bromate $(NaBrO_3)$

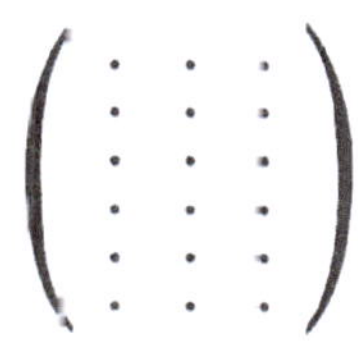

432 (lacks a center of symmetry, but other symmetry elements make all elements zero)

Trigonal

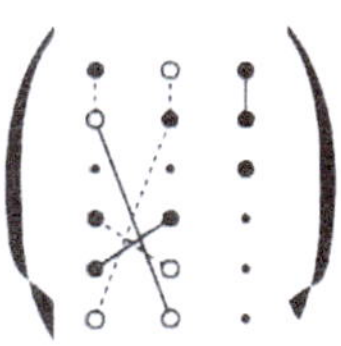

Example: sodium periodate $NaIO_4.3H_2O$

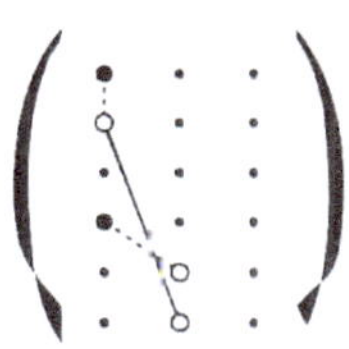

Examples: quartz, Te, mercury sulphide (α–HgS)

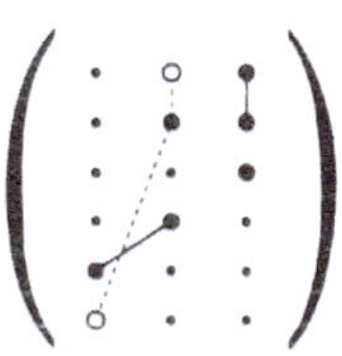

3m (m perpendicular

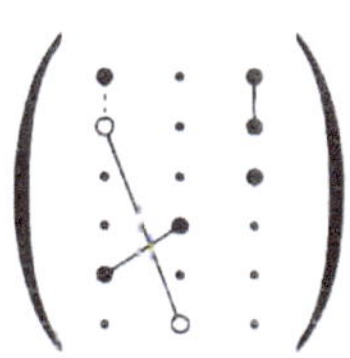

3m (m perpendicular to y)

Examples: $LiNbO_3$, $LiTaO_3$, proustite (Ag_3AsS_3), pyragyrite (Ag_3SbS_3)

(*cont.*)

Table: Electrooptic properties (*cont.*)

Hexagonal

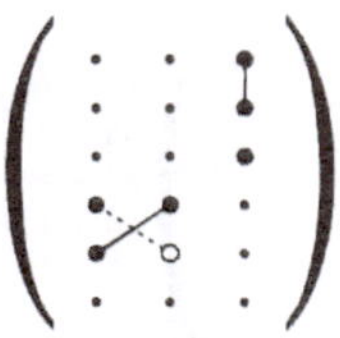

6

Examples:
lithium iodate ($LiIO_3$),
Iodoform (CHI_3)

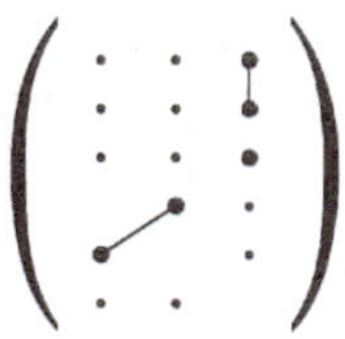

6 mm

Examples:
CdS, CdSe,
ZnS, ZnO

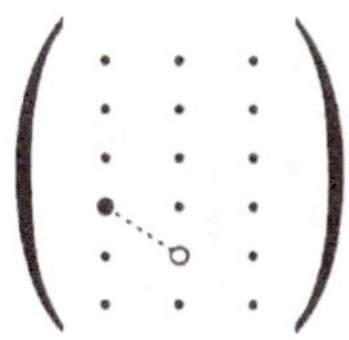

622

Examples:
barium aluminate ($BaAl_2O_4$)
kalsilite ($KAlSiO_4$)

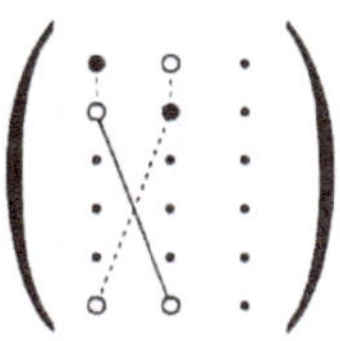

$\bar{6}$

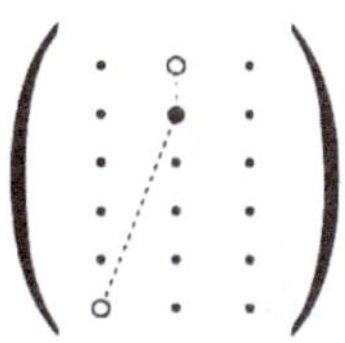

$\bar{6}$ m2 (m perpendicular to x)
standard orientation

Examples:
benitoite ($BaTiSi_3O_9$)

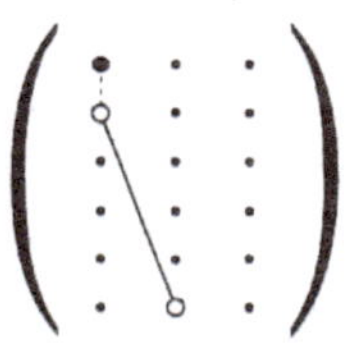

$\bar{6}$ m2 (m perpendicular to y)

Table: Electrooptic properties *Low and high frequency values of elements of the electrooptic tensor for various crystals.*

Material	Symmetry	Electrooptic coefficients (10^{-12} m/V) at specific wavelength (μm) low frequency	 high frequency
KDP	$\bar{4}2m$	r_{63}=-10.5	r_{63}=8.15
		r_{41} = 8.8	
ADP	$\bar{4}2m$	r_{63} = - 8.5	r_{63} = 5.5
		r_{41} = 8.6	
CDA (CsH_2AsO_4)	$\bar{4}2m$	r_{63} = 8.6	
$LiNbO_3$	3m	r_{33} = 32.2(λ = 0.633)	r_{33} = 30.8(λ=0.633)
		r_{13} = 10(λ=0.633)	r_{13} = 8.6(λ=0.633)
		r_{22} = 7(λ=0.633)	r_{22} = 3.4(λ=0.633)
		r_{51} = 32(λ=0.633)	r_{51} = 28(λ=0.633)
$LiTaO_3$	3m		r_{33} = 35.8(λ=0.633)
			r_{13} = 7(λ = 0.633)
			r_{22} ≃ 1 (λ=0.633)
			r_{51} = 20(λ=0.633)
$BaTiO_3$	4mm	r_{51}=1640(λ=0.546)	r_{51} =820(λ=0.546)
			r_{33} = 28(λ=0.633)
			r_{13}=8(λ=0.633)
$PbTiO_3$	4mm		r_{33} = 5.9(λ=0.633)
			r_{13} = 13.8(λ=0.633)
$Ba_2NaNb_5O_{15}$	mm2	r_{33}=48(λ=0.633)	r_{33}=29(λ= 0.633)
		r_{13}=15(λ=0.633)	r_{13}=7(λ=0.633)
		r_{23}=13(λ=0.633)	r_{23}=8(λ=0.633)
		r_{42}=92(λ=0.633)	r_{42} = 75(λ=0.633)
		r_{51}=90(λ=0.633)	r_{51}=88(λ=0.633)
GaAs	$\bar{4}3m$	r_{41}=1.6(λ=10.6)	r_{41}=1.5(λ=10.6)
ZnSe	$\bar{4}3m$	r_{41} = 2.0(λ=0.546)	r_{41}=2.0(λ=0.633)
ZnS	6mm		r_{33}=1.8(λ=0.633)
			r_{13}=0.9(λ=0.633)
SiO_2(quartz)	32	r_{11}=-0.47(λ=0.5)	r_{11}=0.29(λ=0.633)
		r_{41}=0.2	
CuCl	$\bar{4}3m$	r_{41}=3.6(λ=0.633)	r_{41}=2.35(λ=0.633)
		r_{41}=3.2(λ=10.6)	r_{41}=2.2(λ=3.39)

Notation for the properties of materials and abbreviations

a = lattice parameter
ρ = density, g cm^{-3}
α_L = linear expansion coefficient, K^{-1}
c_s = specific heat capacity
κ = thermal conductivity, W m^{-1} K^{-1}
E_g(D) = bandgap (direct)
E_g(I) = bandgap (indirect)
T_m = melting temperature
α_{L} = linear expansion coefficient, K^{-1}
κ = thermal conductivity, W m^{-1} K^{-1}
C_P = molar heat capacity under constant pressure
c_s = specific heat capacity, J g^{-1} K^{-1}
thermal diffusivity α or $D_{\mathrm{th}} = \kappa / C_P \rho$, m^2 s^{-1}
T_{Debye} = Debye temperature K
E or Y = Young's modulus, GPa
G = shear modulus, GPa
K = bulk modulus, GPa
H_{Knoop} = Knoop hardness, kg mm^{-2}
H_{Vickers} = Vickers hardness, kg mm^{-2}
υ = Poisson ratio
ε_r = relative permittivity
q_{ij} = piezo-optic coefficients pPa (picopascal)
μ_e = electron drift mobility
μ_h = hole drift mobility
D_e and D_h = electron and hole diffusion coefficients
n = refractive index
ν_d = Abbe number
e_{ij} = piezoelectric constants
C_{ij} = elastic constants
$m_{hH}{}^*/m_e$ = heavy hole effective mass
$m_{hL}{}^*/m_e$ = light hole effective mass